Zhicai Zhong
Editor

Proceedings of the International Conference on Information Engineering and Applications (IEA) 2012

Volume 5

Editor
Zhicai Zhong
Chongqing
People's Republic of China

ISSN 1876-1100 ISSN 1876-1119 (electronic)
ISBN 978-1-4471-4843-2 ISBN 978-1-4471-4844-9 (eBook)
DOI 10.1007/978-1-4471-4844-9
Springer London Heidelberg New York Dordrecht

Library of Congress Control Number: 2012956302

Printed on acid-free paper

Springer is part of Springer Science+Business Media (www.springer.com)

Preface

Welcome to the Proceedings of the 2nd International Conference on Information Engineering and Applications (IEA 2012), which was held in Chongqing, China, October 26–28, 2012.

As future generation information engineering, information technology, and applications become specialized. Information engineering and applications including computer engineering, electrical engineering, communication technology, information computing, service engineering, business intelligence, information education, intelligent system, and applications are growing with ever increasing scale and heterogeneity, and becoming overly complex. The complexity is getting more critical along with the growing applications. To cope with the growing focus on intelligent, self-manageable, scalable information systems, applications are being created to the maximum extent possible without human intervention or guidance.

Information engineering and applications is the field of study concerned with constructing information computing, intelligent system, mathematical models, numerical solution techniques, and using computers and other electronic devices to analyze and solve natural scientific, social scientific, and engineering problems. In practical use, it is typically the application of computer simulation, intelligent system, Internet, communication technology, information computing, information education, applications, and other forms of information engineering to problems in various scientific disciplines and engineering. Information engineering and applications is an important underpinning for techniques used in information and computational science and there are many unresolved problems worth studying.

The IEA 2012 conference provided a forum for engineers and scientists in academia, industry, and government to address the most innovative research and development including technical challenges and social, legal, political and economic issues, and to present and discuss their ideas, results, work in progress, and experience on all aspects of information engineering and applications.

There was a very large number of paper submissions (1845), and all submissions were reviewed by at least three Program or Technical Committee members or external reviewers. It was extremely difficult to select the presentations for the conference because there were so many excellent and interesting submissions. In order to allocate as many papers as possible and keep the high quality of the conference, we finally decided to accept 542 papers for presentations, reflecting a 29.4 % acceptance rate. We believe that all of these papers and topics not only provided novel ideas, new results, work in progress, and state-of-the-art techniques in this field, but also stimulated the future research activities in the area of information engineering and applications.

The exciting program for this conference was the result of the hard and excellent work of many others, such as Program and Technical Committee members, external reviewers, and Publication Chairs under a very tight schedule. We are also grateful to the members of the Local Organizing Committee for supporting us in handling so many organizational tasks, and to the keynote speakers for accepting to come to the conference with enthusiasm. Last but not the least, we hope you enjoyed the conference program, and the beautiful attractions of Chongqing, China.

October 2012

Yan Ma
Qingsheng Zhu
Shizhong Yang
General and Program Chairs, IEA 2012

Organization

IEA 2012 was organized by Chongqing Normal University, Chongqing Computer Society, Chongqing Copious Prachanda Cultural Exchange Services Company, Chongqing University, Chongqing University of Science and Technology, Yangtze Normal University, Chongqing University of Arts and Sciences, and sponsored by the National Science Foundation of China, Shanghai Jiao Tong University. It was held in cooperation with *Lecture Notes in Electrical Engineering* (LNEE) of Springer.

Executive Committee

Honorary Chair:	Zeyang Zhou, Chongqing Normal University, China
	Xiyue Huang, Chongqing University, China
	Naiyang Deng, China Agricultural University, China
	Shizhong Yang, Chinese Academy of Engineering Chongqing University, China
	Junhao Wen, Chongqing University, China
General Chairs:	Yanchun Zhang, University of Victoria, Australia
	Xinmin Yang, Chongqing Normal University, China
	Yuhui Qiu, Southwest University, China
	Qingsheng Zhu, Chongqing University, China
	Yan Ma, Chongqing Normal University, China
	Lin Yun, Zhejiang Gongshang University, China
Program Chairs:	Zhongshi He, Chongqing University, China
	Huiwen Deng, Southwest University, China
	Jun Peng, Chongqing University of Science and Technology, China
	Wancheng Luo, Chongqing University of Arts and Sciences, China
	Liubai Li, Yangtze Normal College, China
	Guotian He, Chongqing Normal University, China
Local Arrangement Chairs:	Yan Wei, Chongqing Normal University, China
	Lei Lei, Chongqing Normal University, China
Steering Committee:	MaodeMa, Nanyang Technological University, Singapore
	Nadia Nedjah, State University of Rio de Janeiro, Brazil
	Lorna Uden, Staffordshire University, UK

Publicity Chairs:	Shaobo Zhong, Chongqing Normal University, China
	Xilong Qu, Hunan Institute of Engineering, China
	Mingyong Li, Chongqing Normal University, China
Publication Chairs:	Zhicai Zhong, Chongqing Peipu Cultural Services Ltd, China
Financial Chair:	Mingyong Li, Chongqing Normal University, China
Local Arrangement Committee:	Jun Li, Chongqing Normal University, China
	Xiaohong Yang, Chongqing Normal University, China
	Xiaoyan Li, Chongqing Normal University, China
	Shixiang Luo, Chongqing Normal University, China
	Wenshu Duan, Chongqing Normal University, China
	Yong Chen, Chongqing Normal University, China
	Wei Gong, Chongqing Normal University, China
	Yi He, Chongqing Normal University, China
Secretaries:	Jin Zhang, Chongqing Normal University, China
	Wanmei Tang, Chongqing Normal University, China
	Gaoliang Zhang, Chongqing Normal University, China

Program/Technical Committee

Yiming Chen	Yanshan University, China
Hui Wang	University of Evry in France, France
Shuang Cong	University of Science and Technology of China, China
Haining Wang	College of William and Marry, USA
Zengqiang Chen	Nankai University, China
Dumisa Wellington Ngwenya	Illinois State University, USA
Hu Changhua	Xi'an Research Institute of Hi-Tech, China
Juntao Fei	Hohai University, China
Zhao-Hui Jiang	Hiroshima Institute of Technology, Japan
Michael Watts	Lincoln University, New Zealand
Tai-hon Kim	Defense Security Command, Korea
Muhammad Khan	Southwest Jiaotong University, China
Seong Kong	The University of Tennessee, USA
Worap Kreesuradej	King Mongkuts Institute of Technology Ladkrabang, Thailand
Yuan Lin	Norwegian University of Science and Technology, Norway
Yajun Li	Shanghai Jiao Tong University, China
Wenshu Duan	Chongqing Normal University, China
Xiao Li	CINVESTAV-IPN, Mexico
Stefa Lindstaedt	Division Manager Knowledge Management, Austria
Paolo Li	Polytechnic of Bari, Italy
Tashi Kuremoto	Yamaguchi University, Japan
Yong Chen	Chongqing Normal University, China
Wei Gong	Chongqing Normal University, China
Yi He	Chongqing Normal University, China
Chun Lee	Howon University, Korea
Zheng Liu	Nagasaki Institute of Applied Science, Japan
Guotian He	Chongqing Normal University, China
Lei Yu	Chongqing Normal University, China
Michiharu Kurume	National College of Technology, Japan
Sean McLoo	National University of Ireland, Ireland
R. McMenemy	Queens University Belfast, UK
Xiang Mei	The University of Leeds, UK
Cheol Moon	Gwangju University, Korea
Veli Mumcu	Technical University of Yildiz, Turkey
Nin Pang	Auckland University of Technology, New Zealand
Jian-Xin Peng	Queens University of Belfast, UK
Lui Piroddi	Technical University of Milan, Italy
Girij Prasad	University of Ulster, UK
Cent Leung	Victoria University of Technology, Australia
Jams Li	University of Birmingham, UK
Liang Li	University of Sheffield, UK

Hai Qi	University of Tennessee, USA
Wi Richert	University of Paderborn, Germany
Meh Shafiei	Dalhousie University, Canada
Sa Sharma	University of Plymouth, UK
Dong Yue	Huazhong University of Science and Technology, China
YongSheng Ding	Donghua University, China
Yuezhi Zhou	Tsinghua University, China
Yongning Tang	Illinois State University, USA
Jun Cai	University of Manitoba, Canada
Wangmei Tang	Chongqing Normal University, China
Yan Wei	Chongqing Normal University, China
Yuyu Li	Chongqing Normal University, China
Yanxia Wang	Chongqing Normal University, China
Sunil Maharaj Sentech	University of Pretoria, South Africa
Mei Yu	Simula Research Laboratory, Norway
Gui-Rong Xue	Shanghai Jiao Tong University, China
Zhichun Li	Northwestern University, China
Lisong Xu	University of Nebraska-Lincoln, USA
Wang Bin	Chinese Academy of Sciences, China
Yan Zhang	Simula Research Laboratory and University of Oslo, Norway
Ruichun Tang	Ocean University of China, China
Wenbin Jiang	Huazhong University of Science and Technology, China
Xingang Zhang	Nanyang Normal University, China
Qishi Wu	University of Memphis, USA
Jalel Ben-Othman	University of Versailles, France
Lei Lei	Chongqing Normal University, China
Daijiang Chen	Chongqing Normal University, China

Contents

Part V Semantic Grid and Natural Language Processing

Part VII Mathematics and Computation

Part VIII Information Management Systems and Software Engineering

Part I
Business Intelligence and Applications

Chapter 1
Efficient Scheme of Local Colleges Development

Mei-liang Xiao

Abstract Local colleges are the main bodies of Chinese higher education. Their proper and sound development will do well to the development of local economy and society. Compared with famous and key universities, their development is in inferior state and there are lot of problems. This chapter analyses the problems during their development process, then it puts forward relevant strategies so as to provide reference for the development of local colleges.

Keywords Local colleges • Development • Problems • Strategies

1.1 Introduction

With the expanding functions of universities, the traditional universities have been the center of the society instead of "the ivory towers". In China, in the course of popularization of higher education, the reform of management system is going thoroughly [1, 2]. The central government delegates the management power to local government, at the same time; the local governments are running higher education with more initiative so as to further the economic and social development. Now among 1,000 universities, local colleges have take up 70 %. The development of local colleges is good. While compared with the famous and key universities, local colleges are in the dry tree [3]. This article analyzes some development problems of local universities and then puts forward related solutions.

1.2 Analysis of the Problems During the Development

1.2.1 Orientation is Going in for Greatness or Perfection

Nowadays, every nation knows that higher education plays critical role in the development of one's economy and society. The knowledge economy age is

M. Xiao (✉)
School of Foreign Languages, Xiaogan University, Xiaogan 432000, Hubei, China
e-mail: xiaomeiliang@hrsk.net

Z. Zhong (ed.), *Proceedings of the International Conference on Information Engineering and Applications (IEA) 2012*, Lecture Notes in Electrical Engineering 220,
DOI: 10.1007/978-1-4471-4844-9_1,

coming. In such background, people's needs for higher education are more urgent. The demanding standard of talents is gradually raised. People need higher level of education. The shrinking of technical secondary schools and junior colleges and the intense competition of postgraduates examinations are typical examples. In order to survive and develop, the high tide of "upgrading" emerges among local universities: junior colleges are going to be undergraduate colleges or universities, undergraduate colleges are going to start master education. The local colleges all want to be "big" universities. In fact, however, the diversity for talents determines the variety of colleges and universities. Different colleges and universities are on different starting line. It is not necessary for them to develop hand-in-hand. The strategy of "do something and don't do something" is very important. Beijing University is Beijing University itself; Qinghua University is Qinghua University itself. They both have their long history and they have incomparable advantages. If local colleges go in for greatness without thinking of their present situation, they may decentralize their limited resources. Therefore, local colleges must think about their orientation according to their reality.

1.2.2 The Disciplines have No Unique Features

The disciplines are carriers for the colleges and universities to cultivate talents, to carry on scientific research, and to serve the society. The level of disciplines determines the quality of their talents, the standard of their scientific study as well as their capability of serving the society. To some degree, the disciplines determine the survival and downfall of colleges and universities. Thus, modern colleges and universities are going out of their way to start hot specialties in spite of their own reality. From the advertisements of different colleges we can easily find that their disciplines are almost the same as those of famous universities. Local colleges start their disciplines at random, they do not think over the actual needs of the local society. Their disciplines are short of characteristics.

1.2.3 The Norms of Cultivating

Due to the effect of orientation and setting of specialties, many local colleges try their best blindly to keep up with famous universities in cultivating talents. Whatever the famous universities have, they want to have. They pay more attention to the education of record of formal schooling, whereas they neglecting cultivating practical professionals. Recently, the author made visits to several job fair and found that the graduates from local colleges could not be well matched with those from famous universities. Many employing units are unwilling to take the resumes of graduates from local colleges. They have no clear norms of cultivating talents so the graduates are lacking of competitiveness.

1.2.4 *Local Colleges are Short of Initiative of Serving the Society*

Serving the society is one of the functions of colleges and universities. The forms of service are different, such as educational training, technical advice and conduct, cultural service, and so on. Serving the society is the best way of linking theory with practice; it is also the way of checking the level of transferring knowledge to practical force. At the same time, by serving the society they may get financial support. American colleges and universities set up good examples in creating income themselves by serving the society. For example, the percentage of money from service in America in 1990 was 21.7 %, while in China it was not up to 10 %. The percentage of money from service of local colleges is much lower than that of big universities.

1.2.5 *Quality of the Teaching Staff*

In order to seek chances for development, local colleges pay great attention to importing and training talents with high education. However, such talents are more likely to go to regions of advanced economy. What's worse, some of the teachers sent out to study further run off after they got their higher degrees. The outflow of talent intensifies the gap in talents. The quality of the teaching staff is obviously a big problem.

1.3 Developing Strategies for Local Colleges

As mentioned above, local colleges have lots of problems in their developing procedure. In the competitive society, if local colleges want to survive and improve, they first must know themselves. The saying "know yourself as well as the enemy, you can fight a hundred battles with no danger of defeat" is surely the premise for our strategies. The followings are what we suggest.

1.3.1 *Clear Orientation*

There are four meanings of orientation of running schools: hierarchical orientation (status among the higher education system); professional orientation; orientation of standard of talents and characteristic orientation of their schooling. The outline of China's education reform and development has already put forward that related policies should be set for determining the taxonomy of higher education according to different regions, professions, and schools. By doing so, different colleges or universities may have their own clear aims and they can fully show their own features. Usually there are four levels in classifying colleges and universities in China: Research University, teaching and Research University, teaching college, junior college. The need of talents is various, so local colleges should be aware that different colleges have different functions. In hierarchical orientation, they must take "locality" as their basis. The undergraduate education and professional education should be their main roles. In professional orientation and orientation of standard of talents, local colleges may find out the needs of local communities, then start needed professions so as to train useful talents. For instance, Xiaogan city is a district with large population of agriculture. There is great potential in agriculture industrialization and urbanization. As the highest college in Xiaogan, Xiaogan University may focus on develop such professions as agriculture planting, landscape agriculture, city planning, architectural design, and so on. In characteristic orientation of their schooling, local colleges should make their local features distinct to others. Every district has its typical geography, cultural environment, economic condition, and natural resources. From what we have discussed above, in the course of running schools, four principles should be followed: principle of demand, principle of feasibility, principle of locality, and principle of characteristic.

1.3.2 Keeping Local District

Social and economic development of local district and development of higher education stimulate each other. Social and economic development is the base for the development of higher education. Higher education provides the development of social and economic development with talents and technology. Colleges and universities are the storage of knowledge. They should transmit knowledge from generation to generation, cultivate educated talents. Meanwhile they should create new knowledge and make use of them.

Local colleges should also be the research center of culture and the base of talents training. By using their intelligence and science and technology, they may contact and serve the society in many ways. There are three main functions local colleges can perform: cultivating talents, providing advisory and training service and promoting the integration of production, learning and research.

1.3.3 Enhancing the Building of Teaching Staff

Twenty-first century is a very critical period for the strategy of rejuvenating the country through science and education. It is also a critical period for educational reform and development. The key of rejuvenating the country through science and education is talent; the key of cultivating talent is the teacher. Thus, to build a group of teachers with high quality and energy is the most important task in the development of a college.

First, we should lay more stress on the construction of teacher's professional ethics. The notions such as loving the school as dearly as one does one's own home, cherishing posts and devoting wholeheartedly to work should be established. Nowadays the money consciousness occurs among the staff of colleges. Most teachers in local colleges go out after they get higher degree just for money in developed district. To the colleges, harmonious and democratic environment should be built. They should try their best to provide good teaching and researching conditions for teachers and researchers. To the staff, they should consider fame and money properly while admire knowledge more.

Second, we should pay more attention to the construction of discipline and backbone of teachers. The construction of academic pacesetter and backbone of teachers is the main task in the building of teaching staff. Local colleges are short of such academic pacesetters. Now many local colleges take resource sharing as their main ways of building their teaching staff. In fact, it is a way but it is not a reasonable way. We should cultivate our own academic pacesetters.

References

1. Chen XF (2001) Internationalization of higher education: general trend of cross century, vol 23(14). Fujian Education Press, Fuzhou, pp 56–58
2. Xie MB (1999) Higher education, vol 56(6). Higher Education Press, Beijing, pp 78–81
3. Zhu GR (1999) Functions of colleges and universities, vol 45(88). Heilongjiang Education Press, Haerbing, pp 78–90

Chapter 2
Study of Superficies System of Germany and Japan

Cheng-peng Ye

Abstract As an ancient form of property rights, superficies system was originated in Roman law. It effectively resolved the contradictions of ownership land between uses of land and had a high degree of adaptability to the market economy. It was inherited and developed by civil law countries, but later had a huge impact on the property rights legislation of civil law in concept and technology. By combing the evaluative path of the superficies system of continental law legal family, we may find many characteristics in its developing process which include the changed goal of building superficies from "having" to "using", many types of superficies and the strengthened rights of superficies, and so on. These characteristics provide references to further improving China's land-use right system.

Keywords Land of ownership • Builds of ownership • Right to use land

2.1 Introduction

Superficies, as one of the indispensable property rights in modern continental law legal family countries, is granting right on other's land, and hence the owner enjoys the right to transfer or inherit the buildings (including working substance) under or below the land. Also, it is of vital introducing value for China to resolve the problems of land use, public land as well as housing problem.

C. Ye (✉)
Department of Law, Yibin University, Yibin 644007, Sichuann, China
e-mail: yechengpeng@hrsk.net

Z. Zhong (ed.), *Proceedings of the International Conference on Information Engineering and Applications (IEA) 2012*, Lecture Notes in Electrical Engineering 220,
DOI: 10.1007/978-1-4471-4844-9_2,

2.2 The Evolution of Germany Superficies System

2.2.1 *Legislation Reforms*

2.2.1.1 Recognition of BGB on Superficies System

In 1990, the provisions of BGB on superficies system were mainly seen from the general regulations on the real properties in the Part three and the direct regulation on superficies in the Chap. 4, which was the first time to carry out legislation recognition on the superficies with the form of law. As lawmakers despised the dynamic functions of the use of property, it was thought that superficies had no applicable spaces, and the regulations on superficies only had six provisions (from No. 1012–1017). The lawmaker, who was known as the "the loyal recipient of Roman laws", however, did not accept Roman laws inflexibly [1].

2.2.1.2 Intensification of "Superficies Regulations" on Superficies

At the beginning of the twentieth century, Germany who had accomplished industrialization encountered agricultural population urbanization and caused "housing problem". Hence, the problems such as land price rising and land speculation were triggered, and the applicable opportunities of land increased unexpectedly. The superficies system provisioned in the Chapter of real properties superficies of BGB was quite weak, and had large defects and blanks in the content, which was unable to meet the new challenges proposed by times. In order to overcome the land crisis as soon as possible at that time, the lawmakers timely promulgated the "Superficies Regulations" (coming into force on January 15, 1919), which marked the real formation of modern superficies system [2, 3]. This regulation basically abolished the related regulations of BGB on superficies, and adopted 39 articles to reconstruct a relatively perfect superficies system. Compared with BGB, the intensification on the protection of superficies and the promotion on its accommodation were the major two characteristics of new system [4].

2.2.1.3 Improvement of "Housing Ownership & Eternal Residence Right Law" on Superficies

The severe damage of World War II to housing and considerable immigrants from Eastern Europe forced the housing of Western Europe after war to be highly tensional. Also, the land use relation turned into very complex. Then, only relying on changes in the land ownership system and housing lease system could never fundamentally resolve the problem of a land to contain multiple housings. In order to deal with the rapidly increasing housing problem and ease the resulting worse social contradictions, the "Housing Ownership & Eternal Residence Right Law" emerged along with times and came into effect on March 15, 1951. The "housing superficies" was created in the part 4 of Chap. 1 This new type of superficies made limited land capable to carry much more houses, and became the ideal legal means to resolve the housing problem at that time [5]. Housing ownership was the perpetual ownership of multiple land nonownership owners to build houses on others' land, which hence broke the principle of land adsorbing building, establishing the thorough legal separation of buildings and the land ownerships to make them independently exist and independently punished by its ownership owner, strengthening the legal position of building ownership further. As an important law ranking only second to the real property provisions in BGB, the promulgation of "Housing Ownership & Eternal Residence Right Law" promoted the superficies system to be improved well beyond doubts.

2.2.2 Characteristics of Germany Superficies System

2.2.2.1 Property Right

Rome superficies was never the property rights, but was only a land use right with the nature of obligatory right and the owner of superficies could only enjoy the right of transferring and inheriting, while the owner of land ownership still kept the ownership of buildings [6]. Germany civil law clearly stipulated that the superficies was the typical property rights and the owner of superficies possessed the complete ownership on the working substance constructed on the others' land and could also have other limit properties and the claim right of independent performance in addition to transferring, inheriting, and mortgaging rights. In the Germany civil law, superficies had been described as a kind of "similar ownership" or "the right equal to land ownership" [7], which was the intermediary and technological supporting point to promote the separation and combination of land ownership and its building ownership [8].

2.2.2.2 Diversity

Due to the constraint on the plane utilization of land, Rome superficies was only regarded as the right to construct buildings on the lands of others, and the type was too single. In order to meet the needs of contemporary society on the compact utilization of land, Germany superficies types were diversified, which included several types of superficies with different functions besides the traditional superficies: Subordinate superficies (referred to as low-level superficies) provided new legal means for the spaces above and under land to divorce from ground to independently become the object of property rights; the superficies of owners made land ownership owners kept the actual utilization of land when attaining the exchange value of land, mobilizing the initiatives of and ownership owners to transfer land; the housing superficies and partial superficies which were right mainly established on the public land allowed the rights of land ownership owner into a position unable to be performed, exerting an important role in the realization of government housing welfare system and the solution to the housing problem of people with low income.

2.2.2.3 Independence

The superficies in Roman law, was the temporary separation of ownership right, and was only a kind of right restricted on ownership without independence. However, German civil law regarded the superficies independent of ownership, which was a kind of property right set up based on the right contract and had no direct relationship with the powers and functions of ownership. In contemporary German civil law, superficies has a specific registration for immovable property like real property ownership, and can be set up at the first place; when superficies suffers from disturbance or faces up to the danger to be disturbed, superficies will be protected based on the claim right of owner. In Germany, the land ownership is also withered as an "ownership without usufruct" (the right of attribution) because of the restrictions on the superficies; and the superficies changes into ownership to play the same role of ownership with ownership, evolving into "using proprietary" and being applied universally in the daily economic life.

2.3 The Development Changes of Japanese Superficies System

2.3.1 Overview to Japan Superficies System

Since the promulgation of civil law in Meiji 23rd year (1890), there was superficies concept in Japan. After a short while, Japan's legislatures reformulated the civil law (which was successively promulgated from 1896 to 1898 and is being used to now) by imitating the mode of the second edition of Germany civil law draft which was promulgated not long age. The Chap. 4 (article 265–269) of the new law established the superficies system, which introduced or followed the Germany superficies system much more. Since the new lay was implemented, Japan legislature strengthened the superficies system by continuously issuing separate laws. Most noticeably, a series of laws and regulations, such as "Law about superficies" (Meiji 23rd year), "Law about the Protection of Buildings" (Meiji 24th year), "Land Borrow Law" (Taisho 10th year), "Land and House Rent Law" (Taisho 11th year) and "Temporary Treatment Law of Land and House Rent for Disaster City" (Showa 21st year), strengthened the rent right targeting at the ownership of building to make themselves close to superficies and complete the transforming of land rent right into property right when striving for the intensification of their own superficies [9].

2.3.2 *Innovation of Japan Superficies*

2.3.2.1 Changing the Purpose of Formulation

In the view of the essence, Germany superficies influenced by Roman law was the right taking of the working substance on other's lands, focusing on the holding of working substance. In order to achieve the maximum benefit of land, Japan civil law centered on the utilization in the concept of superficies and relied on the combination of land, houses, and other facilities, which was convenient for superficies owners to engage in nonagricultural business operations.

2.3.2.2 Expanding the Scope of Subject Matter

According to the provision of the article 265 of the old Japan civil law, superficies referred to the right to use the other's land (the ground, the space above or beneath the ground) in order to hold working substance on the other's land [10] p 198.

2.3.2.3 Increasing the Types of Superficies

The generation of Germany superficies was only the property consensus between the parties. In Japan, superficies could be generated from the direct provision of law under specific circumstances but not only from the consensus between the parties, which was the legal superficies. The establishment of legal superficies meets the concept of Japanese separating buildings and lands, allowing buildings themselves to own independent transaction. Most noticeably, in 1966 the "Japan Civil Law" added the "Space Superficies" into the article 269 of the superficies chapter in No. 93 Law. The "socialization" of land ownership was a process of the independence of the space interests of land and the gradual confirmation of law on the space right.

2.4 The Evolution of Superficies System in the Continental Law System

2.4.1 *Transformation of Essence*

The modern superficies of Germany, Austria, Switzerland, and French, which succeeded in the concept of Roman superficies, concentrated on the objects above ground and did not think the existence of superficies if no realistic and governable objects were above ground, lied in "having" in essence but not in "using". Modern superficies focused on the utilization of land, and stressed the essence of superficies in "having" but not in "using"; and the existence of working substance did not influence the superficies and the loss of working substance would not result in the perishment of superficies, hence exercising the utilization of land completely.

2.4.2 Enhancement of Rights

Roman superficies was regarded by Roman jurists as the right of superficies owner to build houses and use other's land for a long time and to be enslaved to ownership, which was only a kind of burden of land ownership. From Germany, the superficies of continental law system developed into a typical real property right. This was reflected not only on the establishment of superficies necessary to comply with the basic principle (consensus and registration) of real property right establishment, but also on the aspects of superficies able to transfer, inherit and burden other restricted property rights (subordinate superficies, mortgage, land debt, regular land debt and substance burden) as well as the obligee was able to perform independent claim right, etc.

2.4.3 Increasing of Types

After entering the industrialized society, the development of science and technology expanded the activity space of human beings above or beneath ground, the use of land trended differentiation or spatial pattern, and the types of superficies increased as well. Other than the ordinary type, superficies included subordinate superficies, housing superficies, ownership owner superficies, space superficies, etc. Superficies with a wide variety made full use of various legal means to expand the powers and functions of superficies, meeting the condition of market economy; the real property right and especially the land property right as the basic real property right stepped into the tendency of trading mechanism with the form of multiple levels of right.

Generally speaking, the superficies in continental law system gradually ascended and developed. The superficies system met the social the development tendency of ownership socialization as the burden of ownership. In modern society, the superficies system has become the crucial legal system in the real property right area. Its social and economic values will always be worth our study conscientiously and treasuring carefully.

References

1. Zweigert K, Klotz H (1992) Comparative law, (trans:Pan HD, Mi J, Gao HJ, He WF) vol 78(16). Guizhou People's Press, Guiyang, pp 134–136
2. Sun XZ (1997) Contemporary germany property law, vol 45(7). Law Press China, Beijing, pp 89–90
3. Qian MX (1998) On the object of usufruct. Peking Univ Law J 56(3):45–47
4. Zhou N (1994) Origin theory of roman law, vol 15(78). The Commercial Press, Beijing, pp 456–458
5. Zhao JL, Li SW (1999) On the theories and changes of superficies system—discussing "the construction of china's non-agricultural land use system". The collected papers of 1998 civil law and economic law annual conference, vol 56(3). Xi'an: Shaanxi People's Press, Beijing, pp 45–47

6. Horn R, Klotz H, Les H (1996) Introduction to german civil and commercial law, (trans:Chu J) vol 56(18). China Encyclopedia Press, Beijing, pp 123–125
7. Wo Q, You QX (1999) Japan property law, vol 45(4). Taiwan Wu-Nan Publishing Corporation, Taibei, pp 256–258
8. Shi SK (2000) Introduction to property law, vol 1(55). China University of Politic Science and Law Press, Beijing, pp 24–27
9. Qian MX (2002) On the basic morphology of china's usufruct Yi Jiming. private law, (Vol.2 of Edit. 1) vol 34(56). Peking University Press, Beijing, pp 34–37
10. Zheng YB (1995) Real rights law, vol 178(45). San Min Book Co. Ltd, Taiwan, pp 6–9

Chapter 3
Study on Political Status of Religious Parties in Israel

Xingang Wang

Abstract Israel is a secular and modern democratic state in the western world. However, unlike the other standardized forms of state, the religious parties exert a highly important role in the social life of Israel. In this paper, first of all, an analysis is carried out by the author on the definitions on the most important religious parties in Israel, and then the characteristics of the religious parties are put forward on that basis, and finally a conclusion is made on the highly important roles of the religious parties in the Israel social life in the modern times.

Keywords Religious parties • Religious powers • Social and political life

3.1 Introduction

As is known to all, Israel is the only developed capitalistic state in the Middle East region. At the present time, the parliamentary democracy of the western world, in which the legislative powers, administrative powers and judicial powers are respectively separated but also are constrained and coordinated with each other, is being implemented in this state. Knesset is the supreme authority; the prime minister is the executive head of the government; the leader of the party which receives the most seats in the election of Knesset will hold the post of the prime minister. The court of Israel is divided into the civil court and the religious court, which can enjoy their mutually independent adjudications and simultaneously are not restricted by Knesset and government. From the perspective of the political form, it can be learnt that Israel is a standard modern democratic state in the western world. However, in realities, the Judaism and especially the Orthodox Judaism possess a very powerful strength, and hence the Jewish Synagogue takes up a very influential position in Israel at the present time. Therefore, as the representatives of

X. Wang (✉)
Education Department of Ideological and Political Theories, Hebei Normal University of Science and Technology, Qinhuangdao 066004, Hebei, China
e-mail: xgang24wang@126.com

Z. Zhong (ed.), *Proceedings of the International Conference on Information Engineering and Applications (IEA) 2012*, Lecture Notes in Electrical Engineering 220, DOI: 10.1007/978-1-4471-4844-9_3,

the politics of the different religious powers, the different religious parties exercise a highly important function in the political and social life in Israel.

3.2 The Definitions on the Religious Parties of Israel

It is known to all that the religious parties play a very distinctive function in the actual political life of Israel, and are the third parties grouping system after the Labor Party and the Likud Party in general. In the article which is named the "Israeli Political Parties: Separation, Reunion and the Dividing Line", Chinese scholar An Weihua made a clear definition on the religious parties in Israel. However, seen from the relationship between the religions and the secular political powers, it can be known that what the religious parties often look for actually is the Zionism in general.

3.3 The Characteristics of the Religious Parties in Israel

3.3.1 The Religious Parties are a Relatively Stable Political Power in Israel Politics

Generally speaking, a variety of the political parties within Israel have been in existence, and the unions and separations are also very frequent to happen to the secular political parties. Seen from around the world, it can be learnt that Israel is one of the countries in which the changes in the political parties are the most frequent [1]. However, in comparison with these frequent changes happening in the secular political parties, the religious parties are located in a relatively stable state. Therefore, it can be known that such a stable state of the religious parties in the politics of Israel after a conclusion can be embodied in the four different aspects, which are as shown in the following.

3.3.1.1 The Organizational Forms of Religious Parties are in a Relatively Stable State

It is always said that the organizational forms of religious parties are in a relatively stable state. In the early founding of the state, there are two religious parties to be in existence in Israel. They were the spiritual center party and the Israel orthodoxy party respectively. In the later time, these two parties were separated into the spiritual center workers parties and the Orthodox Judaism parties. In 1956, the spiritual center party and the spiritual center workers party were merged into the national religious party. In addition, the Israel Orthodox Judaism party and the Orthodox

Judaism workers party used to have a great deal of cooperation with each other for a certain period, but there were no unions to be carried out between the two sides ultimately in realities. In 1984, because there were a great number of the differences in opinions, the SHAS party and the Torah flag party respectively were separated from the Israel Orthodox Judaism party successively. In the 1980s, although there were some small parties to be separated from the large religious parties, all the separated small religious parties in addition to the SHAS party all were basically in a disappearance by now. At the present time, a situation of the tripartite confrontation, in which the national religious party, the united torah Judaism and the SHAS party are in co-existence, has coming into being within the Knesset of Israel.

3.3.1.2 The Number of Seats of Religious Parties in Knesset is in a Relatively Stable State

The religious party almost had participated in all the previous parliamentary elections. Also, the number of seats of religious parties in Knesset is in a relatively stable state. In the 15th Knesset of Israel, the religious factions achieved the optimal results in the history and acquired the twenty-fight seats in total. Even through in the most defeated parliamentary election, they gained thirteen seats at the same time. Therefore, seen from the overall situation, it can be known that the strength of the religious parties will have no large changes in the short term. In addition to the 14th and 15th parliamentary elections in the 1990s, the seats of the religious parties were generally kept between 30 seats and 18 seats, and took up thirteen percent of the total parliamentary seats at average.

3.3.1.3 The Political Programs of Religious Parties are in a Relatively Stable State

In the mean time, the relative stability of the religious parties is also embodied in the stability of their political programs. The requirements of the religious parties on the aspects of politics, economies and religions almost have no too large changes for decades. Such a stability of the religious parties is primarily attributed to the relative stability of the voters of all kinds of religious parties.

3.3.2 The Policies of Religious Parties Tends are Inclined to Right-Wing Party and have a Close Relation with the Likud Party

Also, as is known to all, the religious parties are conservative in ideology. In the Israeli-Palestinian issue, it is thought by the religious parties that the West Bank was the land of the Jews in the Bible, resolutely excluded the Palestinians, and was opposite to any suggestions on the autonomy or state founding of the Palestinians. In the domestic politics, the religious parties are more willing to give assistance to the Likud party, and also the Netanyahu praised that the religious parties are their inherent Allies. For example, such a role was presented more apparently in the first president prime minister election of 1996. As the majority of the voters of the religious groups cast a vote according to the instruction of rabbis, both Perez and Netanyahu attempted to attain the personal supports of the rabbis. In the mean time, Netanyahu visited the elder rabbi Itzhak Khadori who was one hundred and six years old before the one night of his last election campaign; Khadori made a prediction on Netanyahu before camera that you would be the prime minister tomorrow. Therefore, it can be known that the personal opinions of Khadori exert a magnificent role in the election.

3.3.3 Religious Parties Pay Main Attention to the Relationship Between State and Religions and Seldom Consider the Issues Unconnected with Religions

Whether the ruling is under the Labor party coalition or the Likud party, the concession of the religious parties on the religious issues is the key factor to absorbing it to participate in the unity government. Furthermore, the party, who can meet the more requirements of the religious parties in the religious issues, has more opportunities to participate in the leadership of government [2]. As the labor party under the leadership of Gurion is with a strong secularism, the strength of the religions is controlled within a certain range. However, within the ruling period of the Likud Party, it attached the highest importance to the senses of Jewfishes and showed the respects to the traditional values and the religions of Israel, and hence formed the compromise and compliment policies to the strengths of the religions, and therefore it can be known that its religious influence is stronger than that in the ruling period of the labor party and will be increasingly more obvious in the future days.

3.4 Roles of the Religious Parties in the Social Life

3.4.1 Frequent Revolts are Launched by the Religious Parties and make the Government of Israel Quire Weak

Today, as the multi-party election and the proportional representation system are implemented in Israel, almost the party in the ruling—Israeli labor party or Likud party has to seek the supports of the religious parties for the purpose of constructing the united government of all parties. Therefore, the religious parties have a major impact on the survival or extinction of the cabinet. With the purpose of gaining the ability to maintain the "survival" of the cabinet, it is necessary for the large party of organizing the cabinet to make a quite large concession on the religious parties, and hence the religious parties often achieve the influences which have gone beyond their own political powers in some policies. For example, the religious parties possess a decision power which is disproportional to their strengths on the survival or extinction of the cabinet [3]. The members and supporters of the religious parties are all the devout jewfishes, and always take up 15–20 % of the total population; the seats of the religious parties usually are 15 or 18 in Knesset. However, the cabinet's crises which are caused by the disputes on the religious issues are the most. According to statistics, during 1948 and 1979, there were about 103 crises in cabinet in Israel. Among them, 35 crises were aroused due to the religious disputes, and almost take up one third of the total number.

3.4.2 *Religious Parties have Religious Privileges in the Social Life*

The privileges of the religions had been formed in the period of the Zionism. After the founding of the state, with the purpose of convincing the religious parties to participate in the united government, the two major ruling parties adopted the compliment polices for the religious parties in some aspects. Therefore, it is not strange for the religious parties to enjoy the privileges in Israel today. These privileges of the religious parties seem to only have the relationship with the life of people, but actually their influences on the political life can't be ignored. For example, during the ruling period of the labor party in 1976, the Lappin government arranged a new made-in-America fighter plane on a Sabbath day to hold the hand over ceremony for the air force of Israel.

3.4.3 *Religious Parties Play an Important Role in Israel Politics, but are still Insufficient to Contend Against the Secular Political Powers*

Before the 14th Knesset election, no matter how these different religious parties are separated and combined, the number of the seats in the previous Knesset is generally maintained between 13 and 18 seats. That is to say, under the best situation, only 15 % of the voters hand over their destiny to the religious parties. Even though the new electoral law was implemented in Israel, the religious parties still achieved a rapid development, and received 23, 27 and 22 seats respectively in the 14–16th parliamentary elections. Therefore, even though the religious parties made the largest victory in history in the fifteenth parliamentary election, they only gained the 22.5 % of the votes all over the country. In recent years, the religious parties left their feet on the politics in all aspects. However, these religious parties still have no abilities to contend against with the secular political parties with the powerful strengths.

For example, the seats of the religious parties in Knesset gained a rapid increase in the 14th parliamentary election, and increased to 23 from the previous 16 seats. This only gives a reflection to that a considerably large number of the secular voters have a serious crisis sense for the progress of peace, the future of the nation as well as their own interests. However, this phenomenon can't suggest that the people in Israel hope the religious powers to get involved in the social life. No increase in the number of the seats of the United Torah Judaism has proved the reality to a great extent. Therefore, it can be known that it is universal for the attitudes of the people are relying on the strengths of the religious parties to intensify the recognition on the cultures of Jewfishes, for the purpose of accomplishing the relative conservations and stabilities of the external and internal policies of the Israel government.

In the mean time, it can be known that it is necessary for people to make clear recognition that the powers of the religious parties can't be kept in the rising tendency all the time, not mention to worrying about the struggles with the secular political powers. For example, in the early 1980s, Israel fell into the dilemmas of the Lebanon war; the domestic economy was depressed, and simultaneously the inflation had jumped into the three digits. Under such a situation, the number of seats of the religious parties in Knesset stayed at 13 (which is a very low level) at two consecutive parliamentary elections. This can give a reflection to that the voters including a considerably large number of the religious people do not hold an expectation on that the religious parties can exert a certain role in the settlements of the problems, which are in existence within the multiple wars and economies.

References

1. Weihua A (2002) Israeli political parties: the separation, reunion and the dividing line. W Asia Afr 4(4):35–37
2. Ruidong Y (1997) Analysis on the politics in Israel, vol 90. Northwestern University Press, Xi'an, pp 47–56
3. Yanan X (2004) Conflict between the religions and the secular political powers in the course of the Israel modernization. Middle East Stud 5(1):42–48

Chapter 4
Standard System and Legal Protection of Food Security of Agricultural Special Products

Zhao Fu-jiang, Zang Chen and Luo Cheng Bing

Abstract By researching the current conditions of standard system and legal protection of food security of agricultural special products in Yanshan mountainous areas, this paper points out that there exist some problems, which are that standard system is not complete, the standard is too old and not proper, protective system of standard is not perfect, the strategy of intellectual property and the system of legal protection are imperfect, etc. According to this, appropriate countermeasures have been proposed to solve these problems.

Keywords Yanshan mountainous areas • Agricultural special products • Food security • Legal protection • Technological barrier

4.1 Introduction

Yanshan Mountainous Areas are located in 39°45′–41°N, 115°40′–119°50′E. The Mountainous Areas have included five cities and 36 counties (city and district), which are Chengde, Tangshan, Zhangjiakou, Qinhuangdao and so on. The gross area is 79.7 thousand square kilometers [1, 2]. The agricultural characteristic industries here are developed in a relatively fast manner, such as the chestnut, grape, almond-apricot, hawthorn, featured mixed fruits and so on. The agricultural characteristic industries here have formed into the industry scale [3, 4]. It is predicted that in the year 2015 the planting area of the exported—oriented base will reach the total amount of 2,700 thousand mu, while the realization value of output will come to 2,200 million. However, what requires paying attention to is that the competition of international trade is becoming more and more serious every

Z. Fu-jiang (✉) · L. C. Bing
Faculty of Literature and Law, Hebei Normal University of Science and Technology, Qinhuangdao, Hebei 066004, China
e-mail: zhaofujiang@cssci.info

Z. Chen
Environmental Management College of China, Qinhuangdao, Hebei 066004, China

Z. Zhong (ed.), *Proceedings of the International Conference on Information Engineering and Applications (IEA) 2012*, Lecture Notes in Electrical Engineering 220,
DOI: 10.1007/978-1-4471-4844-9_4,

day [5, 6]. There are more and more export barriers for the agricultural products in each country. The United Nations Economic Commission for Europe has once pointed out that "about 25 % of the non-tariff barriers are due to the difference of the product standards. In addition, they can as well be caused by applying different checks, tests, product quality guarantees and identification methods." It can be seen that the standard system and the law protection research should be strengthened so as to provide strong and effective guarantees for the development of the agricultural characteristic industries in the Yanshan Mountainous areas.

4.2 Situation Analysis

4.2.1 The Existing Problems

4.2.1.1 The Standard Quantity is Small

The special agricultural industries in the Yanshan District are made up of by the chestnut, grape, almond-apricot, hawthorn, featured mixed fruits and so on. However, on the formulation of relevant standards, the quantities are relatively small. For example, there are only three items for the almond-apricot and Anli pear and there are altogether 13 items for the standards of hawthorn. There are altogether 14 items for the standards of the sweet cherry. Chestnut, as the pillar product of the national geography symbol and the famous brand here, there are only 33 items for its standards. Among the 33 items for the standards, there are 26 items for the regional standards while there are six items for the industry standards.

4.2.1.2 The Standard Renew is lagged in Phase

According to the standard age internationally, the standard division statistic items are usually two to five years. The existing standard ages are relatively long. The standards that are equipped with more than ten years standard age still account for a very large proportion (See Table 4.1).

Table 4.1 Standard ages of five kinds of agricultural special products

Products		Total		Standard age	
	<2		2–5	5–10	≥10
Hawthorn	13	2	1	3	7
Sweet cherry	14	2	3	4	5
Chestnut	33	6	5	9	13
Armeniaca vulgari	3	0	2	1	0
Pyrus ussuriensis maxim	3	0	2	1	0

4.2.1.3 Universally Relatively Low Standard

As for the Positive List System in Japan, there are 74 kinds of limitations standards that are stricter than the existing limitations standards in our country in the aspect of agricultural chemicals. They have involved 247 pieces of limitations standards. There are residue limitations in our country while Japan adopts the "uniform limit". As for the "uniform limit" adopted in Japan, there are 96 kinds of chemicals, involving 350 pieces of limitations standards. As for the Chinese chestnuts of Qianxi, the "export leader" in Yanshan District, their residue limits are added to the amount of 288 from the amount of 52. Among them residue limits standards, the residue limits of phoxim and chlorpyrifos have improved 60 % in the strict degree compared to the existing regional standard DB13/T 728-2005.the residue limit standards for the Omethoate and parathion are not sure. However, the abamectin and imidacloprid in the standard are higher than those in the Positive List System in Japan. The standards do not share the same broad and strict degree. In this way, it will usually become the double "technology tariff". However, there still has not been any revision on the regional standards (See Table 4.2).

4.2.1.4 The Standard Construction has not Formed into a System

The food standard construction of the special agricultural products in Yanshan Mountainous Areas has not formed into a perfect system yet. To take the Chinese chestnut as an example, its standard construction is relatively perfected. The existing standards are only "The Pollution-free Fruits Chinese Chestnut" (DB13/T 728-2005), GB/T 22346-2008 chestnuts quality levels, GH/T 1029-2002 Chinese chestnut, DB1305/T 021-2003 pollution-free agricultural Chinese chestnuts and so on. Altogether there are 33 items. The standards have only involved the following four kinds: Jingdong, Qinglong Chinese chestnut, exported Qinglong, and the Xianxia Super Early Chinese Chestnut.

Table 4.2 MRLs of chestnut about DB13/T 728-2005 and positive list system

Agricultural chemical	DB13/T 728-2005	Positive List System
Phoxim	$\leq$0.05 mg/kg	0.02 mg/kg
Chlorpyrifos	0.5 ppm	0.2 ppm
Omethoate	Absent	0.2 mg/kg
Parathion	Absent	0.05 mg/kg
Abamectin	——	0.02 mg/kg
Imidacloprid	——	0.5 mg/kg

4.2.2 *The Standard Guarantee System is not Perfect*

4.2.2.1 The Detection System is Incomplete

There are altogether 50 and more detection institutions in Hebei Province that are involved in the detection projects for the agricultural products. Among the 50 and more detection institutions, there are 19 institutions at the Level of Province and Department. As for some institutions at the Level of Province and Department, their functions are set up repeatedly. The equipments and facilities are purchased repeatedly. All of these circumstances are unbeneficial to realize the share of social resources. The institutions at the Level of City and County are not perfected. Although there are the detection rules for the Critical Control Point for the threaten analysis of HACCP in the practical operations, the existing standards tend to the testing of the ultimate products. There are imperfect prediction and control process towards the potential food damage [1].

4.2.2.2 The Propaganda and Implementation System is not Deep

The special agricultural products in the Yanshan Mountainous Areas are mostly in the model of traditional agriculture. It is mainly relied on the farmers. Although the government has greatly intensified the leading force and organized the agriculture professional cooperation society as well as supporting the leading enterprises, such causes as the absence of the relevant regulations, the in-depth of the propaganda and implementation, the great amount of investment due to the standardization production of the farmers and the improvement of the product standards and so on, Have resulted in the incomplete implementation of the standards as well as the inactive attitude towards implementation. What is worse, there is a lack of understanding of the international advanced standards and there are not enough acknowledgement to the importance of the standardizations.

4.2.2.3 The Demonstration System is not Broad

It is found through investigation that the special agriculture leading enterprises in the Yanshan Mountainous Areas, some of the relevant enterprises such as the production foundation have not carried out the responsibility of the food safety according to the requirements of the law.

4.2.2.4 The Law Protection System is not Perfected

On the one hand, it has to realize the food safety law insurance in the usual meaning under the framework of "Food safety laws" and "Agricultural Product Quality Safety Law". On the other hand, it is to examine from the perspective of the technology laws and regulations.

4.3 Proposed Scheme

4.3.1 Strengthen the Advanced Standard Construction

Intensify the research and development of the standardization formulation and make up for the absence of standardization. The advanced industry standards and enterprise standards should be formulated connecting with the standards formulated by CAC, IOE, IPPC, ISO, OIV and so on so as to make up for the deficiencies of the standards.

Establish the renewed efficient system for the standards, and strengthen the tracking, study, and transformation of the intentional standards and the advanced standards outside the country so as to improve the overall level of the standardization.

On the matter of standardization limitations, it should use the exportation as the orientation. It should revise the original standards and improve the standardization limitations. In this way, it is able to narrow the distance with the international requirements. It should as well establish and perfect the quality detection system for the agricultural products.

It should formulate the detailed safety production technology specifications and operations procedures for the special agricultural products as soon as possible. Introduce the HACCP, ISO9000 quality management system identification and the ISO14000 environment management system identification so as to construct a perfect food standardization system for the Yanshan Areas [2].

4.3.2 Improve the Detection

4.3.2.1 Improve the Detection System

Use the advanced monitoring system as the guarantee and keep adding detection projects. This usually becomes the technology tariffs for the exportation of the special agricultural products [3].

4.3.2.2 Consturct the Propaganda and Implementation System

The international related standardization information on the special agricultural products should be open to the society in time, making it convenient for the enterprises and the society to inquire so as to improve the standardization level [4].

4.3.2.3 Build High Standard Demonstration System

Food safety law stresses that production enterprise is the first responsible man for the food safety. Promote the industrialization operation mode of the special production in mountainous areas.

4.3.2.4 Perfect the Construction

Hebei province has formulated a lot of laws and regulation. However, the supporting construction for the food safety laws lags behind. It should make clear the important and the less important.

Under the system of WTO, TBT, WTO and SPS are the core rules for the laws problems dealing with technology tariffs of agricultural products [5]. China has entered WTO in 2001-12-11 and shall use it as basis.

Acknowledgments Qinhuangdao City Science and Technology Ministry Soft Science project. The economic strategy research brought out by the law of Food safety implemented in Qinhuangdao City (Project Number: 200901A268). Hebei Province Science and Technology Hall soft science project: the strategy research on the regional diversified planning of grape wine industry development in Changni. Hebei Normal University of Science and Technology humanities and social sciences bidding science research project: the planning research on the product safety of grape wine in Qinhuangdao City (Project Number: SKZB110104).

References

1. Jian-jun Q, Shi-gong Z, Shu-zhong X (2004) Study on promoting great-leap-forward development of the characteristic economies in yanshan hilly areas Hebei provincc. J Agric Sci Technol 5(3):65–72
2. Baldwin RE (1970) Non-tariff distortions of international trade, vol 3(7). The Brookings Institution, Washington, pp 143–148
3. Zhejiang Province Standardization Study Institution (2007) The EU food safety management basic regulations and its study, vol 3. Chinese Standardization Press, Beijing, pp 6–7
4. Yan-bo Z, Shu-yun C (2009) How Jingdong Chinese chestnut has broken through the green barrier. Hebei For 3(3):12–18
5. Hun-lun H, Wen-tao X (2009) Study on the safety evaluation and detection techniques of transgenic foods, vol 3. Science and Technology Press, Beijing, pp 154–161
6. Xiang-wen Z, He-guang W, Xiao L (2010) The influential analysis of the EU technical barriers to the export of the agricultural products of our country. Agric Econ Probl 3(4): 105–110

Chapter 5
Optimization Model of Chinese Enterprise Brands

Xiangdong Zhang

Abstract The world's market has entered into an age in which the enterprise brands are competing with each other highly intensively and the multinational businesses or operations are taking full advantage of the cross-cultural management at the same time. In the mean time, the range of the internationalization of the enterprise brands is getting broader and broader, and there is no exception to the depth of such a kind of internationalization. The differences as well as the conflicts, which exist among the different kinds of the cultures has transformed into one of the most important factors to impose a great number of restrictions on the process of the enterprises getting into internationalization. In this paper, the author makes a detailed analysis on the cultural barriers, which appear in the process of Chinese enterprise brands getting into internationalization. The purpose for the author to making this analysis is attempting to make a choice on an appropriate model from the perspective of the cross-cultural management and also seeking for the new ideas which can be utilized to promote the enterprise brands in China to get into the internationalization with a better effect.

Keywords Brand • Internationalization • Cultural differences • Cross-cultural management

5.1 Proposal of the Problems

During the recent few years, the fact that the enterprise brands in China are getting to the internationalization are being transformed into one of the hottest issues, which receive the wide attentions from the community of world enterprises as well as the field of the theories. In the mean time, the economy in China attains a highly rapid development, which promotes a great number of the enterprises to go out of the gate of the country one after another, and therefore the internationalization of the enterprise

X. Zhang (✉)
Tourism College of Zhejiang, Hangzhou 311231, Zhejiang, China
e-mail: zhangxiangdong@cssci.info

Z. Zhong (ed.), *Proceedings of the International Conference on Information Engineering and Applications (IEA) 2012*, Lecture Notes in Electrical Engineering 220, DOI: 10.1007/978-1-4471-4844-9_5, © Springer-Verlag London 2013

brands of China has transformed into one of the objectives which are sought by a large amount of enterprises as well as the entrepreneurs [1].

However, the internationalization for the brand of an enterprise will be a complex and meticulous systematic task all the time. From the perspective of the multiple existing enterprise brands in the internationalization at the present time, it can be known that almost all of them accomplish such an outstanding results by resting with more than tens or hundreds of hard working accumulation. Therefore, in comparison with their accumulations in the enterprise brands, the internationalization for the brand of a Chinese enterprise will obviously look too much more hasting [2].

At the present time, seen from the development of Chinese large enterprises, Haier, Lenovo, Huawei, Tcl, China Mobile and other Chinese enterprises have walked in front of a great number of Chinese enterprises in the internalization, and therefore can be praised to be the pioneers getting into the internationalization in businesses [3].

However, the process for these large Chinese local enterprises to get into the internationalization can never be even, because there are a large amount of the reasons for such internationalization. Among these reasons, the absence of the cross-cultural management in the process for the brands of Chinese enterprises to get into the internationalization is one extremely important reason.

The effectiveness of the cross-cultural management is one of the highly important factors, which excises an influence on the success and failure of the brands of the enterprises in the process of getting into the internationalization. As is known to all in the community of the world business, the secret for the enterprises which have accomplished a great number of successes in the course of the international business all is that they have the ability to make coordination on all kinds of the resources in the process of the cross-cultural management and even to make them integrated into the pressures caused by different cultures [4].

The so-called brand cultures refers to a kind of cultural phenomenon which can be used to give a reflection to the personification of the brand of an enterprise and also can refer to all of the cultural phenomena of the cultural special characteristics such as the management concept, value, aesthetic concept and other concepts which are accumulated in the brands for a very long time and in the brand management activities, and can also be the sum of the benefit cognition, emotion attribute, cultural tradition and individual image which are presented by the above concepts.

5.2 Influence of the Cross-Cultural Management on the Brand Internationalization

The cross-cultural management means that it is highly necessary to make a research to design an effective organization strategy and structure and to engage in the related management activities by advantaging of the limited human resources under the different kinds of the cultural environments [5].

Besides, for the selection of the cross-cultural management models, it is necessary to make a detailed analysis on the influences of the differences in cultures on the internationalization of the brands [6]. Most importantly, the differences in cultures has the ability to increase the difficulty for the international operation of the brands, and therefore it is highly necessary to select a different cross-cultural management model to push forward the process of the brand of an enterprises to get into the internationalization. These can be embodied in three aspects: first, the differences in cultures can play an influence on the definition as well as recognition on the brands in internationalization, namely, how enterprises will select an appropriate name or logo for their brands; next, the differences in cultures can play an influence on the strategy formulation for the brand communication in the process of getting into internationalization; then, the differences in cultures can play an influence on the brand management efficiency in the process of a brand of getting into internationalization.

As a result, in the process of a brand of getting into internationalization, the simple cloning management model can never be applied, and simultaneously can't enter into the cultural permeating stage at the beginning. Therefore, it can be seen that it is necessary to scientifically and reasonably select the cross-cultural management methods under the different cultural differences in accordance with the actual conditions [7].

The influences exerted by the differences in cultures on the brand of an enterprise of getting into internalization are embodied extremely differently in both developed countries and the developing countries. Within the developing countries, it is necessary to push forward the internationalized operation of the brand of an enterprise by phases and also with a gradual step. Generally speaking, there are different kinds of consumption concepts of the consumers in developing countries and developed countries. The consumers in the countries with developed industries usually possess a higher evaluation on their domestic products.

However, on the contrary, the consumers in the developing countries prefer the imported product too much. Under these circumstances, it is much easier for the internationalization process of the brand of an enterprise in developed countries than that in the developing countries. This situation gives rise to much greater pressures and challenges for the internationalization process of the brand of an enterprise in the developing countries in the aspect of the cross-cultural management [8].

5.3 Cultural Barriers in the Process for Chinese Enterprise Brands to get into Internationalization

5.3.1 Low Brand Awareness of Foreign Consumers on Chinese Products

Through the tens or even hundreds of business management accumulation, European and American brands have been already rooted in the minds of their domestic people and even the people all other countries. Since China made a reform of the market economy, the life span of a great number of Chinese enterprise brands at its domestic markets become quite short, and most of Chinese enterprise brands have not been known by the people in the foreign countries and especially in the developed countries [9]. Even though there is part of the people in the foreign countries to know Chinese product brands, it is never easy to make a change to the consumption habits of the people in the European countries and the Unites States. At the same time, the fake and inferior products which are branded with "Made in China" exert a negative influence on the internationalization process of Chinese enterprise brands to some extent.

What's worse, these fake and inferior products have given rise to the obstacles for Chinese enterprises which have entered or are going to enter the world business stage.

5.3.2 Chinese Brands Lack the Related Cultural Connotation

The cultural connotation absent in Chinese enterprise brands brings about the deletion of the related perceptual values. As a result, Chinese enterprise brands are difficult to fulfil the requirements of the market environment with a high maturity on the connotation and value of brand. The studies conducted by the foreign market investigation companies also have suggested this fact that "by now, similar to Japanese and Korean brands of getting into the European and American markets, "China concept" is still a cheap, low quality impression left on the people of the developed areas and can't provide a powerful value support from the perspective of cultures for Chinese enterprise brands".

5.3.3 Chinese Brands Lack the Ability to Surpass the Geography Cultural Boundary

The investigation and research, which was implemented by Interbrand about the impression on Chinese brands in 2005, can suggest that there was a common belief that the brand experts in all countries thought Chinese brands were more inferior in the aspects of credit, reliability and overall value (key factors) that the leading brands of other countries. At the same time, in the aspect of the brand recognition, there is still a difference between China and the developed countries, which can be reflected from two aspects: one is that cognitive rate of Chinese character brand name is low in the international business stage, and the other is that Chinese character brand name is difficult to be pronounced due to the difference between the pronunciation way and the European and American language family if they are letters. Take HAIER for example: HAIER is regarded as a brand of German at many places as the spelled letters are like German.

5.4 Selecting a Suitable Cross-Cultural Management Model and Promoting the Internationalization Process of Chinese Brands

5.4.1 Holding a Correct Attitude Towards the Influence of Cultural Differences on Brand Internationalization

Each nation has its own unique cultural traditions, customs as well as habits. The national characteristics of a brand culture get increasingly more obvious, and this brand can attract those consumers who like to pursue the fresh senses and the foreign emotional appeals. In the world, China is the only country which owns a continuous civilization history for more than 5,000 of years.

For the above reason, it is necessary for Chinese enterprises to make effort to the exploration on this precious treasure. In other words, it is necessary for Chinese enterprises to reflect traditional Chinese culture in the product name, package as well as the advertisement and promotion.

As a result, it is necessary for Chinese enterprises to pay attention to the tolerance of the brands at the same time. Only the brands possess a cultural permeation, the energy of the values can be released in a real sense.

5.4.2 Selecting a Correct Attitude Towards the Influence of Cultural Differences on Brand Internationalization

The process of the brand of a Chinese enterprise of getting to internationalization is just in the beginning stage. The recognition from the people in the world culture on the internationalization process of the brand of a Chinese enterprise needs to make every effort for a long time as well as the sufficient patience. In the recent 10 years, the international capital input which is under the leadership of the transnational companies receives a new connotation, which refers to the "integration input". In other words, the comprehensive input including concept, standard, technology, system, operation mechanism, management as well as human resources has transformed into a major input form for the international capital input, and is the input of brand in essence. This is a capital input type which is of much greater strategic significance, and is a properly dynamic combination of the internationalization, diversification as well as localization.

Under such a background, it is necessary for the process of the brand of a Chinese enterprise of getting to internationalization to select a "follow the easier part first" permeation-oriented road with a gradual step wisely. For example, Chinese enterprises can select the countries or areas with the relatively small cultural differences to push forward their brand permeation, and subsequently make an expansion on the range of the internationalization of the brands further.

At the same time, Chinese enterprise can take full advantage of the cloning model to push forward the cultural permeation model in the relatively small cultural differences with a slow step. However, in the developed counties or areas with the relatively large cultural differences, it is necessary for Chinese enterprise to select the cultural permeation model and then push forward the unique national cultural model with a gradual step.

5.4.3 Cultivating the Ability of Brand to Surpass the Geography Cultural Boundary

It is necessary for the cultivation on the ability of brand to surpass the geography cultural boundary to be implemented in an all-round way. First of all, it is necessary to make the internationalization of the brand positioning realized. Secondly, it is necessary to build up the concept of the internationalization of the brand design. Furthermore, it is necessary to promote the names of Chinese enterprise brands to get into the internationalization, for the purpose of shortening the distance from the language with the consumers in the foreign countries and therefore taking advantage of the brand transliteration to name it. Third, it is necessary to open up the ideas of the internationalization of a brand strategic planning.

Therefore, it is necessary for Chinese enterprises to fully recognize the local cultures and then make a brand value concept planning, a brand image planning and as well as the brand advertisement planning. In other words, it is necessary to reveal the essence of the brands and ensure the products shown in an all-round way when getting an understanding of the local consumers' key ideologies about the product brands, and also to give expression to the personalities of the products as well as the values contained within the brands, and to highlight the personalized characteristics of the brands.

5.5 Conclusion

As is known to all, it can be known that the brand of an enterprise to get into the internationalization is not only a kind of the economic activities and also is a kind of the cultural behaviors. For this reason, Chinese enterprises only get a real understanding of the influences exerted from the differences in cultures and then push forward the process of the brand of a Chinese enterprise to get into the internationalization by phases and also make a right choice on the cross-cultural management model in the process of the brand internationalization, they will have the ability to step over the barriers from the cultures.

As a result, Chinese enterprise brands will have the ability to create a great number of the highly powerfully global brand in its true sense.

References

1. Yu M et al (2005) Brand communication vol 01. Shanghai Jiaotong University Press, Shanghai, pp 54–62
2. Taylor (2007) Translated by Wang Tiantian vol 05. Google brand strategy. CITIC Publishing House, Britain, pp 05–14
3. Zhu L (2006) Research on the brand strategic cultures, vol 8(02). The Publishing House of Economic Science, pp 67–73
4. Feng G, Shengliang L (2007) Ten problems in Chinese brands, vol 02. China Market Press, China, pp 623–628
5. Yang M (2006) Top international brands-the brand cultural strategy of the luxury products transnational corporations, vol 12. Press of Shanghai University of Finance and Economics, Shanghai, pp 7–13
6. Wei F (2001) Brand internationalization-model selection and measurement. J Tianjin Univ Commer 01(5):73–78
7. Qiao C (2005) Brand cultures, vol 10. Zhongshan University Press, Zhongshan, pp 371–378
8. Zhonghe H (2003) Brand internationalization strategy, vol 09. Fudan University Press, Fudan, pp 34–42
9. Braun (2009) Oversea merger and acquisition-are you ready? vol 10. Sino Foreign Management, Britain, pp 478–484

Chapter 6
Research on Banquet Etiquette of China and America

Shanshan Hu, Yan Zhang, Bingshen Gao and Zhenkai Xie

Abstract With China's development with other countries, no matter inland or on board, it is usual for the Chinese to appear in a banquet. When people take part in a banquet, the decent etiquette can establish people's confidence; what is more, it also can help many friends. Decent etiquette can cultivate people's attainment in work and daily life, especially to ladies, if they project themselves decently, then it may make them look more charming. It is known that the banquet etiquette of modern society assimilates and accede the fruit of the ancient. Some are enacted by people through the social action, and the others are behaviors that come from the history. Both of them have been approbated by people. To comply with the etiquette is a kind of respect to others as well as to people themselves. This paper will stress on introducing and analyzing the differences between China and America in banquet etiquette. After writing this essay, the author has improved the refinement, and also understands that how important the decent etiquette is to a Chinese. Besides, this paper can also offer help to those who plan to take a job concerning foreign affairs.

Keywords Banquet • Etiquette • Comparison • Differences

6.1 Introduction

Since the policy of reformation and opening has been brought into effect in China, this country has more chances to contact with other countries in the area of politics and finance, especially with the western countries. Because of the diversity of

---

S. Hu (✉) · Y. Zhang
Department of Fundamental Courses, Xuzhou AirForce College,
XuZhou 221000, China
e-mail: ss874huw@126.com

B. Gao
China People's Liberation Army Corps of Engineers Command College,
XuZhou 221000, China

Z. Xie
Department of AirForce Finance, Xuzhou AirForce College,
XuZhou 221000, China

Z. Zhong (ed.), *Proceedings of the International Conference on Information Engineering and Applications (IEA) 2012*, Lecture Notes in Electrical Engineering 220,
DOI: 10.1007/978-1-4471-4844-9_6,

background and history, there are a lot of differences among countries, especially between China and America. As a state of ceremonies, China has the fine tradition which is to treat others politely. China has joined the WTO for several years, and it has a more close relationship with other countries. More often than not, this kind of relationship is shown in a banquet. A banquet plays an important role in the social association, and then the etiquette and behaviors in the banquet become crucial.

Since all the countries have different history and culture, they often behave differently when communicating with others. If people do not understand this point, it is easy to make mistakes and become the laughingstock of other countries. For example, in the late Qing dynasty, China closed the country to international intercourse, and it knew little about the west. In a western banquette, a Chinese diplomatic envoy cleaned the knife and the fork with the dinner cloth, just doing it as inside China. He did not know that action meant the knife and fork were not clean, and it was extremely impolite in western countries. When looking at this, the host ordered the waiters to clean all of the table ware, which had made the Chinese envoy embarrassed. There is another example about the banquet etiquette behaviors. Li Hongzhang was invited to a banquet once. Because he did not understand the western banquet etiquette, he drank the water which was used to wash hands after having eaten up the fruits. The foreigners did not understand why the Chinese did so. But in order to avoid the discomfiture, others also drunk the water as the Chinese did. From the situation listed above, it can be concluded that different behaviors of banquet etiquette can lead to huge diversities.

It is known that the Chinese culture and the American culture both come into being and develop under specific condition. They exist with reason, and there is not the problem about which one is best and which one is worse. Therefore, it is necessary to do research on the etiquette and behaviors between China and America. In other words, if the Chinese know little or nothing about the America etiquette and behaviors, they may be in an unfavorable position internationally, no matter in politic or in finance.

This paper will stress on introducing and analyzing the difference between Chinese and American banquet etiquette from the aspect of present in time orientation, seat arrangement, and different tableware. All these will be effective to deal with the Chinese and American banquets.

6.2 A Comparison on Banquet Etiquette Between China and America

6.2.1 Present in Time

Punctuality seems to be an universal concept to everyone. However, just like other issues in cross-cultural communication, the idea and use of time in different countries are different. In other words, cultures vary widely in their conception of time. Even within the same culture, the conception can vary with different activities. Hall, who is well known for his discussion of time across cultures, proposes that cultures organize time in one of two ways: either monochromic (M-time) or polychromic (P-time) [1].

He also points out that M-time is the characteristic of people from English speaking countries, while P-time is the characteristic of people from Asia [2].

According to Hall, P-time (Lateness) cultures deal with time holistically [3]. People in such kind of cultures do not emphasize scheduling by separating time into discrete, fixed segments. They treat time as a less tangible medium so that they can interact with more than one person or do more than one thing at a time. They do not perceive appointments as iron-clad commitments, therefore, in these cultures; personal interaction and relationship development are far more important than making appointments or meeting deadlines.

Since China is a country that falls into the category of P-time, Chinese people are inclined to be a little later than what is scheduled when participating in banquets. Normally, they would be half an hour later or even longer. In order to fill in this "blank" period of time, some entertainment, such as playing cards or chatting with others, etc., is arranged, and tea and some snacks like watermelon seeds or a variety of sweets are served for those who have arrived "earlier" to "kill" the time. Both host and guest get used to that and would not interpret this kind of lateness, whether consciously or unconsciously, as the disrespect to the invitation or an impolite behavior. Sometimes, a host even deliberately set the time earlier, providing more "space" for guests' lateness.

By contrast, people in M-time (Punctuality) cultures like from English speaking countries live their lives quite differently. As a matter of fact, "People in the Western world find little in life exempt from the iron hand of M-time" [4]. Hall writes that, "People of the Western world, particularly Americans, tend to think of time as something fixed in nature, something around us and from which we can not escape; an ever-present part of the environment, just like the air we breathe" [5]. As the word monochromic implies, this approach sees time as lineal, segmented, and manageable. Time is something we must not waste, we must be doing something or we feel guilty [6].

Therefore, appointments and schedules are very important to them. Unlike Chinese people, people in M-time cultures such as Americans tend to follow precise scheduling. Once the time is set, it is rarely changed, and people should take it seriously. Usually everyone is supposed to arrive on time when attending a formal banquet or meeting appointments. Sometimes, it is also acceptable that people can be a little later, but not more than 10 min; otherwise his behavior will be regarded as inappropriate or insulting to both the host and other guests. In west, if someone is late for 20 min, he has to mumble some apologies. And if he is late for more than 30 min he is thought to be impolite or to have met some emergencies. In a word, in M-time cultures one who violates the rule of punctuality shall be punished seriously [7].

6.2.2 Seat Arrangement

Seating arrangement is an important means to show different interpretations on social status or interpersonal relations by making use of space. With regard to seating arrangement in banquet, there exist two major distinctive disparities between China and America [8]. China focused on the direct of south and north and the principle of men first, while America stresses on right and left and the principle of ladies first.

It is a common phenomenon in China that there are many four-character expressions referring to seat arrangement. For example, "mian nan cheng gu", "nan zhou guan mian" and "bei mian cheng chen" and so. These expressions are closely related to Chinese traditional culture. It is known that China is an old country with a long history of more than 5,000 years. There used to be many kingdoms, dynasties, and emperors. Whatever the kingdom or the dynasty was, during the ceremony of the royal inauguration or in the process of discussing governmental affairs with ministers, the emperor was always seated with his face toward south. On the other hand, when ministers met the emperor, their faces were all toward north. Therefore, these phenomenons were reflected in the expressions mentioned above. As time passed by, in China, the seat facing south is usually considered as the most honorable, while the seat facing north is comparatively less important. It can be drawn that south has become the symbolizing of the supreme authority, power, and position in Chinese eyes'. Then comparatively speaking the position of the word "bei" (north) is much lower. This kind of cultural phenomenon can be found even in Chinese other idioms. If the two words "nan" and "bei" coexist in the same idiom, the word "nan" (south) usually appears in front of the word "bei" (north), such as "nan qiang bei diao", "nan yuan bei zhe", "nan zheng bei zhan", "nan lai bei wang", and so on. As a matter of fact, it is just because of this reason that the word "nan" (south) has been gradually elevated into a position relatively high in Chinese people's minds, Therefore, as for dining, the seat facing south undoubtedly is reserved for those who are respectable or powerful. However, in America, the situation is quite different from that in China. The most honorable seat is on the left of the host. In ancient times, people used to hold a dagger in his right hand to assassinate his enemies sitting on his left. Only when the guest was arranged to sit on the left side of the host, was it inconvenient for an assassin to perform the action, but much easier for the host to subdue the assassin and protect the guest [9]. Therefore, the most distinguished guest was arranged to sit on the left side of the host. With the progress of human civilization, this kind of old style of assassination has long become extinct. Nowadays, when Americans arrange seat, they do no longer concern about security, but is from protect of psychological need. Influenced by this, it is widely taken in America that the seat at the right side of the host is much honorable than the one at the left side.

What's more, the phenomenon of "Men is superior to women" can be found in Chinese seat arrangement; while "lady first" can be seen in American seat arrangement. Influenced by the Chinese traditional belief that women should obey to the three obedience and four virtues, men are expected to be dominant in the society and superior to the women. In ancient China, women could hardly appear in the public and they were considered as no morality if they showed their face in the public. Thus, since the image of the female was hardly found in formal occasions, there was no need at all to think about women's seating arrangement. At present, the situation of Chinese women has been greatly improved and their positions have also been remarkably elevated. Today, women in China are getting out of the former subjection to men. Although it has already been widely accepted that their images appear in many formal banquets and the Western principle of "Ladies first" has been gradually acknowledged in China, Chinese women still prefer to sit together on many social occasions. Besides, when the principle of respecting the old contradicts the principle of "Ladies first", the choice of most Chinese people is the former rather than the latter [2].

However, in America, the principle of "Ladies first" is their traditional belief as well as an important social manner. This principle comes from Christianity and it is a dominant religion in America and other Western countries. In America, it is estimated that 86 % of the U.S. population is Christian. Therefore, the Christian tradition is prevalent in this country. In Christianity, the God was named as Jesus. And he was worshiped by the Christians. And meanwhile, his mother, Virgin Mary, is also highly worshiped and deserves much respect. The Christians consider the concept and action of respecting women as a noble virtue. In addition, during the period of twelfth and thirteenth centuries, with the expedition of the Crusaders a special class, the class of knights formed and developed. Because of the great influence of knights and their illustrious positions at that time, they set a series of social manners which were called knighthood or chivalry. These manners were used to guide their behaviors. Then the knighthood was soon spread out among the people and imitated by others as the standard of civilization, from the aristocracy to the folk. Among those chivalry manners, one of the most distinctive features was to respect women, and it has been passed down from generation to generation and still today people like to obey it. Therefore, it is not surprising at all to find the images of women in banquets since the ancient times. Thus, when the seating arrangement for the male and the female is concerned, China and America have to follow different principles.

6.2.3 Tableware Setting

Tableware in Chinese banquet has chopsticks, plate, bowl, spoon and saucer, and so on. There are several things should be paid attention to on the tableware setting. First, the wineglass is put on the upper right. The number of wineglasses should be the same as the variety of the wines. Second, the dinner cloth is fold to flower-kind and put into the water cup, or lay on the plate. When playing host to foreigners, there should be fork and knife, saucer, vinegar and oil, and so on, in addition to chopsticks. What's more, there still should be ashtray and toothpick on the table. Tableware in an American banquet has a fork, knife, spoon, plate, cup and dish, and so on. Compared with Chinese table ware, it is more difficult. For example, the knife can be divided into main course knife, fish knife, butter knife and fruit knife, and so on. And there are main course fork, fish fork, and so on. Both the knife and the fork should be put with the blade and edge inside. There are also soup spoon and tea spoon, tea cup, coffee cup, water cup and wineglass, and so on. Usually, the number of wineglasses should be the same as the variety of the wines.

The table setting for a western dinner is very dainty. The soup plate is put right in front of the guest. The forks are put on the left, and the knives are on the right. The spoons should be put on the plates. Then put the wineglass. From right to left, there is glass for hard drinks, champagne, beer, or water glass. The napkin is put into the water cup, or lay on the plate. The plates for bread and butter are put on the top left corner.

6.3 Conclusions

This paper stresses on introducing and analyzing the difference between China and America on the banquet etiquette. In this paper, the author introduces the etiquette during dinning, including present on time, seat arrangement, and tableware setting, and so on. The author strongly hopes that this paper can help people in both China and America to understand and acquire the differences in order to make association between the two countries proceed smoothly and successfully.

References

1. Hall ET (1990) Beyond culture, vol 21. Anchor Books, New York, pp 387–396
2. Du X (1999) British cultural practices in comparisonm, vol 12. Foreign Language Teaching and Research Press, Beijing, pp 387–392
3. Bi Ji-Wan (1999) Cross-cultural non-verbal communication, vol 33. Foreign Language Teaching and Research Press, Beijing, pp 482–488
4. Bian H (2004) On the differences between Chinese and western food culture. J Nanjing Univ People: Humanities Soc Sci 4(2):78–84
5. Hou X (2002) Utility foreign etiquette, vol 14(3). Xi'an Jiaotong University Press, Xi'an, pp 989–993
6. Samovar LA, Porter RE (2000) Intercultural communication: a reader, vol 18, 9th edn. Wadsworth Publishing Company, USA, pp 86–93
7. Jia Y (1997) Cross-cultural communication, vol 2(3). Shanghai Foreign Language Education Press, Shanghai, pp 134–135
8. Li M (1997) Social customs and culinary culture, vol 4(3). Jinzhou Normal University, Jinzhou, pp 37–43
9. Neuliep JW (2000) Intercultural communication: a context approach, vol 43. Houghton Mifflin Company, Boston, pp 488–494

Chapter 7
New Scheme of Ecological Color Enactment

Ying Huang

Abstract The definition of ecological mode has broken through the traditional definition of "creation". In the traditional manner, creation refers to "a sheer fabrication out of nothing". Something new is produced where nothing should be. The definition of creation is made new from a great amount of subjects, such as hermeneutics, structure, and Taoist philosophy, and so on. The hermeneutics, structure, Taoist philosophy hold the idea that creation can only come from all kinds of different information, life, culture, and elements conversation as well as innovation. That is to say, creation is to create something on the basis of the existing things. However, the thing that is newly created is something where nothing should be, and this is enactment. The new concept of "enactment" has greatly supported the creative education for the human beings. On the basis of these, this chapter has made explorations on the new concepts of ecological color enactment.

Keywords Ecological • Creation • Color enactment • New concepts

7.1 The Explanation of Creation

In ancient Chinese language, "creation" and "fabrication" is used in a respective manner in the first place. In "Explaining Simple and Analyzing Compound Characters", there is some explanations about the word "creation". It refers to "establish, create; knife cut" [1]. It is used in the situation that "Modal particle used at the end of a sentence, expressing assertion and affirmation. Modal particle used at the end of a question for emphasis. Modal particle used at the end of a cause-effect sentence to express explanation. Modal particle used in the middle of a sentence to express a pause". It is to cut damage in somewhere that does not have the damage before. Currently, it has the meaning of damage. In addition to this, there is another meaning. It is to cut the rice. With the knife, it cuts the rice

Y. Huang (✉)
College of Arts, Xi'an University of Architecture and Technology,
Xi'An 710000, Shaanxi, China
e-mail: huangying@cssci.info

Z. Zhong (ed.), *Proceedings of the International Conference on Information Engineering and Applications (IEA) 2012*, Lecture Notes in Electrical Engineering 220,
DOI: 10.1007/978-1-4471-4844-9_7,

that has been mature. Under these circumstances, "creation" has the meaning of harvest. As for human beings, it is to store some rice in order to satisfy the need to survive. It is to make a living out of the difficult circumstances [2].

As for "fabrication", it refers to achievement and goodbye to someone. It has the meaning of making achievements and saying goodbye to everyone. With the evolution of the language, there are some other new meanings that have been fabricated. For example, "the creation is produced at the initial stage of human society" (Han Shu). In addition to this, the "creation" as well refers to the situation that the first one that makes a production. (Tang Yun). Both the Han Shu and the Tang Yun have connected the language meaning of "creation" with the language meaning of "fabrication". In this way, creation has started to be equipped with the connotation of "beginning". Fan hua is the first one in history who connects the meaning of "creation" and "fabrication" and put them into practice. He is a historian in the period of "Southern Dynasty". "The Romance of the Three Kingdoms", one of the most popular classic novels in China, gives a most vivid description of the struggles of this period. The meaning of creation is consistent with the meaning of creation in today. The creation refers to the fabrication or establishment of something that is not there in the previous days.

In the western countries, the word "creation" appropriately first appeared at the ancient Rome age. However, the word "creation" is only used for the fabrication and creation of the city at that period of time. The meaning of "creation" is not as widely used as today. It merely talks about the construction of a city. However, as for the word "creation" that has been popular in the western countries for more than 1,000 years, it refers to the creation of God. The creation has nothing to do with the human beings. Creation is to make something out of nothing, just like the God created the world. Therefore, the word "create" has the following meanings in the "Merriam-Webster Collegiate Dictionary". The explanation of "create" in the "Merriam-Webster Collegiate Dictionary" refers to "the production of something so that the production of that particular thing newly made in the world". At the same time, it is equipped with the meaning that "to make out of nothing". In the later period, the word "creation" is widely used in the western countries. The word "creation" becomes popular. Every human being is equipped with the ability to create without any exception. Nowadays, the human beings start to talk about "enactment" because of the enlightenment of western hermeneutics, structuralistic philosophy, and Chinese Taoist philosophy and Chinese and modern bionomics.

7.2 The Application of Enactment Concept in Artistic Teaching

The new concept of "enactment" has broken through the sense of mystery of creation. In particular, the new concept of "enactment" has served as a forceful support to the creative development of the students in the self potential meaning. In addition to this, the new concept of "enactment" is a great support to the cultivation, and magnification as well as creative education to the students in the level of self potential. Of course, the creative teaching has as well involved a great amount of other problems. For example, the creative teaching has involved such factors as the creation composite factor, the creation level, and the creative thinking. The creative teaching has not only involved such factors as the creation composite factor, the creation level, and the creative thinking, but also been connected with the divergence thinking and concentrated thinking in the creative operations. In addition to this, the creative teaching has something to do with the imaginal thinking and logistic thinking, the intuition thinking and analytical thinking, the conscious thinking and the subconscious thinking as well. However, as for the teachers and the students, they can put the creativity and innovative spirit that we usually mentioned into the practice of teaching only when the teachers and the students have broken through the traditional concepts in the first place and get to know that the creation of human beings can merely come from all kinds of different information, all kinds of different lives, and all kinds of different culture as well as all kinds of different factors and so on. The creation of human beings can only come from the conversations, integration and enlightenment, and so on. However, the creation does not come out of the void. Only after the teachers and the students get to realize this particular point can they put the creativity and creative spirit into the practice of teaching and thus put them into the real life.

Art comes from life. Color acts the same. It is as well originated from life. Colors exist everywhere. In the social life of human beings, color has its way everywhere and nowhere. In addition to this, colors exist in food, clothing, shelter, and means of traveling—the four basic needs of everybody. We usually describe life as the myriads of changes, being infinite in variety, being rich and colorful,being full of sound and color,as well as being vivid and dramatic and of great impression. All of these above descriptions can never be separated from colors.

7.3 The Enactment of Color Ecological Art Teaching

Color enables people to have the emotional connection in an easiest manner. It is able to greatly affect the emotional effects and mental states of the human beings to a very great extent. The decoration sense of beauty has become a kind of enjoyment in the spirit and material life of human beings with a growing increase. On the one hand, human beings have opinions and interests on the colors of their own. On the other hand, people are easily affected by the fashions. They keen to imitate the trends of the fashions. Under these circumstances, the popularities of the colors are produced considering the imitation of fashions. We are able to say for sure that life can never be separated from colors while the innovation of designing colors should come from life more significantly.

Color design and color painting have similarities as well differences.Similarly they lie in the situation that both belong to the manifestation methods of vision art;both are a kind of visible art language; and are able to offer enjoyment by human beings. At the same time, they enable human beings to be equipped with the sense of happiness. However, the differences between the color design and the color painting are shown in the following.

7.3.1 The Observation Methods are Different

The painting color pays special attention to the inherent color, light source color, and environment color as well as the interactive relationship of the visual target. The characteristics of the painting color are based on the color of the objects. As for the color composition, it is outside of the natural color of the objects. The color composition is the exploration on the pure color. The color composition pays special attention to the interactive relationship and the rules between the colors and the colors. The characteristics of the color composition are based on the color itself.

7.3.2 The Manifestations are Different

The painting color is able to show the content of color only by connecting with the objective sharps. The content that is being shown is always specific and true. The color composition has the independent manifestation meaning. The color composition does not rely on the subjective form. Instead of this feature, the color composition expresses the emotions through the precise selection and composition of colors. The content that the color composition expresses is objective, abstract, and has multiple meanings as well as ambiguity.

7.3.3 The Art Styles are Different

As for the painting color, it is built on the foundation of visible target. The painting color has the style of realism. The color image for the color painting is vivid and very realistic. As for color composition, it is built on the foundation of conceptual image. The color composition has the decoration style of simplicity and exaggeration.

7.3.4 The Color Functions are Different

As for the color painting, it belongs to the art of vision. The color painting has a great amount of functions. The functions of the color painting are shown as the three followings: the cognition function, the education function, as well as the esthetic function. The color painting pays more attention to the power of spirit. As for the color composition, it belongs to the works of applied arts. The color composition is not only equipped with the general functions of the art. In addition to the general functions of the color composition, it has a great amount of other functions. For example, the color composition is connected with the materials, and the art technique as well as the products. Moreover, the color composition has the practical value and the economic value.

Color has played a very important role in artistic technique design as the creation activity. In addition to its significant role in the industrial art design, it is one of the major content in designs, which is the same to a great amount of the other factors such as modeling, decorative pattern, and material as well as processing, and so on. The color has accounted a very significant place. There are colors when there are any objects. In the design itself, it has included the color design. His name is enough to strike terror in people's hearts. Colors forestall one's opponent (competitor) by a show of strength. Colors overawe others by displaying one's strength. "70 % colors, 30 % flowers". For example, such things as the fiber pattern design (the textiles pattern design), including sill, carpet, fabric, and so on have looked like flowers in the near side while look like color from a far distance. The importance of color design has been emphasized. When entering a living room with complete designs, the first thing that catches the eyes is the comfortable color tunes. In the similar case, the design of the colors and the decorations play a decisive role. One's move greatly affects the general condition. They have carried a big weight.

The interests in colors have resulted in the esthetic awareness to the colors of human beings. The esthetic awareness is the premise factors for decorations and the beauty of life. Color, effect produced on the eye and its associated nerves by light waves of different wavelength or frequency. Light transmitted from an object to the eye stimulates the different color cones of the retina, thus making possible perception of various colors in the object. Color is able to make human beings to decorate and beautify life. However, the first thing should be equipped with the esthetic awareness. That is to say, how to make advantage of colors in accurate manner and design the colors that meet the rules and specifications. In addition, the colors should be beautiful and be able to be widely accepted by the human beings. The colors should be practical. In addition to these, the colors should be able to play the functions of decorating and beautifying the lives. We are able to draw the conclusion as the followings: it is the fact that we pay special attention to the dual effects of color designs that the practical performance and the esthetic feature are able to be unified. In this case, the double requirements of material and spiritual life can be satisfied.

Color is a property of light that depends on wavelength. When light falls on an object, some of it is absorbed and some is reflected. The apparent color of an opaque object depends on the wavelength of the light that it reflects; e.g., a red object observed in daylight appears red because it reflects only the waves producing red light. The color of a transparent object is determined by the wavelength of the light transmitted by it. An opaque object that reflects all wavelengths appears white; one that absorbs all wavelengths appears black. Black and white is not generally considered true colors; black is said to result from the absence of color, and white from the presence of all colors mixed together.

The connotation of color design has revealed the following arguments to us. Practical and esthetic feature is an inseparable entity that interacts to each other in the design structure. The multiple effects and targets of the color have constructed its application value to the human beings. It has embodied the basic motivations of design. As for the creation of beauty, it has added the existence meaning of practical value as well. The esthetic effects enable the users to have the sense of happiness. In this way, the users will be easier to get to accept the products from the mental part.

The dual characteristics in the essence meaning, the practical and the esthetic feature, have as well become a kind of standard and criterion for the color design advantages and disadvantages. In addition, it has formulated the color creation thinking and design manner so as to restrain the human beings. The decorations of colors include the packing decoration, the book decoration and the decoration arts, and so on. All of these are to express the particular theme image making advantage of the colors. Processing to the next step, it is such things as the furniture, lamps and lanterns, mechanism products, communication tools and environmental design, and so on. Colors should meet special function requirements in these designs. Next, it is to enjoy the vision pleasure. Therefore, the design of color should start from the practical functions, it should have the characteristics. The design colors have the adjustment, symbolic, specialty, identification, and protection. It is not difficult to find out that the functions and the values in the design have exceeded the esthetic area of general arts. "Application" and "beauty" are mutually dependent, and mutual influenced and mutual affected.

7.4 Conclusions

Color design should be depended on the very deep objective ideals of the designers. Only in this way can the sense of beauty for the colors of the products be felt. Only in this way can the sense of beauty be getting rid of the subjective restrictions. Proceeding to the next step, the creativity of the human beings, of the designers should be played in a very free manner when the colors are designed. The designers should make the sensible color and the reasonable color be raised to a higher level and get mutual integration. Creation is to create something on the basis of the existing things. At the same time, the objective vision color beauty should be paid special attention to so as to make a better pursuit. Under these circumstances, catch the beauty of beauty with the creative thinking. It is an important act for the designers to show the beauty of color in their creation process.

References

1. Wu L (2001) On educational value of children literature, vol 1(21). Nanjing University of Arts, Nanjing, pp: 2–9
2. Cui P (2010) Space design foundation teaching research. Central Acad Fine Art 8:96–104

Chapter 8
Research on Rural Development in Henan

Yulian Wang

Abstract The rural development is of great importance for the country. It is to propose several opinions for Henan rural development stabilize national development nature, enhance peasant's confidence to construct new rural area and punish corruption phenomenon through analyzing Henan rural development status and its existing barrier, all of which will provide good development condition for rural development in our country.

Keywords Henan • Rural area • Development strategy • Research

8.1 Introduction

With the rapid development of the economy and technology in our country, the pace of city and countryside integration construction is also quickly advancing [1, 2].

The State of Council is calling for that the whole country carries out the development strategy of industry the industry feeding back the agriculture when the industry develops to certain advanced extend, which also provides good conditions for rapid rural development [3, 4].

As the central plain province of our country, Henan has very important position [5, 6]. As we all known, it enjoys the good reputation of "State's Grain Store", proving that Henan's rural development is relatively fast, especially with vast growing area of wheat, therefore, it is also the strategic province of the state's wheat reserve [7].

Seen from this, the rural area in Henan is very wide and rural development appears lagged status comparing with some advanced industrial provinces.

However, this aspect also proves the great potential of Henan rural development [8]. As a great agricultural province, Henan rural development has certain strategic significance and its development is also a typical case of rural development of the country.

Y. Wang (✉)
Nanyang Institute of Technology, Nanyang 473000, China
e-mail: ylwang873s@126.com

Z. Zhong (ed.), *Proceedings of the International Conference on Information Engineering and Applications (IEA) 2012*, Lecture Notes in Electrical Engineering 220,
DOI: 10.1007/978-1-4471-4844-9_8,

At present, increasingly more people pay attention to agriculture as well as to their livelihood, and also a great number of scholars launch the innovation upsurge of Henan rural agricultural development. Through the analyses on the multiple factors as well as restriction factors in some aspects, these scholars attain a great number of the research experiences and working results on Henan rural agricultural development.

These provide clear directions for the powerful development of Henan rural agricultural development in the future, and hence ensure the agricultural development with high efficiency but not towards the crooked road.

8.2 Henan Rural Development Status and Existing Barrier

Located in the central place of our country, Henan's development has certain strategy, including the West Development and the East aiding the Western areas and other projects that the country has always persisted in implementation, Henan province plays great dominant force and guarantees smooth completion of various national indexes and tasks.

Henan had laid the foundation of agricultural province long before that it has vast plain with Yellow River passing this place, basically providing condition for Henan's agriculture. With continuous development and progress in our country, Henan still assumes the responsibility of important grain growing area, especially the growing of wheat.

The agricultural plant in Henan province is mainly wheat. However, due to working outside of large amount of workforce, bad growing quality and lagged technology, the volume is also subject to restriction. With the continuous development of machinery industry, many rural crop planting modes become fixed planting and form the phenomenon depending on the nature, which is also one factor to influence rural development.

Henan rural development is relatively lagged, including the officer's appointment mode, leader's decision direction and the public's enthusiasm, etc. Due to many factors such as self-interest or wrong decision of various rural leaders, it leads to many peasants not to be able to receive good guidance and explicit advance direction and not to have detailed strategic development outline.

It is a common phenomenon to live day to day. In addition, there is error or disadvantage in introducing agricultural technician in Heman rural area that many agricultural technicians would not like to develop in rural area after seeing the lagged rural development, which is also an important reason leading to shortage of agricultural talents in rural area, thus resulting in development of a series of subsequent questions that the rural area is not able to achieve the promotion of new efficient plant mode and many new agricultural technologies could not be applied so that the rural area has always been in original plant mode and low operation efficiency.

The official corruption may be the biggest obstacle for Henan rural development. The officer should be the model for the public to learn from, have certain popular trust and be able to guide correct advance and development direction for the mass.

However, certain rural officers in Henan give up correct decision direction just for tiny interest. For example, for the land transfer phenomenon present in the rural area, many fundamental leaders bewitch peasants to sell the land and transfer the land ownership to developers to earn certain bonus rather than adopt relevant measures to develop rural economy.

For personal benefit, they transfer all rural lands to developers to implement development projects. At this time, peasant loses the land that they live by. Since many peasants have low education level, they basically lose the retreat route after losing land. They are only dependent on several thousand yuan earned by selling the land; they may be faced with hungry after running up.

This is the impudent phenomenon existed in individual rural areas in Henan. This behavior should be distained by people and it is also an important factor to affect Henan rural development.

8.3 Several Research for Henan Rural Development Strategy

The report in the Party's 17th Congress stressed that it was necessary to strengthen the development balance and make every effort to implement the healthy and quick economic development. This is the general principle of China's economic development and also is the overall idea of China's rural development. From the early 2006 when the socialist new rural construction task, a totally new rural development idea of Chinese government gradually emerged, by which the rural areas will move to the road of the economic development with high quality and quick speed. Agricultural production will be highly industrialized; the dual economic structure will be gradually broken. Henan, as the biggest province of central China, is always one of the main provinces of China to supply the primary agricultural products. Also, the problems in Agriculture, rural areas and peasants will be the barriers for Henan to develop economy in the futures.

8.3.1 Grasp National Development Nature and Stabilize Confidence of Peasants

The national development is at the critical stage while rural development is also one aspect that the country pays close attention to. Among that, the industry feeding back the agriculture does not means that the agriculture makes way for industry to let industry develop casually on their own land thus causing great pollution to the local land.

We are trying to master relevant national law and regulation and the development nature proposed by the Central government and firmly grasp the essence to provide more opportunities and development space for peasant's development.

The agriculture could develop industry but the precondition is to make peasants get maximum interest and a great deal of employment opportunities, guarantee that the living quality of the people could be improved along with the development rather than the so-called land transaction resulting in aggravating their pain to lose the land. Comparing with the land, the money is not omnipotent but only land is the foundation for peasant and rural development.

Henan rural development situation is very optimistic and relevant departments and leaders should pay attention to specific rural development matters proposed by the country.

According to the relevant articles, they should make specific analysis based on the specific questions and propose their own detailed development plan and step.

At the same time, the authority departments of rural development should post and advertise detailed development strategy plan in advance to make joint development of the whole people thus not to take a roundabout course and develop the target with strong force and also make the mass really participate in rural construction that all people make concerted effort to build the hometown and create beautiful future.

Only wide mass have sufficient confidence could rural development plan get smooth implementation and achieve estimated target, so to grasp national development nature and stabilize confidence of peasants is the belief that we shall never give up that they constantly build and develop rural area, laying good foundation for national development.

8.3.2 Punish Rural Official Corruption and Introduce Hi-Tech Agricultural Talents

Since China has always been an official standard country, it is imperative to punish the corruption, especially since the country proposes rural development deployment, the rural development capital will become more and more and the expense and direction of the capital need a account record.

In Henan rural development process, the corrupted officers are ready to go and many are against the relevant national laws.

As the central authority, it has certain difficulty for the management. It must comply with the policy of "the Central leading the local" to severely punish rural corruption officers. The source of most official corruption is the mutual collusion between the town and countryside leaders to promote the relevant national development policy to change the nature when reaching to the local place.

If the cultivation household could get RMB 10,000 of the national development fund to develop their own plant, but the local government only allocates them RMB 3,000. Just because this amount difference, it leads the cultivation household not to develop cultivation course with all their strength and miss good development opportunity thus result in continuous lagged development of rural economy.

Therefore, it is imperative to punish rural officers and moreover the means must be severe. The national central authority could specially send a special fund management group for rural development fund to real-time monitor detailed direction of the capital and directly report to the Central once finding the problem. It is also good for the country to master the flow direction of the fund, thus not to leave chance for greedy official and guarantee full role of special rural development fund.

At present, there is also a serious problem for Henan rural development, which is not able to introduce hi-tech agricultural talents, resulting in not promoting of new rural development mode.

Many knowledge youths after university graduation would not like to develop in rural area that they would like to do other job rather than do their own work in rural area, which is the common phenomenon in the society.

The countryside must show an appearance to develop new times to introduce hi-tech agricultural talents. For example, the countryside could provide powerful materials basis and convenient learning condition for external technology talents to comprehensively achieve learning for use concept.

At the same time, the country should give great help in distribution of technology talents and make talent sending plan before students enroll to the university. West talent plan is an important national decision that to develop the Western region must rely on hi-tech talents.

8.4 Summary and Conclusion

Agriculture is the lifeblood of the world economy. And early civilization of the human society development was mainly the agricultural civilization.

Agriculture not only provided food and clothing resources for the survival of human beings, and also provided the basic material support and protection for the extension and development of human beings.

The national development has closed relation with rural development that rural development lays foundation for comprehensive national development but also provides development condition and opportunity for rural development.

Henan rural development should transfer all sights to three modules of the execution of relevant policies, punishment of corruption phenomenon and hi-tech talent introduction. The policy execution is the only way that the national development is passed.

If without a good policy regulation system and nothing can be accomplished without norms or standards, the development situation will be in a state of disunity. The punishment of corruption is the principle that the State has always insisted.

For the people's livelihood and increase of peasant's living standard, the corruption phenomenon must be removed. The introduction of hi-tech talent is the precondition to determine rural development direction and hi-tech agricultural plant.

Since wide peasants have limited knowledge, those hi-tech agricultural talents are needed to input in the rural construction to bring technology for peasant friend, accurately position agricultural development direction and guarantee efficient and strong rural development.

References

1. Weihua A (2002) Israeli political parties: the separation, reunion and the dividing line. West Asia Afr 4(4):35–37
2. Ruidong Y (1997) Analysis on the politics in Israel, vol 90. Northwestern University Press, Xi'an, pp 87–97
3. Xu Y (2004) Conflict between the religions and the secular political powers in the course of the Israel modernization. Middle East Stud 4(1):42–47
4. Yin G (1996) Analysis on the election of Israel and the prospect. West Asia Afr 4(6):18–19
5. Huang A (2006) Ponderation over development of rural sport under construction of new countryside. J Shanghai Univ Sport 5(06):582–588
6. Ren B, Zhang B (2007) Research on building the socialist new countryside and constructing the harmonious society developing rural sports culture. J Sports Sci 87(01):554–568
7. Zhang G (2007) Building and Consummating the Henan new rural construction policy protection system. Tribune Study 4(09):874–885
8. Wang Z, Qinghua D, Zhou Y (2007) On sports culture construction in China socialism countryside. Sports Cult Guide 4(03):719–726

Chapter 9
Research on Taoism Clean Government Culture

Tao Li, Han-jun Tu and Rong Cheng

Abstract Taoism clean government culture refers to the clean government culture in Taoism thoughts and Taoism culture. The Taoism clean government culture is a kind of new concept and culture that abstracts the honest and clean government in Taoism. "The true object of the Dao is the regulation of the person. Quite subordinate to this is its use in the management of the state and the clan; while the government of the kingdom is but the dust and refuse of it" the core connotation and basic value of Taoism clean government culture lie in the cultivation of one's soul and morality. The cultivation of one's soul and morality is not only the way of running the state, but also the basis of running the state. The leaders should actively learn from the Taoism clean government culture so as to learn to make the state better.

Keywords Taoism clean government culture • Cultivate one's morality

9.1 Introduction

Taoism clean government culture refers to the clean and honest culture in Taoism thoughts and culture. It has contained such contents as ideological and moral cultivation, incorruptible employment, politics and honest education, and so on [1]. "I will do nothing (of purpose), and the people will be transformed of themselves; I will be fond of keeping still, and the people will of themselves become correct. I will take no trouble about it, and the people will of themselves become rich; I will manifest no ambition, and the people will of themselves attain to the primitive simplicity" [2]. "The Perfect man has no (thought of) self; the Spirit-like man, none of merit; the Sagely-minded man, none of fame" [3]. In Taoism, the cultivation of soul and morality is considered as both the way of running the state and the basis of running the state. Therefore, it has a very important profound meaning for the modern clean and honest

T. Li (✉)
School of Public Health Wuhan University, Wuhan 430000, China
e-mail: tao30cli@126.com

T. Li · H. Tu · R. Cheng
Hubei University of Medicine, Shiyan 442000, China

Z. Zhong (ed.), *Proceedings of the International Conference on Information Engineering and Applications (IEA) 2012*, Lecture Notes in Electrical Engineering 220, DOI: 10.1007/978-1-4471-4844-9_9, © Springer-Verlag London 2013

government construction to dig out the essence of the cultivation of one's morality and the basic methods and attitude in the Taoism clean government culture.

9.2 The Connotation and Value of the Taoism Clean Government Culture

As for the Taoism clean government culture, its connotation is the social thoughts that are embodied through the closed combination of modern clean and honest government culture and the traditional Taoism culture; while its value is embodied in the brand new vitality shown by the ancient Taoism culture when it is in the modern clean government construction.

9.2.1 The Major Thoughts of the Taoism Clean Government Culture

Taoism thoughts are an important composite part in the Chinese traditional culture. In the Taoism thoughts, there are a great amount of rich ideological connotations that are related to clean and honest government construction. In the universe there are four that are great, and the (sage) king is one of them. Man takes his law from the Earth; the Earth takes its law from Heaven; Heaven takes its law from the Dao. The law of the Dao is being what it is. It is that the Dao produces (all things), nourishes them, brings them to their full growth, nurses them, completes them, matures them, maintains them, and overspreads them.

It produces them and makes no claim to the possession of them; it carries them through their processes and does not vaunt its ability in doing so; it brings them to maturity and exercises no control over them—this is called its mysterious operation. Keep life and lose those other things. Who cleaves to fame, rejects what is more great. Who loves large stores, gives up the richer state; who is content, needs fear no shame. Who knows to stop, incurs no blame. From danger free, long live shall he [4]. These ideological connotations require people to act according to the rules of "Taoism". In this way, people can improve their own morality.

9.2.2 The Core Connotation of Taoism Clean Government Culture

In Taoism thoughts, "Tao" is regarded as the Mother of all things. When one is about to take an inspiration, he is sure to make a (previous) expiration; when he is going to weaken another, he will first strengthen him; when he is going to overthrow another, he will first have raised him up. The soft overcomes the hard; and the weak the strong [5]. It insists in that the Dao in its regular course does nothing (for the sake of doing it), and so there is nothing which it does not do. "Tao Te Ching" points out that "[The cultivation (of the Dao), and the observation (of its effects)] What (Dao's) skilful planter plants can never be up torn; what his skilful arms enfold, from him can never be borne. Sons shall bring in lengthening line, sacrifices to his shrine. Dao when nursed within one's self, his vigor will make true; and where the family it rules what riches will accrue! The neighborhood where it prevails in thriving will abound; and when it is seen throughout the state, good fortune will be found. Employ it the kingdom over, and men thrive all around. In this way the effect will be seen in the person, by the observation of different cases; in the family; in the neighborhood; in the state; and in the kingdom [6], How do I know that this effect is sure to hold thus all under the sky?". "Zhuangzi" has further pointed out that "men like Yan He do of a truth dislike riches and honors. Hence it is said, 'The true object of the Dao is the regulation of the person. Quite subordinate to this is its use in the management of the state and the clan; while the government of the kingdom is but the dust and refuse of it'. From this we may see that the services of the Dis and Kings are but a surplusage of the work of the sages, and do not contribute to complete the person or nourish the life. Yet the superior men of the present age will, most of them, throw away their lives for the sake of their persons, in pursuing their (material) objects—is it not cause for grief? Whenever a sage is initiating any movement, he is sure to examine the motive which influences him, and what he is about to do. Here, however, is a man, who uses a pearl like that of the marquis of Sui to shoot a bird at a distance of 10,000 feet. All men will laugh at him; and why? Because the thing which he uses is of great value, and what he wishes to get is of little. And is not life of more value than the pearl of the marquis of Sui?"

9.2.3 Basic Values of Taoism Clean Government Culture

Taoism clean government culture is a kind of new concept and new culture that abstracts the honest and clean government in Taoism.

9.2.4 The Medical Meaning of Taoism Clean Government Culture

"Yellow Emperor's Inner Canon" has pointed out that all illnesses are the result of anxiety of hearts. A person who has a lot of things to think about and suffers from a great deal of pressure will have serious illness. "The Dao in its regular course does nothing (for the sake of doing it), and so there is nothing which it does not do. If princes and kings were able to maintain it, all things would be transformed by themselves. If this transformation became to me as an object of desire, I would express the desire by the nameless simplicity. Simplicity without a name is free from all external aim. With no desire, at rest and still, all things go right as of their will".

9.3 Basic Methods and Attitude of Taoism Clean Government Culture

9.3.1 Conform to Nature

In a little state with a small population, I would so order it, that, though there were individuals with the abilities of ten or a hundred men, there should be no employment of them; I would make the people, while looking on death as a grievous thing, yet not remove elsewhere (to avoid it). Though they had boats and carriages, they should have no occasion to ride in them; though they had buff coats and sharp weapons, they should have no occasion to don or use them. I would make the people return to the use of knotted cords (instead of the written characters). They should think their (coarse) food sweet; their (plain) clothes beautiful; their (poor) dwellings places of rest; and their common (simple) ways sources of enjoyment. There should be a neighboring state within sight, and the voices of the fowls and dogs should be heard all the way from it to us, but I would make the people to old age, even to death, not have any intercourse with it. "(The manifestation of simplicity) Sincere words are not fine; fine words are not sincere. Those who are skilled (in the Dao) do not dispute (about it); the disputatious are not skilled in it. Those who know (the Dao) are not extensively learned; the extensively learned do not know it. The sage does not accumulate (for himself). The more that he expends for others, the more does he possess of his own; the more that he gives to others, the more does he have himself. With all the sharpness of the Way of Heaven, it injures not; with all the doing in the way of the sage he does not strive".

9.3.2 A Contented Mind is a Perpetual Feast

The first is that happiness consists in contentment. "The sufficiency of contentment is an enduring and unchanging sufficiency". "We may see who cleaves to fame rejects what is more great; who loves large stores gives up the richer state. Who is content needs fear no shame. Who knows to stop incurs no blame. From danger free long live shall he". The second is to get rid of temptations. The kingdom is a spirit-like thing, and cannot be got by active doing. He who would so win it destroys it; he who would hold it in his grasp loses it. The course and nature of things is such that what was in front is now behind; what warmed anon us freezing find. Strength is of weakness of the spoil; the store in ruins mocks our toil. Hence the sage puts away excessive effort, extravagance, and easy indulgence.

The third is to have a good attitude. "Keep life and lose those other things; keep them and lose your life:—which brings sorrow and pain more near?"; "Favor and disgrace would seem equally to be feared; honor and great calamity, to be regarded as personal conditions (of the same kind). What is meant by speaking thus of favor and disgrace? Disgrace is being in a low position (after the enjoyment of favor) Getting that (favor) leads to the apprehension (of losing it) and losing it leads to the fear of (still greater calamity)—this is what is meant by saying that favor and disgrace would seem equally to be feared".

9.3.3 To be Frugal

"For regulating the human (in our constitution) and rendering the (proper) service to the heavenly, there is nothing like moderation". It is only by this moderation that there is an early return effect (to man's normal state). That early return is what I call the repeated accumulation of the attributes (of the Dao). With that repeated accumulation of those attributes, there comes the subjugation (of every obstacle to such return). Of this subjugation we know not what shall be the limit; and when one knows not what the limit shall be, he may be the ruler of a state. He who possesses the mother of the state may continue long. His case is like that (of the plant) of which we say that its roots are deep and its flower stalks are firm—this is the way to secure that its enduring life shall long be seen.

9.3.4 To Calm the Soul

It is the way of the Dao to act without (thinking of) acting; to conduct affairs without (feeling the) trouble of them; to taste without discerning any flavor; to consider what is small as great, and a few as many; and to recompense injury with kindness. All difficult things in the world are sure to arise from a previous state in which they were easy, and all great things from one in which they were small. Therefore the sage, while he never does what is great, is able to accomplish the greatest things. He who lightly promises is sure to keep but little faith; he who is continually thinking things easy is sure to find them difficult. Therefore, the sage sees difficulty even in what seems easy and so never has any difficulties.

9.3.5 To Have Good Characters

First, he who knows other men is discerning; he who knows himself is intelligent. He who overcomes others is strong; he who overcomes himself is mighty. He who is satisfied with his lot is rich; he who goes on acting with energy has a (firm) will. Second, the Great man abides by what is solid, and eschews what is flimsy; dwells with the fruit and not with the flower. It is thus that he puts away the one and makes choice of the other. Third, there is nothing in the world more soft and weak than water, and yet for attacking things that are firm and strong there is nothing that can take precedence of it;—For there is nothing (so effectual) for which it can be changed. Everyone in the world knows that the soft overcomes the hard, and the weak the strong, but no one is able to carry it out in practice.

9.4 The Modern Leaders Should Maintain Cultivation Thoughts

9.4.1 Cultivate the Great Spirit

It is simply by being pained at (the thought of) having this disease that we are preserved from it. The sage has not the disease. He knows the pain that would be inseparable from it, and therefore he does not have it. I may have been not faithful; whether, in intercourse with friends, I may have been not sincere; whether I may have not mastered and practiced the instructions of my teacher. The (state of) vacancy should be brought to the utmost degree, and that of stillness guarded with unwearying vigor. All things alike go through their processes of activity, and (then) we see them return (to their original state). When things (in the vegetable world) have displayed their luxuriant growth, we see each of them return to its root. This returning to their root is what we call the state of stillness; and that stillness may be called a reporting that they have fulfilled their appointed end. The report of that fulfillment is the regular, unchanging rule. To know that unchanging rule is to be intelligent; not to know it leads to wild movements and evil issues. The knowledge of that unchanging rule produces a (grand) capacity and forbearance, and that capacity and forbearance lead to a community (of feeling with all things). From this community of feeling comes a kingly character; and he who is kinglike goes on to be heaven-like. In that likeness to heaven he possesses the Dao. Possessed of the Dao, he endures long; and to the end of his bodily life, is exempt from all danger of decay.

9.4.2 To be Kind-Hearted

The government must function for the people, show concern for the people and bring benefit to the people, "build a party for the interests of the vast majority; exercise the state power in the interest of the people".

9.4.3 *To Put the Heart into the People*

He who would assist a lord of men in harmony with the Dao will not assert his mastery in the kingdom by force of arms. Such a course is sure to meet with its proper return. Wherever a host is stationed, briars and thorns spring up. In the sequence of great armies there are sure to be bad years. A skilful (commander) strikes a decisive blow, and stops. He does not dare (by continuing his operations) to assert and complete his mastery. He will strike the blow, but will be on his guard against being vain or boastful or arrogant in consequence of it. He strikes it as a matter of necessity; he strikes it, but not from a wish for mastery, When things have attained their strong maturity they become old. This may be not in accordance with the Dao: and it soon comes to an end. Only by this way can our career achieve success.

9.4.4 *To be Indifferent to Fame or Gain*

"Harmony is most precious" should be maintained as the symbol. In the construction process of socialist harmonious society, we should maintain Taoism clean government culture and cultivate in both the physical and mental aspects.

References

1. Wan-qin H, Shi-qi Y, Ping-deng Z (2011) Clean and honest new products in Wudang. Ministry Supervision People's Republic of China 5(4):19–26
2. Huang P (2011) The "Tao Te Ching explained by Huang Pu-min", vol 3. Yuelu Publishing House, Changsha, pp 96–106
3. Yu Q (2006) Modern Chinese National Culture lecture, vol 12. China Changan Press, Xian, pp 374–378
4. Laisen A (2010) Taoism clean culture education base: Wudand Shan Taoism clean culture textbooks. China Fangzheng Press 25(2):14–19
5. Chen X, Yangshen (2010) Principle modern Chinese and western medicine magazine, vol 19(33). pp 65–67
6. Dang H (2009) On Taoism nourishing life. Heilongjiang Hist 12:2349–2350

Chapter 10
Requirement Analysis on Chinese Language of Foreign Students

Su-hong Guo and Ya-jun Yang

Abstract Needs analysis on foreign language has been applied extensively in foreign language teaching, it also provides theoretical support in needs analysis on Chinese language of foreign students in teaching Chinese as a foreign language. This article states the theories and method of needs analysis in teaching Chinese as a foreign language. It analyzes the significance of carrying out needs analysis, we can use needs analysis to identify some problems to improve teaching Chinese effectively as a foreign language.

Keywords Teaching Chinese as a foreign language • Foreign students • Needs analysis

10.1 Introduction

The objective of Needs Analysis is to provide foundation for the course setting of the adults to learn English for specific purposes. That is to say, the objective of needs analysis is to formulate a way that can be used in a special manner so as to adapt to the outline theories of English teaching. It is the ultimate goal of needs analyses [1]. The objective is drawn through the analyses of the objective environment that the language learners are located in. In addition, the objective is as well drawn according to the analyses of the language characteristics that the language learners have [2].

The educational circles in foreign countries have different explanations on the meaning of needs and the meaning of needs analysis [3]. Some work has put forward the concepts of both the Objective Needs and the Subjective Needs [4]. And other work has put forward the concepts of both the Target Needs and the Learning Needs. The work in [5] has made classifications between needs of two kinds of different levels. He considers that needs are of two kinds of different levels. One is "to

S. Guo (✉) · Y. Yang
College of International Exchange, Beijing Union University, Beijing 100101, China
e-mail: guosuhong@guigu.org

Z. Zhong (ed.), *Proceedings of the International Conference on Information Engineering and Applications (IEA) 2012*, Lecture Notes in Electrical Engineering 220, DOI: 10.1007/978-1-4471-4844-9_10,

lay a foundation" and the other is "to develop potentials" [5]. The author of [5] considers that when carrying out the needs analysis with this kind of model, that is to say, when carrying out the needs analysis with the two different kinds of needs, the method that is from bottom to top should be by any means made use of. To speak more specifically, in order to carry out the needs analysis, understanding and analysis of "laying a foundation" should be set forth in. Needs about such aspect as "to lay a foundation" should be made clear and analyzed in the first place. In a gradual process, the needs should be transited to the needs about such aspect as "to develop potential". This process is made use of in order to learn about and make analysis on the learning needs of the learners. As for this kind of process, it is considered as the "needs analysis". From the perspective of the content, needs analysis has mainly included the following two aspects: Target Needs Analysis and Learning Needs Analysis. As for the Target Needs Analysis, it is carried out at the end of the curriculum. It is used to make analysis of the goals that the students should have reached when the courses have finished. The main point for the Target Needs Analysis is that it is a kind of analysis that is carried out focusing on the personal study circumstance of the students, and their learning purpose as well as their foreign language level in the current phrase. As for the Learning Needs Analysis, the main point for this is to get to know the current language circumstance before the class begins and the lectures starts to carry out. It is used in order to investigate the current purpose and levels of the students. In addition, it is used with the purpose to get to know the learning attitude, the learning preference, and the personal wishes as well as the needs and expectations of the students and so on.

Data about all kinds of need for foreign language learners are keeping on perfecting. In addition to this, materials about all kinds of needs are as well being improved day after day. Richterich has put forward the model the needs analysis for the foreign language teaching for the first time [6]. Munby has put forward a model that makes analysis on the needs of the students after a foreign language teaching curriculum comes to an end [7]. The analysis is conducted to get information about the target situation of the students, which is called as Target Situation Analysis(TSA). Munby (1978) [7] has put forward a kind of needs analysis mode that is called TSA, and it is able to be applied as the foundation of the compilation of the outline for the target language course. Although it is the case, this kind of model cannot offer any help when the foreign language teaching activity is carried out in the actual communicating environment. This kind of model fails to be helpful for it does not base on the real communication environment. Some researchers have put forward a series scheme of the model PSA. It has become a kind of supplement of the TSA. PSA is short for Present Situation Analysis. PSA is mainly made use of to get to know the current language circumstance of the students before the foreign language curriculum begins to carry out and the courses start to operate. The PSA is able to use to investigate the advantages and disadvantages of their language learning. There are all kinds of models about the needs analysis. All of the models about the needs analysis are developed recently. They have enlarged the key area for the data collection. In the data collection that is collected by the newly developed models, it has not only included all kinds of subjective data, but also included the learning methods that the foreign language learners have preference to. In addition to this, it has as well included the content related to the following aspects: the requirements and expectations of the learners to the courses that they have been learning as well as the environment that the classroom locates and so on.

Needs analysis has included a great deal of data. It has involved the collection of the subjective information and the objective information. All of the subjective and objective information are related to the teaching activities. They belong to the category of evaluation and research. The key is to find out the distance between the current language application ability of the learners as well as their professional knowledge levels and the degree that they have been eager to reach.

10.2 Needs Analysis

Need analysis is usually carried out according to the following steps: first is to determine the investigate object; and then it is to design the questionnaires; in the third place it is to collect the materials and in the end it is to analyze the information.

Before carrying out the needs analysis, the first thing that should be done is to determine the investigate object well. At the same time, it is necessary to design the research methods for the needs analysis. As for the needs analysis for foreign languages, the most commonly used methods are the Questionnaire and the Interview. As for the method of Questionnaire, it has been mostly applied in the needs analysis of the foreign language. The method of Questionnaire is adaptable to the large-scale investigation of the needs analysis for foreign languages. However, the information reliability and the efficiency control require a great deal of rich experiences when designing the questionnaires. As for the method of the Interview, it has been divided into the following two kinds: the Individual Interview and the Group Interview. The advantages for the both of the interviews are shown as the followings: the amount of information is very large; and the flexibility of interviews is very high and so on. However, there are a lot of disadvantages as well, such as the standards for the interviews do not stay the same. In addition to this, the results of the interviews are difficult to be carried out quantitative study. Under this circumstance, the method of interview is usually combined with the other methods so as to acquire richer information.

The needs analysis of Chinese for foreign students is able to adopt the methods of questionnaire and the interview. According to the needs analysis methods, on the basis of the document analysis, try to collect the investigate items as detailed as possible relevant to the requirements of Chinese by foreign students in the first place. Proceeding to the next step, taking into account the design of questionnaires that was designed by Chuan-bin in the year 2006 [1], the relevant content of needs analysis about Chinese for foreign students should be analyzed in a mutual framework. The relevant content that should be put into the mutual framework is shown as the followings: the objective situation, the objective level, Chinese learning, the application difficulties, the learning motivation, the learning preference, and the language attitude as well as the learning objectives, and so on. In this way, they can be used to determine the main content that needs to be analyzed for the research.

As for the information reliability of the questionnaires, it refers to the consistency of the questionnaires. That is to say, when carrying out the repeat investigation, the same results would be produced. The degree should be the same. As for the efficiency of the questionnaires, it refers to the degree that the questionnaires reflect the content that is to be investigated. Researchers are able to organize a part of the foreign students to take part in the forums in the first place. In addition to this, the researchers are able to make interviews for part of the teachers. They are able to make summary of the information that is offered. Moreover, they can make analysis very carefully and revise every question and selection in the questionnaires. Proceeding to the next step, the method of Test–Retest Reliability can be used in the reliability evaluation for the questionnaires.

After the determination of the questionnaires, it comes to the collection of materials. It then comes to the summarization of the "the consistency of the themes" for the data collected from the interviews. The data analysis applies the Spss 13.0 systematic software. The data of the questionnaires is entered into the computer by the researcher. As for the data collected from the multiple choices, it applies the multiple dichotomy method so as to enter into the computer. The Frequencies analyzed methods can be used to calculate the out of every choice. In addition to this, the Frequencies analyzed methods can be as well used in the calculation of the Pct of Cases of this choice among all of the efficient questionnaires.

After the materials have been collected in a complete manner, what is more important is to make analysis on the materials that are collected. Although there are difficulties in collection and analyzing the data and information about the needs of Chinese by foreign students, it has very important meaning that is hard to imagine. Needs analysis is able to be done in different phrases for the foreign students when they are carrying out their studies. For example, before the opening up of the courses, needs analysis should be done to the target situation of the foreign students. The result can be used in setting the courses. As for the needs analysis for the foreign students in the middle of the course, make analysis on the needs for the current situation. The goal of this is to adjust the content of teaching as well as the teaching manner. After the course has finished, the needs information of the foreign students needs has thus become the foundation for whether the learning objectives have realized or not as well as whether the course setting is reasonable or not.

10.3 Meaning of Needs Analysis

Teaching Chinese as a foreign language is an important branch in the foreign language teaching. As for the core goal of teaching Chinese as a foreign language in the subject system construction, it is "how to let foreigners learn Chinese fast in a short period of time" [2, pp. 12–14]. The subject system of teaching Chinese as a foreign language is made up of by four levels. They are ontology, epistemology,

Methodology of pedagogy, and modern educational technology tool. With the guidance and adjustment of the needs analysis of Chinese language, it will be blind to talk about "what to teach", "how to learn", "how to teach" and "what educational technology and methods are used" and so on in the subject system of teaching Chinese as a foreign language. There cannot be any subject construction of the course that teaching Chinese as a foreign language to speak of, not to mention the perfection of the subject construction [1].

Sun [3, pp. 152–155] considers that needs analysis is an important factor that affects that development of the course teaching Chinese as a foreign language. Needs for Chinese have played a core status in the teaching process of teaching Chinese as a foreign language. Needs analysis has determined and affected the course setting, selection of materials, implementation of course, and teaching evaluation from multiple aspects.

In the first place, needs analysis of Chinese enables Chinese teachers to set the courses, formulated and executes the teaching outline more efficiently. As for the setting of a foreign course, the premise is to do well the needs analysis. Only by fully understanding all kinds of needs of the students during the learning process can the courses be scientifically set. In this way, teaching goals and teaching scheme that meet the practice can be formulated.

In the second place, needs analysis of Chinese pays attention to the learning procedure and process. It pays attention to the actual conditions of the students. This has important meaning for the Chinese teachers to improve their teaching methods and teaching effect. Liu [5, p. 74] makes analysis on the teaching principle that "consider students as the center and teachers as the guidance". "It sets forth in the features and requirements of the students, formulates the teaching scheme, the teaching outline and determines the teaching content, teaching materials and the teaching methods. The learning content should be real and practical and in urgent need of the students. Only by "learning in order to practice" can the learning interests of the students be improved".

10.4 Existed Problems

In the field of teaching Chinese as a foreign language, needs analysis has its own practical value. The domestic scholars have discussed the necessity to develop needs analysis from the early 1990s.

Tang [6, pp. 198–199] stresses that "practical communication needs is a big principle. We should use language theories as guidance so as to arrange the functional items".

Cheng [7, pp. 91–97] considers that "the revise of outline is practical.teh appearance of the first teaching outline symbolizes that the teaching changes from experimental types to scientific types…"

However, there are some weaknesses on the need analysis in the field of teaching Chinese as a foreign language. Till date, there hasbeen not enough and complete specific content on needs analysis. The conclusions drawn are restricted in the study results such as target situation, target level, etc.

Acknowledgments This paper is supported by "study on teaching Chinese as a foreign language with the perspective of the needs of foreign students" by Beijing Union University (number: SK201021X).

References

1. Chuan-bin N (2006) Needs analysis of Chinese as a foreign language, vol 31. Hehai University Press, Nanjing, pp 398–406
2. Lu J (1999) Opinions on the implementation of the basic research of teaching Chinese as a foreign language. Lang Appl 1(4):12–14
3. Sun D (1997) Looking back on the teaching research of teaching Chinese as a foreign language and the records of the research and discussion meeting. Lang Teach Linguist Stud 4(1):152–155
4. Wang ZH (1995) The learning objects and the exterior conditions. Chinese Learn 5(5):45–48
5. Liu X (1997) Brief discussion on the basic principles of teaching Chinese as a second language. Chin Teach World 5(1):68–77
6. Tang C (1999) About functions. Seen Liu X (ed) Brief discussion on teaching Chinese as a foreign language, vol 4(2). Beijing Language and Culture University Press, Beijing, pp 198–199
7. Cheng J (1999) What on earth the grammar outline we need. Chin Teach World 4(3):91–97

Chapter 11
Study on the Rhymed Formulas of China's National Folk Music

Kai Cui

Abstract The song-and-dance duet, as one of the most important national folk music in China, gives a full expression on a highly profound artistic philosophic theory by relying on the easy-to-understand and quite frank languages which are contained within it. Beyond all doubts, this national folk music of China is definitely a conclusion and a quintessence of the practical experiences of the artists who lived in the previous time with an in-depth, and is the essence of the art of the song-and-dance duet in northeast of China. In this paper, by conducting an interpretation and study on the rhymed formula, the artistry values as well as the esthetics values of the song-and-dance duet folk music are brought by the author to light.

Keywords Song-and-dance duet • Rhymed formula • Interpretation • Study

11.1 Introduction

The folk music art-song-and-dance duet is extremely popular in the northeast three provinces (Liaoning province, Heilong province, and Jilin province), the eastern area of northern China as well as the eastern area of Inner Mongolia autonomous province [1]. It is a music art with an artistically unique pattern, which is acted with the performance forms of festively singing and dancing, the cooperation between a man and a woman and simultaneously makes use of the sound rhymes of the dialect in northern China to demonstrate different kinds of loving stories with singing in a performance [2].

The number of the rhymed formula as well as the artistic sayings, which are utilized by all kinds of the artists of song-and-dance due, can be counted by the hundreds. All of them can be put by these artists into the practical application

K. Cui (✉)
Xinmin Normal College of Shenyang University, Xinmin City, Liaoning Province 110300, China
e-mail: cuikai@guigu.org

Z. Zhong (ed.), *Proceedings of the International Conference on Information Engineering and Applications (IEA) 2012*, Lecture Notes in Electrical Engineering 220, DOI: 10.1007/978-1-4471-4844-9_11, © Springer-Verlag London 2013

with ingeniousness. Furthermore, all of them can be said to be the most classic in the national folk music [3]. For example, people often take full advantage of the words that "it would be rather not to eat a meal, but the song-and-dance duet can never be abandoned or forgotten in life" to communicate their deep-hearted infatuation into the art of song-and-dance duet. In the mean time, people also apply the famous quotation that "The south, north, west and east areas of China are characterized respectively by dancing, singing, bantou tunes and juggling clubs", to give a detailed reflection on the characteristics of the song-and-dance duet art at eastern, western, southern and northern China. The art of song-and-dance duet, from the emergence to evolution and from singing tones to singing words, contains the classic rhymed formula everywhere. At the same time, it bears with the enriched contents of the song-and-dance duet in all aspects, and hence can present a dialectical and uniform law.

11.2 Interpretation on Rhymed Formula

11.2.1 Six Characters of Singing Song-and-Dance Duet

First of all, in song-and-dance duet, there is higher importance which is always attached to the six characters and rhymed formula, which are words, sentences, rhyme, allegro, tone, and vigor.

a. Words
   Words are the principal part of a language. Only when the words used in the singing are vivid, people can get a real understanding of the meaning contained in the music. In this point, it is required by the art of song-and-dance duet that the artists have to clearly sing the words and accomplish the state that the words used in the singing can truly give a reflection on life and like an unshaped knife to commit a murder if they are of untruthfulness.
b. Sentences
   Sentences used in the singing of song-and-dance duet mean that the artists are necessary to sing a single meaning in the sentences with accuracy and distinction. Only in this way, the true meaning in the sentences can be conveyed with a better effect, and also the stories and characters can be performed and created more vividly. The artists can really achieve the state that "singing has to give clear evidence to listeners and should be set back if having not made that realized" if only they can get an in-depth understanding of the meanings contained in the singed sentences.
c. Rhyme
   The "rhyme", which is discussed among the artists singing song-and-dance duet, means that the artists have to establish a lasting appeal when singing and performing a song-and-dance duet story. The "rhyme" is the life of the

two artists in transforming their musical voices and also is the core of the requirements asked by people on the standards of musical voices esthetics. It includes not only the cadence but also the especial styles which are rooted in the dialect in the three provinces of northern China.

d. Allegro
The allegro, which frequently appears in traditional Chinese opera or folk music, contains two types of meanings in general. The one is the allegro in meters, and the other is the allegro in rhythm. The allegro in meters refers to the measure in Free rhythm, flow rhythm, 2/6 rhythm, and twinkling rhythm each three times. However, the allegro in rhythm refers to the arrangements of the musical words and their locations. That the importance is attached by the art of song-and-dance duet to sixteen-character rhythm for the musical voices is a very good example.

e. Tone
On the one hand, tone refers to a high tone. This means that it is necessary for the artists singing song-and-dance duet to sing the musical tones accurately and clearly when singing a song-and-dance duet story. The artists can never make their singing deviated from the right tones. Therefore, it is better for them to make a proper adjustment to the musical feeling. On the other hand, tone refers to the melody. This can indicate that the mastering degree of the artists on the tones as well as the names of the tunes and all kinds of functions of the tones.

f. Vigor
Generally , the vigor refers to the intensity of the musical voices. Singing and performing a song-and-dance duet story cannot be like old cattle to climb mountains. Therefore, it is necessary for the artists to make the "edges and corners" in the musical voices of song-and-dance duet distinct and ordered with a cadence. In the process of the practices, the artists, who are in area of the song-and-dance duet art, can be praised to own a solid singing foundation for song-and-dance duet if they can get a real understanding of the six words. In the mean time, the composers can be with might redoubled in the process of creating the musical voices.

11.2.2 Disorder and Negligence are Tabooed Respectively in Free Rhythm and Slow Rhythm

For the rule that the disorder and negligence are tabooed respectively in free and slow rhythm, the author believes that the conclusion on this rhymed formula is obviously full of wisdom. It indicates that the singing with a board in the works can be sung into the "scattered" state, and also the average placement of the sung words as well as no differentiations in length and cadence should never come up. But the rhythms in the musical voices should be arranged in accordance with the contents and meanings which are expressed in the singing words, making the music balanced, cadenced, and applied appropriately.

11.2.3 Female Character is Like a Flower but Relies on the Male Character to Reveal Its Beauteousness

Obviously, this rhymed formula seems to be very easy to read and understand. However, it contains very rich experience in realities. As the main body of the story, the two persons in this story take the harmonious and complete performance as the highest artistic pursuit in a song-and-dance duet.

11.2.4 Men and Women Characters as Partners are the Main Force no Matter a Large Number of Mounted and Foot Enemies Come

In the short and simple art form of the song-and-dance duet, one of the basic characteristics when the Female Character and Male Character in a song-and-dance duet is the cooperation between each other. In the works of the song-and-dance duet for only 30 or 40 minutes, the number of its important characters is usually not more than three or four, but it has a great number of minor characters. All these characters need the cooperation between the major female character and male character to be created in details. In addition, when the artists are conveying something orally or by heart, the performers are required to make a flexible arrangement on the cooperation between characters in accordance with the characteristics of the specific roles, story line and the strong points of the performers.

11.3 Beauteousness Contained in the Rhymed Formula

Through the simple interpretation on the rhymed formula of the song-and-dance duet, it can be felt that the succinct rhymed formula contains multiple factors in the area of esthetics. Furthermore, at today's numerous and complicated social cultural market, the song-and-dance duet popular in northeast China can be in an invincible position, just because it can highlight the characteristics of the folk art forms and the be unique beauteousness in itself.

11.3.1 Distinctive and Conspicuous Common Beauteousness

Common beauteousness is a bridge which is established between the subject and object of the esthetics. The common beauteousness, which is manifested in the art of the song-and-dance duet, is specifically embodied in the distinctive nationalities styles, conspicuous local characteristics, full-bodied local conditions and customs as well as the local folkways and accents. The psychological quality and characters, which are formed from the natural environment, traditional values and the way of life, customs and other factors, will exert an influence on the style features and aesthetic tastes of the folk music in a place. This public aesthetic habit, which is developed from the regional characteristics, living customs, and real social environment, has accumulated into the special hobby and preference of the people to have an appreciation on the works of art. This is just like the rhymed formula that each place has its own way of supporting its own inhabitants, who speak their unique dialect. The art of the song-and-dance due was born in the heated floor of the cold house in northeast China. The creators of the art as well as the main body of appreciating it are the peasants. Therefore, the artistic characteristic of the art is certainly easy to read and understand. Just due to these special charms, the minds of the appreciators can be unrestricted, just like the water in a thoroughly clear brook slowly flowing past the hearts of the appreciators.

11.3.2 Artistic Beauty of the Men Character with Unique Characteristics

In the art of the song-and-dance duet, there are a great number of the rhymed formulas about the female character and the men character. For example, the female character is like the body of a wheel, while the men character is like the roller; a line of the female character is as same as the main part of the singing men character. From these rhymed formulas with the simple images, the artistic position and values of the men character can be felt with an in-depth.

In accordance with the perspective of esthetics, the "men character" is one of the aesthetic attributes in a kind of relatively beautiful objective things. It is not an embodiment of the an "ugly" character, but brings an artistic enjoyment to people with the purpose of satisfying the happiness between the spirits and physical bodies of the audiences. Within the specific works of the song-and-dance duet, the artistic beauty of the men character is embodied incisively and vividly. Particularly, when the works can give an expression to the dialectical relationship among the hardness and softness, dynamics and statics, happiness and sadness, simple and complex, rich and poor, and strong and weak, the men character can better give a reflection on such a relationship, and can reveal the personality, the essence of life, as well as the wise opinions.

11.3.3 Beauty of Short Distance Between Subject and Object

As an art welcomed and loved by a wide range of audiences, the song-and-dance duet has generated a feeling of beauty for people when its influence is exerted in the aesthetic subjects. The generation of such an esthetic feeling, as same as other art forms, conveys the predetermined esthetic effects with its own unique way of communication. The performance of the song-and-dance duet art is not only equipped with the communication ways which are common in the other art forms, and also forms the ways of communication between performers and audiences, performers and characters, characters and characters, and characters and audiences. It mobilizes the emotions of the audiences, and makes a common feeling generated between characters and audiences through the emotional communication. Hence, the best aesthetic state can be achieved.

The unique way of communication in the art of the song-and-dance duet comes into being with a gradual through the understanding of the esthetic attitudes of the audiences in the long-time development. The formation of such a unique way of communication plays a decisive role in the survival and development of the song-and-dance duet in the long run.

From the above, it can be known that the characteristics of the song-and-dance duet in conciseness, easy-to-understand, plainness, accuracy and other aspects receives a full expression in the rhymed formulas of the song-and-dance duet art. Just due to the existence of these unique characteristics from the perspective of esthetics, the art of the song-and-dance duet can still own the power to send out glorious and flowery shine while the other musical forms are suffering from stagnation at the present time. At the same time, the author discovers that the rhymed formulas which are concluded by the artists from the practices are as vast as the smoke and sea and require a wide range of the artistic workers or the people interested in this area to jointly participate in it. Therefore, the author makes a copy of the rhymed formulas which are known by him or felt to have some values in the following, for the purpose of making a common research and discussion with the people within the same industry.

Song is "bone" and speaking is "meat";

Your fluency can be judged when you sing a little;

The pleasance of your singing can be judged by the first sung sentence;

Singing without dancing is not better than selling bean curds; singing with dancing is core;

Wrapped string sound can be heard clearly; woodwind instrument makes the tune profound but not the words; two-stringed bowed instrument makes words and rhymes profound;

The beginning should be joyful and the ending should be hurry; the middle part should be stable and surprising;

Musical voices of the song-and-dance duet seem to be heavy;

The number of the characters is no more than three; and story line is though the works;

It is better to be unfamiliar, but it is precious to be familiar;

Singing can be totally with a warm blood;

Skills lie in hard work; books are necessary to be read, and also skills need practices;

Crying with eyes but not fact, because the crying face cannot reflect the real feeling;

Musical voices should be continuous with vigor.

References

1. Bingchen N, Xiusheng Y (1979) Song-and-dance duet folk music in northeast of China, vol 9(1). Jilin Audio-Visual Publishing House, China, pp 54–59
2. Zhende L, Shukin L, Shuxiao W (2000) The art of song-and-dance duet, vol 9(1). Culture and Art Publishing House, China, pp 328–337
3. Jiyang Z (1992) Analysis on the esthetics characteristics of the song-and-dance duet folk music. Art Studies 1(02):353–359

Part II
Management Science and Engineering

Chapter 12
Accounting Support of Value Management of Listed Bank in Full Circulation of Stocks

Dongfang Qiu and Yiping Liu

Abstract Now a days, the shares of the listed companies are circulated overall. The value management is the core of the management of listed companies, listed banks in China and are facing the competence from home and abroad and the pressure of risk management. We should build the financial decision system focused on economic capital and the added value of economy, the accounting information system centralized in the accounting operation, as well as the information guarantee system which is based on the risk-guided auditing under the guidance of scientific development concept and following the main stream of the value management of listed companies in order to support the value management of listed banks and to realize the sustainable development of listed banks.

Keywords Full circulation • Listed bank • Value management • Accounting support

12.1 Introduction

As the completion of nontradable share reform is approaching, the all-circulation of listed companies stock will realize the pricing function of capital market. Generally, the stock price is the index that reflects the value of listed companies and the property of the stockholders. Those who could not create value in the security market will not get developed and there will also exist the risk of merger and bankrupt. Thus, for listed companies, value management will be a good weapon to improve the value of the company and the property of shareholders. Value management is also an important tool to combine the strategy, the operation, and the

D. Qiu (✉) · Y. Liu
College of Economics and Management, Nanjing University of Aeronautics and Astronautics, Nanjing 211106, China
e-mail: qdf99@163.com

Y. Liu
e-mail: yipingliu@263.net

Z. Zhong (ed.), *Proceedings of the International Conference on Information Engineering and Applications (IEA) 2012*, Lecture Notes in Electrical Engineering 220,
DOI: 10.1007/978-1-4471-4844-9_12, © Springer-Verlag London 2013

market value of a company together. The three main parts of general accounting -financial, accounting, and auditing, can support the value management in trinity. Namely, they can play the decision role of finance, the information supporting role of accounting, and the information guarantee role of auditing.

At the end of the year 2006, China has comprehensively opened the local currency business to foreign banks, and cancelled the restriction of clients and regions and made the commitment to treat foreign banks the same as Chinese banks. It means foreign banks can compete with Chinese banks equally and fairly in China. As the main members in the capital market and under the background of full circulation of the shares, listed banks are facing the competence from interior and exterior. They should establish the business development strategy, adjust the structure and orientation of loan, and accelerate the speed of finance innovation under the guidance of scientific development concept. Only in this way can we improve the ability of risk management and internal control, and be on the way of sustainable and harmonious development. The operational and managerial activities should also follow the main stream of the value management. However, in the listed company which specially operate fund, the supporting role of finance, accounting, and auditing based on value management has its unique content.

12.2 Construct the Financial Decision-Making System Focused on Economic Capital and Economic Added Value

The history of banks in western developed countries demonstrates that, the scientific financial management system is benefit to optimization of the resource configuration; improve efficiency, risk control, and the ability of competence. The economist Simon emphasized that management is the decision, the decision is in the overall management process, and the financial management is all about financial decision.

As commercial bank is of high risk, controlling risk is always its important operation target. Its loss can be divided as the expected loss and unexpected loss. The former can be covered by the financial risk fund while the later can be covered by economic capital originated from the maximization of shareholder value concept. The economic capital is the capital to offset the corresponding unexpected losses which assessed by the management of the commercial banks according to the actual risks of their own business within the range defined by the confidence interval and the prescribed period. Because the clients are uncertain and fluctuating periodically, unexpected loss need to be calculated and defined by economic capital and diminished by its equity capital. Economic capital itself depends on the practical value of the unexpected lose through which we can predict the risk degree of each branch, field and business and effective identification of the level of the risk and profit of each branch. Economic added value (EVA) = net profit on book-the cost of economic capital (the occupation of economic capital × total shareholder return). For commercial banks economic added value reflects the surplus profit when considering the expected lose and unexpected lose, as the former one covered by the bank profit while the later covered by the bank capital. The listed banks should maintain a virtuous cycle of the three important parts-capital, risk, and income to meet the requirement of the maintenance and creation of value. The key point to create value for shareholders effectively is to propel economic capital management in a long term. Economic capital management is the advanced management tool for international banks, and it is also an important means to strengthen the supervision of capital sufficiency and intensify internal capital management and risk management for commercial banks.

To construct the financial management and decision system which is steering for value management and focused on economic capital and economic added value, first, we should import the theory of economic capital and economic added value into finance managing activities of commercial banks, which is the premise of financial management innovation. Second, we should use economic capital to allocate financial resources reasonably, optimize assets structure, and allocate them to business unit, product, client and stuff in order to meet the demand of maximization of the bank value, balance the risk and income through capital allocation, and return to realize a scientific performance evaluation system. We should constantly improve the calculation method of economic added value and the mode connecting to labor cost in aggregate. From the vertical and horizontal dimensions, we should establish the performance evaluation method focused on EVA, with the economic added value and the return on capital adjusted by risk factors being the main index to assess the profit level and evaluate the management from the bank head to the business sites.

Listed banks financial management decision should stick to the principle of the maximization of benefits with the premise of controlling risks and value maximization of the value as the goal, constructing risk control, and return restriction system which focused on economic capital. With the economic capital cover the risk, we should also adjust returns and take the risk-adjusted performance indicator as the core index and establish the long-term incentive and constraint mechanism to guide banks to balance operation risks and income and change business growth mode, accelerate the harmonious development of banks' profit, quality, rate and scope, and improve the whole value of listed banks.

12.3 Intensify the Accounting Information System Centered with Accounting Business Process

The efficient financial decision is related to accounting information system. The accounting information provides important proof for the value management of listed banks.

In the long term, local branches of commercial banks are the basic accounting unit. Every local branch has a set of independent ledger and general ledger.

All commercial banks have five levels of accounting that is street branch, district branch, city branch, province branch, and head bank. So many levels of accounting and such complex levels of management make the delivery of information slower and distorted, which add the risk of accounting information and weaken the function of the accounting management and its inner-control. In the five-level accounting mode, the department is over emphasized to operate business and implement the management. The business processes of commercial banks are depended on their structure which means functional group is formed according to the similarity of their activities. At present, the individual financial department, company business department and accounting department are adopting the typical managing method of department bank, which artificially divides the whole business process. With the reform of listed banks management and the development of information and technology, as well as the practice of unity management of accounting, the flat management and the new system of bank teller, the transformation from "the department bank" to "the process bank" is the difficult task our listed bank face. We must adopt reengineering in accounting operational process and strengthen the information system of accounting operational process in listed banks.

Commercial banks are unique companies whose operating object is capital and accounting object is the movement of capital. The overlap of two objects can easily result in a serious disordered phenomenon of accounting and accountant. Actually, bank tellers participate in the traditional accounting, such as filling in accounting voucher, registering accounting book, formatting and filling in accounting entries, the main duty of bank tellers is still to handle all kinds of business, but not accounting. They face their clients such as depositors and borrowers, and they do not to face those who make use of the accounting report like managers, supervisors, and inventors and so on. Therefore, traditionally people think the accounting of commercial banks is not only accounting but also business, but it actually accounts business. The accounting business process in listed banks mainly means the process of checking, settling, clearing accounts, paying, and receiving money, which is the basis of all commercial banks business and the important information source for management decision. The accounting process reengineering is adopting modern information technology, starting off from personal needs, making the accounting business process reform as the core, and set the goal to improve customer dependence and the response sensitivity to the market. We should pay attention to the circumstances changes for business and redesign the accounting business process and the accounting management model, so that we can make the accounting business process to be a stable one which can add value under the premise of controlling the risk effectively, and improve the marketing and market exploring ability. First, we should actively propel the reform of decoupling of front and back stage and accounting centralized and breakthrough the present management and accounting model of "street branch, district branch, city branch, province branch, head bank" and transfer to the two levels-head bank and first level bank-accounting model. Second, we should reform the organization structure. The management department should not be established in second-level and the lower-level branches and the accounting management sector and the proposal financial sector should be merged to the financial accounting department. At the same time, the bank clerks in branches and the following agencies should be transferred to be the marketing service clerk, which means the front application system does not pitch at accounting system but business handling system so that to help provide wider angle of view and space for business handling system in commercial banks. It not only embodies the exact management mechanism that focuses on clients which is helpful for the organization to innovate and improve the service quality and become more competitive but also simplifies front business process and optimizes the labor structure. Second-level branches assign accountant officer to local networks in order to supervise the accounting business and to accelerate the transfer of networks from the style of accounting dominant to the style of marketing service. Listed banks should rebuild the business process while optimizing the accounting organization structure, propelling the management mode which is focused on business process instead of on department. And we should build the customer-focused vertical and assessment-oriented process with the background centralized processing and the front and background separated but restricted with each other to realize the convenient, automatic, standard process of inner control, and provide punctual accounting information and proof for the value management of listed banks according to the business process achieving the goal of preventing risks.

12.4 Build the Modern Information Security System of Auditing Which is Guided by Risks

The accounting activities are the supervisor of the economic activities in listed banks. The bank auditing is the supervisor of financial and accounting activities and the guarantee of their healthy development.

From the financial crisis born in Asia in 1997 to the American subprime mortgage crisis in 2007 which affected the global economy, all that prove banking as a high-risk industry and the risk has already permeated to all aspects of the operation in a bank. A series of risky cases such as Barings Bank, Daiwa Bank, and Industrial Bank, etc., have already warned us that credit risk, market risk, operation risk and the risk in financial innovation, and the financial derivatives affect each other and influence the safety of banks operation bringing the banks huge lose even bankrupted. Furthermore, it will influence the stability of economic environment and society. Thus, we should build the information security system guided by the modern risk-oriented auditing and incorporate the internal auditing, the certified public accountant auditing and government audit as a whole. First, the function of the modern risk-oriented internal auditing should be brought into play as fully as possible. With the establishing and consummating of the modern company system, the internal auditing which is the important part of the internal administrative of listed banks should learn from the experience of banks in developed companies. We should refresh the concept and establish the risk-oriented internal auditing mechanism to achieve the function of internal auditing like security, liquidity and profitability, and to strengthen the risk management of banks. Second, as financial corporation whose stocks are hold by the public, listed banks should abide by the regulation of capital market, and do the CPA auditing through which to provide reasonable assurance of the legitimacy and fairness of the financial report. Accounting firm and certified public accountant should have the risk-oriented auditing concept based on strategic and systematic view and assess the risk of material misstatement of listed banks from a macroscopic view to medium view and to the microscopic view. Meanwhile we should pay more attention to the auditing in high risk field, adding auditing process and expanding the extent of auditing, which can ultimately control the auditing risk of listed banks and improve the quality of report. Third, as the vast majority of listed banks are the state-controlled shareholding companies, they have the duty to make state-owned assets retaining and increment. At the same time, due to the significance of banks in the national economy it is important for the bank to do the government audit. Therefore, auditing the listed banks is an important part of the government auditing. Audit Commission made financial audit, banking audit, and the audit in state-owned and state holding enterprise as the "three-in-hand" in "Audit Work from 2006 to 2012 Development Plan". Although with the strengthening of the external supervision and the improvement of the listed banks management the capacity to handle unhealthy assets and risk is improving, what listed banks face at present is still the complicated external environment. The risk of management still exists and the environment of government bank auditing is still severe. Therefore, the modern risk-oriented government bank auditing should emphasize the concept of scientific development and build the scientific audit concept with the goal to promote the risk prevention, enhance economic performance and standardize management. It should enhance the auditing to the local branches under the leadership of the head bank, progressively realize the on-line auditing and speed up the construction of auditing data base and the "financial-audit" program. In a word, we should provide the useful information for the listed banks and their stockholders to make decision and improve efficiency and safety for the banks operation through building the risk-oriented internal audit, the CPA audit and the government audit system, which is important part of value management.

Under the environment of full circulation, the value management of listed banks in China depends on the financial decision system which is centered with the economic capital and the economic value added the accounting information system which is centered with the accounting business process, and the information security system which is guided by the modern risk-oriented audit. Though the main part of these works is different, they have the same object which is to realize the value increment and the value management of listed banks and to maintain the sustainable development of listed banks.

Chapter 13
Study of Innovation Legitimacy Effect on Business Model Innovation

Bingcheng Wang, Hao Ding and Xiaona Wang

Abstract The purpose of this study is to test different effects of three components of innovation legitimacy on business model innovation. We selected the full-time employees of 211 enterprises and institutions as samples, built models taking gender, age, education level and rank as the control variables and taking cognitive legitimacy, moral legitimacy, and practical legitimacy as independent variables. This chapter made descriptive statistic analysis, reliability testing, and correlation analysis through SPSS15.0, and tested the relationships between variables using hierarchical regression analysis. The final conclusions: the moral legitimacy shows a significant relationship on business model innovation; while the effect of others on business model innovation does not reach the level of significance.

Keywords Business model innovation • Innovation legitimacy • Hierarchical regression analysis

13.1 Introduction

In the contemporary world, because of globalization, development of information technology and shorter product life cycles, the competitive situation has been changed greatly. Facing the complexity and variability of external environment and intense competitive pressures, companies must find new profit growth point depending on innovation in order to get a long and stable development. Business model innovation is the most original innovation behavior in the development of enterprises' innovation ability. So we can see that the business model innovation is so important to the enterprise. Paul and Tim who are the executives of the Accenture Institute found that the fundamental reason of the failure of a lot of Star Enterprises is excessively focusing on financial indicators but ignoring

---

B. Wang (✉) · H. Ding · X. Wang
China University of Petroleum (East China), Shandong University of Science and Technology, 579, Qianwangang Road, Qingdao 266590, China
e-mail: wxnxyx@126.com

Z. Zhong (ed.), *Proceedings of the International Conference on Information Engineering and Applications (IEA) 2012*, Lecture Notes in Electrical Engineering 220,
DOI: 10.1007/978-1-4471-4844-9_13,

the business model innovation. Thus, the enterprises would easily get into trouble once the existing business model loses the growth potential [1].

The new business model allows the first user to gain a competitive advantage, but it is easy to suffer the blow of the "legitimacy". If enterprises using new business models want to survive and develop, owning only efficiency and performance is not sufficient and the business purpose and the process must be consistent with values, customs and social expectations which is deemed to be legitimate. But the relationship between innovation legitimacy and business model innovation has not been verified. So this paper would divide legitimacy into cognitive legitimacy, moral legitimacy, and practical legitimacy according to the point of [2], and test the specific influence of innovation legitimacy on business model innovation [3, 4].

13.2 Research and Design

13.2.1 The Research Framework

The research framework is shown in Fig. 13.1.

13.2.2 The Research Questionnaire

This study consists of two questionnaires which are about innovation legitimacy and business model innovation.

We design five items for the questionnaire of business model innovation referring [5, 6]. The items include "I will improve the business model initiatively to deal with problems in work", "I will explore the opportunity to improve workflow and services", "I will learn and adopt new business models from other companies", and so on. All items are modified to fit into the 5-point Likert-scale format from 1 (extremely disagree) to 5 (strongly agree).

The questionnaire of innovation legitimacy is designed according to the point of [2] and the questionnaire of organizational legitimacy developed by [6] and it includes a total of eleven items. All items are modified to fit into the 5-point Likert-scale format from 1 (extremely disagree) to 5 (strongly agree).

13.2.3 Method

We use SPSS 15.0 to do descriptive statistics of samples, reliability testing of the various dimensions and the correlation analysis of the various dimensions, test construct validity depending on the average variation of extraction (AVE), and use the method of hierarchical regression analysis to verify related assumptions.

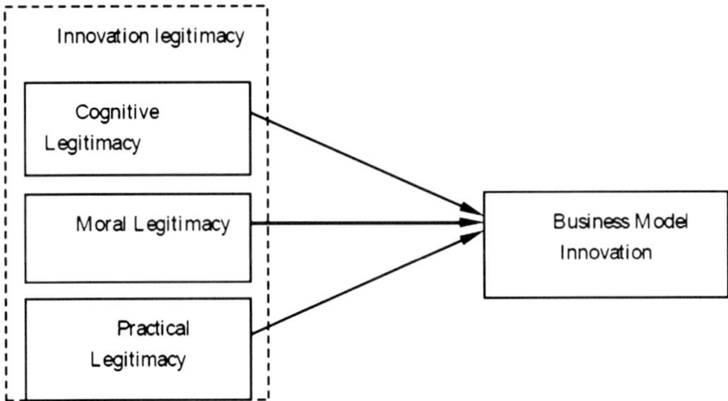

Fig. 13.1 Research framework

Pointed out that the method of hierarchical regression analysis is essentially a confirmatory techniques rather than exploratory approach. During the course of using the method, the decision about inputting the variables in or not will be made by the researchers depending on theory or research needs, so the method is a regression analysis program which has the greatest flexibility, the most theoretical and practical significance [7].

13.3 Results

13.3.1 Descriptive Statistics of the Sample

We delivered 300 questionnaires and received 168 copies with the response rate of 56 %. But there are just a total of 130 valid questionnaires excluding 38 copies which are incomplete or obviously not serious with the valid response rate of 43.3 %. The basic situation of respondents is shown in Table 13.1.

13.3.2 Reliability and Validity Analysis

Because the items of the questionnaire are from the research of scholars at home and abroad, first we test related items appropriately and remove some of the items. As for the standard in the deletion of items, we also use the value of Item-Total correlation in addition to Cronbach's α value, then delete some items whose values are less than 0.3 or negative. At last we test the reliability of each dimension using SPSS15.0, the result is shown in Table 13.2.

Table 13.1 Descriptive statistics of the sample

Level	Effective sample	Proportion (%)	Level	Effective sample	Proportion (%)
Gender			Education level		
Male	76	58.5	Senior high school and below	7	5.4
Female	44	33.8	Junior college	30	23.1
Missing	10	7.7	College	75	57.7
			Master and above	13	10.0
			Missing	7	5.4
Age composition			Range		
<30	53	40.8	General staff	15	11.5
30–40	52	40.0	Line managers	27	20.8
40–50	19	14.6	Department heads	71	54.6
>50	4	3.1	Senior managers	8	6.2
Missing	2	1.5	Missing	9	6.9

Table 13.2 Reliability test of the dimensions

Dimensions		Final items	Cronbach's α	AVE
Innovation legitimacy	Cognitive legitimacy	2	0.627	0.665
	Moral legitimacy	4	0.843	0.611
	Practical legitimacy	3	0.635	0.645
Business model innovation		5	0.838	0.613

Table 13.3 The correlation coefficient between dimensions

Numbers	Dimensions	Cognitive legitimacy	Moral legitimacy	Practical legitimacy	Business model innovation
1	Cognitive legitimacy	(0.816)			
2	Moral legitimacy	0.382^{***}	(0.781)		
3	Practical legitimacy	0.380^{***}	0.509^{***}	(0.803)	
4	Business model innovation	0.198^{*}	−0.019	0.120	(0.783)

Notes * $p < 0.05$, ** $p < 0.01$, *** $p < 0.001$

Table 13.2 shows the reliability coefficients of various dimensions are higher than 0.5, therefore the various dimensions of the questionnaire all meet reliability requirements according to [8] whose recommendations are that Cronbach's α coefficient ranged from 0.5 to 0.7 is credible and ranged from 0.7 to 0.9 is quite credible.

Table 13.2 shows the values of the average variation of extraction calculated of each dimension are higher than 0.5, indicating that each dimension has a convergent validity. Table 13.3 shows that the square roots of the average variation of extraction of each dimension are more than the correlation coefficient between the various dimensions of column, indicating that each dimension has a discriminant validity [9].

13.3.3 The Correlation Analysis

We use SPSS15.0 to conduct the correlation analysis, and the correlation coefficients between the various dimensions are shown in Table 13.3.

13.3.4 Test of the Relationship Between Innovation Legitimacy and Business Model Innovation

We use hierarchical regression analysis to explore the underlying assumptions, make the gender, age, education level, and rank of employees as control variables, innovation legitimacy as an independent variable, the business model innovation as the dependent variable, to test the effect of innovative legitimacy on the business model innovation such as shown in Table 13.4.

In Model 1 (M1), the regression analysis only includes the control variables, Table 13.4 shows $R^2 = 0.030$, so the regression coefficient of each control variable is not significant.

In Model 2 (M2), we add independent variable into regression analysis, Table 13.4 shows $R^2 = 0.096$, $\varDelta R^2 = 0.066$, indicating it explained 9.6 % of the total after adding innovation legitimacy, and the explanatory power increases 6.6 % compared to the M1.Test regression coefficients of each variable, the moral legitimacy: $\beta = -0.234$, $p = 0.044 < 0.05$, reaching a significant level and the maximum value of VIF of each variables: 1.571 < 5, being excluded from the multicollinear factors issues, indicate that the moral legitimacy has a negative impact on business model innovation. The regression coefficients of cognitive legitimacy

Table 13.4 The hierarchical regression analysis of the relationship between innovation legitimacy and business model innovation

	M1				M2			
	β	t	Sig.	VIF	β	t	Sig.	VIF
Control variables								
Gender	0.117	1.168	0.245	1.074	0.138	1.398	0.165	1.094
Age	0.007	0.066	0.947	1.187	0.021	0.202	0.840	1.232
Level of education	0.145	1.487	0.140	1.018	0.127	1.315	0.192	1.041
Range	−0.041	−0.383	0.703	1.217	−0.072	−0.674	0.502	1.289
Independent variables								
Cognitive legitimacy					0.150	1.268	0.208	1.571
Moral legitimacy					−0.234*	−2.038	0.044	1.479
Practical legitimacy					0.188	1.702	0.092	1.363
R^2	0.030				0.096			
Adjusted R^2	−0.007				0.033			
$\angle R^2$	0.030				0.066			

Notes * $p < 0.05$, ** $p < 0.01$, *** $p < 0.001$

and practical legitimacy do not reach significance level, but both coefficients are positive, indicating that cognitive legitimacy and practical legitimacy have a positive effect on business model innovation.

13.4 Conclusions

While previous studies have shown that organization legitimacy hindered the business model innovation to some extent, and also provided a lot of ways in which enterprises can gain legitimacy resources. However, this study shows that not every aspect of the legitimacy has a negative impact on the business model innovation, and just the moral legitimacy has a significant negative impact on business model innovation. This conclusion makes the new business model obtain legal resources more targeted and provides the theoretical basis for business model innovation. In order to obtain the support and endorsement of the holder of resources, enterprises should focus on establishing their own moral legitimacy. Enterprises must abide rules issued by the government and stakeholders to give them a signal that they are consistent with mainstream concept signal; allying with some of the associations than separate action is more legitimate; they can also take the initiative to carry out charitable donations, such as establishing the Hope Primary School to improve their reputation.

References

1. Dai W, Yue L (2011) The business model innovation in post-crisis era. Harvard Business Review 1:72–73
2. Suchman MC (1995) Managing legitimacy: strategic and institutional approaches[J]. Acad Manag Rev 20(3):571–610
3. Dowling J, Pfeffer J (1975) Organizational legitimacy: social values and organizational behavior. Pac Sociol Rev 18:122–136
4. Epstein EM, Votaw D (1978) Rationality, legitimacy, responsibility: search for new directions in business and society, vol 1. Goodyear, Santa Monica, pp 12–15
5. Fornell C, Larcker DF (1981) Evaluating structural equation models with unobservable variables and measurement error. J Mark Res 18:39–50
6. Huang Z, You X (2010) Social network, institutional lawfulness, and performance of globalization by Chinese businesses. Econ Manage J 8:38–47
7. Kleysen RF, Street CT (2001) Toward a multi-dimensional measure of individual innovative behavior. J Intellect Capital 2(3):284–296
8. Nunnally JC (1978) Psychometric theory, 2nd ed. Mcgraw Hill Press, New York
9. Qiu H (2009) Quantitative research and statistical analysis, vol 3. Chongqing Univ Press, Chongqing, pp 12–17

Chapter 14
Identification of High-Tech Enterprises and Management

Shijin Yang

Abstract High-tech enterprises can enjoy a series of preferential tax policy, so that enterprises are active to apply to be high-tech enterprises. In order to meet the conditions of the declaration, many enterprises pursuit quick success, resulting in producing a lot of problems during their application. Therefore, this paper points out the role, emphasis and difficulty of the high-tech enterprises application and analyzes enterprise management problems, and proposes measures to solve the problems. Finally, it draws the conclusion that enterprise should improve its management system with efforts made in the preapplication assessment and planning, organization of implementation and follow-up management.

Keywords High-tech enterprise • Recognition • Enterprise management

14.1 Introduction

It has been a long-term state development strategy to support enterprises in technological upgrading and innovation and to achieve the transfer of the economic growth from extensive mode to intensive mode. Therefore, a series of preferential tax policies have been adopted, which can be traced in the following terms of the latest Corporate Income Tax Law of The People's Republic of China adopted from the year 2008:

1. In a tax year, the portion not exceeding 5 million yuan obtained by a resident enterprise from technological transfer shall be exempted from Enterprise Income Tax (EIT), and the excess shall be taxed at the reduced half rate.
2. The enterprise will enjoy a 50 % additional deduction of research and development expenses incurred from the research and development of new technologies, new products, and new techniques on the basis of the actual deductions.
3. A startup investment enterprise invests, by means of equity investment, in an unlisted small or medium-sized high and new technology enterprise for two or more years, it may set off 70 % of its equity investment against the current

S. Yang (✉)
College of Economics and Management, South China Agricultural University, Guangzhou 510642, China
e-mail: crystal_sxj@yahoo.cn

Z. Zhong (ed.), *Proceedings of the International Conference on Information Engineering and Applications (IEA) 2012*, Lecture Notes in Electrical Engineering 220, DOI: 10.1007/978-1-4471-4844-9_14,

taxable amount of the startup investment enterprise when its equity holding attains to two full years. If the taxable amount is not enough for setoff, the margin may be carried forward to subsequent years for setoff.

4. If the depreciation of fixed assets of an enterprise need to be accelerated due to technological progress or product upgrading, the depreciation period can be shorten or methods of accelerating the depreciation can be adopted.
5. High and new technology enterprises with the key support from the state can enjoy a 15 % deduction of its EIT.

Among the above five preferential tax policies, high-tech enterprises and nonhigh-tech enterprises can enjoy the first four preferential policies. However, only high-tech enterprises can enjoy preferential tax with a reduced rate of 15 %. It shows that concerning the preferential tax policies, being recognized as high-tech enterprises will bring substantial benefits. An enterprise once identified as high-tech enterprises should make full use of these preferential policies to effectively reduce the corporate tax, to upgrade its management mechanism, effectively improve its R&D and develop its core competitiveness thus to lay a solid foundation for its long-term stable and healthy development. These preferential policies cannot be simply understood as general tax incentives.

However, driven by the enormous benefits some enterprises which are not qualified to be recognized as high-tech enterprises take fraud and other illegal means to get through the first stage of the high-tech enterprise qualification recognition. However, this is extremely risky once they cannot pass the review section of qualification recognition. They have to refund the preferential tax, pay for the delay fees and fines, and are not allowed to apply for the identification of high-tech enterprises within five years. Moreover, the central government has been increasing its efforts on the supervision of the recognition of high-tech enterprises. According to the 2011 National Audit Office audit report: among 148 high-tech enterprises in 18 provinces, autonomous regions and municipalities selected as auditing samples, 17 unqualified enterprises are identified as high-tech enterprises by the auditing management organizations at provincial level, which directly led to their refunding of 2.665 billion yuan preferential tax and paying for the delay fees. This brought huge loss to the enterprises. Therefore, any illegal and fraud action to be taken to get through the qualification recognition is risky and costly. Instead, it is feasible for the enterprise to follow the truth and seek ways to improve its R&D management capability, to carry out a reasonable taxation planning, and thus develop itself into a veritable high-tech enterprise.

14.2 Requirements for Being Recognized as High-Tech Enterprises and Materials to Submit

Requirements based on "High-tech Enterprise Recognition Approach" (Ministry of Science and Technology [2008] No. 172) and "High-tech Fields with key supports from the State", Ministry of Science and Technology, Ministry of Finance, State Administration of Taxation jointly issued "High-tech Enterprise Recognition

Guidelines". A high-tech enterprise should meet 6 prerequisites with particular emphasis on the core independent property rights.

Judging from the prerequisites of being recognized as a high-tech enterprise, there are specific qualitative and quantitative indicators [1].

Qualitative indicators are reflected in the following 3 aspects:

1. The enterprise's core intellectual property right must meet specified requirements. There are normally two cases: if it is obtained through the purchase of patents, the filing of the transfer of patent rights in the Intellectual Property Office is needed; if the enterprise owns its intellectual property rights, it needs to go to the State Copyright Bureau to apply for the ownership certificate. This is needed during the submission of materials of RD (R&D project).
2. Ability to transform scientific and technological achievements into production. This indicator is a measure of the enterprise's capability of transforming core technology into the production. The enterprise should get the test report and qualification certificate of products from relevant authorities of the central government. For example, if the product belongs to the software category. The enterprise should obtain the qualification product certificate issued by National Economic and Information Commission. This is required to be included into the materials for the application of PS project (High-tech Products).
3. Research and development organization and management level. It is reflected on the organizational structure of R&D institutions and human resources allocation, reward and punishment system of R&D personnel, R&D project management system and the corresponding financial accounting management approach etc.

The above three aspects of qualitative indicators are not isolated and can be regarded as "piers of a bridge" [2]. The bridge deck is the enterprise's R&D and scientific innovation procedure. Pier No. 1 is the carrier of the R&D project and scientific and technological innovation. Pier No. 2 is the investment in, accounting and auditing of the R&D project and scientific and technological innovation. Pier No. 3 is the achievements (that is, PS). These three aspects are indispensable indicators. If an enterprise cannot meet the requirement of Indicator 1, it shows that it has no capability of conducting R&D projects. In case of Pier 2, it shows that their projects are carried without theoretical and analytic support. If the enterprise cannot meet the requirements of indicator 3, it proves that there is only investment into the R&D projects but with no transformation fruits into production. Therefore, only when an enterprise meet the requirements of the above 3 indicators, can it be recognized as a high-tech enterprise.

Quantitative indicators are: (1) The proportion of total R&D expenses to total sales revenue must meet the specific requirement, that is between 3 and 6 %; (2) High-tech products revenue, namely PS project income accounts for more than 60 % of the total income (including the revenue from the main business and subsidiaries); (3) The total assets and sales revenue each year maintain a certain growth rate (the rate is not specified). The rate reflects the enterprise' s capability of maintaining a sustainable growth and its core competitiveness [3]. Materials to submit. There are 7 documents need to be prepared: Application Form of High-tech Enterprise Qualification Certificate, Enterprise Qualification Certificate, audit reports in the latest 3 years, intellectual property rights certificate, Introduction on R&D personnel, documents to prove the launch of R&D project, materials to prove the enterprise's capability of transforming technological achievements into production.

Judging from the submitted materials, we can see two main characteristics of the requirements to be a high-tech enterprise:

First, requirements on the management of R&D expenses are very strict. The enterprise should fill out the specified information into the application form including (the name of the R&D projects, duration, number of R&D personnels, budget, per year expenditure, organizational structure, core technology, innovation and substantial achievements. For the auditing reports, the enterprise should provide details of the annual R&D expenses including (the amount of internal R & D investment and outsourcing R&D investment, internal R&D investment is divided into R&D personnel salaries and input and depreciation charges, design fees, equipment commissioning charges, amortization of intangible assets and other specific information). The reports need to be rectified and audited by the intermediaries. If the enterprise does not have standardized management of R&D projects, without the launch of project, budget, the arrangement of R&D personnel and appropriate accounting system, it is impossible for an enterprise to provide such data. In addition, without an standard accounting system of the R&D expenses, the enterprise cannot enjoy the preferential tax deduction of 50 % based on the substantial deduction [4].

Second, requirements on the quality of high-tech products (services) are very high, which is reflected in three aspects: (1) Being in the advanced and leading position. The products and services should be included into the high-tech category with the strong support from the state. Enterprises need to fill out the specific form for the application of high-tech products. The form includes the information of key technologies and major technical indicators, its competitive strengthen compared with similar products, intellectual property rights certificates etc. (2) The product sales revenue proportion must meet the requirements: the PS project income accounts for at least 60 % of the total income (including the main business income and income from other sub-business). This information should be indicated in the audit report on the income of the high-tech products (services) by the intermediary. (3) Relevance. High-tech products revenue should be closely related to R&D projects. If the enterprise can provide co-operation contract with universities and research institutions, extra points will be added into the evaluation score.

14.3 Major Challenges and Problems in the Recognition of High-Tech Enterprise

Judging from the requirements and application materials, we can see that the key lies in the input of the enterprises on R&D projects and their capabilities of transforming the R&D achievements into production and the major challenge for the enterprise is to establish an effective management system, including the R&D management, management of the transformation from technology to products, or services and financial management.

The reason is: if the enterprise's R&D capability and technology transformation capability are very strong, but lacking of standard management system, for example, there is no project proposal, no budget and the accounting system is not standard, the enterprise cannot submit all the requirement information, or it is likely it will not pass the review section, which may lead its failure in the recognition of high-tech enterprise. When the enterprise has enough capital, it can improve its R&D capability by cooperating with universities or outsourcing its projects. However, the internal management level can be improved only by the enterprise itself.

At present, the major problems lying in the recognition of high-tech enterprises are: firstly, inadequate preparation, lack of prior assessment and analysis, not having a thorough understanding of application requirements.

Second, senior management pays more attention to the input of R&D and the transformation of achievements, while ignores the establishment of the R&D management system, for example, if there is no incentive measures for those involved into the application, this will directly affect their activeness in the application and if tasks are not well defined among relevant departments, the information cannot flow smoothly, thus causing the waste of resources.

Third, senior management of most enterprises have a bias opinion towards auditing. They treat the auditing as purely financial issues, which they think are the task only for financial department. However, financial department's work can only reflect the input of R&D and the income from high and new technology. If an enterprise has not established a scientific and complete management system to control the planning, project proposal and project operation, its financial department cannot work out a detailed project proposal, budget execution, project progress report required by the application.

14.4 Recommendation of the Application of High-Tech Enterprises

Based on the above analysis, we find that the enterprise should focus on the specific requirements and to improve its management system with efforts made in the preapplication assessment and planning, organization of implementation and follow-up management. The specific suggestion is as follows:

To carry out an effective and efficient pre-application assessment and planning. Before the application, it is imperative for the enterprise to carry out an effective self-assessment, with the focus on the capability of its R&D personnel, the input of capital, the level of the transformation of technical achievements and its competitiveness. The purpose of the assessment is to help the enterprise to know more about itself so as to avoid the waste of resources. Some enterprises want quick success. They make a very rush decision and only until they are submitting the materials, do they realize that they cannot meet some of the basic requirements, which put them into a very embarrassing situation. This step is very important but it is often ignored by the enterprises. The enterprise should compare its real situation with the requirements one by one and try to seek the differences. It can also hire an experienced intermediate organization for their professional guidance so as to lay a solid foundation for its application.

Preapplication assessment is the basic step. Once the planning has been formulated, the enterprise can move on with its application. It shows that assessment ensures a great success of the application.

The major issues need to be considered during the first application are as follows: first, to make sure whether the enterprise needs to restructure itself. According to the requirements of high-tech enterprise, the income from high-tech products (services) and that from nonhigh-tech products (services) should be treated separately. The income from high-tech products and services accounts for at least 60 % of the total income. Therefore, the enterprise should decide whether it needs to restructure itself or not based on its income structure. If it cannot meet the requirement, it is feasible for the senior management to consider to reregister a new enterprise to undertake the business of high-tech products (services) and other related businesses.

Second, the time for application needs to be carefully determined. The enterprise should decide which year is the best time for them to submit its application. Third, it is necessary to establish R&D Department. This will help the enterprise to make a clear difference between the input in R&D and the input in other production areas, so as to establish a standard R&D input accounting system to meet the requirements of strictly controlling of the input in R&D and the preferential tax deduction for the recognition of high-tech enterprise.

Specific measures as follows need to be taken as the application requires the joint efforts from financial sector, R&D department, human resources department, and other relevant departments, it is very important for the enterprise to set up a special group for the coordination and implementation of the application with the head of the company as the group leader to progress the application as one of the key projects.

Second, to focus on strengthening the coordination and communication between R&D sector and the financial sector [5]. R&D department and financial department should strengthen their communication to ensure the correct division of accurate imputation of R&D development costs and accounting, high-tech products and nonhigh-tech products revenues, R&D project budget formulation and control, and so on. R&D department is responsible for the technical aspects, including the management of R&D projects, the identification of the transformation of technological achievements and products, the provision of data and information relating to technology and coordinating with the financial sector in the accounting and auditing of the input in R&D and income from high-tech products (services). Financial department is responsible for standardizing of R&D expense approval process, budget management, accounting, and assisting R&D department in providing financial information of the R&D projects. In summary, the application of high-tech enterprises cannot success without the close cooperation of these two departments. If any party is passive, it will directly affect the result.

If there is not a standardized, scientific research and development management system, the R&D efficiency, the transformation of technological results and the corresponding accounting are bound to be affected. This requires the establishment of management system to control the reviewing of the plan, the launch of the project, and the operation of the project, budget execution and financial statements. Incentive measures and R&D personnel performance appraisal mechanism are also necessary and important. The establishment of the system will facilitate the enterprise for the normalization and standardization of management and to overcome temporary and sudden impact on the management system.

In accounting the R&D expenses, the enterprise should faithfully record the use of R&D expenses, set up detailed items based on different projects and to set up items such as labor costs, direct investment, depreciation charges in accordance with the "High-tech Enterprise Recognition Approach" [6]. The enterprise should finds the relation between the detailed accounting data and required data in the application form and separate the accounting of high-tech revenue and non high-tech income.

The recognition of high-tech enterprises is related with the work of science and technology department, Taxation department and auditing sector [7]. According to the identification procedure, the auditing sector will audit the R&D expenses and income from high-tech products and work out a report. The application group should maintain regular contact with the audit agency, to timely identify the problems and ask for their advice; the recognition of high-tech enterprises is finally determined Ministry of National Science and Technology. Once the enterprise obtains the certificate, it needs to report to the tax authorities in a timely manner. The information should be submitted in accordance with the requirements of the R&D projects. Thus the enterprise can enjoy an additional deduction of research and development expenses on the basis of the actual deductions and a 15 % deduction of its EIT.

The follow-up Management. According to "High-tech Enterprise Recognition Approach", the validate period for the certificate is three years since its issuance. That means the enterprise can enjoy favourable tax policies in the same year when it obtains the certificate. The enterprise needs to reapply in the third year. Therefore, the application is not a once and for all approach. The enterprise should focus on the change of the state polices, improve its core competitiveness, and make full use of the preferential tax policies to lower its operational cost, so as to maintain a sustainable and healthy development [8].

Enterprise management process as shown below:

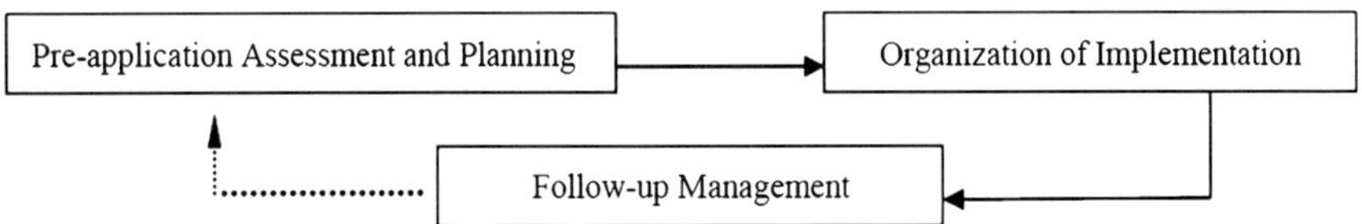

To sum up, a scientific enterprise should seize the opportunities brought by the state strategic adjustment and preferential police, strength is R&D level and its capability of transforming the technical results into production and improve its management mechanism, so as to build itself into a high-tech enterprise with strong core competitiveness, playing a supporting role in the economic and social development.

References

1. Xi C (2011) Some experience of special audit about high-tech enterprise identification, China certified. Public Acc 9:31–32
2. Guanmin N (2011) Discission on the auditing of high-tech enterprise recognition, China certified. Public Acc 5:63–65
3. Guihong S (2010) Identification of high-tech enterprise and its financial accounting and management. Population Econ 3:266–267
4. Jianxi Z (2010) Additional deduction policy application of the innovative enterprises' research and development expenses. Public Acc 10:21–22
5. Fang Z (2009) Interpretation on preferenital tax policy of pre-tax deduction of the cost for research and development. J Shanxi Coll Youth Administrators 4:70–71
6. Datuan M (2011) The discussion of R&D expenses processing in the high-tech enterprise recognition. Enterprise Guide 13:149–150
7. Xiaorong M (2009) The changes and implication of the pre-tax deduction policy of enterprises' research and development expenses. J South China Agric Univ (Soc Sci Ed) 8:80–84
8. Qiyong G (2009) Some experience of the tax policy's practice about additional deduction of research and development expenses. Sci Technol Inf 27:321–323

Chapter 15
Study on Recreational and Leisure Activities Participation

Guang-hui Qiao

Abstract In recent years, many Chinese students went to Korea for study. Leisure has a great meaning for foreign students, through the participation of leisure to spend their time after daily going to school, and then improve the quality of service for life. At the same time, the constraints on leisure participation have received increasing attention in leisure studies in Korea as well as other countries during the past decade [1]. The purpose of this study was to compare the differences of leisure participation activities and leisure constraints between Chinese students and Korean students. A random on-site survey was conducted from March 15th to April 15th 2011 and a systematic sampling approach was used to select 630 respondents (300 Chinese students and 330 Korean students) in Korea Paichai University. This study used factor analysis to select 5 constraint factors on recreational activities participation (Psychological, Time, Accessibility, Partner, and facility and Safety). Seven test analyses also revealed the significant differences from Chinese students and Korean students on the constraint of recreational activities participation. In the end the insufficiency of this study is put forward and recommendations on the future study are also given.

Keywords Constraint • Chinese students • Recreational activities participation

15.1 Introduction

During the past decade, the constraints on recreational and leisure participation have received increasing attention in recreational and leisure studies in Korea as well as other countries [1].

Project Source: Henan Province Department of Education Humanities and Social Research Project, 2012-QN-079; Henan Province Association of Social Sciences Research Project, SKL-2012-3540: Zhengzhou Association of Social Sciences Research Project, Sk1440 Zhengzhou City Soft Science Project, 20121248.

G.-h. Qiao (✉)
School of Management, Henan University of Technology, Zheng zhou, China
e-mail: yangqiao1980@126.com

G.-h. Qiao
Post doctor in Human Geography, Henan University, Kai Feng, China

Z. Zhong (ed.), *Proceedings of the International Conference on Information Engineering and Applications (IEA) 2012*, Lecture Notes in Electrical Engineering 220,
DOI: 10.1007/978-1-4471-4844-9_15,

The statistical data shows that there will be 5.5 million graduates from universities or colleges, and 14.85 million graduates from senior high schools in China. This means the population base of those who have the opportunity to study abroad will increase by about 25 % over 2007. Experts have forecasted that additional factors, including the 2008 Beijing Olympic Games, the inland employment situation, and appreciation of the RMB, and the development of new visa processes may lead to a 30 % increase in students choosing to go abroad. Thanks to the good relations and frequent cultural exchanges between China and Korea, the big influence of "Korean Wave" in China, Korean traditional charming cultures and Korean complete and modern educational system, the number of those Chinese studying abroad in Korea is increasing by 25 % per year. The number of Chinese overseas students in Korea has increased significantly in recent years (http://www.chuguo.cn/news/107927.xhtml). A large number of Chinese students exist in Korea, so a research on their participation to recreational activities and the constraints to participation is very meaningful and will play a role for Korean recreational and leisure tourism industry. As we know leisure has a great meaning for foreign student, through the participation of leisure to spend their time after daily going to school, and then improve the quality of service for life. The leisure participation and experience of leisure constraints among foreign students may provide new insights into the nature of leisure participation on foreign student in Korea. International students encounter many challenges in the process of their adjustment to their host country [2].

The purpose of this study is to investigate the constraints to use of outdoor recreation resources and participation in leisure activities among Chinese students in Korea. The purpose was: to compare the differences of leisure participation activities and leisure constraints between Chinese students and Korean students.

15.2 Literature Review

15.2.1 What is Leisure and Recreation?

Leisure: Consists of your free time and what you enjoy to do. It also involves fun and social acceptance [3].

Recreation: The local part refers to your home area and what you do there [3].

15.2.2 What is Leisure and Recreational Constraints?

Leisure constraint refers to any factor that intervenes between leisure and satisfaction with one's leisure [4].

Several recreational and leisure researchers have developed a variety of conceptual and methodological approaches that serve to explain how constraints on recreational and leisure activities might operate [5]. Crawford and Godbey [6] proposed intrapersonal, interpersonal, and structural constraints. Intrapersonal constrains refer to individuals' psychological states and attributes. Interpersonal constraints refer to interpersonal interaction or the relationship between individuals' characteristics. Structural constraints are related to the intervening factors between leisure preference and participation.

Several theoretical frameworks exist for guiding research concerning those impediments confronting various population groups wishing to pursue leisure and tourism activities [7]. These constraint frameworks have primarily focused on two thematic areas of investigation: activity specific participation barriers, and the impediments facing particular segments of the population [8].

Activity research has centered on identifying the constraints associated with commencing, maintaining, and increasing involvement in particular pursuits [9]. It has also examined the reasons for dropping out of certain activities. Specific activities examined in this regard include hiking; golfing [10]; card playing [6]; camping [11]; tennis [12]; and skiing [13]. Shaw et al. [14] provided support for the negotiation proposition, stating that leisure participation is dependent not on the absence of constraints but on negotiation through them. Recent studies highlighted that constraints were more frequently perceived by participants than non-participants and individuals do negotiate through constraints to start or continue participation in leisure [15]. In other words, higher levels of perceived constraints do not necessarily result in less participation in recreational sport activities, and that individuals do not necessarily react passively to constraints.

15.2.3 Leisure and Recreational Constraints for Foreign Students

During their school years, college students deal with stress from academic life and issues stemming from normal development, such as psychological autonomy, economic independence, and identity formation. Compared to host national counterparts, however, international students experience added difficulty in that they must cope with other forms of stress [16]. This difficulty includes culture shock, language difficulties, adjustment to unfamiliar social norms, eating habits, customs and values, differences in education systems, isolation and loneliness, homesickness, and a loss of established social networks [3].

This paper has done a comparison of recreational and leisure participation and leisure constraints between Chinese university students and local Korean University Students.

15.3 Methods

15.3.1 Questionnaire Development

The questionnaires were designed to measure perception of constraints in recreational and leisure activities, leisure participation pattern, and demographics.

A five point Likert rating scale measuring constraints, ranging from 1 = has not influenced me at all to 5 = has influenced completely, was used. The questionnaires for this study were adapted from the research papers by Crawford et al. [11].

15.3.2 Survey and Data Collection

This study used a random survey. An on-site survey in bi-language Chinese and Korean was administrated to Chinese students and local Korean students from March 15th to April 15th, 2011, in Korea Paichai University. A total of 630 questionnaires were delivered, 624 questionnaires were returned and 602 usable samples were obtained, resulting in a response rate of 95.6 % (298 Chinese students and 304 Korean students). The survey was administered by my classmates and I who understood well the purpose of the study.

15.3.3 Analysis Methods

The collected data were analyzed by Statistical Package for Social Sciences (SPSS) version 15.0. Statistical techniques such as descriptive analysis, factor analysis, and T test were used to achieve the objectives of this study. Descriptive analysis was used to find out the ranking of importance on perception of constraint. Factor analysis was used to reveal the grouping factors of the items of the constraint dimensions. And t test was subsequently conducted on differences of the perception of constraint factors between Chinese students and Korean students.

15.4 Data Analysis and Results

15.4.1 Ranking of Importance on Perception of Constraint

Table 15.1 shows the ranking of importance on perception of constraint for Chinese university students and Korean university students according to the survey. The first five top perceptions' constraints are No physical strength or capability (mean = 5.00); Not feeling fit enough (mean = 4.95); Not interested (mean = 4.94); Not confident (mean = 4.91); Health-related problem (mean = 4.83). So it means when we consider about young students' participation awareness to recreational and leisure activities, we should launch the major strategies according to perception of constraint with different importance.

Table 15.1 Ranking of importance on perception of constraint

Perception of constraint	Mean	Std. deviation	Perception of constraint	Mean	Std. deviation
No physical strength or capability	5.00	1.049	No money	4.62	1.112
Not feeling fit enough	4.95	1.071	Expensive fee	4.53	1.294
Not interested	4.94	1.114	Cost of equipment	4.52	1.245
Not confident	4.91	1.036	No one teach me	4.49	1.279
Health-related problem	4.83	1.036	No one to participate with	4.39	1.437
Did not enjoy before	4.73	1.166	Don't know where to participate	4.39	1.457
Busy life	4.73	1.116	Friends don t have time	4.21	1.516
Work/study to do	4.71	1.272	Not necessary skills	4.08	1.329
No time	4.71	1.125	Inadequate facilities	3.97	1.365
Social commitment	4.68	1.180	Inconvenient facilities	3.96	1.320
Family commitment	4.66	1.121	Afraid of getting hurt	3.93	1.241
			Safety	3.90	1.575

15.4.2 Factor Analysis to the Items of the Constraint

Constraint items (Six Psychological items, Five Time item, Three Accessibility items, Five Partner items, and Four facility and Safety items) have been described by factor analysis with SPSS12.0. In order to improve the results of factor analysis, some constraint factors were given up because of the loading value less than 0.5 and Common Factor Variance is less than 0.4. Finally there are 23 items which can be analyzed by factor analysis. Principal Component Method and Varimax Rotation were used on the extraction of factors rotation so that the common factors can be more satisfactorily explained. According to Eigen-value > 1, extracted five common factors "Psychological factor", "Time factor", "Accessibility factor", "Partner factor", "facility & Safety factor" (Table 15.2).

15.4.3 Test on Differences on the Perception of Constraint Between Chinese and Korean University Students

This study used *t* test to analyze the differences on the perception of constraint between Chinese and Korean university students (Table 15.3).

To these perception of constraint factors: Psychological factor (Chinese students: 4.64, Korean students: 3.96, $p = 0.000$), Time factor (Chinese students: 4.65, Korean students: 3.90, $p = 0.000$), Accessibility factor (Chinese students: 3.20, Korean students: 2.68, $p = 0.041$), partner factor Chinese students: 3.73,

Table 15.2 Results of factor analysis of the items of the constraint dime

Factors and items	Factor loading	Communality	Eigen-value	Variance explained (%)	Reliability coefficient
Factor 1: Psychological			7.736	22.1	0.929
No physical strength or capability	No physical strength or capability	No physical strength or capability			
Not feeling fit enough	Not feeling fit enough	Not feeling fit enough			
Not interested	Not interested	Not interested			
Not confident	Not confident	Not confident			
Health-related problem	Health-related problem	Health-related problem			
Did not enjoy before	Did not enjoy before	Did not enjoy before			
Factor 2: Time			2.075	18.0	0.851
Busy life	Busy life	Busy life			
Work/study to do	Work/study to do	Work/study to do			
No time	No time	No time			
Social commitment	Social commitment	Social commitment			
Family commitment	Family commitment	Family commitment			
Factor 3: Accessibility			1.684	16.0	0.880
No money	No money	No money			
Expensive fee	Expensive fee	Expensive fee			
Cost of equipment	Cost of equipment	Cost of equipment			
Factor 4: Partner					
No one teach me	No one teach me	No one teach me			
No one to participate with	No one to participate with	No one to participate with			
Don't know where to participate	Do no't know where to participate	Don't know where to participate			
Friends do not have time	Friends do nt have time	Friends do no t have time			
Not necessary skills	Not necessary skills	Not necessary skills			
Factor 5: Facility and Safety			1.013	9.8	0.798
Inadequate facilities	Inadequate facilities	Inadequate facilities			
Inconvenient facilities	Inconvenient facilities	Inconvenient facilities			
Afraid of getting hurt	Afraid of getting hurt	Afraid of getting hurt			
Safety	Safety	Safety			
Total variance explained					
Total scale reliability					
KMO = 0.873, Barttlet's test of Sphericity = 2543.590, df = 153, Sig = 0.000					

Table 15.3 Results of T-test for the perception of constraint factors by chinese university students and local korean university students

Category	Psychological	Time	Accessibility	Partner	Facility and Safety
Chinese	4.64	4.65	3.20	3.73	4.49
Korean	3.96	3.90	2.68	4.14	3.77
T–value	5.859	6.015	5.490	4.130	4.869
P value	0.000**	0.000**	0.041*	0.027*	0.030*

P_S: *P < 0.05
**P < 0.01

Korean students: 4.14, $p = 0.027$), facility and safety factor (Chinese students: 4.49, Korean students: 3.77, $p = 0.030$). Chinese students' perception of constraint is higher than Korean students'. It means that Chinese students have higher perception of constraint to recreational and leisure activities' participation.

15.5 Discussion

It can be considered that the recreational and leisure are often ignored in the lives of the university students due to the heavy study. Especially, foreign students are likely to have little recreational and leisure and do not use the same criteria as local students in evaluating their leisure involvement. The present study was an exploratory one aiming to examine the relationships between university students' perception of constraints and nationalities.

In Today's society every country attaches great importance to the development of recreational and leisure tourism and local residents' participation to recreational and leisure activities. There are many previous studies for the constraints of recreational and leisure activities [8, 17]. However, very few documents on the reaction and participation from the foreign students to local recreational and leisure activities, as well as the constraints involved in the analysis and research. However, the foreign students is a very important component in the local tourism industry, they play a very important role in developing local tourism. Their participation, their satisfaction may be a direct impact on local tourism and leisure industry to a higher level of development. In particular the number of Chinese students in Korea and their current situation, it is significant meaningful to have a research on Chinese students to participate in local recreational and leisure activities as well as their perceived constraints. So far, there is no such study which has been done. I preliminarily researched Chinese students in Korea and local university students to participate in recreational and leisure activities as well as the constraints in this study. The results showed that Chinese students in Korea and the local university students are different in the five factors of leisure constraints. Especially the most significant differences are the time factor, followed by the psychological factor, followed by the partner factor, followed by the facility & safety factor, and the last one is accessibility factor. Therefore, the local tourism-related sectors will focus on the five constraints factors to develop local recreational and leisure industry, so that foreign students in Korea, especially Chinese students will be easier to take part in recreational and leisure activities.

Besides, both foreign students and local university students reported a wide range of constraints that either limited or blocked their participation in recreational and leisure activities. It should be noted that university students' constraints to recreational and leisure activities are a function of cultural and social interpretations of nationalities differences. Most of the research on recreational and leisure constraints for individuals does not argue that foreign students have no leisure, but that they have less leisure, or face more constraints, than do local students' counterparts. It is well known that foreign students are constrained to certain leisure participation. Equal opportunity in recreational and leisure is necessary to empower foreign university students in recreational and leisure participation [18].

Finally, I hope this paper will give some considerations to the tourism-related organizations for reducing recreational and leisure participating constraints of foreign students. Regarding the positive association between engagement in extracurricular activities and academic involvement, one possible explanation may be that students' engagement in extracurricular activities and academic involvement are reciprocally influenced by each other [14, 19]. At the same time, taking courses may lead students to strengthen their interests in certain activities. Since the students are able to use the knowledge and skills they have gained in class when they return to their out-of-class activities, they may reap an immediate reward for their learning. This reciprocal relationship, therefore, may encourage students to maintain their levels of motivation to learn in courses, as well as encouraging them to engage in out-of-class activities. The language is possible a constraint as well for participation to leisure and recreational activities. Help the foreign students with Korean language to make their life and social relational activities easier.

Future research needs to be undertaken to examine how university students' perception of constraints are formed. The findings of this research suggest several directions for further investigations of university students' recreational and leisure participation. It implies that the nature of the recreational activity appears to play a vital role in determining the extent to which perceived constraints are effective. Finally, for research to progress, it will be necessary to utilize qualitative methods that facilitate an in-depth understanding of their constraints to recreational and leisure participation. Furthermore, because of this paper is a preliminary research for recreational and leisure activities participation between foreign university students and local university students as well as their different perceptions to the constraint of participation, so in the future, it should study more about:

The differences in recreational and leisure participation between foreign students and local students based on difference gender and age.

The differences in recreational and leisure participation among foreign students based on difference pocket money and length stay in local place.

The correlation between leisure participation and intrapersonal constraints.

References

1. Ahn BW, Yeo IS, Lee CW, Lee JS (2006) The influence of leisure competence on leisure constraint for paragliding participants. The Korean J Phys Educ 45(1):237–248
2. Alexandris K, Tsorbatzoudis C, Grouios G (2002) Perceived constraints on recreational sport participation: investigating their relationship with intrinsic motivation, extrinsic motivation and amotivation. J Leisure Res 34:233–252
3. Zu-nion P (2007) Special interest tourism and research methods, vol 3. PaiChai University Press, Taejon, pp 2–3)
4. Barker M, Child C, Gallois C, Jones E, Collan VJ (1991) Difficulties of overseas students in social and academic situations. Australian J Psychol 43(2):79–84
5. Backman SJ (1991) An investigation of the relationship between activity loyalty and perceived constraints. J Leisure Res 23(4):332–344
6. Crawford DW, Godbey G (1987) Reconceptualizing barriers to family leisure. Leisure Sci 9:119–127
7. Backman SJ, Crompton JL (1990) Di! erentiating between active and passive discontinuers of two leisure activities. J Leisure Res 23:154–161
8. Williams P, Fidgeon PR (1999) Addressing participation constraint: a case study of potential skiers. Tourism Manage 21:379–393
9. Bergin DA (1992) Leisure, activity, motivation, and academic achievement in high school students. J Leisure Res 24(3):225–239
10. Bialeschki MD, Henderson K (1998) Constraints to trail use. J Park and Recreation Manage 6:20–28
11. Crawford DW, Jackson EL, Godbey G (1991) A hierarchical model of leisure constraints. Leisure Sci 13:309–320
12. http://www.chuguo.cn/news/107927.xhtm 1 Admin (2008) The number of students from China studying aboard in 2008 will exceed 200,000 (4):90–98
13. Kay T, Jackson G (1991) Leisure despite constraint: the impact of leisure constraints on leisure participation. J Leisure Res 2:301–313
14. Shaw SM, Bonen A, McCabe JF (1991) Do more constraints mean less leisure? Examining the relationship between constraints and participation. J Leisure Res 23:286–300
15. Leong FTL, Chou EL (1996) Counseling international students. In: Pedersen PB, Draguns JG, Lonner WJ, Trimble JE (eds), Counseling across cultures, Sage Publications, USA (2):210–242
16. McGuire FA, Dottavio FD, O'Leary JT (1986) Constraints to participation in outdoor recreation over the life span: a nation-wide study of limiters and prohibiters. Gerontologist 26:538–544
17. Scott D (1991) The problematic nature of participation in contract bridge: a qualitative study of group related constraints. Leisure Sci 13(4):321–336
18. Park SH, Oh SS (2006) Examining the relationship between perceptions of constraint on recreational sport participation and gender among college students. J Leisure Recreation Stud (7):105, 113
19. Toyokawa T, Toyokawa N (2002) Extracurricular activities and the adjustment of Asian international students: a study of Japanese students. Int J Intercultural Relations 26:363–379

Chapter 16
Research on the Advantages of Informal Finance in Small and Medium Enterprises

Yang Yi and Yan Bailu

Abstract At present, informal financing channel has been a hot topic discussed in the researches on small and medium enterprises (SMEs). As the informal finance owns advantages in the collection of soft information, most of its researches use the information as the complete prerequisite of the assumption. This paper extends this assumed condition, holding that informal credit financing is conducted under the condition of the information asymmetry. It is concluded that informal finance has advantages on the interest rate during the information selection and also on the mechanism of solving moral hazard problems.

Keywords Small and medium enterprises • Informal finance • Financing • Gaming • Incomplete information

16.1 Introduction

Many researches indicate that informal finance is becoming the primary source of small and medium enterprises (SMEs) financing. In the financing of small enterprises, the informal finance plays an alternative role for formal finance to some extent [1]. Informal finance makes a difference to the development of SMEs in Wenzhou and was also widespread. When the financing needs of SMEs was below two millions of RMB, it was difficult for them to obtain supports of bank loans, needing to seek other financing channels [2]. There were needs on the private capitals in Pearl River Delta and Yangtze River delta transferring to China's inlands which severely lack capitals [3]. Main bodies needing finance changed greatly, not only producing different scales of

Y. Yi (✉)
Financial and Economic Department, Guangxi University of Technology, Liuzhou 545006, China
e-mail: xuexi@hrsk.net

Y. Bailu
China Banking Regulatory Commission Jiangxi Office, Nanchang 330008, China

Z. Zhong (ed.), *Proceedings of the International Conference on Information Engineering and Applications (IEA) 2012*, Lecture Notes in Electrical Engineering 220,
DOI: 10.1007/978-1-4471-4844-9_16, © Springer-Verlag London 2013

needs on credit funds, but also bringing financial service needs at various levels. 73 % of the SMEs financing comes from informal financial channels.

We need to first make explanations and regulations on informal finance: The smaller financing needs no mortgage generally; the interest rate can be higher than the formal finance but with various forms such as private free debit, foundation, private banks, pawn shops. There were no fixed rates but the rates often floated greatly, and the contract was not relied on the protection of national laws but with its own restraint mechanism [4]. Executing the contract was mandatory and rigorous [5], and then the default rate was not high. This paper argues that any debit and credit contracts, with their own restraint mechanisms and out of the protection of national laws with the exception of formal finance, belong to the area of the informal finance researches.

16.2 Analysis on the Advantages of the Interest Rate of the Informal Finance

This paper does not intend to deeply analyze the reasons why the informal finance came up, but to analyze the impacts of its micro-operation mechanism on the market equilibrium based on its existence and certain development.

16.2.1 The Assumption on the Clearing of Credit Market

This paper argues that the whole credit market is balanced, namely $D = S$; a balanced interest rate "r" ($r = m_1r_1 + m_2r_2/m_1 + m_2$, a weighted average of two markets' rates) will be obtained if $D = S$, and then SMEs can attain financial loans at the informal credit markets and ultimately achieve a balance of the whole credit market (i.e. $D = S$).

Relevant assumptions on informal financial credit markets:

1. The assumption on interest rates: This paper assumes that the interest rate of informal financial market is higher than that at formal market which is in line with the real market. As the interest rate is less than or equal to informal market, the funds supply stock of informal finance will shift to informal market, which will disappear. A floating rate is applied in informal finance, and the floating rights are in the hands of consignor.
2. The abstraction on the quality of agents: The research on informal financial credits needs a standard to measure different types of agents, but the traditional theory of credit rationing only abstracts the agents to be two types (i.e., high risk and low risk). This paper thinks that the types of informal financial market agents (borrowers) are highly complex with many immature market main bodies and opportunistic behaviors, and hence such a classification is too simple.

This paper abstracts the standards able to measure different types of agents. Namely "quality" is expressed by "q", $q = f(R, \theta)$, in which R means that agents obtain loans for investment or other incomes after other applications, θ is the variable of the project investment risk and q is a continuous function of R and θ ($R \geq 0$, $\theta \geq 0$). Thus, the q with the smaller risk θ and greater R is thought to be a high-quality agent, while the q with greater θ and smaller R is low-quality. In this paper, the agent factor is abandoned to set q, and thus agents' opportunistic behaviors at informal financial markets are common. If there is no effective restraint mechanism, the agents' creditworthiness is only subjective and difficult to be measured, needless to be measured.

3. The distinctions between different types of information: The informal financial sectors have greater advantages than formal financial sectors on the collection of SMEs "soft information" due to its popularity and geography factors (Yifu 2005). This has reached a consensus in current related documents, and information is used as the complete assumed condition to research the balance problems of the whole credit market. We believe that the transaction scope of informal finance goes beyond certain geographical or interpersonal relation when it develops to a period, and inevitably causes information asymmetry in the process of non-mechanism shifting to mechanism. Thus, this paper, taking the information asymmetry of informal financial market as the assumption prerequisite, maintains that the information asymmetry of informal financial market has advantages on mechanisms, such as flexible guarantee mechanism and cooperation mechanism. However, though these mechanisms are flexible, their regional differences are highly large, and also their floating interest rate is a mechanism commonly existing, which are the main reflections of the advantages of the information asymmetry of informal financial market on mechanisms.

16.2.2 Informal Financial Trading in this Paper is Divided into Two Types: Under the Conditions of Complete and Incomplete Information

1. Complete information: It is in line with reality when the trading is completed under the condition of complete information. For example, personal loans are often fully completed under the condition of complete information, and are possible to be impacted by interpersonal and regional factors; due to the geographical advantages of the debit and credit sides, the credit side is easy to collect the "soft information" of the debit side, making information reach complete status. Thus, this paper will not deeply analyze this issue.
2. Incomplete information: Under the condition of information asymmetry, the floating interest rate plays a role in the selection of different-quality agents, hence designing different contracts (r, q). "r" means the level of interest

rate, and q is the "quality" of an agent. Here, four agents q_1, q_2, q_3, q_4 are abstracted; q_1 is high θ and low R; q_2 is low θ and low R; q_3 is high θ and high R; and q_4 is low θ and high R. The distribution probability of q_1, q_2, q_3, q_4 can be expressed as $P(q_1) = P(q_2) = P(q_3) = P(q_4) = 0.25$ (Fig. 16.1).

After the success of work, ER is the expected income, r is the level of interest rate, R and r is in the negative correlation, $U = U(R, r)$;

In the graph of $\mathrm{EU} = U(\mathrm{ER}, r)$, the solid line is the utility curve of agent, dashed line indicates the expected utility line,

$$Uq_4 > Uq_3 > Uq_2 > Uq_1 \tag{16.1}$$

$$\mathrm{EUq}_4 > \mathrm{EUq}_3 > \mathrm{EUq}_2 > \mathrm{EUq}_1 \tag{16.2}$$

Because consignors will raise interest rates as much as possible to achieve their own profit maximization. Then, under the condition of asymmetric information, high interest rates will make q_1 and q_2 out of the market, which is shown as r^* in the graph, while q_3, and q_4 are still left. That is to say, through a rising of interest rate, different agents can be screened from R. But high interest rates does not make the consignors maximize profits ($D = S$), only offering different contracts for different-quality agents, at this moment the consignors will be screened again (Fig. 16.2).

Under the condition of market clearing, interest rate is raised at second time. Due to the raising of the level of interest rate, high-risk agent will exit the market,

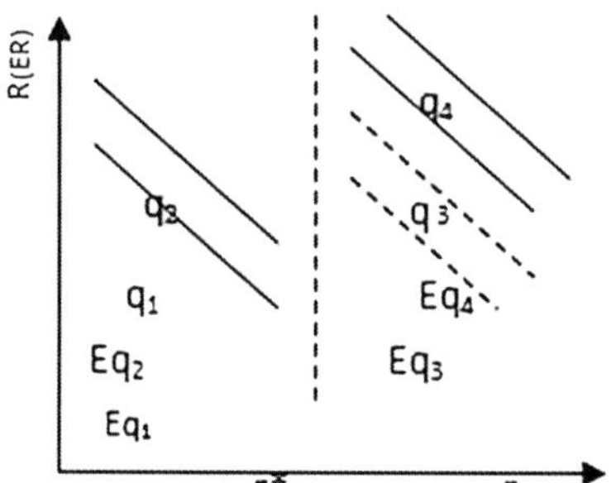

Fig. 16.1 R and ER is differentiated here; R is the income after successful investment

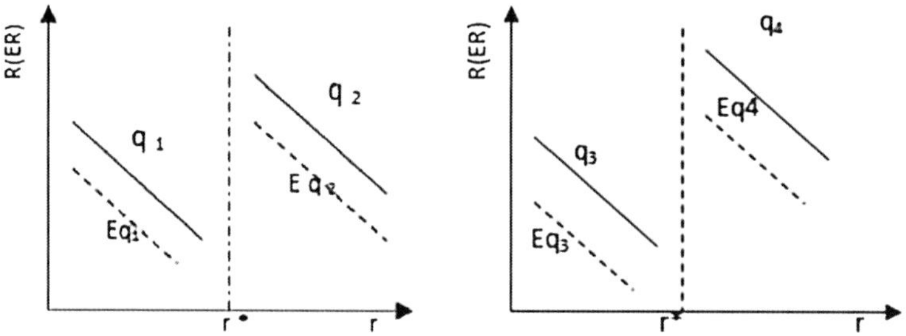

Fig. 16.2 R and ER is differentiated here; R is the income after successful investment

thus achieving differentiation to different-quality agents. Under the condition of market clearing, different-quality agent will obtain different contracts to achieve a balanced agreement (r_1, q_1), (r_2, q_2), (r_3, q_3), (r_4, q_4).

It should be noted that the model mainly describes the roles of floating interest rates in the selection of different types of agents, while real consignor will trade with q_4 with a higher interest rate r, and thus trades with q_3, q_2, q_1 by constantly lowering r, and ultimately achieving market clearing condition. Although the model is different from that of the real world, it is an objective fact that the interest rate plays a role in screening information. Thus, the model is logical, and informal finance has greater advantages than formal finance.

16.3 Analysis on the Advantages of the Informal Finance on the Prevention of Moral Hazard

After informal finance completes the selection of agents by using floating rate, the clearing of the financial market is realized $(D = S)$, producing a market equilibrium interest rate $r^* = m_1 r_1 + m_2 r_2 / m_1 + m_2$ $(r_1 < r_2)$ ("m" means the scale of financing, "r_2" means the level of interest rate of informal financial market), which is the weighted average of the interest rate levels of the four agents (i.e. $r_2 = mq_1 rq_1 + mq_2 rq_2 + mq_3 rq_3 + mq_4 rq_4 / mq_1 + mq_2 + mq_3 + mq_4$ $(mq_1 + mq_2 + mq_3 + mq_4 = m_2$, $rq_1 < rq_2 < rq_3 < rq_4)$. The equilibrium interest achieved at financial markets does not mean that every borrower can obtain loans at the equilibrium interest rate but with interest rate differential which is caused by two types of markets. However, the segmentations of the two types of markets are only short-term. Due to the presence of $r_1 < r_2$, m_2 will be expanded (here it is necessary to explain that $m_1 + m_2$ is not a constant), but m_1 will not necessarily reduce (m_1 means that part of the funds at formal financial market shifts to the informal financial market). When m_1 is invariant, m_2 will increase, suggesting that new capital flows to m_2, and $m1$ will relatively reduce. Thus, informal financial market has a trend towards expansion.

Though the floating rate achieves the selection of agents, it only solves the asymmetric information in advance, and then financial transactions are truly completed only when agent returns the principal and pay the interests. This paper abstracts away the agent's reputation factor, thinks that all the agents' reputations are not different after signing contracts (i.e., there are differences between no q_1, q_2, q_3, q_4, which are defined as H here). Agent will adopt reasonable default behaviors, but whether they perform contracts depends on the validness of consignor's restraint mechanism.

1. Assume there is a borrower (agent) H: H will take opportunistic behaviors (i.e., whether he performs contracts depends on the restraint mechanism or the loss caused by the default of H by the restraint mechanism under the given restraint environment).
2. There are two lenders (consignors) $F1$ (formal finance) and $F2$ (informal finance) sign $C1$ and $C2$ contracts respectively with H.

3. $F1$ and $F2$ will not take default behaviors, that is to say, $F1$ and $F2$ will not take any punitive action. Gaming analysis is conducted: H has two options (performance or default). H will get an income "R" after a project investment, and will pay the principal "m" and the interest "r" when performing the contract.
4. $F1$: If selecting performance, H will pay the principal "m" and the interest "r", so H will obtain indirect income "V" which is consignor's subjective evaluation on the credibility of agent, which is manifested as the easiness of next transaction. Due to the opportunistic behaviors, H will ignore this part of income. Therefore, H's income will be R-m-$r + V$ if he selects performance. And $F1$ will not take punitive actions if H performs the contract, then $F1$'s income is $m + r$.

If selecting the default, H will not have to return the principal "m" and interest "r", but will lose "V". $F1$ has two options: punishment and non-punishment. H's income is R-V and $F1$'s income is -m-r if $F1$ selects nonpunishment; H's income will be R-V-F and $F1$'s income is f-m-r-C if $F1$ selects punishment, in which F means H's loss caused by the F1's punishment and f means F1′s income ($f \neq F$) gained from the punishment on H. Also, F1's income "f" is often less than H's loss "F" caused by the punishment when H selects default (Tables 16.1, 16.2).

F' means the loss of $F2$'s action caused to H, f is the income after $F2$'s action, C' is the action cost of $F2$. The probability of $F1$'s punishment is $P1$ if H selects default, while its punishment probability is $(1-P_1)$; $F2$'s probability is P_2; H will expect the default costs to $F1$ and $F2$.

$$\beta_1 = P_1 (R - V - F) + (1 - P_1)(R - V) = R - V - P_1 F \quad (16.3)$$

$$\beta_2 = P_2 \left(R - V - F'\right) + (1 - P_2)(R - V) = R - V - P_2 F' \quad (16.4)$$

The values of $\beta_1 \sim \beta_2$ depend on the values of P_1F and $P_2 F$, while the values of P_1F and $P_2 F$ depend on f-m-r-C ~ f′-m-r-C′, mainly depending on the values

Table 16.1 C indicates $F1$'s action cost, which can be expressed as the following through the game model

F1	Non-punishment	Punishment
Performance	(R-m-r + V, $m + r$)	(*, *)
Default	(R-V, -m-r)	(R-V-F, f-m-r-C)

Table 16.2 H and $F2$ will conduct the same game model shown in the following

$F2$	Non-punishment	Punishment
Performance	(R-m-r + V, $m + r$)	(*, *)
Default	(R-V, -m-r)	(R-V-F', f'-m-r-C)

of C and C'. The ways of informal finance and formal finance to take punitive actions are different. The informal financial contracts are generally not bound to laws, and formulate severe punishment mechanisms to the performance of debit and credit contracts with the enforcement that may not be legitimate. Moreover, especially unusual means or even the illegal means becomes the mandatory ways for the contract performance (Zhang Qinggeng 2006). Thus, the punishment cost "*C*" of informal finance will be higher, and its illegalness will undertake higher risks. Meanwhile, the cost of formal finance will also be very high because of its agency cost and information cost [6] as well as illegal means and various principal-agent relationships. This paper takes the extreme value $C = C'$ to launch $P_1 = P_2$, and then only one factor F and F' to impact β_1 ~ β_2. F is H's loss after action. The losses caused by illegal and severe means are often higher than legal means. The paper maintains that $F' > F$ can launch $\beta_1 < \beta_2$. If P_1 and P_2 are changed into E (P_1) and E (P_2) (i.e. the expected probability of H on $F1$ and $F2$), and the $F' > F$ holds, then E (P_1) < E (P_2) can come up. Thus, illegal means can make agents subjectively think a greater possibility of informal finance to take actions. If the values of β_1 ~ β_2 depend on E (P_1) F and E (P_2) F', then $\beta_1 < \beta_2$ is launched.

Proposition 1: If $\beta_1 < \beta_2$, facing up to informal financial sectors and formal financial sectors, the expected cost of agent H on defaults in informal financial sectors will be greater than that at formal financial sectors.

Proposition 2: Opportunism commonly exists; if agent tends to break the contract, H will choose default on $F1$ but not on $F2$, forming repayment ration.

Proposition 3: After repeated games, H is easy to establish long-term relationship with $H2$, which can be also called "relationship lending" balanced as (Rm-r + V, $m + r$), but difficult to make long-term cooperation with $F1$.

16.4 Summary

This model assumes that agents have contracts in both markets. H is a theoretical abstraction as well, which can be understood as the rational reflections of agents (SMEs) to incentives under two types of financing environments. Adverse selection and moral hazard often occur in combination. Only the preinformation asymmetry and post-information asymmetry can be effectively resolved, able to make financial transactions actually completed. The selection advantages of informal finance in advance and its later restraint advantages mutually are the integral premise but not independent. Lin Yifu's complete information assumption on informal finance is reasonable from a macroscopic view, because the advantages of informal finance on mechanism make the information cost lower and its limit 0 can be adopted macroscopically, which can be only explained based on a microfoundation. Only introducing a model to the common but ignored informal finance at financial market and conducting in-depth analysis on its micro foundations, reasonable reform programs can be sought from a macroscopic view.

Acknowledgments The Findings at the Research Stage of the Scientific Research Project (201012MS127) of Guangxi Education Department.

References

1. Jie Z, Changfeng S (2006) Capital structure, financing channels and SMEs financing difficulties—an empirical analysis from China Jiangsu. Econ Sci 2(3):111–117
2. Lihong J, Xun Xin (2006) Analysis on the China's SMEs financing particularities and institutional innovation. Shanghai Finance 4(1):774–778
3. Chen H (2006) Ideas on the establishment of SMEs direct information platform. Shanghai Finance 2(2):53–58
4. Yifu L, Xifang S (2005) Information, informal finance and SMEs financing. Econ Res 4(7):876–884
5. Qinggeng Z (2006) Impacts of informal finance on formal finance and countermeasures. Shanghai Finance 5(1):95–102
6. Jie Z (2002) SMEs relationship lending and bank structures. Econ Res 9(6):46–49

Chapter 17
Study on Urban Infrastructure Bottlenecks

Yuhai Chen, Fengxiao Li and Huabiao Wang

Abstract The single investment of the government has always been the main limitation of the urban infrastructure. Broadening the financing channels and doing investment in the market are the effective ways to encourage operation of the urban infrastructure. In this chapter, the author first analyzes the problems lying in the infrastructure construction and urbanization. Solutions will also be provided and discussed. Finally, preliminary study of the policies and measures of urban infrastructure is presented.

Keywords Urban infrastructure • Construction issues • Thinking strategies

17.1 Introduction

The number of infrastructure of urbanization and its quality are two important components to measure the quality of urbanization [1]. Urban infrastructure is made up of electric power, water supply, drainage, transportation, roads, post and telecommunications, communications, environmental protection, disaster prevention, gas supply, heating, and other related engineering facilities [2]. It also includes technology services, culture and education, health and sanitation, business logistics, information, and other nonengineering facilities.

17.2 Problem Analysis

The investment and financing system in the modern urban infrastructure still reserves strong management mode of the planned economy. As a result, there exist a lot of problems in operation.

Y. Chen (✉) · F. Li
Hebei North University, Zhangjiakou 075000, Hebei, China
e-mail: pkohage@163.com

H. Wang
Hebei Institute of Architectural Engineering, Zhangjiakou 075000, Hebei, China

Z. Zhong (ed.), *Proceedings of the International Conference on Information Engineering and Applications (IEA) 2012*, Lecture Notes in Electrical Engineering 220,
DOI: 10.1007/978-1-4471-4844-9_17, © Springer-Verlag London 2013

17.2.1 Lack of Government Investment

Nowadays, the sources of funding of the urban infrastructure are mainly from various kinds of taxes. Although there are many kinds of charges, they still cannot satisfy the funding needs from urban infrastructure. In recent years, urban infrastructure allows local governments to take loans from banks, which eases part of the pressure from shortage of funds in urban infrastructure. However, the loans from commercial banks are mostly short-term, compared with the long-term urban infrastructure that uses large amount of funds. The conflict between these short-term loans and long-term needs make the cost of loaning very high and the repayment pressure increase. The funds central government puts in local government in urban infrastructure are not enough. With limited funds to implement urban infrastructure, the local governments appear to be inadequate.

17.2.2 Blur of Business Entities

The investment and management authority and responsibility in urban infrastructure are unclear and witch hunt. What's worse, their business entities are absent. Governments and authorities are both investors and operators, which is lack of constraint mechanism.

17.2.3 Lack of Long-Term Planning

The operation of urban infrastructure usually adopts executive methods. The determination of projects depends on the amount of money available. As a result, it is lack of long-term planning and continuity of construction. This kind of short-term activities results in problems such as duplication, a waste of funds, and low quality. In this way, the class of urban infrastructure and the taste of construction will still linger in the low level. The irrational layout of small towns and counties as well as industrial parks will be lack of power to support industrialization, urbanization, and the development of information.

17.2.4 Extensive Management

Most of the projects of urban infrastructure have less direct economic benefits, but with better social benefits and macro efficiency. They are lack of clear owners to operate and manage. Due to the reasons that management is operated by many owners and that they take what they need, the state-owned assets are idled, wasted, and lost greatly. Besides, the management is lack of effective integration which results in not enough close convergence and researching guides of the infrastructure such as roads, transportations as well as garbage and sewage treatment in the early, middle, and later stage of the construction.

17.2.5 Limited Bank Credit

Seen from the current situation of our country, we can conclude that although the liability scale of the financing platform in the local governments has reached a healthy level internationally, there exist problems like the platform scale growing too fast of some local government and the operation lacking of regulation. Taking these problems into consideration, our state requires proper management of the financing platform in local governments and processing the debts of some financing platform companies. CBRC has made it clear that commercial banks should complete the clear-up work of local financing platforms comprehensively. At the same time, it requires that any local financing platform projects should be stopped signing the contracts. In that case, the credit from banks to urban construction is limited to a great extent.

17.2.6 Poor Co-Ordination

There is not enough coordination between urban infrastructure and regional economic and social development; among various industries, regions, and department in infrastructure; among the planning, construction and management in infrastructure; between urban infrastructure and lying industries; between urban construction and the use of urban land as well as between key infrastructure projects and supporting projects. They are lack of interrelatedness and coordination.

17.3 Proposed Scheme

17.3.1 Well-Developed Policies and Laws

The first action that should be taken is that we should make up related policies for finance so as to support urban infrastructure. This will provide policies basis for the development of urbanization. Second, the state should introduce some appropriate policies and measures for China Development Bank and Agricultural Development Bank to open up loans with targets. It can also set up some development financial institutions to especially satisfy the loan needs during the course of urbanization infrastructure. Third, the system of related laws and regulations of asset securitization should be built up and improved as soon as possible. The perfection and regulation of the system and building up some necessary financing institutions contribute to the government support and oversight mechanisms in asset securitization. Some related details should be researched intensively. We should also cultivate and regulate the environment of securitization of infrastructure assets, including credit environment and the market environment. Fourth, financial input should be increased. We should focus the expenditure on infrastructure, utilities and public projects. We should also play the function of leverage from the financial investment well by partial investment, discounts, credit guarantee, and taxes relief to attract investment from the whole society. Finally, we can improve the financing and loaning capacity of the urban construction investment companies by increasing their cash flow, fostering the profitability of the infrastructure projects and decreasing the asset-liability ratio with reduction of asset allocation. We should also put aside some amount of funds that are borrowable to devote into debt restructuring.

17.3.2 Reasonably Divide the Layers

17.3.2.1 Reasonably Divide the Layers of Infrastructure Supply

According to the market-oriented situation, there are basically three categories. The first one belongs to the market type, that is, the investment can be recovered and generate accumulation of capitals projects. It should allocate the social resources through the market mechanism on basis of the standard planning and regulation management of the government. The second belongs to the semi-market type which can attract social investment with the help of public policies as well as products and services revolution. The third one belongs to the investment by the government, that is, pure public project or low charges project. Its construction funds are mainly collected from the taxes by the government and from the profits of some operational projects.

17.3.2.2 Scientifically Divide Responsibilities Among Governments in Infrastructure

According to the principle of responsibility symmetry, we should reasonably divide the responsibilities of the planning, constructing, operating, and managing of the infrastructure in cities and counties governments. Municipal governments should implement the great decisions that are made from the state, undertake the forecast of infrastructure needs, overall layout, long-term construction planning and implementation of major projects, and improve the integration and sharing of infrastructure resources. On the other hand, they should build up the interests and mechanism system to encourage the mutual cooperation among county governments, which supervises county governments to make infrastructure construction into the city's master plan. As a result, competition system is arisen. Besides, they should effectively regulate the operators of the infrastructure. As for the strong natural monopoly, its services quality, product prices and public interest protection should be regulated. And for the infrastructure field of its substitutes, they should provide convenience for competition. For the county governments, they should first relate to their own situations to improve the infrastructure planning and properly organize the infrastructure and management in their own regions. Second, according to the principle of graded responsibilities, they should increase the input of public infrastructure in their financial district on basis of the economic development and enhanced financial strength. Third, they should assist the municipal governments to make major infrastructure across regions according to the planning and deployment of the municipal governments.

17.3.3 Solving the Problem of Stable Investment Channels

Financial institutions should identify the entry point and intensify support. On one hand, financial departments should take urban infrastructure into one of the key points of the credit policy and financial supports. The People's Bank should put urban infrastructure supports into the schedule of financial works, fully playing their role of window guidance. Various kinds of monetary policy tools should also be adopted to guide commercial banks and rural credit cooperatives to invest more in infrastructure to small and medium cities. On the other hand, more improved financing channels should also be set up.

17.3.3.1 Select Financial Derivatives

Financial derivatives include protected advance refunded bonds. The establishment and changes of governmental policies, tax law changes, and the fluctuations on exchange rates may all lead to the adverse changes of the cash flow in infrastructure. This will lead to the decrease of economic values in infrastructure projects and bring losses to the bonds holders. A protected advance refunded bond can ensure the holders to return the bond and collect cash with the price set in advance when adverse events occur. In this way, the bond holder not only purchases the bond itself, but also purchases the put option. The second one is convertible bonds with call options. The holders of convertible bonds enjoy the right of turning the bonds into stocks when some events occur in the future. When the infrastructure puts into operation and some special events happens out of people's expectation, the issuers can clear the debts and request redemption. And the holders can turn the bonds into stocks for the sake of sharing more profits. In that case, the original debts can be turned into the company's equity capital.

17.3.3.2 Reasonably Issue Municipal Bonds

Municipal bonds in western countries such as America, Japan, Britain, and France are highly developed and with very high credit rating derivatives. They have a long history of nearly a century. The post-war Japan regarded municipal bonds as a kind of strong financing tool to provide huge amount of capitals for infrastructure. In the 1920s, America has begun to issue municipal bonds to collect huge capitals for infrastructure. Municipal bonds are issued by local public bodies of cities and counties as well as small towns. Their purpose lies in collecting capital needed to develop public facilities. Issuing municipal bonds does help to relieve the capital pressure from urban infrastructure, but there are many issues that should be paid attention to during the course. First, the government issues municipal bonds, aiming at making up the deficit and broadening the public investment. As local bonds, municipal bonds should pay attention to the competition with other kinds of securities when they are contributing to solving the problems above. Second, the issuer should bear high credit rating. In China, the issuer must be the governmental organizations, especially those in relatively developed regions. Third, the scale of issuing municipal bonds is a serious economic issue. Generally speaking, it cannot exceed a certain percentage of the annual revenues of the local governments. And at the same time, it cannot be less than that percentage, either. If the quantity is too small, the bond liquidity will be influenced. If otherwise, the government can't return the principal and pay the interests in the maturity, which reduces the credibility of the government and increases the available size of the local governments.

17.3.3.3 Accept Participation of Investment Trust

The achievement of investment and financing reform of infrastructure projects depend on the coordination of all entities that participate in. They should join hand in hand to design a good financing plan, risk control actions, and revenue sharing program. During this process, financial intermediaries like the investment trusts can not only take advantage of their professional experience and capitals to only participate in the design of financial plans, but they can actively take part in initiating the project companies to become a sponsor in the urban infrastructure. As the investment and financing side, investment trusts accept trust funds from dispersed investors. And there are two main ways—loan trust and equity financing. Loan trust means that the trust companies accept investors' trust by issuing interest warrants to pool funds entrusted and achieve accounting rate management for collection use. By way of project loan financing, they can provide support for the infrastructure projects. Project companies make mortgage in name of the right of operation and machinery equipments. The project sponsors found a project company especially for the financing and operation of the projects. And the project companies collect external loans in the name of their companies instead of the name of the sponsors. They act as both the trustees and the lenders, making loans from companies for the sake of making profits from the projects. The second type is equity financing. The funds of infrastructure are raised by private placement to call on investment companies and industrial enterprises to provide equity support for the infrastructure and the involved enterprises. Besides, capital management and supervision should also be untaken. It can get together the idle funds in society to invest in the infrastructure that has great potential. At the same time, it provides value-added services for the enterprises devoted in infrastructure by way of equity transactions to gain higher investment incomes. With the help of the fund managers who undertake centralized operation and professional management, the fund shareholders can be ensured to gain more benefits than their individual investment.

References

1. Xinglin Y (2004) Promoting urban infrastructure and construction of the finance market with the sense of urban management. City Forum 5(3):73–78
2. Rongjun L (2003) Strategic thinking of promoting the process of urban investment. Econ Reform 6(4):25–29

Chapter 18
Study on the Innovational Function of the Management

Yun Zhang

Abstract Traditional management process summarization should be improved in adapting and instructing the diversified management for modern enterprises, and at present the Chinese enterprises also urgently need the new understanding for the management function. Starting from the cognition and practice of the management, the innovation should be the core function of the management, and it should be embodied in the function of the management.

Keywords Innovation • Function • Management • Process

18.1 Introduction

Four functions such as planning, organizing, leading, and controlling in the management are the mainline in the management theory all along [1, 2]. Traditional management scholars thought that all managers in all departments and all classes in the enterprise would carry out these functions, and effective management is meant to complete these functions successfully [3, 4]. Robbins thought that "the actual management is not so simple like described management function, and there are not simple, limit-clear and pure starts and terminals of planning, organizing, leading and controlling in the real world". We think the innovation runs through the process of the management, and it should be the core function of the management.

18.2 Understanding the Innovational Function

The innovational function of the management can be understood as that the change of exterior environment threatens the survival and running of the system to certain extent, and to avoid the system loss induced by the threat or its extension,

---

Y. Zhang (✉)
Department of Industrial and Commercial Management,
Nanjing Institute of Industry Technology, Nanjing 210046, China
e-mail: g.qun@163.com

Z. Zhong (ed.), *Proceedings of the International Conference on Information Engineering and Applications (IEA) 2012*, Lecture Notes in Electrical Engineering 220,
DOI: 10.1007/978-1-4471-4844-9_18,

the system develops the local or whole adjustment in the interior, or predict the opportunity which is helpful for the system development in the environment when the system observes or experiences the changes in the exterior world, and actively adjust the strategy and technology of the system to actively develop and utilize the opportunity to seek the development of the system. Because of the multiplicity and levity of the threat and opportunity, the innovation should not be institutionalized, and the system has not copy gifted entrepreneurs all along, but the innovational entrepreneur emerges continually. With the development of productivity and the advancement of the technology, the innovational function of the management gradually possesses the natures of recognition and practice.

18.2.1 The Recognition of the Innovational Function

In the management function system, some scholars put forward the hierarchy function structure, and the macro function includes planning, decision making, evaluating, predicting, and other defensive functions, and the basic function includes planning, organizing, controlling, leading, commanding, harmonizing, and encouraging for general management problems, and the micro function includes many subfunctions such as investigating, researching, analyzing, processing, preparing, propagandizing, training, ordering, appointing and removing, inducting, consulting, deploying, implementing, censoring, checking, summarizing, and rewarding and publishing, which is used to prefect and supply the basic function and the macro function. In above three layers including 28 functions, the innovation is not listed. We think that the innovation exists in every function, and especially for the macro function and the basic function, the innovation will decide the result and efficiency of the management to the largest extent.

18.2.2 The Practice of the Innovational Function

Many successful enterprises generally adopt the mode of encouragement management, and form the innovation culture of the enterprise. Toyota Motor Corp alleged that their employees would propose 200 million new ideas every year, and each employee would put forward 35 advices averagely, and 85 % of these advices would be adopted by the company. In the daily management of Microsoft Corporation, the work time could be changed flexibly. Program design is a sort of research work with innovations and highly centralized energies, so the researchers and designers need comfortable feelings to complete their works consciously. Someone may select working in the night, and be on duty at 22 o'clock and off duty at 6 o'clock in the next day. To keep intensive and comfortable work atmosphere, the company divides employees into many group with 5–15 persons to engage in the research and development of the project. Designers all work for a long time, and they input program to the computer ceaselessly all day and print paper pile, and the work rhythm is so quick to make people crazy. Employees consciously delay the work time, and no one wants to be the person who first leaves the office. The work atmosphere forms loose and devoted work principles, and makes the work efficiency of Microsoft Corporation extraordinarily high. The innovational representation of the successful enterprise is the exertion and embodiment of the innovation of the whole enterprise, and the support and perfection of the innovational function is the activator to enhance the competitive force and efficiency of the enterprise.

18.3 The Innovational Function is the Core Function of the Management

18.3.1 Necessary Innovational Function

In the management process, maintaining the order is not enough, and once the management system is closed, the system can but spontaneously go to disorder. As viewed from the entropy theory and the dissipative structure theory, the negative and isolated maintenance is entropy production, and the essential of innovation and reform is negative entropy which is the necessary condition for the system evolvement. Without the innovation but the maintenance, the entropy production in the system can not maintain the order state, and without the maintenance but the innovation, the system will be in the nonbalanced state for ever. Only the management with the optimized combination of maintenance and innovation is the excellent management. Peter Drueker thought that "the deficiency of innovation is the biggest single reason for the ruin of existing organization". In the future, the innovation of the organization and society is same important with the innovation of the technology. Only if the maintenance matches with the innovation and the speed and quality of the innovation is good enough, the enterprise or any human system in the competition could survive and develop healthily.

18.3.2 Innovational Function Represents its Values in Other Functional Activities

Various management functions have their own special representation form. The function of planning is embodied by confirming objective, establishing strategy, and developing the layout, project or plan of the sub-plans. The function of organizing is embodied by the organization structure design and personnel distribution.

The function of leading is embodied by the leaders' instructing and encouragement, and solving conflicts. The function of controlling is embodied by monitoring the activity process to ensure the information feedback and measure correction completed by the plan. If the environment does not change and renovate, the plan will be invariable, and all organizations will sustain good structure design, and the decision making project will be confirmed because of the exact prediction of the result. However, the environment always changes, managers should deal with challenges at any moment, and in each stage and every function of the management activity, they should innovate everywhere at every turn and at any time. The innovational function happens not only in the term of strategic layout decision making, but in the daily management, so it has not special representation form, and it represents its survival and values in all activities of other management functions. In the continually changing environment, the objective, content, form and activity factors of the system activity must continually change and adjust it, or else, the interior factors of the management system will be impacted even eliminated. Therefore, the innovation will be picked out from four functions of the management, and be the core function of the management.

18.4 Innovational Contents in the Management Process

18.4.1 The Innovation in the Planning Function

Some most important strategies in the organization always occur when you least expect them to, which cannot be known by the senior managers before. To fully utilize these strategies, we usually need to identify them and expand their influences. For example, some new purposes found by some salesman occasionally could be turned into the new major operation of the corporation. Planners always look for modes in the failed experiment, seemly random activity or disorder learning, and they usually could find the new methods to solve the problem or consider the problem, for example, finding the new market which does not get in, or understanding the characteristics of the corresponding new product, and so on.

18.4.2 The Innovation in the Organizational Function

Because of increasing competition, the requirement of the enterprise to the innovation is higher and higher, and many enterprises are implementing reforming actively and flow construction, and trying to be learning-type organizations. Modern management theories thought that the organization innovation was the composing of the "technical" advancement and the investment of the management system and structure in the production, distribution and sale was the composing of the total capitals.

18.4.3 The Innovation in the Leading Function

The result of leading is to induce the reform, usually the drastic reform, and form very active reforming potential, which requires that leaders must accept continual challenges and innovations, and make the enterprise be in the advantaged competitive status through active reform.

18.4.4 The Innovation in the Control Function

Joseph Alois Schumpeter emphasized that "the innovation of the enterprise is the drive of the economic development". The intention of the management control is to test whether the system is running normally and developing according to the strategic objective. The innovation in the management control function is also embodied in the project management of the total lifecycle and total system and in the total system and total process quality management and control.

The innovation is not listed as a sort of management function, and it is related to the industrial development level in the initial development stage of the management theory. With the quick development, levity, and uncertain of the scientific technology, every manager will encounter new situation and new problems every day. The conformism can not deal with new challenges, and without the innovation, the enterprise can not sustain the operation and acquire ideal management performance. The drastic competition and challenges of the globalization compel managers to make the innovation in the management process as a sort of core function. In Guntern's book, he pointed out that "the proper combination of sufficient innovation speed and quality is the first key factor to acquire the success". In the further research, we should mainly emphasize the content embodiment of the innovational function and how to embody these contents.

References

1. Guntern G (2000) Challenges of innovational lead, 25(14). Tsinghua University Press, Beijing, pp 29–35
2. Huang R (2007) Contents and methods of the enterprise management innovation. Mod Manag Sci 10(1):79–81
3. Zhang W (2001) New innovational management in new economic times: innovational management of the complex product system. Econ Manag 20(16):69–75
4. Sanduo Z (2000) Management Theory, vol 12(6). Higher Education Press, Beijing, pp 39–47

Chapter 19
Study of Low-Income Housing System in China

Qun Gao

Abstract Low-income housing system in China, provides three forms to Raise funds to building affordable housing and low-cost housing, such as tax rebates and general transfer payments, Special transfer payment, in order to achieve the basic housing needs of low-income groups and to achieve social Fair in the domain a of real estate.

Keywords Low-income housing system • Transfer payment • Safeguard people's livelihood • Social justice

19.1 Introduction

The low-income housing system means a series of system giving priority to the transfer payment system, and the concrete systems that includes the government provides the basic living houses for those low-income families [1, 2], and provides the low-rent house for those low-income families by the power from the country, the local government, and the society, and provides the economically affordable house for low-income families, and provides the price-limit house and the commodity house for the middle and high-income families [3, 4]. The main target of the transfer payment system is the equalization of public services, and the equalization of the real estate means the fair sharing of the special public commodity, i.e., the real estate, which could also help the harmonious development of the real estate economy. Since 2005, the housing policy of China had begun transforming from the market housing to the low-income housing, to solve the housing difficulty of low-income families [5]. In the August of 2007, indicated that the low-income housing policy of China went to be perfected.

Q. Gao (✉)
College of Economic Management, Nanjing Institute of Industry Technology, Nanjing 210046, China
e-mail: gaoqun@hrsk.net

Z. Zhong (ed.), *Proceedings of the International Conference on Information Engineering and Applications (IEA) 2012*, Lecture Notes in Electrical Engineering 220, DOI: 10.1007/978-1-4471-4844-9_19,

19.2 The Current Transfer Payment Policy of China

The finance transfer payment system is based on the tax distribution system of 1994, and it is composed by three parts, i.e., the tax rebates, the general transfer payment, and the special transfer payment, and it is the transfer payment system with Chinese characteristics giving priority to the transfer payment from the central government to the local governments.

Tax Rebates. Tax rebates is the main form of the finance transfer payment of China, and it is the important source of the local finance incomes, so whether the tax rebates is designed reasonably or not determines the reasonability of the whole system. But the tax rebates is still distributed based on the base to maintain the local vested interests, and it embodies the policy that inclines to those regions with strong income ability, and it maintains the vested interests of rich regions, and runs counter to the intention reducing the gap among regions. In addition, the tax rebates belong to the central finance income on paper, but the local finance could determine these capitals finally.

General Transfer Payment. The general transfer payment is the subsidy payout which is arranged by the central finance for local finances to compensate the financial gap of those regions with weak financial strengths. The general transfer payment is the important measure to reduce the financial gap among regions, and it is the main part of the finance transfer payment, and mainly includes many forms such as the common transfer payment, the adjusted wage transfer payment, the national region transfer payment, the rural taxation expenses reform transfer payment, and the year-end balance finance subsidy.

Special Transfer Payment. The special transfer payment is the subsidy capitals established by the central financé to realize special macropolicies and develop strategic targets, and it is mainly used in various public service domains about the people's livelihood. And local finances should use the capitals according to the regulated purpose.

19.3 The Transfer Payment of the Government Financial Capitals

Financial capital is one of the main capital sources of the low-income housing construction. In past local financial budget, the low-income housing construction capitals have not been arranged, and at present, the space to pay the low-income housing capitals in the local financial budget which is originally intense is very limited. For the west regions with difficult financial income, the central government could support the local governments by the forms such as the budgetary investment subsidy and the special subsidy capitals of financial low-rent housing. In addition, the property tax can be collected in the whole country, and part of the taxation could be used for the construction capitals of the low-income housing, which is deserved to be explored.

19.4 The Transfer Payment of the Added Incomes of Housing Fund

According to the "Temporal Methods of the National Social Security Fund Investment Management" of the Ministry of Finance and the Ministry of Labor and Social Security, all net incomes of the social security fund should be brought into the social security fund which should be distributed and invested according to relative regulations of China. The "Methods" regulated that the housing fund was the "long-term housing savings fund", so the housing fund is the guarantee capitals which are specially used for solving the employees' housing problem, with the characteristics such as the specificity, the mutual assistance, and the security. These characteristics of the housing fund also belong to the content of social securities. The housing fund system should realize the mutual benefit, the mutual assistance, and the mutual association among subscribers, i.e., the added income of the housing fund except for the loan risk reserve and the relative management charges is used for the construction of low-rent houses. Since 1998, the construction capitals from the added income of the housing fund have achieved about 10 billions Yuan, which is one form of the mutual assistance, and it could not only effectively solve the problem of difficult housing of low-income families, and accord with the essential attributes of the housing fund such as the mutual assistance and the security, but also embody the intention that China pushed the housing fund system, and promote the health development of the housing fund system.

The added income of the housing fund means the difference between the operation income and the operation payout of the housing fund. The operation income includes (1) the interest income of the personal housing loans of the housing fund, (2) the income of the bank credit, and (3) the interest income of purchasing the national debts. The operation payout mainly includes (1) the agent fees of banks, (2) the deposit interest of the employees' personal housing fund, and (3) other relative management charges. That is to say, the added income of the housing fund is composed of the deposit interest and the interest difference between loans and deposits (i.e., the interest difference between the housing fund deposits and loans). In fact, the added income of the housing fund is generated in the process that the housing fund management center utilizes employees' personal housing funds to satisfy subscribers' demands of purchasing houses and adding values under the support of national policy, and implements many added value activities such as depositing, loaning, and purchasing national debts.

19.5 The Transfer Payment of the Land-Transferring Fees

Land is the basic resource for the human being to survive. The establishment of the land transfer system and the obtainment and distribution of the land transfer income are the basic guarantee for human being to live and develop. The formation of the land price accords with the economic development, the development course of city, and the durative of the land development, and only in this way, the security function of land could be exerted fully. As the important strategic resource of the country, the land is used for the country to obtain the durative profit by the land transfer, which is used for the public establishment service and the public management, and the construction of the low-rent houses. In the Economic Housing Application Procedures of China, the construction lands of the economic houses are provided by the form of administrative allocation, and brought into the yearly land supply plan of the current year, so the land cost of the economic housing construction is almost zero. At the same time, local governments are exempted from various administrative fees and governmental funds of the urban basic establishment expenses for the economic housing projects.

The central government has required local governments to use more land transfer incomes in the housing construction of the low-rent houses and other security houses by the land income transfer payment, which could further solve the difficult housing for middle- and low-income families in the economic regression. The proportion of the low-rent housing guarantee capitals in the new income of land transfer should exceed 10 %, and local governments could properly enhance this proportion according to the actual situations. The sharing of land income is the biggest livelihood, and to realize this sharing, the land price will certainly rise with the maturing of the regions and the land price in the mature regions only ascend and present the tendency of increasing values. In 2009, the income of national funds was 1833.504 billion Yuan, which equaled to 26.8 % of the national finance income in the same period. In the income, the most was the state-owned land sale revenue of use-rights. In this year, the income of national land transfer was 1423.97 billion Yuan, 77.7 % of the income of national funds, and the contribution ratio of the land transfer income to the national finance income was about 21 %. The central government and local governments increased the public finance income by obtaining the lands, and improved the difficult housing of low-income families by the transfer payment.

19.6 Attempt in Enterprises

Because the construction of the low-income housing belongs to thegovernmental function, the construction scale is larger, the "land finance" of the government reduces more, and local governments could let some big enterprises with lands to participate in the low-income housing construction, so some governmental functions could be shared by enterprises, which could not only reduce the burden of the government, but also correspondingly reduce the instant loss of the "land finance". In this way, both the government and the enterprise could obtain their own profits, and at the same time, because the new competitive subject, i.e., the industrial enterprises with lands, will occur in the commercial housing development market, that means the house price may descend properly. According to relative reports, many big enterprises in Shanghai have large number of lands, and the positions of these lands are better than the lands from the governments. It is obvious that these enterprises would use their lands for the commercial housing development and construction, but that was forbidden by the state. On the contrary, without special policies, these enterprises will not use their own lands for the commercial housing development and construction. Though the government has attracted many big enterprises to use their own lands to participate in the low-income housing construction, the policies may be changed. According to the "The Low-income Housing Construction Land Supply Management Implementation Methods of Shanghai", "if the enterprises voluntarily use their lands to apply and participate in the construction of low-income house, the baseline proportion of the low-income housing construction is 50 %". That is to say, the other 50 % of lands could be allowed to construct the commercial houses. If this policy could be implemented, many enterprises would actively participate in the construction of the low-income house.

19.7 Horizontal Transfer Payment Among Governments

In the practice, the horizontal transfer payment of finance has not occurred in the low-income housing among same-level governments. The transfer payment among governments is mainly to balance the gap of the government income because of different geographical environments or economic development levels, and ensure that the local governments could effectively provide housing guarantee for difficult families according to the uniform standard of the state. Generally, the regions with developed economy and rich finance could transfer the finance to the poor regions, for supporting other regions each other, reducing the regional gap, balancing the finance, assisting the society, and supporting the minority. The capitals come from regions with rich finance, and the capital transfer is direct, transparent, and highly effective. With the encouragement of the central government, a nonformula and non-institutionalized transfer payment occurs among provinces and regions.

In addition, the housing funds of different regions are extremely unbalanced, and the management of the housing funds in developed regions is reasonable, but the usably idle funds are less, and on the contrary, the housing funds in middle and west regions are rich, but the management is not standard relatively. The total idle housing funds of China are about 100 or 200 billions Yuan and the funds among different regions are not balanced, and if the uniform housing fund platform could be established in China, the funds in the rich regions could flow to those poor regions, and the low-income housing problem could be solved better.

19.8 Proposed Scheme

19.8.1 Improved Scheme

At present, there are no complete and systematic implementation methods about the finance capitals, land transfer, financial support, and transregional utilization of housing fund of the low-income housing construction, especially the problem how to carry out the capitals lacks in perfect measures. When the planning index has not been confirmed, part local governments negatively implement the financial payouts, which largely reduced the construction of low-rent houses.

The central government should require local governments to institute the low-income housing policies according to their local institutions for low-income families, and strengthen the supervision to local governments for the implementation of policies. At the same time, the construction of laws and regulations about the low-income housing should be quickened, and the local governments' responsibilities about the residence right should be confirmed by the legislation, and the subject and object of the low-income housing should have relative laws and regulations to abide by.

19.8.2 The Construction Capital Source

According to the "Urban Low-cost Housing Management Practices of China", the construction capitals of low-rent house should give priority to the public budget capitals of finance, but only a few cities establish the finance capitals supply plans from the system, and most cities depend on the added incomes of housing funds and the surplus of the low-income housing sales to collect the construction capitals of low-rent houses. This capital proportion is lower, and not stable, and the subsequent capitals could be guaranteed.

Therefore, the central government and the local governments must strengthen the finance capital supply plan of the low-rent houses, fully and reasonably utilize the added income of housing fund, enhance the proportion of the land transfer income in the low-rent housing construction capitals, actively exert the social power, and collect the construction capitals of low-income houses from multiple channels. At the same time, the investment subsidy of the central budget and the special subsidy capitals of the central financial low-rent house should incline to the difficult regions in the middle and west of China.

Acknowledgments The research is supported by the project of the Jiangsu Provincial College Philosophical and Social Science Fund (No. 09SJD790026) (Sponsoring information).

References

1. Gao Q (2006) Brief of the development prospect of traveling property. Constr Econ 15(5):13–15
2. Niu FR, Li JG (2008) Development report of Chinese real estate, vol 45(23). Social Sciences Academic Press, Beijing, pp 167–168
3. Qiao BY, Fan JY, Peng JM (2006) Government transfer payment and local finance endeavor. Manag World 25(25):45–47
4. Yin H, Kang LL, Wang LJ (2007) Financial equalization effect of the government transfer payment: research based on china county-level data. Manag World 13(45):456–457
5. Zeng QF (2010) Causes and countermeasures of the financial risks in chinese real estate. Bus Chin 46(33):24–26

Chapter 20
Effective Managing Approaches in Student Management Work

Qiaoyuan Wei

Abstract Student management work is complicated and also diverse. How to do it well is a problem that must be tackled by student managers. This article states the effective management methods from the aspects of strengthening the daily ideological and political education, the construction of learning style and academic advising, solving the students' practical difficulties, cultivating outstanding student leaders, increasing class cohesion, keeping safety education and maintaining stable operating, emphasizing on home visits, and so on.

Keywords Student management work • Effective • Managing methods

20.1 Introduction

Student management work is complicated and also diverse and that the thoughts of working objects—students are becoming more complex, working enthusiastically, and sense of responsibility are particularly important in order to do a better job in less time and establish a good image in the minds of students [1, 2]. On the one hand, trying to guarantee the investment of certain amount of time and energy to develop good work objectives and fully mobilize and play the subjective initiative of each student, especially student leaders to improve work efficiency; on the other hand, student managers should use every opportunity to strengthen the emotional exchange with students, narrow the psychological distance, actively care about them, be close to them, open their heart to students, and have a heart-to-heart talk and treat them frankly; warmly solve the difficulties for students and timely figure out the emerging questions [3]. Student managers should always adhere to the belief of "everything for the students, for all things of students, for all students", keep the work style of "be aware of the students, relieve students from anxiety,

Q. Wei (✉)
Department of Chemistry and Biological Engineering, Guangxi Normal University for Nationalities, Chongzuo 532200, Guangxi, China
e-mail: weiqiaoyuan@hrsk.net

Z. Zhong (ed.), *Proceedings of the International Conference on Information Engineering and Applications (IEA) 2012*, Lecture Notes in Electrical Engineering 220,
DOI: 10.1007/978-1-4471-4844-9_20,

be heart-warming to students", regard student management work as a cause to sacrifice, pursuing work sincerely, and loving students.

20.2 Strengthen Daily Ideological and Political Education

Student managers should attach importance to the entrance education for new students, teach them how to adapt to college life, how to take care of themselves, organize them to study the school's student management manual seriously by ways like having class meetings, going into the quarters, and having heart-to-heart talks, etc. Educate students to follow school discipline, strictly check on work attendance, make roll, call every sunday night in each dormitory, keep abreast of the situation of students in school, in addition, also go to classrooms to check the class participation constantly, if we found one is absent, then make phone calls or check his dormitory, and find reasons to avoid further absences. Require students to determine learning goals and life goals at the beginning of each semester. Often conducted topic class meetings to learn the Student Management Rules and other related management regulations and actively cooperate with the various aspects of school work. In addition, select the contents from the student handbook which are closely related to student daily life or learning, and then give tests in the form of examination paper. The test not only deepened the understanding of the students, but also can understand some of the ideas of students which can achieve sound effects.

Think highly of the internet ideological and political education on students, establish student personal information files, communicate with or interview the students regularly by means of Fetion, QQ, mail box, telephone, short messages, and so on, listen attentively to heartfelt wishes of students, keep abreast of their thinking dynamic, and guiding them with the utmost care.

Limited ideological and political work of education in the classroom has been far away from meeting the job requirements and the student hostel would also become an important position to carry out the ideological and political work.

In addition, give effective psychological counseling guidance to students who are under psychological distress or mental block respectively and enable them to get rid of obstacles as soon as possible, do self-regulation to improve the mental health, and enhance ability of self-education.

Nowadays, the employment situation is severe and students need to be educated to improve their overall quality and ability. In addition to learn the professional knowledge, students should also actively participate in extracurricular activities.

Pay attention to train the applicants for party membership, cooperate with student party branch secretary and launch students to actively write party membership application and thought reports. Focus on the outstanding students of all aspects in class, encourage and help them, ask them to be hard on themselves according to the standards of party member and endeavor to join the Party at an early date. Promote good department atmosphere and study style through the power of examples.

20.3 The Construction of Learning Style and Academic Advising

Pay adequate attention to class construction and make the best use of the circumstances to mobilize the learning initiative of students; further regulate the daily students management, attach importance to the daily discipline of students, and handle with students who violates disciplines timely, fair and with equitable. At the same time, combined with the construction of a harmonious campus, further regulate students' behavior and promote construction of class atmosphere.

Further improve the system, strengthen the publicity and education on construction of the learning style, with strict management, carry it out seriously, value the exam style, promote study style, check on work attendance well, and regulate test discipline seriously and boost the construction of learning style to a new level. Regularly hold class meetings of various topics and meetings to exchange learning experience and other activities, which emphasize the importance and necessity of learning, so as to form a rich learning atmosphere that students chasing each other in the class. Teach students to learn the specialized knowledge well and read extensively to the best of their ability at the same time, determine the learning orientation, take examinations in school as much as possible for getting some useful qualification certificates, make career planning, and prepare fully for the future employment.

20.4 Advocate Colorful Class Activities

Participate in extracurricular activities organized by classes as often as possible, frequently go deep into students, play down the status of teachers themselves, and let students feel that the teacher is an ordinary member of them. In a relaxed and lively atmosphere, all can easily open up. So encourage and support the class leadership to organize excursion and head the students personally such as picnic in the open air, barbecue, etc., in each semester. These not only let students enjoy the fun of doing it by themselves, but also exercise their survivability; visiting the suburbs not only broaden the horizon of students, but can also enhance their collective consciousness; actively participate in the activities of delicious food street, through buying materials, processing them, and then street hawking all by themselves which make them understand the hardships of making money, and of course appreciate the fun. Carry out the learning from Lei Feng activities in March each year, students can realize the preciousness of love by going to welfare to convey greetings for the old and help in cleaning. The conducting of class volleyball games and English reading competitions that try to let every student to participate which greatly enhances the cohesion of the class.

20.5 Solving the Practical Difficulties for Students

Concerning about the students from poor families, and do thoroughlyinvestigation on poor students in class, separate different levels of poor students, organize the survey and mater the lives and learning of them, and establish poor students file. Talk with them about their physical condition and daily living conditions at ordinary times from time to time. Do poverty assistance job carefully, make a good use of the limited specific funds from the state and school, try to reduce the worries of poor students, and alleviate their life stress caused by their financial difficulties. Strengthen the humanistic concern and psychological counseling to foster their good mental qualities. Help them establish the concept of self-confidence and self-improvement. Poor students often have low self-esteem due to economic reasons, financial assistance to poor students is important, and it is more important to care about their mental health and help them out of inferiority complex. What is particularly important among them is to correct their concept of money, values, and outlook on life, making them aware of whether the economic situation is good or bad does not represent the level of personality, and let them face their own, to explore their potential, knowledge, and abilities as their spiritual food and enrich themselves out of poverty in spirit.

20.6 Cultivating Outstanding Student Leaders, Increasing the Construction of Class Cohesion

In the usual work, counselor or head teacher should be strict on the student leaders and create opportunities for them to establish some credibility in the class. While the time is ripe, boldly let them manage the class, and give full play to their work enthusiasm, initiative, and creativity.

Equality and democracy are the basis for the existence of all the collectives and their healthy development, and class group is also without exception. Head teacher or counselor should set an example and equally treat each student. Everybody has a bright spot, but sometimes overwhelmed by some criteria. Therefore, we must not be cynical to students who have learning or living difficulties, on the contrary, they should be given enough help and encouragement. In this context, the students would not get along with each other with snoblings, making a more harmonious relationship between the students.

20.7 Keeping Safety Education and Maintaining Stable Operating

Strengthen the inspection and supervision of electrical appliances that are out of line by valuing the safety and civilization of the dormitories, and carry out safety and legal education steadily. Often go to student residences, make regular inspection and occasional spot checks, and prevent the "dirty, chaotic, and poor", and so on in the dorms. In addition, strengthen the publicity; create a strong safety culture atmosphere; and enhance student awareness of safety and legal compliance. Through developing educational activities by full using the classroom, broadcast, blackboard newspaper, meetings, and other forms, make safety education on fire protection, traffic, property, and personal safety, and so on. Remind students to keep in mind their own property, personal safety, improve protection awareness during the holidays.

Newly enrolled students came to a new environment, who are not familiar with, can not adapt to or the gap is too great in their imagination, or family financial difficulties, etc., would drop out of school with psychological fluctuations. Students of this type can be kept by adopting the methods of paradoxical intention and giving methodical and patient guidance, etc.

20.8 Emphasizing on Home Visits

Home visit is an important means for school to link and communicate with families. It is also an important element in school education and important complement for teachers to do a good job of their classes. The effectiveness of home visits is directly related to the communication and understanding between the school and the community, teachers and students' parents, and affects the participation and support of the community for schools, thereby affects the overall quality of education in schools. Through home visits, teachers can understand the performance of students learning at home, understand the family background, and know the students living habitual actions. Through home visits, teachers can let parents be aware of their learning, work, and behavior in school. Home visits in the new semester, teachers can learn from parents in all aspects of information of students, such as family status, personality habits, academic performance, etc., but also discuss their existing strengths and weaknesses and noteworthy and so on with parents. All these are very helpful to start the student management work. Before home visits, the route should be worked out, book parents in advance, be familiar with the student records, draws up the conversation content, and other preparatory work.

Only the class teacher or parents, can understand students in-depth, so they can teach students in accordance with their aptitude and educate effectively and with optimization. Home visiting conscientiously is a good way to comprehensively understand students, improve education and teaching, and promote quality education. For the future of our country, the class teachers or counselors are duty-bound, shouldering heavy responsibilities and must strive to do it.

20.9 Conclusion

The highest state of class teacher or counselor is that there is no class teacher or counselor. When our students do not need to tutor in their growing, you would be successful. When students make progress in their learning, happy, with job proficiency and rich in their minds, whether there is teacher or counselor or not is the same. But when students encounter difficulties, or trouble, confused, and can think of the class teacher or counselor around them, and seek help from them, or the sense of security from class teacher or counselor, then, this teacher or counselor is successful.

Experienced a lot from student's management, the joy is continuous. Student management workers should always be with fresh enthusiasm to infect people, warm person with a sincere concern, inspire people with the progressive attitude, shape person with noble spirits, perform the sacred duties continuously, and meet the greater challenge!

References

1. Tong L (2002) Analysis on psychological education in the work of head teacher. In: The eighth national conference proceedings of Chinese mental health association of adolescent mental health professional committee, vol 09(3). pp 27–33
2. Wu YD (2008) Love makes me grow up with students. In: Proceedings of class teacher working forum in Guangxi, vol 28(3). Pp 48–56
3. Li K (2009) Class Cohesion is the primary problem of class construction. Henan Educ 23(3):45–46

Chapter 21
Pricing Strategy of Hotels in Less Developed Regions

Huang Li

Abstract This article has analyzed the influential factors of hotel prices and discussed how to promote the prices potential in less developed regions from the basic law which is based on maximum profit principle so as to execute revenue management pricing strategy. The author puts forward some suggestions that hotels in less developed regions should improve the value of its products and service and guide the consumption demand in order to achieve price discrimination.

Keywords Hotel prices in less developed regions • Influencing factors • Price discrimination • Pricing strategy

21.1 Introduction

The price is an important factor in the management process of hotels, and how to set the scientific and reasonable room price is especially important. Competition of hotels is fierce in less developed regions, as profits of guest room take up more than 50 % of the hotels' revenue more, so the establishment of house prices is the key step of marketing and the important decision for hotels, which is also the key factor affecting the revenue and profit. To achieve maximum profit for hotels in the less developed regions, the establishment of house prices needs to break limitations of the local economic environment, and raise the price to realize the yield management [1, 2].

21.2 Status of the Hotel Rooms Pricing in Less Developed Regions

The hotel rooms pricing is one of the main strategies of the hotel management. The guest room price in less developed regions is affected by variable costs, market structure, capacity of the guest room and changing needs, so the pricing is

H. Li (✉)
Business College, China West Normal University, Nanchong 637002, Sichuan, China
e-mail: huangli@hrsk.net

Z. Zhong (ed.), *Proceedings of the International Conference on Information Engineering and Applications (IEA) 2012*, Lecture Notes in Electrical Engineering 220,
DOI: 10.1007/978-1-4471-4844-9_21,

one-way thinking and takes a one-sided approach to pursuing the short-term profit and lack of overall strategic thinking. What's more, rivals is always competing the price regardless of the cost, so the pricing and the quality of hotel itself do not match, and the price and the discount are both in a state of disorder and confusion. Specifically, these are the following problems:

21.2.1 Insufficient Consumption Demand and Low Room Income

Consumption demand and amount are subjected to regional economic development, while the hotel rooms demand is affected by consumption level, so in less developed regions the main consumption object of hotels focus on the business guests from developed economic or tourists in boom season. In addition to consumption demand and rivals' price, the room price is also affected by the season and some other factors. Consumption demand is obviously insufficient, so the room price is different for even the same star level or brand hotels compared to the hotels in first-tier cities or developed areas. In the sight of the real hotel price in less developed regions, the average price is generally very low, because of the large upfront input and high cost, profits of even some star-rated hotels have reduced greatly for the influence of the room price. The higher the star-rated hotel is, the larger the guest room income gap will be [3, 4].

21.2.2 Pricing is not Systematic and the Room Price is in a State of Disorder and Confusion

Many hotels lack the perfect pricing system and house prices management. Most hotels have not setup a special pricing department, so they cannot predict accurately the guest demand function and cost function, and cannot directly calculate the price of profit maximization, either [5].

21.2.3 Profits of the Room Price has Some Potential to be Dug to Achieve Revenue Management

In the economic environment where competition is more and more fiercer, because of excess production, most hotels are in a aim for running which is also the principle of pricing. As long as the room price can compensate the variable cost and some fixed cost, we can keep the hotel running. So the room price is generally low compared to the same star level or brand hotels in developed areas, and the price has certain potential for rising. In the long run, hotels in less developed regions have to learn how to add value and improve the profit space, which not only needs to raise the room price but also to execute profit management.

21.3 Factors Affecting the Pricing of Hotels in Less Developed Regions

21.3.1 Value of Own Products and Service of Guest Rooms and Variable Cost

The room price is determined by the value of guest room products, which is determined by social necessary labor time for making the product. Labor creating the value of guest room products reflects in the design, construction, decoration, arrangement, and daily service process of the guest room products. There is great difference of support, perfection, and comfortability of facilities. And maintenance, quality assurance, and advanced degree of necessary facilities in first-tier cities are greatly different from those in two three-tire cities, which cost different necessary labor time, therefore the price is greatly different, too. In addition, the hotel price level also embodies the quality of hotel service labor provided by the service personnel. And the operation level of operators in developed areas is higher than that in less developed areas, so the individual labor time may be more than the social necessary labor time. According to the law that the social necessary labor time decides the value and the value decides the supply price, it can reflect that the operation level of operators in developed areas is higher than that in less developed areas [6]. So in less developed areas that the hotel price is low is decided by the value of the social necessary labor time.

21.3.2 Purchasing Power and Consumption Custom of the Consumers

With the development of economy, Chinese people's purchasing power is rising. The rate of business trips among mainland cities and travel and vacation has also raised a lot compared to former years. People also gradually become accustomed to choosing a hotel when travelling across cities. Hotels in less developed regions should cater to this part of people's consumption habits, considering its ability to pay and the demand price, and take some active promotion means like to set a preferential price or to exchange presents with the integral to enhance core competitiveness, so that they can attract a certain number of loyal and fixed customers [7]. At the same time, operators must pay attention to the change of the consumption custom, and people's consumer psychology has changed from the material consumption stage to mental consumption stage even the brand stage. At present, rational consumption idea is grown with the tourist market which is deeply implanted in consumers' consciousness, and people has paid more attention to the comprehensive ratio when choosing a hotel. The hotel room price, room facilities, and comfort levels, and whether traffic is convenient, are the three indexes that the consumers care most. What's more, the hotel also needs to consider spending habits and the ability to pay of consumers and business men in this region in order to determine a reasonable and scientific room price.

21.3.2.1 Competitors and the Competitive Environment

The market competition of hotel industry is unusually fierce, and the price is at a point that both operators and demanders can receive, which is determined by market competition. So the pricing of hotels must synthetically consider the competitor's price level and consumption demand. As a result, before the price for the hotel is established, we need to understand the competition position and environment where the hotel itself locates and research the competitor's price and other marketing factors carefully to establish the price on this basis. At the same time, when hotel rooms are oversupplied, the room price can only reflect operators' survival goal that is just a low supply price. While hotel rooms are in short supply, the room price can only reflect operators' goal of profit maximization.

21.3.2.2 How to Achieve the Optimal Pricing Strategy for Hotels in Less Developed Regions

The basic principle for hotels in less developed regions is still trying to seek profit maximization, so the traditional pricing strategy based on costs is difficult to adapt to the fierce competition and changes of market demand. The precondition of the pricing strategy for hotels in less developed regions is to meet overall operation strategy, and then the hotel should select the optimal pricing strategy which is oriented by demand:

The Pricing Oriented by Consumer Demand

Hotels in the less developed regions should take pricing strategy that is oriented by consumer demand. The consumer-oriented hotel pricing strategy concerns in consumers' demand difference of different segments of the market, and the point of implementation is to identify customers' different demand for products systematically. As the Marginal cost (MC) is the same in the same kind of hotel products. If we want the maximum profit, MR must be the same in different customer groups. But the elasticity of demand is different in various customer groups. According to the calculation formula for marginal benefit that $MR = P\ (1 + E/E)$

(MR is marginal gains, P is the price, E is elasticity of demand), we can deduce that the pricing is absolutely different in different customer groups so that the hotel can gain maximum profits.

To Implement Price Discrimination and Reduce Consumer Surplus

First we can implement price discrimination according to the position of guest rooms and support facilities and then price according to different check-in time. According to international practice, hotel prices are generally calculated with the unit of the room and time. Usually the calculation time for a room/day is from 8 a.m. to 12 at noon next day. But if there is a great difference between the buy time and the check-in time, the customer will feel exchange value is unequal between hotels and them. For example, guests who check in before dawn could feel that exchange value is unequal between hotels and them, and they have just suffered losses. If the hotel can consider that the special buy time will have a special effect on guests' psychology and hotel management, and for guests who check in at this time, hotels can calculate the price by hour prices, which will dissolve the guest's loss psychology and make the guest feel fair and considerate, thereby guests will trust the hotel more and become much more loyal. In additional it is a good way to use preferential price and appropriate discount to stimulate and encourage guest to make full reservations. At last, the hotel can price on the basis of customers' own differences with other marketing methods.

Raise the House Price on the Basis of Promoting Service Value of Products

At present the consensus of hotel prices are mainly the door market (only used for identification), and the real prices that hotels implement are group price, promotion price, contract price, reservation center price, and the individual traveler price. The individual traveler price is for the guests who check in directly without a business contract with the hotel, which is the highest price. The individual tourists can bring greater revenue and profit for the same type rooms. In order to achieve profit maximization, hotels in underdeveloped areas need to develop passenger sources which is mainly based on the tourists from developed regions. But considering the long-term benefits, hotels also need to maintain a batch of faithful target customers and execute the contract price. Even business contract price is signed, it is not unalterable, and the hotel can raise it in proper time according to development of economics. Of course the premise to raise the room price should be based on perfection of the hardware facilities and improvement of service and the added value of corresponding products, and then customers who have signed an agreement would like to pay more. So in addition to trying to satisfy the guests' special needs, hotels in less developed regions also need to master the basic information of guests and the ability to pay, experience the guest potential psychological and physical needs, with sophisticated marketing techniques, to satisfy consumers' more demands through improving the value of products and service so that customers would like to pay more for a deal which can strive for highest possible profit for hotels.

References

1. Li XY, Huang XY (2010) Strategies and Skills for Hotel Prices. Yuncheng Univ J 17(34):99–103
2. Sun MY (2009) Researches on pricing strategy for hotel rooms. Commercial Modernization 78(4):78–80
3. Li YY, Li FS, Han XL (2008) Researches on pricing strategy for hotel rooms in competitive environment. Guide Consumption 9(12):177–180
4. Sha Y (2008) Analysis of pricing strategy for Chinese hotels from management economics. Friends Acc 45(3):6–8
5. Zhang Y (2007) Summary of theoretical researches on pricing methods for hotel rooms. J Tourism 78(5):67–69
6. Li XD (2004) The model construction of price discrimination for chinese star grade hotels. Chongqing Technol Coll J 15(4):3–6
7. Han J (2002) Discussion of application of demand orientation pricing method—take the hotel for example. Guangxi High Commercial Coll J 14(78):3–8

Chapter 22
Performance Evaluation of Special Funds Based on Budget Management in Colleges

Hua Han and Zhong Wei Sa

Abstract This combination of the budget management of the status quo and development trend, the special financial performance evaluation for the strengthening of institutions of higher education budget management role. Through the school performance evaluation, performance index system for the assessment of the establishment and application made a preliminary discussion.

Keywords Budget control • Performance evaluation • Indicator system

22.1 Introduction

College budget is an annual financial revenue and expenditure plan made by colleges and universities based on the career development plans and tasks. It is throughout the whole process of the overall budgeting and budget enforcement, which is also the premise and basis for various economic activities in school. Strengthen the university budget management through performance appraisal and other means has significant meaning in improving the efficient use of funds [1].

22.2 The Status and Development Trend of Budget Management in Colleges and Universities

22.2.1 The Meaning of Budget Management of Colleges and Universities

Budget management is a financial management form widely used within colleges and universities. It is with specific and detailed characteristics compared with

H. Hua (✉) · Z. W. Sa
Finance Department of Beijing, Vocational College of Electronic Science and Technology, Beijing 100029, China
e-mail: dipachof@sina.com

Z. Zhong (ed.), *Proceedings of the International Conference on Information Engineering and Applications (IEA) 2012*, Lecture Notes in Electrical Engineering 220,
DOI: 10.1007/978-1-4471-4844-9_22,

the planning management [2, 3]. And it is carried out to quantify the statistics, distribution, and control for the overall financial revenue and expenditure in school through the whole process of funds collection, distribution, and use.

22.2.2 The Status and Existing Problems in Budget Management of Colleges and Universities

Status and Problems

The economic activities in colleges and universities currently present the trend of diverse and complex. The original internal financial management in colleges and universities cannot meet the new situation, which is mainly reflected in the following [4].

Budgetary institutions are not perfect and the operation process is not standardized.

The way to control the budget enforcement is single. At the present stage, the university budget control is mainly the normative one. There is a phenomenon of "valuing budget and despising management".

The performance appraisal of budget enforcement is not comprehensive enough.

At this stage, the performance appraisal of budget in colleges and universities is mainly through audit inspection. It lays emphasis on normative assessment and ignores performance appraisal in the aspect of assessing the evaluation content. This attaches importance on the compliance and rationality of the use of funds while implementing the budget, but lack of attention on the effective use of funds [5].

22.2.3 The Development Trend of the University Budget Management

Establish a unified budget committee and standardize the budgeting process. With the deepening of budget management in colleges and universities, the awareness of "total involvement and full budget" is growing. Through the establishment of the school budget committee, the relevant functional departments can be unified and integrated, the formation procedure can be standardized, and all these can provide protection to improve the level of budget making.

Intensify the budget management on the special funds and introduce the performance evaluation mechanism. Intensifying the budget management on the special funds particularly in the assessment methods of which the introduction of performance appraisal mechanism has become one of the trends of budget management.

22.3 The Significance of Carrying Out the Performance Appraisal on Special Funds to Strengthen the Budget Management

22.3.1 The Special Funds Management can Directly Impact the Budget Management Standard

With the yearly increasing investment in special funds, the special funds in some colleges and universities have been accounted for over 50 % of the financial allocation. Special funds have characteristics such as a fixed sum is for a fixed purpose, many constraint conditions and so on compared with that of the basic funds. The increasing special funds bring education budget for colleges and universities, and at the same time, add new difficulties to budget management.

22.3.2 The Facilitation of Carrying out Performance Appraisal for Special Funds to Budget Implementation Process

Through performance appraisal to the completed project, the project leader can setup the performance awareness, and know the explicit performance goals. By performance evaluation, the project leader actively involves in budget management, and makes the concept of performance management throughout the entire process of project approval, implementation, and post-management, which can promote the efficient use of special funds.

22.3.3 Performance Appraisal can Enrich the Evaluation Methods in Budget Execution

The purpose of budget management is to improve the efficient use of budget fund. The performance appraisal of the special funds provides an objective evaluation basis for measuring the efficiency in the use of special funds. It is beneficial for the overall assessment and evaluation on university budget management and can greatly promote the level of budget management in colleges and universities.

22.4 The Practice of Performance Appraisal on Special Funds in Colleges and Universities

22.4.1 The Status and Existing Problems of the Special Performance Evaluation in Colleges and Universities

The status of performance appraisal on special funds in colleges and universities,

In recent years, Beijing Municipal Bureau of Finance gradually carried out the comprehensive performance evaluation against the special funds. The proportion of evaluation is gradually enlarged. And it has achieved initial results on strengthening budget management.

Some colleges and universities have also established the internal performance evaluation mechanisms and form the double performance appraisal mechanisms combining both the external and internal performance evaluation. In the practical work of performance appraisal, schools pay attention to the mutual promotion and mutual complementation of external and internal performance evaluation and each performs its own functions. This has played an irreplaceable role on improving the efficient use of funds for the promotion of the school budget management.

The existing problems on performance appraisal of the special funds in colleges and universities, currently, specific performance evaluation is still in its infancy. There are following problems in the process of performance appraisal.

The performance measurement system is single and the refining degree is not enough. And there is an unreasonable phenomenon that a set of indicators assessing all items.

Part of the units lack the initiative on participating in the performance evaluation, and they are in the state of "being evaluated". Some colleges and universities aim at completing the inspection and evaluation by the higher authorities. They are evaluated passively rather than active assessment, causing small evaluation covering surface and the evaluation results cannot completely reflect the condition of overall budget management.

Attach little importance on applying the performance evaluation results. And the results of performance evaluation are not linked to the incentive measures, forming the situation of evaluating superficially and caring little about the results. This gradually fades the importance of performance evaluation and universities and colleges pay little attention to performance evaluation, and the effectiveness in promoting budget management is gradually reduced.

22.4.2 *Method of Performance Evaluation on Special Funds*

22.4.2.1 The Principle of Performance Evaluation

In the 1980s of twentieth century, foreign performance evaluation experts proposed "3E" principle, including Economy, Efficiency, and Effectiveness. Combining the status of the university budget management, we consider that the principle of integrating external evaluation and internal evaluation, the differentiation principle of classification comparison, operational principle, and the principle of combining evaluation results with the incentives should be added in the evaluation principle.

22.4.2.2 The Principle of Combining the External Evaluation and Internal Evaluation

The external evaluation and internal evaluation should be combined. Then, the internal evaluation plays the advantages such as wide coverage, high pertinent, and high instructive of the evaluation results, it also forms a comprehensive evaluation system with the external evaluation. The internal evaluation and external evaluation supplement each other and can promote mutually to fully reflect the general level of university budget management, and give full play to the performance appraisal of promoting development, management, and effectiveness by evaluation.

22.4.2.3 Differentiation Principle of Classification Indicators

Special funds are respectively involving different professional departments and functional management departments. Where funds should be used and how the effectiveness reflects is different in various departments and functional departments. According to different funding sources, patterns, beneficial targets, effectiveness goals, and ways that show the effectiveness and many other features, the performance evaluation index system can be determined respectively. So the evaluation of pertinence and accuracy can be improved.

22.4.2.4 Operational Principle

Special performance evaluation centers on the performance evaluation. It evaluates the effectiveness of special funds at all stages through a series, multi-angle assessment evaluation index, and scientific evaluation methods.

22.4.2.5 The Principle of Integrating the Evaluation Results and Incentives

It combines the results of performance appraisal and reward and punishment mechanism of penalties and rewards, which plays the role of incentives and constraints, is conducive to the application of performance results, and would help to establish the performance management awareness of the administrators and project leaders in colleges and universities.

22.4.2.6 The Implement of Performance Evaluation on Special Funds

Establish a sound internal performance evaluation mechanism from the aspects of institution building, institutional settings, and architectural study, and so on to perfect the internal performance evaluation mechanism and ascertain the organization, personnel division, work processes, and evaluation methods. All special funds will be fully integrated into the coverage of performance evaluation in school. The coverage of performance evaluation would be extended. Through performance evaluation results, the status of the school budget management can be real and fully reflect.

Establish the classified evaluation system. We propose the assessment work plan of "unified management, classification assessment". According to the different using directions of the funds and effectiveness of different expressions, the special projects in school can be broadly divided into four categories which are construction projects of experiment and practice room, education reform classes, information category and infrastructure improvement category. Do performance evaluation respectively and set up different items evaluation index system; all these can fully embody the differentiation principle of classification index. This article takes the most common performance evaluation index system in experimental practice construction category as the object, then analyzes and discusses the performance evaluation index system in colleges and universities.

Standardized performance evaluation processes. Standardized performance evaluation procedure is the important prerequisite to ensure the smooth operation of performance evaluation. Establish a specialized working group on performance evaluation composed by the relevant professionals from the finance department, the state-owned office, audit office, and related functional management sections. The working process includes the choice of evaluation object, evaluation program design, publishing the evaluation results, and so on (See Fig. 22.1).

Strengthen the application of performance evaluation results. First of all, enhance the publicity of the results of performance evaluation; increase its openness and transparency; then, establish the mechanism of combining the evaluated results with the incentives, set up special award fund, or linked to budget indicators and other methods. Associate the performance evaluation results

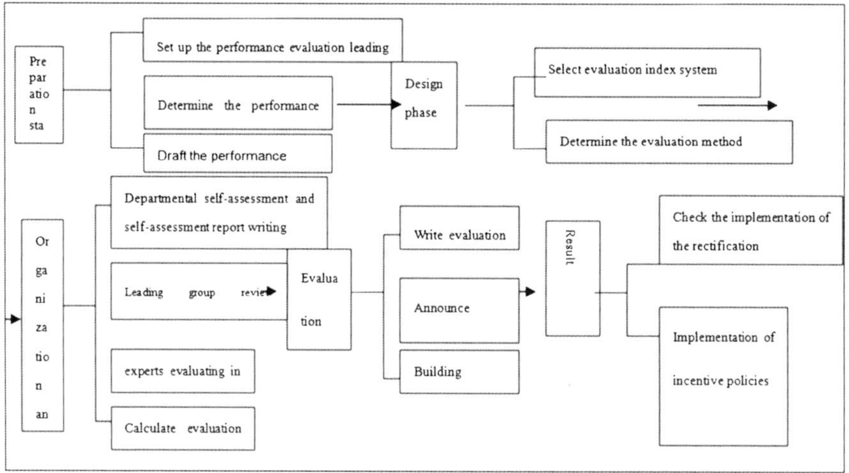

Fig. 22.1 Performance evaluation processes in colleges and universities

with the budgetary allocation department to allocate and control the next year's departmental budget according to performance evaluation results. Then, the colleges and universities, departments, functional management can truly feel the benefits and losses bring the performance evaluation (Table 22.1).

Table 22.1 Performance evaluation index system of construction category of experiment and practice room

First-class index	Second-class index	Third-class index
Operational indicator (80 points)	Project appraisal situation (5 points)	Whether the high professional projects should be proved by the operations specialist (2 points)
		Whether it is proved by its department, its college (3 points)
	Budget declare quality (8 points)	The standard ability of text declaration (3 points)
		The adaptability and specificity of performance objective and the departmental project (3 points)
		The timeliness of budget declaration (2 points)
	Financial appraisal quality (7 points)	The preparation of project appraisal and previewing materials (2 points)
		The timeliness of sending materials to financial evaluation institutions (2 points)
		Project evaluation and subtract situation (3 points)
	Project input service condition (60 points)	Whether the project is completed according to schedule and put into use (15 points)
		Whether to establish the corresponding management system (15 points)
		The cooperation with the school-enterprise after the completion of the project (20 points)
		Project awards and social service (10 points)
Budget management situation (20 points)	Budget management situation (20 points)	The situation of once acceptance and auditing situation (10 points)
		Project expenditure schedule (5 points)
		The compliance of project expenses (adjustment) (5 points)

Establish the performance appraisal rating system. The result of performance evaluation is established based on the analysis of large amounts of data. Data collection is the key link to decide whether the performance evaluation is correct or not. In the current university budget management system, data acquisition would involve a number of departments and many people. Therefore, it is a relatively difficult task to standardize the principle of data collection, collection procedures, and unifying the means of collection. To promote this work, a complete basic database system on performance evaluation, effective information delivery, and feedback system should be established to improve the quality of data collection, thus improve the accuracy of performance evaluation.

Acknowledgments The paper is one of the Beijing Municipal Education Commission Projects, subsidized by "Beijing Vocational Colleges Budget Expenditure Performance Evaluation Index System Research", the project number: PXM2010_014306_108925.

References

1. Hui L, Naibin D (2010) The structure of university project expenditure performance evaluation index system. J Southwest Agric Univ 4:30–32
2. Huanjun C (2010) The status and countermeasures of financial management in colleges and universities. Value Engineering, pp 147–148
3. Lin Y (2010) Analysis on the countermeasures of financial budget management in higher vocational colleges. China's Securities and Futures 10:76–77
4. Zhu Y (2006) Analysis on the construction of performance evaluation index system in colleges and universities. J Chang Sha Tel Tec Vocat Coll 9:72–74
5. Jiangtao C (2010) The concept on building the platform of budget management performance evaluation. J Chengdu Univ 12:365–368

Chapter 23
Management Mode of Public Institutions

Rui-ying Yuan

Abstract As for the management of public institutions in Kaifeng City, it is fully equipped with the color of planning. There are a lot of problems, which are shown as the followings: the management mode falls behind; the management content is vague and general; the management mode is mechanical; the status of a legal person has not been confirmed yet; the personnel administration becomes rigid; and so on. In this case, when choosing the management mode for the public institutions in Kaifeng City, multiple factors should be taken into consideration in a comprehensive manner. There is a great deal of principles that should be taken into consideration, which are taken human beings as the first priority, to comply with the economic and social development, to regard both fairness and efficiency as equally important, and to coordinate in a comprehensive manner. Under these principles, all kinds of modes should be explored and carried out actively.

Keywords Management of public institutions • Management mode • Reform of public institutions • Kaifeng city

23.1 Introduction

Kaifeng City is an area of which the economic development is not so developed. The market-based economy system has not yet been improved and perfected. In Kaifeng, the basic allocation effect of the "market" does not play an apparent role. In the field of economic development, the government still plays a major role. From the perspective of the public institutions, the public institutions that are engaged in the production and operation have apparent reaction to the market signals. In addition to the public institutions that are concerned with production and operation, the public welfare institutions do have some responds to the market changes. The diversified development process of this kind of public institutions makes slow progress considering the backward economic development and the dated ideas. The administrative public institutions have got rapid development due

R. Yuan (✉)
Yellow River Conservancy Technical Institute, Kaifeng 475001, Henan, China
e-mail: yuanruiying@cssci.info

Z. Zhong (ed.), *Proceedings of the International Conference on Information Engineering and Applications (IEA) 2012*, Lecture Notes in Electrical Engineering 220,
DOI: 10.1007/978-1-4471-4844-9_23,

to the reliable resource guarantee and the administrative management and function mode. However, they have a low productivity. The management of the public institutions in Kaifeng City is full of the color of planning.

23.2 The Existing Problems

23.2.1 Basic Conditions

Until the end of 2010, the conditions of the public institutions in Kaifeng City are:

According to the funds supply channel, 4,395 public institutions are divided as the followings: there are 3,658 total supply finance institutions; 241 deficiency payment financial institutions; 496 funds self-management institutions. Among the 1,04,876 compilations, there are 66,091 total supply finance compilations; 18,920 deficiency payment financial compilations; 19,865 funds self-management compilations. Among 1,08,385 staff, there is 72,299 total supply finance staff; 15,965 deficiency payment financial staff; 20,121 funds self-management staff.

According to the types of the public institutions, the 4,395 public institutions are divided into the following types: there are 2,233 education public institutions; 247 culture, news, and electricity public institutions; 280 sanitation public institutions; 362 real estate, urban public usage, social welfare, and economic monitor public institutions; and 900 other public institutions.

23.2.2 The Existing Problems in the City Management

To view the situation as a whole, the development of social programs has a series of problems: the overall arrangement is not reasonable; the productivity is not high; the management is not uniform; and the financial burden is very heavy; and so on.

23.2.2.1 The Management Mode is Very Backward

Until now, it has still been the traditional planned management mode. This has written off the potential of the self-development of public institutions. The traditional planned management mode has restrained the development of public institutions very seriously.

23.2.2.2 The Management Content is Vague and General and the Management Mode is Mechanical

At the moment, the management of the public institutions in Kaifeng City is still in the phrases of primitive institutions management and personnel financial management and so on. The deep goal management and performance assessment have not been carried out yet.

23.2.2.3 The Status of a Legal Person has not Been Confirmed Yet

From 2000, registration of juristic persons for public institutions in Kaifeng City has started to operate. Public institutions act as the secondary organ for the responsible department. The public institutions still belong to the subject of conduction, no matter it is in the daily performance or the financial management or the resources allocation and so on.

23.2.2.4 The Personnel Administration Becomes Rigid

At the moment, the public institutions in Kaifeng City do not have the administrative management authority basically. No matter it is for the leaders in the public institutions or it is for the ordinary personnel, the rights to allocate are still within the department in charge of the government. There are a series of results considering the situations: the staff in public institutions is aging; the structure is in confusion; the development abilities are low; and the development initiatives face frustrations.

23.3 The Factors

At the moment, the field reform of superstructure in our country keeps on moving forward. The reform that has involved the dimension of "social management" is keeping on promoting and perfecting. In addition, it has got great achievements as well. When choosing the management mode for public institutions, places all over the country shall take all kinds of actual circumstances into account comprehensively. They should take such measures are as suitable to local conditions:

23.3.1 The Economic and Social Development Situation

The economic base determines the superstructure. As an important composite part for the superstructure, the reform and management for the public institutions are bound to be restricted by the current conditions of regional economic development. Only by setting off from the current conditions of regional economic development and choose the management mode that adapts to and suitable for the actual conditions of the region, can the public institutions be good managed. The public institutions can play its role and functions fully.

23.3.2 Clear and Definite Goal Orientations

In the first place, there should be a clear and definite goal. Management is only a kind of method. What needs to be explained is that the mental attitude of "the tendency of practice to benefit" is commonly found in the public institutions in the contemporary society. This kind of mental attitude should be changed so as not to hold up the development of social affairs under the drive of "interest".

23.3.3 Fully Motivate the Activity

In the contemporary society, the management of the public institutions has been fully equipped with the color of "planning". As for all kinds of public institutions, their work should be carried out whether under the drive of administrative orders or under the drive of economic benefits. As for the former drive, the public institutions carry out their work out of administrative orders. In this case, their productivity will be low; the resources will be wasted; and there a deathly quiet prevails. As for the latter, the public institutions carry out their work out of economic benefits. In this case, it would result in all kind of difficulties causing high expenses. Under this circumstance, the reform of the public institutions should be taken into full consideration. In addition to the initiatives of the personnel, the productivity should be improved and the activity should be motivated.

23.3.4 The Organic Integration Methods

The 17 national congress of CPC has put forward the rule by law. The 17th CPC National Congress has brought out and carried out the legalization of all kinds of work. This requires us to carry out management according to laws for the public institutions. However, public institutions are a special organization form in our country. As a special organization form, its public welfare benefits may sometimes be "out of work in the market". Under this circumstance, in order to meet the requirements of the public, the government should adopt the administrative methods to carry out the welfare causes or associate the social program development by using economic methods.

23.3.5 Efficient Performance Assessment and System of Accountability

Registration management system for the public institutions has played an important role in the management of the public institutions in our country. The implementation of the registration management system is a milestone for the public institutions management in our country, which has greatly and forcefully promoted the standardized management process of the public institutions. However, the management of the public institutions is still in the simple management phrase of "examination and approval". People have no idea about "how does the public institutions work", "what kinds of effect does it have", and "whether the common people feel satisfied about it" and so on. It has been an urgent need for the management of the public institutions to be equipped with standardized systems. In this case, the foundation of perfect performance assessment system for the public institutions and the system of accountability are in great need.

23.4 The Selection of Management Mode

The 17th National Congress of CPC has put forward the construction requirements of "the service government, the reliable government, the government ruled by law and the economical government". It has made overall arrangement on the management of the public institutions and its reform.

23.4.1 The Overall Management Principles

In the contemporary society, the management of the public institutions in Kaifeng City should comply with the following principles:

23.4.1.1 To Take Human Beings as Essentials

To take human beings as essentials is the fundamental principle in order to develop the Cause of Socialism with Chinese characteristics. That is to say, our public cause comes from the public requirements of the masses of the people. The requirements of the people are the only starting point in order to develop flourishing public course.

23.4.1.2 To Coordinate in a Comprehensive Manner

The social public service is an organic unity entity. It cannot be separated artificially. When formulating the management mode, the general character of the public service should be stood out so as to avoid the interventions of the personal factors as much as possible. The management should be operated by considering the public services of certain region as an entire system.

23.4.1.3 To Regard Both Fairness and Efficiency as Equally Important

It is where the existence of the public institutions and the value of development lie. As for the public services, the fair supply mentioned here refers to the "consumers". The service productivity is the basis for the survival of the public institutions, and the service productivity is as well the core content for the government management departments to strengthen and manage.

23.4.1.4 To Comply with the Economic and Social Development

The development of the social public course should base on the actual circumstances. As for the management of the public institutions, it should comply with the regional economic and social development level. Only in this way can the prosperity and progress of the social programs be made sure. Management that is backward or excessively advanced, all fails to promote the optimum development of the social programs.

23.4.2 *The Basic Mode of the Management of the Public Institutions in Kaifeng City*

It should be set off from the actual conditions of all places in the country. Take into account the relevant way of doing of the foreign public management and the successful experiences of the developed areas when making explorations on the new mode for management of the public institutions in our country. As for the City Kaifeng, the management mode of the public institutions should include the following content:

23.4.2.1 Functional Target Location Management: Nonprofit Institution that Provides Public Products

Take the supply of public goods as the ultimate goal of the public institutions and take the nonprofit property as the inner attribute of the public institutions. Adopt the functional management mode so as to promote the "consolidation" performance of the public institutions.

23.4.2.2 Systematic Volume Control and Dynamic Management

Use the bearing capacity of the regional economic as the basis. Determine the overall planning of the development of the social public services in Kaifeng City in a reasonable manner. Systemize all kinds of public institutions according to the requirement structures and proportions of the public goods of the same kind during the corresponding periods. Found dynamic volume control mode and use it as the basic methods for management. In the end the management departments from the government are able to adjust the volume and structure arrangement of the public institutions at the right moment according to the economic and social development changes.

Diversified Development under the Market-Based System to get completion of the concept that government is the only host. The allocation function of the market should be played fully so as to allocate the resources by the market. Encourage the governmental organization, the folk organization, the legal person from the enterprises, or even natural man to participate in the development and management of the social public service. The government can adopt the method of "compensated purchase" to transfer the production of "public product" so as to attain the maximum benefits.

Corporate Governance System of the Public Institutions that Manages Affairs According to Law Managing affairs according to law is a basic requirement for the future social and public service management. Establish the legal person of the public institutions functions independently and bears corresponding responsibilities according to the laws. Microcosmically, the public institutions can establish the board of directors, and the system of worker representative conference, etc.

Evaluation and Accountability Mechanism that Separating Management and Implementation the separation of management and implementation refers to the separation of the subject of host and the subject of monitoring for the public institutions. In the future, as for the management aspect of the public institution, the main responsibility of the government is to guide and monitor. On this basis, the foundation of perfect evaluation and accountability mechanism that is participated by the masses of people and monitored by the society shall be an important task for a period of time for the management of the public institutions in the future.

Scientific and Reasonable Classified Management there are many public institutions that are equipped with personal features. It is the basic work to make scientific classifications to manage well the public institutions. Only by carrying out different specific measures according to different types of the public institutions the inner activity of all kinds of public institutions can be fully motivated. In this case, they can be promoted to develop flourishingly.

In a word, the management problem for the public institutions is bound to an important subject that needs to be faced by all levels of government for a long time from now on. With the changes of the economic and social environment, its connotations are bound to keep on changing. This certainly will bring a larger and wider exploration field for extensive scholars.

References

1. Kang-zhi Z (2007) Conceptions on serving-oriented government. Urban Manage 4:39–45
2. Zhen-ming C (2007) Understanding of public affairs, vol 3. Peking University Press, Beijing, pp 487–493
3. Jin R (2010) Administrative management public institutions and the reform. Adm Manage Reform 8:37–46
4. Wang W (2008) Opinions on the deepening administrative management system reform adopted in the second plenary meeting in the seventeenth central committee of the communist party of China on 27th, February in the year vol 23, pp 478–485

Chapter 24
Study on the Collected Yugur Ethnic Minority Group

Jun-tao Wang

Abstract This paper has made systematic summary and comprehensive evaluation on different versions of the story "Gesar". The different versions of the story "Gesar is collected which belongs to the Yugur ethnic minority group". It aims at showing a clear outline and grasps the overall situation of the story, which makes people apprehend at a glance.

Keywords The Yugur ethnic minority group • Gesar • Story of different versions • Collection and arrangement

24.1 The Summary and Evaluation of "The Story Gesar" for the Yugur Ethnic Minority Group that is Collected in the 1950s

At the beginning of the foundation of the nation, the nationality language research team in the Chinese Academy of Sciences has done comprehensive and clear popularity work on a great deal of minority groups all over the country. The research work is on a large scale. Among them, the Yugur ethnic minority group today is included. The popularity work at that year has left a great amount of precious information for us today. In addition, it has provided very adorable story information for the story of Gesar when we make research into the Yugur ethnic minority group and the Gesar story.

24.1.1 The Collection of the Story

Title: "historical legend of the Yugur ethnic minority group"

The collection time: 1958-5

J. Wang (✉)
College of Literature, Northwest University for Nationalities, Gansu Lanzhou 730030, China
e-mail: Wangjuntao@cssci.info

Z. Zhong (ed.), *Proceedings of the International Conference on Information Engineering and Applications (IEA) 2012*, Lecture Notes in Electrical Engineering 220, DOI: 10.1007/978-1-4471-4844-9_24,

The collection location: Lianhua Xiang Minghua District the Yugur ethnic minority group autonomous county in Southern Gansu

Narrator: unknown

Collator: Chen Zong-zhen, Tenishev, the nationality language research team in the Chinese Academy of Sciences.

24.1.2 Outline

The "huoling war" is shown as the followings:

Chao tong is very greedy. He envies the fortune of the Yugur ethnic minority group. When Gesar is out of his hometown for 3 years, Chao tong collects with the lord of the Yugur ethnic minority group so as to give away Ke dun, the wife of Gesar. Chao tong acts as the planted agent. With the cooperation of Chao tong, the lord of the Yugur ethnic minority group leads 15,000 army forces to kidnap the wife of Gesar.

When Gesar is back to his hometown, he finds that his wife has been missing. He is so worried that he looks for everywhere. One day, when he is piping, Ke dun hears his voice by accident and becomes very excited. She is piping as well. The couples meet each other and make the plan to get rid of the lord of the Yugur ethnic minority group.

Both sides have a fierce fight. They have fought each other for 3 days and Gesar killed more than 1,00,000 people. Ke dun is back with her husband. Gesar has won the fight and they have a reunion in the end.

24.1.3 Evaluation

As far as it has been known that the earliest version of the story is reflected in this piece of "the Yugur ethnic minority group historical legend". This piece of "historical legend" was recorded as early as the year 1957. It was recorded by the scholar Zong-zhen in our country. In the year 1997, it was lectured as the language information of the Yugur ethnic minority group in the western part when doing the teaching activities [1]. The piece of "historical legend" that is talked about in this paper is one of the examples in it. The passage is numbered No. 2 in the original book. It has research values of multiple dimensions.

24.2 The Summary and Evaluation of the Story "Gesar" for the Yugur Ethnic Minority Group that was Collected During the 1980–1990s

Under the excellent circumstance of the research of "Epic of King Gesar" that is greatly supported by the country, the age of 1980–1990s can be said to be the "golden age" that collects, arranges, and studies the "Gesar" in the Yugur ethnic minority group. Part of the story is shown as the followings.

24.2.1 The Collection of the Story

Title: "the legend of continuous recording of Yang (Ma) an"

The collection time: during the 1980–1990s

The collection location: Mutan Village Xuequan Countryside the Yugur ethnic minority group autonomous county in Southern Gansu

Narrator: Geng dengdaozhibu, En qinzhaolima

Collator: Tian Zi-cheng.

24.2.2 Outline

The content is shown as the followings:

The God has sent Gesar to punish evil. In order to do this, it sent Gesar to be born in the family of Chao tong. The mother and the son were considered as the demons and sent out home.

Chao tong wants to kill both the mother and the son with the heavy storm. He curses at them but fails to kill them. He tries to kill the new born Gesar with poisons as well.

Chao tong sends his fellows to kill Gesar.

A murderous heart under a smiling exterior and a cruel heart under the cover of sugar-coated words: Chao tong receives his grandson back home with hypocritical show of friendship, but he tries to kill him in the way.

Chao tong gives a large crowd of cows to ask Gesar to put out to pasture. In addition, he requires that the cows should be fine and not missing. Gesar is very smart and wins Chao tong again.

Chao tong tries to kill Gesar again with the help of nine lamas from the evil world. However, it is still all in vain. He has no alternative but to send Gesar out of home.

Gesar has special abilities and makes friends with a Bai princess. The princess and Gesar make agreements about marriage privately.

The princess has a bad nightmare while Gesar is not there. Chao tong leads his evil fellows and runs away of the princess. He changes the directions of the mountains and rivers with evil curse.

Gesar is confused and becomes the "man who loses his memory". He gets missed in the Han District and falls in love with a girl who is from Han.

The sister of Gesar talked to Gesar and filled him with wisdom. When Gesar has his sense back, he said goodbye to his wife Han and again turns to kill the evils.

Chao tong is finally killed and the people living in the plain are all very happy. The couples get together in the end. However, Bai princess smiles while crying, for Gesar cuts the weeds and digs up the roots. He has killed her son with the evil.

24.2.3 Evaluation

This piece of story has words of grace. The characters are lifelike and full of vitality. The structure of the story is skillfully arranged. The plots of the story sound very impressive. It is very rare in multiple versions of the "story".

24.3 The Story Summary and Evaluation of the Newly Excavated "Gesar" in the Yugur Ethnic Minority Group During the Present Times (2009–11)

The author has made new collection and arrangements on the story in the culture study for the Yugur ethnic minority group in the Yugur ethnic minority group autonomous county in Gansu Province from the year 2009 to the year 2011, lasting for nearly 3 years. The summary of the story is partly shown as the followings:

24.3.1 The Collection of the Story

Title: "the story of a kaqiaodong"
The collection time: 2009-9
The collection location: Naiman Tribe, Yingpan Village, Huangcheng Country
Narrator: Lan Yu-xiu
Collator: Da Long-dong-zhi, Wang Jun-tao (The author).

24.3.1.1 Outline

The wife of Chao tong is vicious. She treats even the same to her sister-in-law. The brother helps the wicked perpetuate wicked deeds. He helps his wife to bully and oppress his sister. By the name of having a weird baby, he and his wife have driven the sister to the snow mountain.

The uncle has vicious idea that he wants to kill the little baby. Facing the sister's son who has extraordinary power of getting what he wants, all the things he does turn out in vain.

24.3.1.2 Evaluation

This piece of story talks about the things from the birth to the adulthood of Gesar. Compared to the other pieces of story of the same kind, it is relatively in detail. There is no such a piece as the elder brother's wife treats the sister-in-law with vicious ideas.

24.3.2 The Collection of the Story

Title: "the story of Ge shaerzhanu and He erzhanu"
The collection time: 2010–11
The collection location: Dahe Countryside, Southern Gansu
Narrator: Qu mu-da-er
Collator: Zhuo ma, Da long-dong-zhi, Wang Jun-tao

24.3.2.1 Outline

Ge shaerzhanu and He erzhanu had a fight for many years. Neither side has the better of the other. They agree that using the stone to decide who the winner is. It is out of Ge's expectation that he loses and runs away in the end.

24.3.2.2 Evaluation

Although the story is using the stone to decide the winner, the ancient producing tool for the Yugur ethnic minority group has helped him to win.

24.3.3 The Collection of the Story

Title: "the story of Gai Sai-er-han"
The collection time: 2010-10
The collection location: Naiman Tribe, Shi dalong Village
Narrator: Lan Zhi-hou
Collator: Da long-dong-zhi, Wang Jun-tao

24.3.3.1 Outline

Chao tong and his wife do not have a baby for many years. Chao tong asks for a baby from the God.

The God is moved and tells him how to get a baby.

However, Chao tong violates every law of nature. He and his wife tease his sister. His sister has a dream that the son and the moon go into her belly.

The sister is pregnant. Chao tong thinks that his sister has a baby given by the God. He is so envy that he drives her to the snow mountain.

The sister does not die from coldness and her baby is very healthy. The brother then sends his wife to inquire about the information.

The brother gathers together the tribe to tease at the nephew.

The mother and the son are driven to place that is rich and fertile. They live happily.

The mother and the son run into a pride princess. They got the deserved treatments.

Khan attracts son-in-law for his daughter. Shi letuzi who has an ugly face has won nine sets in succession. He becomes a son-in-law of high rank.

Although the princess is married, she feels unhappy and very annoyed, for her husband is dirty and makes her feel vexed.

The husband runs into a Han woman and gets lost and lives several years with her.

The wife of Gesar is robbed by the robber. The wife left some words to her mother-in-law. The lost hero is having back his senses and learns the way to get rid of the evils.

The mother tells everything to her son and the son left his mother. After leaving home, he set out on the long trek to a far distance. The aim of this trip is to look for his missing wife, which has taken a long journey.

The blacksmith helps to make an iron wing. The wife gives a helping hand to her husband and kills the king in the end.

Gesar comes to the hell and takes across sentient beings universally. All living things become mosquitoes and insects and they fly all over the hell. The Han wife becomes the black spider for she kills a lot of lives.

24.3.3.2 Evaluation

This piece of story can be said to be the longest and comprehensive and full of characteristics. It is relatively long, for it has more than ten thousand words. It is relatively comprehensive, for it covers nearly all the "Gesar" story of the Yugur ethnic minority group that can be seen so far. It is relatively full of characteristics, for it is equipped with some plots that the other stories do not be equipped with.

24.4 Conclusions

To sum up, the first step of collection and arrangement of the "Gesar" story of different versions has been basically done. It is basically the "Gesar" stories that have been handed down. All of versions of the story have laid a solid foundation for the future research.

Acknowledgments This paper belongs to phased Achievements of the "Research of 'the Yugur ethnic minority group-the Story of Gesar' in the innovation item for graduate students in Northwest University for Nationalities" Item number (ycx10005).

Reference

1. Jing-wen Z (2002) A piece of study on "historical legend of the Yugur ethnic minority group. J Central Univ Nationalities 2:39–45

Chapter 25
Efficient Accounting Scheme of Currency Capital

Wenbin Wang

Abstract The monetary capital refers to the operation capital in currency status owned by the enterprise, mainly including case, bank deposit, and other currency capitals. The monetary capital plays an important role in enterprise's production and operation. This chapter discusses for accounting of enterprise's monetary capital, having certain reality and guidance significance.

Keywords Enterprise • Monetary capital • Accounting

25.1 Introduction

The monetary capital refers to the operation capital in currency status owned by the enterprise. It is the current asset with the strongest cashability among all enterprise's assets to reflect that the enterprise has the asset for emergency debt, mainly including cash, bank deposit, and other monetary capitals [1].

The monetary capital plays an important role in enterprise's production and operation [2]. The accounting of monetary capital is the accounting for income, expenditure, and balance for monetary capital.

This paper mainly discusses for accounting of enterprise's monetary capital with certain reality and guidance significance.

25.2 Monetary Capital Features

As an important current capital of the enterprise, the monetary capital has its own obvious features:

W. Wang (✉)
Dezhou Occupation and Technical College, Dezhou 253000, China
e-mail: wenb380swang@126.com

Z. Zhong (ed.), *Proceedings of the International Conference on Information Engineering and Applications (IEA) 2012*, Lecture Notes in Electrical Engineering 220,
DOI: 10.1007/978-1-4471-4844-9_25, © Springer-Verlag London 2013

25.2.1 The Monetary Capital has Strong Cash Ability

Generally, the monetary capital could be used to pay for various debts of enterprise and directly purchase various raw materials or labor materials required by the enterprise without realization [3].

25.2.2 The Monetary Capital has Strong Mobility

In the enterprise's production and operation process, many economic businesses are directly involved in receipt and payment of monetary capital, such as purchase of raw materials, sales of the finished products, issuance of staff's salary, etc., all involved in receipt and payment of monetary capital. Therefore, in the enterprise's production and operation, it is especially important to strengthen management and control of monetary capital.

25.2.3 The Income and Expense Amount of Monetary Capital Reflects Enterprise's Production and Operation Scale to Certain Extent

Generally speaking, the bigger the production and operation scale is, the bigger the involved income and expense amount of monetary capital, vice versa.

25.2.4 Whether Income and Expense of Monetary Capital is Prompt Directly Influences Enterprise's Economic Benefit

In its production and operation, the enterprise mainly relies on the currency capital of its business income to pay for various expenses. If the enterprise could not take back its business income timely, its production and operation activity may be interrupted for this reason, even resulting in enterprise's bankruptcy [4].

The monetary capital's feature just explains its important role in enterprise's production and operation activity, so enterprise must pay close attention to monetary capital accounting and must not treat it carelessly, otherwise, it may directly influence enterprise's sustainable development.

25.3 Monetary Capital Accounting Task

In order to enhance management of enterprise's monetary capital and fully play the role of monetary capital in enterprise's production and operation, enterprise must enhance accounting of monetary capital. Its accounting task is mainly including:

Reflect and monitor enterprise's situation to carry out and execute relevant monetary capital management system to ensure enterprise has income and use various monetary capital under the precondition with regulation.

Timely deal with enterprise's income and expense businesses and register on the account; thus to timely reflect income and settlement situation of enterprise's monetary capital so as to ensure the integrity and reasonable use of enterprise's monetary capital.

The accounting of monetary capital is a complicated system project. It must start from many aspects and consider factors in many aspects to do well of this job with purpose.

25.4 How to Correctly Deal with Accounting of Monetary Capital

The monetary capital is mainly including cash and bank deposit, etc., so in conducting monetary capital accounting, it must consider capital accounting in those two aspects to summarize the whole and excellently complete monetary capital accounting work.

25.4.1 Cash Accounting

Cash refers to currency assets used to purchase the required materials and pay for relevant expenses and debts at any time. It is the important composition content in enterprise's monetary capital with the strongest mobility. The use scope is mainly including salary, subsidy, various welfare bonuses or travel expenses of enterprise staff, etc. According to national cash management regulation and settlement system regulation, enterprise must have income and use the cash, enhance cash management and accept supervision of open bank according to "Cash Management Temporary Articles" promulgated by State of Council. Generally, the cash accounting is mainly divided into chronological cash accounting and total classification cash accounting.

25.4.1.1 Chronological Cash Accounting

In order to timely reflect the income and balance situation of enterprise's cash in hand, enterprise must set up dairy cash account; it means that to continuously and daily register cash increase/decrease change situation according to the happening or completion time of cash collection or payment business. This is a kind of original record account, having great role to implement cash's chronological accounting.

The dairy cash account should be chronologically recorded by each item and day according to approved original certificate and case receipt certificate, payment certificate and bank deposit or payment certificate by cashier. In order to timely know and master cash's collection and payment situation and balance amount, the account should be recorded timely and calculated total number and balance of daily cash income and expense at the end of the day as well as comply with the actual deposit amount of cash in the stock.

At the end of the month, the balance in "cash dairy" should be in compliance with balance of "total cash amount" to guarantee settlement daily and monthly and compliance with accounts. If having foreign currency, the enterprise should conduct detailed accounting according to set "dairy cash account" according to RMB cash and foreign currency cash respectively.

25.4.1.2 Total Classification Cash Accounting

In order to comprehensively reflect and monitor collection, payment and balance situation of enterprise's cash in hand, enterprise should also design "cash" total classification account.

Total classification cash accounting should be responsible for registration by accountant without engaged in cashier work. The debit of total classification "cash" accounting should register cash income amount while the credit registers cash expense amount. The balance is in debit to reflect cash's balance in hand.

When enterprise withdraws cash from bank, collects small sales amount below transfer account point, and collects balance of travel expense given back by staff, etc., it should debit "cash" account and credit on relevant account.

When enterprise handles cash expense business according to national cash management system within permissible scope, enterprise should debit relevant accounts, and credit "cash" account according to approved accountant certificate.

The chronological cash accounting and total classification cash accounting are both important composition parts of cash accounting work. Enterprise should often conduct detection for cash and guarantee to have basis thus to ensure orderly conduction of enterprise's production and operation activity.

25.4.2 Bank Deposit Accounting

Bank deposits refer to various deposits that enterprise deposits in bank or other financial institutions. In order to enhance management and control for financial activities, each independent accounting enterprise should apply and open deposit settlement account in local national bank according to regulation.

When enterprise handles settlement business of various accounts for outside, the rest economic businesses must be handled through bank except various businesses in compliance with "Cash Management Article" could be settled with cash.

25.4.2.1 Total Classification Accounting of Collection and Payment of Bank Deposit

In order to conduct control and accounting for accounts deposited in bank, the enterprise generally designs "bank deposit" total classification account.

As the total classification account of calculating enterprise's bank deposit, "bank deposit" should be used as increase, decrease and end balance of bank deposit of the accounting enterprise within certain period. The debit registers increase amount of bank deposit; the credit registers decrease amount of bank deposit while the end balance indicates the enterprise's actual deposit amount in the bank at the end of month.

For enterprises both having RMB deposit and foreign currency deposit in the bank, enterprises should design detailed classification accounts according to currency type in this account and conduct detailed classification accounting for various collection and payment businesses.

In addition, the total classification accounting methods of bank deposit should be varied according to bank's settlement methods. When handling various transfer settlements, enterprise must strictly execute settlement system and abide by settlement discipline.

25.4.2.2 Chronological Accounting of Collection and Payment of Bank Deposit

In order to daily reflect collection, expense, and balance situation of bank deposit in detail, enterprise should design "bank deposit dairy account" (with similar format with cash dairy) according to deposit types in open bank and other financial organizations respectively.

Enterprises having foreign currency deposit also design "bank deposit dairy account" for RMB and various foreign currencies to conduct detailed classification accounting. "Bank deposit dairy account" is generally recorded by cashier according to collection and payment certificate as well as happening sequence of the business.

The balance should be got at the end of the day; the income, total payment amount, and balance amount of this month should be got at the end of month. Also it needs to check accounts with bank regularly.

25.4.2.3 Check Bank Deposit

Enterprise's bank deposit often has various collection and payment businesses in its production and operation process. In order to avoid error in bank deposit, enterprise needs to check accounts with other open banks. Generally speaking, there is once time each month. Its check method is to check each item between "bank deposit dairy account" registered by enterprise and "bank deposit account statement" regularly given to enterprise by its open banks.

If there is any incompliance, enterprise should timely check out the reason and make correction. For non-arrival accounts, enterprise should formulate "bank deposit balance adjustment table" at the end of month and adjust the balance of bank deposit compliance with balance of bank statement. If there is error for accounting record, it needs to timely investigate the reason and adjust accounting record.

25.5 Conclusion

The monetary capital is the enterprise's most powerful current asset that it is easy to become the corruption and misuse object. In addition, as payment tool and debt-repayment mean, the monetary capital is easy to result in a financial strait if its collection and payment does not cooperate well.

So, enterprise must enhance monetary capital management, strengthen monetary capital accounting, and strictly abide by relevant national management system on monetary capital. This will promote enterprise for more beautiful future!

References

1. An W (2002) Israeli political parties: the separation, reunion and the dividing line. West Asia Afr 3(4):35–37
2. Yan R (ed) (1997) Analysis on the politics in israel, vol 90. Northwestern University Press, Xi'an, 1028–1032
3. Xu Y (2004) Conflict between the religions and the secular political powers in the course of the israel modernization. Middle East Stud 3(1):42–46
4. Yin G (1996) Analysis on the election of israel and the prospect. West Asia Afr 4(6):18–19

Part III
Web Science, Engineering and Applications

Chapter 26
An Improved Algorithm Based on Dv-Hop Localization for Wireless Sensor Network

Su Bing and Xue Wei Jie

Abstract DV-Hop is a typical range-free node localization algorithm in Wireless Sensor Networks (WSN); the localization cost and the hardware requirements are not high and localization accuracy of nodes is still low under certain circumstances. Based on the main reasons of the error of DV-Hop algorithm, an improved algorithm is proposed which is based on the average hop distance correction. The algorithm uses error correction value δ to modify the average hop distance, to reduce the deviation between the average hop distance and the true average hop distance. Simulation results show that, the improved algorithm has to more effectively reduce the average localization error of unknown nodes, and to improve the localization accuracy, and not require additional hardware.

Keywords Wireless sensor networks (WSN) • Node localization • Distance vector-hop (DV-Hop) • Average hop distance • Error correction

26.1 Introduction

Node location is one of the key technologies in wireless sensor networks (WSN). In most WSN applications, sensor nodes are randomly distributed, except for some anchor nodes; most of the node positions are unknown. Therefore, only under the premise of the clear nodes' position information, further realization of the locating and tracking of the external target can be achieved. According to the localization mechanism, the existing localization algorithms can be divided into two types: range-based algorithm and range-free algorithm [1]. Compared with range-based

X. W. Jie (✉)
Institute of Information Science and Engineering, Changzhou University,
Changzhou 213164, Jiangsu, China
e-mail: xueweijie2009@126.com

S. Bing (✉)
Key Laboratory of Process Perception and Internet Technology of Changzhou,
Changzhou University, Changzhou 213164, Jiangsu, China
e-mail: subing@cczu.edu.cn

Z. Zhong (ed.), *Proceedings of the International Conference on Information Engineering and Applications (IEA) 2012*, Lecture Notes in Electrical Engineering 220,
DOI: 10.1007/978-1-4471-4844-9_26,

algorithm, range-free algorithm has lower hardware requirements and lower localization cost. This type of algorithm has gained wide attention due to these advantages. Among them, the DV-Hop [2] algorithm is widely used which is one of the range-free localization algorithms. It is a distributed localization algorithm based on range-free proposed by the Niculescu and Nath in the Navigate project. For research on the DV-Hop improved algorithm, the country has also made some achievements, such as literature. These improved algorithms to a certain extent though improve localization accuracy, but also increase the energy consumption and computing cost, etc. However, there is still room for improvement in the algorithms. It uses the product of the average hop distance between the nodes and these hops, to estimate the position of the unknown nodes, so that its localization error is larger. This paper put forward a scheme based on Ref. [3], which is based on average hop distance error correction method to improve the localization accuracy.

26.2 DV-Hop Localization Principle

The basic principle of DV-Hop is that the product of the average hop distance and hop counts between unknown nodes and anchor nodes are used to represent the estimate distance between them. The process implementation of DV-Hop algorithm can be divided into three stages:

26.2.1 *Measure the Minimum Hop Counts to Anchor Nodes*

Each anchor node broadcasts the localization information to the entire work, and its packet includes its position information and hop value initialized to zero. Neighbor nodes receive the packet, save these information, only maintain the packet with minimum hop count on the same nodes, then hop add one and broadcast these message to other neighbor nodes. Through this mechanism, all nodes will receive the minimum hop value distance from each anchor node.

26.2.2 *Calculate and Broadcast the Correction Value*

After an anchor node obtains the minimum hop value distant from the other anchor nodes, we can calculate the distance C_i between two arbitrary anchor nodes $i(x_i, y_i)$ and $j(x_j, y_j)$. The calculation formula is as Eq (26.1) ($hops_{ij}$ is the number of hops between anchor node i and anchor node j):

$$C_i = \frac{\sum_{i \neq j} \sqrt{(x_i - x_j)^2 + (y_i - y_j)^2}}{\sum_{i \neq j} \mathrm{hops_{ij}}} \tag{26.1}$$

Then, using the flooding method broadcast the correction value in the whole network. When each unknown node receives multiple correction values, only retain the average hop distance which was the first received, and discard other received packets. It ensures that most nodes can receive the information about the average hop distance from the closest anchor node. Finally, use the product between the average hop distance C_i and hop counts hops distance of anchor nodes, so as to calculate the estimate distances between unknown nodes and anchor nodes:

$$d_i = C_i \times \text{hops} \tag{26.2}$$

26.2.3 Estimate the Coordinates of Unknown Nodes

When unknown nodes have obtained the estimated distance from three or more anchor nodes, it is the estimated coordinates of unknown nodes that can be calculated by the least square method.

26.3 Related Work

26.3.1 DV-Hop Algorithm Disadvantage

Analyzing the basic principle of DV-hop algorithm, its main error originated from the product of hops and average hop distance, which used the hop distance instead of straight distance. When the estimated value and the actual value of the average hop distance have large deviation, localization error will be increased. In Fig 26.1 gives an example.

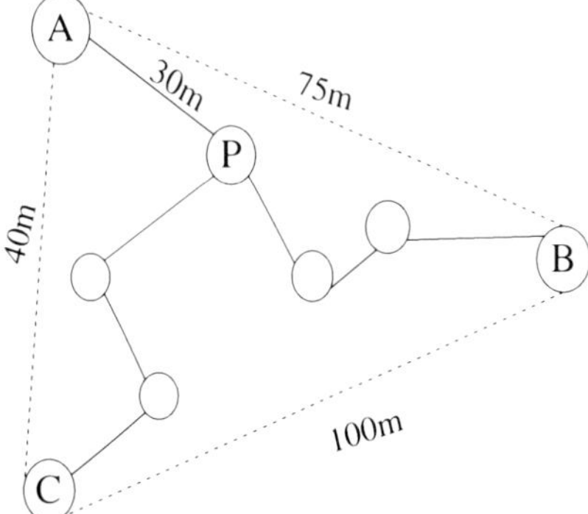

Fig. 26.1 The error schematic of DV-hop algorithm

In Fig. 26.1, A, B, and C are anchor nodes, P is an unknown node. The solid line between the nodes can communicate directly; the dotted line between nodes cannot communicate directly, which is needed for transition of the intermediate nodes. The distance between any two of three anchor nodes are 75, 40, and 100 m. The distance between P and A is 30 m, the smallest hop count of P to A is one, the smallest hop counts of P to B and to C are three hops. The length of the other sides is 20 m. According to DV-Hop algorithm, the calculation method of average hop distance of A, B, and C is:

A: (75 + 40)/(4 + 4) = 14.375;
B: (75 + 100)/(4 + 6) = 17.5;
C: (40 + 100)/(4 + 6) = 14;

We broadcast three correction values in the entire network, but unknown nodes keep only those received from the average hop distance of recent anchor nodes. Therefore, the estimated distance of P to A, B, and C are: 14.735 × 1 = 14.375, 14.735 × 3 = 43.125, and 14.375 × 3 = 43.125; but the actual distance of P and A is 30 m and its error is very large.

26.3.2 The Existing DV-Hop Improved Algorithm

The traditional DV-Hop algorithm does not have high location accuracy. To solve these problems, many scholars from the point of view of reducing average hop distance have proposed better improvement methods [4–8] to reduce error. In [3] was proposed the improved method of the average hop distance of the entire network. The average hop distance of the entire network is defined as cc, see Eq. (26.3):

$$\mathrm{cc} = \frac{\sum C_i}{n} \tag{26.3}$$

Among them, C_i is the average hop distance of a single anchor node, n is the number of anchor nodes.

In [5] the local fixed method is used to consider the closest anchor nodes and other anchor nodes. In solving the average hop distance between unknown nodes and anchor nodes, divide two cases: (1) solution for the distance between unknown nodes and the closest anchor i using Eq. (26.1); (2) solution for the distance between unknown nodes and other anchor nodes. First, calculate respectively the average hop distance d_{ij} between anchor node i and other anchor nodes, then use Eq. (26.5) to calculate the distance between unknown nodes and other anchor nodes.

$$d_{\mathrm{ij}} = \frac{\sqrt{(x_i - x_j)^2 + (y_i - y_j)^2}}{h_{\mathrm{ij}}} \tag{26.4}$$

$$C_{ij} = \frac{(C_i + d_{ij})}{h_{\mathrm{ij}}} \tag{26.5}$$

Among them, node j is another anchor node in the data table of anchor node i, h_{ij} is hop between anchor nodes i and j, d_{ij} is the average hop distance between anchor node i and other anchor nodes.The localization performance of these algorithms has been improved to some extent, but there are still errors, and there is room for improvement.

26.4 Improved DV-Hop Algorithm

This paper is mainly based on the analysis of the traditional DV-Hop algorithm and [3]. Owing to the lack in the literature [3], it puts forward to the improvement scheme, which is to modify the average hop distance, in order to reduce the localization error. First, this algorithm uses Eq. (26.3) to calculate the average hop distance of the entire network cc; then analyzing the deviation between the true distance and the calculating distance between anchor nodes, gets the average hop distance error of the entire network so as to modify the average hop distance of the entire network to improve the precision of localization. Define δ as the average hop distance error of the entire network:

$$\delta = \frac{\sum \{|dt - de|_{\mathrm{ij}} / \mathrm{hops}_{\mathrm{ij}}\}}{n} \tag{26.6}$$

Among them, $|d_t - d_e|_{\mathrm{ij}}$ is the absolute value of the difference of the actual and estimated distance arbitrarily between two anchor nodes i and j. $hops_{ij}$ is arbitrarily the smallest hop count of two anchor nodes i and j, n is the total number of nodes in networks.

In the localization process, the distance on the same path will be counted twice, and will lead to increased computational overhead. Therefore, in order to reduce the amount of computation, to calculate the estimated distance between anchor node i and anchor node j which is on the same path, it is only to calculate de_{ij}, and not to calculate de_{ji}. d_t and d_e show as Eqs. (26.7) and (26.8):

$$d_t = \sqrt{(x_i - x_j)^2 + (y_i - y_j)^2} \tag{26.7}$$

$$d_e = \mathrm{cc} \times \mathrm{hops}_{\mathrm{ij}} \tag{26.8}$$

For example, in Fig. 26.1, A, B, and C are anchor nodes. Calculate the average hop distance of the entire network cc:

$$\mathrm{cc} = \frac{C_A + C_B + C_C}{3} \tag{26.9}$$

Use the hops between the anchor nodes and the average hop distance of the entire network cc to calculate the estimated distance of AB, AC, and BC, using Eq. (26.7) to calculate the actual distance of AB, AC, and BC, and then use Eq. (26.6) to calculate the error correction values δ of the average hop distance of the anchor nodes of the whole network the average hop distance error correction value:

$$\delta = \frac{\left(\frac{|dt-de|_{AB}}{\text{hops}_{AB}} + \frac{|dt-de|_{AC}}{\text{hops}_{AC}} + \frac{|dt-de|_{BC}}{\text{hops}_{BC}}\right)}{3} \tag{26.10}$$

Among them, three is the number of anchor nodes in Fig. 26.1.

The schedule details of the improved algorithm of this paper are as follows:

1. First, all the nodes get the minimum hop and coordinate location away from each anchor node in the networks.
2. In the second stage, use the improvement scheme of this paper is to correct the average hop distance error. First, use Eqs. (26.1) and (26.3) to get the average hop distance of the entire network, then take advantage of Eq. (26.6) to calculate the average hop distance error of all anchor nodes and broadcast δ to the entire network. On the basis of [3] increase a data broadcast stage. After this stage, all unknown nodes and anchor nodes know the average hop distance error of the entire network, so as to get a new equation on the average hop distance of the entire network:

$$\text{HopSize}_{\text{new}} = cc + k\delta \tag{26.11}$$

Among them, k is a variable parameter used to balance on the size of the average hop distance, which is in $[-1\ 1]$. Its value depends on the specific network environment.

Combined with the minimal hop counts between unknown nodes and anchor nodes, solve the estimated distance of unknown nodes to anchor node:

$$d_i = \text{HopSize}_{\text{new}} \times \text{hops} \tag{26.12}$$

3. Get three or more of the estimated distance between unknown nodes and anchor nodes, then use the maximum likelihood estimation method to calculate the coordinators of unknown node.

Assume that the coordinates of n anchor nodes are: $(x_1, y_1), (x_2, y_2) \ldots\ldots (x_n, y_n)$, the coordinates of the unknown node P is (x, y), the distance of each anchor node distance to the unknown node P is $d_1, d_2, d_3 \ldots\ldots d_n$, which can be expressed as Eq. (26.13).

$$\begin{cases} (x_1 - x)^2 + (y_1 - y)^2 = d_1^2 \\ (x_2 - x)^2 + (y_2 - y)^2 = d_2^2 \\ \vdots \\ (x_n - x)^2 + (y_n - y)^2 = d_n^2 \end{cases} \tag{26.13}$$

Equation (26.13) can be written into the form of $AX = b$. Among them, A and b is as follows:

$$A = \begin{bmatrix} 2(x_1 - x_n) & 2(y_1 - y_n) \\ \vdots & \vdots \\ 2(x_{n-1} - x_n) & 2(y_{n-1} - y_n) \end{bmatrix} \quad (26.14)$$

$$b = \begin{bmatrix} x_1^2 - x_n^2 + y_1^2 - y_n^2 + d_n^2 - d_{n-1}^2 \\ \vdots \\ x_{n-1}^2 - x_n^2 + y_{n-1}^2 - y_n^2 + d_n^2 - d_1^2 \end{bmatrix} s \quad (26.15)$$

Using a standard least squares method to solve the equation$AX = b$, the solution is as follows:

$$\widehat{X} = (A^T A)^{-1} A^T b \quad (26.16)$$

26.5 Analysis of Simulation Results

26.5.1 Experiment Environment Setting

In order to verify the performance of the improved algorithm of this paper, using Matlab 7.0 simulation, the simulation results are compared and analyzed. In wireless sensor networks, the N sensor nodes are deployed, the number of anchor nodes and unknown nodes are entered by the user. The simulation area is 100×100 m; these sensor nodes use a randomly distributed way and the experimental results take the average of many experiments, using the localization error formula to evaluate the overall performance of the algorithm. The equation is as follows:

$$\text{error} = \frac{\sqrt{(x_t - x_e)^2 + (y_t - y_e)^2}}{R} \times 100\% \quad (26.17)$$

Among them, (x_t, y_t) is the true coordinator of unknown node, (x_e, y_e) is the estimation coordinator of unknown nodes, R is the communication radius. The sensor nodes are randomly distributed in the 100×100 m square area as shown in Fig. 26.2 ; it has 100 sensor nodes, its communication radius is 20 m, the number of anchor nodes is 20.

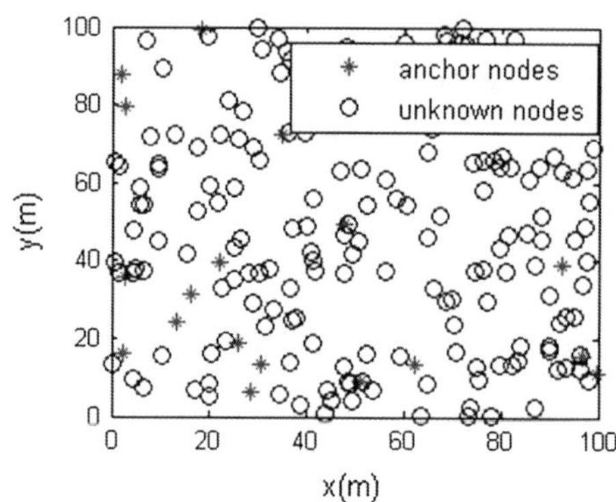

Fig. 26.2 The diagram of the sensor nodes randomly distributed

26.5.2 Experiment Results and Analysis

Figure 26.3 compares the different values of the variables *k*, which has the impact of the localization for the improved algorithm of this paper. 200 sensor nodes are randomly deployed, communication radius is 30 m.

From Fig. 26.3 we can conclude that when the total number of nodes and anchor nodes are constant, the localization error depends on the value of the variable parameters *k*; when the parameters are variable *k* for setting value, the localization error depends on the number of anchor nodes and all nodes. The three simulation curves show that: when the value of *k* is at [0.3 0.7], the algorithm localization error is relatively lower, especially when the value of *k* takes around 0.6. In the following experiments the value of *k* is set as 0.6. But in different networks, the better localization error corresponds to the value of *k* which is also different.

By changing the number of anchor nodes, the impact of the number of anchor nodes on the localization error, random distribution of 100 nodes is studied. The nodes' communication radius is set to 20 m, the number of the anchor nodes is set to 5, 10, 15, 20, 25, 30.

It can be seen from Fig. 26.4 that when the number of nodes and the communication radius of nodes are constant, the localization error of two algorithms

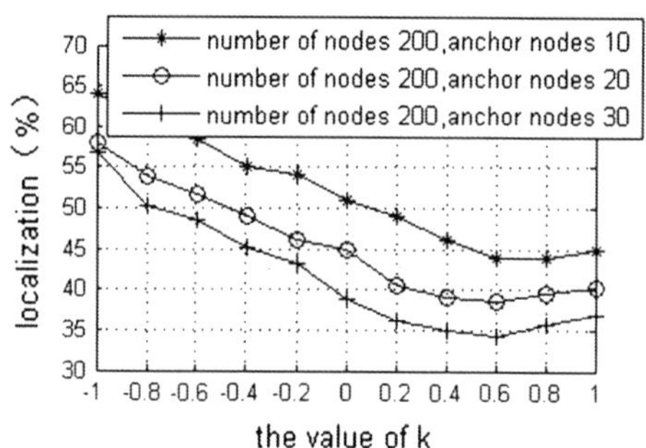

Fig. 26.3 Different value of *k* on the influence of the localization error

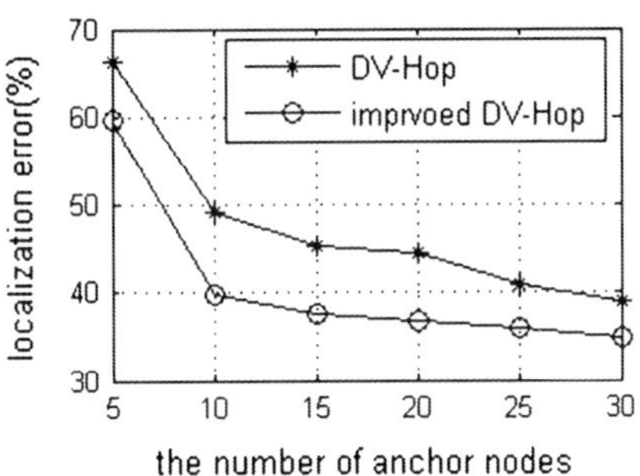

Fig. 26.4 The relationship between the number of anchor nodes and localization error

presents the trend of decreasing, with the increase in the proportion of anchor nodes. But when the anchor node proportion is more than a certain number, the two algorithms' traffic will also increase and the rate of localization error reduction would be slower. Therefore, setting the right number of anchor nodes to some extent can improve the localization precision. Localization accuracy of the algorithm in this paper can be improved by an average of about 5.8 %, compared with the DV-Hop algorithm. Among them, when anchor nodes are 20, localization error of the algorithm of this paper can be reduced to about 8 %.

In order to study the impact of the nodes on the localization error, the number of nodes is set to 100, 150, 200, 250, 300, 350, communication radius is set to 20 m, and the number of anchor nodes is set to 20. As shown in Fig. 26.5, when the number of nodes is the same, the localization error of the algorithm of this paper is lower than the DV-Hop algorithm. In addition, with the increase of nodes, the two algorithms' localization error is reduced to a certain degree. When the number of nodes is at 100, localization error is decreased by about 12 %.

4. In Fig. 26.6, the number of all nodes is set to 100, the anchor node ratio is set as 10 %, communication radius is set to 15, 20, 25, 30, 35, 40 m. Compared with DV-Hop algorithm, the average localization errors of the improved algorithm is decreased by about 6.8 %. When the communication radius is 40 m, the localization error of the DV-Hop algorithm is 30 %, the localization error of the improved algorithm is 23.62 %, the improved algorithm is better than DV-Hop algorithm.

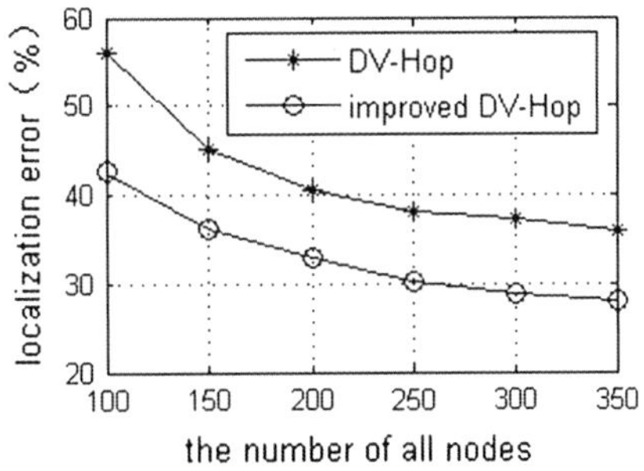

Fig. 26.5 The relationship between the number of nodes and the localization error

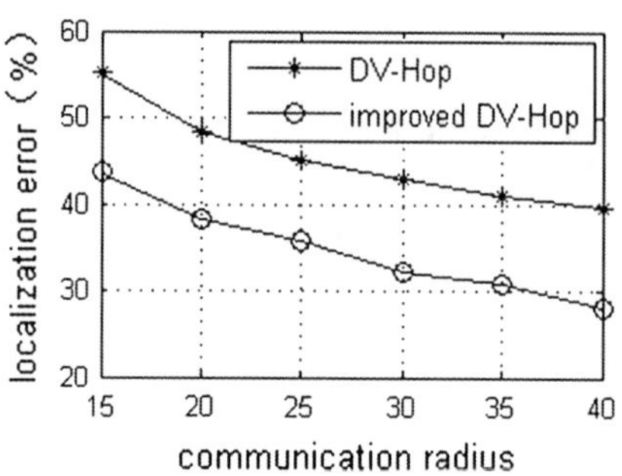

Fig. 26.6 The relationship between communication and the localization error

26.6 Conclusions

Owing to deficiencies in the DV-Hop algorithm, the improved DV-Hop algorithm of this paper uses the error of the real distance and the calculation of the distance between anchor nodes to modify the average hop distance of the entire network, so as to make the average hop distance of the entire network closer to the real value of the average hop distance closer. Eventually the localization accuracy is improved, and also the efficiency of object detecting and tracking. However, the algorithm of this paper in communication and computing cost aspects has increased. Therefore, to reduce the communication and computing cost is the focus of future research.

References

1. He T, Huang C, Blum BM et al (2003) Range-free localization schemes for large scale sensor networks. Proceedings of the 9th Annual Int'l Conference on Mobile Computing and Networking. San Diego: ACM Press, vol 73(38), pp 81–95
2. Lee J, Chung W, Kim E et al (2010) Robust DV-Hop algorithm for localization in wireless sensor network. Int Conf Control Autom Syst 32(26):2506–2509
3. Gang P, Yuanda C, Limin S (2004) Study of localization schemes for wireless sensor networks. Comput Eng Appl 40(35):27–29
4. Long CH, Saeyoung AHN, Sunshin AN (2011) An improved localization algorithm based on DV-Hop for wireless sensor network. IT Converg Srvcs: 333–341
5. Rong Y, Lingkai ZH (2011) Localization algorithm for wireless sensor network based on ACOPSO. Comput Measure Control 19(3):732–735
6. Mingyu SH, Yin ZH (2011) DV-Hop localization algorithm based on improved average hop distance and estimate of distance. Appl Res Comput 28(2):648–650
7. Wei H, Fuyuan X (2012) Research of localization algorithm based on optimal beacon for wireless sensor networks. Comput eng design 33(4):1324–1328
8. Hui L, Shengwu X, Yi L (2011) An improvement of DV-Hop localization algorithm for wireless sensor network. Chin J Sensor Actuat 24(12):1782–1786

Chapter 27
On Journal Citation Based on Network Topology Parameter

Xiuying Du

Abstract To study cited relations, node degree, and betweenness of citied networks of library and information science core journals are studied. Degree index, degree active index, and degree rise index are introduced to quantify the changing direction and speed of the citied frequency of a journal; so are betweenness index, betweenness active index, and betweenness rise index in quantification of the "bridge" role of a journal. $\sum D$, $\sum B$ and $\sum$ are gotten to assess citied ability, bridge role, potential influence, and development foreground of a journal. It is showed that the method is effective in the study of cited relations.

Keywords Journal network • Citation • Topology parameter

27.1 Introduction

References are important information for citation statistics and analysis [1]. Citation analysis is a measurement [2] in study of citation by use of graph theory, fuzzy sets, and other mathematical and statistics methods to reveal the quantitative characteristics so as to evaluate and forecast the development trend of a journal [3]. Citation analysis is useful in exploring subject microstructure [4]. In this paper, the citation network of Chinese 19 core journals of library and information science (JLIS) are studied from their degree and betweenness centralities in consecutive 5 years to assess their influence and future development trend.

X. Du (✉)
Library of Yulin Normal University Yulin, 537000, China
e-mail: yldxy163@163.com

Z. Zhong (ed.), *Proceedings of the International Conference on Information Engineering and Applications (IEA) 2012*, Lecture Notes in Electrical Engineering 220,
DOI: 10.1007/978-1-4471-4844-9_27,

27.2 Main Parameters of a Node

Citation network is a special type of social network, with nodes representing the journals, connections (edges) between the nodes representing the citing and cited of the journals [5]. In a citation network, the importance of a node is equivalent to the significant derived from the connections of the node to others, and called centrality or parameter. In this paper, degree and betweenness [6] of the nodes of the journal network are studied and the influence and the development trends of each journal are assessed. The definition and calculation methods of the two parameters can be found in literature [7].

27.3 Analysis of JLIS Citation Network

27.3.1 Data Source

Data used in the study are collected from CNKI [8]. In that, citing (citied) data of each year of 2006–2010 of 19 JLISs are gathered respectively. The journals are listed in Table 27.1 and each is given a No.

27.3.2 Data Analysis

The degree centrality of each of the journals is directly counted from the data collection, and the betweenness centralities of them are calculated by Ucinet 6 for

Table 27.1 Chinese 19 core journals of library and information science (JLIS)

No.	Journals	No.	Journals	No.	Journals	No.	Journals
1	Journal of library science in china	6	New technology of library and information service	11	Library work and study	16	Library and information
2	Information studies: theory & application	7	Researches on library science	12	Library	17	Journal of the national library of china
3	Document, information & knowledge	8	Library development	13	Journal of academic libraries	18	Information and documentation services
4	Library and information service	9	Library journal	14	Journal of the china society for scientific and technical information	19	Library theory and practice
5	Information science	10	Library tribune	15	Journal of information	–	–

Windows [9]. After that, degree index, degree active index, and degree rise index are to be introduced into to quantify the changing direction and speed of the citied frequency; so are betweenness index, betweenness active index, and betweenness raise index in quantification of the "bridge" role of a journal.

27.3.2.1 Degree Centrality

In degree centrality and relation analysis below, the citation network is directed and with repeated edge in which the citied frequency is considered. Let i ($i = 1$, 2… 19) be a journal and y ($y = 2006, 2007, \ldots, 2010$) be a year. The degree of i in y is the sum of the cited frequencies of i cited by 19 JLISs (including itself) in y. That is $d_{\text{iy}} = \sum_{j=1}^{19} d_{\text{ijy}}$.

Degree Index D

Degrees of different journals are variable in a big scope and not able to reflect their academic position and role, so the degree index is introduced to compare the relative positions of the journals in a citation network.

Definition 1 The average degree of i is the average degree of its 5 year degrees; that is $Ad_i = \sum_{y=2006}^{2010} d_{\text{iy}} / 5$. Ad_y is an average annual degree, which is the average of the degrees of the 19 journals in year y; that is $Ad_y = \sum_{i=1}^{19} d_{\text{iy}} / 19$. The total average degree is the average degree of the 19 journals in 5 years; that is $Ad = \sum_{i=1}^{19} Ad_i / 19 = \sum_{y=2006}^{2010} Ad_y / 5$. Then, the degree index of i is the ratio of its average degree to the total average degree; that is $D_i = Ad_i/Ad$.

The average degree, average annual degree and degree index of 19 JLISs are listed respectively in Table 27.2.

The total average degree is 732.5 ($Ad = 732.5$). $D_i > 1$ is called high degree index, while $D_i \leq 1$ called low degree index. The former is consisted of 1, 4, 5, 7, 10, 12, 13, 15, and their average degree bigger than the total average degree. The latter are 2, 3, 6, 8, 9, 11, 14, 16, 17, 18, 19, and their average degree smaller than the total average degree.

Degree Active Index HD

For the changing of the degree in different year can reflect the development of a journal, the degree active index is introduced to quantify the character.

Definition 2 Let max (d_{iy}), min (d_{iy}) respectively be the maximum and minimum of the degree of i, and $h_i = \max(d_{\text{iy}})/\min(d_{\text{iy}})$ be their ratio;

Table 27.2 Degree index of 19 JLISs

	1	2	3	4	5	6	7	8	9	10	11	12	13	14	15	16	17	18	19	Ad
Ad_i	1,035	716	518	1,645	861	434	953	731	586	985	622	752	784	605	1,021	435	221	556	457	732.5
D_i	1.41	0.98	0.71	2.25	1.18	0.59	1.30	1.00	0.80	1.34	0.85	1.03	1.07	0.83	1.39	0.59	0.30	0.76	0.62	–

$A_{max} = \sum_{i=1}^{19} \max(d_{iy})/19$, $A_{min} = \sum_{i=1}^{19} \min(d_{iy})/19$ be the average of max (d_{iy}), min (d_{iy}) of 19 JLISs respectively, and $A = A_{max}/A_{min}$ be their ratio; then $HD_i = h_i/A*D_i$ is the degree active index of i.

In right column of Table 27.3, there are $A_{max} = 897.89$, $A_{min} = 554.58$, $A = 1.62$ respectively from top to bottom. The journals of $H_i \geq 1$ are 2, 4, 7, 8, 9, 11, 12, 16 while $HD_i < 1$ are 1, 3, 5, 6, 10, 13, 14, 15, 17, 18, 19. The citied frequencies of the former are various in a big scope and their active obviously in the citation network while the latter's vary little and their degree are not active.

Degree Rise Index *RD*

For the degree active index can describe the various extent of a journal but cannot show the direction, the degree rise index is defined to measure the direction and the speed of the variation as followings:

Definition 3 Let $r_{iy} = D_{iy}/D_{iy-1}-1(i = 1, 2, \ldots, 19, y = 2007, 2008, 2009, 2010)$ be the degree rise index of i in year y, and $r_i = \sum r_{iy}$ be the rise extent of i while $Ar = \sum r_i/19$ be the average rise extent of 19 JLISs; then, the degree rise index of i is $RD_i = r_i/Ar*D_i$.

The degree rise index of 19 JLISs are given as Table 27.4.

27.3.2.2 Betweenness Centrality

Betweenness was introduced as a measure for quantifying the control of a human on the communication between other humans in a social network by Linton Freeman [10]. The definitions of betweenness index, betweenness active index and betweenness rise index of 19 JLISs are similar to what have defined in degree centrality above and calculated in Table 27.5 as Bi, HBi, and RBi, respectively.

27.3.3 Discussion

The degree index D, degree active index HD, degree rise index RD and betweenness index B, betweenness active index HB, betweenness rise index RB are summarized to be degree score ($\sum D$), betweenness score ($\sum B$), and total score ($\sum$) in Table 27.5. The observation and analysis as follows:

1. There are several journals with high degree score ($\sum D$) which are 4 (9.35), 7 (5.61), 11 (3.76), 12 (3.59), 15 (3.59) meaning their close relationship with the others, and strong actual strength, powerful potential, and obvious development trend. On the contrary, the others with low degree scores mean their underpowered impetus and unclear prospect.
2. Journals with high betweenness score ($\sum B$) are 3 (8.57), 17 (6.42), 16 (3.77), 11 (3.69), 10 (3.16). They are positioned in the shortcut of so many pairs of the others that they place important betweenness and bridge in citation networks.

Table 27.3 Degree active index of 19 JLISs

I	1	2	3	4	5	6	7	8	9	10	11	12	13	14	15	16	17	18	19	
Max (d_{iy})	1,053	900	575	2,196	991	494	1,532	915	758	1,098	786	1,000	843	684	1,187	558	254	605	631	897.9
Min (d_{iy})	842	553	410	983	685	374	664	485	364	777	330	523	666	522	893	285	188	447	546	554.6
H_i	1.25	1.63	1.40	2.23	1.45	1.32	2.31	1.89	2.08	1.41	2.38	1.91	1.27	1.31	1.33	1.96	1.35	1.35	1.16	1.62
HD_i	1.76	1.59	1.00	5.03	1.71	0.78	3.00	1.89	1.67	1.89	2.02	1.97	1.35	1.09	1.85	1.16	0.41	1.03	0.72	

Table 27.4 Degree rise index of 19 JLISs

I	1	2	3	4	5	6	7	8	9	10	11	12	13	14	15	16	17	18	19	Ar
R_i	0.11	0.48	0.10	0.91	0.30	0.15	1.01	0.58	0.96	0.05	1.05	0.56	0.09	0.13	0.24	0.62	0.29	0.16	0.51	0.44
RD_i	0.16	0.47	0.07	2.05	0.35	0.09	1.31	0.58	0.77	0.07	0.89	0.58	0.10	0.11	0.33	0.37	0.09	0.12	0.32	–

Table 27.5 Degree score ($\sum D$), betweenness score ($\sum B$), and total score ($\sum$) of 19 JLISs *

i	1	2	3	4	5	6	7	8	9	10	11	12	13	14	15	16	17	18	19
D_i	1.41	0.98	0.71	2.25	1.18	0.59	1.30	1.00	0.80	1.34	0.85	1.03	1.07	0.83	1.39	0.59	0.30	0.76	0.62
HD_i	1.76	1.59	1.00	5.03	1.71	0.78	3.00	1.89	1.67	1.89	2.02	1.97	1.35	1.09	1.85	1.16	0.41	1.03	0.72
RD_i	0.16	0.47	0.07	2.07	0.34	0.08	1.31	0.58	0.77	0.07	0.88	0.59	0.11	0.10	0.35	0.37	0.09	0.11	0.32
$\sum D$	3.33	3.05	1.78	9.35	3.23	1.45	5.61	3.47	3.23	3.30	3.76	3.59	2.53	2.02	3.59	2.11	0.79	1.90	1.65
B_i	1.13	1.21	1.04	1.21	1.06	0.83	0.88	0.98	1.15	0.98	1.08	0.96	1.21	0.83	1.25	0.88	0.63	1.04	0.90
HB_i	1.23	1.10	2.82	0.95	0.66	0.75	0.73	1.25	0.80	1.33	1.70	0.87	1.10	0.75	1.13	1.28	2.16	0.65	0.71
RB_i	0.07	0.48	4.70	0.36	0.02	0.73	0.69	0.30	0.22	0.85	0.90	0.31	0.32	0.84	0.46	1.61	3.63	0.02	0.19
$\sum$	2.42	2.79	8.57	2.52	1.75	2.32	2.29	2.53	2.16	3.16	3.69	2.13	2.64	2.43	2.84	3.77	6.42	1.71	1.80
$\sum$	5.75	5.84	10.34	11.87	4.98	3.77	7.90	6.00	5.40	6.46	7.44	5.72	5.17	4.45	6.43	5.88	7.21	3.62	3.45

*$\sum D = Di + HD_i + RD_i$, $\sum B = Bi + HB_i + RB_i$, $\sum = \sum D + \sum B$

3. That some of the journals, such as journal 10 and 11, are found to be higher both in $\sum D$ and $\sum B$ (>3.0, listed from 3rd to 5th), means their citied level and bridging level being all well. So, the degree score ($\sum D$) and betweenness score ($\sum B$) are summed up to be a total score ($\sum$) to assess comprehensively a journal from two parameters as degree and betweenness of a citation network. Journals with high total score ($\sum$) are 4 (11.87), 3 (10.34), 7 (7.90), 11 (7.44), 17 (7.21), meaning their strong overall strength and powerful potential in development.

27.4 Conclusion

This paper mainly studies the node degree and betweenness of 5 consecutive year citing networks of core journals of library and information science based on network topology. In the analysis of node's degree, degree index, degree active index, and degree rise index are defined to quantify the changing direction and speed of a journal's citied frequency; so are betweenness index, betweenness active index, and betweenness rise index defined in the analysis of node's betweenness to quantify the "bridge" role of a journal. As the results, $\sum D$, $\sum B$, and $\sum$ are got as degree score, betweenness score, and total score to assess the citied levels, bridge levels, and potential development prospects of 19 JLISs. As a new method of bibliometrics, the method used in the paper can also be used to study the other problems of documentary metrology.

References

1. Chuanjun S, Yi Y (2010) Libr Inf Serv 54(7):63–67
2. Yin CL (2006) Topological studies of science studies citation network. Dalian Univ Technol 8(8):44–46
3. Zhang S (2011) Bibliometric analysis of grant-funded articles in core journals of library and information science in china. J Intell 5:76–80
4. Tian D (2009) Citing and cited network analysis on journals in library and information science. J Intell 6:48–51
5. Huang D (2011) Research of graduate study team s knowledge sharing network and analysis of influence factors. Inf Sci 10:1539–1544
6. Wu S, Zhang Z (2010) Review on researches of network centrality algorithm. Libr Inf Serv 18:107–110
7. Xiaoqiang G, Xing Z, Naihang T (2009) Validity of journals citation evaluation with centrality indexes of networks. J Acad Libr 5:61–65
8. Weihua Z, Yihong Z (2012) A literature review on information service from library networks in china. Res Libr Sci 2:6–11
9. Liu G, Yang Y et al (2009) Social network about the sharing and running of scientific and technological resources. Libr Inf Serv 14:88–91
10. Freeman L (1977) A set of measures of centrality based upon betweenness. Sociometry 40:35–41

Chapter 28
Community Digital Management System Modeling Mechanism Based on UML

Dan Liu, Chun-Hong Lv, Zu-Hua Guo and Zhi-Wei Tan

Abstract Community Digital Management System is an important means of community digital management. Traditional system analysis design methods have been difficult to guarantee the efficiency and quality of software development. Object-oriented technology can effectively solve this problem. UML is a standard for object-oriented software system modeling tool. This paper introduced UML modeling mechanism. On the basis of analyzing the functions of Community Digital Management System, the paper used of UML to model the Community Digital Management System, which provides a theoretic basis for follow-up development of the system. The model method based on UML can accelerate software development process and improve software quality. At the same time, it can support dynamic business requirements.

Keywords UML • Object-oriented • Community • Digital management • Modelling

28.1 Introduction

With the rapid development of computer technology and the deepening of information needs degree, the production scale of software is increasing. How to develop high-quality software within a reasonable time period is an urgent problem. Object-oriented (OO) technology reduces the difference of the solution

D. Liu (✉) · C.-H. Lv · Z.-H. Guo · Z.-W. Tan
Department of Computer Science, Henan Mechanic and Electric Engineering College, Xinxiang, Henan, China
e-mail: liudan1005@126.com

C.-H. Lv
e-mail: lw3392507@126.com

Z.-H. Guo
e-mail: gzhgg@126.com

Z.-W. Tan
e-mail: 905290825@qq.com

Z. Zhong (ed.), *Proceedings of the International Conference on Information Engineering and Applications (IEA) 2012*, Lecture Notes in Electrical Engineering 220,
DOI: 10.1007/978-1-4471-4844-9_28,

domain and the problem domain, provides a good reuse mechanism and more effectively improves software development efficiency, which provides an effective way for the solution of the above problems.

Unified Modeling Language (UML) is a standard for OO and component-based software system modeling tool. It is a tool which is used to draw the visualization description to the software system model. Its scope is not limited to support OO analysis and design, but also to support the whole process of software development from requirements analysis. It helps developers to draw a clear model for facilitating the exchange.

Taking the analysis and design of the digital community management systems for example, the paper explains how to reduce the difficulty of software development and improve the development efficiency by the UML.

28.2 Introductions of UML Modeling Mechanisms

UML provides with a standard method which is used to observe and describe the various system features from different angles. UML defines five major categories and a total of nine kinds of model diagrams. Table 28.1 lists the UML view and the diagrams included by view and the main concepts related with each diagrams [1, 2].

Table 28.1 UML view and diagram

Primary domains	View	Diagram	Primary concepts
Structure	Static view	Class diagram	Class, correlation, generalization, dependency relationship, realization, interface
	Use case view	Use case diagram	Use case, participant, correlation, extension, inclusion
	Realization view	Component diagram	Component, interface, dependency relationship, realization
	Deployment view	Deployment diagram	Node, component, dependency relationship, position
Dynamic	State machine view	State view	State, event, exchange, action
	Movement view	Active diagram	State, activity, combination
	Interactivity view	Sequence diagram	Interactivity, object, message, active
		Cooperation diagram	Cooperation, interactivity, cooperation, role, message
Model management	Model management view	Class diagram	Subsystem, model
Extendibility	All	All	Constraint, tag value

From the point of view of practical application, when using OO technology to design information system, UML application can be shown in Fig. 28.1. In addition, there are two supporting mechanisms including common mechanism and extension mechanism. In practice, in fact, not all projects need to use all these diagrams. For example, many projects may not use the state diagrams and activity diagrams. A state diagram is used to build life cycle model as an instance of the class. Activity diagram is very useful for the complex process of use case or modeling for the operation of class. In the community information system, the system is mainly focused on data processing, information retrieval and so on. Taking the community digital systems for example, UML modeling analysis will be made.

28.3 UML Modeling Case Analysis of Community Digital Management System

28.3.1 Function Analysis of Community Digital Management System

Digital management [3] refers to the general term of management activities and management practices to quantify management objects and management behavior and achieve the functions such as planning, organization, coordination, and management, innovation through using of computer, communications, networking, and artificial intelligence techniques. Community digital management combines information technology with community management and unified coordinates and mobilizes the various departments, units and the residents to do a good join community work so as to improve the quality of community management and meet the growing material and cultural needs of residents.

Community Digital Management System mainly includes the four functions: community affairs management, community office management, community

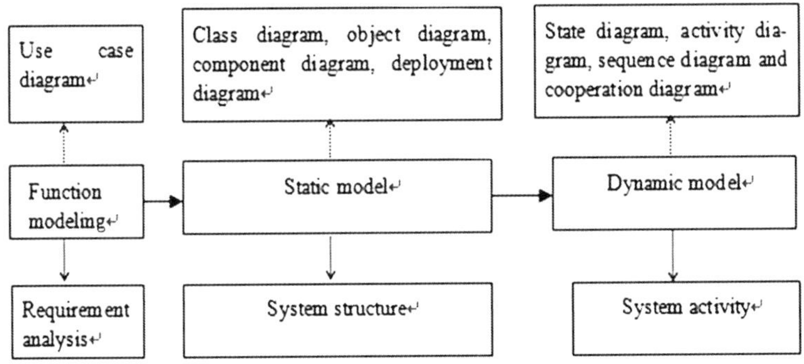

Fig. 28.1 Application position of UML elements in the system design

web-based information services (including online education, e-commerce, online entertainment, etc.), and family intelligent. The key of the digital management is to build a digital management system. And the digital processing modules mainly include:

Data statistics module: community population data (the statistical data such as sex structure, occupational structure, age structure, knowledge structure and so on).

Query help module: it contains the administrative affairs (policy advices such as supporting the military special care, real estate, family planning, etc.), service information (travel, weather, shopping, dining, and entertainment information, etc.), intermediary services (housekeeping, tutoring services, appliance re-pair, employment agencies, home decorating, and other service information advices), emergency services (fire, police, ambulance, emergency housing re-pairs, and other information).

28.3.2 UML Modeling Analysis

28.3.2.1 Building of Static Model

The use case diagram of UML [4, 5] can vividly describe the customers' needs. The main components of the use case model include use cases, roles, and system boundaries. The use case is used to describe the function demand; the role is used to describe external entity related to the system functions which can be user and an entity. It runs the "use cases" in information systems to achieve the purpose and it is the main body of information systems. While the system boundaries are used to define system functional scope. Therefore, UML functional modeling of the system is used to complete the needs analysis. Through use case modeling, the external roles and the systems can be carried out functional modeling. Community Digital Management System is a community service-oriented and it is managed by community officers. By the above functional requirements analysis, use cases and system roles can be defined. The use case diagram is as shown in Fig. 28.2.

Static modeling [6] is a description of the static structure of information systems, which can be achieved by class diagram, object diagram, and package diagram. The class diagram is particularly important, it not only define the classes in the system, but also expresses the links between classes such as association, dependency, generalization, realization, including the internal structure of a class (class attributes and operations). Being different from the data model, it not only shows the structure of the information, but also describes the behavior of the system, which is valid throughout the life cycle of the system. The class diagram technique is the core technology of the OO technology, which used to describe the relationship between the properties of an object and the other object.

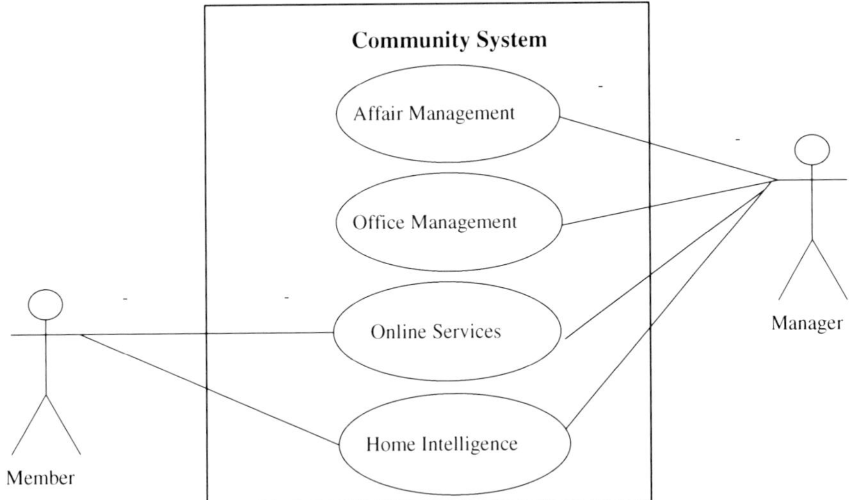

Fig. 28.2 First layer use case diagram of community digital management system

28.3.2.2 Building of Dynamic Model

For Information management system, in addition to a static model, it is more important to analyze various information processing timing and properly control and process this information. So the dynamic modeling can achieve this goal. The behavior diagrams and interaction diagrams describe the cooperation communication mechanism between the objects and the timing of behavior in the cooperation process. Information systems focus on to control information, so the timing diagram is the main description tools of the dynamic modeling of information systems. The online information services subsystems in the community digital management system include online education, e-commerce, online entertainment, etc. Taking online entertainment activities for example, the paper gives its timing diagram. In time of online entertainment, each community member chooses the online entertainment projects according to personal tastes. Through the system interface, it carries out the qualification first to check whether the user is the community member. If he is community member, he can enter, or refused to enter. The timing characteristics depiction of this activity can be shown in Fig. 28.3.

In the analysis phase, the model diagram can describe the static structure and dynamic behavior of the specific area class (or instance). The design period emphases on refining the class function (operations and attributes) extracted in the period of analysis. At the same time, the technology problems such as databases, user interfaces, communications, equipment, and other technical aspects can be solved through adding new classes. The structure design phase mainly defines package (subsystem) and determine the dependencies between the packages and the main communication mechanism. The class diagram pack-age technology can separate logic from the technical logic to reduce the dependence of the class so

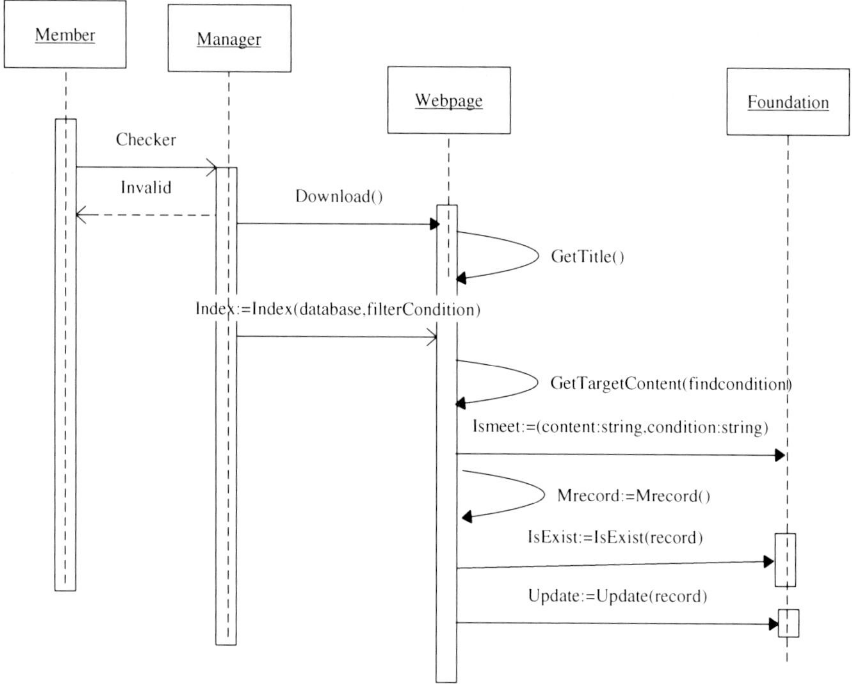

Fig. 28.3 Timing diagram of online entrainment movement

as to achieve high cohesion and low coupling. OO analysis and the design modeling process is not a simple top-down one-way process. In other words, the design model is established not until the analysis model is completed. The iterative approach makes two models staggered growth, so to ensure that the final program can effectively reflect the requirements of the problem.

28.4 Conclusions

The paper introduces the entire development process using UML to model, construct, and test for community digital management system. UML is just a set symbols system for the modeling, rather than means, nor is the software process. But it can be used in a variety of modeling methods and a variety of software process. In addition, it can carry out visualization, detail, structure and documentation for software-intensive systems, products.

In the process of MIS system development, UML can provide with a unified, flexible, and easy to understand expression model to the entire development process, which will help the understanding expansion and maintenance of the software system.

Modeling is to achieve, the software system development method based modeling is MDA, which is Model Driven Architecture. This method is by stepwise refinement of the model, and then the model is a reality. So the product-related documentation can be obtained at the same time. But now the MDA is still in development. If that day arrives, the programmers will no longer use C++, C#, Java to program, but they will use more advanced UML to write programs. This is also one of the reasons using the UML to develop system.

References

1. Cheng-jia D (2007) UML system modeling and analysis design, vol 12. China Machine Press, Beijing, pp 429–431
2. Yi-ping H, Jian-ye Q (2004) Design and realization of community comprehensive service system. Electron Eng 5:208–220
3. Zhi-cheng L (2009) Software engineer and rose modeling case course, vol 34. Dalian University of Technology, Dalian, pp 90–94
4. Xing-peng L, Wei W (2011) Modeling of scientific research management system for universities based on UML. J Hubei Univ Natl Nat Sci Edn) 29:324–326
5. Xiao-ying H (2011) Research and modeling of enterprise logistics information management system based on UML. Coal Technol 30:277–279
6. Cheng Z (2011) Modeling process of book management system based on UML. Jisuanji Yu Xiandaihua 10:48–50

Chapter 29
Research on Network Interactive Services in University Library

Xiang-fei Guo, Mei Zhang and Jian-Jie Du

Abstract The aim of this study was to discuss the research focus on Network Interactive Services in university library in china. We adopt the method of research focus analysis based on the frequency of keyword and g-index, taking database China Academic Journal Database (CAJD) as the data recourse. The results show that development of digital referential consultant service and collaborative model, technology of internet interactive services, knowledge management in the role of the internet interactive service, are the research focus of medical literature retrieval course in the last 7 years.

Keywords Internet interaction services • Academic library • Research focus • G-index

29.1 Introduction

As the network technologies become more sophisticated, the electronic information resources developed, modern library collection vector and collection composition has undergone profound changes. Interaction in the form of the library reader service also accompanied by a profound change the new, interactive network service model will be the future mainstream of the university library services and development trends. Network interaction has become the focus of attention of scholars.

X. Guo (✉) · M. Zhang · J.-J. Du
Library of Hebei United University, Tangshan, China
e-mail: xiangfeig@126.com

Z. Zhong (ed.), *Proceedings of the International Conference on Information Engineering and Applications (IEA) 2012*, Lecture Notes in Electrical Engineering 220, DOI: 10.1007/978-1-4471-4844-9_29,

29.2 Methods

29.2.1 Data Sources

Take the literature published in the "CAJD" as the data resource. The retrieval expression: SU = Library and SU = (Network Interactive+digital reference consultation+online consulting+virtual reference consultation+push service+Blog+BBS+E-mail+OICQ), year from 2005–2011. Obtained a total of 1,800 articles the retrieved articles were grouped according to keywords.

29.2.2 The G-Index

According to the definition of G-index, if the papers are arranged in the de-sending order of their number of citations, $g\psi$ is the largest number such that the summation of the number of citations is at least g^2. In other words, when papers are arranged in descending order of their citations, g-index can be defined as follows [1].

$$g = \max(i) : \sum_i Ci \geq i^2 \tag{29.1}$$

Note that g-index is the largest number iψ such that $\sum_i Ci \geq i^2$

The g-index of key words also could also be calculated like this, we calculated g-index for each keyword the greater the g-index the greater the influence of the keywords [2]. We filtered out 52 high frequency keywords. The key words were arranged by the size of the g-index as shown in Table 29.1.

Table 29.1 High-frequency keywords and g-index of university library network interactive services

Serial number	Keyword	Keyword frequency	Keyword g-index
1	Blog/library blog	201	24
2	Web 2.0	70	24
3	Library	489	22
4	Digital reference consultation	307	21
5	Library 2.0	40	21
6	Viral reference consultation	207	18
7	Information service	140	18
8	University library	232	16
9	Reference consultation	130	16
10	Individualized service	55	16
11	Digital library	90	14
12	RSS	42	14
13	Push service	28	14
14	Knowledge management	48	13
15	Network environment	46	12

(continued)

Table 29.1 (continued)

Serial number	Keyword	Keyword frequency	Keyword g-index
16	Service model	38	12
17	Subject librarian	20	10
18	Cooperative digital reference service	27	9
19	Library service	27	9
20	Comparative study	9	9
21	Service	39	8
22	Reader service	36	8
23	Real-time reference	16	8
24	User	13	8
25	Library website	11	8
26	Knowledge blog	9	8
27	Knowledge service	24	7

Combined with these data, we summed up the research focus of the Network interactive services of library:

1. Present situation research of Network interactive service and cooperation model, Consists of nine keywords: digital reference service, virtual reference service, collaborative digital reference service, the service model, librarians, knowledge services, real-time consultation, and personalized service.
2. Network interactive services related technologies, Consists of five keywords: Web 2.0, Library 2.0, blogs, RSS, Notification service.
3. Knowledge Management of University Library, Consist of two keywords: University library, knowledge management.

29.3 Figures Results Analysis and Conclusion

29.3.1 Situation Research of Network Interactive Service and Collaborative Model

In China, the digital reference consultation service started in quite late years, but the development is quite rapid, there are a growing number of university libraries are carrying out the Digital Reference consultation Service [3]. Digital Reference consultation and the development of advisory services practice, which greatly improve the level of library Network interactive service, also makes the related theoretical studies to become a research focus of network interactive services 2.3 Footnotes.

Depending on the digital reference consultation service in the information interaction and the cooperation way of digital reference consultation services, there are three cooperation models:

1. Asynchronous Digital Reference consultation Service. Generally by using of E-Mail, Web form, BBS, and FAQ

2. Real-time reference service. Generally by using of online chatting room OICQ and UC.
3. Cooperative digital reference service, consist of two or more digital reference Consultation platforms, for example CALIS distributed united Virtual Reference System. Because of the digital reference service is a emergent novelty for the Chinese university, and still at the exploratory and pilot phase, how to make good use of these three service models to carry out more efficient on line reference consulting services and, how to provide truly satisfied information service for wider range of users, is the focus problem for university library at this stage.

29.3.2 Research on Network Interactive Services Related Technologies

Network interactive services related technologies are the security of network interactive services. The web 2.0 technology was mainly talked about. Using web 2.0, university library was impelled to expand information resources, extend library information services, and make the library service level in proved become possible [4–6]. At present, Blog, Wiki, RSS, and Folk snoopy, as the core technique in web 2.0, were widespread used by university library to innovate library information service. University library, as a university information exchange center, how to use these new technologies, new ideas to provide services, have been a very worthy subject. The present study is more about the blog in the digital reference applications and information push service based on RSS.

29.3.2.1 Blog

As a Web log, the blog has been applied in many libraries. University Library can use the blog to build interactive online learning platform, conduct education, and training for users. Using of blog has broken through traditional boundaries of time and space, and makes the object of user socialized and the education more humane. The teachers also can do some tracking and evaluation after class, or offer project or topic, and let the students discuss with each other by blog, While the students through the blog will explore in the learning process, found that share information with other members can improve their collaborative learning, the research ability of study and the independent in learning [7, 8]. The growing popularity of blogs suggests the possibility that some of the works that students need to do in order to read well, respond critically, and write vigorously might be accomplished under circumstances dramatically different from those currently utilized in higher education.

29.3.2.2 RSS

RSS is a technology for syndicating information such as the content of websites. These technologies enable desirable practices such as collaborative content creation, peer assessment, formative evaluation of student work, individual as well as group reflection on learning experiences, and up-to-date information regarding changes in collaborative spaces, and can be used in the development of authentic learning tasks.

The same with the blog, the RSS were also applicator in library, such as museum press released and the thematic guide/navigation. Wuhan University Library is a goof example to use Thematic Guide topics. Information Portal RSS Push Service As more and more libraries started using RSS technology, their forms of service had more innovative application of RSS which can further enhance the library of interactive service capabilities, so that the library network of interactive services become more flexible and proactive [9, 10].

29.3.3 The Application of Knowledge Management in the Network Interactive Services

Today most scholars believe that introducing the concept of knowledge management into the digital library reference consulting services can help to improve the library's operational capability and efficiency of services [11]. The goal of knowledge management is trying to get the most appropriate knowledge in the most appropriate time and pass the knowledge to the most appropriate person; this precisely corresponds to "reference", amity to answer user's questions in digital reference service. Librarians interact with the users through the network means "consultation" and the establishment of the Q&A knowledge base. The advanced nature of the library of knowledge management for digital reference service is mainly reflected in: First, the digital reference source of information organization building that is the production and organization and management of explicit knowledge; The second is the development and externalization of tacit knowledge resources of the Reference Librarian, which will Contribute to the cooperation of the staff, and further improve the provision of new ideas and concepts for Library Digital Reference Service. To help solve the problems encountered in the development process. Knowledge management concept is worthy of further promotion in the network interactive services [12].

References

1. Zhang CT (2010) Relationship of the H-index, G-index, and E-index. J Am Soc Inf Sci Technol 1(3):625–628
2. Zhaoxing , Xiaoqiang G, Ji'an G (2009) Hot spot analysis method based on the theme of word frequency, and g index. Libr Inf Serv 2(2):60–63

3. Jianling Q (2010) Research of digital reference service in university libraries. Mod Inf 3(3):56–59
4. Huaizhu R, Guichun S, Shuangqin C (2004) China's university library virtual reference service analysis. J Inf 4(9):45–49
5. Wenhui Z (2008) Research on the personal knowledge management in the web 2.0 environment. Mod Inf 5(8):121–123
6. Bingsi F, Xiaojing H (2006) Library 2.0: building the new library services. J Acad Libr 6(1):56–59
7. Lingling L (2012) The investigation and research of lib2.0 information service of university library. Libr Work Study 7(1):38–41
8. Lixin X, Shanshan Z (2011) Research of library subject knowledge services based on academic blogs. Libr Tribune 8(6):109–114
9. Xia C (2011) Review of the RSS in the library application. J Libr Sci 9(12):138–140
10. Lingling Z (2011) The constructive application of RSS in the library: take northeast normal university library for example. J Libr Sci 10(10):131–134
11. Yu Q (2010) Library digital reference service based on knowledge management. Libr Work Coll Univ 11(2):55–56
12. Haiying C (2005) Knowledge weblog-the new idea of library knowledge management. J Sichuan Soc Libr Sci 12(5):23–25

Chapter 30
Study of Learning Model Based on Web 2.0

Fuqiang Wang and Lijun Xu

Abstract Compared with web 1.0, the application of BLOG, RSS, Wiki, SNS, P2P, IM technologies provide the learners with a more free, more interactive, more convenience internet website and internetwork platform, which enables the informal learning toward the tendency of personalization and people-oriented. Web 2.0-based informal learning is increasingly becoming a new and hot field of study. On the previous basis of the definition of informal learning, web 2.0 further elaborates and describes it and then summarizes the characteristics and effectiveness of informal learning, constructing a web 2.0-based informal learning model on the basis of corresponding theoretical support.

Keywords Web • Informal • Learning theory • Model

30.1 The Definition and Characteristics of Informal Learning

30.1.1 The Outline of Informal Learning

Pettenati Informal learning is learning that results from work, family, and leisure-related activities of daily living and so on. It has no target, no stipulated time, always treated as experience learning and to some extent as incidental learning [1].

Zeng Lihong thinks that informal learning is purposeful self-directed learning and online learning without strict organizational structure.

Thus, the informal learning occurs at the time and space of informal learning, and transfers and exchanges the knowledge by means of non-academic social interaction and social networking software, a process of exchange the information obtained from outside into their own knowledge.

F. Wang (✉) · L. Xu
Institute of Computer and Information Engineering, Xinxiang University, Xinxiang 453003, Henan, China
e-mail: wfqwang@163.com

Z. Zhong (ed.), *Proceedings of the International Conference on Information Engineering and Applications (IEA) 2012*, Lecture Notes in Electrical Engineering 220, DOI: 10.1007/978-1-4471-4844-9_30,

30.1.2 The Characteristics of Informal Learning

1. Informal learning has no fixed learning goals.
2. Informal learning is completely autonomous study.
3. Informal learning is social activity.
4. Informal learning is not restricted by learning spaces.
5. Informal learning emphasizes communication and collaboration.

30.1.3 High Efficiency of Informal Learning

Jay Cross points out that 75 % of one's knowledge is acquired by informal learning, while formal learning only holds 25 %. Thus, the knowledge gained through informal learning is far greater than the amount of formal learning. Nevertheless, in reality, it takes a lot of time and money of learners on formal learning, who ignore the informal learning. Benefits of learning in this way are low. Compared the informal learning's the inputs and outputs with formal learning's; we can also find high efficiency of informal learning. Formal learning is a high input activity with low output, while in terms of informal learning; the opposite is true, as shown in Fig. 30.1.

30.2 The Theoretical Support of Informal Learning Based on Web 2.0

30.2.1 Social Constructivism

The main research object of social constructivism is the process of interaction and mutual constitution between the individual and society. Especially the reconstruction of the relationship between the individual and society are treated as a primary research purposes and tasks [2]. Social constructivism emphasizes the dialectical relationship between the production process of social and the formation of personal knowledge, and the social constructive role of individual knowledge and

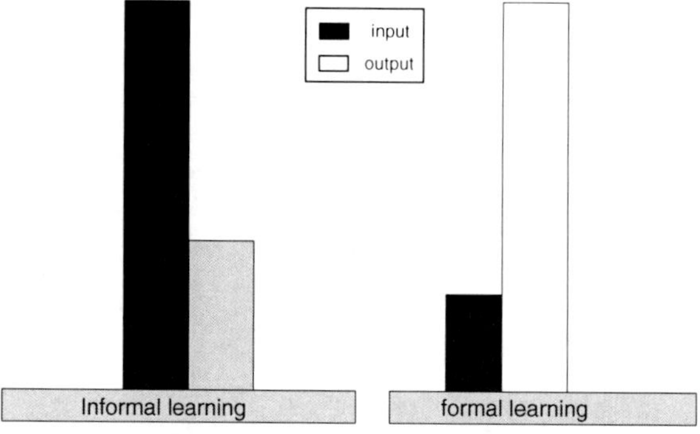

Fig. 30.1 The comparison of input and output between informal learning and formal learning

social knowledge, scientific knowledge. The first step of study is the cultural process to enter some practice community, which can realize the mutual building of society through the participation of social interaction and the organization of the way of culture. In the process of learning, social constructivism pays more attention to how to consulate, collaborate and exchange in order to complete of the construction of personal knowledge in accordance with its natural tendency.

30.2.2 Lifelong Learning Theory

With the development of internet, opening university, and distance education come into being. People begin to realize that learning is not limited to the acquisition of knowledge in classroom and exploratory learning based on internet source also plays an important role. The proposal of lifelong learning makes the concept of learning further expand in people's hearts. And people come to realize that the stagnation in the learning, a lifetime thing, makes progress stagnant. Lifelong learning concept not only changed people's learning concepts but also led learning way to a new field-the field of informal learning. The Web 2.0 provides a better supporting platform and operating environment, pushing the informal learning to a new high.

30.2.3 Learning Theory of Humanism

The educational view of humanist takes human's freedom and full-scale development as the ultimate goal, promoting a relaxed, free learning environment and promoting the new teaching models which fully reflect the cognitive function of student autonomy, designed to cultivate innovator with creative thinking and innovative capability. The combinative use of a variety of Web 2.0-based social software provides the necessary conditions for learners learning by themselves, interactive learning, retrieving and indexing information, as well as broadening the knowledge horizons. Meanwhile, Web 2.0-based learning environment implies a break from the restrictions of the goal and the content of learning. Thus, learners can discover the theme independently, learn initiatively explore knowledge, free from outside interference, which is from the bottom a learner, and learning efficiency is very high.

30.3 Informal Learning Model Based on Web 2.0

The reason why the web 2.0-based informal learning model can meet the needs of most learners is that it brings together a variety of learning software, including Blog, IM, Wiki, RSS, Tag, and so on. Some of these software features a single, while others have a variety of functions. For example: Tag and RSS can only be used as a tool to obtain information, while Blog, Wiki not only can get access to information, organize information but distribute information. Different social software play different role in informal learning, which is combined organically by the

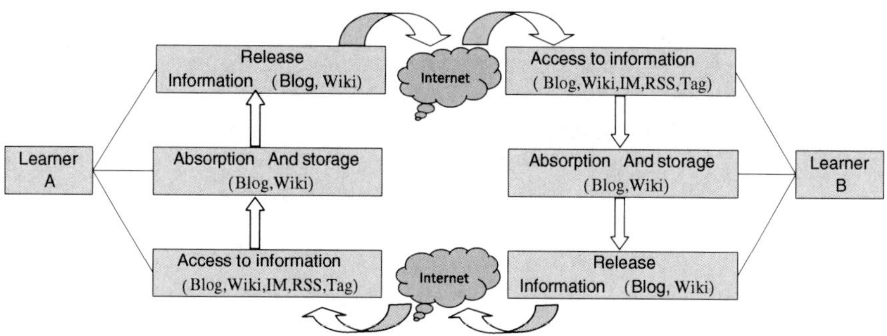

Fig. 30.2 Informal learning model based on the web 2.0

Web 2.0-based informal learning model. As shown in Fig. 30.2, to A and B as an example of the two learners in informal learning based on Web 2.0.

In this learning mode, learners A and B, respectively in their own Web 2.0 learning environment, can choose places for learning according to their preferences, such as in their own home, schools, or even the workplaces, as long as the place with the computer and the network can become their learning environment. As the leader, learner who has great interest on learning handles a variety of social software based on Web 2.0, and dominates the theme and content of their own learning. On the basis of web 2.0, the informal model of study, which meets the needs of learners to acquire, absorb and store, publish the information freely, set all kinds of powerful social software in one with low technology.

30.3.1 Acquire Information Get Access to Information

Web 2.0-based learning mode, learners can not only subscribe to their favorite channel through the RSS technology to keep track of update resources at any time, but also share related learning resources distributed by other members directly from the various subjects and many community of practice groups. Blog and Wiki with the function of filtering the information, learners can be directly connected from the Blog and Wiki for group collaboration in the group to get the latest, most valuable and professional knowledge combination, Quick and easy access to information for learners to reduce a lot of unnecessary trouble of filtering the information, but also to avoid the interference of redundant information, greatly improving the learner's learning efficiency.

30.3.2 Absorb and Store Knowledge

Though obtaining knowledge is important, absorbing and storing knowledge are more significant. It is a relatively complex process to turn the information into his own knowledge. However, the web 2.0-based Blog and IM' software simplify the process of absorbing and storing knowledge. First of all, any outside knowledge can fuse with his own knowledge system in the way of thinking independently and exchanging. As social communicative tool, Blog and IM make it possible between the learners and learners, between learners and teachers that spark of their views ideas collide, and they exploit their different attitudes toward one thing, even form the consensus fusing into their knowledge system. Second, the temporary obtaining of information does not mean long-term memory [3]. Especially in the informal learning we encounter knowledge by chance, if not recorded in time we would forget them easily. So the timely recording of new information and learning experience appears to particularly important. Blog, in the form of log, documents recent learned lesson, and their understanding of knowledge, which are the crystallization of the wisdom of the learner, their own system of knowledge base by constructing.

30.3.3 Distribute Information

Traditional network put learners at a passive position in receiving information. The emergence of Web 2.0 makes it convenience that the learner can easily distribute their own information. Applications elements of Web 2.0 information lower the threshold for the individuals to participate in the creation of information. So they do not have to master the technology of procedural language and Dreamweaver, and anyone can set up their own online blogs, participate in wiki writing, recommend books, or comment, to realize the personal information dissemination. It is necessary to obtain the information before distribute it. Search engines, such as network pick, RSS, Blog, and other social software provide learners with very timely information and access to screening tools. Accumulation of substantial information will inevitably bring about the release of information. Blog as a platform for information dissemination for learners play a role that cannot be ignored.

In summary, Web 2.0-based informal learning reflects features of people-oriented, active construction, open innovation, collaborative interaction, aggregation and sharing. And it is these highlight advantages that meet needs of the characteristics of the network learning, that will become the most popular learning model in the new age.

References

1. Pettenati MC (2010) Informal learning theories and tools to support knowledge management in distributed COPs. http://ftp.informatik.rwth-aachen.de/Publications/CEUR-WS/Vol-213/paper47.pdf
2. Jia-ning X (2008) The process and characters of informal scholarly communication based on web 2.0. Inf Sci 1:53–59
3. Jing Y, Chengxin L (2007) Web 2.0 and postmodernism turning of learning mode. Mod Educ Technol 1:69–71

Chapter 31
Network Based on the Complex Task State-Level Workflow Modeling

Erxi Zhu, Min Xu and Hong-Lei Lu

Abstract The conceptual model complex workflow tasks, on this basis, the use of colored Petri nets for the task interface, the task of the state network, composite version of the task described in detail, the level of proposed workflow task state network. Compared to the current workflow modeling method, the use of network-level task state, with levels of skills in support of the task version of configuration tasks by explicitly express the advantages of the different states. Task state-level network can be used to model complex workflows, to meet the complex business environment under the conditions of the workflow needs of the model.

Keywords Petri net • Business process management • Workflow • Business process modeling

31.1 Introduction

With the acceleration of economic globalization increasing competition in the market era the enterprise business environment in which great changes have taken place, giving companies an enormous challenge [1, 2]. Faced with this grim situation, governments, scholars have put a lot of money and manpower, and manufacturing processes, automation technology, production patterns and management techniques, and other point of view, put forward many useful enterprises to improve competitiveness and overall strength methods [3, 4]. It is recognized that the problem lies not only in the specific mode of manufacturing technology and production itself, but also because they are still confined to the traditional mode of

E. Zhu (✉) · H.-L. Lu
Department of Computer Engineering, Jiangsu College of Information Technology, Wuxi 214153, China
e-mail: chenwb_33@126.com

M. Xu
Department of Electronic Information Engineering, Jiangsu College of Information Technology, Wuxi 214153, China

Z. Zhang (ed.), *Proceedings of the International Conference on Information Engineering and Applications (IEA) 2012*, Lecture Notes in Electrical Engineering 220, DOI: 10.1007/978-1-4471-4844-9_31,

production and management of the old framework, recognizing that the business process optimization and management of the necessity and importance. Therefore, since the twentieth century, 90 years later, business process management (BPM) technology has been more and more attention researchers [5, 6].

BPM based on computer network technology, workflow technology, enterprise application integration, and XML technology, business process from the perspective of a full range of enterprise management, and support continuous improvement of business processes, the core idea is for the enterprise and corporate between the various business processes to provide a unified modeling, execution and monitoring environment. Currently, the process involved in BPM modeling, operation, analysis, monitoring and other aspects of carrying out a more in-depth research, but also made some research results. But the face of the increasingly complex and dynamic BPM practices, there are still many problems to be solved.

This paper studies the complex modeling of the workflow model, designed to address changes in business work flow of the work to achieve just the dynamic change of workflow process changes in, without affecting the operation of business processes across the enterprise approach so that the work flow system under different conditions with different levels of processing changes and error correction function, while changes in self-search to make adaptive changes in corporate management to reduce vulnerability.

31.2 Business Process Management and Workflow

31.2.1 Business Process Management

Business process (BP), also known as BPM in order to achieve certain purposes of the implementation of the activities of a series of logically related set of business processes to meet the needs of the market output is the product or service. That is, in part or all of the participation of organizations and personnel, the use of corporate resources, in accordance with pre-defined business rules, the participants and documents between organizations, information, and transfer tasks and carry out daily tasks processing and management decision-making, in order to achieve the intended business objectives.

BPM is the process of transformation from the related business areas such as business process improvement business process improvement (BPI), business process reengineering in the developed.

BPM, also known as business process management. From a management theory or strategic level, BPM is an internal event in an environment and external events, by a group of interdependent business processes proceed to describe the business, understanding that, organization and maintenance. Perspective from the specific implementation, BPM can be divided into process analysis, process definition and redefinition, resource allocation, scheduling, process management, process quality and efficiency measurement, process optimization.

31.2.2 The Definition of the Workflow

Established in 1993, gives the Workflow Management Coalition defines workflow: Workflow is a kind of fully or partially automate the business process, a process it according to the rules, documents, information or tasks to perform in between the different for delivery and implementation. In order to achieve the different interactions between workflow products, has given the workflow management system definition: workflow management system is a software system, it has completed the workflow definition and management, and in accordance with pre-defined in the computer work flow logic to promote the implementation of the workflow instance. At the same time, workflow management system's main function was to illustrate in detail:

1. Process design and definition: analysis by business process modeling, use of the actual business process modeling tools to complete formal definition of a computerized handle the conversion.
2. Run the control phase: execution services through the workflow software, repeated instances of the process and control.
3. Human Computer Interaction stage: implementation through the application software interface between the user and the interactive application tools.

31.2.3 4-Level Task State Net Workflow Model

The current Petri net workflow models are too simple; such as is usually the work of the nets, first of all it does not explicitly express the status of the task changes, it is difficult to distinguish the status of a task; Second, workflow net is a classic Petri net, it does not have modular capabilities, and therefore can not express the hierarchical structure; Finally, the workflow net is a static description, workflow task difficult version of the configuration. The task of state-level Petri nets network combines an intuitive and mathematical rigor, a strong expression of the advantages of the rules. It is a special type of hierarchical colored Petri nets. Task state-level network supports both the description of the workflow hierarchy and multi-tasking version of configuration, but also through the existing theory to structural analysis and verification.

31.2.4 CPN Description

CPN is defined as follows:

CPN is a multi-group

$$\left(\sum P, T, A, N, C, G, E, I\right) \tag{31.1}$$

where $\sum$ is a finite non-empty set of colors; P is a finite set of non-empty place; T is a finite set of non-empty translation; A is a finite non-empty arc set; and

$$P \cap T = P \cap A = T \cap A = \varphi \tag{31.2}$$

N is a node function,

$$N : A \rightarrow P \times T \cup T \times P \tag{31.3}$$

that

$$\forall a \in A,\ N(a) = (Source,\ Destination) \tag{31.4}$$

where

$$(S, D) \in P \times T \cup T \times P \tag{31.5}$$

C is the color of the function P,

$$C : P \rightarrow \sum \tag{31.6}$$

The guard function G is T;

$$\forall t \in T : \left[Type(G(t)) = B \wedge Type(Var(g(t)) \in \sum \right] \tag{31.7}$$

$$B = \{true,\ false\} \tag{31.8}$$

$Type$ is the type of expression or parameter, Var is the expression contains the parameters; E is the arc expression function,

$$\forall a \in A : \left[Type(E(a)) = C(p(a)) \wedge Type(Var(E(a)) \in \sum \right] \tag{31.9}$$

where $p(a)$ is $N(a)$ the place; I is the CPN's initialization function,

$$I : P \rightarrow Exp,\quad \forall p \in P : [Type(I(p)) = C(p)] \tag{31.10}$$

31.2.5 Description of Task State Net

Use of the expression of colored Petri nets to the task interface, to form a task state network. Can be described from two perspectives, the external interface out the details hidden within their own, the effect is like a change, the interface between it and other task dependencies expressed through various library; task describes the internal structure of the interface to the internal state of dependence on external control and data state transition under the action of the process Fig. 31.1.

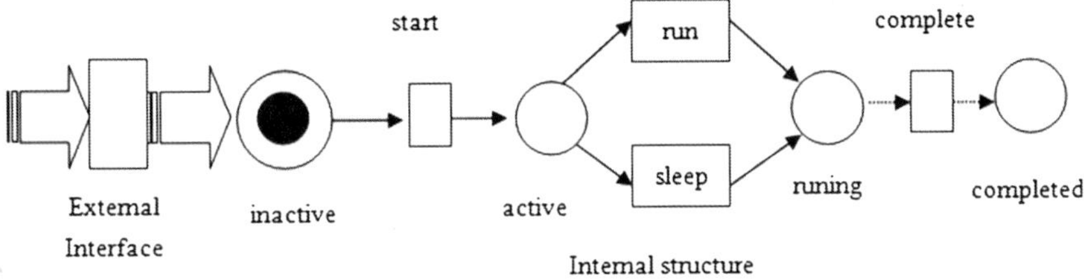

Fig. 31.1 Task state net

31.2.6 Version of the CPN Description of Complex Tasks

In the workflow model, the complex task to achieve the corresponding versions of multiple tasks, each task version of the interface and the composition of multiple tasks, these tasks are complex interfaces between the control flow and data flow dependence Contact (task version of the difference reflects constitute the interface between the task and rely on two different links).

The expression of workflow patterns: work flow model that is abstracted from the real business out of the basic work flow control structure, it is independent of specific modeling language. Workflow model more accurately support more workflow patterns, the model's ability to express the stronger. In fact, every workflow model is a combination of multiple control flow dependence; we only select the model given the CPN workflow model described in Fig. 31.2.

Choice Pattern:

$$TI1Cho(TI2, TI3) \equiv (TI1\underrightarrow{cp- > st}TI2 \vee TI3) \cup (TI1\underrightarrow{sk- > sk}TI2 \vee TI3) \quad (31.11)$$

if the $TI1$ end. The translation in activation of λ_1 or λ_1, This $TI2$ and $TI3$ the inactive place get started only one token, other tokens can only be skipped, That

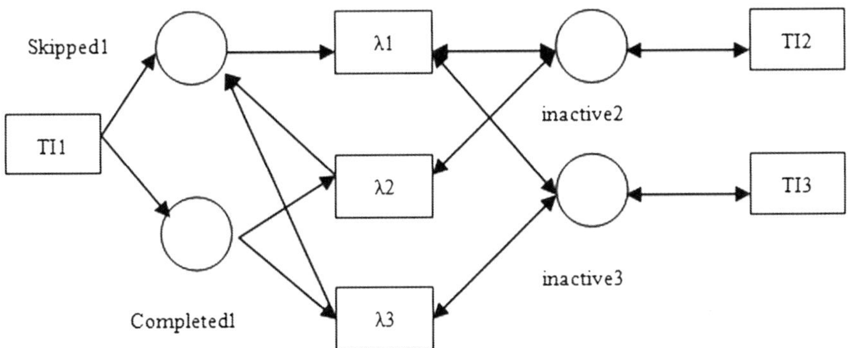

Fig. 31.2 Choice pattern of CPN descriptions

only one implementation of $T12$ and $T13$; skip if $T11$, $T12$ and $T13$ the inactive place that are given a skip token, so $T12$ and $T13$ are skipped.

31.3 Conclusion

With the deepening of workflow applications and business environment more complex, workflow model should have more description and validation analysis. This paper introduces the workflow modeling the basic concepts, the concept of workflow tasks and complex models are analyzed in this paper, the use of colored Petri nets task interface, the task state network, the task version of the compound described in detail raised the level of the workflow task state network. Compared to the current workflow modeling method, the use of the advantages of network-level task state is: expressive power with a level of support tasks version configuration, explicitly expressed through the different states of the task. The model can express complex control dependence; the use of state-level task nets can model complex workflow.

References

1. van der Aalst WMP, Kumar A (2001) A reference model for team-enabled workflow management systems. Data Knowl Eng 38(3):335–363
2. Mentzas G, Halaris C, Kavadias S (2001) Modeling business processes with workflow systems: an evaluation of alternative approaches. Int J Inf Manage 21(2):123–135
3. Chiplunkar C, Deshmukh SG, Chattopadhyay R (2003) Application of principles of event related open systems to business process reengineering. Comput Ind Eng 45(3):347–374
4. Sheng Liu D, Min Wang J, Chan SCF, Guang Sun J, Zhang L (2002) Modeling workflow processes with colored Petri nets. Comput Ind 49(3):267–281
5. Beckjord ES, Cunninggham MA, Murphy JA (1993) Probabilistic safety assessment development in the United States 1972–1990. Reliab Eng Syst Saf 39(3):159–170
6. Murata T (1989) Petri nets: properties, analysis, and application. Proc IEEE 77(4):541–580

Chapter 32
Trusted Network Model and Repair

Qian Xu

Abstract The credibility of the proposal and development process of the network and how it works, in the traditional network model, based on credible combination of life sciences to the behavioral characteristics of the data contained in the main line, build a credible model of network theory, in this proposed based on a credible fix the overall network design and a detailed discussion of the trusted network repair services technology, given the credibility of the model repair and restoration of network services, work flow, provide a credible rehabilitation program. Resources in ensuring the restoration of the program transmission of fast, reliable based on the performance with a certain extension.

Keywords Trusted network • Trusted remediation • Network security

32.1 Introduction

Security is defined as: the state or character away from danger, to guard against espionage or sabotage, crime, attack or escape measures [1]. And then specifically related to computer security, we use the international organization for standardization (ISO) definition: for data processing system to establish and adopt the technology and management of security, protection of computer hardware, software, and data is not due to accidental and malicious destruction of reasons, change, and disclosure. Physical security includes the protection of related computer equipment, software data logic level include the integrity, confidentiality, availability, and so on.

In the computer field, "credibility" is mainly derived from the development of fault-tolerant computing, 2000 IEEE international conference fault-tolerant computing (FTCS) with the international federation for information processing (IFIP)

Q. Xu (✉)
Electrical & Information Engineering College, Shaanxi University of Science and Technology, Xi'an 710021, Shaanxi, China
e-mail: okohag@126.com

Z. Zhong (ed.), *Proceedings of the International Conference on Information Engineering and Applications (IEA) 2012*, Lecture Notes in Electrical Engineering 220,
DOI: 10.1007/978-1-4471-4844-9_32,

10.4 trusted computing working group working meeting critical applications combined and renamed trusted systems and networks for the IEEE international conference on (ICDSN). In the fault-tolerant computing, reliability is defined as a computer system the nature of the services it provides the user can be an act of perception, and can prove its reliability, and users are able to interact with another system (human or physical system) [2].

Trusted Computing Alliance trusted computing group (TCG) on "credible" is defined as [3]: "an entity in achieving a given target, if its behaviour is always as expected, then the entity is credible." The "act in line with expectations" in the different scenarios may be different, according to this definition, TCG trusted network interconnection trusted network connect (TNC) since the working group put the TNC from V1.0 targeted to: define a open solution architecture that allows network administrators to apply for access to the network terminal according to the identity and security status to determine their network access.

32.2 Trusted Computing Platform

Traditional security architecture of the terminal to the core of TCB, TCB itself tamper-resistant requirements, self-protection, but we all know, and pure software system is difficult to guarantee the safety of TCB, the TCB after an attack, can not ensure the integrity of their own from destruction. So people start looking for new ways to solve security problems TCB, trusted computing came into being.

People will trust in the trusted computing roots into the TCB, the TCB as the core, and then passed through the chain of trust mechanisms, a measure of level, a certification level, and gradually establish and expand trust chain, in order to achieve the boundary gradual expansion, protect its integrity. Visible terminal is the source of a security incident, the terminal should be a high level of security implementation, so that insecurity from the terminal to be controlled. The terminal security thinking has been valued by the people, and people have gradually realized that a software system can not guarantee the information from the fundamental safety of the system, so in this context the establishment of trusted computing platform.

IBM, Microsoft, and so well-known enterprises set up in 1999, trusted computing platform alliance TCG [4], reorganized in mid-2003 TCG. The core idea of trusted computing is "transitive trust", that is, if from an initial "trusted root" starting platform computing environment in every conversion, this credibility be maintained by way of transfer is not destroyed, then the platform of the computing environment to always be trusted. Trusted computing platform to support cryptography to secure the operating system as the core of its basic structure shown in Fig. 32.1, this level of measurement level, a certification level of trust chain, the trust relationship extended from the underlying hardware platform to the operating system, and then expanded to the upper application, and ultimately build a credible application platform.

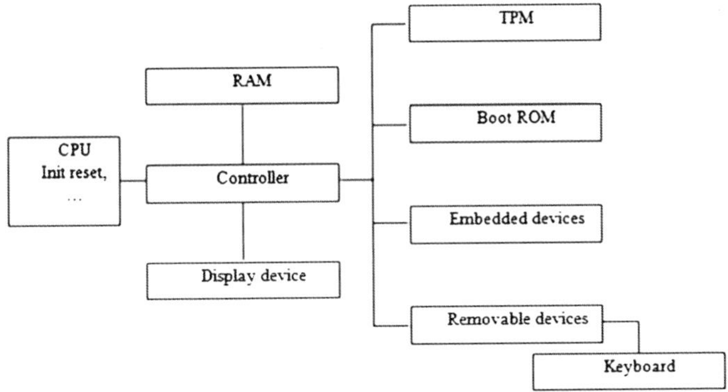

Fig. 32.1 Structure of trusted computing platform

TCG aims to develop in line for a variety of terminal platform design of trusted computing industry norms, thus taking into account the universal platform, which is defined as a trusted computing platform, a common reference structure. Including building a "trusted root" of the key modules, Trusted computing platform by the board and build on it CPUT, memory and some peripherals, and embedded firmware, BIOS,

Trusted building blocks (TBB) of common composition. The main features of trusted computing platform are embedded on the motherboard to have credible building blocks, the trusted computing platform building blocks of trust is the trusted root.

32.3 Trusted Network Model

Comprehensively improve the defensive capabilities of the network is reliable overall objective of the network; it must be done to achieve the goal of full integration of security resources and management, security, survivability, and manageability organic bound together. Have a global consciousness, from a global perspective to risk analysis assessment and management, and then customize the global scope and distribution strategy, and timely follow changes in environmental conditions and intelligent adjustment, Trusted network from the range of perspective, to be able to build a clear border security, and automatically according to changing conditions and make the appropriate extension and convergence. This form of diversification in the current network networking increasingly blurred boundaries of the network has more significance today.

Trusted Network System in the internal flow of data (data flow) above, Management of the data stream contains the knowledge needed to trusted network, business data, implicit in the data stream in user behavior. For knowledge, content, behavior management, and updates are the core of the trusted network.

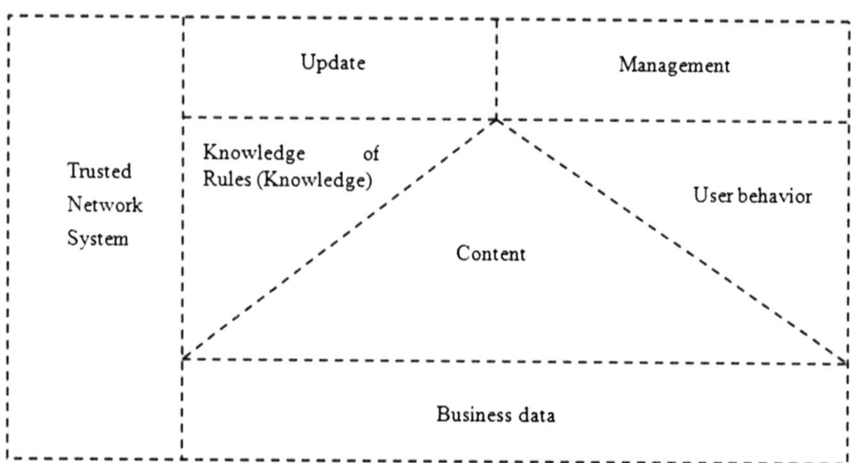

Fig. 32.2 Trusted network model diagram

Trusted network terminal equipment need to access the internal network, the first access switch via 802.1X protocol level access terminal authentication, the terminal if you do not meet the data access switch can not forwards. If the access terminal problem, the attack can not be extended to the whole range within the internal network.

Trusted network based on the concept and characteristics of the model we can obtain the following diagram, shown in Fig. 32.2.

32.4 Trusted Repair Service Workflow

Credible framework for fixed networks is based on the TNC to develop and design modules. Repair in the trusted network, if the access terminal in the access authentication phases, because the property is not credible and can not access network complete, trusted repair in consultation with the user client is a credible fix. Repaired by a trusted access terminal, then re-access authentication and access to trusted networks.

In the trusted network access process, evaluation, segregation, and repair are the three basic stages of this process. Assessment phase is mainly based on credible information authentication module integrity of local policy on the terminal evaluation, when needed credible information indicating a credible measurement module repair; isolation stage is if a trusted access network at the access terminal integrity verification does not pass, then the PDP will be directed to the terminal PEP redirected to an isolated environment, where the conduct and integrity-related updates; repair phase is the data access terminal for the process of repair, making the access terminal PDP provided to meet the requirements to

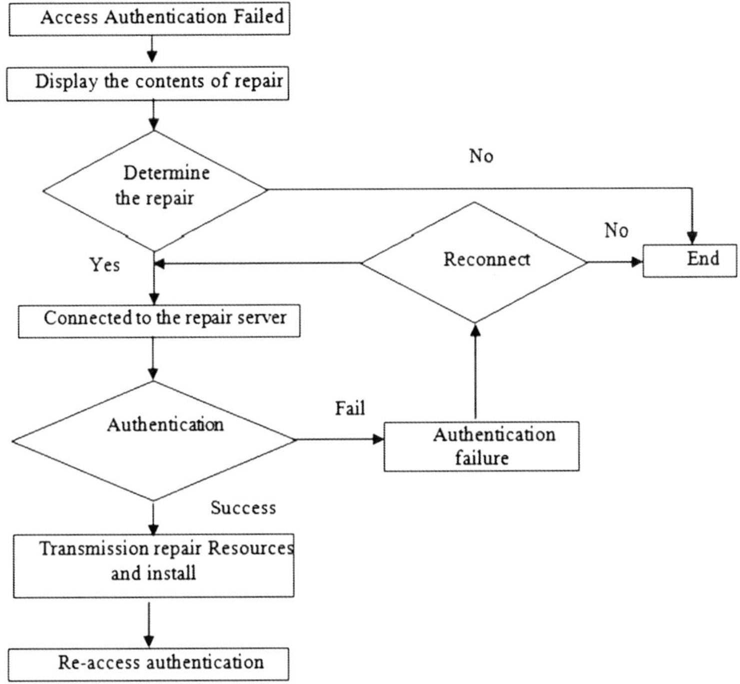

Fig. 32.3 Flowchart trusted repair service

access the network. The action isolation and repair is repair services, repair services can be seen in the importance of reliable network access process. If the lack of credibility of the process of repairing the access network link, then the integrity verification fails for the access terminal will not know what reason is due to the protected network can not access or can not be trusted through a secure channel for access to repair data thus unable to gain access to the protected network access. Trusted repair service process shown in Fig. 32.3.

32.5 Conclusion

Through the course of development of network security with the current edge of the network concept of credible research, introduced the proposed trusted network and its working principle and development process, in the traditional network model, based on trusted with the life sciences, the behavior of the data contained in characteristics of the main line, build a credible model of network theory, on this basis, the proposed network's overall credibility of repair design, and discussed in detail trusted network repair services technology, given the credibility of the model and the fixed network repair service workflow and provide a credible rehabilitation program. Through this analysis we can see the original trusted network on the basis of network security is more emphasis on behavior management, and behavior-based management is bound to extend intelligence to judge, so the network structure and more tend to change the physical structure of intelligent life.

At the same time due to the current development of network technology makes the network perimeter continues to fade, but gradually a clear and credible network of trusted zones. Therefore, network expansion and integration is becoming a trend, coupled with the data in a wider range of efficient transmission of the demand, making this trend more apparent. Trusted network at a higher level of security is worth further study.

Acknowledgments This work is partially supported by Shaanxi Provincial Science and Technology Department: Computer Software and Security Technology (2009JM8003).

References

1. Meifeng S, Jian G (2004) The research of formal specification of intrusion detection rules. In: Proceedings-conference on local computer networks, vol 25(15). LCN, Tampa, pp 429–430
2. Harn L, Hsin WJ, Mehta M (2005) Authenticated Diffic-Hellman key agreement protocol using a single cryptographic assumption. IEEE Proc Commu 12(10):404–410
3. Beth T, Borcherding M, Klein B (1994) Valuation of trust in open network. In: Gollmann D (ed) Proceedings of the European Symposium on research in security, vol 7(2). Springer, Brighton, pp 3–18
4. Suganuma T, Yasue T, Kawahito M, Komatsu H, Nakatani T (2005) Design and evaluation of dynamic optimizations for a java just-in-time compiler. ACM Trans Program Lang Syst 27(15):732–785

Chapter 33
Research Websites of Vocational Colleges

Peng Jun

Abstract This article provides the example of the construction and management of the research website—Science News Profile in Sichuan Business Vocational College. By analyzing the status and problems existing in the research websites construction in Sichuan Business Vocational College, this article proposes solutions to the related problems mentioned.

Keywords Vocational college • Research websites • Problems • Strategies

33.1 Introduction

Campus network is one of the most important parts of the construction in education information [1, 2]. It is also the reflection of modernization in educational technology equipment. What's more, it is one of the symbols of the modern education. The application of campus network can update the management concepts of school administrators, enhance the quality and efficiency of management and achieve the communication function inside and outside schools conveniently. It can enrich the vision of the teachers, which changes the traditional teaching methods and concepts of the teacher, making it possible to improve the educational concepts [3, 4]. Besides, the campus network is beneficial to the development of the creative thinking of teachers, which can improve their ability to obtain, analyze, and deal with information, making them much easier adapt to the modern society. Based on the campus network in schools, the staffs in the Guidance Department of Research and Industry Cooperation organize the construction of research websites.

The website of Science News Profile—http://scswkyc.mynet.cn in Sichuan Business Vocational College was built in 2006. It is the platform for the Guidance Department of Research and Industry Cooperation to release

P. Jun (✉)
Guidance Department, Sichuan Business Vocational College,
Research and Industry Cooperation, Chengdu 611131, Sichuan, China
e-mail: iaejlxd@sina.com

Z. Zhong (ed.), *Proceedings of the International Conference on Information Engineering and Applications (IEA) 2012*, Lecture Notes in Electrical Engineering 220,
DOI: 10.1007/978-1-4471-4844-9_33,

news to the public. Besides, it is also the main format for the internal staffs to declare and manage their research projects, register and examine their research achievements and display the school-enterprise cooperation projects. This research website in the college has been the most effective and convenient way and one of the interactive platforms for the students, teachers as well as the outside to know the Guidance Department of Research and Industry Cooperation. Science News Profile has experienced three revisions in succession since it was set up.

33.2 Existing Problems in Research Websites

With the rapid development of the Internet and the wide spread of network technology, the number of visits to the research websites in vocational college has increased increasingly. And this situation is growing more and more apparent. During 4 years operation of the website, four problems have been found out.

33.2.1 Insufficient Updates of the Contents in the Website

In the early stage of the construction of the research website, the website builders were full of passion. They were also ambitious about the future development of the website, getting together, and commutating with each other about the updates. As a result, they did a good job. However, they are not well prepared for the long-term maintenance and operation for the website. Their momentum was often enough, while they are tired of updating the website in later period. The construction of the website is lack of staffs and ensured funding, which often depends on the professional teachers to build it after school. However, the professional skills and energy are relatively limited in schools, in which the slow updates of the website in the long run are caused. What's worse, once the website is not updated in time, the visitors' interest in the website will decrease dramatically. As a result, the number of visits to the website keeps small.

33.2.2 Propagation Conditions are Under Improvement

Nowadays, the access conditions of the staffs in vocational colleges are comparatively good. However, the differences between the hardware configuration and software environment are very large. For the construction of the research website in colleges, the shortage of the hardware and software especially that of the application software is very obvious.

In the late period of the website construction development, small amount of funds is invested. Free server was put into use in the early construction of the website when there was quantity limit for monthly upload capacities. As a result, a lot of files could not be uploaded. It was a very difficult issue to maintain the website without outside investments. And here only refers to the maintenance of personal sites. Long-term development requires the input of contents and operation. The input here refers not only to the input of more personal energy, but also to more funds and relationships.

33.2.3 Insufficient Website Applications

Since the early development of the research website, there are columns of Introduction of the Department, Science News, Assembly Systems, Research Achievements, and Proposed Suggestions in its homepage. When using, the visitors can find out that the set rate is very high in the news columns like Science News. During recent years, more than 230 pieces of news are released which include 83 pieces of news relating to the leading inspection and various kinds of conferences. The news photos are basically about the conferences. However, information related to the activities carried out by the students and teachers is not enough. Therefore, we can see that the research website in our college focuses mainly on information release and college promotion, while the teaching management, office automation, educational resources, and communications between students and teachers are seldom dug into.

33.2.4 Poor Openness and Narrow Participation

In the process of the website construction, building, updating, and maintenance are all depend on individual builders, which makes narrow participation of the students and teachers. An excellent website can not only depend on the technical staffs, but also depend on the positive attitude of the students and teachers during the course of website construction.

The main reason resulting in the narrow participation is the poor openness of the website. On one hand, the right of releasing information in the website concentrates on the network managers, which obstructs the passion of the students and teacher to participate in the website construction. On the other hand, the website doesn't take the users' participation into consideration in its technical features. And the poor openness of the website results from many elements—the builders lacking of related concepts, technical staffs lacking of technical standards as well as network safety problems brought by the openness.

33.3 Reasons Analyses of the Existing Problems from the Website

33.3.1 Own Problems from the Website

First, the own problems from the website are very fundamental. They prevent the research websites from developing. The purpose of website construction is unclear. Generally speaking, website construction requires clear purpose before everything is started. Long-term market and feasibility researches are performed before action is taken. However, research websites are different as mentioned previously. The author planned to fulfill his ideas by constructing the website in the very beginning when its future existence was not regarded as the primary factor. Without clear purpose, the website was self-sufficient in the early stage of the development, lacking of essential conditions to develop hardware and software environment where the lacking of application software was especially apparent. When the website was finished, the feeling was just like finishing an experiment. Websites without clear purpose and features are naturally not popular among public. They will fail in the fierce competition. The positioning of the website functions is not accurate and the contents imitate the large portals, which foreshadowed the single template and insufficient iterativeness.

33.3.2 Lacking of Power of the Comprehensively Large Websites and Large Production Team

The research website of our college was constructed similar to the comprehensively large websites, which is lack of its own characteristics. It does not catch the psychological needs of the visitors. Its contents cannot meet visitors' needs, neither, or they are not creative. They may be reprinted from some authority sites, lowering the level of our research website.

33.3.3 Lack of Financial Support

Technically speaking, a successful website depends on the policies and financial supports. Both software and hardware can support the long-term operation of the website. This could not make the case in the early operation of the research website. It also resulted in the low technology level indirectly. The low and unreasonable input of software and hardware makes it easy to build a website but hard to maintain, let alone good operation. Nowadays, research websites that have good communication effects and win attention from students, teachers, and the colleges are those with good software and hardware conditions and are supported by the colleges in economy. However, the general research websites are built and developed by the passion and ambition of some staffs in the early stage. The staff community and individuals input limited amount of funds and energy. But without support of the follow-up funds in the website development, software and hardware supports are lost. At that time, the construction of the website will become an impossible mission just like the saying that one cannot make bricks without straw.

33.3.4 Lack of Effective Management

A relatively healthy boot system assessment can be also lack of related monitoring and incentive systems as well as staffs training. Maintenance and management of the website and collection and release of information both require managers of the websites. The management staffs should not only have certain information skills, but the ability to grasp the words. The constructors and defenders are all teaching staffs. To build an all-rounded website and make it perform positive functions is very hard to achieve. School administrators and ideological and political education advocates are of great important to guide and train them.

In reality, a successful website exists in the long run and has influences with its applications, management, maintenance, and development. For research website in the college, it is basically healthy based on the healthy mechanism of the college, institutional setup, financial support, as well as the management and motivation. All these elements will accelerate the speed of the updates of the website construction, promoting the development of potential sites, and preventing the situation of stagnation.

33.4 Strategies of Solving Problems Existing in Research Websites

33.4.1 Adhere to the Characteristics and Reasonable Positioning

In the early stage of their construction, college research websites should be positioned properly according to their own characteristics in contents. Besides, the purpose of their construction should also be clear, such as Introduction of the Department, Research News, Resources Sharing, School-enterprise Cooperation, and Online Study. After the target is set, some setting of the columns should be emphasized. Some related contents should also be dug into deeply. Staffs should be organized to collect information, too. The graphic design and information details should be paid more attention to at the same time. In this way, the whole positioning of the website is clear and the forecast target can be achieved. The emphasis of the construction of the website can be selected based on the characteristics of research targets in the college. If improper module is chosen, the result would be counter productive.

33.4.2 *Rational Division and Sharing of Labor*

Rational sharing is the core in the construction of excellent websites. Construction of the research website is not the patent for R&D staffs and teachers whose main responsibilities are providing elements, building technical platforms, and offering technical support. Constructing a website cannot be finished overnight. On the contrary, it needs the process of accumulation. The construction of a website aims at achieving resources and information sharing and bringing convenience to the users. As for the campus network, users include students, teachers, parents, education administrators, and enterprises.

When setting the columns, constructors of the website should take the vitality of the column into consideration. There are at least two conditions for the column to become a vivid one. First, there should be clear security for the daily maintenance and updates of the column contents. Second, there should be a number of visits to the column. Otherwise, the column will be lack of vitality. The existence of that column should be reconsidered, because it is meaningless to construct a website whose contents cannot be updated and number of visits is very small.

33.4.3 *Technical Support and Open Platform*

The level of technical skills of the website makers affects directly the openness and vitality of the website. The college should encourage and support related teachers to improve the skills for constructing websites in the system, which can ensure relatively strong technical security and develop technical platform where the majority of teachers and students can take part in. If the skills are really insufficient, colleges can entrust outside power to help to develop the website in the early stage. But they should provide plans for the construction in details. Besides, the columns needed and solid technology required should also be provided to the developer.

33.5 Effective Management

33.5.1 *Improve Website Quality and Content Management*

Website quality is the core of the brand. Without quality assurance, the brand is just like desk decoration. Compared to the pure management of website construction, the operation of the research websites pays more attention to the image modeling and promotion planning. Website quality and content are the basis of its development. To improve the affinity and vitality of the website, high quality as a support is the most important thing. Improving the management and supervision of the research website from its quality and contents will lower the negative effects and play a positive role actually.

33.5.2 Improve the Management Method with the Technical Support

First, strong technical support is the basis of enriching the contents of the website. Information and technology develops rapidly. Studying, mastering, and using the new technology and the new system can seize the initiative and attract more attention. If the research website in colleges makes full use the advance technology, the comprehensive advantage can be ensured to be brought into play. At the same time, good hardware and software inputs are the foundation of the website construction. Many of the influential websites in colleges are those secondary sites built based on the campus networks. Although the campus networks provide a strong main server covering the whole campus and proving pretty good environment for the existence of the student networks, the development and construction of websites still need a great amount of elements such as material, energy, offices, and facilities that offer proper conditions of software and hardware environment. In this way, developers can maximum their imagination to a great extent. The comprehensive function of the website and expected benefits can also be achieved. All of these can't be provided by general research website developers. With the support of the college, research and development of the website will be more standardized.

Acknowledgments This dissertation is the research result of the topic—Research and Development of Application Software of the Management System in Vocational Colleges Research (No. 10YKC05) from Sichuan Business Vocational College.

References

1. Chen R (2004) Practical guide of website construction and management, vol 12(6). Anhui Cultural and Audiovisual Press, Hefei Anhui Province, pp 64–69
2. Tao D (2002) Project management of website construction, vol 53(23). Post and Telecom Press, Beijing, pp 589–597
3. Yu W (2003) Application of information architecture theory in information organization. Libr Dev 10(5):734–741
4. Yang J, Yu Y, Gao R (2003) Campus network and education library resource construction, vol 8(1). The Open University of China Press, Beijing, pp 84–89

Chapter 34
Rule Base Design of Ipv6 Network Intrusion Detection System

Feng Yu and Wei Liu

Abstract In order to improve the IPv4/IPv6 network intrusion detection system efficiency, the rules of intrusion detection system, feature library has been designed, gives a detailed list of rules and regulations define the structure, the rule base using three-dimensional linked list data structure, rules and their options are stored in the list, the scan to find the list to find the first match with the rules of data packets, in the case of matching, pattern matching, or search using a test plug to match. The design can improve the alarm system, intrusion detection accuracy, and efficiency.

Keywords Intrusion detection • IPv6 • Rules • Agreements

34.1 Introduction

Intrusion detection technology as a proactive security technologies in a timely manner all kinds of malicious intrusion detection and network response when compromised, it is traditional security technologies such as firewalls and reasonable supplement [1], is an emerging network security technology, but also The current theoretical study of computer network security, a hot spot.

Next generation Internet Protocol IPv6, network management, control, security, performance, and many more powerful than IPv4, and efficient. Therefore, in the next generation Internet protocol IPv6 environment, the establishment of an efficient, real-time network intrusion detection system is of great significance. Intrusion detection is to take isolated network behavior and unsafe behavior characteristics of known process of matching rules; the rule base is the need to guard against all known attacks on the characteristics of libraries [2]. Captured from the network data packets and each has been defined rules match, if found, a packet matches the rule, it

F. Yu (✉) · W. Liu
Network Center of Shenyang Jianzhu University, Shenyang 110168, Liaoning, China
e-mail: yufeng@hrsk.net

Z. Zhong (ed.), *Proceedings of the International Conference on Information Engineering and Applications (IEA) 2012*, Lecture Notes in Electrical Engineering 220,
DOI: 10.1007/978-1-4471-4844-9_34,

means an attack is detected, and then act in accordance with the rules of processing (such as sending warning etc.); if you search for after all the rules when no match is found, it means the message is normal. It can be seen, the rules for the design and organization of the alarm system accuracy and efficiency of the system has a great impact, the intrusion detection system is mainly based on rule matching intrusion detection, therefore, according to rules file (also known as the rule base) establish a good structure is essential rule match, the only way to detect the engine processing module can be the basis for this rule base intrusion detection.

34.2 Intrusion Detection and the Working Principle of the Rule Base

As shown in Fig. 34.1, among them, the main source of information system audit data or the original network data packets [3]. Data pre-processing refers to the pre-treatment data collected will be transformed into models for the detection of the received data format, including the removal of redundant information, i.e., data reduction. This is the key field of intrusion detection. Detection model is built based on various detection algorithms model testing. Since the detection rate of single-detection model less than ideal, often require multiple parallel test model analysis and processing, and test results of these data fusion in order to achieve satisfactory results. Security policy is set according to security policy requirements. Response to treatment mainly refers to the results of comprehensive security strategy and the response made by the process, including the generation of test reports, notify the administrator, disconnect from the network or firewall configuration changes to active defense measures.

The rules of the system is text-based files [4], at boot time, read all the rules file, rules file processing time required to store the rules in some way so that the face of the packet after the match. The system uses a three-dimensional (3D) linked list data structure, the rules and their options are stored in the list, the scan to find the list to find the first match with the rules of the packet, and then, in the case of matching, finding patterns, or using a test plug-in match to match.

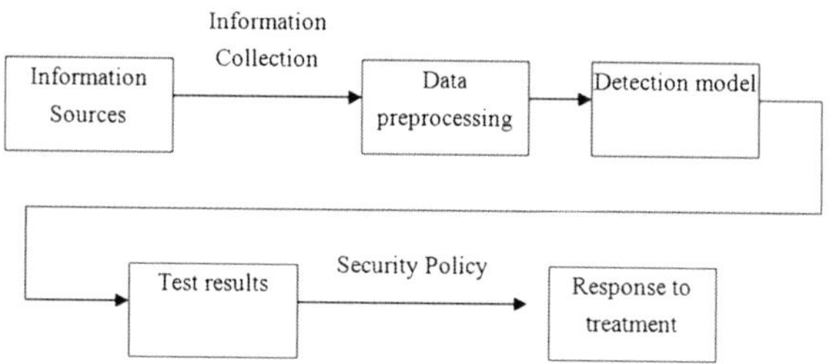

Fig. 34.1 The working principle of intrusion detection

Its working principle is described as follows: when a source address in the unit of time (this time set the threshold to their own) constantly monitored segment to send packets, then use this address that the user scan the network section, need to record some of the address information and scanning behavior, but also can send feedback to the address, to remind each other that we have detected his misconduct. Furthermore, if a source address in the unit time to stop some of the host or a host of different TCP and UDP port to send data packets, the same users that the use of the address of the host system scan. Of course, in the process, but also to detect some sign of TCP header flags, and other special segments of data packets.

34.3 Definition of the Rules

Rules define the methods used by intruders. In the intrusion detection system, the rule contains the contents of two parts: rules and rules of the first option [5].

34.3.1 Rules of Head

First divided into rules of behavior rules, agreements, direction descriptor, source information, the purpose of information and other 5 parts, by these restrictions, rules to deal with the amount of data can be effectively reduced.

Rules of behavior: the definition of what to do when the rules are matched, such as Alert (alarm), Log (record), Pass (ignored), dynamic, and activate and so on.

Agreement: that the rule which corresponds to a protocol such as TCP, IP, UDP, and ICMP and so on.

Direction descriptor: refers to "->" or "<-" through which to understand which engine is "source" information, which is "objective" information.

Source information: source IP address can include, network, ports, and other information.

Objective information: IP address can include the purpose, network, ports, and other information.

For example, the regular first alert tcp $EXTERNAL_NET any ->$HOME_NET 21, where rules of behavior are alert; protocol field is TCP; sources of information including: source IP address portion—$EXTERNAL_NET, refers to any address; source port—any, refers to any port; -> arrow indicates the direction of the session, here is the $EXTERNAL_NET $HOME_NET any port to port 21 on; the purpose of information, including: $HOME_NET, which is defined in the rules similar to the beginning of the file NetNumen.conf $ var the variable to use throughout the rule, which is defined as the internal network; 21 attacks, said the purpose of the port.

34.3.2 Rule Options

Rules of the keys used by a number of options, the system provides a variety of options for keywords, the description of the option syntax rules are as follows:

Msg "FTP EXPLOIT wu-ftpd 2.6.0 site exec format string overflow Linux." this is a warning;

Flow to_server, established, the rules of the system options section contains links to test plug-in keywords. Flow option refers to the list of the third dimension that of a client-server testing plug-in points to a pointer, TCP client-server plug-in link to the session state maintenance and reorganization of the module to detect if a packet is included in the established session.

Content, the message load in the search for a pattern, this field is an important feature of this system. Comparable data can contain binary data. This will use the Boyer–Moore search algorithm for content matching.

Reference, the keywords include third-party information on the interpretation of the attack.

Classtype, attacks tend to be classified so that the user quickly identifies the attack and its priority. Priority single digits: 1 for high, and 2, 3 that low.

Sid, the only signs to the rules of the system.

Rev, the version number of amendments, always sid MS.

In addition to the above examples some of the option keyword and more keywords that are not involved are described below:

Logto, records comply with this rule for all messages to the user specified file instead of the standard output file.
ttl, check the IP packet's TTL field value.
tos, check the IP packets TOS field value.
id, check the IP packet fragmentation ID field value.
ipoption, check the IP packet's option field value.
fragbits, check the IP packet fragmentation bit value.
dsize, check the IP packet length value of the load.
flags, check the TCP-flags field.
seq, check the TCP sequence number field value.
ack, check the TCP acknowledgment field value.
hype, check the ICMP protocol type field value.
icode, check the ICMP code field value of the agreement.
icmp_id, check the ICMP response to the message ID field value.
icmp_seq, check the serial number of ICMP response message field value.
content-list, the message load in the search for a pattern collection.
Offset, adjust the content option, and set the pattern matches the beginning of the offset.
Depth, adjust the content option, set the maximum search length of pattern matching.
nocase, the contents match the string is not case sensitive.
session, a session record for a specified application layer data.

rpc, review the RPC service on an application or process the call.

resp, activation response (such as: close the connection, etc.). Convenient option to achieve this response and its format is a list of close connection.

React, activation response (such as: access blocked sites, etc.), which is a more advanced response mechanisms.

Priority, specify the severity level of this rule, each type of attack has a default level.

urlcontent, allows a URL in the search request matches the specific format, It and similar content, but only in the URL in the search.

tag, the keyword allow rules not only record the packets triggering term rules, but also the subsequent network data can be recorded to facilitate later analysis.

ip_porto, test IP header Protocol field, content is the / etc / protocols in the protocol name.

sameip, used to check whether the source address and destination address the same.

stateless, and TCP state maintenance module used in conjunction with the reorganization of the session, allowing the rules of the match does not take into account the state of connection, the format is stateless.

regax, the keywords specified in the content can contain wildcards.

In fact, the above options for each rule in the source code corresponding to the various processing plug-ins.

34.3.3 Rules File

In this system, the rules are initially present in the form of text file, called the rules of this document file, it exists in the program directory or subdirectory of the system and classified according to different groups, for example, includes the contents of FTP attacks rules file ftp.rules, also contains a TCP content of the rules file tcp.rules attack and so on.

Moreover, in the rules file can also define and use variables, the specific syntax is:

Var <variable name> <variable value>

Variables are included in the rules file, generally used to indicate IP addresses or networks, the following example, the representative of the variable HTTP_PORTS HTTP port number 8081.

When the system starts, it will read all the rules file, and create a multidimensional list.

34.4 Rule List

The rules of the intrusion detection system organized into a 3D linked list structure, which is divided into two parts: the list head (Chain) and the list options (Chain Options). Included in the list is the first rule of the total number of attributes, and different test attributes option is included in the list of different options. For example, if a rule files in the specified CGI-BIN number of the rules of exploration activities, and they all have the same source / destination IP address and port number. To speed up the rate of detection of these common attributes will be compressed into a single list head, while the detection of each of the different attributes associated with the header structure of the various options to save the list.

The data structure is mainly involved in rules.h file, one of the most important data structure have the following: the first function pointer list of rules; rule options function pointer list; response function pointer list; rules option structure; Rules head structure; activation list; IP address structure; rule list structure.

34.4.1 Rules of the First Function Pointer List

The first function pointer list rules, rules, options function pointer list pointer to the list of three structural response functions were similar, the following list of rules the first function pointer as an example.

```
typedef struct_RuleFpList
{
Void * context;
int (* RuleHeadFunc) (Packet *, struct_RuleTreeNode *, struct_RuleFpList *);
/ * First check the rules of function pointer * /
struct_RuleFpList * next; / * function pointer to the next rule node * /
}
RuleFpList;
```

34.4.2 The Data Structure Header

Header data structure in the form of clear, in accordance with the protocol type, the internal pointer to the first list of different rules, specifically as follows:

```
typedef struct_ListHead
{
RuleTreeNode * IpList; / * point to the IP protocol list * /
RuleTreeNode * TcpList; / * point to the TCP protocol list * /
RuleTreeNode * UdpList; / * point to the UDP protocol list * /
RuleTreeNode * IcmpList; / * point to the ICMP protocol list * /
# Ifdef IPV6
RuleTreeNode * Icmp6List; / * point to ICMP6 protocol list * /
# Endif
struct_OutputFuncNode * LogList; / * LOG function function list * /
```

```
struct_OutputFuncNode * AlertList; / * ALERT function function list * /
struct_RuleListNode * ruleListNode; / * IP rules list * /
}
ListHead;
```

34.5 Conclusion

Intrusion detection technology to protect one of the main means of information security, intrusion detection system design and organization for the rules of the alarm system accuracy and efficiency of the system has a great impact, the paper intrusion detection system works in the rule base, rule definitions, rules list structures were discussed in detail. This design uses a 3D linked list data structure, rules and their options will be stored in the list, the scan to find the list to find the first match with the rules of the packet, and then, in the case of matching, pattern matching to find or using a test plug to match. But how to establish a good rule base matches the structure of the rules will be the focus of future research.

References

1. Morton D (1997) Understanding IPv6. PC Netw Advisor 20(5):43–48
2. Spafford EH, Zamboni D (2000) Intrusion detection using autonoumous agents. Comput Netw 34(17):547–570
3. Mukherjee B, Heberlein LT, Levitt KN (1994) Network intrusion detection. IEEE Network 5(2):26–41
4. Tidwell T, Larson R, Fitch K et al (2001) Modeling internet attacks. In: Proceedings of the 2001 IEEE workshop on information assurance and security, vol 37(23). pp 54–59
5. Lee W (1999) A data mining framework for building intrusion detection model. IEEE Symp Secur Priv 29(14):120–132

Part IV
Grid Computing and Cloud Computing

Chapter 35
Iron and Steel Industry Confirmer Wharf Financing Mode

Quan Yuan and Rong Wang

Abstract Given China's iron and steel industry current situation, large in scale and technological backwardness, it is low on profit, lacks innovation, and competitiveness. This paper analyzes the application of using confirmer wharf financing model in iron and steel industry, and further on the procedure the Quartet (the bank, the iron and steel industry, distributors, logistics enterprises involves in the confirmation warehouse financing process). It concludes that the confirmer wharf can play a valuable role in promoting the supply chain flow and logistics and information flow in the industry, and gives iron and steel manufacturers constructive recommendations to avoid the risks related to confirmer wharf.

Keywords Confirmer wharf financing • Iron and steel industry • Risk avoidance

35.1 Introduction

In the "Twelfth Five-Year Plan" period, structural adjustment and industrial upgrading of China's iron and steel industry will become the main melody of the development [1, 2]. Iron and steel industry merger and reorganization will accelerate. The iron and steel industry in China will face three challenges: the domestic economic growth will be reduced; the cost of production factors has a gradually rising trend; the restriction of resource and environment will further hasten [3, 4]. Although China has a large share of iron and steel products in the global market, but the industry's technological backwardness has become a bottleneck. General speaking, iron and steel products made in china are of low quality, and lack of competitiveness, which causes the anxiety and frustrates the entire steel industry [5].

Q. Yuan (✉) · R. Wang
Department of Management Science, Hubei University of Technology Engineering and Technology College, Wuhan, China
e-mail: eminy_yq@hotmail.com

Z. Zhong (ed.), *Proceedings of the International Conference on Information Engineering and Applications (IEA) 2012*, Lecture Notes in Electrical Engineering 220, DOI: 10.1007/978-1-4471-4844-9_35,

35.2 Related Work

Given the data envelopment analysis conducted on China's major large and medium-sized iron and steel enterprises in terms of scale efficiency and relative efficiency, they claimed some small and medium-sized iron and steel enterprise has a higher economic efficiency, many large steel enterprise has a relatively low efficiency. Some work has done Grainger causality test on China's Baosteel, TISCO, Anshan Iron and steel, and WISCO. The results shows the scale effects are not obvious in the industry in this country.

From the iron and steel industry point of view, the author addresses the current iron and steel industry of small and medium-sized enterprises financing problem by using confirmer wharf financing, as well as enhance the entire steel industry supply chain business flow, logistics, and information flow.

35.3 Iron and Steel Industry Confirmer Wharf Financing Research

Given iron and steel industry current situation, it is not difficult to find out that the change on external logistics mode results in the higher steel trade requirements, the root cause is the internal factors has not been resolved. The ways to enhance the iron and steel industry's competitiveness are the optimization of product-line structure, the increase of the trading scale, form a supply, and demand network chain structure mode, however, it needs a lot of capital support. Only financing in large scale can realize iron and steel industry system optimization, therefore improve the iron and steel industry market competitiveness. The effective application of confirmer wharf financing can resolve the financing difficulties in the iron and steel industry.

35.3.1 Confirmer Wharf Operation Procedure of Iron and Steel Industry

Confirmer wharf is characterized by a first ticket goods, i.e. in the small and medium-sized enterprise (financing) pay a certain margin, financial institutions issued acceptance, meaning to the supply chain upstream core enterprise (Supplier), acceptance is a financial institution, After receiving Bank acceptance draft, upstream core enterprise delivers goods to logistics enterprise warehouse which the bank designated. Goods entering the warehouse to warehouse receipt pledge after. According to the bank's instruction, logistics enterprises sent goods to small and medium enterprises in batches. Specific processes as shown in Fig. 35.1.

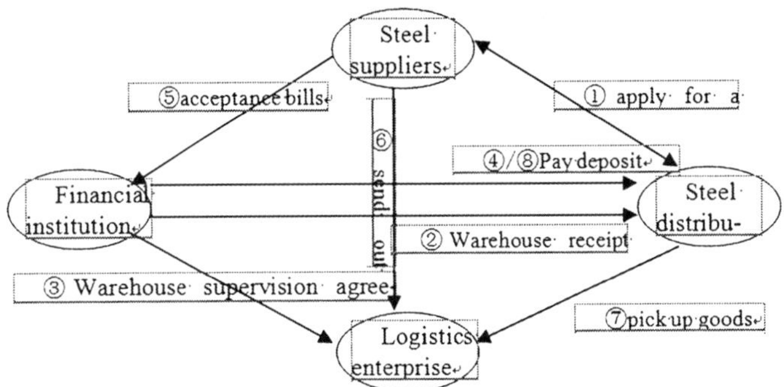

Fig. 35.1 Confirmed warehouse financing mode

Small and medium enterprise as a distributor of steel and upstream core enterprise as steel suppliers jointly sign the sales contract, and negotiate to apply for loan to financial institutions (such as a bank) by the small-sized enterprise. Banking and steel suppliers, buyers, steel logistics enterprises signed a "Quartet cooperation agreement".

Bank approved a certain amount of credit line for steel distributors, and cleared the first margin ratio. Steel distributor with a sales contract in the bank obtain predetermined loan amount of warehouse receipt pledge, and bank acceptance bills which steel suppliers as the payee are issued by the buyer, and then bank acceptance, especially for payment for the upstream supplier.

If the core enterprise commit to repurchase and credit status examined by the bank, then the bank will sign "cooperation agreement", "account supervision agreement" "quality assurance agreement". The bank commission the logistics enterprises to provide warehousing regulatory services, and sign the "irrevocable bank to help exercise of its guarantee," "pledge goods replacement notice". Steel suppliers, logistics enterprises and banks signed a "cooperation agreement", "warehouse supervision confirmed warehouse agreement," "comprehensive credit agreement".

Steel distributor pays to the bank acceptance fee and acceptance margin; bank acceptance, meaning for steel suppliers; according to "Quartet cooperation agreement" provisions, the iron and steel distributors deposit a certain margin to the bank, and steel suppliers will be equal to the bank designated goods logistics warehouse after receiving the bank acceptance bills; The bank informs logistics enterprises to release the corresponding proportion of cargo to the steel distributors through the "Notice of shipment", and then steel distributors extract the corresponding amount of goods from warehouse; after the sale, steel distributors pay a guarantee deposit again.

From the down payment to achieve sales, it cycles until the down payment account equal to the amount of the draft, and steel distributors fetched all goods. At the same time, the pledge contract and repurchase agreement accordingly cancel. The agreement book summary is shown in Table 35.1.

Table 35.1 Confirmed warehouse protocol summary table application in the iron and steel industry

Protocol Name	Agreement Shall
Quartet cooperation agreement	Steel supplier Steel distributors Bank Logistics enterprise
Sales contract	Steel supplier Steel distributors
Cooperation agreement; Account supervision agreement; Quality assurance agreement	Steel supplier Bank
Cooperation agreement; Warehouse supervision confirmed warehouse agreement; Comprehensive credit agreement	Steel supplier Bank Logistics enterprise
Irrevocable bank to help exercise of its guarantee; Pledge goods replacement notice	Bank Logistics enterprise

35.3.2 Significance of Confirmer Wharf Financing Model in the Iron and Steel Industry

Confirmer wharf financing business model reduces financial pressure of the trading parties, and allows the market to achieve win—win situation. So confirmer wharf financing operations are in the profit position of the Quartet.

From steel suppliers perspective, they gain greater business profits, obtain competitive advantage of industry chain, reduce the cost of capital, reduce the accounts receivable occupation, and guarantee the collection; from steel distributors perspective, it solves the full purchase capital difficulty, through large quantities of orders for suppliers to give preferential prices, reduce cost of sales, and on the basis of seasonal differences makes it is possible to obtain higher profits; as for the banks, it has advantages achieving chain marketing, meet all aspects of customer needs, and has a risk control advantages, at the same time get rich service fees and possible discounting a bill of costs, mastered the right of taking delivery of logistics; on logistics enterprises, they can obtain relevant proceeds, proceeds from the store and enterprise management of goods to the buyer for a fee, and valuation and impawning supervision for banking intermediary services charge a percentage of the fee.

35.3.3 Special Risks Related to the Confirmer Wharf Financing in the Iron and Steel Industry

The shortage of funds, financing difficulty has a vital impact on the capital-intensive iron and steel industry. But it is the one that iron and steel industry must encounter and resolve, there is no way around it. For the distributors, risk exists mainly in the line of credit is superfluous, supply exceeds demand, sales increased risk; for the Bank's risk, there is a possibility that producer and distributor may fraud the loan, despite the dealers have to pay the bank deposits, balance of manufacturer's warranty, but for the most importantly, delivery is currently defined as a real right is the right to dispute claims of fierce. In addition, the iron and steel enterprises are not responsible for the guaranty liability, commitment to product buy-back, this makes the distributors face difficulties in capital returns.

Confirmer wharf financing in the iron and steel industry has some special risks, namely steel enterprises as a supplier of steel iron do not assume the nature of guaranty liability, nor promise of product buy-back; the biggest risk is the supply enterprises to bear the responsibility of refund; while for steel suppliers, it eliminates risk which the distributors bear in the mismanagement case.

On the steel industry upstream by coal, logistics, such as prices and labor cost rise, downstream by the construction, automobile, shipbuilding, machinery, household appliances and other major steel industry demand growth is blunted affect, iron and steel enterprise has a very low profit. For the security of the capital, in accordance with the requirements of banks, the iron and steel enterprises as a producer must take a series of responsibility and obligation. If a slight negligence is occurred in the process of operating, fulfil responsibility is not in place, the iron and steel enterprises must assume capital compensation and financing bank guarantee responsibility.

35.3.4 Confirmer Wharf Business Risk Avoidance for the Iron and Steel Manufacturer

In order to ensure the safety of bank funds, iron and steel enterprises as a manufacturer in the confirmer wharf business cooperation undertake the following duties and obligations: Acceptance and custody of the bank's capital in accordance with the requirements of the bank, handle orders for distributors, delivery of steel to keep or distributors, returning the fund balance to bank according to the confirmed warehouse cooperative agreements. From the manufacturer perspective, to avoid risks in Confirmer wharf business one should pay attention to:

First, when Confirmer wharf cooperation agreements drafting, iron and steel enterprises should define the responsibilities and obligations; In product sales industry, it insist on not provide external guarantee principle; After completion of sales, it should insist no commitment to repurchase obligation principle for achieved sales of product.

Second, a careful analysis, summarized various confirmed warehouse cooperative business operation process and risk control points, generally can be divided into the following four aspects of the basic measures to control risk:

(1) In the collection business, bank capital must be recognized and received.
(2) In ordering business, one must place an order according to the strict agreement. First, keep and use received funds in line with agreements; second, to provide contract details in the delivery address and the consignee.
(3) In the product delivery business, strictly supply delivery steel according to the contract.
(4) To fulfill obligation to refund the balance of the funds to the bank according to the agreement by the parties.

Finally, iron and steel enterprises should standardize the operation against the warehouse business cooperation. Iron and steel enterprises improve the bank on the dealer's financial support through standardized operation against the warehouse business cooperation, control risk effectively, and safeguard financial security. And through the bank's financial support, it can better solve the dealer financing needs, so that dealers can continue to expand the distribution quantity, improve distribution capacity, ultimately to achieve the promotion of iron and steel enterprise product sales and market share.

35.4 Conclusion and Prospects

In the past company always take loan directly from the bank, through confirmer wharf iron and steel manufacturers can either seal market sales, define production plan, and can seek indirect financial support from the dealers. Iron and steel enterprises confirmer wharf application make the bank, enterprise three parties can greatly meet the needs for development. But the research process inevitably exist some disadvantages. For confirmer wharf business risks, this paper analysis only from the iron and steel manufacturers point of view of risk avoidance, not from the bank, steel distributors perspective, the author will aim at these problems in later more in-depth study.

References

1. Yanfeng L, Xia L (2008) Logistics financial. Sci Press 7:87–90
2. Xin S, Zhenghua L, Li Y (2004) A newsboy model with budget cost constraints. J Jilin Univ 8:45–49
3. Jingjing Z (2007) Logistics finance theory research on commercial bank credit risk management in commercial bank operation and management, vol 3. pp 90–99
4. Quan Y (2012) Supply chain analysis and prospect of financing problem of small and medium sized enterprises, logistics engineering and management, vol 5. pp 56–59
5. Ming L (2007) On the national electric marketing strategy, vol 7. pp 12–18

Chapter 36
Green Suppliers Selecting Based on Analytic Hierarchy Process for Biotechnology Industry

Xueling Nie

Abstract The objectives of the research are twofold: (1) to establish collaborative evaluation criteria of green suppliers utilizing AHP; and (2) to construct evaluation processes according to the aforementioned set of criteria. In order to fulfil the two goals, an investigation via utilization of the analytic hierarchy process (AHP) is made to the selection process for green suppliers in the biotechnology industry. The results show that the major concerns in terms of green supplier selection for biotechnology companies are currently CGMP certification, established environmental policies, and product acknowledgement. Also, an evaluation form consisting of green criteria and weights is constructed to facilitate the selection process.

Keywords Green supply chain • Green supplier selection • Analytic hierarchy process • Biotechnology industry

36.1 Introduction

In recent years, the hi-tech industry has been greatly supported by governments, and a great deal of talent and funds have been absorbed into several relative fields. The biotechnology industry is a kind of sunrise industry, as it requires relatively low energy consumption and high knowledge intension, and offers high added value. According to a forecasting report of the Industrial Economics and Knowledge Center at the industrial technology and research institute (ITRI), the global biotech market grows at 18 % per annum on average [1].

Supporting enterprises with technology needed for green supply chains is one way to raise industry competence. With the looming of a green industry

X. Nie (✉)
School of Business Administration, University of Science and Technology Liaoning, Anshan 114051, China
e-mail: nxlok@163.com

Z. Zhong (ed.), *Proceedings of the International Conference on Information Engineering and Applications (IEA) 2012*, Lecture Notes in Electrical Engineering 220, DOI: 10.1007/978-1-4471-4844-9_36,

revolution, enterprises must be environmental-friendly, and must collaborate with both upstream and downstream supply chain partners.

The choice of an appropriate cooperative partner is at the heart of supply chain management, and material or product/service providers are the most closely associated with an enterprise [2]. If we can find a supplier that complies with industry characteristics and meets supply chain requirements, then supply chain competence can be enhanced.

The objective of this study is to develop a green supplier evaluation model for the biotechnology industry. Specifically, the objective is twofold: (1) to establish collaborative evaluation criteria of green suppliers utilizing AHP; and (2) to construct evaluation processes according to the set of aforementioned criteria.

36.2 Research Methodology

This study mainly adopted AHP to analyze the importance of various green supplier evaluation criteria.

AHP is a multi-attribute evaluation method developed in 1971 by Prof. Thomas L. Saaty of Pittsburg University, and it is primarily used to solve decision problems in uncertain situations and with multiple evaluation criteria [3]. Combined with expert discussion, the AHP hierarchy structure can be generated from confirmed criteria. To avoid excessive industry discrepancies between the experts, all experts chosen for this study were senior executives of purchasing or environment engineering departments within the industry. This study included five such experts, who were all senior executives of listed or over-the-counter pharmaceutical firms with years of industry experience, and who all had unique viewpoints on subject of this study. Eisenhardt (1989) pointed out that, when conducting a case study, the case number should normally be between four and ten, as fewer cases hamper theory construction, and more cases become hard to analyze due to the relatively large amount of data. Therefore, this study included senior executives of five firms as interview subjects.

The above procedure can be shown as follows: (1) obtain criteria from literature, and filter them by expert interview to form AHP hierarchy structure; (2) design AHP questionnaire and assess comparative values; (3) compute weight of each criterion, and construct evaluation form as well as operational process.

36.3 Results and Discussion

36.3.1 Step 1: Build Criterion Hierarchy

Using multiple criteria, AHP is a simple evaluation method to determine precedence. This study used AHP to confirm the five categories and 14 criteria (CGMP certification, environmental protection policy and objectives, executive

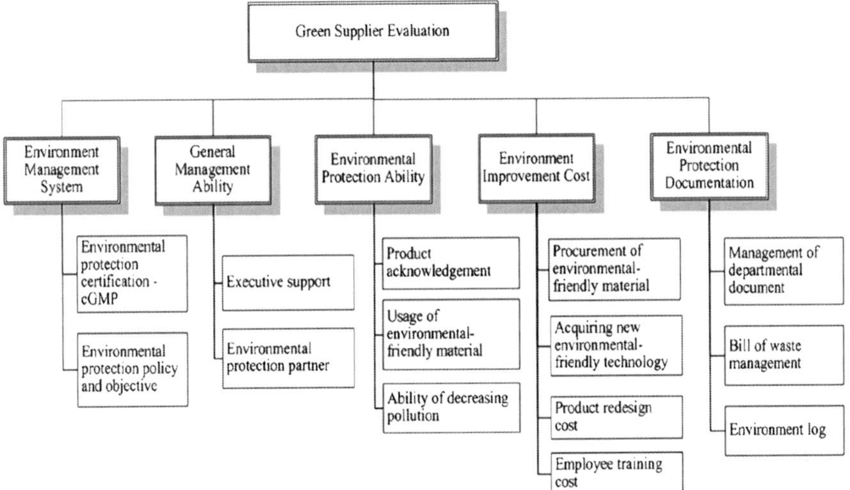

Fig. 36.1 Criterion hierarchy of AHP

support, environmental protection partner, product acknowledgement, usage of environmental-friendly materials, ability to decrease pollution, procurement of environmental-friendly materials, acquiring new environmental-friendly technology, product redesign, employee training cost, management of departmental documents, bill of waste management, environment log). The hierarchy structure was built and is shown in Fig. 36.1.

36.3.2 Step 2: Calculate the Weight of Each Criterion

This step was divided into four stages: performing interviews based on the questionnaire; establishing a pairwise comparison matrix; computing criterion weights; and calculating consistency.

This study completed the AHP expert questionnaire on the basis of the above criteria. In total, ten experts were interviewed and eight questionnaires were valid, i.e., consistency index, CI < 0.1, and consistency ration, CR < 0.1. If the number of effective questionnaires had amounted to three, it would have indicated compliance with the AHP hypothesis (Lin, 2005). The experts interviewed were primarily senior pharmacy executives with decisive power, and they were also knowledgeable regarding environment protection issues. Thus, they were able to express precise and insightful opinions.

The comparison matrix was then established. Once the pairwise comparison matrix was established, the precedence of each criterion could be calculated. After normalization process, the normalized pairwise comparison matrix can be obtained, as shown in Table 36.1. The criterion score was calculated by averaging the normalized values in each row; these were in turn used to compute consistency measures and a consistency index, and finally the consistency ratio.

36.3.3 Step 3: Summarize the AHP Questionnaire

We processed the questionnaires into the pairwise comparison matrix one by one, and calculated precedence according to the above steps. Every column in Table 36.2 represents the weight of each criterion rated by each expert (e.g., the first expert rated environment management systems with a weight of 0.557); every row represents a criterion of each hierarchy in the AHP structure.

AHP uses a consistency ratio to check the pairwise comparison consistency, and if this ratio exceeds 0.1, it indicates inconsistent judgment. In this case, the decision maker had to correct the original values of the pairwise comparison matrix. It was also necessary to check the CI and CR values—if the consistency ratio was less than 0.1, then the pairwise comparison consistency was deemed to be a reasonable level, e.g., for the first expert, $CI = 0.085 < 0.1$, $CR = 0.076 < 0.1$.

36.3.4 Compute the Average Weights

A total of ten experts attended this questionnaire investigation. Two questionnaires with high inconsistency were omitted; hence there were eight effective

Table 36.1 Normalized pairwise matrix for main criterion

	Environment management system	General management ability	Environmental protection ability	Environment improvement cost	Environmental protection documentation	Criterion score	Consistency measure
Environment management system	0.121951	0.078647	0.122804	0.241371	0.1851852	0.149992	9.519056
General management ability	0.121951	0.393236	0.1228049	0.241371	0.2592593	0.227725	11.830840
Environmental protection ability	0.609756	0.393236	0.6140243	0.241371	0.2592593	0.423529	20.230784
Environment improvement cost	0.121951	0.078647	0.1228049	0.241371	0.2592593	0.164807	14.056875
Environmental protection documentation	0.024390	0.056233	0.0175611	0.034516	0.0370370	0.033947	0.755350

Table 36.2 Weights of criterion categories and CR values

No.	Environment management system	General management ability	Environmental protection ability	Environment improvement cost	Environmental protection documentation	CI	CR
1	0.557	0.147	0.154	0.090	0.049	0.085	0.076
2	0.567	0.161	0.154	0.075	0.041	0.083	0.074
3	0.386	0.166	0.343	0.051	0.051	0.070	0.060
4	0.388	0.159	0.338	0.071	0.042	0.097	0.086
5	0.149	0.423	0.164	0.227	0.033	0.090	0.080
6	0.276	0.155	0.155	0.184	0.227	0.090	0.080
7	0.348	0.177	0.177	0.222	0.073	0.040	0.040
8	0.557	0.130	0.130	0.130	0.051	0.010	0.010
Total	3.228	1.518	1.665	1.050	0.567	–	–
Average	0.403	0.189	0.208	0.131	0.070	–	–
Rank	1	3	2	4	5	–	–

questionnaires in total. Every hierarchy in each questionnaire has its own weight and consistency ratio. To integrate the questionnaire weights as given by these experts, this study calculated the weight of every criterion in all questionnaires using a weighted average. For instance, the weights of the environment management system criteria as rated by expert one thru expert eight were 0.557, 0.567, 0.386, 0.388, 0.149, 0.276, 0.348, and 0.557 respectively. These eight weights add up to 3.228, and the average is 0.403. Calculation results of all weights in all categories by all experts are shown in Table 36.2. It was found that the criterion precedence order was environment management systems, environmental protection ability, general management ability, environmental improvement costs, and environmental protection documentation.

36.3.5 Analyze the Weight of Each Criterion

The overall weight analysis was performed, and the results are tabulated in Table 36.3. It was found that pharmaceutical manufacturers put a high priority on certification (0.611) and environment protection policies (0.388), which is also why the government has played an active role in tutoring manufacturers to obtain their environmental protection certification; further, pharmaceutical manufacturers insist that environmental protection ability is more important than general management ability, because general management ability can be obtained through training and education.

It also can be seen that although the criterion weight of executive support (0.512) was greater than the weight of environmental protection partner (0.487),

Table 36.3 Green supplier evaluation form

Main criterion (weight)	Sub criterion (weight)	Rank	Overall weight (a)	Score (b)	Weighted score (a*b)	Sub-total
Environment management System (0.403)	Environmental protection certification—cGMP (0.611)	1	0.246			
	Environmental protection policy and objective (0.388)	2	0.156			
General management ability (0.189)	Executive support (0.512)	4	0.097			
	Environmental protection partner (0.487)	5	0.092			
Environmental protection ability (0.208)	Product acknowledgement (0.550)	3	0.114			
	Usage of environmental-friendly material (0.186)	8	0.039			
	Ability of decreasing pollution (0.262)	7	0.054			
Environment improvement cost (0.131)	Procurement of environmental-friendly material (0.201)	11	0.026			
	Acquiring new environmental-friendly technology (0.441)	6	0.058			
	Product redesign cost (0.238)	10	0.031			
	Employee training cost (0.118)	13	0.015			
Environmental protection documentation (0.070)	Management of departmental document (0.556)	8	0.039			
	Bill of waste management (0.291)	12	0.020			
	Environment log (0.148)	14	0.010			
SUM =						
A. Excellent ($90 \leq SUM \leq 100$)	Final decision:					
B. Very good ($80 \leq SUM < 90$)	Ready for placing order (A, B)					
C. Good ($70 \leq SUM < 80$)	Re-evaluation needed after improvement (C, D)					
D. Fair ($60 \leq SUM < 70$)	Avoid placing order (F)					
E. Fail ($0 \leq SUM < 60$)						

the difference was small. This indicates the importance of executive support in a company, as well as becoming an environmental protection partner along with a supply chain member. In this way, the general management ability of a pharmaceutical factory can be elevated, and consequently the pharmaceutical factory can respond effectively to rising environmental protection pressure, and also acquire more advantages to facilitate overall development.

The weight of product acknowledgement was 0.550, which was the most important among the three criteria in the Environment Protection Ability category. The customer impressions of the purchased product influence the overall image of the company. Therefore, experts demand that product R&D, manufacturing processes, product packaging and final delivery must not threaten the environment; hence experts put high emphasis on product acknowledgement. Pharmaceutical factory attention to using environmental-friendly materials and lowering pollution was not as relatively significant, because CGMP certification mandates manufacturers to have regulations related to environmental protection. In this study, experts thought that if products could comply with regulations during the design process, then no additional costs would be needed.

As to pharmaceutical factories, the primary criterion regarding environmental protection related documentation was how each department deals with the related documents (0.556), as well as how they deal with purchasing department categorized related waste or purchased green products, in terms of documenting them for supervisors or CGMP certification teams to audit. Manufacturers paid more attention to bills of waste management because related pollutants are discharged in the manufacturing process; therefore, environmental protection engineers must dispose of the waste effectively and document the process for auditing.

After the weight analysis for each criterion category and each criterion, an overall weight for each criterion was calculated by multiplying the weight of the criterion category and the weight of each criterion. Taking CGMP certification as an example, the overall weight was calculated by multiplying the weight of the criterion category (environment management system), 0.403, by the CGMP certification weight, 0.611, which yielded an overall weight of 0.246. The overall weight and its rank, as shown in Table 36.3, represent the importance of each criterion in the whole supplier evaluation model.

36.3.6 Construct a Green Suppliers Evaluation Form

Based on the evaluation criteria and their weights, a green supplier evaluation form was proposed as shown in Table 36.3. By rating each supplier based on the proposed criteria and weights indicated in Table 36.3, the suppliers can be effectively evaluated. The supplier evaluation results were divided into groups A, B, C, D, and E, according to the summary of weighted scores. Thus, the final decision of whether or not to place an order can be made according to the supplier level: (1) A or B: adequate to place orders, (2) C or D: hold until the supplier makes improvements on environmental protection issues, and (3) E: avoid placing orders.

36.4 Conclusion

The results are summarized as: (1) the way pharmaceutical factories currently evaluate green suppliers is mainly based on checking whether or not the supplier has passed the CGMP certification, it has established environmental protection policies, and it has product acknowledgements. (2) Based on the expert interviews and literature analysis, this study provides several criteria for biotechnology manufacturers to use when evaluating green suppliers. (3) A model was constructed that details the process of evaluating upstream green suppliers for biotechnology manufacturers. Based on the evaluation criteria and the procedures provided in this study, relative weights for the criteria were calculated, and a green supplier evaluation form was proposed, which can help decision makers to evaluate green suppliers in a more systematic manner.

References

1. Banerjee SB (2001) Corporate environmental strategies and actions. J Manage Decis 39(1):36–44
2. Bansal P, Roth K (2000) Why companies go green: a model of ecological responsiveness. J Acad Manage J 43(4):717–736
3. Curkovic S, Sroufe R, Melnyk S (2005) Identifying the factors which affect the decision to attain ISO 14,000. J Energ 30(8):1387–1407

Chapter 37
XAEDI: A XML-Based Adaptable Event-Driven Integration Framework

Zhihong Liang, Hongwei Kang, Yong Shen, Xingping Sun, Qing Duan and ShengLin Yang

Abstract After combining content release, subscription middleware features, and the traditional event-based software integration together, focus on the problems in the existing system has established an adaptable and efficient events-driven integration framework with XML as the message interaction basis by means of core models defining and filter testing designs such as architecture design and event models.

Keywords Event-driven • Integration framework • Events model

37.1 Introduction

Currently, quantities of prototype systems based on content release and subscription have emerged from both home and abroad, including ENS (event notification service) system and SDI (selective dissemination of message) system. However, there exists a common problem in these systems that as the middleware based on content release and subscription, event service or message agent, they mainly take the support for the loose coupling communication among the systems instead of their integrated features such as efficiency and adaptability into consideration. Besides, the existing methods either only focus on the single event mapping or lack the adaptability for the composite processing of events. There will be brief introduction for some influential studies in the following part.

Nowadays, systems like Gryphon, Siena, and Elvin are the famous ones in term of the content-based Pub/Sub middleware; the event expression forms taken by which are similar to subscription languages with the single semantic constraint as

Z. Liang · H. Kang (✉) · Y. Shen · X. Sun · Q. Duan · S. Yang
School of Software, Yunnan University, No. 2 North Cuihu Road, Kunming, Yunnan, People's Republic of China
e-mail: hwkang.yn@gmail.com

Key Laboratory for Software Engineering of Yunnan Province, Kunming, Peoples' Republic of China

Z. Zhong (ed.), *Proceedings of the International Conference on Information Engineering and Applications (IEA) 2012*, Lecture Notes in Electrical Engineering 220, DOI: 10.1007/978-1-4471-4844-9_37, © Springer-Verlag London 2013

their basic construction and forward single event [1, 2], object to the composite events required by integration and take the flat pattern as event basis, which means that the event content is the set of "attribute equalling value" pairs, the subscription terms are the linkage among simple languages based on various properties and poor expression capability [3].

The XML-based Pub/Sub system is mainly made up of X Filter [4], Y Filter [5], XTrie [6], and Web Filter [7] which belongs to the message retrieval field as well named as SDI system specifically speaking. They are applied for the XML message released through the Internet; most take X Path as the selective language of XML and are superior to the message acquisition method based on flat pattern in term of message selecting intensity and expression capability. However, these systems fail to support the composite procession of multievent documents and the custom generation of new XML documents, which are the indispensible features for the integration framework.

Studies of the sequential logic of composite events mainly focus on the active database. For example, the SAMOS database applies Petri Nets to detect composite events, capable of expressing parallel behaviors and managing complex data with parameters; however, it only executes several kinds of simple operations of composite events and fails to cater to the requirements made by integration for composite events. Some work has designed a composite relation and sequential expression of language support events called Snoop and has detected the composite events through tree structure, while it shares the common problems with SAMOS. Currently, there are not many studies on the sequential logic of composite events in the Pub/Sub system. Some work has discussed the composite events detection under distribution environment and focused more on the topology distribution structure. The above two both fail to support the composite events modes such as the split and reconstruction of the XML documents required by integration.

37.2 Establishing Design Structure of XML-Based Event-Driven Integration Framework

37.2.1 Design Structure of Integration Framework

The XML-Based Adaptable Event-Driven Integration Framework is short for XAEDI. In the XAEDI, the integrated software modules or systems are called participants which carry out indirect interaction by sending or receiving XML-form messages representing events happening instead of direct interaction. Events refer to the changes or increase happing in a certain integrated system and composite events refer to various logical combinations of different events, both of which can be distributed to interested integrated systems through XAEDI. The loose coupling integration of software systems can be realized by means of selective transmission of event messages among different software modules or systems.

In XAEDI, participants who send out event messages are called informers and participants who subscribe and receive the interested event messages are called listeners. A participant can be the informer and the listener at the same time. Adaptable change (such as code modification) is a must for integrated software to be an interactive participant of XAEDI and such kind of adaptable process can be executed skillfully, which is out of the study scope of this paper.

XAEDI mainly consists of Event Message Answer Components, Subscription Query Answer Components, Registration Message Answer Components, Filter Engine, and Event Message Announcement Components.

Event Message Answer Components: waiting for receiving, perpetuating the event message from informers, later analyzing and sending to the filter engine for processing.

Subscription Query Answer Components: receiving and storing the subscription query terms from listeners, later analyzing and sending to the filter engine for processing.

Registration Message Answer Components: waiting for receiving, storing, and forwarding the registration message from participants.

Filter Components: they are the core as well the research key point of this paper. They mainly consist of the Participants Registration Components, Primitive Event Filter, Composite Event Filter, Output Message Transformation, Adaptable analyzing, and optimizing Rules. Participants Registration Components are applied by participants to send registration, release, or subscription intentions to XAEDI and only those successfully registered participants can execute actual message interaction with XAEDI. Besides, XAEDI can store participants' event message types through their registration and release all the available subscription event message types to subscriber. Primitive Event Filter is applied to filter the event message and all the primitive subscription terms sets. Composite Event Filter is applied to compositely detect composite subscription terms. Output Message Transformation is utilized to map the event message to reconstruct and transfer according to output formats. Adaptable analyzing and optimizing rules focus on the integrated adaptability and will cause practical optimization under certain conditions.

Event Message Announcement Components: distributing event message successfully mapped and execute adaptable change to interested listeners.

37.2.2 Formal Representation of XAEDI

The system made up of XAEDI and the integrated participants can be presented as a Triple (Π, Θ, and Σ), with $\Pi = \{p_1, p_2, \ldots, p_m\}$ representing the participate sets, $\Theta = \{F_1, F_2, \ldots, F_n\}$ as the example sets of XAEDI, and $\Sigma = \left\{O_1, O_2, \ldots, O_p\right\}$ as the interactive operation sets between participants and XAEDI. The distributivity of the framework is not researched in this paper.

XAEDI can be represented as XAEDI = (ET, SB, CT, FT, NT), with *ET* (Event) representing event model which has described the basic event message model and event composite operation model, *SB* (Subscription) representing subscription model which has described the event subscription model, *CT* (Component) representing component model which has described the construction components and their interactive relations in the XAEDI, *FT* (Filter) representing filter model which has described the registration patterns of participants, filter algorithm of basic XML event message, detection methods of composite events, and adaptable change methods of output message, and *NT* (Notification) representing the announcement model which has described the announcement methods of event message and route strategies.

37.3 Establishing Core Models

Based upon set theory, event algebra theory, and research achievements of related fields, the basic event model describing the specifications of basic events, sequential logic model describing sequential relations under distribution environment, event composite model describing the multievent composition, forwarding and trigger operation, and subscription model describing subscription languages choice and methods are established.

37.3.1 Sequential Logic Model

In the event-driven software integration framework, events happening at certain time and place refer to messages sent by participants and characterized by instantaneity and atomicity; therefore, each event is of a related time stamp attribute to show its happening time. Time stamp attribute is necessary for both the sequential basic events and the multievents ranked by sequential relations. Supposing there exist a partial order $<$ of time stamp, representing that events happen according to complete sequence order and a total ordering order $\prec$ representing that events happen according to partial sequence order, with $\cup$ representing the relation of time merging, all of which consist of sequential logic relations.

Under distribution environment, when event releasing time fails to equal the event receiving time of integration framework because of network delay and so on, the time intervals rather than single time points are applied, which can not only maintain consistency with the physical time order of events, but also can represent the time intervals of composite events.

$t = \left[t^l, t^h\right]$ is applied to represent the time interval of an event, t^l the start time, t^h the end time, and t_1, t_2 the time interval of e_1 and e_2 , respectively. The sequential logic relations cater to the following equivalent conditions:

$$t_1 < t_2, t_1^l < t_2^h \tag{37.1}$$

$$t_1 \prec t_2,\ \left(t_1^h < t_2^h\right) \vee \left(t_1^h = t_2^h \wedge t_1^l < t_2^l\right) \tag{37.2}$$

$$t_1 \cup t_2,\ \left[\min\left(t_1^l,\ t_2^l\right),\ \max\left(t_1^h,\ t_2^h\right)\right] \tag{37.3}$$

37.3.2 Basic Event Model

Message releasers publish messages by means of events, and subscribers subscribe events with subscription language. Events and subscription languages are an interactive whole, with the event patterns deciding the basic patterns of subscription languages and vice versa. As for XML event, it differs from events of other formats with its content organized by Hierarchical Tree, including structure messages apart from elements and attribute messages.

Event definition: in this integration framework, event represents the XML message caused by the state change of some time point and released by participants. XML event is a triple: XML-Event equaling (element, attribute, and structure). Element and attribute they themselves are of names, types, and values, with attribute relying on certain elements and structure restricting the structure relations among elements including four kinds of relations such as relations between father and son, brothers, ancestors, and descendents.

Structure relation figure of event types. The definitions of Event Space, Composition Contribution, Event Query, Event Class, Event Trace, Profile, and Event-Profile Matching are established orderly, which are the foundation of event models.

Event processing model: it mainly establishes a complete process from the releasing, filtering, and to informing the XML event message so as to indicate the flow direction and processing details of XML message flow in the integration framework.

37.3.3 Event Composite Model

Through executing sequential constraint extension of algebra theory (constructed by nonempty sets and a set of operations acting on them) and combining the operation semantics of active database, this paper has defined a set of event composite operations acting on event sets, including several binary operations (conjunction, extraction, and timing sequence), several unary operations (selecting and denying), and several multiple operations (timing and arbitrariness), proving that event and space shut these operations and sets out. Based on the above, the event algebra is equipped with parameters and transformed into parameterized event algebra which endows the composite operations with different semantics according to different parameters so as to cater to the adaptabilities of event composition for different integrations. Eventually, the formal definitions of relevant composite operations are provided based on parameterized event algebra.

E_1, E_2 are two different event types belonging to event space E i.e., $E_1 \subset E, E_2 \subset E \neg e_1, e_2 E_1, E_2$, are event examples belonging to E_1, E_2 , respectively, i.e., $e_1 \in E_1$, $e_2 \in E_2$, nonformal definitions of various composite operations are as follows:

Extraction event ∇: $E_1 \nabla E_2$ happens only when $e_1 \in E_1$ happens or $e_2 \in E_2$ happens

Conjunction event Δ: $(E_1 \nabla E_2)_T$ happens only when $e_1 \in E_1$ and $e_2 \in E_2$ happen $(e_3) := \max \{t(e_1), t(e_2)\}$

Sequence event: $(E_1; E_2)_T$ happens only when $e_1 \in E_1$ happen in advance, later in the T interval $e_2 \in E_2$ happens, $t(e_3) : = t(e_2)$

Timing event $\perp$: $(E_1 \perp E_2)\, E_3$ happens only within certain intervals between $e_1 \in E_1$ and $e_2 \in E_2$ whether $e_3 \in E_3$ will happen, $t(e_1) \leq t(e_3) \leq t(e_2)$ Denying event $\bar{E}_T$ happens only within certain intervals, without $e \in E$ happening

Selecting $\bar{E}_T$ event $[i]$: $E^{[i]}$ happens only when the ith, even of E sequence $e \in E$ happens

Adding event example selecting parameters to single construction event, adding event example consumption parameters to composite event and endowing composite operations with different semantics by setting different parameters in order to adjust to different practical contexts are regarded as the context rules for composite event parameters. Currently, taking the following conditions into consideration: $first\_Dupfirst\_Dup(E)$ representing the first event of E sequence, $last\_Duplast\_Dup(E)$ representing the latest event of E sequence, $all\_Dupall\_Dup(E)$ representing all the events of E sequence, $i\_Dupi\_Dup(E)$ representing the ith event of E sequence, *all pairs*, *unique pairs*, and *repeated pairs* representing all the pairs, the only pair and repeatable pairs of consumption event sequence respectively. The following is a composite example imposed composite operations and additional parameters: $\left((all_{dup(E_1)};\ first\_dup(E_2))_{7days}\right)_{all\_pairs}$.

37.3.4 Subscription Model

The selecting of subscription languages is the core of subscription model. The subscription and query language catering to semistructured XML documents take in some features of query language SQL catering to the structured database. However, the different query objects of them lead to great difference in them. Nowadays, plenty of XML query languages catering to the W3C standards have been proposed, such as X Path, X query, XML-RL, and XML-QL. According to the advantages XML-QL shows in the data extraction of XML documents, integration of different documents, and the formatting level structure of new documents, XML-QL has been applied as the subscription language of our framework.

However, there are the following disadvantages for XML-QL in expressing event subscription, especially in composite event subscription:

The even-driven integration framework regard content matching as its identification basis; therefore, it fails to identify the names of XML events in advance;

In XML-QL, if in sentences are ignored, the XML matching sentences of different objects cannot be distinguished.

Without the introduction of time concepts, XML-QL fails to support the judging of sequential logic and event composition.

Therefore, currently adding sequence operators and composite operators are considered being applied to extend XML-QL by replacing "in sentences" with "block variables" and the extended XML-QL is called XML-EQL. XML-EQL equals block variables set ∪, composite operations set ∪ parameter, and context rules set ∪ XML-QL. Based on the above, grammar definition EBNF of XML-QL will be extended to achieve the grammar definition XML-EQL.

37.4 Conclusions

The loose coupling and extensible integration capability provided by middleware services is a must for establishing large-scale software integration, and the Pub/Sub system based on content is a kind of this middleware. Subscribing the needed events on the Pub/Sub system is what required to integrate the scattering software systems through Pub/Sub system and the number change of the integrated systems has no impact on the whole integration system. Second, the capability for the integrated systems to process XML documents is what required when XML is taken as the expression forms of events and the message interaction forms among integrated systems, without any extra requirements for the system construction, according to which it can be concluded that Pub/Sub system is of sound integrated extendibility.

The integration framework driven by adaptable events proposed in this paper is a newly design framework featured by exchanging message by XML, possessing certain integration adaptability, mapping according to content, being feasible for the multievent composite detection of different integration sources and the adaptable change for XML events, and executing efficient filter by optimized strategies, which has combined content release, subscription middleware features, and the traditional event-based software integration together, and focused on the problems in the existing system and taken the requirements and adaptability needed by software integration into consideration.

Acknowledgments This work, a study on XML-based Adaptable Event-Driven Integrated Software Framework is funded by Yunnan Science Foundation under Grant No. 2009CD009, by the Open Foundation of Key Laboratory of Software Engineering of Yunnan Province under Grant No. 2010KS04, 2011SE11, and 2011SE12, and by the research funding for teacher with a doctorate under Grant No. XT412004.

References

1. Banavar G, Chandra T, Mukherjee B (1999) Matching events in a content-based subscription system. In: Proceedings of the 18th ACM symposium on principles of distributed computing (PODC, 99). Atlanta, USA (5):53–61
2. Carzaniga A, Wolf AL, Rosenblum DS (2001) Design and evaluation of a wide-area event notification service. ACM Trans Comput Syst 19(3):332–383
3. Gough J, Smtih G (1995) Efficient recognition of events in a distributed system. In: Proceedings of the 18th Australian computer science conference (ACSC18). Glenolg South Australia (4):55–65
4. Ahinel M, Franklin MJ (2000) Efficient filtering of XML document for selective dissemination of information. In: Proceedings of the 26th International conference on very large data bases. Cairo (05):53–64
5. DiaoY, Fisher P, Franklin MJ, To R (2002) Yfilter Efficient and scalable filtering of XML documents. In: Proceedings of the 18th International Conference on Data Engineering (ICDE, 02). San Jose (7):341–342
6. Chan CY, Felber P, Garofalakis M, Rastogi R (2002) Efficient filtering of XML documents with X path expressions. VLDB J 11(4):354–379
7. Pereira J, Fabret F, Llirbat F, Jacobsen HA, Shasha D (2001) Web Filter a high throughput XML-based publish and subscribe system. In: Proceedings of the 27th international conference on very large data bases. Roma (7):723–724

Chapter 38
Library Resource Construction Crises and Strategies in the Digital Environment

Sheng Chun Shao, Jing Xu and Shuang Qi Yang

Abstract In order to respond to the various crises in the process of library resources construction in the digital environment, and to reduce the negative impact of the crisis to the library, this chapter discussed the basic types of crises at first, which included the resources type crisis, the resources structure crisis, the resources proportion crisis, the resources layout crisis, and the resource preservation crisis; then it pointed out the main causes of the crises, at the same time analyzed the mechanisms which cause the crises, at last this chapter pointed out strategies to copy with all kinds of crises immediately.

Keywords Digital environment • Library • Resource construction crisis • Strategy

38.1 Introduction

Library resource construction crisis refers to the threat factors or problems which have negative impacts on library resources, and it is uncertain and harmfulness. It includes the resource type crisis, the resource construction crisis, the resource preservation crisis, and the resource structure crisis [1]. The library is affected by the social information environment, the most remarkable change which the library faces in current information environment is the digitalization and networking of the information exchange and it caused a lot of contradictions between the library and readers, the library and the publishers, and the library and information providers. All these contradictions may trigger library resource construction crises [2]. The types of library resource construction crises mainly include the following five types.

S. C. Shao (✉) · J. Xu · S. Q. Yang
Library of Hebei United University, Tangshan, China
e-mail: shaoshengchun@126.com

Z. Zhong (ed.), *Proceedings of the International Conference on Information Engineering and Applications (IEA) 2012*, Lecture Notes in Electrical Engineering 220,
DOI: 10.1007/978-1-4471-4844-9_38,

38.1.1 The Resources Type Crisis

In the face of the sharp rise of digital resources and the convenience of use, more and more users tend to use the digital products, especially the academic journals, conference papers, dissertations, and digital resources which have become a preferred, accustomed, and reliant use for scientific researchers. Therefore, many schools began to order a large number of topics or comprehensive full-text databases of periodicals, while they try to stop a corresponding printed journal. Such as the library of Tsinghai University since 2005, with the increasing number of database order and the decreasing number of printed edition of the journal, in 2008, they only ordered about 198 printed journals. What's more, authoritative report predicts that in 2016 science and technology journals will mainly be typed of digital version as the basic form of publication [3]. Under the E-science environment, engaging in scientific research activities needs to use a variety of types and multiple channels of information in addition to traditional printed resources, journal articles, books, and other resources. It also includes new digital resource types, such as preprints, laboratory reports, personal communication, especially speech, memos, data sheet, and other new resources type [4]. Therefore, the decision of library collection resource type will become more complex and resource acquisition work will be more challenging.

38.1.2 The Resources Structure Crisis

It is mainly the aging of paper resources. Using SPSS statistical regression, paper resources can be obtained and the aging rule can be expressed as a formula;

$$M = Ke^{-\mathrm{bt}} \tag{38.1}$$

K and b are constants, t is a variable, t equals Circulation Statistics minus the year of publication.

In order to effectively express paper resources in the aging speed, we can use directly in half-life index to express different attenuation velocity, a calculation formula is given below:

$$T_{1/2} = In\, 2/b \tag{38.2}$$

In order to statistically define the readers' utilization level, we can also use the following two formulas:

$$t = 0 - \infty \tag{38.3}$$

or

$$S = K/b \tag{38.4}$$

S is the statistical sample, it can express the coefficient of utilization of various types of books in different statistical years clearly [5]. So the aging of paper resources results in the resource structure crisis directly.

38.1.3 The Resources Proportion Crisis

Now the blindness, randomness, and the lack of the sense of quality in the digital literature resources construction have caused the proportion imbalance of collection resources. Along with the rapid development of digital and network, the digital resources in library take a growing proportion. Library resources include paper resources, electronic resources, and cyber source. A paper resource includes paper books and paper periodicals. Electronic resources and cyber source include electronic books, electronic journals, databases, and a variety of network information resources. According to the survey of university library, paper resources account for 70 % in the entire document resources, and electronic resources account for 30 %. How to make the reasonable proportion of the paper resources and electronic resources exerts the use in knowledge and information dissemination of library, which reflects the modernization degree of library collection.

38.1.4 The Resources Layout Crisis

The library's resource layout is dividing library information into a number of sections that correspond to the function of the spatial structure, so that the reader can easily use and it will give full play to the library information resources and value of the model. Because of the modern architecture of the library, it results that three layout cannot play out its features, which brings crisis about resources layout impressions [6].

38.1.5 The Resource Preservation Crisis

The library is a growing organism. Only the metabolism and embracing new ideas can make it maintain its vitality. But there is a widespread phenomenon in the library's collection resources construction. The Library's collection building pays more attentions on number and ignores the resources quality. For example, in the universities and colleges, the library holdings are used as a standard measure. So that quality colleges and universities have had to pursue library holdings and not to tick. This makes a number of outdated, obsolete, useless literatures with a value of literature in the midst of the occupied space, which leads to the waste of library material and human crisis of resource preservation.

38.2 Mechanism Analysis of Resources Construction Crisis

The mechanism of library collection resources crisis is the library elements structure is unreasonable, and forms that collection resources structure cannot balance, and results that collection resources does not adapt environment.

38.2.1 The Digitization of the Literature Resources

Digital library changed the traditional library books of the static characteristics of the literature services, realized the multimedia access, remote network transmission, intelligent retrieval, cross library seamless links, create trade surplus space and time of the information service of the new state. Therefore, the literature resources digitization makes the traditional collection resources face new information environment, the users' demand, and the challenge of the market competition. On the one hand, the literature resources digitization makes the whole information of any content of space to be characterized, analysed, linked, interactive, and blended together, so people can from all levels and many angles to the flexibility to analysis, organization, performance, make use of the information content. On the other hand, when the user can access a large number of documents easily, the major migration process of obtaining information constraining bottleneck will occur. Users will be required to find further analysis of the literature. mining in the literature of the internal hidden knowledge content and its mutual logical relationship, find themselves do not know the structure and laws [7]. At the same time, publishers in digital network constantly enrich service based on content, change the way of service, not only providing digital full-text documents, and will be a variety of digital resources, web services, and even the library catalog is linked together, forming a new digital information service platform. Therefore, the literature information resources digitization will make the traditional collection resources construction system changed.

38.2.2 The Rapid Development of Information Technology

The rapid information technology changes have proposed a new challenge to the construction of the literature resources crisis. Along with the development of information technology and the impact of the digital wave, give the reader brought new experience and to the pressure from the library. From the library developmental view , modern library is seen as a result of the application of information technology; this trend since the 1990s have become powerful and prosperous increasingly. Computer technology, especially the communication technology and Internet technology, rapidly develop the appearance of the Web 2.0 technology, in order to emphasize the reader experience and social features and greatly accelerate the progress of the library. Web 2.0 is the development and integration of Web 1.0 is a series of new thinking in the concrete embodiment of digital library with distinct characteristics of open, transparent, and common participation, self organization, and social interaction and sharing. Web 2.0 is to let users from network information receive into information maker and communicators, from the audience to main body, from the individual to the new Internet service model of community. And is to readers-oriented to turn, the Internet more than ever, the rich and colorful and it

Table 38.1 Parson of digital resource quantity in Tsinghai University library from 2001 to 2008 [10]

Type of resource	2001	2008	Growth rate %
Database (numbers)	187	400	214
Digital journals (kinds)	14,430	44,000	304
Digital books (kinds)	2,04,000	11,00,000	539

Table 38.2 Comparison of digital resource quantity in Jia Xing University library from 2001 to 2008

Type of resource	2001	2008	Growth rate %
Database (numbers)	2	22	214
Digital journals (kinds)	6,000	31,550	304
Digital books (kinds)	23,000	5,76,000	2,504

has strong attraction, the readers can't only the literature retrieval and utilization of information resources, but is evaluation of information resources and creator.

38.2.3 The Stereotyped Reading Behavior

The traditional reading behavior brought pressure to the library resources construction. From the Internet, since its appearance, readers making use of the library in the overall trend abate, and borrowing rate of the printed literature significantly reduced. From Tables 38.1 and 38.2, data can reflect the trend. Today, many readers think Internet search engine can provide high quality information faster and more abundant than library; according to the characteristics of the knowledge to identify, screen, fuse and knowledge innovation, the reader has not only met the use of information search, but still hope to organize, assess, to create, interact with others, and share information and expect the vast multitude of digital information knowledge mining from information acquisition development to knowledge discovery.,

38.3 Library Resource Management Strategies in the Digital Environment

Crisis management discusses the mechanism causing crises and works out the crisis management according to the mechanism. The strategies of library crisis management mainly include the following aspects:

38.3.1 Strengthen the Organization and Management of the Collection of Information Resources

Today in the rapid development of the information environment, library information resources not only include all the library entity collection resources, but also include the virtual collection which do not have the ownership and the library needs to pay, remote access, signed by online, which build the library information resources system.

The established and development of the science in collection of information resource system not only depends on the library information of resources to carry on the long-term complement and accumulation, but also depends on organization and management to getting information resources or scientific and reasonable [8].

Therefore, library information resources organization and management of library is an important link of the construction of the information resources.

38.3.2 Strengthening the Build and Sharing of the Library Resources

Along with the rapid development of information technology, not only all forms of information were in the sharp growth, but also the production, storage, and transmission way of the information had the profound changes. In the digital age, the traditional library and digital library coexist complementarily, especially the increasing and expanding digital resources, the shape and the carrier and diversification of the access method of the resources, make any single library collection construction independently, at the same time, just rely on library reader demand model has no longer apply. Therefore, information resource sharing has become an inevitable trend of the development of the library.

38.3.3 Strengthen Library Resources Security System Construction

The ultimate goal of the construction of the information resources is to build an information resource guarantee system which can satisfy the demand of information society furthest. This is an entity system, including the reserve system and service system of the information resources [9].

The security system makes limited information to satisfy social need furthest and make it convenient to develop fully and use efficient, which has the resources network system of hierarchical structure science and reasonable space layout for material foundation and the social target of information resource sharing. This security system also makes the literature information socialization as organization form and the network technology electronic computer communication as the means. Therefore, information resource guarantee system occupies very important position on the library literature resources construction.

38.3.4 Perfect Digital Information Resources Construction

Along with the rapid development of information technology and network environment better and better, the digital information resources not only become more and more large in the proportion of the number, but also increasingly play an incomparable role in the services which is provided by the library than the printing document resources. Digital information resources have different characteristics with the printed document in production, storage, transfer, and carrier and so on. So far digital information resources also have completely different content and method with the printed document resources in collection, organization, processing, development, and utilization. Compared with the traditional printing literature, the use of digital resources has no time and space limitation, and the database has more retrieval ways and has more flexible method, and the inspection rate is high, and has faster contents updates, and according to need users can selectively download, print, leading to deeply popularity. Therefore, perfecting digital information resources construction is the indispensable part of the library literature resources construction system.

References

1. Yi H, Fei Y (2009) Research on library crisis and crisis management under digital network context. Library inf serv 1(4):13–16
2. Zhao WH (2010) Discussion on library crisis management under digital environment. J Mod Inf 2(5):95–98
3. (2009)The China education and research computer. Trend of retaining data management strategy for information integration in colleges and universities upgraded network. http://www.edu.cn/sj-6538/20090706/t20090706-389057.shtml
4. Cai YC, Huang G, Weiguo M (2009) Tactics on crisis of library collection development. Library Inf Serv 3(3):26–29
5. Wang Y (2009) The study on the aging regularity of the library books and its application. J Intell 4(6):49–53
6. Zhang HP (2010) Arrangement of multi-dimensional body's library information resource and construction of characteristic reading room. Library 5(8):112–116
7. Zhu H, Cao Z (2010) Effect of digital document information on theory and practice of collection structure of library. Library Tribune 6(6):87–89
8. Wu XF, Li DM, GW Li (2011) Network grey literature resources organization and management. J Library Sci 8(13):151–154
9. Xi ming X (2008) Information resources construction. In: She C (ed) Wuhan University Press, Wuhan 9(15):244–248
10. Wang JY (2009) The changing document resources construction in university libraries. Inf Stud Theor Appl 7(12):1543–1547

Chapter 39
Information Construction and Sharing Strategy Based on Regional Library Alliance

Yun Hong Lv

Abstract Library alliance is the library consortium which is organized to share resources and mutually benefit by common agreement and contract restrict. At present our country university libraries, public libraries all appear library consortium. It has important significance to realize the information resources sharing. But as a result of long term since the influence of traditional concepts, the existence of funds is nervous and area development is lopsided. In view of this, this chapter puts forward the co-construction and sharing of information resources dive lomenta strategy.

Keywords Library alliance • Information resource • Co-construction and sharing

39.1 Introduction

The library as an important part of culture and education, the highest realm is knowledge building and sharing. In the era of information and knowledge, material resources, information resources, human resources, and energy resources constitute the modern social resources support, especially the information resources have attracted much attention, has become a national strategic resources, each country put into an unprecedented funds and manpower to increase on the construction of information resources. But facing the reality environment, academic prosperity, all kinds of media information resources increase demand for information and readers toward a refined, deep, sharp, specifically, new, and prospective demand direction as more and more users tend to remote access to library information resources and services. In the world, no one information institution can meet the diverse information needs of readers by itself, so in order to plan as

Y. H. Lv (✉)
Hebei United University, LibraryJianshe Road 57, Tangshan, Hebei, China
e-mail: lvyunhong@126.com

Z. Zhong (ed.), *Proceedings of the International Conference on Information Engineering and Applications (IEA) 2012*, Lecture Notes in Electrical Engineering 220,
DOI: 10.1007/978-1-4471-4844-9_39,

a whole, different types and regions of the library information resources allocation, to better carry out the co-construction and sharing of information resources, effectively eliminating information division, increase the exchange of information between the libraries, the system and region more efficient, high-quality information service of regional library alliance emerge as the times require. Library alliance is organized to achieve the sharing of resources and mutual benefit by common agreement and contract restrict library consortium [1]. Regional library consortium is the area-centered library cooperation organization, which is defined by the geographic position adjacent to the two or more than two libraries on the basis of equality, voluntary, which makes relevant articles of association rules and the signing of a cooperation agreement, consisting of regional alliance [2].

39.2 Analysis of the Mode of Regional Library Alliance

Regional library consortium carried out mainly printed resource sharing in the early, interlibrary loan, document copy preference literature entity sharing cooperation as well as the union catalog, union list of periodicals resource sharing cooperation project [3]. Later, it become joint purchase of document resources, the combined reference and consultation, automation technology services coordinated resource sharing and library construction. Along with the development of computer network and communication technology and the deepening cooperation between members of the library alliance, service function is also in constant expansion and extension of electronic resources, such as joint procurement, automation system and network maintenance, and building local characteristic database. At present, our country colleges and universities, scientific research, and public library appear to be in library alliance sharing mode which is mainly divided into the following two.

39.2.1 National Library Alliance

China Academic Library and Information System, referred to as CALIS, is approved by the State Council of higher education in China "project 211" Nine Five "fifteen" overall plan the three public service systems. The purpose of the CALIS is, under the lead of the Ministry of Education, make the national invest in modern library ideas, advanced technological means, and the integration of rich literature resources and human resources and build the China Academic Digital Library as the core of education literature united security system, to realize information resources co-construction and sharing in common knowledge and to maximize the social and economic benefits for China's higher education. The CALIS management center located in Peking University, is set up by arts and science, engineering, agriculture, medicine, the four national literature information service centers, and the seven regional literature information service centers China Northeast, China Southeast, Central, Northwest, Southwest, Northeast of Southern China, and a Northeast regional defense literature information service center.

39.2.2 Regional Library Consortium

The predecessor of Shanghai literature resources co-construction and sharing network was the Shanghai literature information resource network. In 1994, Shanghai area public, scientific research, universities, intelligence the four systems of 19 libraries and information institutions held a signing ceremony in Shanghai library about the Shanghai literature information resource network. In May 13 1999, the meeting for co-construction and sharing of the literature resources was held in Shanghai, the conference puts forward the Shanghai literature resources sharing plan in 3 years to realize the public, university, scientific research system library networking in Shanghai and at the same time, the conference identified the information platform, the procurement coordination, and implementation of information services and personnel training in three specific schemes [4, 5].

39.3 Regional Library Consortium Construction Problems

39.3.1 Ideas Behind

Long term since, our country information institutions used to be closed to international intercourse, provide for one, each unit thought to meet the needs by itself and lacks resource sharing initiative. On the resource construction pursuit "big and complete, small and complete" in cooperation and it emphasizes on the self interest. On the surface, co-construction and sharing of resources has become the consensus, but the actual operation of the resource sharing consciousness is still weak. As a university literature resource sharing service network for example, this school library is both from paper document and electronic database, are relatively more full, professional, and strong features, and other regional college libraries are relatively weak a lot. The library within the alliance is not in a level, without full-time service personnel, therefore, the library exists between although nominal alliance, but the actual communication and cooperation is very little, since the project has been established a formal cooperation and share the project can be counted on one's fingers, causing the regional alliances in a form without content, a name without a performance state of nothingness.

39.3.2 The Imbalance of Regional Development

The relevant literature shows the scope of the whole country (Hong Kong, Macao, Taiwan area colleges and universities are not within the scope of statistics) Library

Table 39.1 Table of basic conditions of library alliance

Alliance name	Set up	Number	Network address	Administrative unit	Management center
Beijing network library	2001	39	http://www.netlib.edu.cn	Beijing municipal commission education	Capital normal university
Tianjin university digital library	2004	20	http://www.tjdl.cn	Tianjin municipal commission education	Tianjin polytechnic university
Shanghai education network library	2000	152	http://www.shelib.edu.cn	Shanghai municipal commission education	Shanghai jiao tong university
Jiangsu university digital library	1997	240	http://www.jalis.org.cn/	Jiangsu municipal commission education	Nanjing university

Union coverage is not high still, and there exists the imbalance of regional development. Due to the influence of unbalanced economic development between regions, the digital library alliance in the regional development is not balanced at present. The regional library alliance construction is mainly concentrated in Beijing, Tianjin, Shanghai, Guangdong, and other eastern coastal areas and the Midwest construction number is little. As shown in Table 39.1.

39.3.3 Insufficient Funds

From the present building union perspective, its main source of funding is government financial support and union funds. In our country, west region economic development is relatively backward; funds became the bottleneck of the local library alliance construction and development [6]. Foreign library alliance in addition to government funding, has a lot from a large number of social, business, and other aspects, this deserves our reference and study. To solve the problem of insufficient funding, to obtain stable funding, realize the source of funds of the plurality of channels.

39.3.4 Resource Integration Degree Shallow, Services Need to Improve

Regional library information service is less, the service is single, cooperative digital reference service did not become mainstream. For University Library Alliance as an example, through the network survey it has been carried out on the online

Table 39.2 Tables of library alliance co-construction and sharing

Alliance name	Union catalog	Reference	Subject navigation	Document delivery	Characteristic database
Beijing network library	University network bibliographic database	Online reference	Key subject resources	CALIS	No
Tianjin university digital library	Tianjin university union catalog	Virtual reference service (construction)	Page navigation	CALIS	15
Shanghai education network library	Chinese bibliographic database, foreign periodical database	CALIS virtual reference system	Key disciplines of core journals	CALIS	No
Jiangsu university digital library	Books union catalog database	CALIS virtual reference system	Navigation database	CALIS	22

reference services of union of six, in addition to Beijing network library and digital library of University of Hubei province to carry out cooperative digital reference service, the rest are directly connected to the Union Center Library Reference for service. As shown in Table 39.2.

39.4 Regional Library Consortium in Information Resource Co-Construction and Sharing Strategy

39.4.1 Strengthening Communication, Optimize the Allocation of Resources

The establishment of regional library alliance, the literature resources, electronic resources purchase, each member of library overall plan, according to China's library consortia construction practice and the hall itself, can be taken to the province as a unit; the establishment of provincial League Committee, by the union committee according to the members of the subject characteristics, set up the resource development of long-term planning on the basis of unified planning, unified layout, unified management to carry out the overall construction. Division of labor must be clear, each doing his own job, but also give full play to their respective characteristics, to play the overall benefit and joint security advantages, to prevent the waste of resources and repeated construction through the union

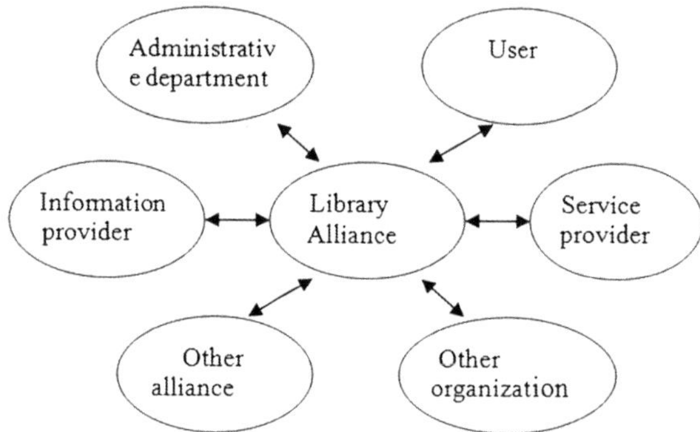

Fig. 39.1 Charts of library alliance communication and coordination

internal, external communication and coordination, and to achieve the long-term development of Library Alliance [7] and Communication diagram, as shown in Fig. 39.1.

39.4.2 Strengthening the Standardization, Establishing Information Communication Platform

Library alliance standardization construction is to realize the co-construction and sharing of information resources on the premise. Library alliance is an important aspect of the members through the network to share data. Library is different between the format of the data mark, and cataloging standards are not unified. This is bound to affect the construction of the shared literature information resources.. Along with the digital information resource utilization rate rise ceaselessly, only to strengthen standardization construction, can realize the sharing of information resources. On construction, we should first consider on customer information using in specific circumstances, set out from the fact, think readers with high-quality service as the goal, to realize standard construction [8]. Through the establishment of information communication platform, as shown in Fig. 39.2, unified

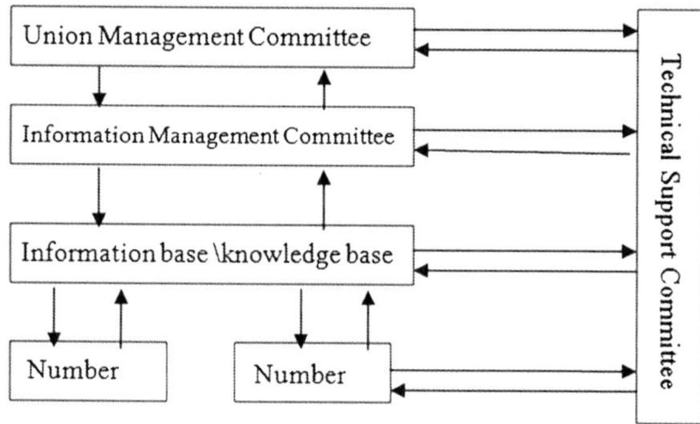

Fig. 39.2 Chart of library alliance information communication

release operating standards and processes, achieve union internal communication and coordination and the standard consistency and information of the smooth operator.

39.4.3 To Strengthen Cooperation and Coordination, Establishment of Performance Evaluation Mechanism

On the basis of communication and coordination of regional digital library alliance, we must establish a set of regular, ongoing performance evaluation system on the alliance cooperation project inspection, both on the alliance activities and tasks, capital and resource configuration of a supervision and management, but also assessment and evaluation on the alliance operation mode through establishing perfect and applicable performance evaluation strategies and evaluation index system of alliances, scientific system, comprehensive evaluation, to promote better development [9, 10].

References

1. Liang X (2012) Mobile Digital Library Federation: The Trend of University Information Resources Sharing. J Mod Inf 1(12):63–66
2. Chen H (2007) Discussion on foreign regional library consortiums. J Inf 3(4):126–128
3. Li C (2011) Research on the library alliance inspiring to Guangdong University library alliance construction. Res Library Sci 4(4):85–88
4. http://www.th.superlib.net/news/showNews.Action?id=871&flag=logo&cataid=863
5. Lin H (2010) Investigation and analysis of current situation of regional universities digital library federation. Library Hunan Univ Commerce 6(4):61–63
6. Lijun E, Xu Z (2012) An investigation and analysis on the construction situation of regional library alliance in our country. Library 7(1):62–65
7. Yafang F, Lang G (2009) Research on the operation mechanism of regional library alliance in China. J Library Work Res 8(3):8–11
8. He L, Wu L (2010) Regional digital library alliance construction status and development strategy in China research on libraries. Sci 9(12):13–16
9. Yuan J (2010) On the Strategy of Improving the Service Quality of Library Consortium. Inf Doc Serv 10(5):74–77
10. Wang RF (2010) Establish regional library consortium to develop local literature. J Library Inf Sci Agric 2(2):52–54

Chapter 40
Study of H-index in Evaluation of Library and Information Science Journals

Jian Jie Du, Xiang Fei Guo and Mei Zhang

Abstract This study aims to make an empirical research of h-index in the evaluation of library and information science journals. The evaluation objects are 19 kinds of library and information science core journals from the Journal of Library Science in China using citation, h-index, and relative h-index. It proved that this journal has a top h-index, relative h-index, citation, and highest citation in the library and information science core journals, and high quality authors of highest citation. The papers in the journal have a high quality and excellent academic influence.

Keywords Citation • H-index • Relative h-index • Periodical evaluation • Library and information science

40.1 Introduction

Journal evaluation is an important part of the bibliometric study. It can reveal the number of disciplinary literature in the distribution of journals through quantitative analysis by the law of development and growth trends of the journal. Provide important reference for optimizing the use of academic journals, while improving the intrinsic quality of academic journals [1].

This study aims to make an empirical research of h-index in the evaluation of library and information science journals. The evaluation objects are 19 kinds of library and information science journals from A Guide to the Core Chinese Periodical [2] and the authors from Journal of Library Science in China.

40.1.1 Evaluation Index

The 19 kinds of library and information science journals from A Guide to the Core Chinese Periodical and the authors from Journal of Library Science In China were evaluated using citation, h-index, and relative h-index.

J. J. Du (✉) · X. F. Guo · M. Zhang
Library of Hebe United University, Tangshan, China
e-mail: dujianjie@126.com

Z. Zhong (ed.), *Proceedings of the International Conference on Information Engineering and Applications (IEA) 2012*, Lecture Notes in Electrical Engineering 220,
DOI: 10.1007/978-1-4471-4844-9_40,

The journal paper citation frequency is the cumulative number of times the paper was published after a certain time domain in all kinds of literature references; it can be used to evaluate the quality of journals and academic achievement level of the scientific research staff of the academic units. The citation frequency is higher, if the search results generated by the social, economic, and ecological benefits are more. Obviously, highest frequency has been cited is ranked first in the sequence of the cited papers in a journal are cited from large to small arrangement of papers cited. Within the scope of this study, papers' citation frequency ≥100, in cites qualitative change caused by the quantitative area, provide a basis for judge the papers frontier level of activity. This article defines the papers' citation frequency ≥80 as high citation frequency.

The h-index, a new scientific measurement evaluation, was invented by J. E. Hirsch in 2005. Individual scientists h-index is defined as and only if h articles each of Nap papers published by a scientist to receive at least h times the number of citations, each won at least h times the number of citations, the rest of Nap-h papers' citation rate, this value of h is the h-index of the scientists [3], it can be defined as follows.

$$h = \max(I) : \text{cit.} \geq I \tag{40.1}$$

H-index reflected and practiced the "quality" and "quantity" to both; it is a new evaluation concept of seeking the number on the basis of the stress on quality [4]. Compared to traditional bible metric indicators which only reflects the number, it has obvious advantages [5]. H-index can be used in academic influence of the measurement and evaluation of individual scientists, measurement of the research collective academic influence, the evaluation study of the influence of academic journals [6], and also used in evaluation of universities academic strength [7] or disciplines [8].

Related scholars have proposed a relative indicator of the h-index to exclude the influence of amount of journal papers to h-index, which is the result of h-index divided by the volume of journal papers, called journals relative h-index, it can be relatively objective and reasonable evaluation of the size of academic influence [9].

H-index is mainly applied to the evaluation of higher level of academic research staff, but when evaluating the lower level of academic research staff, to get the desired results, it has to be complemented by other evaluations, such as "total citation frequency", "total articles" etc [10]. So in this paper, these three indicators were used to evaluate.

40.1.2 Data Sources

Using the name of the 19 kinds of library and information science journals as "literature published sources", retrieval domain is limited to 1999–2008, retrieval led the "China Academic Literature Network Publishing Database" (Inter database retrieval, retrieval date: 2010/09/03). Using the citation frequency as "Sequencing Analysis Method", retrieval led the journals' total articles, h-index, total number of papers of citation frequency ≥80 and ≥100, the highest citation frequency, comparative study the research object.

Using the high citation frequency authors of Journal of Library Science In China as an evaluation object, retrieve h-index, total number of papers, and his total number of papers on Journal of Library Science In China, and total citation frequency, evaluate the authors' academic influence.

40.2 Statistical Result

The 19 kinds of library and information science journals' h-index, citation frequency, and the high citation frequency authors' academic influence of Journal of Library Science In China are shown in Table 40.1.

Table 40.1 The authors' academic influence of journal of library science in China

Serial number	Author	H-index	The quantity of articles	Total citation frequency	The quantity of articles of this journal	Citation frequency of this journal
1	Jiang Yong fu	28	76	2,156	14	706
2	Qiu Jun ping	28	197	3,241	17	655
3	Zhang Xiao lin	24	55	2,619	8	908
4	Hu Chang ping	18	58	880	14	336
5	Fan Bing si	18	49	1,592	8	332
6	Wu Wei ci	16	43	888	8	376
7	Li Jia qing	16	34	743	3	170
8	Chu Jing li	15	39	941	4	402
9	Cheng Yan an	14	34	922	5	324
10	Ma Wen feng	14	27	825	3	245
11	Suo Chuan jun	13	34	801	11	282
12	Zhao Ji hai	10	18	607	3	277
13	Wu Jian zhong	10	25	776	3	241
14	Li Wu	10	14	494	1	134
15	Tan Xiang jin	9	21	375	6	308
16	Gao Bo	9	23	359	5	272
17	Sun Jian jun	8	15	474	4	187
18	Lu Gong ping	8	13	388	2	120
19	Zou Zhong min	6	8	295	1	155
20	Lin Ping zhong	5	8	210	3	152
21	Lv Jun sheng	5	8	219	1	141
22	Guo Tai min	5	4	210	1	122
23	Xu Wen bo	5	9	351	3	101

40.3 Discussion

40.3.1 Analysis of Citation Frequency and H-Index

We can see that the indicators of Journal of Library Science in China ranked first in the 19 core journals.

Among the 19 core journals, there are 85 articles' citation frequency ≥100, the Journal of Library Science in China accounting for 28 (32.9 %), there are 178 articles' citation frequency ≥80, the Journal of Library Science in China accounting for 53 (29.8 %), the proportion was significantly higher than other journals. This shows that the journal is the overall level of higher and stable development of the papers authors, and published more academic and influential papers.

The h-index of this journal is 67, ahead of other journals in the core journals in Library and Information, the h-index is high because it published a series of authoritative, theory, practice, the effectiveness of the research in the field of library and information in China Formation of a strong influence resulting in larger social and economic benefits.

Furthermore, the relative h-index of this journal is on the top of the 19 core journals.

40.3.2 Analysis of Authors

High quality author group is to produce the basic elements of the prestigious journal; high citation frequency authors play a key role to improve the h-index of the journal.

The authors of Journal of Library Science in China whose citation frequency ≥100 were 23. The h-index, quantity of articles, and total citation frequency were shown in Table 40.1.

From the Table 40.1, we can make a conclusion that high citation frequency of authors posting number of articles ranging from dozens to more than 200 papers; they have the potential to become a high h-index of authors. Quid Jumping, Jiang Yong fu, Hu Chang ping, whose articles more than 10, are high yield and core authors. And, the total citation frequency of Jiang Yong fu, Qiu Jun ping, Zhang Xiao lin, Hu Chang ping, Fan Bing, Wu Wei chi are >300. The h-index of Jiang Yong fu, Qiu Jun ping were 28, Zhang Xiao lin, Hu Chang ping, Fan Bing were 24/18/18. Hirsch account that h-index can even reach 20 for a successful scientist. So, authors' h-index can reach 20 or more were powerful.

40.4 Conclusion

From this study, we can make a conclusion that the Journal of Library Science in China has a top h-index, elative h-index, citation and highest citation in the library and information science journals, and high quality authors of highest citation. The papers in the journal has a high quality and excellent academic influence.

H-index is a useful auxiliary which evaluates papers' quantity and quality of innovative practice enriches the bible metric tools [11]. If used in conjunction with other bible metric indicators, make the journal more comprehensive and impartial evaluation [12].

References

1. Chen W, Li WL (2010) Research on relevance of h-index, rates of web-downloaded and impact factor–core journals of library and information in CSSCI as example. Inf Sci 1(12):1832–1836
2. Zhu Q, Dai LJ, Cai RH (2008) A guide to the core chinese periodical, vol 2(5), Beijing University Press, Beijing, pp 112–116
3. Hirsh JE (2005) An index to quantify an individual's scientific output. In: Proceedings of the national academy of sciences of the United States of America, vol 3(46), pp 165–169
4. Wu JZ (2011) The application of h-index to research level evaluation of discipline specialty-based on national literature analysis in phytonology. Library Inf Serv 4(20):28–31
5. Zhao JM, Qiu JP, Huang K (2008) A new scent o metric indicators-h-index and its application review. Nat Nat Sci Found 5(1):23–32
6. Zeng JJ (2011) A study of institutions evaluation with h-index and its extended index. Library Inf Serv 6(10):65–68
7. Zhou ZF, Wan RG, Yu SW (2010) Analysis of the academic level of universities based on h-index vision. J Intell 7(3):71–74
8. Nie C, Wei ZF (2010) Empirical study on improvement of h-index based on academic influence difference. J Intell 8(5):89–91
9. Jin ZY, Zhao DQ (2011) Correlation of the h rich-index with evaluation system of academic ranking. Library Inf Serv 9(4):33–38
10. Zhang Xue mei (2007) Evaluation of Library and Information academia of the h index Library and Information Service 10(8): 48-51
11. Shi J, Yuan TF (2009) Analysis on applying h-index to evaluate the quality of academic journals. J Med Inf 11(8):46–49
12. Ou Q (2010) A case study on h-index of the core journals in agricultural engineering. J Library Inf Sci Agric 12(4):200–202

Chapter 41
Efficient Characteristics Database Construction Scheme

Ying Zhao and Yun-peng Zhang

Abstract The recent research priorities and directions were reviewed and evaluated based on the survey and statistical analysis of the literatures published from 1996 to 2009 in the characteristic database field, which aimed at exploring the principle of building a database of the University Library Database, and proposing the copyright issues which should be noted in the characteristic database construction.

Keywords Characteristic database • Databases of special subject • Characteristic special subject database • Library

41.1 Introduction

Characteristic database is an information resource library which relies on the collection of information resources, aiming at the user's information needs. It collects, analysis, evaluates, treatments, and stores a subject or a topic for the value of the unique information, and these features are, in accordance with certain standards and norms, resources for digital processing to meet the needs of individual users. The characteristics of the database fully reflect the unit characteristics with counterparts in the literature data resources; the library is set up on taking the full advantage of characteristics of museum collections on the basis of shared information database. It has to reflect the collection of information resources characteristics, to provide users with personalized information services and construction is made in accordance with certain standards and norms which can be shared.

Y. Zhao (✉)
Library of Hebei United University, Tangshan, Hebei, China
e-mail: zhaoying@126.com

Y. Zhang
Jitang College of Hebei United University, Tangshan, Hebei, China

Z. Zhong (ed.), *Proceedings of the International Conference on Information Engineering and Applications (IEA) 2012*, Lecture Notes in Electrical Engineering 220,
DOI: 10.1007/978-1-4471-4844-9_41,

41.2 The Necessity of Building Characteristics Database

The construction of characteristics database is the need for developing distinctive collection resources and the sharing of resources. A distinct theme and prominent feature is the essence of literature information. Database development for the University Library, is one of the main task of the construction of its resources [1]. Library should be based on the audience and key disciplines, the requirements of the university's survey research discipline and the law of professional teaching standards, self-development to create the characteristic database for the teaching, research, and information resources support, relying on the professional features to enhance library resources utilization, and enhance the influence of the discipline at home and abroad. Meanwhile, the construction of characteristics database is to improve the precious books utilization needs. Each library has ancient books and documents, and other characteristics of library resources to protect the ancient resources and to play their library urgent need to address the problem [2]. The digital help protect rare and fragile documents, at the same time meet the readers demand to access these documents.

41.3 The Status of the Construction of Characteristics Database

By using the National Knowledge Infrastructure database as data source, we searched the 450 articles to related literature in 1996–2009. Through the literature on the identification, eliminating the repeated and unrelated literature, finally we got 448 articles Income data processing after order into Excel, and the method of bibliog metrics in statistical analysis.

Statistics a subject, the research papers published, we can from the concept of time on the subject, special understanding of the development of research in the statistical analysis. The characteristic database of research papers distribution is shown in Fig. 41.1. Posting of the characteristics of database research, from 1996 to 2009 was the gradual upward trend, indicating that the characteristic database development research have more and more attention by people. Posting the amount

Years	1996	1997	1998	1999	2000	2001	2002	2003
paper (piece)	1	0	2	1	5	20	19	34
Percent (%)	0.22	0	0.45	0.22	1.12	4.46	4.24	7.59
Years	2004	2005	2006	2007	2008	2009	total	
paper (piece)	35	47	62	61	80	81	448	
Percent (%)	7.81	10.49	13.8	13.62	17.9	18.1	100	

Fig. 41.1 The age distribution of the characteristics of database research papers

of these three stages divided featured database for 13 years: The year 1996–2000 is the beginning of the characteristic database research, from 2001 to 2006 for rapid development phase, and from 2006 to 2009 for the continued development stage.

The literature research topic is the main characteristics of the contents of the documents. Through to the topic of thesis of statistical analysis, we can reveal the characteristic database research status, direction, and characteristics, and the knowledge of the existing advantage and disadvantage, and the research emphasis in the future clearly and its development trend. Through the research of the characteristic database of statistical analysis of the subject, the characteristic database research subject distribution is concluded, see Fig. 41.2.

Statistical results from subjects distribution look, database established research papers several for 237 articles, accounting for 52.9 % of the total paper, is nearly 14 years of research contents next is the database technology research papers for 51 article number, accounting for 11.4 % of the total paper, show that the theory discussion has gradually turned to the application practice research. And intellectual property rights, the resource sharing, the service and management of the number of research papers are under 5 %; therefore, we need to strengthen the experts and scholars.

The author conducted a preliminary investigation on the construction of characteristic database in the University Libraries. The statistics show that currently a number of university libraries have established the relevant characteristics database according to the advantages of the school subjects, the collection, geographical etc. Such as the basic education literature database of the Library of East China Normal University, the Quant characteristics database of Xi'an Transportation University Library, the robot information database of the Shanghai Transportation University Library, the Basho figure research topics character library of Sichuan University Library, and the Hue Studies and Hub Shi feature database of Anhui University. These school libraries have also established teaching results database, such as doctors and masters' thesis database.

Topic	Papers(piece)	Percent (%)
multi-monograph research papers	38	8.5
database established research papers	237	52.9
the database technology research papers	51	11.4
Present situation and the development research papers	46	10.3
a general introduction to database research papers	27	6.0
Intellectual property research papers	14	3.1
standard and quality research papers	13	2.8
Resource sharing papers	11	2.5
The service and management of the number of papers	11	2.5

Fig. 41.2 Topic distributions of characteristics of the database research papers

Although some of the University Libraries on the characteristics of database construction had been made some progress, the whole of the National University Library is still in its infancy, far from being a construction system; there are also some problems of the building of sharing system and service system. Mainly manifested in the following aspects: a single database type, the professional features is flat and monotonous; mainly are the catalogs; bibliographic database, while lack of the multi-based; full text; graphics library to abstracts. The literature of deep-level processing features database is not enough, and the degree of standardization is not high; indexing is not standardized; nonstandard format. In addition, special database projects are the lack of a nationwide unified planning, layout, and division of labor, decentralized building a database, serious low-level redundant development of the phenomenon.

41.4 The Principles of Characteristics Database Development

41.4.1 The Unique Principle

The unique of characteristic database construction is the first prerequisite, it should have distinct characteristics, characteristics means that valuable content [3]. Build unique content, can avoid redundant construction of the literature information resources. So we should do in-depth research topics in the collection based on fully reflect the subject characteristics and geographic characteristics, special features, otherwise, the database construction will lose value.

41.4.2 Standardization and Normalization Principles

Standardization and regulation is an important guarantee for building a high-quality database, it is a prerequisite for the implementation of digital resource sharing system expansion on the basis of standardization and normalization of the characteristics of database that has a living space [4]. Characteristic database construction must comply with international, domestic common data recorded in standard data format standards, data indexing standards, norms and control standards [5].

41.4.3 Scalability Principles

The construction of characteristic database is a long-term work, not easy, it need the accumulation of the literature, and gradually build a perfect process. Characteristics of the database should have good scalability; upgrade capabilities, compatible with different devices, systems, and network platforms, the database as technology development, business expansion and the expansion of the scope of services, to achieve a smooth upgrade and transition [6].

41.4.4 Sharing Principles

Features a database of construction needs a lot of manpower, financial, and material, a single department building a database will be subject to the constraints of technology, capital, information resources, it is difficult to guarantee the quality of the database, and also incompatible with the scale of development of database construction, more important because of their different functions, different types of library-oriented readership of different species in the collection of literature resources, number, version, and features different emphases [7]. Therefore, starting from the advancing network resources to build, sharing between interlibrary library and research groups should break boundaries, play groups, forces, complementary advantages, the road of joint construction and sharing.

41.4.5 Continuity and Systematic Principle

The characteristic database information acquisition and database building process must pay attention to data collection of comprehensive, diversity, system, and continuity. After the database to determine topic selection, information material acquisition around topics that apply to complete and systematical as much as possible on the one hand, the library will need about this characteristic literature by printing document converted into digital document. On the other hand, the library also widely collects information resources of the topic selection. To this end, we need to open up the source documents [8].

41.5 Copyright Issues Should be Noted in the Construction of Characteristic Database

41.5.1 The Copyright of Collection Digital of Information Resources

The digitization of the existing collection of document resources is the ways and means of the construction of the characteristics database [9]. China's intellectual property protection provisions have not yet clearly defined the law of the behavior of the digital qualitative, and academics generally believe Digitization is copyright to copy, on the grounds there are two main reasons, the first, in 1996, the World Intellectual Property Organization proposed "The protection of the substantive provisions of the treaty of Literary and Artistic Works" and the U.S. national information infrastructure to promote the White Paper published by the working Group on "Intellectual Property and national information infrastructure" are clearly identified works digitized are copied; the second one is according to China's newly revised "Copyright Law" Article 10 (5) provides that the "Copy refers to printing, photocopying, lithographing, sound recordings, video, rip, copy, etc. will work to produce one or more acts" Although this definition does not included digitization, the properties of replication behavior (subjective purpose, a certain way, labor characteristics) point of view, digitization of behavior fully comply with such a definition. Collection Digitization of ownership of copyright in reference to the relevant provisions of the reproduction rights to perform, the other person without the permission of the copyright, not the implementation of the original works of digital behavior, otherwise it will result in infringement.

On the basis of China's "Copyright Law" to divide the library's collection of literature, it can be divided into two categories: one is in the public domain literature; the other one is nonpublic areas of the literature. The involved copyright issues in these two types of literature digital are different. "Copyright Law" requires public domain works, including works (such as ancient books, rare books, etc.) over the term of protection (50 years), are not in the copyright protection within the geographical scope of works and NA of the copyrighted work (such as laws, regulations, current events, etc.), the reproduction rights of this part of the literature is no longer dominated by the copyright holder, no longer protection, so libraries need to be digitized without causing infringement [10].

41.5.2 The Copyright of Database Resources

Copyright issues involved in the integrated development of literature resources in the database should be taken to different solutions for different types of databases [11]. In general; the database can be divided into three types: one is not entitled to the copyright file database. For this type of database, database developers just own the copyright, but have no rights on the database of information materials. When the library collects database characteristics of database construction, it must obtain the permission of the copyright owner, by paying the necessary compensation [12]; the second is a database of copyrighted files. For this type of database, database developers just have the copyright of the database, the Library database developers have to pay for labor and remuneration, and the database of information resources as possible. The third is a hybrid database of copyrighted files and files do not have rights. For this type of database, it is necessary for the library to combine the above two cases during using it, using of different approaches to solve the copyright issues.

41.5.3 The Copyright of Network Resource Utilization

Network of information resources with rich, dynamic, and timeliness of the characteristics of the library collection of network information resources, it is difficult to determine whether to make use of information resources protected by copyright [13]. At present, China is yet to enact intellectual property laws related to the use of network resources; libraries can usually adopt the following two approaches: first, for the social public information, unless the authors make a special statement, otherwise the library can be freely used. Second, the online works protected by copyright, as long as any copyright works, whether it is part use or full use of the library must be authorized, and pay the corresponding remuneration.

The twenty-first century, with the development of science and technology, human society is gradually stepped into the era of information technology, computer technology, modern communications technology, and network technology as the main feature of modern information technology has been developed by leaps and bounds, for literature information resources The share of construction provide a reliable technical support. We have to face the contradiction between the personalization of user information needs and the surge in the number of information resources. This force information services to face reality, to analyze the situation, the construction and sharing of the distinctive feature of information resources on the agenda, and this is the inevitable choice for the Information Society environment. Take full advantage of information technology and a rich collection of resources and building characteristics database for teaching and research services for economic construction and social progress should actively explore the work of University Libraries.

References

1. Liu BW (2008) Discussion on the creation of specialized characteristic database in university library. Sci Tech Inf Dev Econ 1(14):65–66
2. Xu JX (2008) On the establishment of characteristic database of university library. J Anhui Univ Technol (Social Sciences) 2(3):162–163
3. Sun SM (2011) Discussion on the principles and standards of the construction of university library's featured database. Sci Tech Inf Dev Econ 3(27):39–41
4. Tang HP (2009) Research and implementation of the self-building characterized database based on B/S. J Mod Inf 4(2):109–111
5. Dong B (2010) The application of DC metadata in special subject databases–take "multinational corporations research special subject database" of Nankai university library as example. Library Work Study 5(4):42–44
6. Tang ZH (2009) Analysis on the principles of the construction of characteristic databases in universities. Sci Tech Inf Dev Econ 6(12):19–21
7. Lin F (2007) Research on the metadata architecture of subject database. Library J 7(5):68–72
8. Liu Y (2012) Analysis of the present situation of characteristic database construction in academic library—take Hunan agricultural university library for example. Library 8(4):141–143
9. Lu XF (2012) An investigation and analysis of the construction of characteristic databases of university libraries in fujian province. Res Library Sci 9(3):58–61
10. Lu XF (2012) The copyright issue of the copyright in the construction of library's characteristic database and strategies. Sci Tech Inf Dev Econ 10(5):3–5
11. Chen HD, Zhang CY, Chang J (2011) Investigation and analysis of the construction of characteristic database in university libraries of Gansu province. J Mod Inf 11(10):61–63
12. Sun SM (2011) Discussion on the principles and standards of the construction of university library's featured database. Sci Tech Inf Dev Econ 12(27):39–41
13. Zhou Y, Ma SK (2010) The study on progress of the research on characteristic database construction of China's universities. Sci Tech Inf Dev Econ 13(20):156–159

Chapter 42
Hardtop Platform of University Knowledge Sharing

Xia Zhang, Xiong Hu and Xiang Hua Ruan

Abstract Knowledge sharing is the process of being aware of knowledge needs and making knowledge available to others by constructing and providing technical and systematic infrastructure. The paper analyzes the contents of teachers' knowledge and the characteristics of Knowledge Sharing, and then under the hardtop and eclipse technology, we designed and built a platform in University Knowledge Sharing. The small-scale experiment proved that it could reduce expenses and promote the university knowledge sharing.

Keywords Knowledge sharing • Hardtop platform • University • Teachers

42.1 Introduction

With development and wide application of the network, the information of Internet has changed pay more and more attention. Now most of the knowledge-based management and management strategy are very genera, this situation goes against the development of knowledge sharing. College as a knowledge center, the knowledge sharing has its own characteristics, sharing platform as a means of promoting knowledge sharing of the university, it was very practical meaning that Knowledge Sharing's level has been improved.

42.2 Present Situation of Knowledge Sharing

College as a knowledge center, the knowledge sharing situation was not very good [1]. In the present situation, investigation found that at present various colleges and universities generally had set up their own campus network, but few have teaching

X. Zhang (✉) · X. H. Ruan
Library of Hubei United University, Tangshan, Hubei, China
e-mail: ZhangXia@126.com

X. Hu
Department of Information Engineering of Tangshan College, Tangshan, Hubei, China

Z. Zhong (ed.), *Proceedings of the International Conference on Information Engineering and Applications (IEA) 2012*, Lecture Notes in Electrical Engineering 220,
DOI: 10.1007/978-1-4471-4844-9_42,

class	subject knowledge	Specialized Knowledge for Study-oriented	knowledge of Learners and education environment
content	substantial knowledge, conceptual knowledge	Associative teaching	the knowledge about students and deficiency about student knowledge
data	Teaching KB, teaching resources (teaching plan, Courseware, Case, item bank ,etc)	Syllabus, teaching schedule, class management, class organization, the reward and punishment, etc	educational theory, teaching thought, Experience in Teaching Medical, Insight, Intuition, mental model, teaching style

Fig. 42.1 Contents of teachers' knowledge

and scientific research personnel. Trends of research and research results of this kind of information are published, but lack in dynamic teaching results to exchange this Shared mode. With the department of basic research activity in stagnation, the teaching research and teaching situation of communication are almost nonexistent [2]. The teacher many knowledge sharing by osmosis, such as kirks way, but in our country's colleges and universities teaching system and communion between teaching tutor system is not perfect, all kinds of corresponding incentive mechanism is very few, the communication between teachers are few and far between. In short, the focus of this paper is to consider the use of advanced technology, for the integration of existing resources, construction of an advanced and efficient sharing platform and ensuring the effective implementation of knowledge sharing.

42.3 Content Analysis of Teacher's Culture of Knowledge Sharing

Teachers as mentors and promoters, master knowledge of various types. According to teachers' working, university teachers' knowledge could be divided into 3 types: subject knowledge, specialized Knowledge for study-oriented, knowledge of learners and education environment [3]. References for Wagner and Novak, we can obtain the scale of teachers' knowledge (Fig. 42.1).

42.4 Content Analysis of Teacher's Culture of Knowledge Sharing

Knowledge sharing is to point to realize the demand of knowledge, technology, and system through the infrastructure construction, the knowledge is provided to other those who need it. Kim Songhua, according to statistical analysis, shows that affect university teachers' knowledge sharing has two levels, the six key factors: relation dimension (cognitive, trust, and open communication and cooperation) and structure size factors (based on their IT infrastructure reward system and communication channel). Cognitive is the most important factor, the direct influence of teachers' knowledge sharing [4]. Incentive system is the second most important factors, the direct impact of the campus material sharing.

University teachers' job to a great extent have the independent side (in the process of teaching, teaching and learning the knowledge transfer the unidirectional), so the sharing is in existence without the support of one side or hesitation; so in the design process of the open should be considered reasonable dimension, progressive cognitive support, effective incentives, and support trust environment. This is the platform which should be considered in planning.

42.5 Solution of Network Configuration Based on Hardtop

42.5.1 Hardtop

Apache Hardtop is a framework for running applications on large cluster built of commodity hardware. The Hardtop framework transparently provides applications both reliability and data motion [5]. Hardtop implements a computational paradigm named Map/Reduce, where the application is divided into many small fragments of work, each of which may be executed or re-executed on any node in the cluster. In addition, it provides a distributed file system (HDFS) that stores data on the compute nodes, providing very high aggregate bandwidth across the cluster. Both Map Reduce and the Hardtop Distributed File System are designed so that node failures are automatically handled by the framework.

42.5.2 Solution of Network Configuration Based on Hardtop

Hardtop is Master/Slave Structure Master: The Master (Name Node) manages the file system namespace operations like opening, closing, and renaming files and directories and determines the mapping of blocks to Data Nodes along with regulating access to files by clients [6]. (Data Nodes) are responsible for serving read and write requests from the file system's clients along with perform block creation, deletion, and replication upon instruction from the Master (Name Node).

In the experimental process using a machine as a Master, responsible for Name Node and Job-Tracker work, two machines as a Slave, responsible for Data Node and Task Tracker work, a machine as a development test environment, database has ubuntu2 computer. Detailed configuration see Fig. 42.2.

42.5.3 Configuration Hardtop Environment

Establish Master to each one of the Slave by SSH letter certificate. Due to the Master will start by SSH Slave all the Hardtop, so need to build a one-way or two-way certificate guarantee command not need to input password [7]. In the Master and all the Slave machines on execution:

Host name	Hardtop role	IP	Hardtop jpg
Unbent	Master slaves	192.168.20.212	Name Node Job Tracker
Ubuntu1	slaves	192.168.20.214	Data Node Task Tracker
Ubuntu2	slaves	192.168.20.237	Data Node1 Task Tracker
Windows	Development test environment	192.168.20.240	

Fig. 42.2 Environment of configuration

#ssh-keygen-trash-p“” –f ~/.sash/id dash
#cat ~/.sash/id_dsa.pub >> ~/.sash/authorized keys
#chimed 755 ~/.sash/
#chimed 644 ~/.sash/authorized keys
Then will every machine ~ /. SSH/id dash.
/.sash/authorized keys

42.5.4 Modification of Files

Hardtop Site Xml
<Property>
<name>fs.default.name</name>
<Value>hdfs://localhost:9100</value>
</property>
<Property>
<Name>mapped. Job Tracker</name>
<Value>localhost: 9101</value>
</property>
<Property>
<Name>doffs. Replication</name>
<Value>1</value>
</property>
Conf/core-site.xml
<Configuration>
<Property>
<name>fs.default.name</name>

<Value>huffs: // local host: 9000/</value>
<Name>hardtop tmp.dir</name>
<value>/hadoop-0.20.2/tamp</value>
</property>
</configuration>
Conf/mapred-site.xml
<Configuration>
<Property>
<Name>mapped. Job Tracker</name>
<Value>local host: 9001</value>
</property>
</configuration>

42.5.5 SSH Access

The husker user on the master (aka husker@master) must be able to connect (a) to its own user account on the master–i.e. sash master in this context and not necessarily sash local host—and (b) to the husker user account on the slave (aka husker@slave) via a password-less SSH login [8, 9]. If you follow my single-node cluster tutorial, you just have to add the husker@master's public SSH key (which should be in $HOME/.sash/id_rsa.pub) to the authorized keys file of husker@slave (in this user's $HOME/.sash/authorized keys). You can do this manually or use the following SSH command:

Hauser @master: ~$ sash-copy-id -I $HOME/.sash/id_rsa.pub husker@slave

On your master node, try to sash again (as the hardtop user) to your local host and if you are still getting a password prompt then.

$ chimed go-w $HOME $HOME/.sash
$ chimed 600 $HOME/.sash/authorized keys
$ chow whom $HOME/.sash/authorized keys.

42.5.6 How to Build and Install the Plug-In

When compiling Hardtop, the Eclipse plug-in will be built if it founds the Eclipse environment path in the ant property "eclipse home". The build framework looks for this property in ${hardtop-sac-root}/sac/contrib./eclipse-plug-in/build. Properties and in $HOME/eclipse-plug-in build properties.

A typical $HOME/eclipse-plug-in build. Properties file would contain the following entry: eclipse. Home=/path/to/eclipse.

Then the plug-in should be built when compiling Hardtop: ant clean package (from the ${hardtop-sac-root} directory), which will produce {hardtop-sac-root}/build/contrib./eclipse-plug-in/hardtop-${version}-eclipse-plugin.jar.

42.5.7 Key Code

Public class DBA chess Write public static void main (String rags)

Job Conf = new Job Conf (DBA chess Write class);
Conf set in put Format (DB In put Format class);
Conf set out put Format (DB Out put Format class);
DB Con figuration configures DB (conf, "com I jobs Driver" "jobs: mysql ://192.168.19.237:3306/test", "root", "123456");
String fields = {"id", "title", "content", "urn"};
DB In put Format set Input (conf, Blog class, "blog", null, null, fields);
DB Out put Format set out put (conf, "blog", fields);
Conf set Map per Class (DB Map per Reader class);
Conf set Reducer Class (DB Map per Write class);
Job Client runs Job (conf);
Public class DBA chess Reader
Public static void main (String rags)
Job Conf = new Job Conf (DBA chess Reader class);
Conf set out put Key Class (Long Writable class);
Conf set out put Value Class (Text class);
Conf set in put Format (DB In put Format class);
File Out put Format set Out put Path (conf, new Path ("doubt put")); DB Con figuration con figure DB (conf "com myself jobs Driver" jobs: mysql://192.168.19.237/test, "root", "123456");
String fields = {"id", "title", "content", "urn"};
DB In put Format set Input (conf, Blog class "blog", "id = 1", "id", fields);
Conf set Map per Class (DB Map per Reader class);
Conf set Reducer Class (Identity Reducer class);
Job Client runs Job (conf);

Acknowledgments The paper is sponsored by the Subject technology in Hubei.

References

1. Peter H, Candy S (2011) Viewing libraries from the perspective of multiple stakeholders. Library Inf Sci Res 1(3):101–102
2. Counts S, Fisher KE (2010) Mobile social networking as information ground: A case study. Library Inf Sci Res 2(12):98–115
3. Gerhard X (2011) A lattice matrix method for hyper spectral image unfixing. Inf Sci 3(10):1787–1803
4. Cheng, HY, Wang WJ (2011) Reversible. Steganography based on side match and hit pattern for VQ-compressed images. Inf Sci 4(11):2218–2230
5. Quant KP (2011) A discrete artificial bee colony algorithm for the lot-streaming flow shop scheduling problem. Inf Sci 5(8):455–468

6. Ahmad A Abu (2010) Information security governance in saudi organizations: an empirical study. Inf Manage Comput Secur 7(18):4–7
7. Choy N, Kim D (2010) Knowing is doing: an empirical validation of the relationship between managerial information security awareness and action. Inf Manage Comput Secur 8(4):84–89
8. Archon V (2010) Science data scandals put spotlight on info practices. Inf World Rev 6(3):1–6
9. Vacant WR (2012) E.H.E-government implementation strategies in developed and transition economies: a comparative study. Int J Inf Manage 9(6):70–74

Chapter 43
Study of Music Design Based on Computer Fractal Technology

Xiaomei Qi

Abstract Fractal technology is not only used in art image formation, but also can be used for image compression. It is an important application in the computer industry, chemical industry, and even in the economic aspects. For the creation of the music domain, it also has the fractal characteristics. Therefore, this chapter has introduced the fractal technology based on the perspective of the computer fractal technology. The automatically generated music, image processing, compression editing, test, and other aspects are analyzed; it can be better applied in the field of music, so as to achieve a high quality, high definition music effect.

Keywords Fractal technology • Algorithms in music composition • Music fractal • MIDI

43.1 Introduction

With the progress of society and the rapid development of information, music production changes from the simple and nature sound to rich clear, and colorful pictures with high quality sound, video, audio, and other effects in the fusion. Faced with further strengthen in the music generating technology, we have proposed an application based on the modern information technique in this paper. The computer fractal technology can be fully utilized in refining the fractal music [1]. The music becomes multiple iterative music effect based on fractal technology from the initial note scale with a time of fluctuation. It can realize the macroscopic and microscopic real music in very vivid simulation creation way.

X. Qi (✉)
Dongying Vocational College, Dongying 257091, China
e-mail: xiaomei_qi21@yeah.net

Z. Zhong (ed.), *Proceedings of the International Conference on Information Engineering and Applications (IEA) 2012*, Lecture Notes in Electrical Engineering 220,
DOI: 10.1007/978-1-4471-4844-9_43,

43.2 Concept of Fractal Technology

Fractal technology is a special way based on the whole and local relationship between processing, and uses the space of symmetric structure and its similarity [2]. It has created out of view and rich picture with full of vigor and vitality and imagination artistic realm and charm by a variety of image simulation.

At present, the most famous analysis technology is Koch-Kurve (Koch curve). It is actually a mathematical curve that has been divided into three equal segments [3]. It is an equilateral triangle in the middle of the three divisions for the replacement, so that repeated replacement can obtain an infinite endless repetition of results. Koch curve tends to infinity based on the length of angle. Because the curve is processed in many repetitions that increase by 4/3, after n times it becomes $(4/3)^n$. The curve is shown in Fig. 43.1. When the n is infinite, then curve length approaches infinity. Fractal dimension's calculated result is 1.26.

Through the study of Julia set C (constant) has added into another calculation method. It closes to the answer, and will have a Nova's fractal graphics. It is shown in Figs. 43.2 and 43.3.

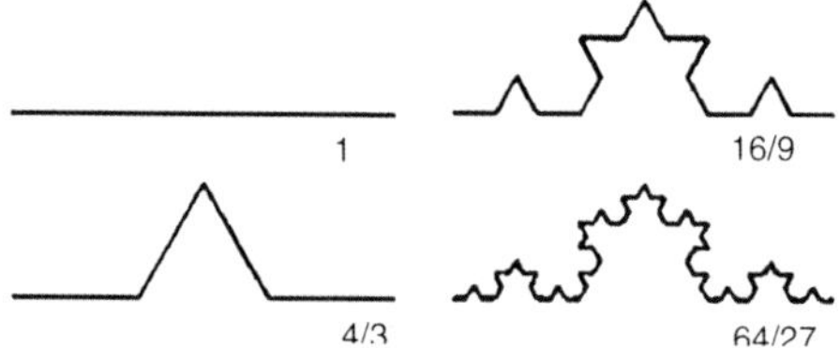

Fig. 43.1 Koch curve

Fig. 43.2 Newton fractal

Fig. 43.3 Newton fractal

Application of fractal geometry in musicology is a favorable tool for music production and processing. It is a major direction of music course [4]. With the help of computer fractal calculation, simulation of making beautiful music data comes from the fractal calculation.

43.3 Music Elements in Fractal Mapping

Image fractal is a mapping relationship that has calculated values and the color value based on fractal mathematics. That is composing method and note value through the algorithm [5]. Due to the different mapping methods, the algorithm obtains different mapping formula. In 128 bit, 0–127 are MIDI file in note value range. Central C represents 60. Due to a number of particularly high or low uncommonly used notes, we do not use numerical method for note value method. 29 between 73 are the main sound symbol that is a set of stored data. By using the calculation method of fractal dimension, the results of index number numerically determine the play notes.

In fractal image production process, the classical algorithm of Mandelbrot is used. It is a collection of graphics polynomial (f (2) = z (2) + c). Through the iteration in the parameter plane image, iterative formula is shown as follows [6]:

$$Z_{n+1} = Z_n^2 + c, \; z, \; c \in C \tag{43.1}$$

Decomposition is shown as follows:

$$x_{n+1} = x_n^2 - y_n^2 + a \tag{43.2}$$

$$y_{n+1} = 2x_n y_n + b \tag{43.3}$$

Through a fixed initial point, the starting point is *C*. After several iterations, numerical results are: (1) number is infinite, and does not belong to the *M* set; (2) the iterative results are in a range of areas, the initial point is always within the collection. In the process of calculation, the iterative calculation of plane point runs *K* times. Then *C* is colored and generated image branch. Based on this method, fractal music is increased. The music is the one-dimensional number; it can be obtained in the straight line for iterative calculation of *K* value. Then the fractal music is determined by *K* value.

43.4 Music Production Based on Mandelbrot Fractal Music Algorithm

Notes are identified by random range of complex set operations. Based on Mandelbrot fractal set, complex categories are defined before playing the music; and operational method is provided. Then a music player and a voice signature array are defined, and by the same time a point set of operations are defined. Through the numerical iteration, the real and imaginary parts of the plural of numerical selection index get value array notes. By using the Internet Explorer browser, the fractal music is processed in Microsoft XP computer operation system.

Fractal music playback code procedures are shown as follows [7]:

```
Public void play [//Fractal music playback mode]
Try Random c=New Random (system. Current time millis; // According to the
computer system, it gets time random parameters
Complex r, z, zold; // Zn+1, Zn and R were defined.
Double nonmr, normi; // the real and imaginary part of the plural
Imt pitchr, pitchi;
While (stop); // Plural range is [−2, 2]
C=New complex (nextdouble*3.0–2.0)
Z=new complex;
Zold=z;
Repeetold=−2;
For (int k=1;k<iterations;k++); // Iterative operation
Pitchold=−1;
If (modulus(z)>3) [break;]//Normalization is [0, 1]
z=(z.times (x)). plus(c);
nom=(real()+2.0)/3; // Tone index number
bitchr=(int) (non*(scale2_length)); // Playing a tone
Thread. skep(200);
Catch (exception)
Print stack;
```

43.5 Application in the Music Based on Fractal Technology

Through the analysis of technical study notes, beating motion constitutes rules of fractal; but a periodic motion can also constitute the random fractal. Such as the butterfly paragraph in "Butterfly Lovers", the melody is identical; and the bass melody and rhythm change at the same time (Figs. 43.4 and 43.5).

Fig. 43.4 Score note changes the *line* chart

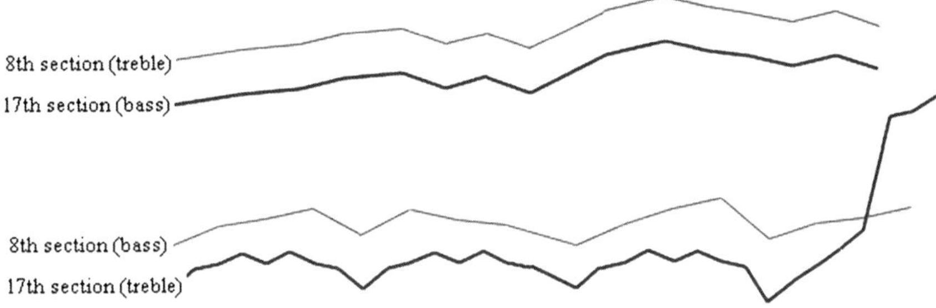

Fig. 43.5 The soprano and bass line of music section

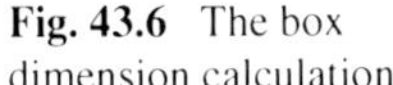

Fig. 43.6 The box dimension calculation

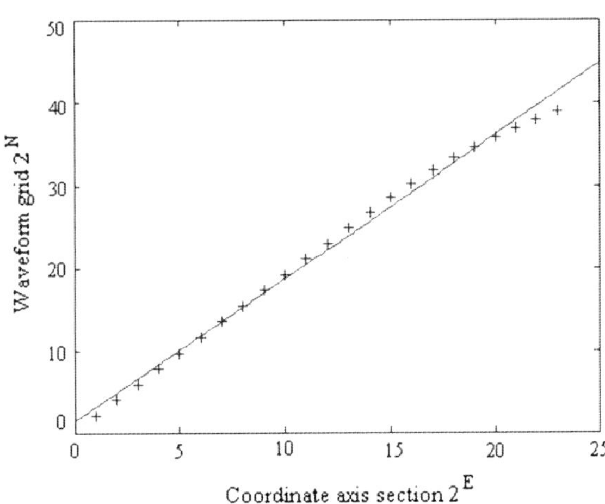

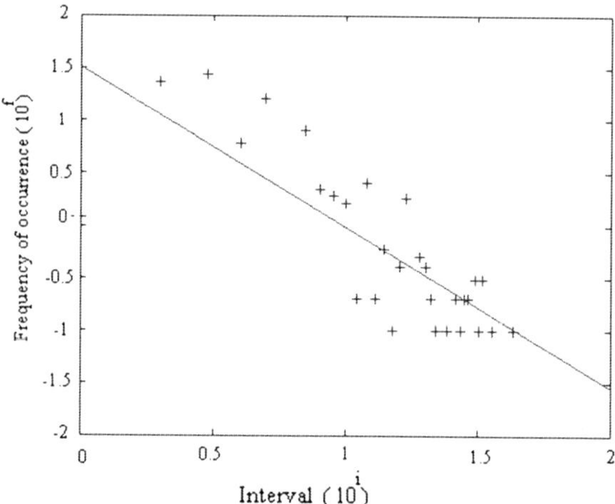

Fig. 43.7 The score calculation

Through comparing the treble line, part of treble has the same shape, but there are great differences in the bass; but it is not similar with fractal form. Because its form is very complex, and the notes of the audio signal using the box dimension mathematics calculation method get the following image [8] (Fig. 43.6):

The map shows that slope by calculation of the linear image, then get the fractal dimension was: $D = 1.5362$. Then, we compute the new statistics of the music. Through a linear fitting, straight line is shown as Fig. 43.7.

The fractal dimension is: $D = 1.5276$

According to the comparison and analysis of its fractal dimension, the difference between notes is very small. All tones show an expression of grief and two types of emotional color depend on the fractal dimension value.

43.6 Conclusion

Music is a combination of modern art and science, and fractal technology is a new technology of the computer. This article has analyzed the music creation process, the relationship between the two aspects and the music fractal characteristics from the computer fractal technology perspective. According to the image data of examples, computer fractal technology greatly enriched the modern music production. Through this technology, we can realize the vivid music, vivid effect, and achieve high grade, and high quality music effect. Fractal technology is not only the fully displayed natural scenery, but also shows remarkable art visual effects with sight and dynamic effective combination.

References

1. Qi D (2009) Fractal and computer generation, vol 03. Science Press, Beijing, pp 563–566
2. Jiang W (2010) Chaos, fractal and music. The new voice of yuefu 5(2):8–17

3. Yang R (2009) Music analysis and composition, vol 3. People Music Press, Beijing, pp 313–315
4. Yu A, Yu CJ, Yu SS, Zhou J (2011) WFA algorithm and its application in image compression based on the fractal geometry. Comput Appl 3(1):117–123
5. Zhang Y (2010) 1/f fluctuation theory and 1/f music. J Liaoning Educ Administration Inst 2(1):201–209
6. Guo D (2011) Fractal geometry in computer graphics applications. Mech Des 3:55–60
7. Bulmer M (2010) Music from fractal noise. J Univ Queensland 01:5–9
8. Hsu KJ, Hsu AJ (2010) Fractal geometry of music. Proc Natl Acad Sci 4:938–942

Chapter 44
Data Acquisition and Processing on Environmental Geophysics

Zhizheng Zhou

Abstract With the development of geophysics in China, the further development of environmental geophysics made great progress in our country and it has made great achievements. However, current environmental geophysics has not summed up the unique method with characteristics. This chapter first analyzes the environmental geophysics data acquisition and processing method, based on the characteristics of environmental geophysics, and then puts forward the methods to improve environmental geophysics data acquisition accuracy and precision of the method of preliminary results.

Keywords Environmental geophysics • Data acquisition • Data processing

44.1 Introduction

Environmental science has been mixed with the geophysics after a long time of developing separately. The first time people merged these two is that when some American scientists try to study the Environmental by geophysics methods [1]. And in 1985, American society of exploration geophysicists (SEG) listed the Environmental Geophysics as an independent subject when they analyzed their work annually. And then, more and more related academic discussions, conferences, and papers came out after that time [2]. All of these are the signal of development of Environmental Geophysics. The similar study in China has started since the end of the last century. Even though it is a little bit later, a lot of related work and great achievements have been done in the last 20 years, such as collected papers and specific seminars [3].

Z. Zhou (✉)
School of Computer and Information Engineering, ShanDong University of Finance and Economics, Jinan 250014, China
e-mail: yangzz_13@163.com

Z. Zhou
School of Geophysics and Information Technology, China University of Geosciences, Beijing 100081, China

Z. Zhong (ed.), *Proceedings of the International Conference on Information Engineering and Applications (IEA) 2012*, Lecture Notes in Electrical Engineering 220, DOI: 10.1007/978-1-4471-4844-9_44, © Springer-Verlag London 2013

The faster development of economic and more concern of environment quality, more important the Environmental Science and the Geophysics are. This forces the Environmental Geophysics to keep moving. In terms of the world's development, Environmental Geophysics will play a more important and global role in the urban construction, infrastructure construction, environmental protection and control, and national security sectors [4].

The economy grows rapidly since the economic open policy in our country. And we proposed the strategy of sustainable development many years ago; however, the fact now is better economy, worse environment. All of the environmental pollution, are becoming even worse, and endangering people's lives; including water resources pollution (city rural groundwater, and surface water, and inland river, lake, and offshore water resources), city air pollution, regional ecological environment damage (include grassland sandy, and soil sanitization), electromagnetic pollution, and industrial and the city waste pollution. Although the related departments are aware of the seriousness of the problem and try to do some adjustments, the pollution is much worse than we imagined, and we cannot eliminate the effects overnight, especially some hidden pollution [5]. Generally, the negative effects may cause huge economic losses, and wide area harm for environment. Recently, environmental authorities have publicized the "33211project" and "a single control with dual aims" strategy to improve the environment. However, the specific method is needed if we want to have some knowledge of regional ecological problems, especially when we study the pollution in the deep of the earth [6].

44.1.1 Environmental Geophysics Tasks

Environmental geophysics is a subject that crosses the Environmental science and Geophysics, and geophysics is the major part of it. Environmental geophysics measures the earth's formation, dynamics, and other geological phenomena by quantitative physical methods, which covers the study of the Earth's hydrosphere and atmosphere, (for example, the ionosphere Dynamo, aurora electro jets, and magnetopause current system), and even the physical properties of other planets and their satellites. Environmental science is to study the environment, and the solution of environmental problems, which incorporates more of the social sciences for understanding human relationships, perceptions, and policies toward the environment [7]. Environmental engineering focuses on design and technology for improving environmental quality [8].

Environmental geophysics is using the basic theory and method to study the relationship between physical characteristics of earth and human surviving environment (include natural and artificial environment). This relationship includes the effects on human surviving environment by geophysical field, and the artificial environment of changes due to natural or human actives [9], In General, geophysics, especially exploration geophysics (or applied geophysics), whose essence is to find the variety of mineral resources from the lithosphere. In contrast, the major task of environmental geophysics is to track the destination and consequences of the developed resources on the Earth (especially the distribution within the biosphere). Thus, we may believe that Environmental Geophysics is the development of exploration geophysics [10].

On the Environmental Geophysics seminar, which was held by Department of Earth Sciences in October, 1997, the committee recommended some research areas based on the urgent problems of Environmental Geophysics:

(1) Environmental effect of geophysical field;
(2) Geophysical methods of Environmental pollution (Especially water pollution) Effective monitoring;
(3) Geophysical methods on ecological environment and geological hazard monitoring and forecasting;
(4) Geophysical study of global change.

In addition, Jiang, Ornubu, and Cui Lin Pei, also put forward their point of view. Summarizing these ideas, environmental geophysics research contents should include:

(1) Monitoring of Environmental disasters caused by natural forces, as well as the observations of environmental change.
(2) Monitoring and investigations of the atmosphere, hydrosphere, and lithosphere (Soil) pollution raised by human life.
(3) Study and measurement of radiation, noise, and vibration pollution caused by economy development and traffic-building.
(4) Global changes caused by anthropogenic emissions.

44.1.2 Current Achievement of Environmental Geophysics

In recent years, environmental geophysicists in the world have done a deal of fruitful work in regional eco-environment, water treatment, and landfill monitoring using geophysical techniques [11]. All of the work is related to the urban construction, infrastructure construction, environmental protection, national security, and so on. With the constantly expanding application area, the content of environmental geophysics covers engineering construction, water resources, geothermal resources exploration, evaluation and forecast of disaster, and environmental pollution detection and monitoring, construction quality detection, and archeological research. It even extends to the ecological agricultural and biological growth monitoring. Environmental geophysics is more and more close to daily life and national development. For these existing difficult problems in environment and engineering Earth physical exploration, many of them can be better solved, such as reservoir dam seepage problems, detection of sandwich wall, advanced prediction of tunnel issues and active fault detection [12].

44.1.3 Advantage of Environmental Geophysical Methods

There are a number of ways to detect and resolve environmental issues. Just like the theory and numerous works instances showing that, compared with other methods, the environmental geophysical methods have the following advantages:

(1) *Low investment and quick, high economic benefit*. Especially as a means of long-term monitoring, real time, and short cycle time [13].
(2) *No or minimize secondary pollution*. Especially compared to conventional drilling method, this can keep pollutants away from going deep as drilling continues, to avoid more pollution.
(3) *Wide range of applications*. Geophysical methods can speed up the first class environmental issues, including calculating the range of seismic activity, area and speed of ground subsiding, scope and distribution of groundwater pollution and seawater intrusion, early-warning and prediction of the Earth's crust. In addition, geophysical methods have unique advantages for research in marine environment and the pole. And for the subjects which are not allowed of destructive detection, we can do the nondestructive detection by Geophysical methods.
(4) *Predictability*. We can predict the direction and speed of pollutants' moving in the media, and providing scientific and effective environmental assessment data.

44.2 Methods and Basic Principles of Environmental Geophysics

Compared to the subjects' traditional Geophysics, the subjects of Environmental Geophysics are usually on the ground surface, more concentrated in the industrial centers and large cities, and few physical differences with the surrounding objects. There are several commonly used methods of Environmental Geophysics, like magnetic methods, electrical methods (including DC electrical method, electromagnetic method), and using ground-penetrating radar. The effective combination of different approaches or methods is chosen according to the different investigation targets.

44.2.1 Magnetic Prospecting

Magnetic exploration is one of the important methods of geophysical exploration. All kinds of rocks and ores have different magnetic nature which may cause geomagnetic anomalies in its local area. Magnetic prospecting is to discover and study these magnetic anomalies in turn looking for magnetic ore, studying the geological structure, and solving environmental problems. The high-precision magnetic prospecting is a new technology developed in recent years, the sensitivity of which is greatly improved. For example, the sensitivity of G858 cesium magnetometer has reached 0.01 nT.

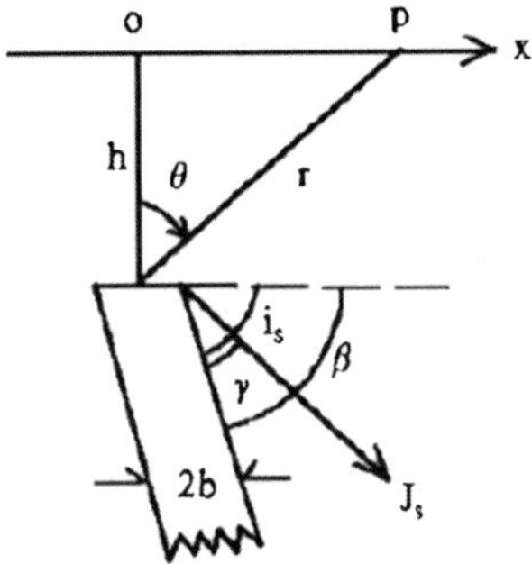

Fig. 44.1 Method of infinite extension of deep-sheet bias magnetization

The following is the method of infinite extension of deep-sheet bias magnetization solving the magnetic field of the magnetic body (Fig. 44.1).

$$Z_a = \frac{2m}{x^2 + h^2}(h \cos \gamma - x \sin \gamma) \tag{44.1}$$

$$H_a = \frac{2m}{x^2 + h^2}(x \cos \gamma - h \sin \gamma) \tag{44.2}$$

$$\Delta T = \frac{2m}{x^2 + h^2}\{(h \cos \gamma - x \sin \gamma) \sin I_0 - (x \cos \gamma + h \sin \gamma) \cos I_0 \cos \delta\} \tag{44.3}$$

In the formula: $m = 2bJ_s\sin\beta$, $\gamma = \beta - i_s$, I_0 is the geomagnetic declination, δ is the geomagnetic inclination.

44.2.2 Electrical Prospecting

Electrical prospecting is a method that analyzes and interprets the characteristics and principals of artificial and natural electric field, or alternating magnetic field, based on the different electromagnetic and electrochemical properties between rock and ore. Electrical prospecting is widely used. High-density resistivity method, three dimensional DC method, and transient electromagnetic method are some common methods. As a new array exploration method, high-density electrical method is widely used in environmental monitoring recently, especially in the investigation of geological disaster, crack, and hidden danger of reservoir dam.

The following is the theoretical basis of the high-density electrical: Manually applying DC to the underground, underground electric field distribution can be observed by instrument on the surface. The electric field distribution to satisfy the differential equation

$$\nabla^2 = \frac{-I}{\sigma}\delta(x - x_0)\,\delta(y - y_0)\,\delta(z - z_0) \tag{44.4}$$

In (44.4): x_0, y_0, and z_0 are coordinates of the Power supply; x, y, *and* z are coordinates of field point.

The method to get the resistivity of high-density electrical prospecting is by applying DC via A, B pole and measuring potential difference between M and N pole, the apparent resistivity values of x point can be figured out using the formula 44.5:

$$\rho_s = \frac{\nabla V}{I} \cdot K \tag{44.5}$$

K is device coefficient: $K = \frac{2\pi}{\frac{1}{AM} - \frac{1}{AN}\frac{1}{BM} + \frac{1}{BN}}$

44.2.3 Ground-Penetrating Radar

Ground-penetrating radar is an efficient shallow detection method, which detects objects by analyzing the amplitude, frequency, and shape of the echo of the high-frequency pulse wave based on the electrical differences between the media. Ground-penetrating radar is used to detect subsurface or nonmetal structure of the building by using high-frequency electromagnetic waves. Internationally, use of GPR should date back to earlier 1970s, twentieth century; it was mainly used in nonpermafrost ice environment research at that time. Compared with the traditional ground-penetrating radar, the phased-array ground-penetrating radar can detect in much deeper, and has better signal noise rate and higher resolution. Figure 44.2 shows the working diagram of the geological radar profile method.

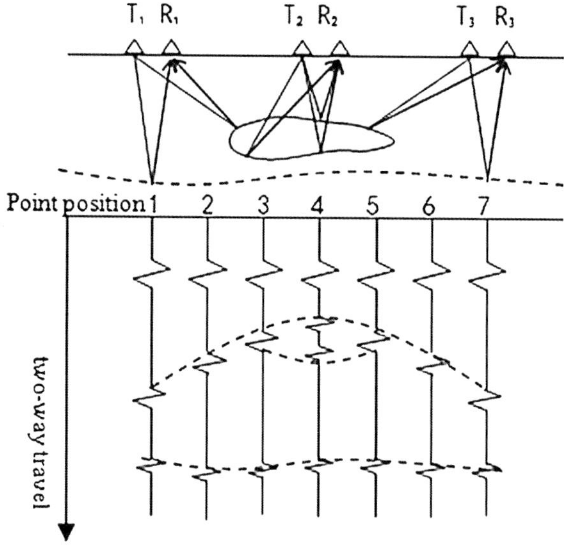

Fig. 44.2 Working diagram of the geological radar profile method

44.2.4 Radioactive Methods

There are various radioactive substances in our life, such as stones in home decoration materials, ground water, and radon in the soil, phosphate fertilizer production, and nuclear waste. The existence of these radioactive materials will cause some pollution to our environment.

Radioactive method is to study the environment through capturing ray of radioactive material and detecting the strength. According to the ion captured in the process, radioactive method can be divided into alpha ion detection method, beta ray method, gamma ray method, neutron detection method, Gamma ray method and neutron detection method are more common.

In addition, gravity method, acoustic methods, and other methods are playing an important role in the environmental Geophysics, and yielded fruitful results.

44.2.4.1 Status of Environmental Geophysical Data Acquisition and Processing

Data is the core of scientific research. Data collection, preprocessing, and output the results, each step determines the final result of environmental Geophysics work.

Quality of data collection, which relate to the exploration equipment, collection methods, and so on, is very important in the whole process. The higher the signal noise rate of the original collected data, the better stack for data processing section.

Existing environmental geophysical data acquisition devices, adopt embedded technology and GPS technology, and now no matter what method the exploration equipment is using, the computer technology is used more or less, especially later in the data acquisition and data processing. However, single board computers are currently used on the data acquisition device, especially on the sensor port, which is not good in terms of processing speed, operating frequency, and response. To a certain extent, this affects the accuracy of data collection, even sometimes poses a near-end crosstalk. Although we can reduce the effects of interference factors in the data collection in some ways, we cannot completely eliminate these distractions. Therefore, the study on how to achieve high signal-to-noise ratio and high resolution results in data acquisition has become even more urgent. This requires adjusting measures corresponding to the local conditions, to provide effective exploration data. With the development of environmental Geophysics, we must improve the quality of acquisition.

Including Environmental Geophysics, geophysical data are often in huge amount. With the increasingly high demands on accuracy of data collection and data volumes increasing, also the data processing should be updated.

44.3 Development and Prospect of Geophysical Data Acquisition and Processing

Environmental geophysics mainly works in the shallow surface, so the data collection is different from mineral exploration. Mineral exploration has to go to the underground to dozens of meters or more than 10 km deep in a wide area, therefore, the related collection data are in large amount. Data acquisition of Environmental geophysics is often limited to a range of areas, and with much smaller depth, however, higher accuracy is expected, and more interference, especially the interference factors near the ground, compared with the mineral exploration. Therefore, strengthening data acquisition and processing needs to be done in the following aspects:

(1) Enhancing the study of new methods and technologies. When pollution has just begun, physicochemical characteristics of polluted and unpolluted media are too similar to distinguish them by regular geophysical methods. Therefore, we need to improve the sensitivity of some geophysical methods, present new arguments, and new research methods, such as wavelet analysis, artificial neural networks, high-order statistics, inversion, and artificial neural networks, and so on.
(2) Establishing a database of petrophysical parameters, which can be shared effectively between research institutions by some kind of mechanism. Study on physical properties of various pollutants, and establish related models and database.
(3) Strengthening the development of new special equipment detector, which can improve antijamming ability, signal acquisition, and resolution capabilities.
(4) Using an integrated Environment Geophysical exploration method. Field investigation involves with multiple disciplines, for some area, due to the complexity of surface geological conditions, single physical method is often hard to succeed, and multiple methods should be taken at this time (For example, heavy magnetic electric quake joint inversion, takes the advantage of different Geophysical methods.
(5) Strengthening program efficiency and fault-tolerant technique of computer software.
(6) Taking the watchdog's design into account, strengthen self-recovery ability of data acquisition equipment, reasonable distribution of upper and lower computer function.
(7) Establishing quality control and evaluation system of Environmental Geophysics data acquisition under different environmental conditions.
(8) Making good use of latest computer technology, taking advantage of the high-performance computing and parallel processing, using high-performance processor in the data acquisition equipment to improve the accuracy of the data acquisition and the efficiency of data processing.

44.4 Summary

Environmental Geophysics is an emerging discipline, and I have only a preliminary understanding of the whole environmental geophysical system. I focused on related areas of embedded technologies and distributed system in the earlier time. In order to get better understanding of Environmental Geophysics, I will try to figure out how to combine professional experience and Environmental Geophysics as my future work.

References

1. Wang H, Chen J (2010) Studies on geophysical software integration environment. Oil Geophys Prospect 2:234–236
2. Zhang H, Sun Z (2012) Geophysical methods in environmental monitoring application status. Northern Environ 1:56–59
3. Wei J, Liu Y, Qiu Z, Yue C, Shiming P (2009) Development of data platform for integrated seismic data interpretation. Geophys Prospect Petrol 2:121–124
4. Wang J, Chao C, Xia J (2004) Progress in environmental and engineering geophysics. Proceedings of the international conference on environmental and engineering geophysics (ICEEG), vol 2. Science Press, China, pp 376–379
5. Glistening J (2004) A session of the international conference on environmental and engineering geophysics held in Wuhan. Adv Geophys 19(1):295–300
6. Kong Y (2009) Research on the GeoVideo data model and its application development. Geogr Geo-Info Sci 5:87–89
7. Cast H (2009) Development of geophysical science report, vol 3. China Science and technology press, Beijing, pp 246–249
8. Cheng Y, Yang J, Zhao Z (2007) Present situation and development of environmental Geophysics, vol 4. Progress in Earth Science, Beijing, pp 363–365
9. Zhang Z (1989) Of Earth to learn a new field of application of geophysical survey-environment. Adv Geophys 4:25–28
10. Jiang H (1997) Environment and geophysics, vol 3. Seismologic Press, Beijing, pp 61–67
11. Chen Y, Baixun X (2005) On the present situation and development of ground-penetrating radar. J Eng Geophys 4:561–563
12. Hong TQ, Fang F (2007) Research and development of portable instrument for measuring gamabased on hardware platform of ARM9 Hedianzixue Yu Tance Jishu, vol 3, pp 31–33
13. Cheng Y, Yang J (2005) Introduction to environmental Geophysics. Geological Publishing House, Beijing, pp 2132–134

Chapter 45
Detecting Approximately Duplicate Records in Database

Xingrui Liu and Lijun Xu

Abstract The existing database system data quantity is huge, many of which are repeated data. Using the traditional approach for detecting approximately duplicate records to find similar duplicate records in the database will involve very large time complexity and space complexity, unable to obtain very good results. This chapter presents a method based on improved genetic neural network approach for detecting approximately duplicate records, using genetic algorithm to optimize the network's initial weights; and then using the BP algorithm to train the detection data to obtain network model. The experimental results show that this method can effectively solve the huge amount of approximately duplicate record data detection problem.

Keywords Approximately duplicate records • BP algorithm • Neural network • Initial weights

45.1 Introduction

Due to the rapid development of IT, the system integration has been widely used in various fields. Many types of dirty data caused by problems exist in the existent database systems: duplicate records, data entry errors, missing values, different units of measurement, and outdated coding and so on. An important issue integrated by different data sources is that the same or similar records in grammar may represent the same record in the real world, in that approximate duplicate records caused by the merging of multiple data sources is a key issue; therefore, repeat information detection and elimination also becomes a research hotspot.

Nowadays, Similar duplicate record detection is largely based on the thinking of the "sort+merge" [1]. Document uses the sliding window method to detect the

---

X. Liu (✉) · L. Xu
Institute of Computer and Information Engineering, Xinxiang University, Xinxiang 453003, Henan, China
e-mail: liuxingrui@xxu.edu.cn

Z. Zhong (ed.), *Proceedings of the International Conference on Information Engineering and Applications (IEA) 2012*, Lecture Notes in Electrical Engineering 220, DOI: 10.1007/978-1-4471-4844-9_45, © Springer-Verlag London 2013

similar duplicate record; Document regards each record as a string, sorts the records, and scans sequentially all the sorted records by using the priority queue with the fixed-size and then clusters them; Document puts forward an approach based on the N-gram to repeat the similar record; Document decomposes each field value into a number of the tokens, and to sort the tokens, then to sort the records, using a sliding window way to compare records in the approaching range. Document, respectively, uses the method of supporting vector machine to detect duplicate records. Finding similar duplicate records in large databases by using traditional similar duplicate record detection will involve a lot of time complexity and space complexity, and because of characters position sensitivity, it cannot guarantee a similar record at the neighboring position when sorting the records, which leads to that algorithm based on the q-gram or "sliding window" clustering or "priority queue clustering" cannot be achieved very good results [2]. On the other hand, these methods have a higher efficiency relatively to deal with the small data sets. But with the expansion of data size, efficiency often cannot further enhance [3]. This paper presents a duplicated records detection method based on genetic neural network, making full use of the nonlinear mapping of the neural network and global optimization features of genetic algorithm. The thinking based on the learning and the evolution will be applied to the detection of duplicate records, which avoids the traditional method to calculate the weights of attributes. The experiments show that this method can effectively solve the large amount of problems of large data sets' duplicate record detection, not only have good detection accuracy, but have a good time efficiency.

45.2 Related Concept Introduction

45.2.1 Genetic Algorithm

Genetic Algorithm is a new way that during simulated biological evolution human beings deals with the complex problems about principle of survival of the fittest and the mechanism of chromosomes' exchanged information among groups. It uses a kind of code to represent the complex structure and names every code as an individual or a chromosome. The population is a muster of codes about the fixed number kept by Algorithm. Human beings could get the high qualified codes by operating each individual in population to imitate the biological evolution. GA' operation of inheritance include cross, variation, and selection. Variation is changing the position of gene according to probability and imitating the variation of hereditary material; the cross is the accidental combination between two individuals according to probability and imitating the chromosomes' exchanging process during sexual reproduction; selection is imitating the survival of the fittest in nature. GA has been widely used in every optimized field and become the focus for the interdisciplinary research [4]. However, there are still some shortcomings and limitations with GA, which are the numbers of iteration, the slowness of convergence, easiness to reduce to the limitation, advance of convergence and so on. There are two questions that should be solved during the research and application of GA: how to overcome the pressure of competition during selection and how to keep the variety of population about GA by crossing variation to improve the searching quality of GA [5]?

45.2.2 Neural Network

The neural network is based on the simplest way to abstract and simulation of the human brain, involving some disciplines such as biology, computer, electronics, and physics, which is a cross-discipline, developed in recent years and has broad application prospects. Neural networks simulating biological nervous system makes interaction response to real world. The broad parallel interconnection networks are composed of some adaptive simple units, constituting the neural network through a large number of connecting neurons [6]. Its basic unit group is neurons and is also the simulation and simplification of biological neurons. Neural network will combine storage and processing of information, eliminating bottlenecks between the traditional calculation of storage and computing and neural network have a strong fault tolerance and robustness, good generalization, analogy, association, promotion, and not having a big impact on overall because of local damage. Neural networks consist of simple processing units of a large number of parallel and distributed processors, this processor has the natural characteristics of the storage and application of empirical knowledge, and this characteristic can be summarized as two aspects, first, the knowledge gained through the weights stored in the neural network can be extracted for use; second, gaining knowledge from the external environment through the learning process using neural network which also has a self-learning ability, and can constantly improve themselves through learning in the new environment and adapt to environmental changes. Therefore, the neural network has a very wide application prospects on many aspects, such as function intended, target recognition, and automatic control.

45.3 Improved Genetic Neural Network for Detecting Approximately Duplicate Database Records

In practical applications, the records in large database are enormous; so we use neural networks to solve record matching problem, which requires a flexible approach. First of all, we trained neural network according to the sample data of the matching records and recorded matches, that is, first calculate all matching records in the sample data and nonmatching records in similarity of the common properties, and then regarding matching records and nonmatching records in similarity of respective property as input of neural network to train. After training, bring the records to be matched in the database in similarity of respective property into neural networks, then judge the records whether they are matched or not according to the outputs. Record matching method based on neural network, its main advantage is that do not need to estimate the property weights directly, but done by the inherent relationship among properties of neural network learning, as the environment changes with strong adaptive capacity. But due to the neural network of received go slow and go into a local optimum easily. This article uses the advantages of genetic algorithms to overcome these problems, and combination of genetic algorithm and BP algorithm, also solve the problem that it takes too long to find close to the optimal solution because of using GA separately. So our training section of the BP neural network is divided into two parts: first using genetic algorithm to optimize the network of initial weights, and then using BP algorithm to detect data to get network models.

Algorithm specific steps are as follows: Suppose there are three layers BP network, WIH_{ij} is the connection weights for the input layer of *i*-node and hidden layer nodes of *j*-node; WHO_{ji} is the connection weights for the hidden layer *j*-node connection and *i*-node of the output. Hi is the hidden layer i-node output; Ii is the i-contact output of the input layer; O_i is the output layer *i*-node output.

(1) Initial population *P*, including the mutation probability Pm as well as for any WIH_{ij} and WHO_{ji} initialization.
(2) Calculating the evaluation function and sort; you can select individual network according to the probability values:

$$p_s = f_i / \sum_{i=1}^{N} f_i \tag{45.1}$$

The f_i is the fitness value for individual *i*, you can measure it by the sum of squared errors *E*, that is

$$f(i) = {}^{1}\!/_{E(i)} \cdots E(i) = \sum_{p} \sum_{k} (V_k - T_k)^2 \tag{45.2}$$

$k = 1,\ldots, n$ is the output layer node; $p = 1,\ldots, n$ is the number of training samples; $i = 1,\ldots, N$ is the chromosome number; T_k is a teacher signal.

(3) The use of probability Pm mutations produces the new individual G_j of G_j'.
(4) Put the new individual into the population *P*, and calculate the individual's evaluation function.
(5) Calculate the error square sum of ANN, if it reaches a predetermined value εGA, then exit, else go to (3), continue to the genetic manipulation.
(6) In the initial value of genetic GA optimization as initialized weights, with BP algorithm training network, until the designated precision εBP (εBP < εGA).

For better applying to different types of character sets, algorithm adopts the convenient internal code sequence value of character to sort handle the record. In order to improve the detection efficiency, according to internal code sequence value of character, using genetic neural network thought, first to optimize the processing of large data sets, and gather the large data sets into more small data sets, then detect separately the similar duplicate records. In order to improve the detection accuracy, according to the important degree of fields, using rating method to calculate every field's weights; besides, by multiple times detection technology, resolving part of the similar duplicate records problem effectively. The general basic ideas of the algorithms are as follows:

(1) According to the actual important fields, the user to the extent specified field level, the system according to the rules of computing the field level finally unification, and finally unification level into the corresponding field weights.
(2) Select key fields or field some place, calculation of field code value selection sequence, and to the entire data set records for sorting, again according to the character encoding sequence value, will be large data set into many small optimization data set.
(3) In the small data using field on matching and effective weighted weights are repeated record similar strategies of detection, and to test out similar marking the repeat record deal with.
(4) Select other key fields or some bits of the fields, and repeat the two steps above. Then you should make the data gathering better newly. After making the data gathering better, you should check the similar and repeated record, and set the repeated mark, avoid that you ignore checking.
(5) Accord to the actual situation, you should divide the every recorded property value into the gathering of having no common in atomies pile, and inherit the coding. While when you divide the value, you should concern about some problems, for example, you should notice the Chinese and western or Chinese and western mixed situation. In order to minimize the record for inputting to bring the spelling errors or repeat the abbreviation for the influence of the inspection records.
(6) Account the similarity of every record in group and represent record property value, and then you should work out the similarity of the two records. When you are accounting, you should compare the two records according to the property value coding from the first step.
(7) Set threshold, and identify all the records of the computing results that go beyond the threshold as the approximately duplicate records. Delete them and repeat the above-mentioned steps until all the records are detected out.

Please note that the first line of text that follows a heading is not indented ("pla" style).

The first lines of all subsequent paragraphs are ("Normal" style).

45.4 Experiment and Results Analysis

We made eight experiments of comparison on these two algorithms, respectively. The experimental data is one student's information databank. It includes 1,000, 2,000, 5,000, 10,000, 50,000, 100,000, 500,000, and 1,000,000 records, and the neural network structure is 12. 6. 3. The environment of the experiment is PC CPU Pentium 2.4, Windows XP, and RAM 512 M.

The experiment compared two kinds of methods of detection of duplicated records obtained experimental results can be seen that the proposed algorithm have better test results. Be seen from Fig. 45.1 that the method of detection time was significantly lower than the BP neural network method for large record sets. From Figs. 45.2, 45.3 can be seen, the BP neural network method in the data set size the

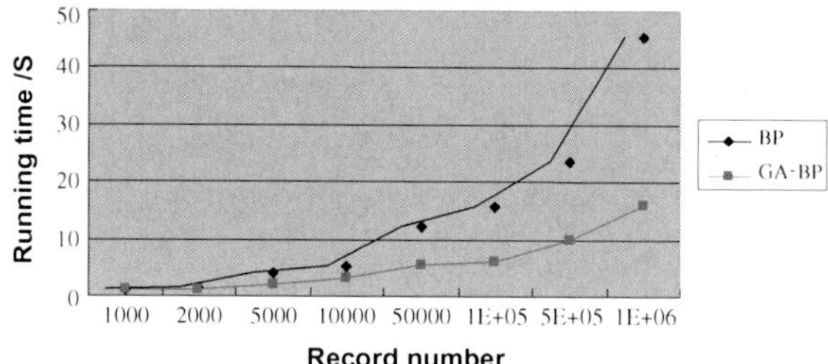

Fig. 45.1 Running time comparison of the two algorithms

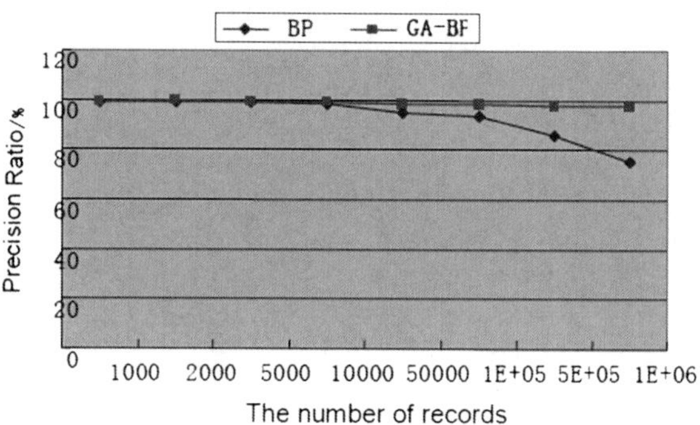

Fig. 45.2 Two comparisons of the precision ratio

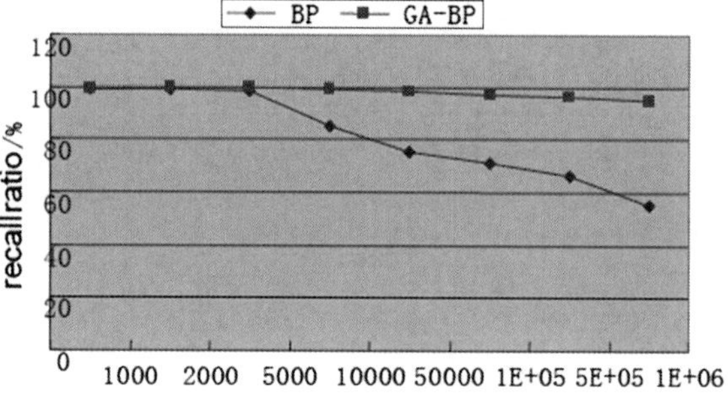

Fig. 45.3 Two comparisons of the check full rate

case can be higher recall rate and precision rate, but the effect is not very good in the case of large-scale data sets. In the case of the detection of large-scale data sets, checking the full rate and precision rate to maintain a relatively high value, indicating that the proposed algorithm has good scalability. From the experimental results it can be seen that while using the duplicate records to train the neural network, it can be avoided not to repeat the test data interference in the case, result accuracy is very satisfactory [7]. BP neural network to solve record duplication, there are still different input on the BP neural network gives the same output, thus an error mapping, affecting the accuracy of the duplicate record results.

From the comprehensive experimental results, we can further improve the precision and recall ratio of duplicate records, by using the same training data to train the neural network repeatedly in the circumstances of the different connection weights and thresholds and recording repeatedly multiple network output to make the intersection be the final outcome of the duplicate records. The experimental results show that we can solve the problem that the different neural network inputs may correspond to the same output and get the error mapping results. Furthermore, the method can effectively solve recognition problem of similar duplicated records of the large amount of data.

45.5 Conclusion

In order to improve the quality of the data sources that have been excavated, how to eliminate duplication of information in the data source is a hot topic in the study of data cleaning. In this paper, the duplicated records detection method, based on genetic neural network, make full use of the nonlinear mapping of the neural network and genetic algorithm global optimization features. It will be based on the idea of learning and the evolution of thought application to the detection of duplicate records, to avoid the traditional method to calculate the attribute weights. The experiments show that this method can effectively solve the large amount of data similar to the duplicate record detection; it not only has good detection accuracy, and has a good time efficiency.

References

1. Elmagarmid AK, Ipeirotis PG, Verykios VS (2007) Duplicate record detection: a survey. IEEE Trans Knowl Data Eng 19(1):1–6
2. Huang L, Jin H, Yuan P (2008) Duplicate records cleansing with length filtering and dynamic weighting. Fourth Int Conf Semant Knowl Grid 8:95–100
3. Hernandez M, Stolfo S (1995) The merge purge problem for large databases, vol 5. ACM Press, New York, pp 127–130
4. Monge AE, Elkan CR (1997) An efficient domain-independent algorithm for detecting approximately duplicate database records. Proceedings of workshop on research issues on data mining and knowledge discovery, vol 7. Tucson, pp 23–29

5. Gravano L, Ipeirotis PG (2001) Using Q-grams in DBMS for approximate string processing. IEEE Data Eng Bull 24(4):28–34
6. Lee ML, Lu H, Ling TW et al (1999) Cleansing data for mining and warehousing. Proceedings of the 10th international conference on database and expert systems applications, vol 5. Florence, pp 751–756
7. Goldberg DE (1989) Genetic algorithms in search, optimization and machine learning, vol 1. Addison-Wesley, MA, pp 1–3

Chapter 46
URP Based on Cloud Computing

Lijun Xu and Fuqiang Wang

Abstract Research university in the URP digital frame construction, introduces the definition of cloud computing features and key technologies, was proposed based on the architecture of cloud computing URP. That should be based on URP cloud information to establish a unified standard, providing a platform and interface specification, and the various application systems to integrate loosely coupled way to achieve resource sharing and exchange of information for the user interface provides a unified and personalized access to services.

Keywords Cloud computing • Digital campus • Application system

46.1 Introduction

Digital Campus is a web based use of advanced information technology means and tools to achieve from the environment, resources, and to the activities of all digital, on the basis of the traditional campus to build a digital space to expand Xianshi dimensions of time and space on campus so as to enhance the efficiency of the traditional campus, expansion of traditional school function, and ultimately the overall educational process information [1]. Education and information has become the top priority of the information age. The development of universities, colleges and universities, have also made information about teaching different levels of development. However, due to investment in education is not the consistency and continuity of information technology is not, plus all units to maintain relative independence, resulting in the building of campus information resources, conflicts, lack of unified planning resulting in the campus resources fragmentation, serious waste of funds, resources not used effectively [2].

L. Xu (✉) · F. Wang
Institute of Computer and Information Engineering, Xinxiang University,
Xinxiang, Henan, China
e-mail: xljkeyan@126.com

Z. Zhong (ed.), *Proceedings of the International Conference on Information Engineering and Applications (IEA) 2012*, Lecture Notes in Electrical Engineering 220,
DOI: 10.1007/978-1-4471-4844-9_46,

General information management system applications by functional departments according to their needs, from departments, grassroots perspective developed from research and development at different times, in the development process, the information did not follow uniform standards, software development platform And data formats are not the same, the system cannot be achieved between effective data sharing, artificially created one after another information island [3]. Lack of unity between the application access interface, resulting in a lack of campus network application systems integration, lack of effective management of information resources sharing. A variety of applications using different software platforms, the lack of a unified application access interface, the lack of integration between applications cannot directly access each other's data and functions that need human treatment, Lack of a unified user interface, the user for different applications require different identity-by-visit, access to resources and lack of a unified application interface [4].

To solve these problems, we take advantage of cloud computing technology to build a digital campus architecture, cloud computing is distributed processing, parallel processing, and the development of grid computing, data storage in the cloud on it, software and services placed in the cloud, built on standards and protocols, can be obtained through a variety of devices. In the cloud computing model, users need applications, data storage does not run and stored in the user's personal computers, mobile phones, and other terminal equipment, but on the run and stored in large-scale Internet server cluster. The cloud model can be limited to the use, integration of existing hardware and software resources, reduce procurement costs, in conditions of limited funds and time to expedite the construction of the campus network of information resources [5].

46.2 Cloud Computing

46.2.1 Definition and Characteristics of Cloud Computing

Cloud computing the concept was first proposed by Google the company, "cloud" refers to the computer cluster, each group includes hundreds of thousands or even millions of computers [6]. "Cloud computing" is a web application model is a distributed computing technology, the basic concept of network computing will be a huge program automatically split into numerous small subroutine, and then to the more than Huge system composed of servers, the search, the results after calculation and analysis back to the user [7]. With this technology, network service providers can in seconds, completed billions of dollars and even tens of millions of information, to receive and "super computer" the same powerful performance of network services [8]. To this end, the narrow cloud computing refers to the delivery of IT infrastructure and usage patterns, refer to on-demand through the network, and scalable way to obtain the necessary resources; generalized cloud computing refers to the service delivery and usage patterns, to demand that the network, Easy way to obtain the necessary expansion of services. This service can be IT and software, Internet-related, but also other services. Cloud Computing is a grid computing, distributed computing, parallel computing, utility computing, network storage, virtualization, Load balancing, and other traditional fusion of computer and network technology product.

46.2.2 The Key Technology of Cloud Computing

Cloud computing is a new type of super-computing model, virtualization, data storage, data management, programming models, and many unique and advanced technology.

Virtualization technology, Virtual machine, that server virtualization is the underlying architecture of cloud computing an important cornerstone. In server virtualization, virtualization software, the hardware required to implement the abstract, resource allocation, scheduling and management, virtual machine and host operating system, and the isolation between multiple virtual machines and other functions, the current typical implementation (basically become the de facto standard) have Citrix Xen, VMware ESX Server, and Microsoft Hype-V and so on.

Data storage technology, To ensure high availability, high reliability and economy, a distributed storage cloud to store data using redundant storage so as to ensure the reliability of data, high reliability software to compensate for the unreliability of the hardware, providing low-cost reliable system. In addition, the cloud computing system needs to meet the needs of large numbers of users, in parallel to a large number of users. Therefore, the cloud must have a distributed data storage technology, high throughput and high transmission rate characteristics. The current data storage technology are Google's Google File System (GFS, nonopen source) and HDFS (Hadoop Distributed File System, Open Source), these two technologies have become standard.

Data management technology, Cloud computing system for processing large data sets, analysis, to provide efficient services. Therefore, the data management technology must be able to efficiently manage large data sets. Second, how to find the huge specific data, cloud computing is calculated from data that must be addressed. In addition, the characteristics of cloud computing is the mass storage of data, after reading a lot of analysis, how to improve the data update rate. and to further enhance the random read rate of data management technology of the future must be resolved. Data management technology of cloud computing is Google, the most famous BigTable data management technology, while the development team is working on similar Hadoop open source BigTable data management module.

To make it easier to bring the enjoyment of cloud computing services, allowing users to make use of the programming model to write simple programs to accomplish a specific purpose, the Cloud to be very simple programming model. Must ensure that the background of complex parallel execution and scheduling transparent to users and programmers. At present the IT vendor's "cloud" program programming tools are based on Map-Reduce programming model.

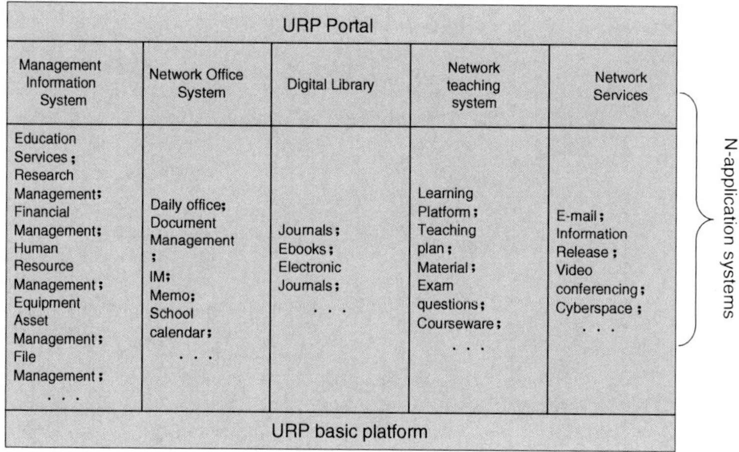

Fig. 46.1 URP functional block diagram

46.3 URP Based on Cloud Computing

URP can be summarized as: a portal, N application systems, and a basic platform. URP is a portal personal portal is the access of the digital campus entry points, including the application of the aggregation and display, personal desktop customization, single sign-on, and application of roaming, available resources, search and query functions. N application systems is a direct and URP campus teaching, management and other business-related management information systems, office network, digital libraries, online teaching and service-related network applications. A basic platform is a basis for the underlying URP platform is the key to the entire program, including application management, user management and authentication, rights management, data exchange, and other services modules. URP architecture is shown in Fig. 46.1.

URP digital campus portal is the entry point for access to the user to provide personalized user interface, users enter the portal, in addition to public information can be seen, but can only see the services commensurate with its status as the user's personalized Web portal, personalized web portal functions are as follows.

(1) *Unified entrance*: Unified user management, unified authentication, single sign-on, and access control management.
(2) *Application integration and navigation*: The ability to follow the application of an integrated system management interface specification, so follow the standard application of the resources and capabilities to better display; to the properties under a variety of applications to guide users to more efficiently and quickly using the good in all applications.
(3) *Customization*: Users can customize their desktop as the same preferences to customize the portal according to their own user interface.
(4) *Resource retrieval and query*: The ability to allow users in numerous digital campus, they need to quickly find the resources and information.

URP the middle of the N application system is used for teaching, research, management, and service a variety of applications, users can meet all kinds of information management and information services on demand is the main University information content.

(1) *Management Information Systems*: Including the academic, research, finance, human resources, equipment, assets, archives, and other management information systems;
(2) *Network office*: To provide users with a network of office environment, including the school calendar, address book, notes, instant messaging, discussion forums, document management, and Web Space Management; the various aspects of the majority of users will move to online office, and online reporting, online approval, online supervision, and truly online office and management.
(3) *Digital Library*: The University of Journals, books, papers, and other digital library resources to uniform standards and norms to manage, through the intelligent search technology and broadband high-speed network will be a variety of multimedia information to users, support Teaching, and research knowledge center.
(4) *Online teaching*: A support to provide users with online lesson planning, courseware, material construction, network classes, online communication, online self-study, examination, and other teaching process network teaching platform.

URP public platform is the basis for the integration of the various application systems provide a range of basic services, so the application can be achieved between the data exchange and sharing, mutual visits between applications, unified user interface, and the dissemination of information access interface to achieve plug and play application.

(1) *Application Management*: In order to set the uniform application of external interface specification, interface specification is divided into the core set of extensions and optional set of interface specifications to support different levels of integration of different applications, so follow the standard application can easily Integration.
(2) *User management and authentication*: To provide a unified user management and authentication, and provides users single sign-on and application support for roaming between, to achieve the authority of the electronic identity management and services.
(3) *Rights Management*: Achieving school institutions, jobs, and positions in the digital campus maps for users to access applications, and applications provide unified access between the rights management specification enables new applications can be automatically discovered and users.
(4) *Data exchange*: To provide a unified view of information and standard data exchange service, making the standardization of data exchange between applications.

University of resource planning resources that you can configure teachers (professional, title, education, participation in research projects), teaching resources (electronic lesson plans, electronic courseware, e-books, audio and video data, and network test systems), laboratory equipment (uses, status, and Number), classroom (curriculum, number of seats, location, and timing), capital (number, source, and support the project), and staff (skills, age). However, the University can also be seen as a special kind of business, the company's products are of high quality graduates and research of high value, the University's resource allocation and management is carried out around the two, the pursuit of information sharing, application integration, and unified access interface. Therefore, URP system uses a unified database for centralized management of information, where unity is the unity of logic, either in a distributed database can also be used the way the data center. URP system to different users with different levels of access to the database, the database server receives the user request access to the database, to establish a connection with the authentication center, the identification of user identity authentication center. Certification provides the key to the server can decrypt the user requests information.

46.4 URP Implementation Based on Cloud Computing

Due to URP of web service is to make it possible to exchange data between different application systems in educational internal institutions so that these data can be mutually cooperated to meet the new demand. All the data from different database application systems, such as: educational management system, research management system, school management system, the wage query system, financial management systems and so on. The data storage structure and data structure may be different, they are integrated by loosely coupled web service and are visited through the application of the system at different access as well as are transferred through the HTTP protocol and SOAP after getting data, which can break through the restrictions between the firewall and the system and ensure that data between different systems can be easily separated and integrated.

Digital Campus education resources platform from the logic functions can be divided into four layers. They can be seen from top to bottom: the school portal, information service layer, applications and database systems, and network infrastructure platforms. For users, the system has a "shielding effect", they could not feel the data obtained is extracted from multiple systems and processed.

The application system and database system layer should be pretreated before building a framework of web service platform. This is because different applications in the design time is generally not considered the problem of data sharing, which led different system database to be duplicated. For example: school management system and the elective system all have the basic database information which also could bring some potential problems to Web service:

(1) Too much redundant data in the distributed system is inconvenient to manage.
(2) Information service layer need to spend more resources on redundant data processing and calculation.

(3) Maybe the different systems exist the same information, which could result in data sync.

Therefore, we must integrate these systems that may have duplicate data. Two methods we can use like this:

(1) *Streamlining the application system.* Allows the system to achieve low coupling and function asynchronous not relevant between the different systems.
(2) If you cannot do the first step, consider to make a database to synchronize database synchronization and minimize the possibility of data synchronized.

How to build a web service-based education resources platform framework? We divided into the following steps:

(1) *System requirements analysis.* To determine the services provided by web service and the application systems involved.
(2) *System design.* To determine the division of the system modules and functions.
(3) According to the first two modules of the interface between data types and parameters, to determine the interface description between the module.
(4) *Design method.* To design out method the interface access.
(5) *Test code.* Provide a unified authentication campus information portal, including integrating all application systems applications. Information portal is not just the integration of information, but also the integration of the database.

46.5 Conclusion

Comprehensive information society in the twenty-first century, the trend has been further development of education reform, corporate governance has moved from material requirements planning (MRP) to develop to the ERP, so colleges and universities must follow the historical trends, seize opportunities, the use of "cloud computing" technology Bold exploration and testing. According to the core idea of cloud computing, cloud services, content, cloud computing will be applied to the construction of Digital Campus, Building Digital Campus will be the basic idea of the contents of construction, platform construction, databases, etc., to bring new ideas. University through the construction of URP, to some extent, can reflect the level of university information management, such as books, research, and student administration information management, information technology and network level, it will fundamentally determine the management of colleges and universities and school efficiency; or even as a measure of the development of a college-level indicator of the level and growth.

This article is based on the status of higher education information technology and issues analysis, applied to cloud computing in the digital construction of colleges and universities, was proposed based on the architecture of cloud computing

URP. This makes full use of university infrastructure, digital resources can be fully shared, the university offers a variety of online education model, teaching reform and accelerates the pace of building information, and promotes information technology throughout the education process. In short, with cloud computing technology and application development, universities take advantage of cloud computing will bring great economic benefits, management efficiency, and social benefits.

References

1. Dongxing J, Shi K, Chu H et al (2004) University resource planning scheme, vol 4. Tsinghua University (Natural Science), Beijing, pp 151–156
2. Hong ZJ, Song D (2005) University resource planning (URP) of air engineering, vol 22(4). Shenyang Polytechnic University, China, pp 263–265
3. Li X (2006) Colleges and Universities University of resource planning (URP) related issues, vol 2. Tianjin University, Tianjin, pp 78–81
4. Zhang J, Wang Y (2008) URP digital campus system construction, vol 10(1). Beijing University of Posts and Telecommunications, Beijing, pp 103–105
5. Hongwei S, Xuning L (2010) Cloud computing in the campus network resources management study, vol 12(6). Shijiazhuang University, Hebei, pp 33–35
6. GI LES J. Google to TP translati on ranking (2009) http://www.nature.com/news/2006/061106/full/news06110626.html
7. Wikipedia. Cloud computing (2009) http://en.wikipedia.org/wiki Cloud_computing
8. China Cloud computing net. What is Cloud Computing? (2009) http://www.cl oudcomputing2 china. cn/Article/ShowArticle. as p?Article ID = 1

Chapter 47
Uncertainty Reasoning in Education Evaluation Forecast

Bo Wu and Tuo Ji

Abstract Education Evaluation is an important area of educational research and is an important part of educational activities. By the limit of education methods and educational resources, it is difficult to forecast the results of Education Evaluation. Uncertain reasoning is often used not strict enough, but the reasoning results consistent with the intuition of human experts, the probability can also be given some explanation. Uncertainty reasoning method is often used than subjective bayesian approach. Using subjective bayesian approach, it can achieve some forecast Education Evaluation, such as forecast of individual learning, class examination pass rate of Forewarning Analysis, etc.

Keywords Uncertain reasoning • Subjective bayesian approach • Education evaluation forecast

47.1 Introduction

In education activities, we often use examinations for education evaluation. For the limit of education method and education resources, it is very hard to intersperse the education activities with exams [1, 2]. Therefore, it is difficult to predict the results of the examination; it means that it is difficult to forecast the results of Education Evaluation [3, 4]. In this chapter, we will take the university curriculum teaching activities as an example, and forecast the results of Education Evaluation [5, 6]. The basic idea is that by using the factors that affect the examination results

B. Wu (✉)
Department of Education Ocean, University of China, Qingdao 266101, China
e-mail: wubo@hrsk.net

T. Ji
Computing Center Ocean University of China, Qingdao 266101, China
e-mail: jituo@hrsk.net

Z. Zhong (ed.), *Proceedings of the International Conference on Information Engineering and Applications (IEA) 2012*, Lecture Notes in Electrical Engineering 220, DOI: 10.1007/978-1-4471-4844-9_47, © Springer-Verlag London 2013

such as Midterm data etc., reason for the final exam, and make predictions on the final examination. In the reasoning process, uncertainty reasoning method will be used to predict Education Evaluation.

47.2 Uncertainty Reasoning Method Introduction

Because of the imprecise and incomplete information, analysis and prediction system often have to deal with a lot of uncertainty. Reasoning is often used in uncertain reasoning methods of nonstandard logic. Uncertainty comes from the knowledge of objective reality and subjective knowledge level of awareness. Uncertain reasoning is often used not stringent enough, but the reasoning results consistent with the intuition of human experts, it can also be given some explanation by the probability.

Uncertainty comes from the knowledge of objective reality and subjective knowledge level of awareness. In the thinking process, uncertainty is often the emergence of a state of mind. Uncertainty reasoning method is proposed and researched from 1970s of last century. Uncertain reasoning is often used not strict enough, but the reasoning results consistent with the intuition of human experts, the probability can also be given some explanation.

Uncertain problem model must explain the representation, Computing and semantics of uncertain knowledge. Expressed refers to the description of the uncertainty of the methods adopted, which is a key step to solve uncertain reasoning. Calculation mainly refers to the spread and updates of the uncertainty. Calculation of the uncertainty mainly refers to the spread and updates. Semantic refers to the representation and calculation of what is meant, that is, to interpret them.

47.3 Forecast of Individual Learning

"ACCESS programming" is a required course for noncomputer majors. Statistics from the historical examination of the course final exam pass rate was 85 %, that is, the prior probability of failing was 0.15. Analysis by the teacher of the course, There may be events A1…A20 (listed in Table 47.1 in the evidence column) which have an impact on the event B (failed the exam). The introduction of each of the two measurements of evidence A LN and LN are defined as follows:

Sufficiency measurement LS reflect the emergence of A to support B:

$LS = \frac{P(A|B)}{P(A|\neg B)}$, when LS > 1, that the emergence of A supports B;

Necessity measure LN reflects the nonoccurrence of A to support B:

$LN = \frac{P(\neg A|B)}{P(\neg A|\neg B)}$, when LN = 1, that the nonoccurrence of A do not impact B;

LS value for each evidence obtained by the statistical or historical data given by the experts, and LN values is set to 1 is based on the considerations: some evidence does not appear, will not affect the results. Establish probability function: $O(X) = \frac{P(X)}{1-P(X)}$, the $P(X)$ of the [0, 1] to zoom into values in [0, ∞] of O (x).

Is not difficult to verify:

$$O(B|A) = LS \cdot O(B)$$

$$O(B|\neg A) = LS \cdot O(B)$$

Existing student A, the basic conditions listed in Table 47.1, an evidence of the "fit" column value is 1 indicates that the evidence occurred. When there is evidence A1, A2…AK inevitable, calculate the probability P(B) of him failing in a final exam of change. By the rules subjective bayesian approach of reasoning,

(1) From P(B) = 0.15, known
O(B) = 0.1765

(2) O (B|A1) = LS·O(B) = 1.95·0.1765 = 0.430
P (B|A1) = 0.301

(3) O (B|A1A2) = LS·O (B|A1) = 2.68·0.430 = 1.153
P (B|A1A2) = 0.536

Table 47.1 Student A's individual learning forecast

Evidence		LS	NS	Fit	O(B)	P(B)
A1	Midterm grades <60	1.95	1	1	0.176	0.150
A2	Nonenglish languages	2.48	1	0	0.344	0.256
A3	Midterm grades >80	0.75	1	0	0.344	0.256
A4	60 <= Midterm <= 80	1.20	1	0	0.344	0.256
A5	First year students	1.17	1	0	0.344	0.256
A6	Sophomore	1.07	1	1	0.344	0.256
A7	Three students	1.09	1	0	0.368	0.269
A8	Fourth grade students	1.95	1	0	0.442	0.306
A9	Extension students	1.10	1	0	0.442	0.306
A10	Rehabilitation of student	1.85	1	0	0.442	0.306
A11	Girls	0.99	1	1	0.442	0.306
A12	Retaking students	1.02	1	0	0.437	0.304
A13	Liberal arts students	1.15	1	1	0.437	0.304
A14	Elective students	1.46	1	0	0.503	0.335
A15	Prerequisite failed	1.69	1	0	0.503	0.335
A16	Job situation is not good	1.81	1	1	0.503	0.335
A17	Attendance frequency <60 %	1.55	1	1	0.911	0.477
A18	Minority	1.00	1	0	1.411	0.585
A19	Student leaders	0.99	1	1	1.411	0.585
A20	Sports specialty students	1.55	1	0	1.397	0.583
Probability						58.3 %

(4) Similarly can be drawn from:

$$O(B|A1 \cap A2 \cap \ldots \cap \mathrm{AK}) = \prod_{i=1}^{k} \frac{\mathrm{O\,(B|Ai)}}{\mathrm{O(B)}} \cdot O(B) = 33.992$$

By $P(B) = \frac{O(B)}{1-O(B)}$, student A's probability of failing the final exam is 58.3 %.

47.4 Class Examination Pass Rate of Forewarning Analysis

As indicated above, the course "ACCESS programming" final exam pass rate was 85 %. By analyzing the historical examination of the course data, and summarizing the proportion (A') of class A, all kinds of situations are listed in Table 47.2. Now, we use the uncertainty reasoning to forecast the final examination pass rate of the class.

Table 47.2 The proportion of the class situation of the various data

Evidence		LS	NS	Historical statistics (probability) (%)	Case A' of class A (%)
A1	Midterm grades <60	1.95	1	29	35
A2	Nonenglish languages	2.48	1	4	20
A3	Midterm grades >80	0.75	1	8	7
A4	60 <= midterm <= 80	1.20	1	63	58
A5	First year students	1.17	1	10	8
A6	Sophomore	1.07	1	62	47
A7	Three students	1.09	1	12	12
A8	Fourth grade students	1.95	1	3	25
A9	Extension students	1.10	1	13	8
A10	Rehabilitation of student	1.85	1	45	42
A11	Girls	0.99	1	50	54
A12	Retaking students	1.02	1	40	45
A13	Liberal arts students	1.15	1	10	13
A14	Elective students	1.46	1	14	15
A15	Prerequisite failed	1.69	1	15	18
A16	Job situation is not good	1.81	1	9	10
A17	Attendance frequency <60 %	1.55	1	14	21
A18	Minority	1.00	1	15	18
A19	Student Leaders	0.99	1	7	10
A20	Sports specialty students	1.55	1	1	2

First calculated evidence "midterm grades < 60" (A1) this index. From the historical data analysis, LS value of "midterm grades <60" is 1.95. For the group of "midterm grades <60", their final exam probability P (B | A1) can be calculated by the following:

$$O(B) = \frac{0.15}{1 - 0.15} = 0.1765$$

$$\mathrm{O}\,(\mathrm{B}|\mathrm{A1}) = \mathrm{LS} \cdot \mathrm{O} \cdot (\mathrm{B}) = 1.95 \cdot 0.1765 = 0.344$$

$$P(B) = \frac{0.4302}{1 + 0.4302} = 0.256$$

This means that 25.6 % of the "midterm grades <60" (A1) students will not pass the final exam. But how to analyze the class A, 35 % in the "midterm grades <60" probability of occurrence, its final exam on how much influence it? Accordance with the subjective Bayesian approach, use the following derivation, obtain the interpolation Figure.

(1) By condition, P (A1 | A1′) = 0.35
(2) If P (A1 | A1′) = 1, then

$$P(B|A1) = \frac{LS \cdot P(B)}{(LS - 1) \cdot P(B) + 1} = \frac{1.95 \cdot 0.15}{(1.95 - 1) \cdot 0.15 + 1} = 0.256$$

(3) The known P (A1 | A1′) = P (A1) = 0.29, with P(B) = 0.15. Interpolation Figure (Fig. 47.1) can be obtained:

$$P(B|A') = 0.15 + \left(\frac{0.256 - 0.15}{1 - 0.29}\right) \cdot (0.35 - 0.29) = 0.159$$

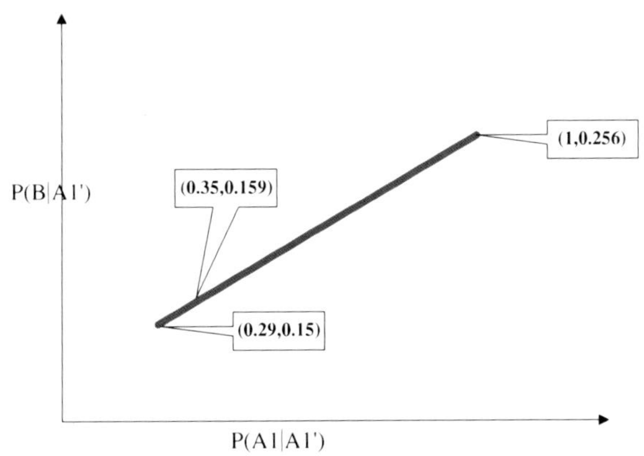

Fig. 47.1 Calculation of P (B | A ') of the interpolation figure

Table 47.3 Class A of the final examination pass rate

Evidence		LS′	NS	O(B)	P(B)
A1	Midterm grades <60	1.060	1	0.187	0.158
A2	Nonenglish languages	1.169	1	0.219	0.179
A3	Midterm grades >80	1.002	1	0.219	0.180
A4	60 <= midterm <= 80	0.979	1	0.214	0.177
A5	First year students	0.997	1	0.214	0.176
A6	Sophomore	0.978	1	0.209	0.173
A7	Three students	1.000	1	0.209	0.173
A8	Fourth grade students	1.153	1	0.241	0.194
A9	Extension students	0.995	1	0.240	0.193
A10	Rehabilitation of student	0.968	1	0.232	0.188
A11	Girls	0.999	1	0.232	0.188
A12	Retaking students	1.001	1	0.232	0.188
A13	Liberal arts students	1.004	1	0.233	0.189
A14	Elective students	1.004	1	0.234	0.190
A15	Prerequisite failed	1.017	1	0.238	0.192
A16	Job situation is not good	1.006	1	0.240	0.193
A17	Attendance frequency <60 %	1.033	1	0.247	0.198
A18	Minority	1.000	1	0.247	0.198
A19	Student Leaders	1.000	1	0.247	0.198
A20	Sports specialty students	1.004	1	0.248	0.199
Probability					19.9 %

(4) For Class A, from the Eq. (3) the derivation can draw the ratio of "Midterm grades < 60″ makes the probability of failure rate rose to 0.159. With this results update "midterm grades < 60″ LS 'values:

$$LS' = \frac{P\left(B|A1'\right)}{P\left(B|A1\right)} = \frac{0.159}{0.15} = 1.060$$

Similarly, get LS 'value of A2, A3… A20 in the list below: (Table 47.3).

Derived from the above know, forecast of the class A final examination failure rate was 19.9 %.

47.5 Summary

Education Evaluation is an important area of educational research; it needs to study the problem of so many factors involved. Education Evaluation mode are also different. In this chapter, from the perspective of uncertain reasoning, we propose the thought of the realization for Education Evaluation Forecast.

References

1. Shi Etc CY (2002) Artificial intelligence theory, vol 11(2). Tsinghua University Press, Beijing, pp 55–56
2. Cai ZX, Xu GY (2003) Artificial intelligence and applications, vol 34(16), 3rd edn. Tsinghua University Press, Beijing, pp 178–279
3. Tan PN, Steinbach M, Kumar V (2006) Introduction to data mining, vol 34(55). People Post Press, China, pp 345–346
4. SZ An Etc (2005) Data warehouse and data mining, vol 56(8). Tsinghua University Press, Beijing, pp 88–89
5. Han JW, Kamber Kamber M, Fanming W, Xiao FM (2006) Translation data mining concepts and techniques, vol 44(5). China Machine Press, China, pp 78–79
6. Shi BZX (2009) Write office 2007 office applications, vol 67(6). China Tsinghua University Press, Beijing, pp 78–79

Part V
Semantic Grid and Natural Language Processing

Chapter 48
Computation of Optimal Embedding Dimension and Time Lag in Reconstruction of Phase Space Based on Correlation Integral

Sen Xia

Abstract Many characteristic invariants including Lyapunov exponent, fractal dimension, and Kolmogorov entropy can be used to measure the chaotic degree of dynamics system. A statistical computation method to determine the optimal embedding dimension and time lag parameter in reconstruction of phase space via the method of delay coordinates status space reconstruction to calculate the correlation function was introduced in this paper. Some scenarios for research on chaotic time series analysis have been also put forward.

Keywords Correlation integral • Time series • Embedding dimension • Time lag • Dynamics system

48.1 Introduction

Continuous dynamics system differential equation and discrete dynamics system difference equation can be effectively used to describe the deterministic systems. But, just often only the discrete experimental data sampled can be recorded to form a discrete time series. The data series evolved with time accordingly usually are derived from evolution of some component variable in the process of dynamics system measurement. For the reason that strong correlation and coupling exist among all the individual components, all other unknown component series information are already enriched in the componential time series sampled and recorded in experiment. Measures to process the times series is helpful to draw important useful information from dynamics system even then make it realizable to predict the future tendency and development of dynamics system.

S. Xia (✉)
Faculty of Computer Engineering, Huaiyin Institute of Technology, Huaian, Jiangsu 223003, China
e-mail: archlinuxandroid@gmail.com

Z. Zhong (ed.), *Proceedings of the International Conference on Information Engineering and Applications (IEA) 2012*, Lecture Notes in Electrical Engineering 220, DOI: 10.1007/978-1-4471-4844-9_48,

In the case of higher dimension and complicated dynamics system, to restore and to find the real features of strange attractor is sometimes incredibly difficult. The predicting accuracy will decline rapidly [1]. During the process of predicting the future behavior of times series, a key procedure is to reconstruct the phase space with the help of delay coordinates. And the most vital one in the reconstruction of phase space is the reasonable selection of embedding dimension and time lag in embedding space.

48.2 Diagnostic and Analytical Methods of Chaotic Dynamic System

Whether a time series is chaotic or not and the degree of chaos existed in it cannot be easily concluded before scientific analysis and critical computation. The dynamics system is chaotic when and just only under the condition that the orbit and form of attractor in the phase space reconstructed behaves between stable periodic motion and completely random movement. Only the one which is infinitely extended and folded and never intersect itself at the same time can be referred to as strange attractor.

There are some methods to diagnose the chaotic degree of dynamic system including numeric computation, geometric, and statistical methods. And in one of the statistical method, the mean value of some system variable is calculated and analyzed to judge or measure the chaotic movement. The fractal dimension, Lyapunov exponent, and entropy are the remarkable indexes of statistical characteristics of dynamics system [2]. The characterization of dynamics system attractor can be dealt with from the perspectives of both the macro features and micro characteristics. In the former perspective, all the attractor or infinite orbit are averaged normally and then they obtained characteristic variables are individually called Hausdorff dimension, Lyapunov exponent, and entropy. And, in the latter one, it means the numbers, the types and eigenvalues of unstable periodical orbits.

48.3 Reconstruction of Phase Space

48.3.1 Theoretical Foundation

Phase space is a very important and valid means used to analyze dynamic systems. And, at least two or more variable time series data are needed to form the phase space. But, just only one component variable times series data are usually detected. As we know, the evolutionary development of independent component variable with time was decided or controlled by the whole system movement; so, the time series formed by independent component variable must carry the whole system's rule of movement even the useful information is hidden in the component sampled time series. According to the basic principles of nonlinear dynamic system, to calculate the first derivative and the second derivative and more higher-order derivatives is a good choice to form the phase space. But, when the noise hidden in the signal data set, the error will be greater, which can make the credibility of phase space analysis have a greatly reduced quality.

A convenient and valid method was proposed and used to reconstruct the phase space from single variable time series [3]. Namely, the delay coordinates of some variable in the original system can be effectively utilized to reconstruct the phase space.

48.3.2 Statistical Computation

The original time series sampled is supposed and obtained as the following:

$$X_1, X_2, X_3, \cdots, X_N \tag{48.1}$$

Suppose the time delay is represented by the symbol of Γ, namely the time interval of the resampled time series chosen from the original time series.

Other resampled time series are:

$$Y_1 = (x_i, x_{i+\Gamma}, x_{i+2\Gamma}, \ldots, x_{i+(m-1)\Gamma}), i = 1, 2, \ldots, M \tag{48.2}$$

And, M is the numbers of phase point in the reconstructed phase space.

$$M = N - (m - 1)\,\Gamma \tag{48.3}$$

Here, m is the value of embedding dimension, and Γ is the time lag.

Then, the reconstructed phase space can be expressed like the following vector forms:

$$Y = [\,Y_1,\ Y_2,\ Y_3,\ \ldots,\ Y_M]^T \tag{48.4}$$

The correlation integral is illustrated like the following equation:

$$C\,(m, N, r, t) = 2\left[\sum H\left(r - |Y_i - Y_j|\right)\right] / \left[M\,(M - 1)\right], 1 \le i \le j \le M \tag{48.5}$$

The function, namely $|Y_i - Y_j|$ is used to calculate the distance between Y_i and Y_j.

$$\begin{aligned} H\,(x) &= 1,\ \text{when}\, x \ge 0 \\ H\,(x) &= 0,\ \text{when}\, x < 0 \end{aligned} \tag{48.6}$$

The statistical index variable S is expressed as:

$$S\,(m, N, r, t) = C\,(m, N, r, t) - C^m\,(1, N, r, t) \tag{48.7}$$

And $S\,(m, N, r, t)$ is used to represent and illustrate the correlative characteristics. The time lag Γ and embedding dimension m is determined by the statistical index variable $S\,(m, N, r, t)$.

In order to investigate the nonlinear correlative features and eliminate the false time correlation, the original time series $\{x_i\}$ is divided into t different disjoint subtime series. And $S\,(m, N, r, t)$ can be calculated by these disjoint sub series.

When $t = 1$, just only one time series existed:

$$\{x\,(k_1)\,, x\,(k_2)\,, x\,(k_3)\,, \ldots, x\,(k_N)\} \tag{48.8}$$

$$S\,(m, N, r, 1) = C\,(m, N, r, 1) - C^m\,(1, N, r, 1) \tag{48.9}$$

When $t = 2$, it exists two different sub time series:
The first subone is:

$$\{x\,(k_1)\,,\; x\,(k_3)\,,\; x\,(k_5)\,,\; \ldots,\; x\,(k_{N-1})\} \tag{48.10}$$

And the second subone is:

$$\{x\,(k_2)\,, x\,(k_4)\,, x\,(k_6)\,, \ldots, x\,(k_N)\} \tag{48.11}$$

The two subtime series are not intersected to each other. And their length is the same, namely equal to $N/2$.

Then, work out the mean value of their $S\,(m, N, r, 2)$:

$$S(m, N, r, 2) = \left\{ \begin{array}{l} [C_1(m, N/2, r, 2) - C_1^m(1, N/2, r, 2)] \\ +[C_2(m, N/2, r, 2) - C_2^m(1, N/2, r, 2)] \end{array} \right\} /2 \tag{48.12}$$

To take into the universality of forms of $S\,(m, N, r, t)$ into consideration,

$$S\,(m, N, r, t) = \left\{\sum \left[C_s\,(m, N/t, r, t) - C_s^m\,(1, N/t, r, t)\right]\right\} /t, s = 1, 2, \ldots, t \tag{48.13}$$

When $t \to \infty$, the following equation can be obtained:

$$S\,(m, r, t) = \left\{\sum \left[C_s\,(m, r, t) - C_s^m\,(1, r, t)\right]\right\} /t,\ s = 1, 2, \ldots, t \tag{48.14}$$

To measure the degree of $S\,(m, r, t)$ changing with the independent variable t, a new difference function of $\Delta S\,(m, t)$ can be defined as the following and some selected r is represented by symbol of r_j:

$$\Delta S\,(m, t) = \max\left\{S\left(m, r_j, t\right)\right\} - \min\left\{S\left(m, r_j, t\right)\right\} \tag{48.15}$$

Normally, the range of N, m, r should be suitable and reasonable, when the following condition is met with the requirements:

$$\sigma/2 \le r \le 2\sigma \tag{48.16}$$

Where, σ represents the mean square root of statistical variables. And the mean value of $S(m, r, t)$ is symbolized by $\check{S}(t)$:

$$\check{S}(t) = \left[\sum\sum S\left(m, r_j, t\right)\right] / (||m|| \cdot ||j||) \tag{48.17}$$

Here, $||m||$ is the numbers of embedding dimension and $||j||$ is the numbers of r_j.

At the same time, to calculate the mean value of $\Delta S(m, t)$, the following can be obtained:

$$\Delta\check{S}(t) = \left[\sum \Delta S(m, t)\right] / ||m|| \tag{48.18}$$

Then, the following statistical variable of $S_{cor}(t)$ can be given just like:

$$S_{cor}(t) = \check{S}(t) + \Delta\check{S}(t) \tag{48.19}$$

And the minimum value of $S_{cor}(t)$ is used as the optimal value of time windows Γ_w, we can write out the optimal time lag Γ:

$$\Gamma_w = \min\{S_{cor}(t)\} \tag{48.20}$$

Then, the optimal embedding dimension m takes the form of the following accordingly:

$$m = (\Gamma_w/\Gamma) + 1 \tag{48.21}$$

48.4 Scenarios for Research on Chaotic Time Series

Some concrete dynamic systems such as Logistic mapping, Lorenz system, are good examples to apply the computation method introduced in this chapter into the practical environment. Combining the thinking of delay coordinates status, space reconstruction and the method of reconstruction of phase space, to decide the optimal embedding dimension and time lag with the help of programming via computer. And then go forward, to compare with other computation methods of phase space reconstruction, then observe the difference among them. It will be believable that the advantages and even disadvantages can be found interesting. After the observation of known dynamic systems have been experimented and some ideal result meets with the computation method, other unknown sampled time series is then the good examples sampled from real world.

48.5 Summary

The statistical computation can be easily implemented by programs with the help of programing. It possesses some obvious advantages such as easily operated, small-scale computation amount, dependable for small data set, and strong antinoise ability. Although the method introduced in this chapter has no very profound theoretical basis, it is at least one of the valid methods to calculate the optimal embedding dimension and time lag helpfully used to reconstruct the phase space. And then, it makes it possible to predict the future behavior of dynamic system. During the social activities, it can be helpful for the prediction of chaotic economy tendency and even helpful for the subprocess of realizing and improving the accuracy of weather forecast.

References

1. Lv JH, Lu JA, Chen SH (2002) Analysis of chaotic time series and application. Wuhan University Press, Wuhan, pp 93–104
2. Kantz H, Schreiber T (1997) Nonlinear time series analysis, vol 8(2). Cambridge University Press, Cambridge, pp 12–19
3. Takens F (2003) Detecting strange attractors in turbulence. Lect Notes in Math 8(3):34–39

Chapter 49
Research on the Financial Input of S&T in Minority Areas Based on the Elasticity Analysis

Hong-Ling Li and Chao Fang

Abstract By the method of elasticity analysis, this chapter researches the financial input of S&T in the minority areas during 2000–2008. It shows that the scape of input is not stable but changes dramatically and a part of the index is negative. This shows that the local government has not realized the importance of S&T enough. But during this period, all the elasticity indexes are becoming more and more centralized. This shows the input mechanism is getting better. This chapter claims that it is necessary for the local administrators to realize the importance of S&T; to establish rules or laws for the S&T input, to widen the way of getting S&T fund, and to improve the efficiency of using the financial input of S&T.

Keywords Ethnic minority areas • Elasticity index • Financial input of S&T • Gross domestic product (GDP)

49.1 Introduction

In recent years, although the performance of economic construction in minority areas is outstanding, the whole situation is still lower than the public expect. Compared with majority areas, it shows a huge discrepancy in each index such as region GDP and per capita disposable income. For example, nine regions of main minority areas occupied 7 of last 10 positions again, on the ranking list of GDP of all parts of country at 2009. Furthermore, the monitoring index of comprehensive S&T advancement level in minority areas is amid 30–40 % over a long period of time, its well below the

H.-L. Li (✉)
College of Public Management, South-Central University for Nationalities, Institute of Rural Development in Minority Areas of SCUN, Wuhan 430074, Hubei, China
e-mail: lihl723@163.com

C. Fang
College of Public Management, South-Central University for Nationalities, Wuhan 430074, Hubei, China

Z. Zhong (ed.), *Proceedings of the International Conference on Information Engineering and Applications (IEA) 2012*, Lecture Notes in Electrical Engineering 220, DOI: 10.1007/978-1-4471-4844-9_49,

same index of developed areas. This situation goes against the healthy development of regional economies obviously.

The financial input of S&T is the premise to undertake S&T activities, only by establishing a mechanism to ensure stable growth of government investment in S&T it can play an important role in economic development. In summary, this chapter takes time as a clue, with nine regions of ethnic minority areas as object, to research the scientificity and stability of S&T input, to help the ethnic minority areas establish a healthy and stable mechanism of S&T input, then make it served for S&T development and social progress in the minority areas.

49.2 Review on the Pertinent Literature

From the literature that author owned to see, researchers did lots of work on financial input of S&T at present time, the main aspects from the content followed:

The first one is the studies on noumenon of S&T input. Studies such as these focus on its importance, characteristics, and structure, etc. Figured that the financial input of S&T has a series of output effect, such as trigger effect and ripple effects; hence, it will affect the modifying on correlative S&T policy and the adjusting on solution configuration of S&T input by government and enterprises [1].

The second one is the studies on efficiency of S&T input. Studies such as these combine later output of S&T with former input of S&T for the purpose of heightening the ratios between two [2]. Found that the whole efficiency of S&T input in our country was on the low side, but the efficiency of S&T input in developed areas was higher from 2004 to 2006, by dividing our country into six regions then using (Data Envelopment Analysis) DEA. Lots of studies on allocative efficiency of S&T resource fall into this class too [3].

The third one is the studies on the economic benefits of S&T input. This class is a special case of studies on efficiency of S&T input, which focus on reviewing the relation between S&T and economic growth.

Zhu Chun-kui (2004) tested the Co-integration Relationship between economic growth and financial input of S&T from 1978 to 2000, the result showed that the financial input of S&T was strongly related to the economic growth, there was a long-term stable relationship between the two. Found that from 1987 to 2009, the contribution rate of S&T input in Guangdong Province to economic growth was 60.8 % though the analysis and causality test [4]. By conducting an empirical interprovincial panel-data analysis, it was found that the financial input of S&T in 31 provinces (cities) of our country benefited to economic growth. As the local financial input of S&T increases every 1 %, the local GDP increased 0.14 %.

Some researchers have different findings as well. Fan Bo-Nai (2004) took the data of S&T input and the economic growth as sample to study, and found that despite there is a clear unidirectional causality between them, but it is not a simple linear relation, the contribution of S&T input to economic growth is very limited, it is also influenced by the input structure and the level of resource allocation. The research result by Shi Ping (2010) shows that, the government financial input of S&T and the internal expenditures of S&T activities certainly promoted the economic growth, but by long-see, the contribution and effect to the economic growth of the former was much higher than the latter obviously. They long-term equilibrium analysis and the dynamic analysis of financial input of S&T and economic growth was done during 1978–2008; the result shows that the financial input of S&T promoted the economic growth on one hand and there also existed a time-lag effect and the phenomenon of diminishing marginal utility on the other hand [5].

In a word, this research found that current research has the following characteristics though the literature review:

First, the relationship between the financial input of S&T and the economic growth has been substantiated in many times. Researchers are concerned about input of S&T from early times, with the development of knowledge economy in recent years; the correlative researches have been improved on both amount and dynamics of study. In the previous researches, whatever method the researchers took, the research conclusion affirmed that S&T input has played a key role in economic growth and the correlation between them.

Second, the input of S&T in the ethnic minority areas has not been a hot research topic yet; the developed areas have attracted more researchers' attention. For instance, Dong Cui-ling (2010) calculated the economic contribution rate of S&T input with Guangdong province as research object; Han Dong-ling (2010) referenced the S&T input in Jiangsu, Zhejiang, and Shanghai, which are located in east China, to do comparative analysis of S&T input in Anhwei province, etc. Researchers do structural and efficiency analysis of S&T input in the ethnic minority areas occasionally (Liu Jin-song, 2009). But generally speaking, the research quantity was not enough, and from the research content to see, researchers focused on the measurement of result, few of them did the policy explanation of S&T input itself [6, 7].

At last, researchers seemed to pay much attention to the innovation of research methods currently. While the research models have become more and more complicated, few people concern the outstanding realistic problems; the research conclusion was also lack of new ideas. Therefore, this chapter reviewed the system of S&T input in the ethnic minority areas which was clearly insufficient in recent times, by the simple method of elasticity analysis; strive for the melioration of current situation [8].

49.3 GDP Elasticity Analysis of the Financial Input of S&T in Ethnic Minority Areas

49.3.1 The Basic Method of Analysis and Thoughts of Analysis

Elasticity is a frequently-used concept in economics. Generally speaking, if only there exists a functional relationship between two variables, we can use elasticity to express the sensitivity of the dependent variable's reaction percentage to the independent variable. Specifically, it is such a number, which tell us, when an economic variable varied by 1 %, the percentage of the other one varied that caused by it. In economics, the general formula of elasticity is:

Elasticity coefficient = Variation proportion of dependent variable/Variation proportion of independent variable

This formula expressed: when the independent variable varied, how Y the dependent variable will vary accordingly. According to the elasticity coefficient $e > 1$, $e = 1$ or $e < 1$, we called these full of elasticity, unitary elasticity, and lack of elasticity.

At the present time, elasticity analysis has already been used in various fields of quantitative analysis. For example, in the field of energy technology, electric elasticity coefficient and energy consumption elasticity coefficient have been international indexes that can be used to measure the matching degree between energy development and economy development in one country, which have become a hot research topic in recent years. They reflect the ratio of the increment speed of energy consumption to the increment speed of (gross domestic product) GDP in a period of time, which could be used in adjusting the relationship between energy and economic growth from macro perspectives. In 1953, the Party Central Committee proposed the policy "the electric power industry is antecedence", the elutriation has become an important target of economy development; thus, the electric elasticity coefficient of our country >1 for a long time, just like many developing countries namely the energy production was in advance to the economic growth. But, with the development of science and technology, the fuel efficiency kept rising, the structure of national economy was also changing constantly, therefore, the energy consumption elasticity coefficient has descended. In our country, with the construction of two-oriented society, the awareness of energy conservation and emission reduction constantly reinforced, and the energy consumption elasticity coefficient was downtrend. It is reported that the electric elasticity coefficient was 1.36 during the tenth 5 year plan, and it dropped to 0.9 during the 11th 5 year plan.

This chapter reviewed the correlation between financial input of S&T and economy development in the minority areas, to see whether the minority areas have established a correspondingly secure mechanism of S&T input when the economic conditions are limited. Especially in recent years, on the conditions that economic environment is constantly improved, whether the S&T input in the minority areas was optimized. This chapter adopted the method of elasticity analysis, with nine regions of ethnic minority areas as research object, to calculate the sensitivity of the S&T input to the speed of economic growth each year from 2000 to 2008. The reasons for abandoning linear-regression analysis are the following two points: first, research has found that it is not a simple linear relation between the S&T input and economic growth. Second, the regression analysis can estimate the relationship between the two in a long period of time only, but the elasticity analysis could shorten the period of analysis and depict the relationship in each year.

49.3.2 The Analysis Result

According to the formula of elasticity coefficient and the purpose of this research, the elasticity formula of regional GDP of the financial input of S&T in ethnic minority areas is:

E = (ΔY/ΔX)*(X/Y) = annual variation proportion of the S&T input indicator/annual variation proportion of the economic indicator.

Here into, the S&T input indicator was represented by local financial allocation for S&T, and the economic indicator was represented by local GDP. It is noteworthy that the reasons for using 'local financial allocation for S&T to represent S&T input followed: On one hand, the research purpose of this article was investigating the S&T awareness of local government. On the other hand, although our country has establish ed the diversified frame of S&T input which include both government and enterprises, but limited by the scale and ability of enterprises, the S&T input of government will still play a key role in it in a long period of time, especially in the minority areas.

According to the measurement result above, we can make the line graph as Fig. 49.1:

The measurement result was on the Table 49.1

49.3.3 Unscrambling of the Analysis Result

It is easy to see from the statistical data shown in 2000–2008, the region's economic indicators of all ethnic groups have been in a good upward trend, and the GDP increased steadily. But at the same time, the situation of its financial input of S&T changes are more complicated. Combined with its flexibility and elasticity coefficient table line chart, we can see the following characteristics:

First, the elasticity coefficient was unstable, even showing a huge shock situation.

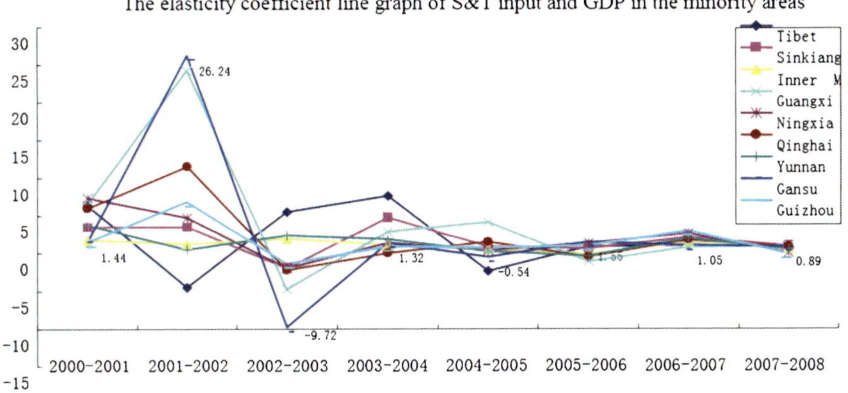

Fig. 49.1 The elasticity coefficient line graph of S&T input and GDP in the minority areas from 2000 to 2008

Table 49.1 The elasticity coefficient chart of S&T input and GDP in the minority areas

Elasticity coefficient	Tibet	Sinkiang	Inner mongolia	Guangxi	Ningxia	Qinghai	Yunnan	Gansu	Guizhou
$E_{2000-2001}$	6.33	3.45	1.74	6.45	7.32	5.98	3.73	1.44	1.48
$E_{2001-2002}$	−4.46	3.49	1.19	24.24	4.74	11.44	0.51	26.24	6.87
$E_{2002-2003}$	5.52	−1.99	1.88	−4.71	−2.04	−2.23	2.33	−9.72	−1.43
$E_{2003-2004}$	7.62	4.75	1.13	2.84	1.41	0	1.88	1.32	0.86
$E_{2004-2005}$	−2.35	0.82	0.47	4.09	0.39	1.53	0.21	−0.54	0.7
$E_{2005-2006}$	0.88	0.57	−0.07	−1.09	1.34	−0.47	−0.42	1.55	1.06
$E_{2006-2007}$	1.02	2.11	1.43	0.79	2.6	1.81	1.97	1.05	2.94
$E_{2007-2008}$	0.79	1.05	0.56	1.04	0	0.85	0.56	0.89	−0.17

The stable elasticity coefficient indicates that the region has a more mature, 'living within our means' the S&T input system. It shows that the region will adjust its financial input of S&T according to the change of economic conditions timely. But seen from Graph 1, in the 8 years with data analysis, the elasticity coefficient of each minority areas has significant jump, especially in Gansu and Guangxi, the volatility of its two is large. That is, when its GDP rose 1 %, its financial technology investment grew 26.24 %. But can this show that the local government attaches great importance to science and technology development and economic conditions in the premise is still lack of adequate S&T input to ensure it? Which, although in this year, its GDP rose by 1 %, its local financial technology funding compared to last year was reduced nearly 10 %; Guangxi Zhuang Autonomous Region have taken place a similar change in its elasticity fell from 24.24 at the same time in 2001–2002 to −4.71 in 2002–2004.

Second, the situation of negative elasticity coefficient appears frequently.

The elasticity coefficient is negative, meaning that GDP changes in direction and funding of local financial technology changes in the opposite direction. In this study of the 72 elastic coefficients, 14 coefficients are negative and the share is up to 19.4 %. As the selected sample data in each region's GDP each year are increased, therefore, the negative elasticity indicates that the region in the case of economic indicators to improve its science and technology investment appeared to reduce. The most typical is the previously mentioned Gansu Province, in the region when the GDP rose from 105.288 billion RMB to 112.537 billion RMB, the cost of technology reduced from 730 million to 240 million, and the elasticity is −9.72. Similar situation also occurred in the same period of Guangxi Zhuang Autonomous Region. In Tibet, Qinghai, Gansu, Guizhou, and Guangxi, there are two cases of economic growth with the phenomenon of technology spending cuts in the period of 8 years [9].

But at the same time, the results also showed that: In the 8 years of data analysis, the elasticity of minority areas in general appears a trend "discrete first and then centralize".

Elasticity coefficient in the minority areas in the line graph representing the nine ethnic regions to nine lines, both showed different levels of 'first discrete focus on' trends. Meanwhile, the data also show that in 2004, the financial technology investment of the ethnic majority areas to GDP reflect changes in sensitivity, that elasticity generally greater than 1, showing the elastic (up to 26.24); but after 2004, the elasticity of all regions in general decline in concentrations between 2007 and 2008 units of elastic in the vicinity.

The all above indicates, before 2004, the S&T input system in minority areas is relatively unstable, and be lack of procedural. But as time goes by, the S&T input system will get a certain degree of curing, and the kind of "freestyle" S&T input will be reduced effectively. To some extent, it shows that the mechanism of S&T input in the minority areas is progressing.

49.4 Suggestion on Perfecting the Financial Input of S&T System in Minority Areas

Science and technology are the key role in regional economic development and social progress. Since the new economic growth theory, which is led by Romer, undertaking the biochemistry inside the S&T progress, more and more countries and regions increase science and technology investment, hopefully to avoid the diminishing marginal returns rule of capital and realize the sustainable development of economy. For instance, in 2008, although the international financial crisis has serious costs for Indian economy, the public investment in S&T of India improves significantly. At the 95th annual science meeting, Prime Minister Singh said the government will improve the investment in science and technology, in the coming 5 years, they will improve the research and development input from 1 to 2 %. In November of the same year, in order to slow down the influence of the international crisis to biotechnology, the India government releases "The Biotechnology Industry Partners Plan" which costs 3.5 billion rupees. The plan will prevent the biotechnology research and development activities from affecting heavily because of the financial crisis.

In recent years, science and technology development in minority areas in China has made remarkable achievements, the output value of hi-tech industries, patent application, and so on various data are rising. But in the mean time, the Financial Input of S&T system in minority regions still needs to be perfected. Although the elasticity analysis in this research is much limiting, according to the result of the above analysis, we can conclude that there is not a promising Financial Input of S&T system in minority regions. Therefore, this study proposes the following suggestions:

49.4.1 Improving the Local Ethnic Government's Consciousness of S&T

It is the basic premise of setting up a sound system of the Financial Input of S&T in minority region. Our ethnic minority regions are almost spread over remote areas, due to long-term influence of nature condition, culture factors, and the level of economic and social developments in those areas is lagging significantly; therefore, the understanding of technology is comparatively low, the manager may pay much more attention to education, medical treatment, and agricultural development and so on. These all will result in the unstable state of financial input of S&T; behind the seriously vibrating elasticity, there will be the randomness problems of science and technology investment decision. Obviously, it is no good to ethnical technology's rapid development. So we must enhance the technology consciousness of ethnical region.

There are many approaches to enhance the technology consciousness of the governmental managers in minority areas. First, in cadre selection and retention stage, carefully choosing the candidate who own strong awareness of science and technology and excellent cultural level; second, through education-training, especially as holding on-the-job-training, promoting the scientific literacy of the cadres. Finally, the government in minority areas may realize the importance of increasing the Financial Input of S&T and develop the technology characteristic of minority areas practically.

49.4.2 Setting up the Institutional Financial Input of S&T in Minority Areas

That is the security protection of cultivating stabilized the financial input of S&T system. From the above analysis, we can conclude that the Financial Input of S&T in minority areas has not accordingly been increased as well as the economic indicators' substantial growth, even not any regularly variation; on the contrary, it will be in large vibration. Which indicates that in sometimes the Financial Input of S&T in minority areas is not completely lack of financial support, but rather lacks institutional guarantee, eventually, the investment scale cannot get judicial protection, so the Financial Input of S&T loses stability.

Therefore, we advise that we should institutionalize the financial input of S&T through establishing the legal system, for instance, in the 12th Five Year Plan, the government in minority areas can set goals of relative quantity and elastic coefficient to the Financial Input of S&T, and then take it down through (management by objectives) MBO, there will be accordingly indicators, which are regarded as the key indicators of the performance evaluation on local science and technology management departments, finally, there are laws to go by the financial input of S&T in minority areas, both the incentive system and the restraint mechanism.

49.4.3 Broadening the Channels of S&T Financing on Condition that Ensuring the Government's Financial Input of S&T

It is the necessary support for local government's input of S&T. Healthy and stable mechanism of S&T input is based on the whole society's participation, especially in the case of the unsteady financial resources input of S&T, social forces seems more important. Currently, insufficient funds are the crucial reason that is restraining science and technology development in minority areas; therefore, broadening the channels of S&T financing can alleviate the bottleneck pressure, providing the necessary supplement for the financial resources input of S&T. For example, we can practice the form of "investment grants", "tax preference", and so on, encouraging enterprises' science and technology input and ensuring the quantity of the financial resources input of S&T; strengthening the cooperation of science and technology between the regions, so we can fix the nature resource, characteristic resources in minority areas, and the technical resources and financial resources in majority areas together to achieve the goal of increasing the financial input of S&T together, guiding the investment areas of civilian capital, setting up multiple channels and multilevel system of the financial input of S&T.

49.4.4 Improving the Level of Scientific and Technological Management and Optimizing the Allocation of Resources

Currently, although the fiscal solvency in minority areas is limited, many social undertakings need to be developed. So on the premise of limited resources, we must improve the level of scientific and technological management so that we can adopt the financial input of S&T to the most appropriate field, earning the maximization of the social and economy benefits. Meanwhile, the local government in minority areas may improve the level of scientific and technological management through personnel selection and training and so on by setting up the cooperative mechanism between the government and university, government and government through various training and practice, achieving the goal of improving the cadres' allocation level of resources of S&T. Finally, all these will promote the progress of S&T and social development in minority areas.

References

1. Peng HT, Wang F (2010) Study on effects of financial input and output for S&T. Forum Sci Technol China (7):5–8
2. Sun XJ, Zeng GP (2009) Is there stabilized proportion of "optimal" R&D investments pattern. Stud Sci Sci (6):1815–1821
3. Zhao LY, Shi P (2009) Research on metrics for Spillover effect of input of basic research. Forum Sci Technol China (6):30–35
4. Zhang QR (2009) Empirical study of relative efficiency of regional S&T input based on DEA model. J Dalian Univ Technol (Soc Sci) (6):75–78
5. Wei SH, Wu GS (2005) Research on the efficiency of regional science and technology (S&T) resource allocation. Stud Sci Sci (4):467–473
6. Xiao M, Xie FJ (2009) Analysis of provincial allocation Efficiency of R&D resources and the determinants in China. Soft Sci (7):1–5

7. Zhu CK (2004) Study on dynamic equilibrium relation between economic growth and financial input of S&T. Sci Sci Manage S& T (8):29–33
8. Dong Cui L (2010) Measure on economic contribution rate of S&T input at Guangdong province. Bus Economy (9):5–7
9. Li HJ, Zhao JM, Ma YS (2010) Study on relationship between local financial input of S&T and economic growth based on interprovincial panel-data model science and technology progress and policy (7):44–51

Chapter 50
Solution of Two-Dimensional Groundwater Flow Steady

Bo Chao Qu, Wen Jing Zhao, Lin An Shi and Li Li

Abstract In this paper, we introduced the relevant theories of Radial Basis Functions and the Generalized Hermit Interpolation. And on that basis, we will construct the symmetric form of collocation methods using radial basis functions, and it will be applied to two-dimensional groundwater flow problems. It is a real mesh less method to solve differential equations, which does not need to divide the meshes and avoids the complex mesh generation process when it solves numerical discretization. When we adopt important radial basis function with symmetric collocation methods to solve groundwater problems, the simulation results are much better satisfied with easier operation, higher computational accuracy, and fewer errors.

Keywords Symmetric • Radial basis function • Collocation method • Two-dimensional groundwater flow steady

50.1 Introduction

50.1.1 The Radial Basis Functions

Definition 50.1 A function $\Phi: \mathbb{R}^s \rightarrow \mathbb{R}$ is called radial provided there exists a univariate function $\varphi: [0, +\infty] \rightarrow \mathbb{R}$ such that
$\Phi(x) = \varphi(r)$, where $r = \|x\|$, and $\|\cdot\|$ is some norm on $\mathbb{R}^s$—usually the Euclidean norm [1].
We usually use the RBF such as

1. Gaussian RBF: $\varphi(r) = e^{-(\varepsilon r)^2}$
2. Inverse multiquadric (IMQ) RBF: $\varphi(r) = \frac{1}{\sqrt{1+(\varepsilon r)^2}}$
3. Multiquadric (MQ) RBF [2]: $\varphi(r) = \sqrt{1 + (\varepsilon r)^2}$

B. C. Qu (✉) · W. J. Zhao · L. A. Shi · L. Li
Qing Gong College, Hubei United University, Tangshan, China
e-mail: qubochao@126.com

Z. Zhong (ed.), *Proceedings of the International Conference on Information Engineering and Applications (IEA) 2012*, Lecture Notes in Electrical Engineering 220,
DOI: 10.1007/978-1-4471-4844-9_50, © Springer-Verlag London 2013

We now use a radial basis function expansion to solve the scattered data interpolation problem in $\mathbb{R}^s$ by assuming

$$P_f(x) = \sum_{k=1}^{N} c_k \varphi(\|x - x_k\|_2), x \in \mathbb{R}^s.$$

The coefficients c_k are found by enforcing the interpolation conditions, and thus solving the linear system

$$\begin{bmatrix} \varphi(\|x_1 - x_1\|_2) & \varphi(\|x_1 - x_2\|_2) & \cdots & \varphi(\|x_1 - x_N\|_2) \\ \varphi(\|x_2 - x_1\|_2) & \varphi(\|x_2 - x_2\|_2) & \cdots & \varphi(\|x_2 - x_N\|_2) \\ \vdots & \vdots & \ddots & \vdots \\ \varphi(\|x_N - x_1\|_2) & \varphi(\|x_N - x_2\|_2) & \cdots & \varphi(\|x_N - x_N\|_2) \end{bmatrix} \begin{bmatrix} c_1 \\ c_2 \\ \vdots \\ c_N \end{bmatrix} = \begin{bmatrix} f(x_1) \\ f(x_2) \\ \vdots \\ f(x_N) \end{bmatrix}$$

50.1.2 The Generalized Hermit Interpolation Problem

We now consider data $\{x_i, \lambda_i f\},\ i = 1, \ldots, N,\ x_i \in \mathbb{R}^s$, where $\Lambda = \{\lambda_1, \ldots, \lambda_N\}$ is a linearly independent set of continuous linear functional and is some data function. For example, λ_i could denote point x_i and thus yield a Lagrange interpolation condition, or it could denote evaluation of some derivative at the point x_i. However, we allow the set Λ to contain more general functional such as, e.g., local integrals. This kind of problem was recently studied in Beaton and Langton [3].

We try to find an interplant of the form [4]

$$P_f(x) = \sum_{j=1}^{N} c_j \psi_j(\|x\|),\ x \in \mathbb{R}^s$$

With appropriate (radial) basis functions so that P_f satisfies the generalized interpolation conditions

$$\lambda_i P_f = \lambda_i f,\ i = 1, \ldots, N$$

To keep the discussion that follows as transparent as possible we now introduce the notation $\xi_1, \ldots, \xi_N$ for the centers of the radial basis functions. They will usually be selected to coincide with the data sites $\chi = \{x_1, \ldots x_N\}$. However, the following is clearer if we formally distinguish between centers ξ_i and data sites x_i. It is natural to let

$$\psi_j(\|x\|) = \lambda_j^{\xi} \varphi(\|x - \xi\|)$$

With the same functional λ_j that generated the data and φ one of the usual radial basic functions. However, the notation λ^{ξ} indicates that the functional λ

now acts on φ viewed as a function of its second argument ξ. We will not add any superscript if λ acts on a single variable function or on the kernel φ as a function of its first variable. Therefore, we assume the generalized Hermit interplant to be of the form

$P_f(x) = \sum_{j=1}^{N} c_j \lambda_j^{\xi} \varphi(\|x - \xi\|)$, $x \in \mathbb{R}^s$ and require it to satisfy

$\lambda_i P_f = \lambda_i f, \; i = 1, \ldots, N$

Then, we can linear system $Ac = f_\lambda$ which arises in this case has matrix entries $A_{ij} = \lambda_i \lambda_j^{\xi} \varphi, \; i, j = 1, \ldots, N$, and right-hand side $f_\lambda = [\lambda_1 f, \ldots, \lambda_N f]^T$.

50.2 Basic Principles

Considering the problem of Two-dimensional groundwater flow steady, i.e. the points in the aquifer head values do not change against the time, the equation is as follows [5]

$$LH = -\nabla(K\nabla H) = f, \; (x, y) \in \Omega \tag{50.1}$$

$$L_2 H = \frac{\partial H}{\partial n}\Big|_{\Gamma_2} = q \tag{50.2}$$

$$H\Big|_{\Gamma_1} = H_b \tag{50.3}$$

In the equation, $K = K(x, y)$ is the groundwater permeability coefficient, $f(x, y)$ is the source and sink terms, L and L_2 are the differential operator used in the differential equation, and Ω is the research area.

$$LH = -\nabla(K\nabla H) = -\left[\frac{\partial}{\partial x}\left(K\frac{\partial H}{\partial x}\right) + \frac{\partial}{\partial y}\left(K\frac{\partial H}{\partial y}\right)\right]$$

In order to be able to apply the results from generalized Hermit interpolation that will ensure the nonsingularity of collocation matrix, we propose the following symmetric radial basis function collocation method is applied to the problem of flow steady of groundwater; we can construct the following expansion for the function of the head value

$$\begin{aligned}\hat{H}(x) &= \sum_{j=1}^{N_0} c_j L^{\xi} \varphi(\|x - \xi\|)\,|_{\xi=\xi_j} + \sum_{j=N_0+1}^{N_0+N_1} c_j L_2^{\xi} \varphi(\|x - \xi\|)\,|_{\xi=\xi_j} \\ &+ \sum_{j=N_0+N_1+1}^{N_0+N_1+N_2} c_j \varphi\left(\|x - \xi_j\|\right)\end{aligned} \tag{50.4}$$

As it were $N = N_0 + N_1 + N_2$. As per Eqs. (50.1), (50.2), (50.3) we can get the discrete linear equations $Ac = F$. i.e.,

$$\begin{bmatrix} \hat{A}_{LL^{\xi}} & \hat{A}_{LL_2^{\xi}} & \hat{A}_L \\ \hat{A}_{L_2L^{\xi}} & \hat{A}_{L_2L_2^{\xi}} & \hat{A}_{L_2} \\ \hat{A}_{L^{\xi}} & \hat{A}_{L_2^{\xi}} & \hat{A} \end{bmatrix} \begin{bmatrix} c(x_i) \\ c(x_j) \\ c(x_k) \end{bmatrix} = \begin{bmatrix} -f(x_i) \\ q(x_j) \\ H_b(x_k) \end{bmatrix}$$

In the above

$i = 1, \ldots, N_0,\ j = N_0 + 1, \ldots, N_0 + N_1,\ k = N_0 + N_1, \ldots, N_0 + N_1 + N_2$

$$\left(\hat{A}_{LL^{\xi}}\right)_{ij} = LL^{\xi}\varphi\left(\|x - \xi\|\right)\Big|_{x=x_i,\xi=\xi_j}, x_i, \xi_j \in I$$

$$\left(\hat{A}_{LL_2^{\xi}}\right)_{ij} = LL_2^{\xi}\varphi\left(\|x - \xi\|\right)\Big|_{x=x_i,\xi=\xi_j}, x_i, \xi_j \in I$$

$$\left(\hat{A}_L\right)_{ij} = L\varphi\left(\|x - \xi_j\|\right)\Big|_{x=x_i}, x_i \in I, \xi_j \in B$$

$$\left(\hat{A}_{L_2L^{\xi}}\right)_{ij} = L_2L^{\xi}\varphi\left(\|x - \xi\|\right)\Big|_{x=x_i,\xi=\xi_j}, x_i, \xi_j \in I$$

$$\left(\hat{A}_{L_2L_2^{\xi}}\right)_{ij} = L_2L_2^{\xi}\varphi\left(\|x - \xi\|\right)\Big|_{x=x_i,\xi=\xi_j}, x_i, \xi_j \in I$$

$$\left(\hat{A}_{L_2}\right)_{ij} = L_2\varphi\left(\|x - \xi_j\|\right)\Big|_{x=x_i}, x_i \in I, \xi_j \in B$$

$$\left(\hat{A}_{L^{\xi}}\right)_{ij} = L^{\xi}\varphi\left(\|x_i - \xi\|\right)\Big|_{\xi=\xi_j}, x_i \in B, \xi_j \in I$$

$$\left(\hat{A}_{L_2^{\xi}}\right)_{ij} = L_2^{\xi}\varphi\left(\|x_i - \xi\|\right)\Big|_{\xi=\xi_j}, x_i \in B, \xi_j \in I$$

$$\hat{A}_{ij} = \varphi\left(\|x_i - \xi_j\|\right), x_i, \xi_j \in B$$

If we apply the abovementioned differential to the selected radial basis function, to solve the undetermined coefficients c_j and apply the results back to the formula Eq. (50.4), we can find the approximate solution [6–9].

50.3 The Application of Two-Dimensional Groundwater Flow Problems

50.3.1 The Two-Dimensional Steady Flow Problem with Continuous Changing Factors

We assume the study area Ω as the square area, and inside the area it meets [10]

$$\frac{\partial}{\partial x}\left(K\frac{\partial H}{\partial x}\right)+\frac{\partial}{\partial y}\left(K\frac{\partial H}{\partial y}\right)=0 \tag{50.5}$$

In the above, $K(x, y) = x^2$.
The Exact solution is

$$H = x^2 - 3y^2 \tag{50.6}$$

When the first boundary condition is satisfied on the boundary of the square area of Ω, we can calculate the specific value by Eq. (50.6).

If the corresponding interval of x, y of the research area Ω is $[5, 20]$, the head value of any point in the research area Ω can be calculated based on the formula of Eq. (50.6), and the result as per Fig. 50.1. We apply the aforementioned Radial Basis Function of IMQ, and the Symmetric collocation method in 1.3.1 to solve the approximation of the head value, and the calculated results shown as Fig. 50.2. We further compare the absolute error of the analytical solution and the approximate solution of head value as Fig. 50.3.

Conclusion: from Figs. 50.1 and 50.2, it can be found that almost no significant differences exist between the exact solution and approximate solution, and the graphics are very similar, and almost coincide. The Fig. 50.3 shows very small absolute error exists between exact solution and approximate solution. Therefore,

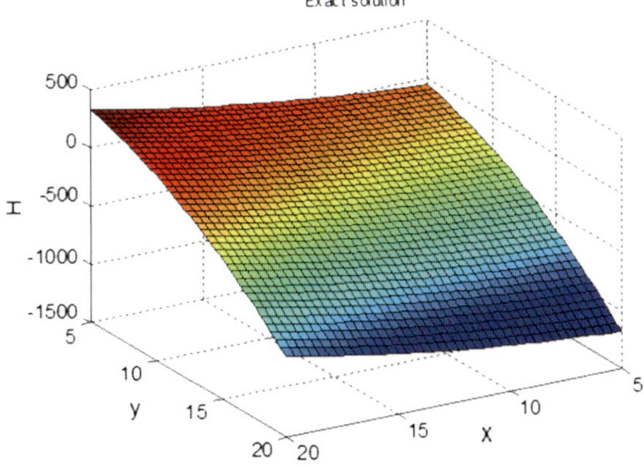

Fig. 50.1 Exact solution

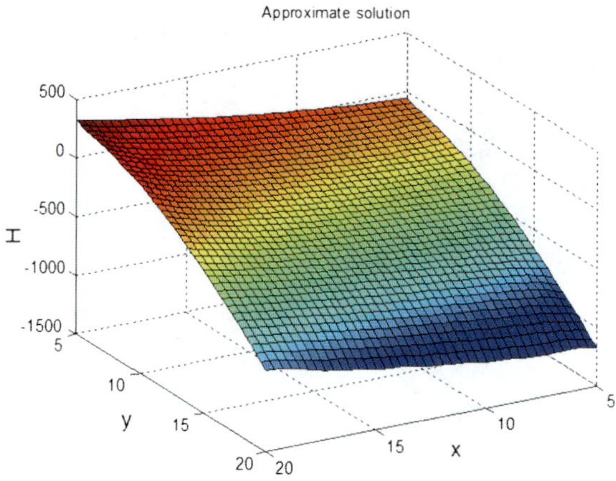

Fig. 50.2 Approximate solution

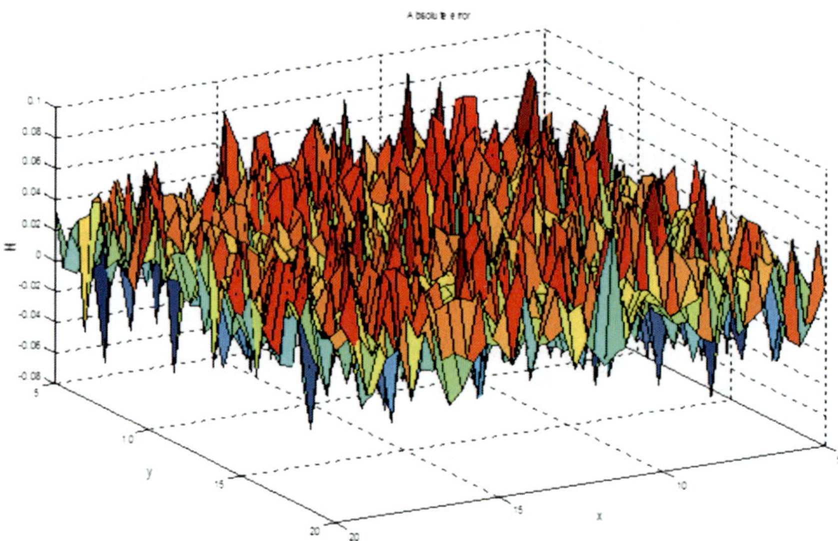

Fig. 50.3 The absolute error comparison of exact solution and approximate solution

through the specific practice, it can be found that the accuracy of symmetric collocation method for the numerical simulation of two-dimensional steady flow is much higher and can be much better simulate the actual problems.

References

1. Stein EM, Weiss G (1971) Introduction to fourier analysis on Euclidean spaces princeton, vol 1(14). Princeton University Press, New Jersey, pp 133–137

2. Hardy RL (1971) Multiquadric equations of topography and irregular surfaces. Geophys Res 2(11):905–912
3. Beaton RK, Langton MK (2009) Integral interpolation, in algorithms for approximation, vol 3(12). In: Isle VA, Levelly J (eds) Springer, Heidelberg, pp 199–218
4. France R, Hagen H, Nielson GM (1995) Repeated knots in least squares multiquadric functions, in Geometric Modelling-Dagstuh, vol 4(8). In: Hagen H, Farin G, Noltemeier H (eds) Springer, Berlin, pp 177–187
5. Yu qun X, Chun hong X (2008) Numerical simulation for Groundwater, vol 5(100). Science Press, Beijing, pp 134–137
6. France C, Schaback R (1998) Convergence order estimates of mesh less collocation methods using radial basis functions. Adv Comput Math 6(4):381–399
7. Fishier GE (2009) Solving partial differential equations by collocation with radial basis function. Surface Fitting Multiresolut Methods 7(5):412–416
8. Wend land H (2008).Scattered data approximation, vol 8(9). Cambridge University Press, Cambridge, pp 511–515
9. Jonathon B, Amine S, Chen K (2008) The Hermit collocation method using radial basis functions. Eng Anal Boundary Elem 9(24):607–611
10. Yu Qun X, Shu Jun Y, Chun Hong X, Yun Z (2004) Application of multistage finite element method to the simulations of groundwater flow. J Hydraul Eng 10(7):7–13

Chapter 51
Intersection of Triangulated Surfaces Based on Background Grid

Lian Qing Shu

Abstract An approach for the intersection of triangulated surfaces based on background grid is proposed. Our algorithm uses the bounding box of the triangulated surfaces to filter out the triangles that do not possibly have any intersection; then, a background grid is introduced to determine pairs of the triangles which are possible to intersect. Finally, our approach calculates out the intersection line segment between triangles. Experiments showed that the proposed approach can quickly and efficiently calculate the intersection line between triangulated surfaces.

Keywords Triangulated surface • Intersection • Bounding box • Background grid

51.1 Introduction

Computation of the intersection line between surfaces is an important task in computer-aided design. Many industrial applications such as solid modeling of assembly operation, robot path planning, generation of tool paths for NC machining, and generation of manufacturing data for composites, need to calculate surfaces intersection line. Generally, there are two ways to represent a surface in computer. One is using B-splices patches, Coon's patches, or nubs patches to represent surfaces. The other is adopting discrete data structures such as tessellate facets or triangular facets to represent surfaces. Utilizing triangular facets to represent surfaces have many strong points in engineering applications such as simple mathematical description, versatility, and flexibility; so the computation of triangulated surfaces' intersection lines is a hot research issue in computer graphics.

The naive way of computing intersection line between two triangular surfaces consists of calculating intersection segment between each triangle in one surface and each triangle on the other surface, then connecting each intersection segment end by

L. Q. Shu (✉)
Zhejiang University of Media and Communications, Hangzhou, China
e-mail: shulianqing@163.com

Z. Zhong (ed.), *Proceedings of the International Conference on Information Engineering and Applications (IEA) 2012*, Lecture Notes in Electrical Engineering 220,
DOI: 10.1007/978-1-4471-4844-9_51, © Springer-Verlag London 2013

end. However, there are a lot of triangles that do not have intersection segments, it will greatly reduce the speed of computing intersection line if we calculate for each couple of triangles including some that has no intersection segment. Currently, the algorithms of calculating triangulated surfaces' intersection line mainly include: (1) intersection between implicit surfaces based on Oct-tree [1], these algorithms are accurate and reliable, but they depend on the selection of Oct-tree root cell; (2) intersection between triangular meshes based on axis-aligned bounding box or bounding ball [2, 3], these methods are easy to implement and do intersection test, but the compactness is poor and the speed is slow; and (3) intersection between triangulated surface based on oriented bounding box tree [4], these methods have good compactness and fast speed, but they are complex to implement and do intersection test.

A new method of computing intersection line between triangulated surfaces is proposed in this chapter. We first adopt bounding box of triangulated surface to filter out some triangular patches, which have no intersection segments; then, a background grid is introduced to determine pairs of triangles which is possible to intersect. This method improves the speed of computing intersection line between triangulated surfaces greatly; experiments showed that the proposed approach can quickly and efficiently calculate the intersection line.

51.2 Related Concept

51.2.1 Triangulated Surface and its Data Structure

For each triangular patch on the triangulated surface, it generally has three adjacent triangular patches. If two triangular patches have a common edge, then they are neighbors to each other. As Fig. 51.1 shows, the edge V_2V_3 is a common edge between $T3$ and $T7$, so the triangular patch $T3$ is a neighbor of triangular patch $T7$ about edge V_2V_3. In Fig. 51.1, edge V_1V_9 is an edge of $T1$, but $T1$ has no neighbor about V_1V_9, so V_1V_9 is the boundary edge of the triangulated surface. We also record each triangular patch's neighbor about vertex. For example, triangular patch $T7$ is the neighbor of triangular patch $T3$ about V_{10}.

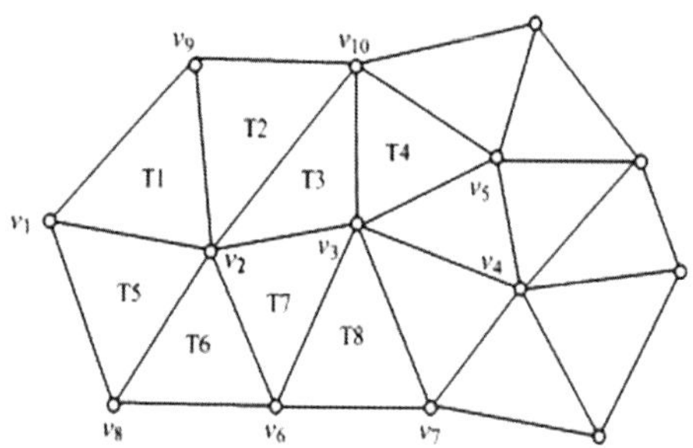

Fig. 51.1 Triangulated surface

We adopt a simple data structure [5] to record the topological information of triangular patches. The data structure is shown as follows:

Triangular patch List

{*TK*; // the kith triangular patch on the triangular patch list;
$V_1{}^k$; // the first vertex of *TK*;
$V_2{}^k$; // the second vertex of *TK*;
$V_3{}^k$; // the third vertex of *TK*;
$R_1{}^k$; // the neighbor triangular patch of *TK* about vertex $V_1{}^k$;
$R_2{}^k$; // the neighbor triangular patch of *TK* about vertex $V_2{}^k$;
$R_3{}^k$; // the neighbor triangular patch of *TK* about vertex $V_3{}^k$;}

51.2.2 Bounding Box

51.2.2.1 Bounding Box of Triangulated Surface

We calculate the bounding box *C* of a triangulated surface *S* with all vertices on *S*. We use $[X_{\min}, X_{\max}]$, $[Y_{\min}, Y_{\max}]$, and $[Z_{\min}, Z_{\max}]$ denote the bounding box of triangulated surface. $X_{\min}$, $Y_{\min}$, and $Z_{\min}$ are the minimum coordinate values of all vertices on *S*, respectively; $X_{\max}$, $Y_{\max}$, and $Z_{\max}$ are the maximum coordinate values of all vertices on *S*, respectively. As Fig. 51.2 shows, the triangulated surface is totally inside bounding box *C*.

51.2.2.2 Bounding Box of Triangular Patch

If the three vertices of the triangular patch *T* is V_1, V_2, V_3, the coordinate value is $V_1 = (a_1, b_1, c_1)$, $V_2 = (a_2, b_2, c_2)$, and $V_3 = (a_3, b_3, c_3)$, then the bounding box of the triangular patch *T* is $[x_{\min}, x_{\max}]$, $[y_{\min}, y_{\max}]$, $[z_{\min}, z_{\max}]$, in which

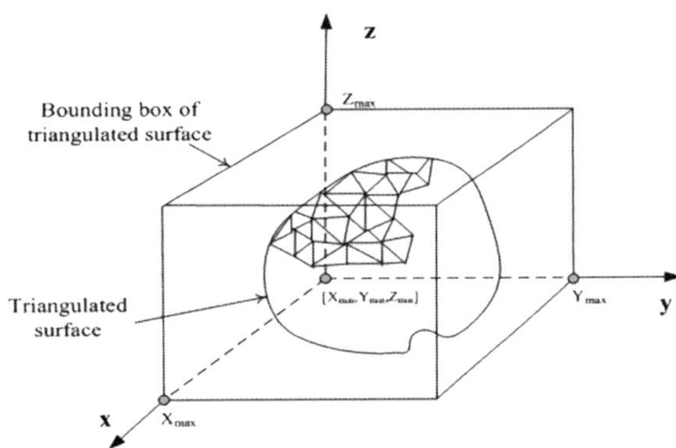

Fig. 51.2 Bounding box of triangulated surface

$$x_{min} = \min(a_1, a_2, a_3), x_{max} = \max(a_1, a_2, a_3) \quad (51.1)$$

$$y_{min} = \min(b_1, b_2, b_3), y_{max} = \max(b_1, b_2, b_3) \quad (51.2)$$

$$z_{min} = \min(c_1, c_2, c_3), z_{max} = \max(c_1, c_2, c_3) \quad (51.3)$$

51.2.3 Background Grid

The efficiency is low if we calculate intersection line between triangulated surfaces *S*1 and *S*2 by considering all the triangular patches on *S*1 and *S*2. As only a little part of triangular patches on *S*1 and *S*2 have intersection segment, it is necessary to filter out some triangular patches which have no intersection segment. We introduce background grid, by which a lot of triangular patches are filtered out.

A more sophisticated spatial subdivision with cells of variable sizes may be necessary for triangular patches of great difference in size. Here, we only discuss the background grid with same size. Followed by the sequence of background grid cells, we record the triangular patches that intersect with the cells. If there is at least one triangular patch from each group of the triangular surfaces in a same background grid cell, then there maybe intersecting lines in this background grid cell. If all the triangular patches from one triangular surface are not in the background grid cell, there will be no intersection segment. Hence, this condition will be used to check possible intersection as the first step.

51.3 Intersection of Triangulated Surfaces

51.3.1 Filtering Triangular Patches

Two triangular surfaces are denoted by *S*1 and *S*2, and the bounding box of the triangular surface *S*1 and *S*2 are denoted as *C*1 and *C*2. If *C*1 and *C*2 are disjoint, then *S*1 and *S*2 cannot intersect; if *C*1 and *C*2 intersect, then the triangular patches of *S*1 and *S*2 which is within *C*1 and *C*2 intersection public area may intersect (As Fig. 51.3 shows), any outside the public areas of *C*1 and *C*2 is impossible intersect. Thus, we can filter out some triangular patches that impossible intersect by bounding box [6].

The steps of filtering out some triangular patches by bounding box are as follows:

To compute out the bounding box of triangular surface *S*1 and recorded as *C*1; the bounding box of surface *S*2 recorded as *C*2; and to calculate the common part of *C*1 and *C*2, recorded as *C*.

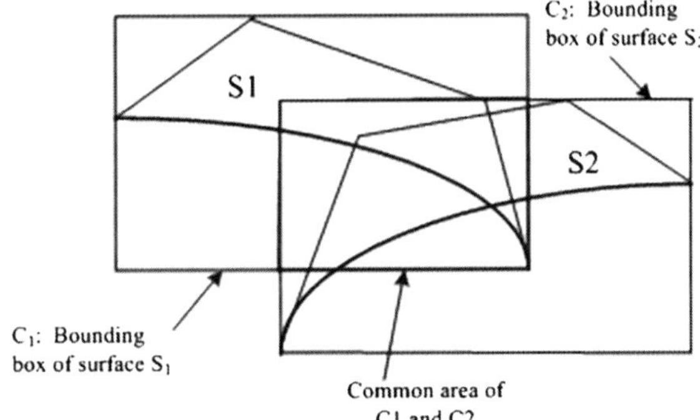

Fig. 51.3 The common area of bounding boxes

For each triangular patch on $S1$, determine if it is in C. If the triangular patch is in C; record it in possible intersecting queue $Q1$.

For each triangular patch on $S2$, determine if it is in C. If the triangular patch is in C; record it in possible intersecting queue $Q2$.

51.3.2 Pick out All Possible Intersecting Triangle Pairs

On the public area C, we define a background grid. The size of background grid cell is given as: $d_x = l_x / N_x, d_y = l_y / N_y$, and $d_z = l_z / N_z$, where l_x, l_y, l_z is the length, width, and height of public area C. The quantity of background grid cells in C is $N = N_x \times N_y \times N_z$. Here, the edge length of background grid cell is 1.2 times the average edge length of all triangular patches on $S1$ and $S2$ [7]. The steps of picking out all possible intersecting pairs of triangular patches by background grid cell are as follow:

Define a background grid on public area C of $C1$ and $C2$. Each background grid cell is recorded as C_i

For each triangular patch on possible intersecting triangular patch queue $Q1$. Determine whether triangular patch $T_k (T_k \in Q1)$ and background grid cell C_i intersect; In background grid cell C_i record the triangular patch TK if it intersects with C_i; End.

For each triangular patch on possible intersecting triangular patch queue $Q2$; determine whether triangular patch $F_j (F_j \in Q2)$ and background grid cell C_i intersect; in background grid cell C_i, record the triangular patch F_j if it intersects with C_i; End.

For each background grid cell C_i (from 1 to N); Exam each background grid cell C_i in turn, cell will be ignored if either TK from $S1$ or F_j from $S2$ is missing; End.

51.3.3 Calculating Intersection Line Segment

For each retained background grid cell C_i in which there have been possible intersecting pairs of triangular patches, the intersection segment for each pair of triangular patches is calculated. We use the following method [5]: assume that calculating intersection line segment between triangular patch *DEG* and triangular patch *ABC*, there are four cases of triangular patches' location [8].

In order to get the crossing point between triangular patch *DEG* and *ABC*, we must consider if the edge *DE*, *EG*, and *GD* have crossing points with triangular patch *DEG*. As Fig. 51.4a shows, edge *DE* will have a crossing point with triangular patch *ABC* if it meets the following two conditions:

Point *E* and point *D* are located in the opposite side of the plan, where the triangular patch *ABC* is in.

Crossing point *P* is inside the triangular patch *ABC*.

We adopt formula (51.4) to determine whether condition 1 is satisfied.

$$(AD \cdot \vec{N})(AE \cdot \vec{N}) \leq 0 \tag{51.4}$$

In which, $\vec{N}$ is the normal vector of triangular patch *ABC*, and it can get by formula (51.5);

$$\vec{N} = AB \times AC \tag{51.5}$$

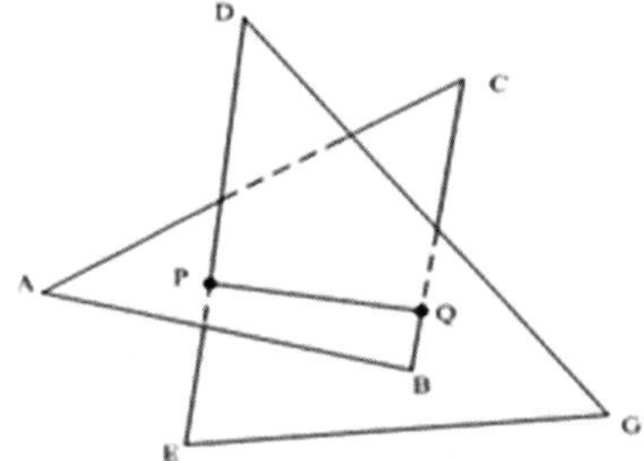

(a) crossing point *P* is inside the triangle

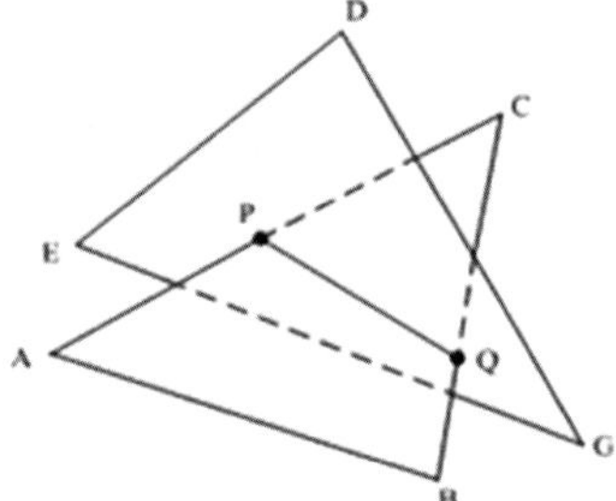

(b) crossing point *P* is on the edge of triangle

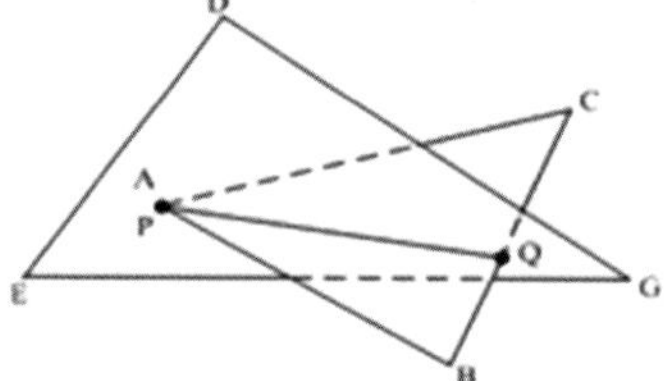

(c) crossing point *P* is on the vertex of triangle

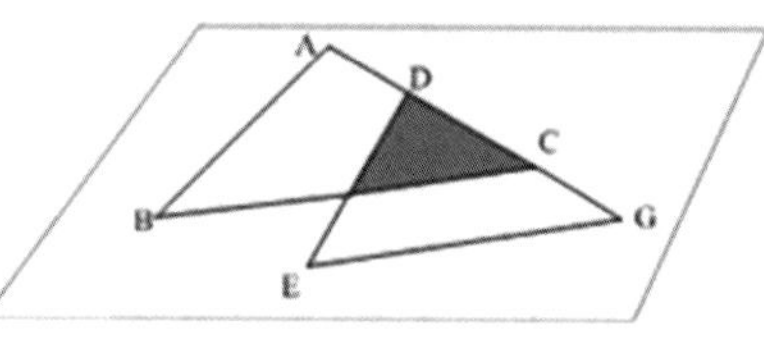

(d) two triangles is on the same plane

Fig. 51.4 The illustration for the position of two triangles

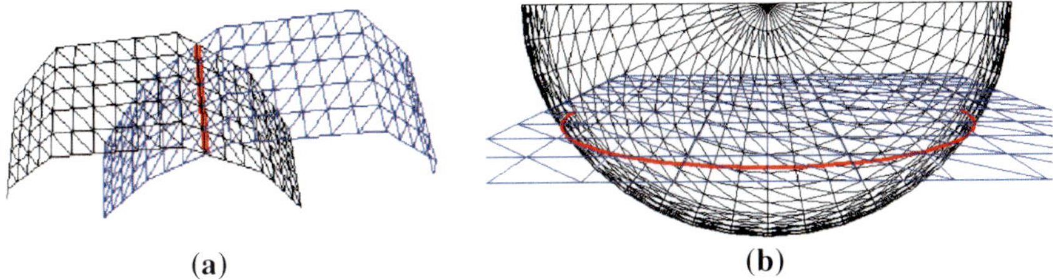

Fig. 51.5 Intersection results of two triangular surfaces

If the above condition 1 is satisfied, we calculate crossing point P by formula (51.6)

$$P = tE + (1 - t)D \tag{51.6}$$

In formula (51.6), $t = {}^{d}/_{(d-e)}$, in which $d = AD \cdot \vec{N}, e = AE \cdot \vec{N}$. Finally, if formula (51.7), (51.8), and (51.9) are satisfied, then P is inside the triangular patch ABC.

$$(AB \times AP) \cdot \vec{N} \geq 0 \tag{51.7}$$

$$(BC \times BP) \cdot \vec{N} \geq 0 \tag{51.8}$$

$$(CA \times CP) \cdot \vec{N} \geq 0 \tag{51.9}$$

Similarly, the second crossing point of triangular patch *ABC* and *DEG* can be obtained by considering if the edge *AB*, *BC*, and *CA* have crossing point with triangular patch *DEG* [9]. The second crossing point is denoted by *Q*, and then the line segment *PQ* is the intersection line segment of triangular patch *ABC* and *DEG*.

In this way, we can obtain all intersection line segments of triangular patch pairs. By connecting this intersection line segments end to end, the intersection line between the two triangular surfaces is obtained. The intersection line may be closed; it also can be open.

51.4 Experimental Result

Based on the above ideas, the algorithm has been implemented in Visual C++ 2003 platform. Figure 51.5a shows two triangular surfaces, on each of which include 200 triangular patches; Fig. 51.5b shows one spherical surface which includes 800 triangular patches and one plane which includes 200 triangular patches. Experiments show that the proposed algorithm can quickly and efficiently calculate out the intersection line between triangular surfaces.

51.5 Conclusions

In the basic way to compute the intersection line between two triangular surfaces, the efficiency is relatively low. In order to improve the speed of computing intersection line between triangular surfaces, we proposed an approach for the intersection of triangular surfaces based on background grid in this chapter. Experiments show that the proposed approach can quickly and efficiently calculate the intersection line between triangular surfaces. Especially, when two triangular surfaces only have a small part in their common area, our approach has significant advantages.

References

1. Yu Z-S, Wu Q-D, Li Q-Y (2001) Intersection between implicit surfaces based on Oct-tree. J Tong Gig Univ 2(5):567–570
2. Huber E (1998) Intersecting general parametric surfaces using bounding volumes. In: Proceedings of the 10th canadian conference on computational geometry, vol 3(6). Montreal, Canada, pp 151–155
3. Figures redo LH (1996) Surface intersection using affine arithmetic. In: Proceeding of conference on graphics interface, vol 4(8). Toronto, Ontario, Canada, pp 168–175
4. Gottschalk S, Manchu D (1996) Obtrude: a hierarchical structure for rapid interference detection. In: Proceedings of SIGGRAPH'96, vol 5(3). New Orleans, Louisiana, pp 171–180
5. Lo SH, Wang WX (2004) A fast robust algorithm for the intersection of triangulated surfaces. Eng Comput 7(20):11–21
6. Zhu X-X (2000) Free curve and surface modeling techniques, vol 8. Science Press, Beijing, pp 131–135
7. Jiang Q-P, Tang J, Yuan C-F (2008) Fast triangle mesh intersection algorithm based on uniform grid. Comput Eng 6(19):1123–1127
8. Philip J, Schneider H, David H (2005) Eerily computer graphics algorithm, vol 9. Publishing House of Electronics Industry, Beijing, pp 396–399
9. Peng Q-S (1984) An algorithm for finding the intersection lines between two b-spine surfaces. J Zhejiang Univ 10(5):85–98

Chapter 52
Research on Hospital Medical Expense Based on BP Neural Network

Feng Yin Su, Jian Hui Wu, Chao Chen and Dong Wang

Abstract Data on hospitalization expense for coronary heart disease from 2009 to 2010 in one tertiary hospital were taken as examples, so as to construct fitting models for hospitalization expense based on BP neural network. Sensitivity analysis of influence factors was executed to evaluate the effect degrees of influence factors on hospitalization expense. The purposes of this study was to evaluate application value of BP neural network used in hospitalization expense study, explore and analyze the evaluation method fitting for the construction and character of hospitalization expense, and use medical costs rationally and control irrational increase of hospitalization expense. The results showed that the main influence factors of hospitalization expense were hospitalization days, treatment outcome, times of rescuing, and age, the comprehensive influences of which were 0.83101, 0.76113, 0.73227, and 0.44537, respectively.

Keywords Hospitalization expense • BP Neural network • Influence factors • Sensitivity analysis

52.1 Introduction

Artificial neural network is a type of mathematical models which simulates biological neural network and is connected by massive simple neurons. Artificial neural network is a complicated network system that can in parallel process and nonlinear transfer information [1]. The application of artificial neural network has infiltrated into medical area recently. As one of the fullest developed and most used network models, BP neural network, which was most fully developed and widest applicator, was named after its special algorithm which used error back propagation algorithm according to the adjusted regulations of network weight [2].

F. Y. Su (✉) · J. H. Wu · C. Chen · D. Wang
Hubei Province Key Laboratory of Occupational Health and Safety for Coal Industry, Department of Epidemiology and Health Statistics, Hubei United University, Tangshan, China
e-mail: sufengyin@126.com

Z. Zhong (ed.), *Proceedings of the International Conference on Information Engineering and Applications (IEA) 2012*, Lecture Notes in Electrical Engineering 220, DOI: 10.1007/978-1-4471-4844-9_52, © Springer-Verlag London 2013

Nowadays, study of hospitalization expense has become the hot spot issue of our society [3]. As the distribution of data on hospitalization expense always displayed is skewed and can be affected by many complicated factors which are related to each other, there exists some limits for traditional statistical methods to study hospitalization expense. As BP neural network has no requirements for type or distribution of the material, has the ability of fault tolerance, and is able to define the complex mapping relation between input variables and output variables, BP neural network has supported a brand new method for processing data with complicated relationships [4].

Our study took coronary heart disease for instance, constructed the BP neural network for hospitalization expense of hospitalized cases to figure out the main influence factors of hospitalization expense, in order to explore and analyze the evaluation method fitting for the construction and character of hospitalization expense, and could support basis for taking target-oriented measures to use the medical resource rationally and control the growth of hospitalization expense.

52.2 Basic Theory of BP Neural Network

52.2.1 Basic Thought of BP Neural Network

BP neural network, which was first proposed by a research group led by Rumelhart and McClelland in 1986 [5], is a kind of multilayer forward propagation networks and one of the most used neural networks currently. The classical network topology structure of BP neural network is composed by input layer, hidden layer, and output layer. It has been theoretically proved that function approximation of any accuracy from N dimensional to M dimensional can be implemented by a BP neural network with one hidden layer for any continuous function on a closed interval. Compared with traditional statistical methods, advantages of BP neural network are as followed: has no requirements for type or distribution of the material, is able to define the complex mapping relation between input variables and output variables by self-study and self-organization, and has much better ability to deal with nonlinear problems than traditional statistical methods.

52.2.2 Algorithm and Principle of BP Neural Network

BP neural network is based on gradient descent algorithm; the training of the whole network is composed by forward propagation and back propagation. According to forward process, sample signals are put into input layer, output from output layer after processing by weight, threshold, and transfer functions of neuron. If the error between output value and expected value is over expected error, the error back propagation stage is initiated to modify, error signals are sent back through the former connecting passage, error signals are minimized by modifying weights of neuron of each layer. The process to constantly revise weights is the process of network training. The loop ends until the output error has been reduced to allowable value or has reached the fixed training number.

52.2.3 Basic Steps of BP Modeling

Simply speaking, there are three steps to construct BP neural network: (1) network initialization; (2) training; and (3). simulation. Network initialization is to select and fix some network parameters, such as number of network layers, number of neurons in each layer, transmission function of each layer, speed of training, weights, threshold, and algorithm of training. Make sure to avoid over fitting during training [6].

52.2.4 Select the Optimization Algorithm

As standard BP algorithm can easily be immerged in partial minimum and convergence speed is too slow, the training time might be very long [7]. Many optimization algorithms have appeared nowadays, such as LM algorithm and Newton algorithm. The convergence speeds of these new algorithms were greatly improved [8].

52.2.5 Sensitivity Analysis

By changing some part of network input to observe the corresponding change of network output, the significance of this part for predicting output can be determined. The concrete methods are to change the recorded variables of samples in succession, then record the maximum and minimum output during the changing process, and calculate the percentage of the difference value of maximum and minimum output in maximum output, and the sensitivity is the mean of all the recorded percentages [9, 10].

52.3 Application

Matlab 7.1 developed by Math Works Company was used to construct BP neural network model in our study, and influence degrees of influence factors on hospitalization expense were evaluated by editing programs of sensitivity analysis based on Matlab.

52.3.1 Data Sources

Data in our study come from front pages of medical record with coronary heart disease in one tertiary hospital in Tangshan city from 2009 to 2010, and number of total cases was 2,437.

52.3.2 Data Initialization

All of the data were recorded by computers; cases with missing values and illogical cases were rejected. The number of valid cases was 2,126, and proportion of valid cases in all cases was 87.26 %.

52.3.3 Database Division

Quasi-Newton Method (OSS algorithm) was used in our study. Sequence database, by numbers of medical records, selected numbers 3, 8, 13, 18, … (in equal intervals) in succession as testing set, the others as training set. Make sure database was divided into training set (80 %) and testing set (20 %). Training set was used to construct BP neural network model, and testing set was used to test network and its generalization ability (Fig. 52.1).

Factors	Code	Quantized methods
Gender	x1	Male=1, Female=2
Marital status	x2	In marriage=1, Others=2
Ages	x3	Years
Admission times	x4	Once=1, Twice and more=2
Rescue	x5	Yes=0, No=1
Payment pattern	x6	Insured=1, uninsured=2
Hospitalization days	x7	days
Treatment outcome	x8	Cured=1, Improved=2, Uncured=3,

Fig. 52.1 Quantized methods for influence factors

Training times	5		15		20	
	Testing set	Training set	Testing set	Training set	Testing set	Training set
1	0.7460	0.7162	0.7009	0.7188	0.7059	0.7173
2	0.7455	0.7171	0.7068	0.7149	0.6978	0.7180
3	0.7214	0.6938	0.7003	0.7183	0.6997	0.7174
...	...	...	...	...	...	...
98	0.7451	0.7159	0.7008	0.7182	0.6981	0.7188
99	0.7483	0.7188	0.7009	0.7186	0.6999	0.7184
100	0.7465	0.7179	0.6991	0.7178	0.7015	0.7184

Fig. 52.2 Performance outcomes of different training networks

52.3.4 Performance Results of Training Network with Different Numbers of Neurons in Hidden Layer

Networks with different numbers of neurons in hidden layer were randomly tested 100 times in our study; the results were shown in (Fig. 52.2).

52.3.5 Results of Hospitalization Expense Modeling

A network model with both satisfactory generalization ability and fitting ability was selected after multiple training. The BP neural network model on hospitalization expense and the influence factors of hospitalized patients with coronary heart disease were finally constructed (Fig. 52.3).

Network structure parameters	Network training parameters	Simulation results of testing set	Fitting results of training set
hidden layers: one	Training algorithm: OSS algorithm:	R=0.86209	R=0.84331
neurons in hidden layer: 15	A total cessation of training iterations: 20	R^2=0.74321	R^2=0.71118
neurons in input layer: 8	Leaning speed: 0.01	$R^2_{adj} = 0.41534$	$R^2_{adj} = 0.52435$
neurons in output layer: 1	Performance function: *SSE*	SSE=5.2389e+009	SSE=2.081e+010
	Stop training *SSE**=1.22463	MSE=1.6738e+007 RMSE=4091.2	MSE=1.7327e+007 RMSE=4162.6

* SEE was for normalized data and RMSE was for reverse normalized data.

Fig. 52.3 Parameters of BP neural network on hospitalization expense

Rank	Influence factor	Sensitivity
1	Hospitalization days	0.83101
2	Treatment outcomes	0.76113
3	Times of rescuing	0.73227
4	Age	0.44537
5	Admission times	0.41421
6	Marital status	0.40873
7	Payment pattern	0.33751
8	Gender	0.07391

Fig. 52.4 Ranking of influence degrees of influence factors for hospitalization expense

52.3.6 *Results of Sensitivity Analysis on Influence Factors of Hospitalization Expense*

Sensitivities of influence factors were analyzed in our study, the results were as followed (Fig. 52.4).

As can be seen from Fig. 52.4, the factor with the largest influence degree was hospitalization days, which was followed by treatment outcomes, times of rescuing, age, admission times and so on. Factors with less influence degree were marital status, payment pattern, and gender. The results were concordant with theories, actual situation, and the other literature reports.

52.4 Conclusion

The theories of artificial neural network have been gradually improved nowadays. As one of the most used models, the medical applications of BP neural network become more and more popular. BP neural network was based on developed computer technology and has abandoned traditional statistic analysis methods which hypothesis should be put forward before verified. As no hypothesis was required for the study problems, the application prospect of BP neural network for exploring etiology was outstanding, especially when the relationships of variables are unknown. The application has shown that the applicability and prospects of BP neural network for hospitalization expense and influence factors were satisfactory.

References

1. Ge ZX, Sun ZQ (2007) Neural network theory and Matlab 2007 simulation, vol 1(2). Electrics Industry, Beijing 45–49

2. Liu B (2006) Research of model construction for hospitalization expense based on BP neural network, vol 2(6). Zhejiang University, Hangzhou, pp 101–109
3. Yan yan J, Jie Y (2006) The study of reasons and countermeasures for rapid rise of medical service costs. Cost Theor Appl 3(7):81–86
4. Wang J, Li M, Hu YT (2009) The analysis on the influencing factors of hospitalization expenses of the patients with gastric cancer by using bp neural network. Chin J Health stat, 4(14):99–103
5. Rumelart DE, James L (2009) McClelland and the PDP research group, ends. Parallel Diatribe 5(5):123–126
6. TED Processing (1986) Voles 1 and 2, vol 6(12). The M. I. T. Press, Cambridge, pp 171–176
7. SPSS (2009) Inc. neural network algorithms. Chicago Ill, 7(15):17–23
8. Xuan Jun D, Wei Na J (2010) Predictive efficiency comparison of ARIMA-time series and BP neural net model on infectious diseases. Mod Pract Med 8(3):142–145
9. Liu HB, Yang YL, Duan ZW (2009) Based on the neural network model to predict the future of coal workers pneumoconiosis hazard research. Chin J Health Stat 9(4):617–618
10. Zhang Q, Tian J (2009) Research on hybrid prediction methods of silicosis based on BP neural. Network 10(4):264–267

Chapter 53
Risk Assessment of Cement-Soil Pile Composite Foundation Using Rough Set

Hongqi Wang, Yiwen Zhu, Junsheng Chen and Shanshan He

Abstract Rough set theory was used to analyze the stability level and risk grade of 30 site tests of cement-soil pile in Chaozhou Water Supply Control Project. And the confidence level and risk of cement-soil pile composite foundation was evaluated to determine the stability of whole Chaozhou Water Supply Control Project. With the refined deduce of rough set analysis theory, the results according to five steps are shown in this paper. The results indicate that the contrast test between rough set analysis results and the site test results are consistent. The contrast results prove that the risk assessment method of rough set theory is effective for analysis cement-soil pile composite foundation.

Keywords Rough set theory • Cement-soil pile • Risk assessment • Reliability analysis

53.1 Introduction

At present, geotechnical engineering risk assessment theory in our country is still in the stage of development. Mature risk assessment system has not been formed. Only in nuclear power station, tunnel, oil drilling platform, large group of underground cavern, metro tunnel, and other large civil engineering construction, risk assessment is widely used, while risk assessment of cement-soil pile composite foundation is almost a blank.

There are certain aspects of the research results on reliability analysis of cement-soil pile composite foundation. A lot of scholars did some relation calculation of the reliability analysis and engineering test, and got some regularity

H. Wang · Y. Zhu · S. He
School of Civil Engineering and Architecture, Wuhan University, Wuhan, China

J. Chen (✉)
School of Civil Engineering and Transportation, South China University of Technology, Guangzhou, China
e-mail: jschen@scut.edu.cn

Z. Zhong (ed.), *Proceedings of the International Conference on Information Engineering and Applications (IEA) 2012*, Lecture Notes in Electrical Engineering 220,
DOI: 10.1007/978-1-4471-4844-9_53,

and beneficial conclusion. Since the 1950s, geotechnical engineering risk analysis research also has made some achievements. But the achievements are mainly in the establishment of the concept and qualitative research, and quantitative research often troubled in the calculation of reliability. Present findings are not much on how to further the combination of technology and economic indicators, and are lack of qualitative and quantitative adaptable risk analysis and evaluation methods.

No special risk assessment method is present for cement-soil pile composite foundation. Only the reliability of cement-soil pile composite foundation is being researched. So, it is still needed to build a complete risk analysis system and model.

53.2 Guangdong Chaozhou Water Supply Control Project Survey

General Situation Guangdong Chaozhou Water Supply Control Project is located in Chaozhou, Guangdong Province of China, which is close to Chaozhou city. The project is used for rational allocation of water resource between town, industrial, and agriculture. And it is a pivotal project for water power, shipping, and water environmental protection. Engineering level is 1 and the main buildings are grade I level.

Geology Situation Field geological survey data show that there are 70–100 m thick quaternary strata in the dam site, among thick soft plastic to plastic shape silt or mucky soil. On the dam axis, soft soil thickness of Dongxi is 11.3–18.9 m, and level elevation is −5.0 to 0.0 m; soft soil thickness of Xixi is 12.9–18.9 m, and level elevation is from −2.89 to 0.72 m.

Construction Deep mixing cement-soil pile was adopted to strengthen the 20 m thick silt. In the preliminary design, cement-mixed ratio of cement-soil pile composite foundation is 15 %, pile diameter is 500 mm, center to center spacing of piles is 1.2 m, the pile length 18 m, and the layout pattern of pile is Mei flower arrangement. Bearing capacity of composite foundation is 130 kPa, bearing capacity of single pile is 120 kN, and unconfined compressive strength of pile is larger than 1,500 kPa.

53.3 Rough Set Theory

Rough set theory, which was proposed by Polish mathematician Z. Pawlak in 1982, is a new mathematical tool to solve fuzzy and uncertainty problems [1, 2]. The main idea is to keep the unchangeable classification ability, reduce the knowledge, and export the decision-making or classification rules.

The Index Data Acquisition Based on Rough Set Theory Based on the established evaluation index, and on the basis of composite foundation field tests of Chaozhou Water Supply Control Project in Guangdong province, integrating with the eight factors of failure criterion of cement-soil pile, basic data of evaluation index are obtained. Part of data is acquired from test image, which would bring slight numerical difference. Part of index data is general value due to no special test for pile body (for example, deformation modulus of pile is average value 80 MPa due to no pile body test). On the other hand, there are tiny differences between some data and the real value. Because index data will be discrete during rough set data processing, these data disadvantages on the overall risk assessment can be ignored.

Data Discretization Rough set theory cannot treat continuous attributes directly and effectively. Therefore, before carrying out the operation, all data must be discretely processed. Due to the general characteristics of composite foundation, different data of the same evaluation index has no obviously difference. So in this article, adaptive discrete method is used to process the data. According to the three given breakpoint parameters, all data of each evaluation index of 30 groups of test data are divided into four discontinuous levels, and the attribute values are combined to reduce the number of attribute value and improve the fitness of knowledge.

Stability evaluation index of cement-soil pile composite foundation are divided into four levels, including grade 1–4, which separately mean poor stability, general stability, good stability, and excellent stability. Basic data of evaluation index used for cement-soil pile composite foundation is scattered, which constitute knowledge representation system.

The Index Reduction and Weight Determination according to calculation and analysis, each index is necessary with no redundancy index [3]. It means that eight indexes reduction result is itself.

According to the calculation results, weights of eight evaluation indexes are displayed in Table 53.1 Obviously, the weight of eight indexes is not much different due to any very different geometry properties of cement-soil pile composite foundation.

Determination of Single Factor Membership Take single factor basic data about middle Sect. 53.1 of switch into the single factor membership functions [4] to middle Sect. 53.1 of switch, get the single factor risk evaluation sets V = {v1, v2, v3, v4} = {poor stability, general stability, good stability, and excellent stability} membership matrix is:

$$R_1 = \left\{ \begin{matrix} 0.000 & 0.000 & 0.833 & 0.617 \\ 0.617 & 0.617 & 0.000 & 0.000 \\ 0.000 & 0.000 & 0.200 & 0.800 \\ 0.000 & 0.667 & 0.333 & 0.000 \\ 0.000 & 0.000 & 0.413 & 0.587 \\ 0.000 & 0.500 & 0.500 & 0.000 \\ 0.000 & 0.000 & 0.300 & 0.700 \\ 0.071 & 0.929 & 0.000 & 0.000 \end{matrix} \right. \tag{53.1}$$

Table 53.1 The weight of evaluation index

Index X	E_0	φ	Pile spacing	Length	Replacement rate	Cement content	Content of organic matter	Age
Weights σ	0.124	0.129	0.120	0.129	0.134	0.124	0.124	0.115

According to this method, single factor membership matrix of each single pile (30 groups of single pile data) of cement-soil pile composite foundation can be calculated.

Comprehensive Feudatory Degree Evaluation of the corresponding weight vector of risk assessment factors of cement-soil pile overall stability is U:

$$\begin{aligned}\omega &= [\delta 1 \cdots \delta 2 \cdots \delta 3 \cdots \delta 4 \cdots \delta 5 \cdots \delta 6 \cdots \delta 7 \cdots \delta 8] \\ &= [0.124 \cdots 0.129 \cdots 0.120 \cdots 0.129 \cdots 0.134 \cdots 0.124 \cdots 0.124 \cdots 0.115]\end{aligned} \tag{53.2}$$

Use $M(\cdot, \oplus)$ type operation combines the weight vectors and fuzzy matrix R [5] and acquires comprehensive membership B. For the middle Sect. 53.1 of dam, its comprehensive membership $B1$ [6]:

$$B1 = [0.0297 \cdots 0.3623 \cdots 0.3248 \cdots 0.2822] \tag{53.3}$$

According to $b_M = \max\limits_{1 \le j \le n} \{b_j\}$ and the maximum membership degree principle, maximum membership degree $B1$ equals to 0.3623, and the corresponding risk assessment rating is V2, which means stability level of the middle Sect. 53.1 of dam is general.

According to this method, comprehensive feudatory matrix of cement-soil pile composite foundation can be calculated.

53.4 Results and Analysis

53.4.1 Analysis of the Test Results

Through comprehensive evaluation results of 30 groups of test data, it is easy to find that 30 group areas of the pile body stability are in general stability and good stability level, and 90 % of the 30 groups of test data are in good stability level. It means that 30 groups of pile data are mainly in good stability level, without major risk. In addition, the group data of membership which is located in the risk level of good stability is 0.5, and some data is even more than 0.7. It means that 27 pile bodies have good stability. On the other hand, 3 pile bodies have still poor stability and data of membership is nearly 0.3, which shows that the stability of 30 groups of test pile body is good, but 30 % of pile bodies are in high risk (poor stability). The poor stability piles mean risk potential [7], and the relevant departments should strengthen inspection and monitoring, find hidden danger in time, eliminate hidden dangers, and control accident.

Calculation and analysis on 30 groups of test data indicate that there are no poor stability piles and 90 % piles have good stability. According to site test of cement-soil pile composite foundation, some conclusions can be drawn as the follows:

The ultimate bearing capacity of pile is larger than 260 kPa. The design load is 130 kPa. The average subsidence under 260 kPa loads is about 5 mm.

Under different replacement ratio, the maximum settlement is not more than 6.6 mm.

Ultimate bearing capacity of single pile obtained by plate loading test is larger than 430 kN.

Damage loadings of pile body obtained by unconfined compressive test are, respectively, 1650, 2250, and 2000 kPa.

Generally, cement in all test piles is even, setting state is good, and compressive strength is high.

The contrast above indicates the site test conclusions and fuzzy comprehensive evaluation are similar which means rough set theory is applicable for cement-soil pile composite foundation to evaluate comprehensive risk assessment. The rough set theory provides the real-time control analytical methods to understand the risk status.

53.4.2 Counter-Example Comparative Analysis

In order to further prove the validity of the rough set model, based on positive verification, counter-example comparative analysis is performed combining with the damage test data to verify the rough set model from reverse side.

At present, Chaozhou Water Supply Control Project is in normal operation, and there are no damage examples of cement-soil pile composite foundation. Therefore, based on site test of single pile plate loading test, counter-example comparative analysis is used. According to ultimate loads of pile 2, pile 21, and pile 22, the corresponding deformation modulus of Pile 2, Pile 21 and Pile 22 are, respectively, 16.97, 21.892, and 23.703 MPa. Only the deformation modulus changes with load increasing during the single pile damage test, and other seven soil parameters are unchanged. So the basic data of counter-example contrast can be obtained (Table 53.2).

According to site test and analysis, only the deformation modulus changes with load increasing during the single pile damage test and other seven soil parameters are unchanged. Although all eight indexes need to be considered to analyse

Table 53.2 Basic data of opposite verifying test

Test point	E_0 (MPa)	φ (o)	Pile spacing (m)	Pile length (m)	Replacement rate (%)	Cement content (%)	Age (d)
Pile 2#	16.97	31	1.2	18	24.2	15	33
Pile 21#	21.892	31	1.2	18	24.2	15	34
Pile 22#	23.703	31	1.2	18	24.2	15	32

the pile stability, the relative importance is different to rough set theory analysis mentioned above. So, eight indexes cannot be equally weight computing. Expert evaluation method is adopted to evaluate the weight of eight indexes under pile damage situation. Through five experts' independent evaluation, new weights of eight indexes are given. Considering the common influence of macroscopically comprehensive analysis of the index, other seven indexes are unchanged, so partition area of risk level is the same as rough set analysis (Table 53.3). Therefore, when selecting fuzzy mathematics evaluation, the comprehensive membership of seven other indexes of Pile 2#, Pile 21#, and Pile 22# is unchanged. According to Table 53.2, under the maximum loads of three groups of test pile, three groups of comprehensive test pile membership degree matrix results are shown in Table 53.4:

According to the principle of maximum membership level, calculation results of site tests of Pile 2#, Pile 21#, and Pile 22# belong to first level, which is poor stability. The result is consistent with the site tests. The site tests show that when deformation modulus are, respectively 16.97, 21.892, and 23.703, Pile 2#, Pile 21#, and Pile 22# are damaged.

It should be noted that treating only three groups of site test as counter-example analysis may be not very persuasive. But through this method, one counter-example can achieve the desired effect. If permitted, abundant counter-examples for verification are much better.

Through contrast model calculation results and site tests between positive and negative aspects, it is proved that model calculation results and site tests are highly consistent, which means the rough set model is applicable for Chaozhou Water

Table 53.3 Stability evaluation index level of cement-soil pile composite foundation

Evaluation index	Poor stability	General stability	Good stability	Excellent stability
Deformation modulus (MPa)	<25	25–40	40–100	>100
φ (0°)	<5	5–20	20–40	>40
Pile spacing (m)	>6	3–6	1–3	<1
Pile length(m)	>30	18–30	10–18	<10
Replacement rate (%)	<5	5–10	10–20	>20
Cement content (%)	<5	5–15	15–20	>20
Content of organic matter (%)	>15	5–15	1–5	<1
Age (d)	<7	7–14	14–90	>90

Table 53.4 Matrix statistical table of comprehensive membership of damage test

Test point	V1	V2	V3	V4	Max membership	Risk level
Pile 2#	0.661	0.339	0.000	0.000	0.661	V1
Pile 21#	0.562	0.438	0.000	0.000	0.562	V1
Pile 22#	0.526	0.474	0.000	0.000	0.526	V1

Supply Control Project. It is mainly because that fuzzy mathematical evaluation method [8] is proper to solve uncertain problems and the running status of cement-soil pile composite foundation is an unknown fuzzy situation. Simultaneously, a lot of problems or factors are difficult to be defined clearly. Hence, it is accurate to use fuzzy evaluation method conclusions in this paper, and fuzzy evaluation method can be used for the whole pivotal project.

On the other hand, traditionally, when most scholars use fuzzy mathematics to evaluate object, experts scoring method [9, 10] is usually used to determine weight, which may contain subjective personal thought. It may deny the objectivity of research object to a certain extent. Rough set theory used in this paper for determining weight is completely from the characteristics of data itself. Through the data analysis and integration, the interactions among different factors are determined. Interactions between different factors and relative importance of different factors are known. Rough set theory analyzes the research object in objective point of view which can reflect the actual situation of research objects. In this paper, risk assessment using rough set theory to determine the weight is accurate. It proved that evaluation method is correct and the method to determine the weights of evaluation is suitable as well.

53.5 Conclusions

This paper first introduces the risk assessment method and theory, and then uses site test data of whole water supply control project to calculate and analyze to acquire the stability level of 30 single pile and membership degree of each risk level. Through the contrast verification with site test, the theoretical calculation and site test results are consistent with each other, which prove that the method is suitable for cement-soil pile composite foundation.

References

1. Han ZX, Zhang Q, Wen FS (1999) Rough set theory and its application. Control Theor Appl 2(4):17–20
2. Xu HF, Wei LH, Lu JF (2011) Tunnel safety risk assessment study in Meihuaqing. Control Subgrade Eng 5(3):153–155
3. Wang GY, Cui HL, Li Q (2009) The study of determining weight factors method in the slope stability assessment based on rough set theory. Rock Soil Mech 8(13):2418–2422
4. Chen YS, Fang J, Tong ZG (2011) The application of the fuzzy mathematics method in water distribution network of the water quality evaluation. Water Supply Drainage Southwest 33(22):15–18
5. Zhang YJ, Cao WG, Zhao MH (2011) The interval fuzzy evaluation analysis method about the karst area highway subgrade's stability. Chin J Geotech Eng 33(14):38–44

6. Chen Y, Zhao MH, Cao WG (2004) Fuzzy evaluation of the end of pile in highway and bridge's karst stability. In: Eighth national geotechnical engineering and process research papers, vol 13(6). pp 310–315
7. Fan F (2010) Five kinds of risk control on the process of project implementation. Railway Eng Enterp Manag 53(22):19–20
8. Wu JP, Cheng Q (2011) Celerity estimation method on cost of railway station building engineering based on fuzzy method. Railway Transp Econ 33(8):1255–1259
9. Cong RG, Xiong C, Wu YN, Zhang YL (2010) The project evaluation method and application of adjustment based on the target of the logical relation. Proj Manag Technol 8(16):833–838
10. Huang XP (2010) The structure state evaluation of subway tunnel used the expert scoring evaluation system. Undergr Eng Tunnels 34(21):350–361

Chapter 54
Electronic Commerce Innovation and Development Strategy of Travel Agency Under Low-Carbon Economy

Wei Huang and Shaojian Zhou

Abstract With the development of informatization, travel agency e-commerce has become an inevitable trend. Comparing with the traditional travel agency, the advantages of low cost and high efficiency are in line with the idea of low-carbon economy. This study starts from low-carbon economy development trend, combines with the present situation of travel agency e-commerce development, and proposes the innovation and development strategy of travel agency e-commerce.

Keywords Low-carbon economy • Ravel agency • Commerce • Development strategy

54.1 Introduction

With the constant growth of global population and economic scale, the energy application brings about series of environmental problems and hazards. Low-carbon economic model based on low energy consumption and low pollution, low emission is a major progress of human society following agriculture and industrial civilizations [1, 2]. Travel agency e-commerce directly applies innovation technology and innovation mechanism of the new century through the low-carbon economy mode and low-carbon lifestyle, realizes the sustainable development of the society which is an important part of e-commerce industry, and also is a kind of trend of the tourism economic development.

W. Huang (✉) · S. Zhou
Modern Service Industry Institute, Zhejiang Shuren University,
Hangzhou 310015, Zhejiang, China
e-mail: cindy112919@163.com

S. Zhou
e-mail: zsj7501@sina.com

Z. Zhong (ed.), *Proceedings of the International Conference on Information Engineering and Applications (IEA) 2012*, Lecture Notes in Electrical Engineering 220,
DOI: 10.1007/978-1-4471-4844-9_54, © Springer-Verlag London 2013

54.2 The Concept Generation of Low-Carbon Economy

Low-carbon economy refers to the guidance of the concept of sustainable development, through technological innovation, system innovation, industrial transformation, new energy development, and other means [3], as far as possible to reduce the high carbon energy consumption of coal and oil, reduce greenhouse gas emissions, and to achieve a win–win economic development form of economic and social development and ecological environment protection. Developing low-carbon economy reduced the impact of economic development to the earth's ecological cycle, adjusted the economic structure and improved energy efficiency, and realized the dynamic balance between man and nature, economy society, and earth ecology [4].

54.3 Travel Agency-Commerce Development Status

Travel agency e-commerce refers to business system which takes the network as the main body, takes tourism information database, electrification commerce bank as the foundation, and uses the most advanced electronic means to operate travel agency and its distribution system. China's tourism e-commerce website appeared in 1996 [5]. At present, more than 5,000 websites have tourism information. More than 300 professional travel websites among them are divided into three categories: the regional website, professional website, and the portals travel channel. But now, the development speed of travel agency e-commerce is not quickly enough, and exist many problems.

54.3.1 Travel Agency-Commerce Consciousnesses is Weak Concepts are Backward

China's travel agency has experienced sustainable development for more than 20 years, the start and development of travel agency website is only experienced less than 10 years, its constant development and adjustment are under the application of network technology in the tourism industry. The enterprises that realized the importance of the travel agency e-commerce development are not many; a lot of enterprises are still standing and waiting [6]. Comparing with the traditional travel agency business, the travel agency e-commerce development depends on the modern technology means, high quality employees, and relative large investment. A few travel agencies using of e-commerce go forefront. Due to the e-commerce macro environment is not mature and reluctant to emphasize the business planning and management on the e-commerce, only few travel agencies are really aware of the concept change and the huge impact to tourism industry brought by e-commerce. Most of the travel agencies still simply take e-commerce as a tool coordinating with the current operation and reducing the management cost, few travel agencies are actively looking for technical solutions, so as to change their position in the travel market.

54.3.2 *Shortage of Travel Agency e-Commerce Talent Lack of Development Potential*

Travel agency e-commerce talent should have tourism professional knowledge and e-commerce practical experience; students with major in tourism are difficult to grasp the technology concept and operate e-commerce essentials, and students with major in e-commerce have technology but no tourism expertise. At present, tourism professionals put forward the requirements of product design, range of services, the website's content and form from the characteristics of the tourism industry and the perspective of business strategy; after the completion of the website, the information is collected and analyzed. According to the requirements of the tourism professionals [7], the network technical staff can design, build websites, provide technical support for the normal operation of the website, but also continue to tap the e-commerce functions, so the website can give full play to its advantages. The development of new technologies and the new world economy need travel agency professionals with the travel agency knowledge and network technology knowledge. In e-commerce, they should understand the traditional travel agency business strategy, but they can also integrate the new technology application of the travel agency. However, at present China's travel agencies lack of compound talent that is familiar with network technology and proficient in the tourism business which results in the thought that tourism business and network technology cannot be a good combination. The majority of the professionals of travel websites are major in computer; few have comprehensive knowledge of the tourism industry, so it is far short of the needs of travel agencies e-commerce development.

54.3.3 *Poor Travel Agency Website Operability Small Amount of Visitor*

At present, travel agencies have not formed real networks, 50 % of domestic travel agencies still remain at small workshop stage. When searching for "certain travel agency in a city", "certain travel agency" in Baidu, you can enter only few several travel agency websites, or you have to enter the public comment website to search website information of the travel agencies [8], but the travel agency home page is difficult to enter. Some travel agency websites ranking rearward in search engines, it has to spend long time to access; and some travel agency websites is difficult to find on the Internet. Through analysis of the travel agency websites, we find that many websites still remain at the traditional tourism business model; website content is not distinctive and attractive, and too simple, or even completely blank in some columns. In addition, their services are more simple, tourist routes, scenic spot introduction, online booking, reservation, car rental, and other modules have almost become fixed content, and less involved in the design of tourist routes, self-service travel arrangements. Ticketing inquiries, visa processing, and other functions cannot be used in some travel agency website. Moreover, the website information update is slow, some routes price are outdated, and no reference value. A number of factors cause less visitor access travel agency website, so the online trading is cold, and cannot attract tourists.

54.3.4 *Network Reputation Cannot Guarantee Poor Transactions Safety*

Because of the rapid popularity of Internet, travel agency e-commerce caused wide attention; it is considered as the most potential growth point of IT industry. However, when we deal trade in the opening network, one of the most important factor of e-commerce popularity is how to guarantee the safety of the data transmission. Network vitality and hacker make visitor difficult to feeling of tourism products and increase purchasing risk, network credit cannot assure. Despite that there is Internet booking and payment, but the online "promises" cannot cash; one of the characteristics of tourism product is off-site consumption. Product quality, service level, and the spirit level to tourists will be evaluated after the completion of consumption, this make the social credit becoming a very outstanding problem in the travel agency e-commerce development. Most Internet users are used to obtain travel information from the Internet, then contract with travel agency, they wish to negotiate prices, and face-to-face contact. Relative simple website marketing, market positioning are not clear prospects; when asked why they do not want to buy online, the majority of people answer is that they are worried about the hacker attacks and lead to credit card information lost and tourism product quality is not guaranteed.

54.3.5 *Relative Simple Website Marketing Market Positioning is not Clear*

The trade requirement of travel agency to consumer (B to C): travel agency website marketing should include advertising, market research, product design and business promotion, online sales, online consulting, and customer relationship management. But most travel agency websites are very attention for information dissemination, travel agency and product brief introduction, the most important part is the company profile, tourist routes, and message. According to the survey, 37.52 % of the travel websites are only tourist information dissemination platforms, without online booking, only 61.84 % E-mail booking in travel website provide online booking services, and 7 % travel website can pay online. As for network research, new product network development, online consulting, network and customer relationship management and marketing are not involved. On the other hand, travel agency website lack of clear market segmentation, their target customers are business travelers or self-service tourists, including all the domestic consumers and tourists. Their information and product are for "all ages". Disregard of consumers at different levels of market segmentation, it will fail to achieve seamless connection between travel agency websites supply and consumer personal demand.

54.4 Combination of Low-Carbon Economy and Travel Agency Ecommerce

While e-commerce applied to the travel agency industry is short, its development momentum is very strong. E-commerce has become a new mode of travel transactions of the information age. Especially the development of low-carbon economy is becoming the trend of the world economic and social change circumstances, travel agency e-commerce with its reduction of cross-regional exchange of information costs, improved the efficiency advantages of cross-regional information exchange which in line with the direction of low-carbon economic development. Tourism is the typical industry of cross-space operation, the operation of the tourism inter-space takes information cross-space as a precondition, and the convenience of the travel agency e-commerce brought by international tourism is an important choice for the development of low-carbon economy.

54.4.1 Reduce the Cost of Information Exchange

The cost of the information exchange is proportional to the distance, the travel industry associates with catering, lodging, travel, shopping, and entertainment, which needs constant communication with related enterprises around the world, and needs to travel around the world, wastes manpower and financial resources, but you can solve these problems through the Internet to reduce communication costs. Such as traditional telephone and fax could increase information transmission cost, but the average cost and marginal cost of multinational information exchange will be extremely low if you use the Internet. A website, no matter how many people visit, its production and maintenance costs are the same, so greater the distance, the greater cost savings.

54.4.2 Save the Tourism Market Transaction Costs

Travel agency e-commerce reduced intermediate segment, the producers and consumers may deal directly, and the virtual market of the travel agency e-commerce has replaced the traditional market, brokers, agents, and wholesalers, retailers, saved the cost of market intermediaries. The travel agency e-commerce reduced the consumers' information searching costs, so that the tourists can get more and more useful information from the tourist destinations and tourism enterprises and the tourists can have more choices and save their time.

54.4.3 Improve the Efficiency of Information Exchange

Tourism enterprises of different countries transmit information and deal with tourism business directly through travel agency e-commerce Internet; it saved the cost and shortened the time, accelerated cash flow, and saved interest expense.

54.5 Travel Agent e-Commerce Innovation and Development Strategies in Low-Carbon Economy

54.5.1 Enhance the Awareness of Travel Agency e-Commerce Increase the Integration of the Intensity of Inter-Industry

Tourism enterprises changing their traditional marketing concepts and Internet high degree integration is a trend. The travel agency should in line with good tourist network, widen business ideas. For example, currently Ctrip occupies more than 50 % of Chinese online travel market share. Ctrip supplies hotel reservations, air ticket booking, vacation reservation, business travel management, high-speed train ticket purchasing, travel information, and a full range of travel services to over 50 million registered members. Service scale and the resource scale is the core advantage of Ctrip. Ctrip has established a set of modern service system, including: customer management system, room amount management system, call queuing system, order processing system, the E-Booking air ticket reservation system, and service quality monitoring system. Rely on these advanced services and management systems, Ctrip provide more convenient and efficient service to its members. Therefore travel agency e-commerce is not a subversive to traditional tourism, but the convergence of network and travel. Travel agency is tourism brokers, they should be skilled to involve in a network business, carry out e-commerce, take network platform as an office, engage in information processing, transmission, and transform the price difference profit into “virtual business” information profit.

54.5.2 Segment of the Tourism Market Development of Personalized Products

Different tourism market segments have different information inquiry and tourism product distribution channels. Such as: frequent travelled self-service market is likely to use online service to arrange travel routes, and ultimately, they reserved travel products. Therefore, travel agency should as far as possible process market segment and choose the right target market, transform the reservation agency into tourism manager and consultant who can increase the value of tourism products. Transform the sales center into one-stop service travel agency who can provide advice ad develop special services. Lvmamma.com was founded in 2008; it has become China's leading new B2C travel e-commerce websites, and China's largest self-service booking and information service platform. Lvmamma.com positions itself at self-tourist market, takes discounted ticket, free travel, special hotel as core service, they also supply bus free travel with the group, long-term travel, outbound travel, and provide one-stop service. Travel agency uses the website to provide various of tourism information to tourists, and necessary guidance services, according to the wishes of the tourist, the trip will meet individual needs, and save money and time for tourist, thus to win the heart of the tourists.

54.5.3 Pay Attention to the Travel Agency Website Management Update the Travel Agency Website Information Egmont of the Tourism Market Development of Personalized Products

The design of travel agency e-commerce website and home page should be illustrated lively and attractive. Tourist route, attraction picture, instruction, and price should be listed in detail, and pay attention to fashion culture, meet market demand, and avoid the product simple concoction. In addition, it should update website information and website designs in a timely manner, so that to attract tourists and make the website more abundant. It also should increase website practicality, increase services, such as tourist route design, self-service travel arrangements, and online virtual tourism. To some extent, taking above services can increase the attraction to tourists, and increase traffic to your website. Reasonable selection of high efficiency, free or fee-engine, registration of travel agency website keywords, website name, address, and website's index, making the website ranking more forward in search engine as far as possible, dissemination of the travel website.

54.5.4 Improve the Quality of the Travel Agency Staff Cultivate Compound Talents of the Travel Agency e-commerce

China's tourism enterprises are toward development road of internationalization, standardization, market-oriented, and opening up, travel agency e-commerce talent is not only the computer network professional, but also they should have network technology, foreign languages, marketing, and tourism knowledge. Employees of tourism enterprise must learn e-commerce, master e-commerce operation method, and enhance ability of deal with tourism business and enterprise management by e-commerce. On the other hand, the travel agency should attach great importance to cultivate the younger generation, make them not only understand e-commerce technology, but also understand the tourism business knowledge.

References

1. Liangang C (2010) Low-carbon economy creates the best mode of e-commerce services e-commerce 12. J Comm Manage (4):344–349
2. Fei F (2003) Study of Chinese B2C travel e-commerce profit model—Ctrip and springtour.com. Tourism Sci 18(4):70–75
3. Desheng L (2009) E-commerce development trend analysis in current world Global. Econ Outlook 24(6):31–38
4. Shengli L (2009) Status mode and prospect of China's tourism e-commerce development. Economist 25(24):244–248
5. Ming L (2010) Analysis of development path of China's traditional large-scale travel agency in the context of informatization. Hubei Social Sci 25(3):93–98
6. Guiyang Z, Ying C, Lei Z (2010) Disputes of low-carbon economy and e-commerce. Money China 12:36–37
7. Haixia W (2011) Development of Chinese tourism electronic commerce analysis. Securities Futures China 25(10):144–145
8. Caixia L (2011) The electronic commerce development strategy China Business and Trade. J Bus Res 25(30):118–119

Chapter 55
Text Information Processing of the English-Literature Discourse

Jian Zhong

Abstract The main objective of this paper is to convert written English discourse into machine-generated comprehensive conversation. This paper is proposed to provide a complete speech synthesis for English text information. The main application of the Text-To-Voice (TTV) system is to help people with comprehension through having the text read to them by computer software. The TTV system will help in retrieving data from sites that contain information in different language styles. The system has been successfully developed for English-literature discourse. A voice-based series connection was used, which reduced the con-chaining points and hence minimized error. The synthesis was tested on expert listeners to ascertain its quality.

Keywords Speech synthesis • English-literature • Machine generated

55.1 Introduction

Speech information processing is the artificial production of human speech. A software system used for this purpose is called a speech synthesis, and can be implemented in software or hardware. A Text-To-Voice (TTV) system converts normal language text into speech; other systems render symbolic linguistic representations like phonetic transcriptions into speech. Synthesized speech can be created by concatenating pieces of recorded speech that are stored in a database. Systems differ in the size of the stored speech units. The speech units can be phonemes, voices, etc. For specific usage domains, the storage of entire words or sentences allows for high-quality output.

The paper is proposed to specifically address the issues of building synthesis voices for the different text Information. The techniques employed for synthesizing

J. Zhong (✉)
Guangdong Polytechnic Normal University, Guangzhou 510665, China
e-mail: jianzhong_3@hotmail.com

Z. Zhong (ed.), *Proceedings of the International Conference on Information Engineering and Applications (IEA) 2012*, Lecture Notes in Electrical Engineering 220,
DOI: 10.1007/978-1-4471-4844-9_55,

speech from text may be broadly classified into three categories: (1) Formant-based, (2) Parameter-based and (3) Series connection-based. The three subcategories of Series connection-based are (1) Unit Selection Information processing, (2) Voice processing, and (3) Domain-Specific Information processing. We use unit selection information processing [1] in which, the prerecorded words are split into voices and stored with the corresponding voice names and maintained in a database. Based on the input, the voice units are selected from the database and connective to generate the speech.

To make the speech voice more natural Rhythm and Smoothing is applied at right places. Various techniques available for smoothing includes Spectral Smoothing, Best group, Waveform Interpolation, LP Techniques, and Pitch Synchronous Overlap Add (PSOLA) [2]. We use a technique based on Best group algorithm.

The most important qualities of a speech information processing system are naturalness and Intelligibility. Naturalness describes how closely the output voices like human speech, while intelligibility is the ease with which the output is understood. The ideal speech synthesis is both natural and intelligible. Speech information processing systems usually try to maximize both characteristics. The technologies for generating comprehensive conversation waveforms are listed below.

Serial information processing
Unit selection information processing
Voice information processing
Domain-specific information processing
Formant information processing
Segmental information processing
MM-based information processing
Sine wave information processing.

Each technology has its own strengths and weaknesses, and the intended uses of an information processing system will typically determine which approach is used. In Hudson Text-to-Speech Information processing System [3], Serial information processing approach is used where natural speech is connective to give the resulting speech output. Formant-based speech synthesis employing half-voice series connection [4], involves identifying and extracting the formants from an actual speech signal (labeled to identify approximate half-voice areas) and then using this information to construct half-voice segments each represented by a set of filter parameters and a source signal waveform. In Markov Model (MM) based speech information processing system [5] the speech waveform is generated from MM themselves, and applies it to English speech information processing using the general speech information processing architecture of Festival Framework.

In TTV system [6], they use phoneme con-chaining technology with an attempt to cover all types of language styles under a single framework. This system can detect the languages and it dispatches the text to the corresponding phonetic synthesis. Here, rhythm is not given much importance and also the quality of the speech can be improved to some extent. We have proposed a Speech synthesis to overcome the drawbacks of these systems to the extent possible.

The paper is organized into three sections. In Sect. 55.2, we discuss the proposed system with the description of modules. In Sect. 55.3, the results are discussed.

55.2 Speech Synthesis TTV System

The TTV system comprises modules for Preprocessing, Encoding Conversion, Segmentation, Series connection, Rhythm, and Smoothing.

55.2.1 Preprocessing

The first stage of a TTV system is the pre-processing module, called tokenization. It converts the input text into a sequence of words and symbols to be processed by the rest of the system. It identifies and makes decisions on what to do with punctuation marks and other non-alphabetic textual symbols (e.g. parentheses), identifies

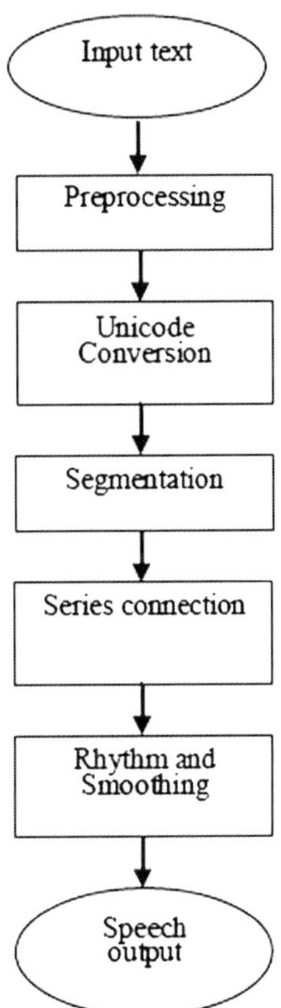

Fig. 55.1 Pre-processing

and expands abbreviations, acronyms, and numbers to full-blown orthographic strings. Each input line is scanned and each recognized construct (word, number, symbol, etc.) is converted into an appropriate word or sequence of words as shown in Fig. 55.1.

Preprocessing includes:

The removal of single and double quotes from the start and end of the token;

Expansion of abbreviations, acronyms, and numbers;

In the case of English information input, if there is any English word, it is replaced by an equivalent word in the corresponding language style.

55.2.2 Encoding Conversion

Encoding is an industrial standard allowing software to consistently represent and manipulate text expressed in most of the world's writing systems. In our work, we have used Encoding conversion to accept the input text in multiple fonts. The logic behind Encoding conversion is to identify the font in which the given text is encoded and applying Encoding to the given font.

Since the input may be in TAM, TAB, or TSCII, the encoding techniques for TAM, TAB, and TSCII are explored and for each and every character in the input, Encoding is applied to convert the given text into encoding. Similarly, the encoding technique of phrase font is explored and Encoding en-coding is applied to the input. In case of speech synthesis input, the text to be given is already converted to encoding and saved as a text file containing inputs from all the English-literature text. If the input is going to be in a single language style like English-literature discourse then the Encoding conversion will be carried out internally.

55.2.3 Segmentation

Speech in English-literature discourse is based on basic voice units which are inherently voice units made from V, CV and kinds of combinations of C and V, where C is a consonant and V is a vowel. From perceptual results, it is observed that from different choices of speech units, the voice unit performs well.

55.2.4 Series Connection

In serial speech information processing, the segments of recorded speech are connective to produce the desired output [7]. Generally, this technique produces the most natural-voicing synthesized speech since the number of concatenating points is less. There are three main subtypes of serial speech information processing, namely Unit selection information processing and Voice information processing. In our work, we use unit selection information processing in which, the prerecorded words are split into voices and are stored with the corresponding voice names and maintained in a database. In case of English dialect and Spoken English embedded in English-literature discourse, the voice units are named using English letters based on the pronunciation in the corresponding language style.

Based on the input, the voice units are selected from the database and connective using MATLAB simulator integrated with java to create complete utterances.

55.2.5 Rhythm and Smoothing

55.2.5.1 Rhythm

Rhythm is the rhythm, stress, and intonation of speech. Rhythm may reflect the emotional state of a speaker; whether an utterance is a statement, a question, or a command; whether the speaker is being ironic or sarcastic; emphasis, contrast and focus; and other elements of language which may not be encoded by grammar. Hence, applying the concept of rhythm to the TTV system makes the information processing speech voices more like human speech. There are several techniques used for rhythm [8]. We make use of punctuation marks like '!', '?', ';' to give stress. In this method, if the words in the input text are followed by any of these special symbols they are read with intonations. For this method working, the voices which are to be read with intonation are maintained as a separate database and these voice units are connective when intonation is to be given.

55.2.5.2 Smoothing

Smoothing is used to smooth the transition between segments in order to produce continuous output as that produced by human speech.

There are several techniques used for smoothing like Spectral Smoothing, Best group, Waveform Interpolation, LP Techniques and PSOLA. Among all these techniques best group algorithm is easy to implement at low cost.

It is common in serial information processing that the boundaries of speech segments are fixed, but the best group technique allows the boundaries to move to provide the best fit with adjacent segments. A measure of mismatch is tested at a number of possible segment boundaries until the closest match is found. While any form of measure may be used, for the sake of improving spectral quality, using a spectral discontinuity measure is appropriate. Measures considered include honey Mel-Frequency Phrase Coefficients (MFPC) and the Auditory-Neural Based Measure (ANBM). It is not necessary to implement best group to perform spectral smoothing, but it does provide some improvement at a small cost.

We proposed a new technique for Smoothing based on Best group algorithm. We initially set the boundaries to search for the optimal concatenating point. If the given voice unit is the beginning of the word then the boundary is set from 5n/6th position to n where, n is the length of the voice unit. Otherwise the boundary is set from the beginning to n/3rd position. Then the Hamming windows are formed to process on the selected portions to find the concatenating point. The zero-crossing counts are found for all the hamming windows and the point in the area containing the minimum count is chosen as the optimal concatenating point. Finally the voices are connective at the chosen points to generate the smoothened speech output.

55.3 Results and Analysis

The results are analyzed using two parameters, namely MFPC and Mean Score (MS).

55.3.1 Mel-Frequency Phrase Coefficient

MFPC are coefficients that collectively make up an Mel-Frequency spectrum (MFC). They are derived from a type of phrase representation of the audio clip. The difference between the frequency spectrum and the MFC is that in the MFC, the frequency bands are equally spaced on the honey scale, which approximates the human auditory system's response more closely than the linearly spaced frequency bands used in the normal frequency spectrum. This frequency warping can allow for better representation of voice, for example, in audio compression. In voice processing, the MFC is a representation of the short-term power spectrum of a voice, based on a linear cosine transform of a log power spectrum on a nonlinear honey scale of frequency.

MFPC are commonly derived as follows:

Take the Fourier transform of (a windowed excerpt of) a signal.

Map the powers of the spectrum obtained above onto the honey scale, using triangular overlapping windows.

Take the logs of the powers at each of the honey frequencies.

Take the discrete cosine transform of the list of honey log powers, as if it were a signal.

The MFPC are the amplitudes of the resulting spectrum.

MFPC are commonly used as features in speech recognition systems, such as the systems which can automatically recognize numbers spoken into a telephone.

They are also common in speaker recognition, which is the task of recognizing people from their voices. The MFPC values are calculated for a set of original words say, and the connective words. The Euclidean distance between the values of P and Q are calculated using the formula,

$$\sqrt{(p_1 - q_2)^2 + (p_2 - q_2)^2 + \cdots + (p_n - q_n)^2} = \sqrt{\sum_{i=1}^{n} (p_i - q_i)^2} \tag{55.1}$$

The deviation of the connective words with the original words in Information based on their MFPC features. The average deviation of MFPC value for connective English words is 1.42, for English dialect 1.45, and for Spoken English 1.47. Though smoothing technique works well for all the English text minimum deviation is obtained for English.

55.3.2 Mean Score

The MS provides a numerical indication of the perceived quality of received media after compression and/or transmission. The MS is expressed as a single number in the range 1–5, where 1 is lowest perceived audio quality, and 5 is the highest perceived audio quality measurement. The MS is generated by averaging the results of a set of standard, subjective tests where a number of listeners rate the heard audio quality of test sentences read aloud by both male and female speakers over the communications medium being tested. A listener is required to give each sentence a rating using the rating scheme and MS is the arithmetic mean of all the individual scores, and can range from 1 (worst) to 5 (best). We took objective measurements from listener tests. Sixty expert listeners were asked to indicate their preferences in terms of naturalness and intelligibility for different words and phrases.

The listeners ranked the smoothened speech produced after series connection as compared with natural speech. The results taken using these parameters signify that the speech is generated with minimal error.

55.4 Conclusions

The system has been successfully developed for the text Information of English-literature discourses. Voice-based series connection was used which reduced the con-chaining points and hence minimal error. Best group technique has been used for Smoothing which resulted in natural voicing speech. The synthesis was tested for its quality through expert listeners. In the proposed work we process only domain restricted text. In future the Synthesis has to be enhanced to process unrestricted text and the Smoothing technique needs to be explored further to find the optimal concatenating point.

References

1. Maedche A, Staab S (2000) Mining non-hierarchical conceptual relations from network information. In: Proceedings of knowledge engineering and knowledge manage me methods, models and tools 12th international conference, vol 23(12). Springer, Berlin, pp 189–202
2. Sanchez D, Moreno A (2008) Learning non-hierarchical relationships from web documents for domain concept construction, data and knowledge engineering UdoUH0Ui:10.1016/j.datak.2007.10.001U vol 64(3), pp 600–623
3. Allen J (1995) Relational language understanding, vol 24(8). The Benjamin/Cummings Publishing Company, Redwood City, pp 342–345
4. Lehmann J, Hitzler P (2007) A refinement operator based learning algorithm for the ALC description logic. In: Proceedings of international conference on inductive logic programming, vol 13(11). Springer, Corvallis Berlin, pp 147–160
5. Villaverde J, Persson A, Godoy D, Amandi A (2009) Supporting the discovery and labeling of non-hierarchical relationships in concept learning, expert system applications H1Udoi:10.1016/j.eswa. 2009.01.048U vol 36(7), pp 10288–10294
6. Bontcheva K, Cunningham H (2003) The semantic web: a new opportunity and challenge for human language technology. In: Proceedings of the workshop on human language technology for the semantic web and web services, vol 15(24). Sanibel Island, pp 1144–1148
7. Marinho L, Buza K, Schmidt-Thieme L (2008) Folksonomy-based collabulary learning. In: Proceedings of international semantic web conference, Springer, Karlsruhe, Berlin 6(14):261–276
8. Guarino V, Masolo C, Vetere C (1999) Ontoseek: content-based access to the web. IEEE Intell Syst. H2Udoi:10.1109/5254.769887U vol 14(3), pp 70–80

Chapter 56
Grid Management Based on Intelligent Agent Cooperation

Haoyue Zhu

Abstract In this paper we described some details towards the use of collaborating Intelligent Agents for the management of grid environments. We outline how collections of collaborating intelligent agents in grid can be a step towards better management of grids. We showed how rules can be used at different levels of the hierarchy to facilitate the cooperation among intelligent agents. The results of three different schemes show the importance of cooperation between agents at different levels and how this cooperation can help to increase efficiency of current management. We also showed how the communication messages can be inferred automatically from rules and get generated on the fly.

Keywords Intelligent management • Cooperation • Rule based • Grid management

56.1 Introduction

Grid computing environments often depend on virtualization technology where client applications can run on separate operating machine logic (ML), particularly for providers of Infrastructure as Operating. Such environments can consist of many different host computers each of which might run multiple MLs. As the number of hosts, ML, and client applications grows, management of the environment becomes much more complicated. The grid provider must worry about ensuring that client operating level agreements (OLA) are met, must be concerned about minimizing the hosts involved, and minimizing power consumption. Our focus is on how to better manage the virtual machine and system infrastructure of the grid provider.

In recent years there has been a lot of research into "Intelligent Computing" [1], especially about how to build intelligent elements and agents [2]. Intelligent agents try to monitor and manage resources in real time by building systems that are

H. Zhu (✉)
College of Mathematics and Computer Engineering, Xi'an University of Arts and Science, Xi'an 710065, China
e-mail: braightmoon@sina.com

Z. Zhong (ed.), *Proceedings of the International Conference on Information Engineering and Applications (IEA) 2012*, Lecture Notes in Electrical Engineering 220, DOI: 10.1007/978-1-4471-4844-9_56,

self-configuring, self-optimizing, self-healing, and self-protecting. In the broader vision of intelligent computing, large complex systems will consist of numerous intelligent agents handling systems, applications, and collections of operating [3].

We consider the use of rule-based agents in addressing this problem and with an initial focus on a hierarchy of intelligent agents where rules are used at each level to help agents decide when and how to communicate with each other as well as using polices to provide operational requirements.

56.2 Grid Management

56.2.1 Architecture and Machine Logic

The infrastructure, such as Amazon EC2, is typically composed of data centers with thousands of physical machines organized into multiple groups or groups. To have a better understanding of the grid provider environment and architecture, we take a closer look at an open-source infrastructure for the implementation of grid computing on computer groups. There are five elements that form the grid infrastructure: grid controller (GC), Web storage controller (WS3), elastic block storage controller (EBS) [4], controller of group (CG), Terminal Controller (TC).

These elements can physically locate on one single machine to form a small grid but each one has a different role in forming the grid infrastructure.

56.3 Approach

Based on the previous discussions, we propose to use a number of different intelligent agents [5]. For example, an IA for an Apache Web server should only focus on the behavior of the Web server and not the relationship that it might have with a database server and the TC IA should only focus on the behavior of the MLs inside that terminal and the general performance of that terminal.

The hierarchy of agents can be expanded dynamically into more levels as required. A good example of splitting and combining elements in the hierarchy is illustrated in the work of [6] to improve the scalability of the hierarchical approach; we have not considered this in this paper.

56.3.1 IA Requirements

In this section we explain the assumptions we have for a general intelligent agent to work in cooperation with other agents. Although IAs are heterogeneous and can belong to different vendors, they should all follow some specifications to make the cooperation possible.

We assume that inside each IA there is an event handling mechanism for processing, generating events, and notifying the interested parties inside the IA. For example, there could be an event bus and different subscribers to certain events (within the IA) and upon raising those events any subscribers will get notified. This event handling mechanism is useful for handling event, condition, action rules, and also for communication between agents (explained below).

56.3.2 Communication Model

In the previous work, we suggested the use of a message-based type of communication between IAs. Several different types of messages were proposed as sufficient for communication between agents:

Msg = <Type, Info>
Type = NOTIFY|UPDATE_REQ|INFO
Info = Metrics|Details
Metrics = {<m,v>|m is the metric name, v is the metric value}
Details = <T, Metrics>
T = <HelpReq, OLAViolation ...>

By using a message "Type", we introduce the possibility of different types of relationships between agents (e.g. request, response) and based on the type of message, one agent can expect the kind of information that would be available in the Info section of the message. The Info can be the latest metrics of elements managed by a particular local agent or could be details on some event that has happened. Having a small set of different types of messages also makes it easy to define the operation of each IA.

The form of each of these types of messages is as follows:

M1 = <NOTIFY, Details>: When one agent wants to raise an event in another agent it can be encapsulated inside a notify message. The type and content (Metrics) of the event is very system specific and can both be defined in the Details portion of the message. Possible events would be a "help request event", "rule violation event", "system restart event", "value update event", etc. We illustrate this type of message in the following section. When an agent receives a notify message from another agent, it will raise an event inside the event bus and deliver it to interested subscribers (e.g. evaluate proper rules).

M2 = <UPDATE_REQ, Metrics>: This is a message asking for the status of the metrics declared in Metrics. Another manger can respond to this message by sending an INFO message back. These Metrics can be specifically declared in rules that used for a communication or it can be inferred automatically from rules. The Metrics are very dependent on the nature of the system and can be different from one system or application to another. Examples of such information include CPU utilization, memory utilization, number of requests/second, number of transactions, available buffer space, packets per second, etc.

M3 = <INFO, Metrics>: This is a message that provides information about metric values, which can help the process of decision making in the higher level agent. This message is usually sent in response to the UPDATE_REQ message from a higher level agent (e.g. M2 explained before).

The UPDATE_REQ message is sent from higher level agents to lower level ones. INFO messages are sent in response to the UPDATE_REQ message and NOTIFY messages are sent from one agent to another based on the need. We will explain in more detail how we can use rules to generate these messages for communication among IAs based on demand.

56.3.3 Inferring Messages from Rules

In order to better illustrate the problem and approach, we will show several examples of rules that can be used at different levels of a hierarchy and how these rules can influence the relationship between agents.

Assume that on each ML there is a LAMP (Linux Apache MySQL-PHP) stack that hosts the Web application and that one IA is managing the applications inside that ML. We use event, condition, action (ECA [6]) rules to specify operational requirements, including requirements from OLAs, and we also use rules to identify and react to important events.

At IA startup, each managed object will get configured with proper values for properties and then at run time the IA can automatically figure out metrics that need to be refreshed and generate the proper message for updating them. This message will look like:

Msg = <UPDATE_REQ, {<cpuUtilization, null>, <memoryUtilization, null>}>

Then, upon receipt of this message by the IMML1 and IMML2, a reply INFO message is automatically generated to be sent back. If for any reason, these IAs cannot calculate these values then they can send an INFO message back with null values which show that there was a problem in getting values for the requested metrics. The general form of the INFO message is:

M3 = <INFO, Metrics>

Metrics = {<m1, v1>, <m2, v2> ...} and an example of the message to be returned would look like:

Msg = <INFO, {<cpuUtilization, 65>, <memoryUtilization, 78>}>

There can be separate rules that specify the need for sending these messages but the important point is that it is not necessary. Rather, these messages can be automatically inferred from what is defined in the rules.

56.4 Prototype

In order to evaluate our approach, we used two MLs running on a single server with LAMP installed on them and a two-tier Web application based on an online store was configured to run on the MLs. There was also a privileged intelligent agent running in the physical server and its job is to manage (optimize based on rules) the behavior of that server by collaborating with the agents running inside MLs.

We used KML virtualization with a distribution to build the guest MLs. "Domain 0" is the first guest operating system that boots automatically and has special management privileges with direct access to all physical hardware by default. The agent running inside Domain 0 has the authority to change the configuration of other MLs such as allocated memory, allocated CPU cores, etc.

We implemented the intelligent agent using the Ponder2 system and used Ponder Talk for communication between agents.

56.4.1 Experiments and Results

We used an open source online store called "Taobao" to measure the response time of Apache Web server running on ML1. We used Agent to generate loads to this virtual store and measured the response time of Apache in three schemes. The ultimate goal of the whole system is to keep the response time under a certain threshold (e.g. 510 ms) that we assumed was defined in an OLA.

56.4.2 Scheme

No cooperation. In the first scheme we disabled all communications between agents. In this case, only the local agents tried to optimize the system based on rules that they had. In this case, when the load increases the local agent tries to adjust the Web server by allocating more resources.

As a result, the system will face more OLA violations and the response will get worse. Thus, the load is more than what this system can handle alone. This also causes a long-term violation of the OLA which could mean more penalties for the operating provider.

We calculate two measures of the performance of the system and agents in this case: the total time that the system could not meet the OLA (T) and the percentage of time that the system spent in a "violation" (V). For these experiments each time interval was 1 s. Therefore, the results for Scheme 1 are:

T1 = 21 s
S1 = Total time = 27 s
V1 = T1/S1 = 0.78 = 78%

One Level Cooperation. In the second scheme, we consider the situation when the local agent can request help. When the local agent can no longer make adjustments to the system, it requests help from the higher level agent. This is specified in the rules of IA1 and IA2, as mentioned in the previous section with the exception that in this case memory limits can change. For example, a rule of IA2 would be:

Limit = 510.
On Event: HelpRequestEvent
If (ML1::memoryUtil > 86 & ML1::cpuUtil > 95)
ML1: increaseMem: +52 max: limit

The current limit for increasing memory is set to a default value (e.g. 510 MB), but it can change over time based on the changes in the system. We will see an example of this in Scheme 3.

There are still subsequent instances where there are occurrences of heavy load and occasional OLA violations still happen. In these cases, IA1 still sends the help request to IA2, but since IA2 has allocated all available memory to ML1 (as per its rule), it cannot do more and simply ignores these requests. To solve this problem, we add another level of management to the system.

Based on the output for this scheme, we calculated the same measures of performance:

T2 = 10.5 s
S2 = 27 s
V2 = T2/S2 = 0.39 = 39 %

As is evident it showed that the time that the system spends in "violation" of the OLA is much less.

Two-Level cooperation. In the third and final scheme, we use another level of management to help reduce the occasional OLA violations happened in Scheme 2. Like the previous schemes, the local agent (IA1) tries to adjust the Web server at points A, B, C, and D. At points E and F, IA2 assigns 52 more megabytes to ML1 to solve the stress. At point G there is another OLA violation. At this point, IA1 asks for help from IA2 but since IA2 already assigned all the available memory as per its rule, it cannot provide more help and automatically creates a help request which it sends to its parent.

IA4, running at the group control level, has a global view [7] of all physical servers and finds the least busy server. It then tells the IA2 to migrate one of the MLs to that server.

The calculation of our measures for this scheme is as follows:

In this case, after migration, there is more memory available at the IA2 level and the memory limit is increased. Therefore, at point H when the load is getting higher and another OLA violation happens, IA1 asks for help and IA2 responds by adding 52 more megabytes to ML1. The same process happens at point I where IA2 adds another 52 MB to ML1 and after that the response time stays below the OLA threshold although the load is still very high.

T3 = 11.8 s
S3 = 43.5 s
V3 = T3/S3 = 0.27 = 27 %

In this case, even with the migration of one of the MLs, the percentage of time in a "violated" state is much less than in Scheme 2.

Table 56.1 Time and violation rate in three scenarios

	T (seconds)	V (%)
Scheme 1	21	78
Scheme 2	10.5	39
Scheme 3	11.8	27

56.4.3 Discussion

Table 56.1 summarizes the time and percentage in a "violated" state for the three schemes. Not surprisingly, having more IAs making changes to the system and components decreased the impact of violations. Most importantly, this happened automatically without administrator intervention and without adding any new hardware which means improvement in the current system efficiency.

The results show that there is definitely an advantage when IAs can cooperate. A single intelligent agent cannot solve all performance problems just by itself because it has only a local view of the system with some limited authority to change things. Thus, the current infrastructure can be used more efficiently and provide better operating with less chance of violating OLAs without adding new computational resources.

56.5 Conclusions

In this paper we described some details towards the use of collaborating intelligent agents for the management of grid environments. We showed how rules can be used at different levels of the hierarchy to facilitate the cooperation among intelligent agents. We also showed how the communication messages can be inferred automatically from rules and get generated on the fly.

In this work we assumed that rules are defined and delivered to agents by system administrators, but as a future work we are planning to make this process more automated.

We then implemented these ideas in a prototype and showed how this cooperation can be useful to preserve the response time of a Web server under a certain threshold (defined in OLA).

Further work on this approach can lead to more automated management of grid environments enabling more efficient use of the grid infrastructure as well as meeting OLA requirements while using fewer resources.

References

1. Gensym SP, Kumar R, Sangal R (2002) A data-driven information processing approach for Indian languages uses voice as basic unit. In: international conference on natural language processing (ICON) vol 15(7), 311–316

2. Chappell DT, Hansen JHL (2002) Spectral smoothing for speech segment series connection. Speech Commun 36(13):3–4
3. Jayavardhana GL Rama, Ramakrishnan AG, Vijay Venkatesh M, Murali Shankar R Hudson (2001) A text- to-speech information processing system. Paper presented in the Chinese Internet 2201 Conference and Exhibition vol 21(6), pp 325–329
4. Schötz S (2006) Data-driven formant information processing of speaker ag. In: Ambrazaitis G, Schötz S (eds.) Lund working papers 52, proceedings of Fonetik vol 26(19). Lund, pp 105–108
5. Tokuda K, Zen H, Black AW (2002) An MM-based speech information processing system applied to English. Paper Presented in the proceedings of IEEE speech information processing workshop vol 14(32), pp 425–429
6. Kisiten A (2001) http://dhvani.sourgeforge.net. (Alwani-TTV System for Indian Languages) vol 535, pp 62–66
7. Geetha TV (2007) Language models for Chinese speech recognition. In: Publication in IETE special issue on spoken language processing vol 24(5), pp 378–383

Chapter 57
Petri Net-Based Workflow Model of Multi-Node Cycle

Chaodong Lu and Zhimei Zhao

Abstract In order to address complex application environments workflow modeling, using Petri nets as a workflow modeling tool, Petri nets based on multi-node cycle workflow model the complexity analysis. The results show that when the number of nodes in the workflow model is small, it can be addressed by increasing the monitoring place circulation problems and improve the complex environment of the efficiency of workflow modeling.

Keywords Petri nets • Workflow • Process model

57.1 Introduction

A workflow is a class of full or partial automates of the business process, which according to a series of process rules, documents, information, or tasks,can be different between the implementation of transfer and implementation [1, 2]. The purpose of the workflow is broken down into good by the work of the task, role, according to certain rules and procedures to perform these tasks, and gain monitoring and management, so as to improve efficiency, reduce costs, and enhance the level of production and operation management objectives [3, 4].

With the increasingly wide range of workflow applications, the research on workflow technology is also being carried out extensively. The main research direction of workflow modeling can be divided into two areas with the run time, modeling studies to investigate issues related to the workflow model, and running study focused on how to improve the performance and reliability [5]. A workflow is a class of full or partial automates of the business process, which according to a series of process rules, documents, information, or tasks passed between the different actors and implementation. Workflow model is a special kind of

C. Lu (✉)
Wuhan University of Technology, Wuhan 430070, Hubei, China
e-mail: luchaodong@hrsk.net

Z. Zhao
Henan Institute of Engineering Zhengzhou, 451191, Henan, China

Z. Zhong (ed.), *Proceedings of the International Conference on Information Engineering and Applications (IEA) 2012*, Lecture Notes in Electrical Engineering 220,
DOI: 10.1007/978-1-4471-4844-9_57, © Springer-Verlag London 2013

process model, which not only describes the process of composition of the logical relationship between activities and events, with the general model of the basic characteristics of the process, but can perform by the workflow management system, thus achieving enterprise business process automation.

Petri net was discovered by a German scholar called Carl Adam Petri in 1962 as a discrete event dynamic system (DEDS) modeling and analysis; it has both intuitive graphical representation means and the strict mathematical basis, which is widely of all ages. Petri net is a powerful graphical tool for describing the process on the basis of strict mathematical derivation, but it is a completely formal tool that allows you to visually describe a workflow process. Petri nets take into account the strict language of the two aspects of meaning and graphics, and are the ideal workflow modeling methods. Compared with other workflow models, workflow based on Petri net structure of a simple model to describe the characteristics of ability, is the ideal workflow modeling method.

57.2 Petri Net Theory and the Basic Rules

Petri net structural elements include the place, translation, and the arc. The dynamic characteristics of the Petri net system are by using token identity expressed as a token of the node containing the place dots; they are in the place of the dynamic translation in the different states of that system.

Petri net definition: the classic Petri net is a six-tuple

$$\sum = (P, T; F, K, W) \tag{57.1}$$

where:

$$P = \{P_i : i = 1, 2, \ldots, |P|\} \tag{57.2}$$

Is a limited place collection in which $|P|$ is the number of the place;

$$T = \{t_i : i = 1, 2, \ldots, |T|\} \tag{57.3}$$

is a limited transition collection, in which $|T|$ is the number of the transition.

$(P \cap T = \varphi)$; $F\,(P \times T) \cup (T \times P)$ is a set of nodes in flow relation; $W : F \to \{1, 2, \ldots\}$ is the arc weight function, $W = 1$ Default; $M : F \to \{0, 1, \ldots\}$ is the marking; $M_0 : F \to \{0, 1, \ldots\}$ is the initial marking;

For a translation $t \in T$, ${}^{*}t = \{p \in P\ (p, t) \in F\}$ is the set of all input places $t^{*} = \{p \in P\ (t, p) \in F\}$ is the set of all output places.

The place and translation's predecessor and successor are:

$$\forall t \in T,$$

$$\cdot t = \{p\,|p \in P \wedge (p,t) \subseteq F\} \tag{57.4}$$

$$t\cdot = \{p\,|p \in P \wedge (t, p) \subseteq F\} \tag{57.5}$$

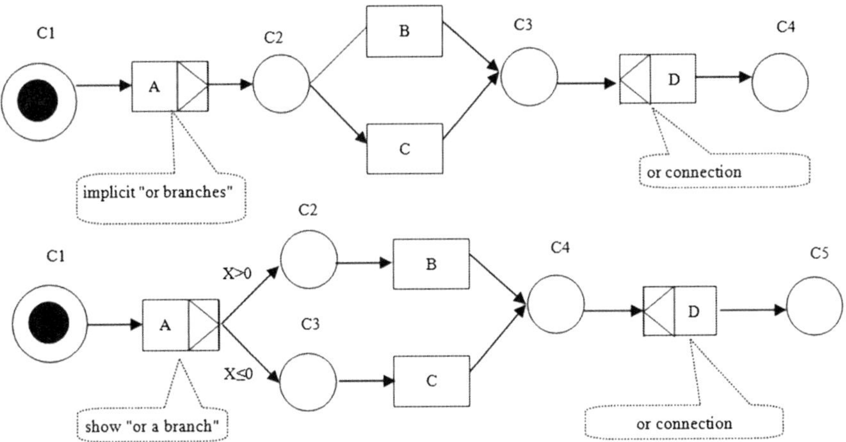

Fig. 57.1 Petri net diagram of the selection of conditions between

$\forall p \in P,$

$$\cdot p = \{t \,|t \in T \wedge (t, p) \subseteq F\} \tag{57.6}$$

$$p \cdot = \{t \,|t \in T \wedge (p, t) \subseteq F\} \tag{57.7}$$

For example, Petri nets work to define the relationship between flow conditions selected according to certain conditions, a branch of the decision process into activities that require workflow with the implementation of two basic primitives: "or connection", "or branch." Or connection is expressed as a multiple input arcs place; when activity B or C is finished, the process can continue with activities in D. "Or branch" is divided into two types, implicitly "or branches", and the show "or a branch" (Fig. 57.1).

57.3 Petri Net Workflow Model Expressions

Petri net model have standard semantics in many workflow management systems. The process modeling methods used are based on events, for example, the most common activities of the network method. These methods to transfer between the activities and events are clearly defined, and for the activities of the state are not clearly reflected in the model, compared to, Petri net being a state-based modeling method, which clearly defines the state of the model elements, and its evolution is driven by the state. Therefore, they are based on Petri net workflow process that also has very clear and strict definitions. Workflow Management System Workflow provides the six original

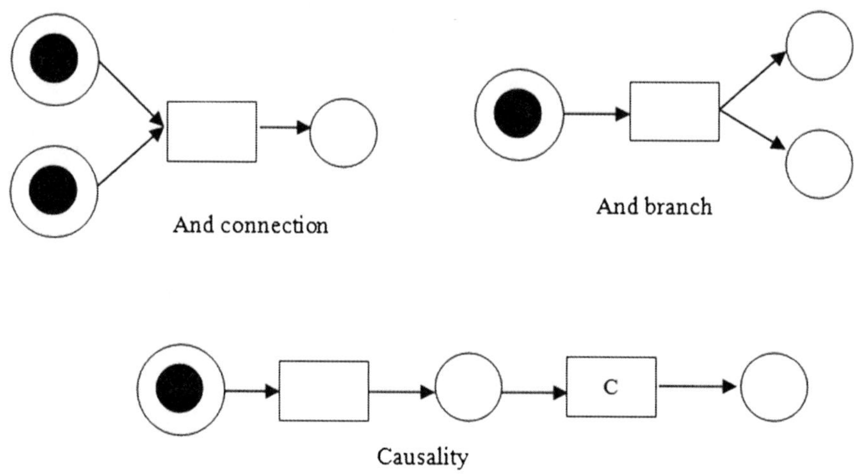

Fig. 57.2 Workflow primitive conversions to petri nets

languages: and, or a connection, circulation, and the branch, or branches, causality on which Petri nets can be supported. Figure 57.2 shows three of them converted; these primitives are unambiguous mapping into Petri nets form of expression.

57.4 Petri Net-Based Workflow Model of Multi-Node Cycle

Petri net is a suitable model for describing the concurrent characteristics of the system. It has become the ideal workflow modeling tool. However, in practice, the use of the basic Petri net models too often face complex problems. There is the understanding that the workflow process is not intuitive, not clear, or even an error process. Especially in the circular flow model, as is the case with the actual type, it is too complex, often an infinite loop situation. Modeling to solve problems in circular flow model, we monitor the place by increasing the problem solving cycle; this model is less suitable for the case of the node.

For n nodes, assume the following models:

$$\sum = (P, T; F, M_T, M_0) \tag{57.8}$$

where

$$\begin{aligned} P = \{&P_0, P_{n+1}, P_1, P_2, \ldots, P_n, P_{1\_1}, P_{1\_2}, \ldots, P_{n-1\_n-1}, P_{i-j}, \\ &1 \le i \le n, 1 \le j \le n, i \ne j\} \end{aligned} \tag{57.9}$$

$P_1, P_2, \ldots, P_n$ states that the same level processing node n;

$$P_{1\_1}, P_{1\_2}, \ldots, P_{n-1\_n-1} \tag{57.10}$$

represents the corresponding location of each surveillance;
$M(P_{i\_i}) > 0$ states its corresponding node P_i has processed the event, the event is not turning.

P_{i_j} represents the corresponding node has already processed the event P_i, $M(P_{i\_j}) > 0$ node P_i that the corresponding event is being processed and can be transferred to other nodes or directly P_j;

$$T = \{t_{0\_1}, t_{0\_2}, \ldots, t_{0\_n}, t_{0\_n+1}, t_{i\_2}, \ldots, t_{i\_\frac{n+1}{2}(C_{n-1}^{0}+C_{n-1}^{1}+C_{n-1}^{2}+\cdots+C_{n-1}^{n-2}+C_{n-1}^{n-1})}, 1 \le i \le n\} \tag{57.11}$$

$$F = P \times T \cup T \times P \tag{57.12}$$

$$M_0 = (1, 0, \underbrace{0, \ldots, 0}_{n}, \underbrace{1, 1, \ldots, 1}_{n-1}, \underbrace{0, \ldots, 0}_{n\cdot(n-1)}) \tag{57.13}$$

$$M_T \subseteq [M_0\rangle \tag{57.14}$$

Can be calculated by the model number of the place:

$$|P| = n = 2 + (n-1) + n\cdot(n-1) = n^2 + n + 1 \tag{57.15}$$

Under the rules, all of the translations in P_0 and $P_1, P_2, \cdots, P_n$ by the place issue, since $P_1, P_2, \cdots, P_n$ is the same for each type of P_i, and that no two libraries will trigger the same translations that an event cannot be handled by multiple nodes; hence in the calculation of translations in this part of $P_1, P_2, \ldots, P_n$ when the trigger can be triggered by a place translations in calculated P_i, multiplied by n, can be triggered by $P_1, P_2, \ldots, P_n$ were all translations. Together with all the vicissitudes of P_0 trigger, you can get the total number of translations.

Arc in a calculation to define the trigger for each translation of $P_1, P_2, \ldots, P_n$, months into the arc n and months arc out to the arc n, to ensure that each translation can be triggered unique. Coupled with translations in P_0 include a trigger to the arc, the arc can be a total of:

$$\sum|F| = 2\cdot n\cdot\sum|T| - 2n + 2 \tag{57.16}$$

Analyzing the above calculation results, the model number of the place is proportional to the square with the growth in the number of nodes. When the number of nodes is small, you can monitor the place by increasing the number of cycles to solve the problem.

57.5 Conclusions

The process of Petri nets, a strictly formal system compared to other non-formal diagram technology, avoids the ambiguity, uncertainty, and contradiction and is an ideal tool for workflow modeling. In the increasingly complex application environments of multi-node cycle and workflow modeling, Petri net has become the model of key issues. In this paper, in the Petri net modeling of circulating flow of the complexity of the work carried out draws the following conclusions: when the number of nodes is small, it can be addressed by increasing the monitoring place circulation problems; but more the nodes, the model place, translation geometric increase in number, but not the application of the modeling method.

References

1. Murata T (1989) Petri nets: properties, analysis, and application. Proc IEEE 77(4):541–580
2. Mentzas G, Kavadias S (2001) Modeling business process with workflow system: an evaluation of alternative approaches. Int J Inf Manage 21(14):123–135
3. Sadiq W, Orlowska ME (2003) Analyzing process models using graph reduction techniques. Inf Syst 25(2):117–134
4. Park1 C, Choi I (2004) Management of business process constraints using BPT rigger. Comput Ind 55(2):29–51
5. Osborn S (2002) Integrating role graphs: a tool for security integration. Data Knowl Eng 43(3):317–333

Chapter 58
A Novel Isolated Speech Recognition Method Based on Neural Network

Guojiang Fu

Abstract The Radial Basis Function Neural Network architecture has been shown to be suitable for the recognition of isolated words. Recognition of words is carried out in speaker-dependent mode. In this mode the tested data presented to the network are the same as the trained data. The 16 Linear Predictive Cepstral Coefficients with 16 parameters from each frame improves a good feature extraction method for the spoken words, since the first 16 in the cepstrum represent most of the formant information. It is found that the performance of radial basis function neural network (RBF) classifier is superior to MLP classifier. It is found that speaker 6 average performances is the best performance in training MLP classifier and speaker 2 average performances is the best performance in training RBF classifier. It is found that average speaker 4 performances is the best performance in testing MLP classifier and speaker 1 average performance is the best performance in testing RBF classifier.

Keywords Neural network • Speech recognition • Multilayer perception

58.1 Introduction

Automatic speech recognition has been an active research topic for 50 years. Along with digital computing and signal processing, the speech recognition problem puts forward the comparison system of further research, clear [1, 2]. The possible application range include controlled electrical appliances, on fully featured have to-text software, automatic operator-assisted service, and voice recognition AIDS for the handicapped [3, 4].

G. Fu (✉)
Information and Control Engineering Institute, Shenyang Jianzhu University, Shenyang, China
e-mail: fuguojiang@guigu.org

Z. Zhong (ed.), *Proceedings of the International Conference on Information Engineering and Applications (IEA) 2012*, Lecture Notes in Electrical Engineering 220, DOI: 10.1007/978-1-4471-4844-9_58,

They are mainly divided into four trends: acoustic-phonetic method, pattern recognition method, artificial intelligence method, and neural network to be realized [5–7].

Speech recognition has a great potential to become an important factor of interaction between the human and the computer in the near future [8–10]. The success of a speech recognition system must not only determine the characteristics that exist in input mode in a point in time, but also has input mode changing over time article [1]. On the contrary, radial basis function neural network (RBF) networks without the need for special adjustment and the training time becomes short about delay neural network.

Dataset The vocabulary set is composed of six words: "passion", "galaxy", "marvellous", "manifestation", "almighty", and "pardon". Six different speakers (two male and four female) are allowed to utter the above words. For uttering each word six times speech databases were recorded in wave files. So there are 216 wave files. Each of these wave files are trained and tested.

Preprocessing The speech signals are recorded in a low noise environment with good quality recording equipment. The signals are samples at 11 kHz. Reasonable results can be achieved in isolated digit recognition when the input data is surrounded by silence.

Sampling Rate 150 samples are chosen with sampling rate 11 kHz, which is adequate to represent all speech sounds.

Windowing In order to avoid discontinuities at the end of speech segments the signal should be tapered to zero or near zero and hence reduce the mismatch.

58.2 Feature Extraction

The goal is for feature extraction of speech signals through finite number of measures signals. This is because the whole information acoustic signal is the process of many, but not all of the information is about a specific task. In the present speech recognition system, the feature extraction method usually finds a relatively stable different example, although such speech sound differences or different environment characteristics and, at the same time, the spokesman for part of the relatively complete information on behalf of speech signal. Linear forecast coding (LPC) is a tool mainly used in audio signal processing and speech processing. The spectrum of the envelope on behalf of a digital signal is a compressed form of speech by using the linear forecasting information model. This is one of the most powerful voice analysis techniques, and one of the most powerful methods for quality good speech coding in low bit rate that give very accurate estimate parameters of the lecture. The LPC speech signal is analyzed through the forecast, and eliminates the effects of the format from the speech signal intensity and frequency estimation of the rest of the buzz. The process of removing the filter is called the resonant; the rest of the filtered signal after subtraction analog signal is called the living. The description of the digital frequency, intensity buzz, and the resonant signals can be live stored or transmitted to other places. LPC comprehensive speech signal process create a source signal with reversing buzz parameters and residual. The resonant used to create a filter (tube) represents the source and runs through the filter, leading to speech. Because speech signal, in the process, with time on doing a lot of short speech signal, generally includes 30–50 frames per second .

The spread directly filter coefficients is not recommended, because they are very sensitive mistakes. In other words, a very small error may distort the whole spectrum, or worse, a small mistake may make the prediction filter unstable.

Generally used for voice LPC is analysis and secondary synthesis. As a kind of speech compress in telephone companies, such as in the GSM standards, It can also be used to secure wireless, where the voice must be digital, encryption, and transmit speech narrow channel.

In the LPC analysis one tries to predict $x_n$on the basis of the p previous samples,

$$x_n' = \sum a_k x_{n-k} \tag{58.1}$$

Then $\{a_1, a_2, \ldots, a_p\}$ can be chosen to minimize the prediction power Q_p

where $Q_p = E\left[\left|x_n - x_n'\right|^2\right]$

Linear Predictive Coding is used to extract the LPCC coefficients from the speech tokens. The LPCC coefficients are then converted into cepstral coefficients. The cepstral coefficients are normalized between −1 and 1. The speech is blocked into overlapping frames of 20 ms every 10 ms using Hamming window. LPCC is implemented using the autocorrelation method. A drawback of LPCC estimates is their high sensitivity to quantization noise. After converting LPCC coefficients into cepstral coefficients where the cepstral order is the LPCC order and decreasing the sensitivity of high and low-order cepstral coefficients to noise, the obtained cepstral coefficients are then weighted. Sixteen Linear Predictive Cepstral Coefficients are considered for windowing. Linear Predictive Coding analysis of speech is based on human perception experiments. Sample the signal with 11 kHz. Number of frames is obtained for each utterance from LPC coefficients.

58.3 Recognition Methodology

In the present case as model, each classifier trying to determine the set characteristic vector and input from current signal belongs to a specific type of digital or incomplete class. As a sample, not as a specific professional class is a random choice.

58.4 Classifiers

Several classifiers are tested for the mentioned dataset. The structures of successful classifiers in recognition are described in the following subsections.

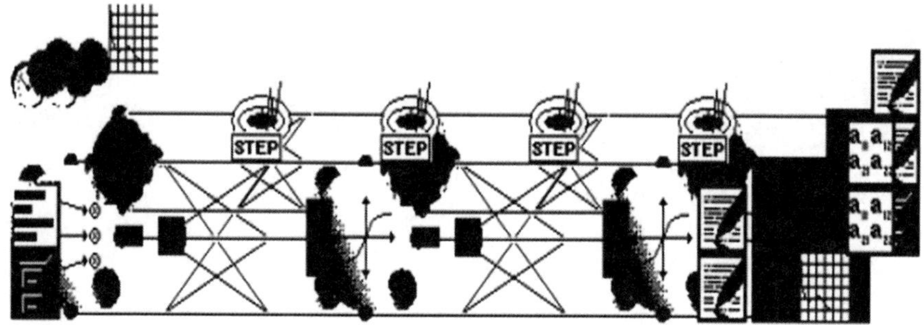

Fig. 58.1 MLP network architecture with step learning rule

58.4.1 Multilayer Perception

As shown in Fig. 58.1, the network has an input layer (on the left) with three neurons, one hidden layer (in the middle) with three neurons, and an output layer (on the right) with three neurons.

There is one neuron in the input layer for each predictor variable. In the case of categorical variables, N-1 neurons are used to represent the N categories of the variable.

Input Layer—a vector of predictor variable values ($x1 \cdots xp$) is presented to the input layer. The input layer (or processing before the input layer) standardizes these values so that the range of each variable is -1–1. The input layer distributes the values to each of the neurons in the hidden layer. In addition to the predictor variables, there is a constant input of 1.0, called the bias that is fed to each of the hidden layers; the bias is multiplied by a weight and added to the sum going into the neuron.

Hidden Layer—arriving at a neuron in the hidden layer, the value from each input neuron is multiplied by a weight (wji), and the resulting weighted values are added together producing a combined value uj. The weighted sum (uj) is fed into a transfer function, which outputs a value hj. The outputs from the hidden layer are distributed to the output layer.

Output Layer—Arriving at a neuron in the output layer, the value from each hidden layer neuron is multiplied by a weight (wkj), and the resulting weighted values are added together producing a combined value vj. The weighted sum (vj) is fed into a transfer function, σ, which outputs a value yk. The y values are the outputs of the network.

If a regression analysis is being performed with a continuous target variable, then there is a single neuron in the output layer, and it generates a single y value. For classification problems with categorical target variables, there are N neurons in the output layer producing N values, one for each of the N categories of the target variable.

58.4.2 Radial Basis Neural Networks

The core of a speech recognition system is the recognition engine. The one chosen in this paper is the RBF. This is a static two-neuron layer feedforward network with the first layer L1, called the hidden layer and the second layer, L2, called the output layer. L1 consists of kernel nodes that compute a localized and radically symmetric basis functions.

The pattern recognition approach avoids explicit segmentation and labeling of speech. Instead, the recognizer used the patterns directly. It is based on comparing a given speech pattern with previously stored ones. The way speech patterns are formulated in the reference database affects the performance of the recognizer. In general, there are two common representations:

The output y of an input vector x to an RBF neural network with H nodes in the hidden layer is governed by:

$$y = \sum_{h=0}^{H-1} w_h \phi_h(x) \tag{58.2}$$

where w_h linear weights are ϕ_h are the radial symmetric basis functions. Each one of these functions is characterized by its center c_h and by its spread or width σ_h. The range of each of these functions is [0, 1].

Once the input vector x is presented to the network, each neuron in the layer L1 will output values according to how close the input vector is to its weight vector. The more similar the input is to the neuron's weight vector, the closer to 1 is the neuron's output and vice versa. If a neuron has an output 1, then its output weights in the second layer L2 pass their values to the neurons of L2. The similarity between the input and the weights is usually measured by a basis function in the hidden nodes. One popular such function is the Gaussian function that uses the Euclidean norm. It measures the distance between the input vector x and the node center c_h. It is defined as:

$$\phi_h = \exp\left(\|x - c_h\| / 2\sigma_h^2\right) \tag{58.3}$$

58.5 Training Phase

The networks are usually trained to perform tasks such as pattern recognition, decision-making, and motor control. The original idea was to teach them to process speech or vision, similar to the tasks of the human brain. Nowadays tasks such as optimization and function approximation are common. Training of the units is accomplished by adjusting the weight and threshold to achieve a classification. The adjustment is handled with a learning rule from which a training algorithm for a specific task can be derived. The Multilayer Perceptron and Radial Basis Function Neural Networks are trained for spoken words for six

Table 58.1 RESULTS for training MLP (%)

	Passion (%)	Galaxy (%)	Marvellous (%)	Manifest-aion (%)	Almighty (%)	Pardon (%)
Speaker 1	95	96	98	88	95	97
Speaker 2	97	96	97	92	97	98
Speaker 3	96	98	96	97	95	96
Speaker 4	97	95	97	87	94	97
Speaker 5	95	95	97	98	96	89
Speaker 6	95	96	98	97	97	96

Table 58.2 Results for training RBF (%)

	Passion (%)	Galaxy (%)	Marvellous (%)	Manifest-aion (%)	Almighty (%)	Pardon (%)
Speaker 1	98	98	99	97	99	99
Speaker 2	99	99	99	99	99	99
Speaker 3	99	98	99	99	98	99
Speaker 4	99	99	99	99	97	99
Speaker 5	98	99	98	97	98	95
Speaker 6	97	98	98	97	98	96

Table 58.3 Overall performance averages

Classifier (%)	Overall performance average (%)
MLP	95.47
RBF	98.21

speakers. The learning rate is taken as 0.01and momentum rate is taken as 0.3; the number of epochs is taken as 100. The Random Gaussian Method is chosen for initialization.

The performance of MLP and RBF classifiers for each speaker have been computed and presented in Tables 58.1 and 58.2 respectively. The overall performance average for both classifiers MLP and RBF have been computed and presented in Table 58.3.

58.6 Testing Phase

The same Multilayer Perceptron and Radial Basis Function Neural Networks are trained for spoken digits for six speakers. The learning rate, momentum rate, and the number of epochs chosen are the same as in the training phase. The initialization chosen is also the same as that of training phase.

The performance of MLP and RBF classifiers for each speaker have been computed and presented in Tables 58.4 and 58.5 respectively. The overall

Table 58.4 Results for testing MLP (%)

	Passion (%)	Galaxy (%)	Marvellous (%)	Manifest-aion (%)	Almighty (%)	Pardon (%)
Speaker 1	98	97	97	98	88	98
Speaker 2	97	97	97	97	96	88
Speaker 3	96	96	98	96	97	89
Speaker 4	98	99	96	97	98	97
Speaker 5	97	95	97	97	97	89
Speaker 6	96	95	98	98	95	98

Table 58.5 Results for testing RBF (%)

	Passion (%)	Galaxy (%)	Marvellous (%)	Manifest-aion (%)	Almighty (%)	Pardon (%)
Speaker 1	100	99	99	99	100	100
Speaker 2	99	100	98	99	99	100
Speaker 3	98	99	99	100	99	99
Speaker 4	99	99	99	99	98	99
Speaker 5	99	99	98	99	97	96
Speaker 6	97	98	98	98	98	97

Table 58.6 Overall performance averages

Classifier	Overall performance average (%)
MLP	96
RBF	98.69

performance average for both classifiers MLP and RBF have been computed and presented in Table 58.6.

58.7 Conclusion

RBF neural network has become an increasingly popular neural network and the different applications are likely to be the primary competitors of multilayer perception. Most of the inspiration comes from traditional RBF network the statistical pattern recognition technology. The unique feature is a process of radial basis function neural network of hidden layer. The idea is that of input space patterns. If these clusters of cluster center are known, then the cluster centre distance can be measured. In addition, the distance measure is nonlinear, so that if a pattern is close to cluster centre it provides a value close to 1. In statistical neural network learning mechanism are not biologically plausible–not occupied the researchers insist on biological analogy.

This is to become an increasingly popular neural network and a different application, which is likely to be the main rival multilayer perceptron RBF network.

References

1. Benyettou A (1995) Acoustic phonetic recognition in the arabex system. Int Work Shop Robot Hum Commun ATIP95 44:34–45
2. Al-Alaoui MA, Mouci R, Mansour MM, Ferzli R (2002) A cloning approach to classifier training. IEEE Trans Syst Man Cybern Part A Syst Hum 32(6):746–752
3. Picton P (2000) Neural networks, vol 35. Palgrave, NY, pp 145–157
4. Tan Lee, Ching PC, Chan LW (1998) Isolated word recognition using modular recurrent neural networks. Pattern Recogn 31(6):751–760
5. Gurney K (1997) An introduction to neural networks, UCL Press, University of Sheffield, Sheffield
6. Berthold MR (1994) A time delay radial basis function for phoneme recognition. In: Proceedings of international conference on neural network, vol 33. Orlando, USA, pp 102–109
7. Rabiner L, Juang B–H (1993) Fundamentals of speech recognition, vol 213. PTR Prentice Hall, San Francisco, pp 378–385
8. Kandil N, Sood VK, Khorasani K, Patel RV (1992) Fault identification in an AC–DC transmission system using neural networks. IEEE Trans Power Syst 7(2):812–819
9. Morgan D, Scolfield C (1991) Neural networks and speech processing, vol 9. Kluwer Academic Publishers, Berlin, pp 46–53
10. Park DC (1991) M A El-Sharakawi and Ri Marks II electric load forecasting using artificial neural networks. IEEE Trans Power Syst 6(2):442–449

Chapter 59
Evaluation on Enterprise Resource Planning Project Based on Fuzzy-AHP

Chang-shan Li

Abstract In this study, an enterprise resource planning (ERP) implementation of the project risk analysis to deal with suppliers, to determine the risk standard benefit from the literature, experts and ERP software users, the problem solving fuzzy treatment using AHP methodology, and then the sensitivity analysis to evaluate our risk analysis result is verified, specifically for our portfolio companies.

Keywords ERP project • Risk analysis • Fuzzy-AHP method

59.1 Introduction

Enterprise resource planning (ERP) is a function module into system application software management resources that coordinate the activities of the boundary. ERP is a key information technology (IT) application used to capture, store, and disseminate information to improve efficiency and visibility in dealing with the actual goods that move within the firm [1]. In the world of globalization, the market is more competitive, seeks new business opportunities for the organization, and enhances competitiveness. It is generally accepted that it has been used to fundamentally improve business. Most enterprises, therefore, strengthen the competition for the use of advanced ERP system, such as [2]. It is the implementation of a challenging, time-consuming, and expensive process that requires notable company time and resources. Despite this fact, in the past few years, many companies have implemented ERP system, because the ERP is a key ingredient to gain competitive advantage and improve operation of a "lean" manufacturing system [3].

C. Li (✉)
Jilin Business And Technology College, Changchun 130062, Jinlin, China
e-mail: lichangshan@cssci.info

Z. Zhong (ed.), *Proceedings of the International Conference on Information Engineering and Applications (IEA) 2012*, Lecture Notes in Electrical Engineering 220,
DOI: 10.1007/978-1-4471-4844-9_59,

59.2 The Numerical Example

There are many authors who proposed the fuzzy-AHP method. Decision makers usually find that it is confidence interval value judgment than to judge [4, 5]. This is because he/she usually cannot clear his/her choice because of the comparison in fuzziness process.

In this study, we prefer long degree analysis method, the method of steps because than other fuzzy-AHP it is the easy way out. Fuzzy sets and fuzzy-AHP are not detailed here for the application of the famous [6, 7].

The methodology of fuzzy-AHP used here for ERP project risk analysis was applied in a Turkey Company for an auto industry for operations. Experts and ERP software users determine the risk standard benefit from the literature. These standards are shown in Fig. 59.1. This paper aims to evaluate the risk level at the surface in the ERP project execution and validates the results specifically for our portfolio companies.

First of all, in the fuzzy-AHP standard the weight of the importance of the choice must do more. For this reason, we must have a language terms and their equivalent fuzzy number is-measures. Language in terms and their equivalent fuzzy number is considered in this paper, see Table 59.1. Therefore, the main goal of the evaluation standard, and according to the scheme, these standards must realize evaluation. Then, after all of this evaluation program, the weights of the choice are calculated.

The weights' calculation details are given below. Because the other calculations are similar for each comparison matrix they are not given here and can be done simply according to the computations below, by using MS Excel 2007. The value of fuzzy synthetic extent with respect to the ith object (i = 1, 2,..., 8) is calculated as:

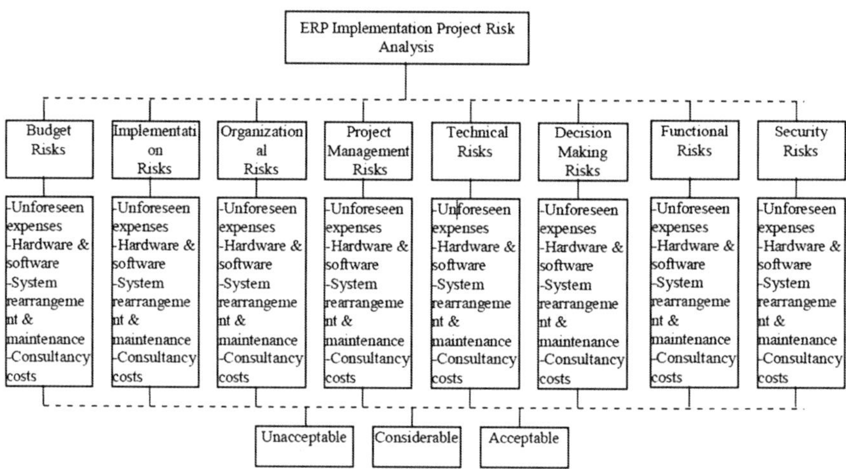

Fig. 59.1 ERP implementation project risk analysis

Table 59.1 Fuzzy comparison measures

Linguistic terms	Triangular fuzzy numbers
Absolute	(7/2, 4, 9/2)
Very good	(5/2, 3, 7/2)
Good	(3/2, 2, 5/2)
Weak advantage	(2/3, 1, 3/2)
Equal	(1, 1, 1)

$S_{BR} = (6.5, 10, 13.5) \otimes (0.01036, 0.015625, 0.0222) = (0.067, 0.156, 0.3)$
$S_{IR} = (5.167, 6.667, 9.5) \otimes (0.01036, 0.015625, 0.0222) = (0.053, 0.104, 0.211)$
$S_{OR} = (5, 6.667, 11) \otimes (0.01036, 0.015625, 0.0222) = (0.051, 0.104, 0.244)$
$S_{PMR} = (5.167, 6.667, 9.5) \otimes (0.01036, 0.015625, 0.0222) = (0.053, 0.104, 0.211)$
$S_{TR} = (5.5, 8.5, 13.5) \otimes (0.01036, 0.015625, 0.0222) = (0.056, 0.132, 0.3)$
$S_{DMR} = (4.66, 7, 10.5) \otimes (0.01036, 0.015625, 0.0222) = (0.048, 0.109, 0.233)$
$S_{FR} = (6.333, 8.5, 14) \otimes (0.01036, 0.015625, 0.0222) = (0.065, 0.132, 0.311)$
$S_{SR} = (6.667, 10, 14) \otimes (0.01036, 0.015625, 0.0222) = (0.069, 0.155, 0.311)$

Also, $V(M \geq M1)$ probability levels:

$V(S_{BR} \geq S_{IR}) = 1.00$, $V(S_{BR} \geq S_{OR}) = 1.00$,
$V(S_{BR} \geq S_{PMR}) = 1.00$, $V(S_{BR} \geq S_{TR}) = 1.00$,
$V(S_{BR} \geq S_{DMR}) = 1.00$, $V(S_{BR} \geq S_{FR}) = 1.00$,
$V(S_{BR} \geq S_{SR}) = 1.00$
$V(S_{IR} \geq S_{BR}) = 0.73$, $V(S_{IR} \geq S_{OR}) = 1.00$,
$V(S_{IR} \geq S_{PMR}) = 1.00$, $V(S_{IR} \geq S_{TR}) = 0.85$,
$V(S_{IR} \geq S_{DMR}) = 0.97$, $V(S_{IR} \geq S_{FR}) = 0.83$,
$V(S_{IR} \geq S_{SR}) = 0.73$
$V(S_{OR} \geq S_{BR}) = 0.77$, $V(S_{OR} \geq S_{IR}) = 1.00$,
$V(S_{OR} \geq S_{PMR}) = 1.00$, $V(S_{OR} \geq S_{TR}) = 0.87$,
$V(S_{OR} \geq S_{DMR}) = 0.97$, $V(S_{OR} \geq S_{FR}) = 0.86$
$V(S_{OR} \geq S_{SR}) = 0.77$
$V(S_{PMR} \geq S_{BR}) = 0.77$, $V(S_{PMR} \geq S_{IR}) = 1.00$,
$V(S_{PMR} \geq S_{OR}) = 1.00$, $V(S_{PMR} \geq S_{TR}) = 0.85$,
$V(S_{PMR} \geq S_{DMR}) = 0.95$, $V(S_{PMR} \geq S_{FR}) = 0.88$,
$V(S_{PMR} \geq S_{SR}) = 0.77$
$V(S_{TR} \geq S_{BR}) = 0.90$, $V(S_{TR} \geq S_{IR}) = 1.00$,
$V(S_{TR} \geq S_{OR}) = 1.00$, $V(S_{TR} \geq S_{PMR}) = 1.00$,
$V(S_{TR} \geq S_{DMR}) = 1.00$, $V(S_{TR} \geq S_{FR}) = 1.00$,
$V(S_{TR} \geq S_{SR}) = 0.90$
$V(S_{DMR} \geq S_{BR}) = 0.78$, $V(S_{DMR} \geq S_{IR}) = 1.00$,
$V(S_{DMR} \geq S_{OR}) = 1.00$, $V(S_{DMR} \geq S_{PMR}) = 1.00$,
$V(S_{DMR} \geq STR) = 0.88$, $V(S_{DMR} \geq S_{FR}) = 0.87$,
$V(S_{DMR} \geq S_{SR}) = 0.77$
$V(S_{FR} \geq S_{BR}) = 0.91$, $V(S_{FR} \geq S_{IR}) = 1.00$,
$V(S_{FR} \geq S_{OR}) = 1.00$, $V(S_{FR} \geq S_{PM}R) = 1.00$,
$V(S_{FR} \geq S_{TR}) = 1.00$, $V(S_{FR} \geq S_{DMR}) = 1.00$,
$V(S_{FR} \geq S_{SR}) = 0.92$

$V(S_{SR} \geq S_{BR}) = 1.00$, $V(S_{SR} \geq S_{IR}) = 1.00$,
$V(S_{SR} \geq S_{OR}) = 1.00$, $V(S_{SR} \geq S_{PMR}) = 1.00$,
$V(S_{SR} \geq S_{TR}) = 1.00$, $V(S_{SR} \geq S_{DMR}) = 1.00$,
$V(S_{SR} \geq S_{FR}) = 1.00$

The weights representing the importance levels of the criteria of the main criteria are shown in Fig. 59.2. Now, the ERP implementation project risk levels must be evaluated with respect to each criterion. All comparison tables are not shown here. But, all weights calculated by pairwise comparisons of main criteria for the risk levels are given aggregately in Fig. 59.3 (Table 59.2).

The results and finally the risk ranks are given in Fig. 59.4. According to the fuzzy-AHP algorithm results based on our standard level, the level of risk is high. ERP for our case companies is not acceptable.

We use here, sensitivity analysis to evaluate risk analysis method. Therefore, the obtained fuzzy-AHP to two standard weight change, organization and project risk management, and others are the same. This cannot be accepted as the level of risk is very high according to the present standards. We note that the risk is that two standard levels drop and is zero, in the implementation of the project risk.

Sensitivity analysis of the results can be seen from the graph in Fig. 59.5. If the risk level in the organization and project management process is reduced to zero, then in the considered ERP project, the level of risk is determined to be acceptable. The enterprise must take action to eliminate these risks especially in process.

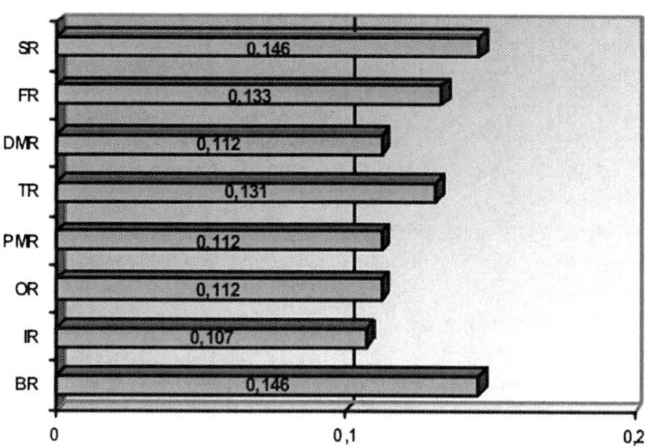

Fig. 59.2 The weights of the main criteria

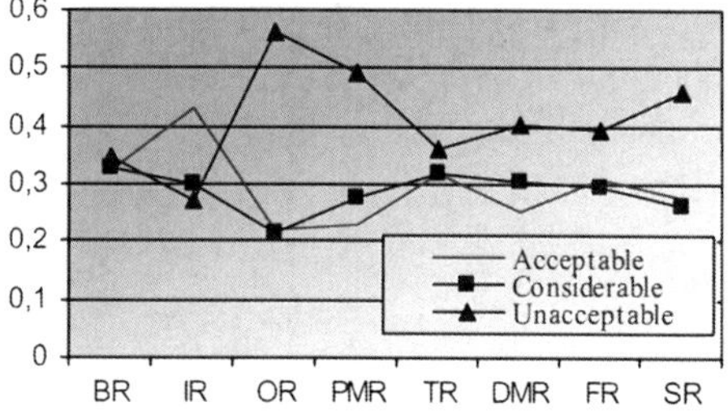

Fig. 59.3 The risk level weights according to the main criteria

Table 59.2 The main criteria comparison matrix

	BR	IR	OR	PMR
BR	(1, 1, 1)	(1, 3/2, 2)	(1, 3/2, 2)	(1, 3/2, 2)
IR	(1/2, 2/3, 1)	(1, 1, 1)	(1/2, 1, 3/2)	(1, 1, 1)
OR	(1/2, 2/3, 1)	(2/3, 1, 2)	(1, 1, 1)	(2/3, 1, 2)
PMR	(1/2, 2/3, 1)	(1, 1, 1)	(1/2, 1, 3/2)	(1, 1, 1)
TR	(2/3, 1, 2)	(1/2, 1, 3/2)	(1, 3/2, 2)	(1/2, 1, 3/2)
DMR	(1/2, 2/3, 1)	(1/2, 1, 3/2)	(1/2, 1, 3/2)	(1/2, 1, 3/2)
FR	(2/3, 1, 2)	(1, 3/2, 2)	(1, 3/2, 2)	(1, 3/2, 2)
SR	(2/3, 1, 2)	(1, 3/2, 2)	(1, 3/2, 2)	(1, 3/2, 2)
	TR	DMR	FR	SR
BR	(1/2, 1, 3/2)	(1, 3/2, 2)	(1/2, 1, 3/2)	(1/2, 1, 3/2
IR	(1/2, 2/3, 1)	(2/3, 1, 2)	(1/2, 2/3, 1)	(1/2, 2/3, 1
OR	(1/2, 2/3, 1)	(2/3, 1, 2)	(1/2, 2/3, 1)	(1/2, 2/3, 1
PMR	(1/2, 2/3, 1)	(2/3, 1, 2)	(1/2, 2/3, 1)	(1/2, 2/3, 1
TR	(1, 1, 1)	(1/2, 1, 3/2	(2/3, 1, 2)	(2/3, 1, 2)
DMR	(2/3, 1, 2)	(1, 1, 1)	(1/2, 2/3, 1)	(1/2, 2/3, 1
FR	(1/2, 1, 3/2)	(1/2, 1, 3/2	(1, 1, 1)	(2/3, 1, 2)
SR	(1/2, 1, 3/2)	(1, 3/2, 2)	(1/2, 1, 3/2)	(1, 1, 1)

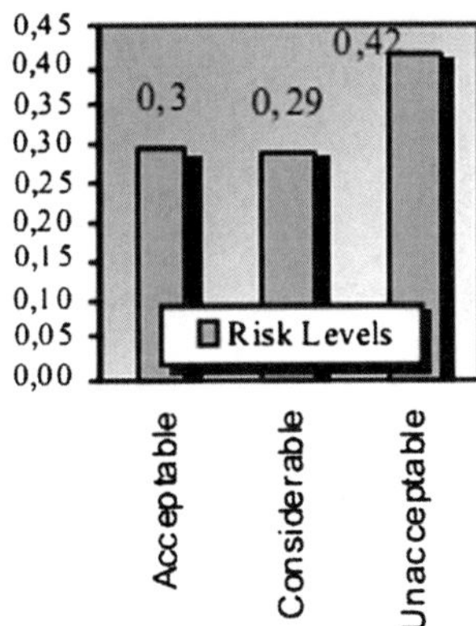

Fig. 59.4 The final weights of risk levels gained via fuzzy-AHP methodology

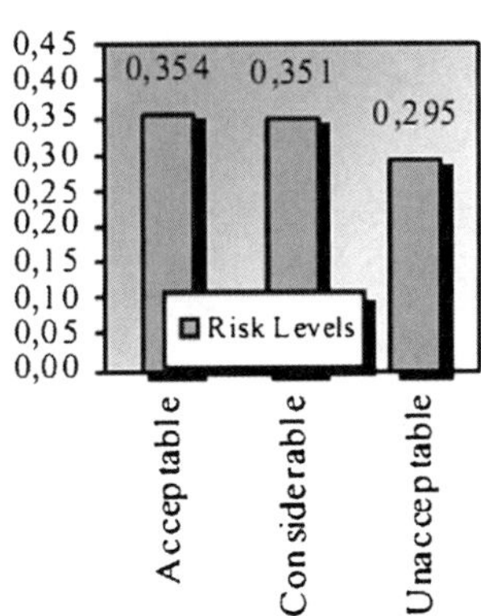

Fig. 59.5 Model result change by sensitivity analysis

59.3 Conclusions

The ERP project risk analysis processing method, fuzzy-AHP is processed in order to solve this problem. Then an example is given that shows the method of applicability and performance. At the same time, the sensitivity analysis is discussed and explains the results of the method.

References

1. Chen G, Wang J (2010) Analysis on performance evaluation system of ERP implementation. Int Conf Inf Sci Manage Eng 3:185–188
2. Grabski SV, Leech SA (2007) Complementary controls and ERP implementation success. Int J Account Inf Syst 8:17–39
3. Verville J, Palanisamy R, Bernadas C, Halingten A (2007) ERP acquisition planning: a critical dimension for making the right choice. Long Range Plan 40:45–63
4. Markus ML, Tanis C, Van Fenema PC (2000) Multisite ERP Implementations. Commun ACM 43(4):42–46
5. Wu L-C, Ong C-S, Hsu Y-W (2008) Active ERP implementation management: a real options perspective. J Syst Softw 81:1039–1050
6. Sarker S, Lee AS (2003) Using a case study to test the role of three key social enablers in ERP implementation. Inf Manage 40:813–829
7. Sheng YP, Pearson M, Crosby L (2003) Organizational culture and employees' computer self-efficacy. Inf Resour Manage J 16(3):42–58

Part VI
Computer Graphics and Image Processing

Chapter 60
SAR Image Automatic Registration Based on PCA-SIFT and Mahalanobis Distance

Jianxun Zhang, Kaiwen Zhang, Wenbin Niu and Jinghua Huang

Abstract In order to realize the registration between the overlapping areas of the adjacent frames in SAR image sequence, a SAR image automatic registration algorithm based on PCA-SIFT and Mahalanobis distance was proposed. First, combine the improved PCA-SIFT and Mahalanobis distance to obtain the feature points and complete the original matching; next, apply the improved RANSAC algorithm to remove the mismatching, and estimate the transformation matrix; then, accomplish the mapping between the images according to the obtained matrix, and do the resample on the mapping. Finally, SAR image automatic registration is realized. Experiments show that the improved algorithm can achieve image registration with effectiveness and robustness.

Keywords PCA-SIFT • Mahalanobis distance • RANSAC • SAR image registration

60.1 Introduction

SAR image registration is the process to seek the space geometry transformation of two (or more) images of the same scene taken from different viewpoints, at different times, and/or by different sensors [1, 2]. It is the step prior to SAR image fusion and SAR image mosaics, and it is also the key step of motion tracking and target recognition.

There are mainly two kinds of SAR image registration methods: the method based on block and the method based on features. In the former, the registration is completed based on the gray value and it cannot handle such conditions as image distortion [3] and illumination change in the image. As the image distortion and uneven illumination exist in the SAR image, it is difficult to complete the correct matching using the

J. Zhang · K. Zhang (✉) · W. Niu · J. Huang
School of Computer Science and Engineering, Chongqing University of Technology, Chongqing 400054, China
e-mail: xfkw5258@126.com

Z. Zhong (ed.), *Proceedings of the International Conference on Information Engineering and Applications (IEA) 2012*, Lecture Notes in Electrical Engineering 220, DOI: 10.1007/978-1-4471-4844-9_60, © Springer-Verlag London 2013

method based on the block. The geometric matching is completed according to the common characteristics among the images in the method based on features [4, 5]. Hence, the image registration method based on the features is suitable for SAR image.

In view of the above, according to the features of SAR image, a registration algorithm based on PCA-SIFT and Mahalanobis distance was proposed. The experiment results show that the registration is well completed and the proposed algorithm is effective and robust.

60.2 The Related Algorithms

60.2.1 Feature Extraction Algorithms

According to Ref. [1], feature extraction algorithms can be divided into the following three kinds: area feature extraction algorithm; line feature extraction algorithm; point feature extraction algorithm.

Among the point feature extraction algorithms, the SIFT feature proposed by David G. lowe in 2004 and its later generations are the new developments. Ref. [6] shows that SIFT and its extended algorithms perform very well among the many feature extraction algorithms under performances such as scale invariance, rotation, affine transformation, and illumination variation.

PCA-SIFT, a SIFT extended algorithm, was proposed by Ke and Sukthankar [7]. Compared to SIFT, PCA-SIFT perform much better in illumination invariance [8].

An improvement point based on PCA-SIFT is proposed according to the features of SAR image in this paper.

60.2.2 Similarity Measure

The matching is completed according to the feature similarity in the registration algorithms based on features, and the similarities of the features include Euclidean distance, block distance, Mahalanobis distance, and Hausdorff distance.

Affine transformation invariance can be used for matching the feature points. And as a result of the affine invariance of Mahalanobis distance, the results of PCA-SIFT algorithm are reprocessed using Mahalanobis distance and the new matched points are obtained.

60.2.3 Remove the Mismatch

The matched points that fit into the transformation model between the reference image and the registration image are called "inlier" and the points that do not fit into the model are called "outlier". The estimation of the model parameters is greatly influenced by the outlier and the outlier must be removed.

Random sample consensus (RANSAC) was first proposed by Fishler and Bolles in 1981. The basic idea of the RANSAC is that, first, design a search engine and use it to iteratively remove those input data which are not conforming to the estimated parameters, and then use the correct input data to estimate the parameters.

An improved RANSAC was adopted to remove the mismatch so as to obtain the accurate matched points and improve the accuracy of the registration.

60.2.4 Image Transformation

The image geometric transformation between registration image and reference image can be divided into direct transformation method and resampling method[3].
Direct transformation method

The corresponding position (ξ, η) in the reference image of each feature point (X_1, Y_1) in registration image is calculated according to the column order, and the gray value of point (ξ, η) is calculated by interpolation and the gay value of (ξ, η) is equal to the gray value of point (X_1, Y_1).
Resampling method

The corresponding position in the reference image of each feature point (ξ, η) in registration image is calculated according to the column order, and the gray value of point (X_2, Y_2) is calculated by interpolation and the gay value of (X_2, Y_2) is equal to the gray value of point (ξ, η).

60.3 SAR Image Automatic Registration Based on PCA-SIFT and Mahalanobis Distance

Based on the above, a SAR image automatic registration algorithm based on PCA-SIFT and Mahalanobis distance was proposed. The flowchart is as follows:

The implementation steps were as follows (Fig. 60.1):

60.3.1 Feature Point Extraction Based on the Improved PCA-SIFT

Detect the extreme value points in the scale space, and primarily determine the position and scale of the key points.

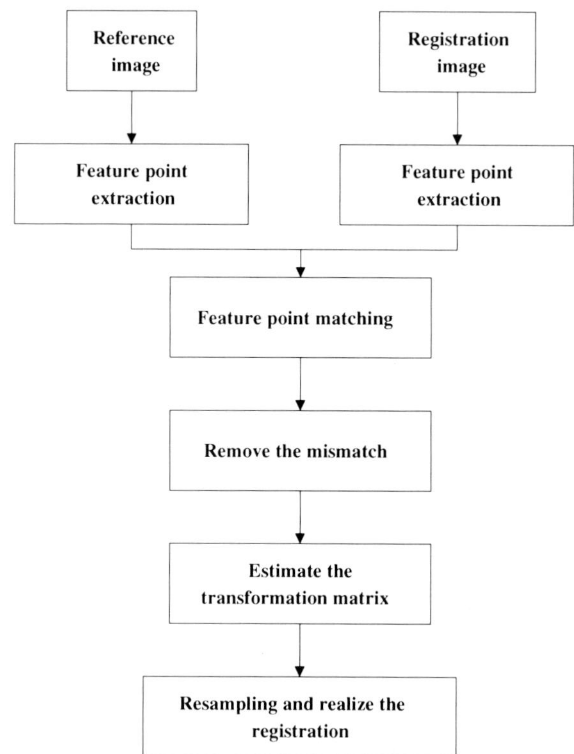

Fig. 60.1 The process of the proposed registration algorithm

A two-dimensional image was described by $I(x, y)$. Comparing each pixel of the middle layer (except the bottom and the top layer) in the DOG Pyramid scale space with its 26 adjacent pixels, the point with the maximum or the smallest extreme value was found out to be the initial feature points; then the location and its corresponding scale were recorded.

An improvement point based on PCA-SIFT was that the layer of DOG Pyramid was reduced from 4 to 2. The reason to do this was as follows:

The transform scaling did not exist in the SAR images in the project.

The scale invariance of PCA-SIFT was realized through building the pyramids, and the experiment proved that the building of pyramid took up the maximum time of the whole algorithm.

The layer of the pyramid was reduced from 4 to 1 in turn, and when the layer was 2, the result was the best.

$$D(x, y, \sigma) = (G(x, y, k\sigma) - G(x, y, \sigma)) \times I(x, y) = L(x, y, k\sigma) - L(x, y, \sigma) \tag{60.1}$$

there, k represents a constant of the two adjacent scale space magnification, and $G(x, y, \sigma)$ represents Gauss kernel functions and σ is the scale factor.

Determine the position and scale of the key points, and remove the key points with low contrast and unstable edge points, so as to enhance the stability and improve the anti-noise ability.

Specify the direction parameters for each key point to make the operator with the rotation invariance.

In order to make the detected feature points with the rotation invariance, the gradient direction of the pixel around the key point was used to specify the direction parameters for every key point.

$$\begin{cases} \theta(x,y) = \arctan((L(x,y+1) - L(x,y-1))/(L(x+1,y) - L(x-1,y))) \\ m(x,y) = \sqrt{(L(x+1,y) - L(x-1,y))^2 + (L(x,y+1) - L(x,y-1))^2} \end{cases} \tag{60.2}$$

Equation (60.2) represents the amplitude and direction of the gradient of pixel (x, y).

Generate PCA-SIFT characteristic vector.

It needed three steps to calculate the PCA-SIFT feature point descriptor. First, import a PCA-SIFT projection matrix; second, detect the feature points; third, obtain the characteristic vector of feature point dimension reduction.

60.3.2 Calculate the Mahalanobis Distance

For sample space $X = \left\{(x_1, y_1)^T, (x_2, y_2)^T, \ldots, (x_n, y_n)^T\right\}$ made up with n points, the Mahalanobis distance between any point $X_i = (x_i, y_i)^T$ and the sample mean is as follows:

$$d_i = \sqrt{(X_i - \mu)^T C^{-1} (X_i - \mu)} \tag{60.3}$$

There, C^{-1} represents the inverse matrix of C. C represents the convariance matrix. μ represents sample mean.

In the algorithm, set the standard deviation threshold of the Mahalanobis distance to be h. Then the flow of obtaining the final matched points is as follows:

Among the original matched points, taking four point pairs as a group, the Mahalanobis distance of the four point pairs $d_1 = \{d_{1(1)}, d_{1(2)}, d_{1(3)}, d_{1(4)}\}$ and $d_2 = \{d_{2(1)}, d_{2(2)}, d_{2(3)}, d_{2(4)}\}$ were calculated based on the above equation. The corresponding standard deviation of Mahlanobis distance $S(m = 4)$ was calculated and the pair where S was smaller than h was reserved. Finally the final matched points set were obtained.

60.3.3 Eliminate the Mismatch Using the Improved RANSAC

The registration image and the reference image are respectively described as $I_1(x_1, y_1)$ and $I_2(x_2, y_2)$, and the affine variable relationship between them is as follows:

$$\begin{pmatrix} x_1 \\ y_1 \\ 1 \end{pmatrix} = \begin{pmatrix} t_0 & t_1 & t_2 \\ t_3 & t_4 & t_5 \\ 0 & 0 & 1 \end{pmatrix} \cdot \begin{pmatrix} x_2 \\ y_2 \\ 1 \end{pmatrix} \tag{60.4}$$

Or $X_1 = TX_2$.

There, $X_1 = (x_1, y_1, 1)^T, X_1 = (x_2, y_2, 1)^T$ respectively represent the homogeneous coordinates of the registration image and the reference image. The matrix T can be estimated using at least three matched point pairs[4].

The procedure of the improved RANSAC algorithm is as follows:

First of all, set the number of the current best estimated interior point to be 0.

Repeat the random sampling for N times. Three matched point pairs were needed to estimate the matrix T. Set P_1 as the probability which showed that a pair was inter points, and $\varepsilon = 1 - P_1$ as the probability which represented that a pair was outer points. So, when the number of sampling was N, then: $(1 - P_1^4)^N = 1 - P$ and $N = \frac{\log(1-P)}{\log(1-(1-\varepsilon)^n)}$. n was the number of the matched point pairs that were used to estimate the matrix T. Here, $n = 3$.The situation that the three points were not in line in the three point pairs should be guaranteed, and if this situation occurred, the three point pairs must be selected again.

Calculate the transformation matrix T according to the three matched point pairs.

Calculate the Euclidean distance $d(x_{1(i)}, Hx_{2(i)})$.

Set up the distance threshold t, and take the matched points satisfying $d < t$ as interior points;

If the current number of interior point was larger than U, T and the interior point set were taken as the current best estimation and U was updated. If the number was equal to U, the one with the small number of registration was selected to be the current best estimation. At the same time, the number of remaining matching N was dynamically estimated. If the current iteration times reached N, T and the interior set were reserved and the interior was stopped.

The DLT algorithm was applied to estimate T on all the matching points in the current interior set.

60.3.4 Image Transformation

After determining the geometry transformation parameters, the registration image would be transformed into the coordinate system of the reference image. The position of the pixel in the registration image was obtained through resampling method. As the gray value was probably a floating number, the final gray value of the position in the SAR image was obtained though bilinear interpolation in this paper.

60.4 Experimental Result

The data in this paper are obtained from a certain institute. Figures 60.2 and 60.3 are the overlapping areas of the adjacent frames in SAR image sequence. Figure 60.4 shows the matched feature points and Fig. 60.5 is the final SAR image after registration. The experiment results are as follows:

Fig. 60.2 Reference image

Fig. 60.3 Registration image

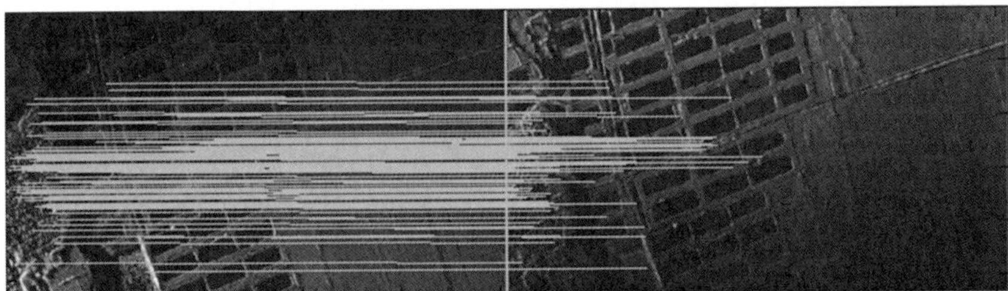

Fig. 60.4 The matched feature points

Fig. 60.5 The image after registration

As shown in Fig. 60.4, 660 feature points were detected in the reference image and 705 points were detected in the registration image, and 99 pairs of feature points were matched. The whole process consumed 0.0288462s.

Both subjectively and objectively, the results show that the proposed algorithm performs very well and satisfies the automatic registration requirements.

60.5 Conclusion

A SAR images automatic registration method based on the PCA-SIFT and Mahalanobis distance was proposed in this paper, and the steps were as follows: First, obtain feature points and primary matched points by combining the improved PCA-SIFT and Mahalanobis distance; second, remove the mismatches using the improved RANSAC algorithm, and estimate the transformation matrix; third, complete the mapping between images, and complete resampling. Finally the SAR registration is realized. Experiments demonstrate the effectiveness and robustness of the proposed algorithm.

Acknowledgments The authors wish to thank Xu Kai and a certain institute. This work was supported in part by a grant from Chongqing Education Commission, and the number is KJ080623.

References

1. Gao C (2007) Research on feature-based automated registration method for aerial image sequence stitching, master degree paper of national university of defense technology
2. Fonseca LMG, Costa MHM, Castellari SP, (2002) Automatic registration of radar image, IGRA SS
3. Zhang H (2006) Study on remote sensing image mosaics, master degree paper of Xi'an electronic science and technology university
4. Wang J, Li Q (2008) Research on SAR image registration based on homograph matrix. J CAEIT 3(6):557–660
5. Zhang J, Liu Y (2010) The medical image registration based on multi-resolution image cone and Hausdorff distance. J Chongqing Univ Technol Nat Sci Edition 24(1):59–64
6. Mlkolajczyx K, Schmid C (2005) A performance evaluation of local description. IEEE Trans Pattern Anal Mach Intell 27(10):1615–1630
7. Ye K, Sukthankar R (2004) PCA-SIFT, a more distinctive representation for local image descriptors. In: Proceedings of the conference on computer vision and pattern recognition, IEEE, Washington, vol 1. pp 511–517
8. Luo J, Gwun O (2009) A comparison of sifts Pca-Sift and SURF. Int J Image Process IJIP 3(4):143–152

Chapter 61
Improved Image Fusion Algorithm Based on Wavelet Transform

Jianxun Zhang, Wenbin Niu and Kaiwen Zhang

Abstract An improved image fusion algorithm based on wavelet transform is proposed. First, wavelet transform was applied in fusion image to obtain low frequency and high frequency. Second, low frequency and high frequency were fused respectively by different methods. For low frequency, a weighted average fusion method according to correlation coefficient was proposed. For high frequency, a fusion method based on the biggest local square difference was proposed. Finally, the performance of the image fusion was evaluated using criteria like entropy and average gradient. Simulation experiments show that this algorithm improves entropy and average gradient and enriches the detailed information of images besides keeping information about original images.

Keywords Image fusion • Wavelet transform • Correlation coefficient • Entropy • Average gradient

61.1 Introduction

Currently, with the quick development of multi-media technology and the wide application of image sensor, the scope of image application is becoming wider and wider. The traditional image processing technology cannot meet the need for increasing accuracy, while new image fusion technology is effective on improving image quality and obtains more information. The study on image fusion has been increasing since 1990s, and the scope extends to remote sensing image processing, medical image processing, robot, intelligent manufacturing, military application, etc.

J. Zhang (✉) · W. Niu · K. Zhang
School of Computer Science and Engineering, Chongqing University of Technology, Chongqing 400054, China
e-mail: zjx@cqut.edu.cn

W. Niu
e-mail: nwb522412857@163.com

Z. Zhong (ed.), *Proceedings of the International Conference on Information Engineering and Applications (IEA) 2012*, Lecture Notes in Electrical Engineering 220, DOI: 10.1007/978-1-4471-4844-9_61,

Image fusion, images or image sequences of some abstract sense obtained by two or over two sensors simultaneously (or not simultaneously) are integrated to produce a new explanation [1] about the sense, which cannot be achieved by the information from a single sensor. The aim of image fusion is to reduce uncertainty.

The methods of image fusion mainly includes three kinds [2, 3]: pixel level image fusion, feature level image fusion, and decision level image fusion, of which pixel level image fusion is used most widely. Its main methods [4] lie in: weighted average method, logic filter method, statistical optimization method, wavelet method based on wavelet transform, fusion method based on neural network, etc. Wavelet transform is one of the most widely used methods nowadays. However, traditional wavelet fusion only puts stress on high frequency fusion rules and neglects the function of low frequency fusion rules on the overall image fusion effect. In this paper, an improved image fusion algorithm based on wavelet transform, aiming at the low frequency component of image fusion, is proposed. It does not only take measures of maximum quality, minimum quality, and weighted average, but also makes threshold according to correlation coefficient and has filed fusion based on standard deviation weighted average. Aiming at high frequency component, it fuses images according to local maximum.

61.2 Image Fusion Based on Wavelet Transform

Images to be fused are disassembled by two-dimensional wavelet. Low frequency component, level high frequency component, vertical high frequency component, and diagonal high frequency component are obtained. For example, what is shown in Fig. 61.1 is the two–level disassembly diagram of two-dimensional wavelet. The low frequency component of image, LLi (i = 1,2), is the approximate image of the original image. It contains most energy information about the image, while the high frequency component, is the detailed expression of the original image, LHi, HLi, HHi (i = 1,2), including the texture and margin of the image.

Fig. 61.1 Two-level disassembly diagram of two-dimensional wavelet

LL_2	HL_2	HL_1
LH_2	HH_2	
LH_1		HH_1

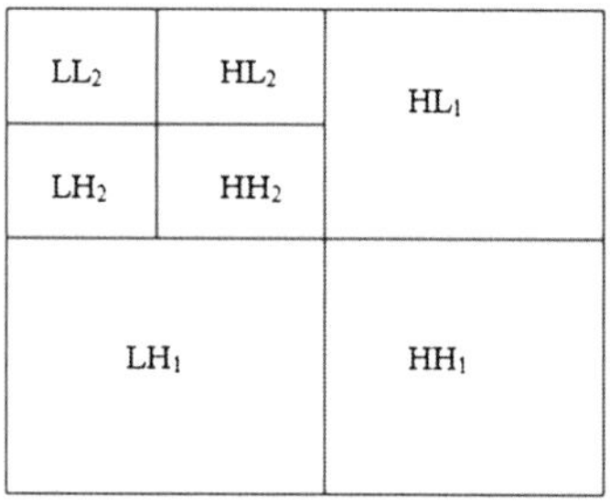

61.2.1 The Principle of Wavelet Transform Fusion

Take two images for instance, and the fusion of multi images can be concluded according to this example. Suppose A and B are two images to be fused after registration and C is the image after fusion; the concrete steps of the algorithm are as follows:

Images A and B are wavelet transformed after registration, and the low frequency component and high frequency component of A and B are separated.

Aiming at low frequency component and high frequency component of different disassembly levels, different fusion algorithms are applied.

Low frequency component and high frequency component after fusion are wavelet transformed inversely and reconstructed to form Image C after fusion.

61.2.2 The Traditional Rule of Wavelet Fusion

From the principle of wavelet transform fusion algorithm, we know the choice of fusion principle is very important to both the quality of image fusion and the algorithm. Low frequency component and high frequency component after wavelet decomposition have different meanings; therefore, fusion rules are different.

Low frequency component represents the approximation of images, and usual fusion methods for it include weighted average method and maximum quality method [5]. Of them the latter means choosing the pixel of bigger grey value from two images to be fused as the pixel of fusion image. Weighted average method means weighting and summing up corresponding pixels of two images. Suppose $A_L(x, y)$ and $B_L(x, y)$ are low frequency components after wavelet decomposition, and $F_L(x, y)$ is the image after fusion, then:

$$F_L(x, y) = W_1 \times A_L(x, y) + W_2 \times B_L(x, y) \tag{61.1}$$

In the above equation, $W^1 + W^2 = 1$, W^1 and W^2 are weighting factors. When $W^1 = 0.\ 5$, $W^2 = 0.5$, it is called average method.

61.2.3 Improved Wavelet Low Frequency Fusion Strategy

Traditional low frequency fusion of wavelet algorithm chooses easy weighted average method. Its advantage is easy for algorithm and realizing and suitable for real-time dealing. When grey differences are not big, it may be effective. However, when there is a great difference in the fusion image, the patch trace is very obvious, and the contrast of the fusion image will decrease.

In this paper, a frequency fusion method, in which correlation coefficient is regarded as threshold and weighted average processes according to standard deviation, is proposed. Correlation coefficient is the standard of variables. In the image disposing, correlation coefficient reflects the correlation of two images. When correlation coefficient approaches 1 most two images approach each other best, as follows:

$$rxy = \frac{\sum_{i=1}^{N} (X_i - \overline{X})(Yi - \overline{Y})}{\sqrt{\sum_{i=1}^{N} (X_i - \overline{X})^2 \sum_{i=1}^{N} (Y_i - \overline{Y})^2}} \tag{61.2}$$

According to mathematics theory, in common situations, when correlation coefficient is over 0.8, it is considered that two variables have strong linear relationship, thus, correlation coefficient is taken as the threshold parameter in this paper, and low frequency component is fused by the fusion method in the area of standard deviation weighted average. The detailed steps of low frequency fusion strategy are as follows:

Suppose A and B are low frequency component images of two original images after wavelet composition, LA(i,j), LB(i,j) are the pixel values of two low frequency images at the point (x,y), LC(i,j) is the pixel value of the image after fusion at the point (x,y), then:

1. Calculate the correlation coefficient values of A and B in the neighboring area of 5×5 (or 3×3, 7×7), of which the center is the point (x, y).

$$C(A,B) = \frac{\sum_{i,j} [(LA(i,j) - \overline{LA}) \times (LB(i,j) - \overline{LB})]}{\sqrt{\sum_{i,j} [(LA(i,j) - \overline{LA})^2] \sum_{i,j} [(LB(i,j) - \overline{LB})^2]}} \tag{61.3}$$

In the above equation, $\overline{LA}$ and $\overline{LB}$ are average values of the pixels in the neighboring area.

If $C(A,B) \geq \sigma$, then after fusion, the pixel value of the point (x,y) is LC(x,y) = (LA(x,y) + LB(x,y))/2.

If $C(A,B) < \sigma$,the pixel value of the point corresponding to point (x,y) is LC(x,y) = K1 × LA(x,y) + $K2$ × LB(x,y).

In the above equations, $K_1 = \frac{\sigma_{LA}}{\sigma_{LA} + \sigma_{LB}}$, $K_2 = 1 - K_1$,σ_{LA} and σLB are standard deviations of area LA and area LB respectively, σ is threshold and prescribed as 0.8, then the standard deviation can be concluded from:

$$\sigma = \sqrt{\frac{\sum_{i=1}^{n} (Xi - \overline{X})^2}{n - 1}} \tag{61.4}$$

61.2.4 The Fusion Strategy of High Frequency Components

As for high frequency components, which includes detailed information such as margins and is very active, we take the fusion measures based on regional variance. Suppose TA, TB are high frequency components obtained from two images after wavelet transform. TF being the high frequency component after fusion, the contract steps of this method are as follows:

Calculate regional variances of the neighboring area 5×5 (or 3×3, 7×7) of TA(x, y) and TB (x, y), which takes point (x, y) as the center, and the result is named var1 and var2. Calculate:

$$\text{cor var} = \frac{2 \times \text{var1} \times \text{var2}}{\text{var1}^2 + \text{var2}^2} \tag{61.5}$$

If cor var $\geq T$, usually $(0.5 < \mathrm{T} < 1)$, then:

$$\mathrm{TF}(x, y) = W_1 \times \mathrm{TA}(x, y) + W_2 \times \mathrm{TB}(x, y) \tag{61.6}$$

In the above equation: $\mathrm{W1} + \mathrm{W}_2 = 1$, $\mathrm{W}_1 > 0$, $\mathrm{W}_2 > 0$.
If corvar $< T$, then:

$$\mathrm{TF}(x, y) = \begin{cases} \mathrm{TA}(x, y), & \text{var1} \geq \text{var2} \\ \mathrm{TB}(x, y), & \text{var1} < \text{var2} \end{cases} \tag{61.7}$$

61.3 The Evaluation Standard of Image Fusion

The evaluation of fusion result is an important step of image fusion. In many applications of image fusion, the final user is the human being; therefore, visual specialty is one of the most important factors to be considered. However, during human evaluation of image fusion quality, many subjective factors will exert influence on the result [6]. Aiming at the disadvantage of subjective factors, objective evaluation standard must be introduced. In this paper, entropy and average gradient are applied to evaluate the quality of image fusion.

61.3.1 Entropy

Entropy reflects information richness degree directly. The bigger entropy is, the more information there are in images, and vice versa. The equation is as follows:

$$H = -\sum_{i=0}^{L-1} P_i \log P_i \tag{61.8}$$

In the above equation, H is the entropy of images, L is the gray series of image, and P_i is the ratio of the pixel whose gray value is I and the general pixel of image. The bigger the entropy value is, the more average information image includes.

61.3.2 Average Gradient

Average gradient is the definition of image. It reflects the image expression ability of detail contrast. The equation of average gradient is as follows:

$$Ave_V = \frac{\sum_{i=0}^{n-2}\sum_{j=0}^{m-2}|F(i+1,j) - F(i,j)|}{(n-1)\times m} \tag{61.9}$$

$$Ave_H = \frac{\sum_{i=0}^{n-2}\sum_{j=0}^{m-2}|F(i,j+1) - F(i,j)|}{n\times(m-1)} \tag{61.10}$$

$$AG = \sqrt{Ave_v{}^2 + Ave_H{}^2} \tag{61.11}$$

In the above equations, AveV, AveH are average gradient values of vertical and level directions respectively of corresponding images. AG represents the average gradient value of image. The bigger the value is, the better the definition of the image.

61.4 The Result and Analysis of Emulation Experiment

The data of emulation experiment are from multi-focus clock images, whose resolution is 256 × 256. Figure 61.2 and 61.3 are original images taking by left-focus and right-focus respectively. Images to be fused must be registered. Whether to register or not will influence fusion quality directly. The data in this paper are from registered images. Wavelet transform method, DBSS, is applied. In order to validate the effect of the algorithm in this paper, the algorithm of this paper in experiments is contrasted with the maximum pixel value, minimum pixel value, and

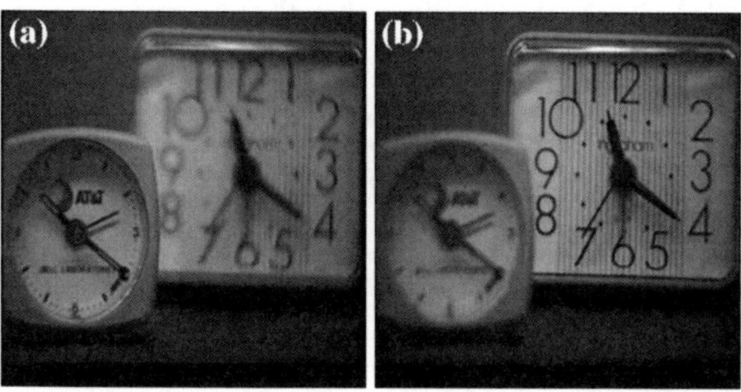

Fig. 61.2 **a** Original image of left-focus **b** Original image of right-focus

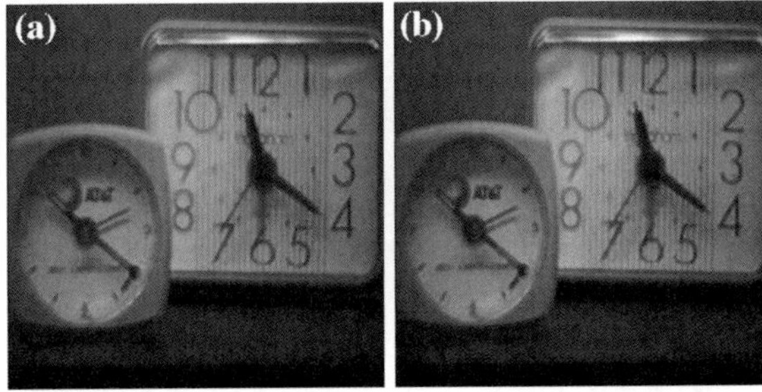

Fig. 61.3 **a** The maximum of pixel value **b** the minimum of pixel value

Fig. 61.4 **a** Traditional wavelet transforms **b** the algorithm of this paper

Table 61.1 The contrast of entropy value and average gradient value before and after image fusion

	Fig. 61.4	Fig. 61.2	Fig. 61.2	Fig. 61.3
Entropy	7.3132	7.2109	7.3433	7.4518
Average gradient	4.0746	4.9739	6.2512	6.3936

traditional wavelet transform. The platform of experiments is MATLAB 7.1, and the results are as follows:

From Table 61.1, the entropy of the image fused by the algorithm proposed in this paper is bigger than that of the maximum pixel value, the minimum pixel value, and traditional wavelet transform. It also includes more information. The average gradient is improved in comparison to the above algorithms. The algorithms proposed in this paper enhance the quality of image fusion effectively, and the efficiency is improved dramatically in the platform MATLAB.

61.5 Conclusion

The algorithm of image fusion is one of the focus of studies in recent years, and is widely applied in the military and medical areas. An improved wavelet transform fusion algorithm is proposed in this paper. For low frequency component after wavelet decomposition, weighted average fusion strategy based on correlation coefficient is applied; for high frequency component, fusion strategy based on regional variance is applied. The experiments prove that the algorithm effectively improves the quality of definition of fusion image.

References

1. Chang CI, Chen K, Wang JW (1994) A relative entropy-based approach to image thresholding. Pattern Recognit 27(9):1275–1289
2. Huang XY, Zhou LW, Gao ZY (1999) Multispectral image fusion using wavelet transform. SPIE 6:28–36
3. Gang H, Liu Zhe X, Xiaoping GR (2006) Research and recent development of Image fusion at pixel level. Pattern Recognit Lett 27(11):121–127
4. Luorc Y (2002) Multisensor fusion and intergration: approaches, applications, and future research directions. IEEE Sensor J 2(2):107–109
5. Varshney PK (1997) Multi-sensor data fusion. Electron Commun Eng J 9(12):245–253
6. Huang W, Liangjing Z (2007) Evaluation of focus measures in multi-focus image fusion. Pattern Recognit Lett 28(4):493–500

Chapter 62
Tree Modeling Based on Two Orthogonal Images

Shibo Yu, Fuyan Liu and Hao Li

Abstract This paper introduces a method that is based on two orthogonal images, which only need the front and side images of the tree to create a 3D tree model that is similar to the real model from any angle. First, we get the front and side images of the tree, then combined with user interaction to extract the front and side silhouette of the tree, and extract the main trunk. Second, we establish a visual hull according to the silhouette of tree. Third, system uses a visual hull and the built-in subbranches to generate three-dimensional branch structure of tree. Finally, we add the leaves to the branch, and then we get a complete three-dimensional tree model. In this paper, we implement a system to demonstrate out method.

Keywords Image-based modeling • Tree modeling • Visual hull

62.1 Introduction

Tree model is widespread in a variety of game scenarios, simulation systems, and film scene. Tree modeling is a complex and time-consuming work, because of its variety of species and the complex structure. Hence, establishing a realistic tree model in a fast, convenient, and real-time way is important.

Traditional modeling approach is the manual way, which requires the user to describe every details of tree. Only images needed to establish three-dimensional model, image-based modeling approach make tree modeling more automatic. Tree

S. Yu (✉) · F. Liu · H. Li
School of Computer Engineering and Science, Shanghai University,
Shanghai 200072, China
e-mail: su47yuwenshu@gmail.com

F. Liu
e-mail: lfy@shu.edu.cn

H. Li
e-mail: 67658993@163.com

Z. Zhong (ed.), *Proceedings of the International Conference on Information Engineering and Applications (IEA) 2012*, Lecture Notes in Electrical Engineering 220,
DOI: 10.1007/978-1-4471-4844-9_62,

model is not only easily found in the real world, but also easy to take a photo from the realistic model; hence, image-based modeling techniques have become a useful approach for tree modeling .

Inspired by Tan [1], we come up with a new method that applies image-based modeling to the tree modeling, which is based on two orthogonal images of the tree. This method only requires the front and side images of the tree, combined with user interaction, and then automatically generate a complete tree structure. This method is easy to use, need little user interaction, and can establish a realistic tree model.

62.2 Related Works

Tree modeling methods can be divided into rule-based modeling, image-based modeling, and other methods. In 1968, Lindenmayer [2] create character rewriting system—L system, which is useful for the simulation of cell interaction. L system use a series of character substitution rules to replace the original string and get a result string, then parse the string to get results. Prusinkiewicz [3] and Lindenmayer use L system for the construction of the tree model. As the self-similarity feature of the tree, which means the subtree structure is similar to the trunk, Oppenheimer [4] use fractal algorithms to create a tree model according to this feature. Weber and Penn [5] used geometric rules to define the tree model; the relative trunk angle and trunk length ratio and other parameters to define the trunk structure. But such a system is difficult to use, which requires a certain expertise, especially in the rules written and parameters set, it is not suitable for the amateur or new users.

In recent years, image-based tree modeling has become a research focus. Quan [6] and Tan [7] used multiple images to modeling a tree, which is a point's cloud-based method. This method uses an ordinary camera take to multiple pictures around the tree, and then use the three-dimensional reconstruction techniques to obtain the points cloud. It restores the main branches according to the points cloud, and then use the graphical interface, allowing users to edit the branches to meet the real tree. It adds leaves to the branch to complete tree. The advantage of this approach is we can get an almost real tree model that can be viewed in all directions; the disadvantage is we cannot always get enough multiangle images. Tan [8] used a single image to model a tree, ours idea come from Tan.

The sketch-based tree modeling method is proposed by Okabe [9]. It simply drew two-dimensional branches structure sketch by the user, and then quickly generates 3D geometric model. The advantage of this modeling approach is that users do not need any professional knowledge to just draw the sketch of tree branch; the disadvantage is that the result is an unreal modeling. Reeves [10] and Neubert [11] bring particle system into the tree modeling to get a real-like tree model. There are methods that also use the laser scanner to obtain tree modeling, such as XU [12].

62.3 Tree Modeling Method Based on Two Orthogonal Images

Our method is divided into the following steps, capture images of the front and side images, extract the silhouette of the tree, extract the main trunk, construct a visual hull, automatically generate the branches structure, and add leaves. The first step is to obtain the front and side of the tree. It can be captured by an ordinary digital camera. Because we use the uncalibrated camera, we should capture the image as orthogonal as possible, and camera distance from the tree as equal as possible, and the tree should be in the center of image. Compare with using one image method, using two orthogonal images could build a high authenticity visual hull. The next step is to extract the silhouette of the tree from the front and side images, which requires user interaction. This step requires the user to set the region of interest area, background area, and tree area, and then automatically extract the tree silhouette. The next step is to construct a visual hull, which we can cut branch that outside of it and make the generated tree shape similar to the real tree model. The next step is to generate the structure of the branches. We use the visual hull and built-in sub-tree structure to automatically generate a complete structure of the branches. This step is useful for new subtree to replace the old branches. The key step is the replace strategy and growth algorithms. Finally, add leaves to the branches structure, and we get a complete tree model.

62.3.1 Extract Tree Silhouette and Trunk

Our method use the "GrabCut" that proposed by Rother [13] for image segmentation. The reason why we select GrabCut is that we do not require very fine extraction of the tree silhouette, we just get physical characteristics, and GrabCut support user interaction very well. Tan [8] also obtained good result by using the Gaussian mixture model (GMM) to partition the image. GrabCut needs to set the image area, foreground, and background. GrabCut is an iterative algorithm; it can use the user interaction and the result of previous step to get a better result, as shown in Fig. 62.1.

Fig. 62.1 Silhouette and trunk extraction

Basic user interaction:

1. Set the interesting area for the tree, use green color. The area containing the tree silhouette must be included.
2. Set the background of the tree; use black color, the area outside of the tree.
3. Set the real silhouette for the tree, use red color. This part must be tree silhouette.
4. Set the trunk; use blue color, main trunk of the tree.

62.3.2 Construct Visual Hull

We need a 3D hull that makes sure branches always in it when we automatically build branches structure. This chapter uses the Visual Hull propose by Laurentini [14]. This method reconstructed 3D surface based on multiple angle image. Okabe [9] and Tan [8] used only one image; they get a two-dimensional silhouette, and then rotated it to get a three-dimensional shell. Compare with the method of Okabe and Tan's method, visual hull is more closer to the real tree model shape, whether it is positive or side. This chapter is based on two orthogonal images modeling, we do not have the conditions to get all angles image, and therefore we simplify the method to build Visual Hull. We use front and side silhouette to cut the space, then we get an enclosing visual hull. As shown below Fig. 62.2.

62.3.3 Generate Branches Structure

This chapter uses the automatic method that is proposed by Tan [8] to generate tree. This is a nonparameter tree growth system, it uses visual hull and 3D points to control the growth of the branches. The growth processes is continuing use the new subtrees instead of the original branches or add new branches to the sub-tree. This chapter uses the built-in subtrees structure to synthesize the complete structure of the branches. The subtree structure is divided into two categories: Type I, suitable for the branches of

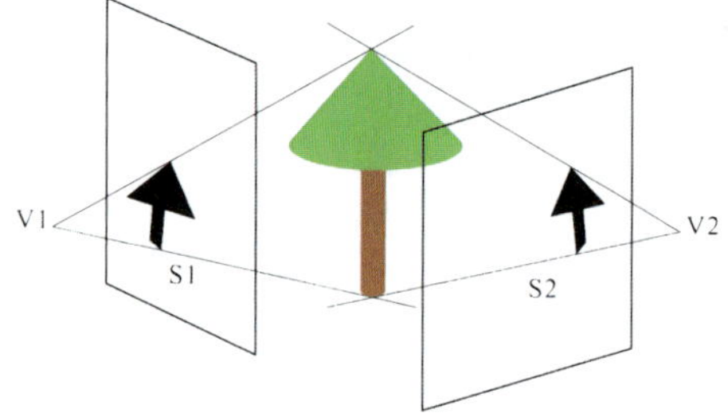

Fig. 62.2 Visual hull

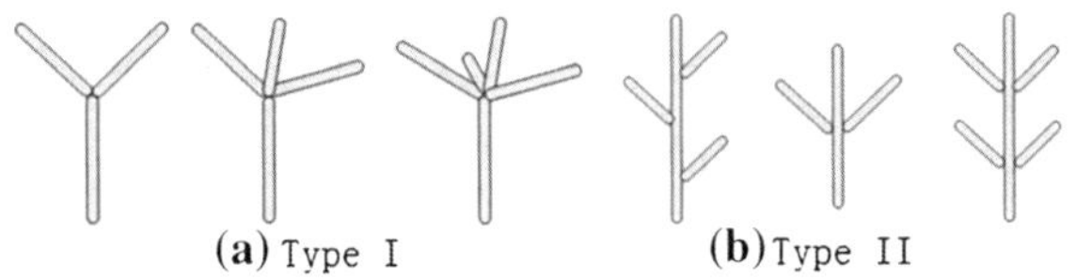

Fig. 62.3 Built-in sub-tree structures, **a** Type I, **b** Type II

the ordinary type structure; Type II, suitable for the structure of A-shaped branches, as shown in Fig. 62.3.

62.3.3.1 The Process of Branch Growth

Our growth algorithm as same as Tan, description as follows:

1. Convert previous 2D branches structure into 3D structure that use as main trunk. The approach used in the text is extracts blue marker as part of the main trunk.
2. Choose an alternative branch. The select strategy is to select the largest radius of branch from the existing branches.
3. Select a sub-tree structure to replace the current branch. The strategy is to select the system built-in sub-trees structure one by one and rotate it, and then calculate the minimum cost of sub-tree. We use this sub-tree to replace the current branch. The method of calculate cost will be described below.
4. Branch pruning. We cut the branch that outside of the visual hull.

62.3.3.2 Point-Based Branch Growth Control

We need to control the growth of branch. We must avoid branches structure concentrated or unnatural in particular side. In particular, branches structure should not only close to 2D silhouette, but also close to the 3D visual hull. Okabe [9] use the method that rotate 90 ° of the sub-tree as same as Tan [8] used. We use two control factors, control factors based on two-dimensional silhouette, control factor base on three-dimensional visual hull.

Red point P is the sampling points on tree branch. Yellow point S is the sampling points on tree silhouette or visual hull.

Two-dimensional control factor is used to control the branches structure close to the two-dimensional silhouette. We uniform sample points S from silhouette, as shown in Fig. 62.4. P is the sampling points on the three-dimensional branches, and then projected onto a 2D plane. Both S and P are two-dimensional points. We calculate the minimum distance between S and P to determine which subtree to be selected. Same as calculate minimum $E^{2D}(T)$.

$$E^{2D}(T) = \sum_i \mathrm{dist}^{2D}(s_i, T) \tag{62.1}$$

Fig. 62.4 Control points

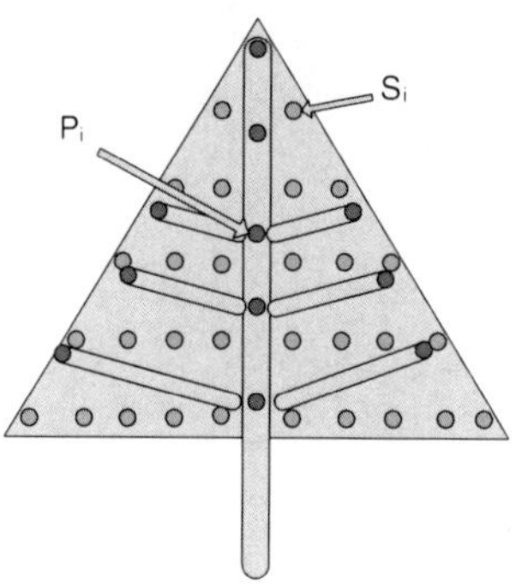

$$\mathrm{dist}^{2\mathrm{D}}(Si,T)=\min_{j}(d^{2\mathrm{D}}(Si,Pj)) \quad (62.2)$$

$$d^{2\mathrm{D}}(Si,Pj)=|\mathrm{Si}\text{ - }\mathrm{Pi}| \quad (62.3)$$

Equation (62.1) is total distance between P and T, (62.2) is minimum distance between Si and P and (62.3) is two-dimensional Euclidean distance between *Si* and *Pi*.

Three-dimensional control factor is used to control the branches structure close to the visual hull. We uniform sample points S from visual hull, as show in Fig. 62 4. P is the sampling points on the three-dimensional branches. Both S and P are three-dimensional points. We calculate the minimum distance between S and P to determine which subtree to be selected. Same as calculate minimum $E^{3D}(T)$

$$E^{3D}(T)=\sum\nolimits_{i}\mathrm{dist}^{3\mathrm{D}}(d_i,T) \quad (62.4)$$

$$\mathrm{dist}^{3\mathrm{D}}(Si,T)=\min_{j}(d^{3\mathrm{D}}(Si,Pj)) \quad (62.5)$$

$$d^{3\mathrm{D}}(Si,Pj)=|Si-Pi| \quad (62.6)$$

Equation (62.4) is total distance between P and T,(62.5) is minimum distance between Si and P and (62.6) is three-dimensional Euclidean distance between Si and Pi.

We alternante calculate two-dimensional distance and three-dimensional distance. This not only ensure the branches structure close to the two-dimensional tree silhouette, but also close to three-dimensional visual hull.

62.3.4 Add the Leaves

After generate branch structure, the next step is to add the leaves. Our method is adds leaves base on branch length and the level of branch. If a branch is short or on the last level of the tree, it will be added a large amount of leaves.

Table 62.1 Result of tree modeling

	Num of branches	Num of leaves	Time
Model (Fig. 62.5c)	1,001	20,140	8 min 35 s

Fig. 62.5 Result of tree modeling, **a** Front, **b** Side, **c** Tree model

62.4 Results

This article uses the following tree image, the result tree model look-like a real tree, as shown in Table 62.1.

As shown in Fig. 62.5a is front image of tree, b is side image of tree, and c is the result model. The result demonstrate out method is available, and our system is easy use and generate look-like tree model.

62.5 Conclusion and Further Work

Our method can easily create a tree model, that is real look-like, but it cannot do real-time modeling. The next step is to accelerate the modeling process, to reduce modeling time. In the process of building a tree model, many steps can be carried out in parallel. We can use parallel technology to accelerate the modeling process, such as CUDA, OpenCL parallel computing platform.

References

1. Ping T et al (2008) Single image tree modeling. Siggraph Asia '08 Acm Siggraph Asia 04:3231–3236
2. Lindenmayer A (1968) Mathematical models for cellular interactions in development I&II. J Theor Biol 22:280–315

3. Prusinkiewicz P, Lindenmayer A (1990) The algorithmic beauty of plants, vol 24. Springer, New York, pp 145–146
4. Oppenheimer P (1986) Real time design and animation of fractal plants and trees, vol 20. Computer Graphics (SIGGRAPH'86 Conference Proceedings) pp 55–64
5. Weber J,Penn J (1995) Creation and rendering of realistic trees. Proceedings of SIGGRAPH'95, pp 119–128
6. Quan L, Tan P, Zeng G, Yuan L, Wang J, Kang SB (2006) Image-based plant modeling. Siggraph 08:599–604
7. Tan P, Zeng G, Wang J, Kang SB, Quan L (2007) Image-based tree modeling. In Siggraph 08:87–91
8. Tan P, Fang T, Xiao J, Zhao P, Quan L (2008) Single image tree modeling. In Acm Siggraph Asia 12(02):345–348
9. Okabe M, Owada S, Igarashi T (2005) Interactive design of botanical trees using freehand sketches and example-based editing, vol 24(3). Computer Graphics Forum, Proceedings of Eurographics, pp 487–496
10. Reeves W, Blau R (1985) Approximate and probabilistic algorithms for shading and rendering structured particle systems. Comput Graph 19:313–322
11. Neubert B, Franken T, Deussen O (2007) Approximate image-based tree-modeling using particle flows. ACM Trans Graph 26:03–12
12. Xu H, Gossett N, Chen B (2005) Knowledge-based modeling of laser-scanned trees, vol 31. In Siggraph'05 Sketches, pp 47–50
13. Rother C, Kolmogorov V, Blake A (2004) Grab cut: interactive foreground extraction using iterated graph cuts. ACM Trans Graph 23:309–312
14. Laurentini A (1994) The visual hull concept for silhouette based image understanding. IEEE PaMI 16(02):150–162

Chapter 63
Remote Sensing Image Retrieval Based on Color and Texture

Qinkun Xiao, Mina Liu and Song Gao

Abstract Remote sensing image retrieval based on texture features and color features is proposed, for meeting the limitations of single features of remote sensing image retrieval and the computational cost of traditional retrieval methods. On the analysis of the existing remote sensing image retrieval, the general frame of remote sensing image retrieval which is based on the color and Gabor wavelet texture features are established. According to the optimization of filter parameters which designed a group of multi-scale and multi-directional filters, the two image feature fusions are based on texture feature and color feature. Then image retrieval prototype system is designed and implemented based on color and texture features. The obtained color and texture features are used to retrieve image database. The experimental results show that the proposed method is efficient.

Keywords Remote sensing image retrieval • Texture feature • Color feature • Gabor

63.1 Introduction

With the increasing of the remote sensing image database, more and more remote sensing images need to be retrieved quickly and efficiently. In recent years, more and more people are concerned with the content-based remote sensing image retrieval technology.

Content-based remote sensing image retrieval which is through the analysis of remote sensing image, first, extracts the visual information such as color, shape or texture features, or the combination of these features. Second, similarity calculation with the image feature vector of image library. Based on the size of similarity output, the retrieval results are obtained.

Q. Xiao · M. Liu (✉) · S. Gao
Xi'an Technological Uinversity, No. 4, JinHuaBeiLu, Xian City, Shaanxi Province, People's Republic of China
e-mail: 294041616@qq.com

Z. Zhong (ed.), *Proceedings of the International Conference on Information Engineering and Applications (IEA) 2012*, Lecture Notes in Electrical Engineering 220, DOI: 10.1007/978-1-4471-4844-9_63, © Springer-Verlag London 2013

In recent years, based on color, texture and shape, space hierarchical relationships and other single visual feature algorithm research have made some achievements. However, the algorithm is only part of the image of a single visual feature property [1]; the retrieval is poor in universality, so the retrieval efficiency is not very satisfactory.

63.2 Texture Feature Extraction

The texture which has nothing to do with the color and brightness of the image reflected in the visual features of homogeneous phenomenon reflects the intrinsic properties of the total surface. Because of containing a large number of the texture information in remote sensing image, the texture feature extraction is already becoming an important area of research, which is based on the content of remote sensing image retrieval.

There are four texture feature extraction methods: statistical analysis [2], structural analysis [3], model analysis [4], and spectrum analysis [5]. Gabor wavelet used in this issue is based on transform analysis of a texture feature extraction method. This method is in a wavelet transform and uses Gabor basis function to achieve same scale and direction feature extraction.

63.2.1 Texture Feature Extraction

Gabor wavelet transform can be identified as a mathematical tool for extracting information on the multi-directional and multi-scale. Gabor filter is used by Gabor function as the unit impulse response function of band-pass filter with good filtering performance. The output can be seen as the Gabor wavelet transform input signal [6]. The two-dimensional Gabor function formula is:

$$g(x, y) = \frac{1}{2\pi\sigma_x\sigma_y} exp\left[-\frac{1}{2}\left(\frac{x^2}{\sigma_x^2} + \frac{y^2}{\sigma_y^2}\right) + 2\pi jWx\right] \tag{63.1}$$

where W is the complex Gaussian function modulation frequency. In Fig. 63.1, (a) shows the two-dimensional Gabor function of the real part of the surface distribution, (b) shows two-dimensional Gabor function of the imaginary part of the surface distribution, (c) shows the two-dimensional Gabor function of frequency distribution.

$$G(u, v) = \exp\left\{-\frac{1}{2}\left[\frac{(u - W)^2}{\sigma_u^2} + \frac{v^2}{\sigma_v^2}\right]\right\} \sigma_u = \frac{1}{2\pi\sigma_x} \sigma_v = \frac{1}{2\pi\sigma_y} \tag{63.2}$$

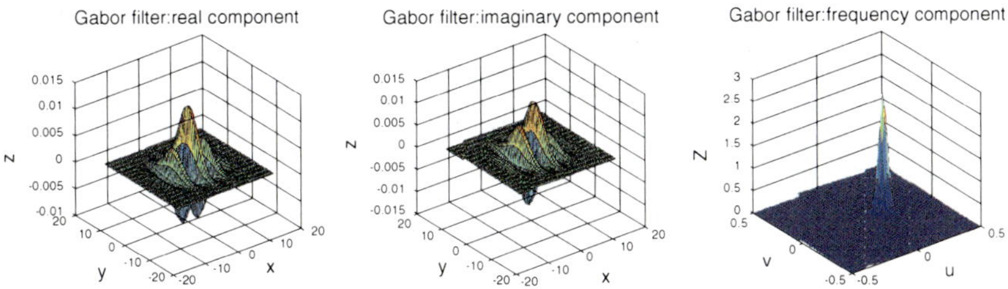

Fig. 63.1 The surface distribution of the two-dimensional Gabor. **a** Real part of figure. **b** Imaginary part of figure. **c** Frequency part of figure

Using different scales of filter can detect the local features of images in different scales; during the texture feature is extracted by basis function of Gabor. $g(x, y)$ as the mother wave function which with moderate expansion and rotation changes, get a group of self-similar filters, called Gabor wavelet [7].

$$\begin{aligned} g_{mn}(x, y) &= a^{-m} g\left(x', y'\right), a > 1, m, n \in Z \\ x' &= a^{-m}(x \cos\theta + y \sin\theta) \\ y' &= a^{-m}\left(-x \sin\theta + y \cos\theta\right); \theta = \frac{n\pi}{K} \end{aligned} \tag{63.3}$$

The entire space is divided into 12 × 5 bins. So the feature information of the pixel pi is a log-polar histogram of the coordinates of the rest points, setting measured by the reference point as the origin. The outline of any object can be indicated by a matrix size of $n \times 60$. a^{-m} is the factor of scale; K is the number of direction, get a set of direction and scale are different filter groups through the change of m and n value. Assume that S stands for scale, U_k and U_L represent the high and low center frequency; this will get m, n of the filter parameters σ_u, σ_v

$$a = \left(\frac{U_h}{U_l}\right)^{S-1}, W = U_h, \quad \sigma_u = \frac{(a-1)\,U_h}{(a+1)\sqrt{2\ln 2}} \tag{63.4}$$

$$\sigma_v = \tan\left(\frac{\pi}{2k}\right)\left[U_h - 2\ln\left(\frac{\sigma_u^2}{U_h}\right)\right]\left[2\ln 2 - \frac{(2\ln 2)^2\sigma_u^2}{U_h^2}\right]^{-\frac{1}{2}} \tag{63.5}$$

63.2.2 Gabor Texture Feature Extraction

This part based on Shape Context feature information assumes an image as $I(x, y)$ whose size scale of $w \times h$, with Gabor filter can be expressed as:

$$W_{mn}(x, y) = \sum_{x_0}\sum_{y_0} I(x - x_0, y - y_0) g^*_{mn}(x_0, y_0) \tag{63.6}$$

In this formula, $g^*_{mn}(x_0, y_0)$ is $g(x_0, y_0)$ complex conjugate function: x_0, y_0 variables which is the size of template filter; μ_{mn}, σ_{mn}. Then we can find the center and boundary information of the tracked target and determine the target. Represent an image texture features extracted respectively, by the formula

$$\mu_{mn} = \frac{1}{wh} \sum \sum |W_{mn}(x, y)| \tag{63.7}$$

$$\sigma_{mn} = \frac{1}{wh} \sqrt{\sum \sum [|W_{mn}(x, y)| - \mu_{mn}]^2} \tag{63.8}$$

The features of remote sensing image are accurately expressed by Gabor texture features. Set $K = 6$, $S = 4$ is the filter's scale and direction parameter, the highest frequency and lowest frequencies were set to the highest frequency and lowest frequencies $U_k = 0.4$, $U_l = 0.03$. Set the parameters shown in Fig. 63.2 (the first deputy of the original picture). Assuming that the image *I*, *K*, calculate the similarity distance:

$$d_{mn}(I, K) = \sum_m \sum_n \left(\left| \frac{\mu^I_{mn} - \mu^K_{mn}}{\sigma(\mu_{mn})} \right| \right) + \left(\left| \frac{\sigma^I_{mn} - \sigma^K_{mn}}{\sigma(\sigma_{mn})} \right| \right) \tag{63.9}$$

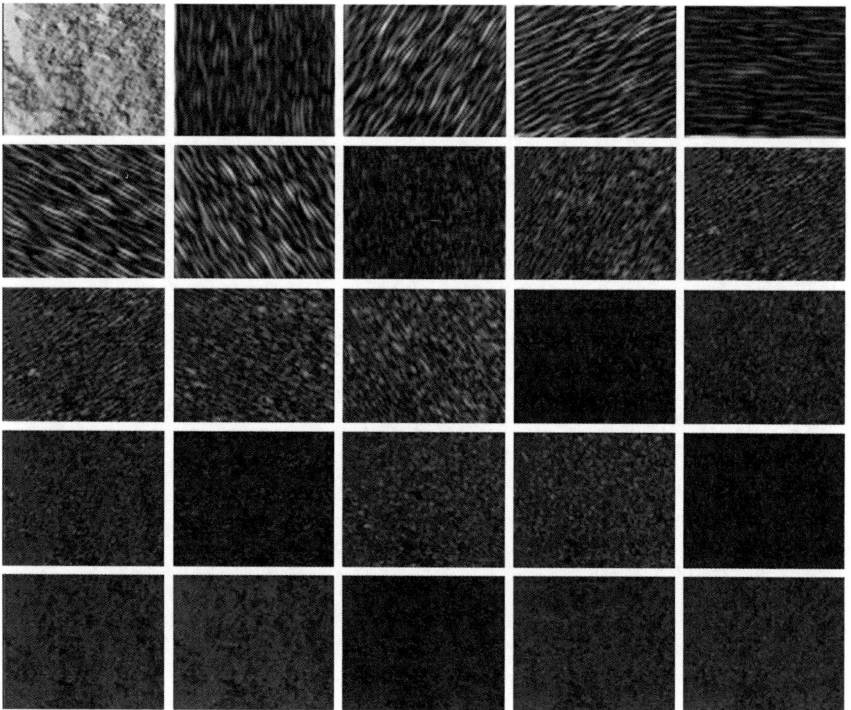

Fig. 63.2 $K = 6$, $S = 4$ filtered images

In this formula $\sigma\ (\mu_{mn})$ and $\sigma\ (\sigma_{mn})$ represent standard deviation of the mean features and standard deviation of the features of the library.

63.3 Color Feature Extraction

Color features in image retrieval are the most significant features. The traditional RGB color model can distinguish the color difference to be nonlinear and there is no direct sense, so it is not a good color description system. In the color representation methods, we choose to comply with the visual sense of the HSV model. HSV color model is closer to human visual characteristics. It is composed of three components, hue H, saturation S, and brightness V, which are based on a perception of the color model. In order to reduce the computational cost, decrease in the image quality case does not obviously result in some representative said color image be extracted, so as to reduce the storage space and improve the purpose of processing speed. HSV color space is properly quantified and then the histogram is calculated, which is calculated to be much less. According to experience data, can H, S, V components be quantized interval by the color of the human perception, etc., the type (1.10) to HSV space quantification.

$$H = \begin{cases} 0 & H \in [0,35] \\ 1 & H \in [36,71] \\ 2 & H \in [72,107] \\ 3 & H \in [108,143] \\ 4 & H \in [144,179] \\ 5 & H \in [180,215] \\ 6 & H \in [216,251] \\ 7 & H \in [252,287] \\ 8 & H \in [288,323] \\ 9 & H \in [324,359] \end{cases} \quad S = \begin{cases} 0 & S \in [0,0.1] \\ 1 & S \in (0.1,0.2] \\ 2 & S \in (0.2,0.3] \\ 3 & S \in (0.3,0.4] \\ 4 & S \in (0.4,0.5] \\ 5 & S \in (0.5,0.6] \\ 6 & S \in (0.6,0.7] \\ 7 & S \in (0.7,0.8] \\ 8 & S \in (0.8,0.9] \\ 9 & S \in (0.9,1.0] \end{cases} \quad V = \begin{cases} 0 & V \in [0,0.1] \\ 1 & V \in (0.1,0.2] \\ 2 & V \in (0.2,0.3] \\ 3 & V \in (0.3,0.4] \\ 4 & V \in (0.4,0.5] \\ 5 & V \in (0.5,0.6] \\ 6 & V \in (0.6,0.7] \\ 7 & V \in (0.7,0.8] \\ 8 & V \in (0.8,0.9] \\ 9 & V \in (0.9,1.0] \end{cases} \tag{63.10}$$

The HSV color space is divided into $L_H \times L_V \times L_S$ zone after quantitative picture, L, m, n representative of the three HSV quantitative series respectively $L_H = 10\,L_V = 10\,L_S = 10$ which is color space divided into 1,000 similar color intervals, H, S, V components are combined for a one-dimensional feature vector $L = L_S \times L_V \times H \times S + V$, according to the value of the above three components can be $L = 10H + 10V + 10S$. In this way, the quantized image statistics are one-dimensional histograms of the 1,000 handle. The color histogram is an important method to extract color features, which has inherent rotational invariance, scale invariance, and translation invariance. Statistic histogram of image is a one-dimensional discrete function.

$$h_k = \frac{n_k}{n}, K = 0, 1, \cdots, L - 1 \tag{63.11}$$

K represents the features of the color; L is the number of features, in this paper assume 1,000 to get the color histogram $H = h_1, h_2, \cdots, h_{1,000}$ of the image P, which has been the color feature vector.

63.4 Experiment

With matlab programming, a GUI interface is built and implemented based on color and texture features for image retrieval prototype system. The core module of this system consists of three parts:

Color histogram was used to extract the feature of image library, to retrieve the picture of the feature extraction.

Texture feature is extracted by Gabor wavelet, followed by the image library to retrieve the pictures of texture feature extraction.

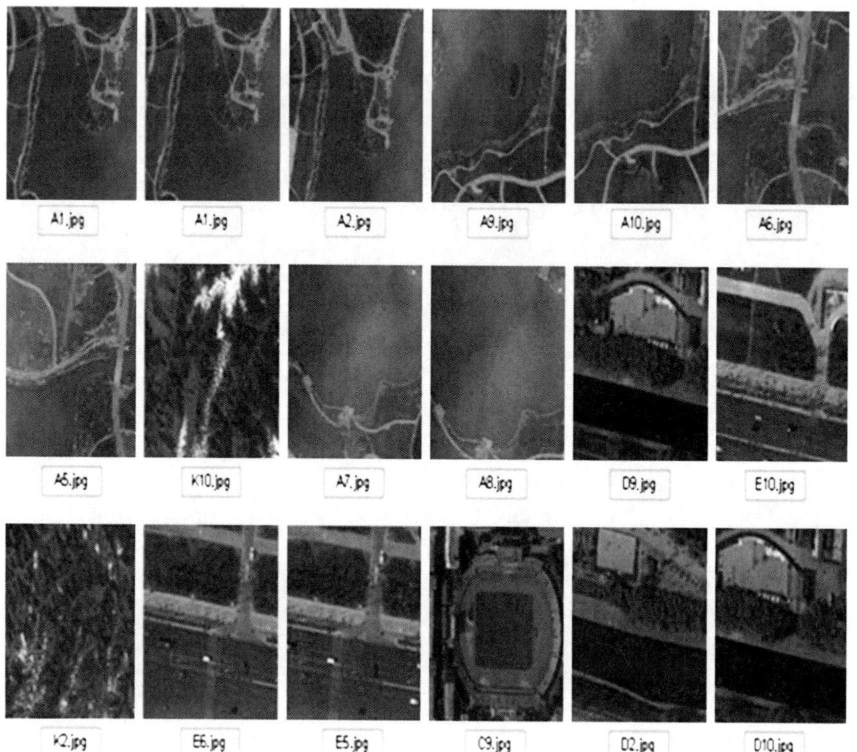

Fig. 63.3 *A*1 search results

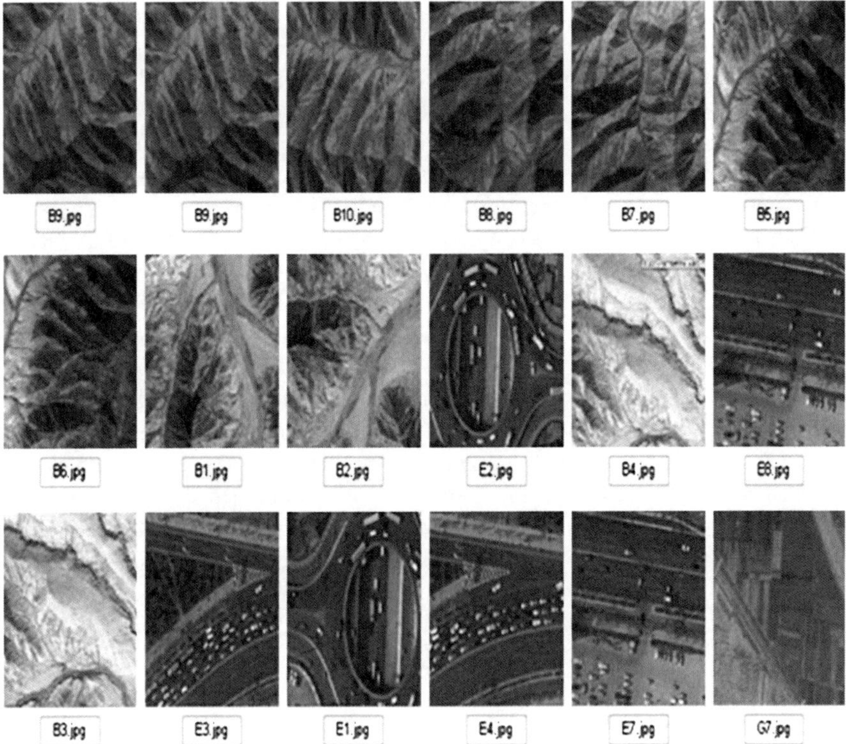

Fig. 63.4 *B*9 search results

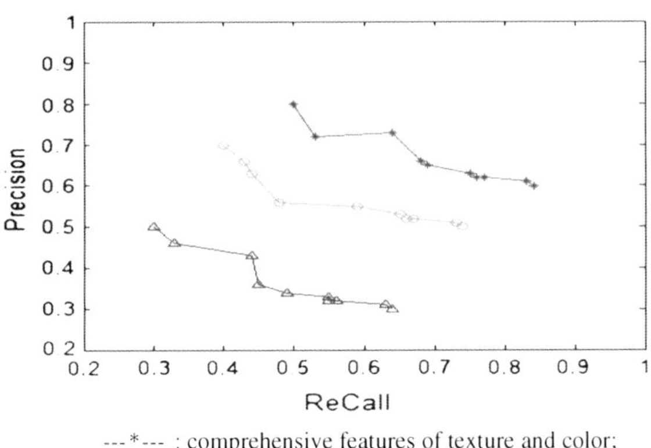

Fig. 63.5 Recall-precision curve, *astic* comprehensive features of texture and color; *circle* color feature; *triangle* texture feature

To retrieve the picture features and picture library all the features with the Euclidean distance and the similarity calculation.

Downloading remote sensing images in this experiment to establish a 10 set of libraries, each group of the library consists of 10 similar pictures. The following picture shows the GUI interface with matlab programming to build an image retrieval system. Figures 63.3 and 63.4 is $A1$, $B9$ search results: the first diagram for the input image, after 17 pictures for the similar image retrieval.

From Figs. 63.3 and 63.4 it can be clearly seen that color and texture features are used to express the content of image information.

Figure 63.5 depicts a single feature of recall precision curves and a single feature after the comprehensive recall precision curve, as shown below:

This experiment may prove that the results retrieved by the comprehensive utilization of color and texture feature is more in line with human visual requirements than the single color or texture feature both in its color and distribution. The retrieval effect accuracy improved clearly in mountains and architecture of the complex texture image. The average precision of retrieval algorithm reached more than 70 %, compared with the single texture algorithm, which was approximately by 20–30 %.

63.5 Conclusion

It can be inferred from the experiment that the image retrieval methods based on color and texture features for performance are relatively good; especially compared with applications in color and texture features of image retrieval, the merit is obvious. In experiment or practical application, it is essential to select the appropriate scale and direction, according to the need to limit the number of sequencing planning results in the output, for very accurate search results. In addition, it is needed to limit its output results sort planning number to get very correct retrieval results.

References

1. Ye Y (2006) Based on the information fusion of image retrieval research. Huaqiao Univ 2:32–64
2. Tumara H, Mori S, Yama W (1987) Texture features corresponding to visual perception. IEEE Trans Syst, Man Cyber 8(6):460–473
3. Young DC, Sang YS, Ck N (2003) Image retrieval using BDIP and BVLC moments. IEEE Trans Circ Syst Video Technol 13(9):951–957
4. Khot A, Hernandez OJ (2003) Color image retrieval using multispectral random field texture model and color content features. Pattern Recognit 36(8):1679–1694
5. Manjunath BS, Ma WY (1996) Texture features for brow sing and retrieval of image data. IEEE Trans Pattern Anal Machine Intell 18(8):837–842
6. Liang Z, Yang J (2008) Gabor wavelets transform based texture image retrieval research. Chin Sci Pap Online 35:42–45
7. Mingzhong Z (2011) Multi-scale gabor wavelet transforms in image retrieval applications. Electron Sci Technol 24(8):121–125

Chapter 64
Synchronous Monitoring Method of Ballistocardiogram and Electrocardiograph Signals

Dan Yang, Linlin Ye, Xu Wang and Ning Ye

Abstract Heart working processing involve electrophysiological and pump function. Traditional cardiac function can often be assessed by detecting ECG signals, which only reflect the electrical activity of the heart. In this paper, a measurement device is designed to monitor the heart activity state comprehensively. The proposed device monitors two kinds of physiological signals: ECG and BCG. Ballistocardiogram (BCG) signal reflect cardiac systolic and diastolic activity of heart pumping process. According to its measurement principle and signal properties, the detecting device is designed and realized. For BCG signal, sensor circuit is by resistance strain weighing sensors fitting in home-use electronic weight scale and ECG signal is recorded through the electrodes posting on the body surface. Signal processing circuit with high gain and low noise is composed of former amplifier with feedback circuit, filter circuit, and main amplifier. Real-time display module displays the signal through MATLAB GUI. The results show that the device is suitable for measuring BCG and ECG signals. In practical testing, the proposed device properly monitors cardiac activity, so the design is reasonable.

Keywords Electrophysiological and pump function • ECG • BCG • MATLAB GUI

64.1 Introduction

Traditionally, Electrocardiogram (ECG) is often used to provide useful cardiology information about the heart function state, which are often measured with electrodes placed on the skin of patients. It causes restraint and produces some

D. Yang (✉) · L. Ye · X. Wang · N. Ye
School of Information Science and Engineering, Northeastern University, Shenyang 110819, China
e-mail: yangdan@ise.neu.edu.cn

Z. Zhong (ed.), *Proceedings of the International Conference on Information Engineering and Applications (IEA) 2012*, Lecture Notes in Electrical Engineering 220,
DOI: 10.1007/978-1-4471-4844-9_64,

psychological burdens to the human body. Recently, Ballistocardiogram (BCG) used to represent the weak body vibration caused by the heart pumping out blood was discovered [1]. It has been known since the late eighteenth century but was seldom used in medical practice because of technological limitations [2]. BCG signals comprise components attributable to cardiac activity, respiration, and body movement, which can be measured without any electrodes attached to the subjects [3]. Therefore, it is a potential application for monitoring heart condition at home.

Many studies have been done on the development of BCG applications for physiological statue monitoring. BCG was early proposed by Starr and Wood in 1961 [4]. Because of the limited condition, it was substituted by ECG at that time. In 1991, Ben H. Jansen et al. designed the electrostatic charge sensitive mattress to monitor BCG signal [5]. In 1996, Yu Meng-sun et al. designed the micro movement sensitive mattress to record BCG complex. But it takes so much space that the sensor's size must match the bed. The sitting posture takes a long time in a person's one-day activity. Because of these reasons, Sakari Junnila and Alireza Akhbardeh et al. designed the BCG chair based on EMFi sensor, in 2006 [6]. Although these researchers did a lot of work on the BCG acquisition system, there are still many important parameters for design needing to be improved [7].

This paper presents a synchronous monitoring system of ECG and BCG signals, which is composed of detecting acquisition module and signal processing module. This makes it possible to monitor the heart status more comprehensively and evaluate the condition of the patient's heart even at home. The device can help doctors more accurately understand cardiac function, predict and diagnose heart diseases, evaluate cycles and the performance of a heart drug, monitor the quality of sleep, etc.

64.2 Synchronous Monitoring System Design

Figure 64.1 shows the system architecture of the proposed system. The BCG signal measuring module senses the BCG signals via the four pressure sensors composed of full-bridge circuit, and the vibration is transformed into electrical signal by the pressure sensors. The ECG signal measuring module senses the ECG signals via the electrodes placed on the skin of people. Then these two modules respectively perform corresponding signals pre-processing to amplify the signals. Later, it filters out noises. To transfer BCG signals (or ECG signals) to the MCU board, the analogy BCG signals (or ECG signals) are first converted into digital signals with analogy-to-digital converter. Then the MCU board can process the BCG signals (or ECG signals) and transmit these data to PC.

64.2.1 System Hardware Design

Figure 64.2 shows a circuit schematic representation of the diverse stage followed in order to obtain the BCG signal. BCG signals from the four pressure sensors

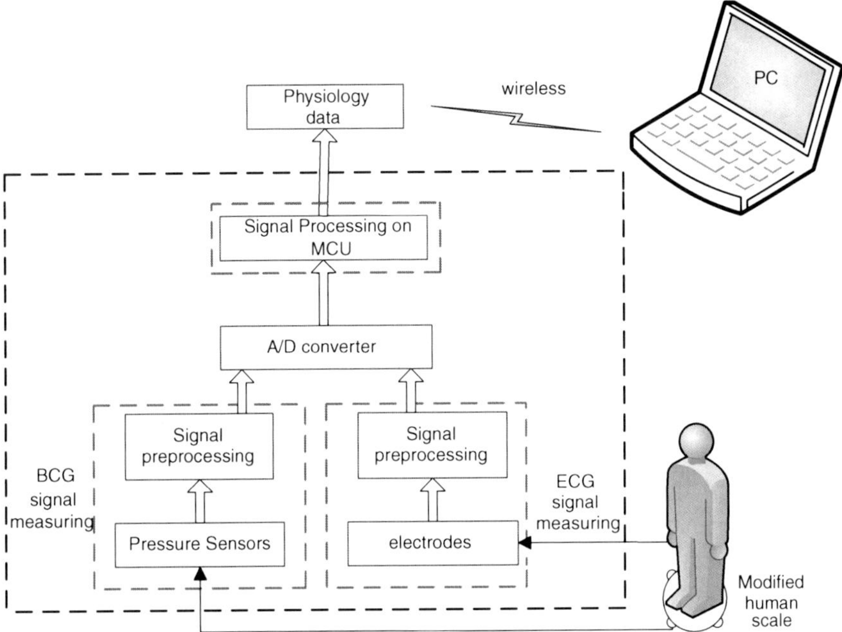

Fig. 64.1 System architecture of the proposed system

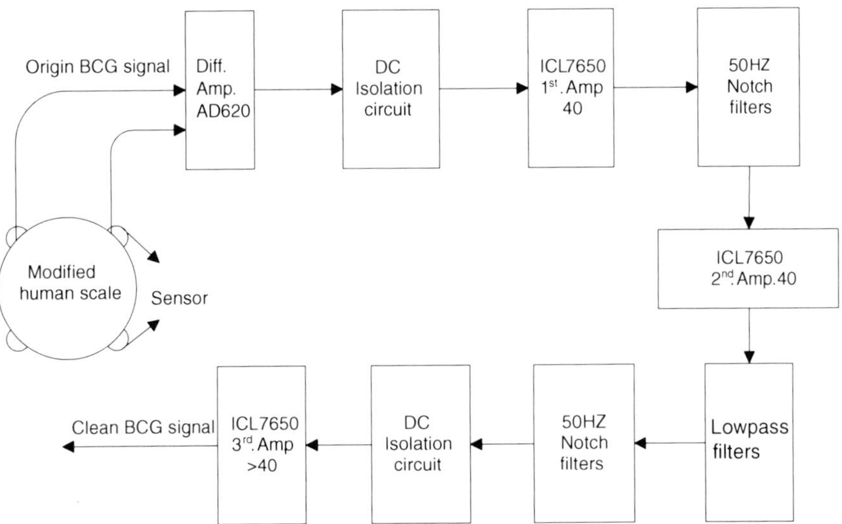

Fig. 64.2 Schematic of the proposed BCG detecting module

composed of full-bridge circuit first pass through a differential amplifier with gain of 10, avoiding the noises overriding the BCG signals, implemented by an instrument amplifier (AD620). For BCG signals peak-to-peak typically are 1 uv, an amplification of 106 is necessary to collect useful physiological parameters. This is performed by three grade amplifiers (ICL7650). Two of them are used to amplify the signal with a gain of 40, the third is used to adjust amplification by the practical needs. ECG signals are 1 mV peak-to-peak typically, so they do not need as many amplifier series as BCG signals. They just need a front differential amplifier with a gain of 10 and an inverting amplifier with a gain of 40. For physiological signals, filter circuits are designed necessarily. A fifth order Chebyshev low pass active filter is used to restrict the BCG signals in a band of 0.1–20 Hz after the first stage of amplification, the pass-band cut-off frequency, stop-band cut-off frequency and stop-band attenuation are 20, 50 Hz, 45 dB respectively. The amplifier UAF42 is selected for the lower pass active filter. For avoiding 50 Hz components of BCG signals, 50 Hz notch filter is used twice. Finally, the DC isolation circuit is applied in the design to improve the capacity of DC-resistance. The design of the circuits in the ECG signal detecting module is the same as the BCG signal, so we do not repeat them here.

64.2.2 System Software Design

System software design consists of ED (End Device) and AP (Access Device). ED is a terminal equipment node composed of wireless transceiver chip CC2500 and main control chip MSP430 that complete signals collection and distribution. AP is also composed of wireless transceiver chip CC2500 and main control chip MSP430 that are used for receiving data and communicating with PC.

The principal idea of the software design is: first, AP data center equipment completes the initialization of BSP and MSP430 to realize communication with the PC, while waiting for the ED terminal equipment to join the node. When the pre-processed ECG signals or BCG signals are sent to ED, ED will transmit these signals to A/D for converting and then send them to AP through CC2500 wireless. After these data are received by AP, they are sent to the PC via the serial port and stored in the buffer of MATLAB. MATLAB processes these data and displays them through MATLAB GUI. Finally, it achieves an expected effect of real-time synchronous monitoring ECG and BCG signals.

64.3 Experiments and Analysis

In the experiment, we recorded the BCG signals of the person standing on the body scale and the ECG signals via the electrodes placed on the skin of the person. First, we detect ECG and BCG signals through an oscilloscope. Figure 64.3 shows the waveform of synchronous ECG and SCG signals.

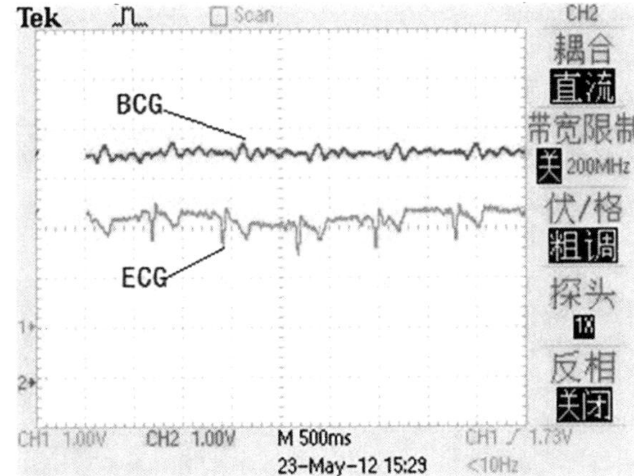

Fig. 64.3 The waveform of synchronous ECG and SCG signals

Where CH1 is BCG signal waveform and CH2 is ECG signal waveform. From the waveform chart, we note that BCG and ECG signals are synchronous, and BCG signal lags behind the ECG signal. For BCG signal, its waveform is clear, *W* wave group is obvious; the value of *J* wave can be clearly identified. For ECG signal, its QRS wave, *P* wave and *T* wave are obvious. The collected BCG signal waveform and ECG signal waveform are all cyclical.

During the system modules testing, first we run the MATLAB and open the designed the MATLAB GUI program. Then we set the communication parameters: port number: com3, baud rate: 9,600, transferring bit number: 8, stop bit number: 1. Finally we click the mouse and open the serial port. The program running results is shown as Fig. 64.4.

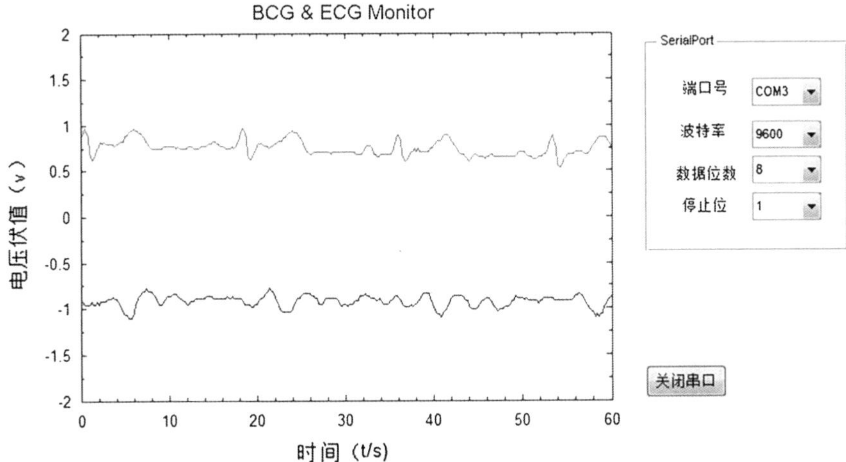

Fig. 64.4 The waveform of synchronous ECG and SCG signals in MATLAB GUI

64.4 Conclusion

Generation principles and synchronous acquisition method of ECG and BCG were described in this paper. From the analysis and results above, we can obtain that BCG and ECG signals are consistent at a time. BCG signals have a certain corresponding relation with ECG signal in each cycle, and BCG signal lags behind ECG signal. The synchronous monitoring of BCG and ECG signals could detect heart activities' status comprehensively and accurately. Currently, the device of synchronous monitor BCG signal and ECG signal does not exist. Thus, the design research in this paper is of worth. The results show that the proposed method would be helpful in developing a tool for monitoring people's health status. In the future, we will make more efforts to develop wireless embedded equipment. This research laid the foundation for developing physiological parameters monitoring embedded systems of the future.

Acknowledgments This work is supported by the Fundamental Research Funds for the central universities (N100304008, N110316001, and N110304002).

References

1. Pollock P (1986) Ballistocardiography: a clinical review. In: MAJ, Ottawa, vol 7(6). Canada, pp 778–783
2. Akhbardeh A, Kaminska B, Tavakolian K (2007) A modified blind segmentation method for ballistocardiogram cycle extraction. In: Proceeding of the 29th IEEE EMBS annual international conference, vol 8(6). Lyon, pp 896–1899
3. Pinheiro EC, Postolache OA, Girao PS (2010) Automatic wavelet detrending benefits to the analysis of cardiac signals acquired in a moving wheelchair. In: Proceeding of the 32nd IEEE EMBS annual international conference, vol 8(2). Buenos Aires, pp 602–605
4. Mack DC, Patrie JT, Suratt PM et al (2009) Development and preliminary validation of heart rate and breathing rate detection using a passive, ballistocardiography-based sleep monitoring system. IEEE Trans Inf Technol Biomed 13(1):111–120
5. Postolache OA, Silva Girao PMB, Mendes J et al (2010) Physiological parameters measurement ased on wheelchair embedded sensors and advanced signal processing. IEEE Trans Instrum Meas 5(9):2564–2574
6. Sakari J, Alireza A, Laurentiu C, et al (2006) A wireless ballistocardiographic chair. In: Proceedings of the 28th IEEE EMBS annual international conference, vol 9(1). USA, pp 5932–5936
7. Bruser C, Stadlthanner K, Brauers A, et al (2010) Applying machine learning to detect individual heart beats in ballistocardiograms In: Proceeding of the 32nd IEEE EMBS annual international conference, vol 9(7). Buenos Aires, pp 1926–1929

Chapter 65
Color Image Segmentation Algorithm of Rapid Level Sets Based on HSV Color Space

Yue Zhao, Xin Xu, Chao Chen and Dan Yang

Abstract In order to solve the problem on how to segment the color image, we put forward the color image segmentation of rapid level sets based on HSV color space. First, we adopt the method of rapid level sets based on CV mode to improve the segmentation speed. Next, the segmentation of color image can be made with the introduction of the concept of color difference under HSV color mode. In this case, the segmentation can be made exactly even the gray color is very close since the color information is fully employed. The experiment result shows this method mentioned here is more effective in segmentation than the traditional segmentation method of level sets.

Keywords Level set • C–V model • Image segmentation • Color difference • HSV color space

65.1 Introduction

Image segmentation as the basic link of the image processing has been a hot and difficult problems in image engineering. For decades, researchers continue to explore new image segmentation method, so that solving the problem is more close to practical application. Level set model was first proposed by Osher, and Sethian for tracing the interfaces among different phases of fluid flows [1]. Level set method as a new image processing method based on partial differential equations, with free topological transformation as well as easy integration a variety of features. Or more of the advantages of the method have recently been the concern of many scholars in the field of image segmentation. However, because of it built

Y. Zhao (✉) · X. Xu · C. Chen · D. Yang
Teaching and Research Institute of College Computer, Bohai University, Jinzhou 121013, China
e-mail: zy_ky7777@126.com

Z. Zhong (ed.), *Proceedings of the International Conference on Information Engineering and Applications (IEA) 2012*, Lecture Notes in Electrical Engineering 220, DOI: 10.1007/978-1-4471-4844-9_65, © Springer-Verlag London 2013

on the image gray level information on the basis of the color image segmentation is less effective, and split the slower, to the practical application difficult.

Chan and Vase [2] combined with level set ideas and Mumford-Shah model, the CV level set model. CV level set model to extract the object boundary does not depend on the gradient of the image, and it is applied to the gradient meaningless or fuzzy edge image. Ref. [3] proposed a fast level set algorithm, which greatly improves the segmentation speed, but the algorithm provides only two fast segmentation model, based on threshold and scope of these two models based on the gradient of the fast model. Ref. [4] based on the color gradient of the level set method, which uses the HSV color model, combined with the use of the Bayesian classification model (use the Bayesian classification model), and integration of the color of statistical characteristics of the region to determine the contour of the target area boundary, but the split speed. The color level set model in [5] gives a new convex nature of the variation formulation and its rapid calculation method, the new model can truly reflect the characteristics of the color image boundary detection function duality principle, but the color the use of information is not sufficient.

How fast segmentation of color images, we propose a fast level set for color image segmentation algorithm based on HSV color space. First, in order to improve the segmentation speed, we use fast level set method to evolve. Second, for color images, we introduce the concept of color difference in the HSV color model space to segment image.

65.2 Fast Level Set Segmentation Method

65.2.1 Level Set Method

The level set method is a method to solve the active contour model, initially applied to the fluid mechanics field, was widely used in various curve evolution of scientific research and engineering. Compared with the parametric active contour model, the level set segmentation model free topological transformation, solving without parameters, etc.

Level set method, the plane closed curve implicit expression for the surface of the three-dimensional continuous function $\varphi(x, y)$, one has the same function value with the value curve $\varphi(x, y)$, called the level set function, embedded closed curve as zero level set $\{\varphi = 0\}$.

The initial level set function $\phi\,(x, y, 0)$ usually in accordance with the distance function is defined as:

$$\varphi(x, y, 0) = \begin{cases} d(x, y, C), if\ x \text{ is inside } C \\ 0, f\ x \text{ is on } C \\ -d(x, y, C), if\ x \text{ is outside } C \end{cases} \tag{65.1}$$

where $d\ (x,\ y,\ C)$ is on behalf of the distance which from the point $(x,\ y)$ to the curve C.

65.2.2 C–V Model

Mumford-shah model is an ideal image segmentation model, but in the specific solving more difficult. Chan and Vase proposed the C–V model is the level set based thinking and the Mumford-shah model. It does not use gradient information, but to minimize the energy functional square evolution curve. On Mumford-shah model Another difference is that the CV level set model in the energy function, an increase of the area (inside (C)) and the Mumford-shah model of the original length of the Length (C) promote the evolution curve reaches the object boundary.

Join the area and length, the CV method, the energy functional is expressed as:

$$\begin{aligned} \varepsilon\left(u_1, u_2, C\right) &= \lambda_1 \int_{\text{inside}(C)} \left|I(x, y) - C_1\right|^2 \mathrm{d}x\mathrm{d}y + \\ &\lambda_2 \int_{\text{outside}(C)} \left|I(x, y) - C_2\right|^2 \mathrm{d}x\mathrm{d}y + v \cdot \text{Length}(C) + \mu \cdot \text{Area}(C) \end{aligned} \tag{65.2}$$

where, C is the boundary of the region Ω1 and Ω2. C_1, C_2 is a constant value, respectively, the average gray scale of the internal region of the curve C and the outer region. Contour C in the actual boundary, that is, $\mathrm{C} = \mathrm{C}_0$ (C_0 is the boundary of the segmented regions), the energy functional $\varepsilon(\varphi)$ to obtain the minimum.

Joining the Heaviside function, it is the following:

$$\begin{aligned} \varepsilon(\varphi) =& \lambda_1 \int_\Omega \left|I - C_1\right|^2 H(\varphi)\mathrm{d}x\mathrm{d}y + \lambda_2 \int_\Omega \left|I - C_2\right|^2 (1 - H(\varphi))\mathrm{d}x\mathrm{d}y \\ &+ \mu \int_\Omega \delta(\varphi) \left|\nabla\varphi\right| \mathrm{d}x\mathrm{d}y + v \int_\Omega H(\varphi)\mathrm{d}x\mathrm{d}y \end{aligned} \tag{65.3}$$

$$\Omega_1 = \{x, y : \varphi(x, y) <; 0\}, \text{meanvalue}(\Omega_1) = C_1 \tag{65.4}$$

$$\Omega_2 = \{x, y : \varphi(x, y) > 0\}, \text{meanvalue}(\Omega_2) = C_2 \tag{65.5}$$

$$\Omega = \Omega_1 + \Omega_2 + C \tag{65.6}$$

Using the variation method and gradient descent flow Eq. (65.3) to minimize operator can obtain the following form:

$$\frac{\partial\varphi}{\partial t} = \delta(\varphi)\left[\mu\nabla\cdot\left[\frac{\nabla\varphi}{|\nabla\varphi|}\right] - \lambda_1(I - C_1)^2 + \lambda_2(I - C_2)^2 - v\right] \tag{65.7}$$

65.2.3 Improved C–V Model

Shi et al. [3], a fast real-time level set segmentation method, greatly improved the segmentation speed. But its speed is based on threshold and gradient, thus affecting the use of the algorithm. So that the algorithm has to improve the general applicability of Goal Orientation and other people in the literature [7], creating a new CV speed, as follows:

$$F(x,y) = |I(x,y) - c_{out}| - |I(x,y) - c_{in}| \tag{65.8}$$

Among them: the c_{out}, c_{in} respectively, the average gray value of the contour line of external and internal, $I(x, y)$ to point (x, y) at the gray-scale value. The average gray value of the difference between where the two points gray values and contour the outer region and inner region, In order to speed power curve evolution Fast Level Set as CV-based fast level set segmentation algorithm can be described as follows:

The level set function is discredited into four integer values:

$$\phi(x) = \begin{cases} 3 & \text{if } x \text{ is an exterior point} \\ 1 & \text{if } x \in L_{\text{out}} \\ -1 & \text{if } x \in L_{\text{in}} \\ -3 & \text{if } x \text{ is an interior point} \end{cases} \tag{65.9}$$

Among them, the level set value is negative, the area within the contour area, marked as Ω_{in}, the level set for the positive region is the external area of the contour, denoted by Ω_{out}. The region of level set is -3 called the internal region, the region of value of $+3$ called the outer region. Internal lists and external linked list can be defined as follows:

$$L_{\text{in}} = \{x \mid x \in \Omega_{\text{in}} \text{ and } \exists\ y \in N(x) \text{ such that } y \in \Omega_{\text{out}}\} \tag{65.10}$$

$$L_{\text{out}} = \{x \mid x \in \Omega_{\text{out}} \text{ and } \exists\ y \in N(x) \text{ such that } y \in \Omega_{\text{in}}\} \tag{65.11}$$

where, $N(x)$ is the point of the neighbors around the point x the upper and lower's domain? Through the exchange of elements in the internal lists and external list achieve the evolution of the curve. The linked list elements of the exchange the switch in and switch out two-step operation to complete the linked list L_{out} of the switch in operating point x exchange to the linked list in L_{in}, will no longer belong to the elements of the internal linked list L_{in} delete, switching to the internal region, the operating curve outward expansion; contrary switch out the operation will be the point of the linked list L_{in} exchange to the linked list L_{out}, the operating curve inward contraction.

Two lists of elements in the exchange to be decided by the speed F of positive and negative. Point L_{out}, when F is greater than 0 the switch in operation, the point of exchange within the contour; point L_{in} to F is less than 0 perform switch out operation, the point of exchange to the outside of the contour. Fast Level Set Segmentation Method based on CV without solving partial differential equations, without re-initialization of the signed distance function, only width of 2 pixels in the narrow-band region is calculated, so the split speed has greatly improved.

65.3 Level Set Segmentation Algorithm Based on HSV Color Space

65.3.1 The Color Space Selection

Traditional level set segmentation drawback is that the segmentation of color images is not satisfactory. The curve evolution is to rely on gray-scale information-driven. Therefore, when the object and the background gray level information is difficult to distinguish, the power of curve evolution is not enough, leading to the split failure. The gray value of the target and background the same size but different color segmentation must not be correctly segmented target, using the CV model. Therefore, we introduced the color space for color image segmentation will be more effective.

Select the appropriate color space is effectively split the image, but also relates to methods and strategies used. Common RGB space all colors as a combination of three primary colors, but there is a strong correlation between the three-component, and thus not suitable for directly used for the independent operation of three components-based image segmentation. Space close to the human eye color perception of hue, saturation, and brightness (hue, saturation, value, and HSV) space, which, hue and saturation are collectively referred to as chrome, not only describes the color wavelength component distribution, but also that the depth of shade of colored light, Luminance component and color information. The image processing and computer vision algorithms can be in the HSV color space, easy to use, they can be dealt with separately and independent of each other. Therefore, in the HSV color space can greatly simplify the image analysis and processing workload. HSV color space and RGB color space is just a different representation of the same physical quantity, and thus exists between them, the conversion relations.

65.3.2 Conversion of RGB Space to HSV Color Space

Point was transformed into the HSV color space from RGB space that can be defined as:

$$V = \max(R, G, B) \tag{65.12}$$

$$S = (V - \min(R,G,B))/V(\text{当时}, V = 0 \;\; S = 0) \tag{65.13}$$

$$H = \left\{ \begin{array}{ll} 60\,(G - B)\big/(V - \min\,(R, G, B)) & \text{if}\, V = R \\ 120 + 60\,(B - R)\big/(V - \min\,(R, G, B)) & \text{if}\, V = G \\ 240 + 60\,(R - G)\big/(V - \min\,(R, G, B)) & \text{if}\, V = B \end{array} \right\} \tag{65.14}$$

If the proceeds of $H < 0$, then $H = H + 360$. Such hue information H in the range of 0–360°. V and S ranges from 0 to 1.

65.3.3 The Color Difference on HSV Space

Effective segmentation of color images, the color difference in the HSV color space to further the introduction of CV-based fast level set. HSV color space, H denotes the hue (hue), S means the saturation (saturation), and V represents the brightness (value).

Remember HSV space color values for $P_1 = (H_1, S_1, V_1)^T$ and $P_2 = (H_2, S_2, V_2)^T$, the color difference is defined as:

$$\Delta_{\text{HSV}}(P_1, P_2) = \sqrt{(\Delta_I)^2 + (\Delta_C)^2} \tag{65.15}$$

The color difference in (65.15) instead of (65.8) gray, to get fast level set based on the HSV color model speed:

$$F(x, y) = \nabla_{\text{HSV}}(P(x, y), V_{\text{out}}) - \nabla_{\text{HSV}}(P(x, y), V_{\text{in}}) \tag{65.16}$$

Where V (x, y) point (x, y) at the color value, V_{out} and V_{in} were the average of the external area of the contour and the internal region of the color value. Color level set segmentation model to make full use of the image information, considering the image's hue, saturation and intensity, so you can get a better segmentation results.

65.4 Experimental Results

In order to verify the feasibility and effectiveness of the proposed method, the real image data were simulated. In the dual-core 2.0 GHz Pentium processor, we used matlab2010b to achieve the proposed algorithm. We denote by CV model and the proposed fast level set segmentation algorithm based on the HSV color space to test contrast.

It can be seen from the results Figs. 65.1 and 65.2 classic CV model of the level set method to split the color image, part of the foreground objects were mistakenly split into the background, goals and background intensity close to background as foreground segmentation.

Fig. 65.1 Red flower

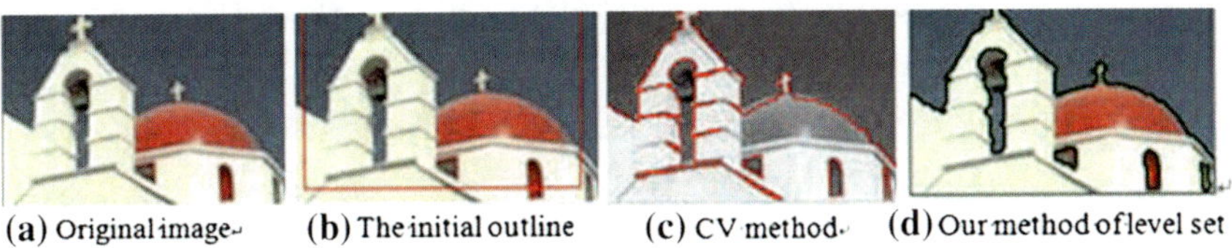

(a) Original image (b) The initial outline (c) CV method (d) Our method of level set

Fig. 65.2 Church

65.5 Conclusions

This paper presents a fast level set for color image segmentation algorithm based on HSV color space. First, in order to improve the segmentation speed, fast level set method based on CV model evolution. Second, the segmentation of color image can be made with the introduction of the concept of color difference under HSV color mode. In this case, the segmentation can be made exactly even the gray color is very close since the color information is fully employed. The experimental results show that compared with the traditional level set segmentation, our approach has better color image segmentation.

Acknowledgments This work is supported by the Fundamental Research Funds for the Central Universities (N100304008, N110316001 and N110304002).

References

1. Osher S, Sethian J (1988) Fronts propagating with curvature-depen-dent speed: algorithms based on hamilton-jacobi formulations. J Comput Phys 79(1):12–49
2. Chan FT, Vase L (2001) Active contours without edges. IEEE Trans Image Process 10(2):266–277
3. Yongang Shi, Clem Karl William (2008) A Real-Time Algorithm for the Approximation of Level-Set-Based Curve Evolution. IEEE Trans Image Process 17(5):645–656
4. Xu J, Wu J, Ye F et al (2008) A level set method for color image segmentation based on bayesian classifier. In: International conference on computer science and software engineering, CSSE 2008, vol 2(5), Wuhan, pp 886–890
5. Jingwen Z, Yuantao C, Jiaying W (2011) New color images segmentation algorithm based on level set. J Inf Comput Sci 8(14):3107–3114
6. Jun L, Xin Y, Peng-Fei S (2002) A fast level set approach to image segmentation based on mumford-shah model. Chin J Comput 25(11):1175–1183
7. Xiao-fei W, Cui-rong Y, Yong Y et al (2012) Fast level-set segmentation based on color difference in HSI color space. Appl Res Comput 29(3):1135–1137

Chapter 66
Election Path Based on Image Pressing and Mathematical Modeling

Jianneng Zhong

Abstract In order to have better awareness and understanding of each election method's connotation and role, it is the aim of this chapter using image analysis, mathematical modeling, the literature, and other research methods, to study the elections that have been polled, the similarities and differences of direct and indirect elections in two ways, respectively, to carry model analysis for these two election methods. This is the first time that image analysis method is used to compare the pros and cons of direct election and indirect elections, and thus provide a reference for the development of democracy. This paper aims to establish an electoral model method, studies the form of direct and indirect elections, and the electoral path optimization analysis which can obtain whether direct election is better than indirect election, to provide a theoretical basis for electoral activities.

Keywords Image processing • Mathematical modeling method • Election model • Model analysis • Optimal control

66.1 Introduction

As a political practice, elections are the base of the modern democratic system. The inner electoral is the development of a fundamental system of inner democracy, which is also to measure the important sign of the development of inner democracy. At different stages of the life of democracy, each election has different roles because awareness and the value of its democratic function are not the same [1]. For better knowledge and understanding of these election forms, to build the model based on direct elections and indirect elections, through an intuitive model for these elections' two forms to carry on the comparative analysis, further to find out their basic democratic function, and the intrinsic relationship with each other,

J. Zhong (✉)
Business School, Central South University, Changsha 410083, China
e-mail: zhongjn96@163.com

Z. Zhong (ed.), *Proceedings of the International Conference on Information Engineering and Applications (IEA) 2012*, Lecture Notes in Electrical Engineering 220,
DOI: 10.1007/978-1-4471-4844-9_66, © Springer-Verlag London 2013

in order to provide a reference for the development of democracy [2, 3] are the aims of this paper. The commonly used election methods are direct and indirect elections.

Direct election is defined by the voters directly vote for state representative institutions and representatives of national public official's election, and refers to the electors directly participating in the election to exercise their voting rights in the mode of election [4]. It allows each voter who has a chance to choose their own most trusted authority in the eyes, and can directly reflect the people will, realize the voters will, to better mobilize citizens to participate in the management of state affairs enthusiasm, help to strengthen the voters and elected the relation, it has more democracy, effectively restricts government organizations control or election attempts, thereby avoiding the indirect elections in electoral violations will happen [5, 6]. The results of direct election can be used to the majority rule to explain. The so-called majority rule refers to a bill or decision that must be approved by voters in favor of 1/2 or more to pass a plurality voting rules. In operation, majority rule can be divided into the simple majority rule and the proportion of the majority rule [7]. According to the simple majority rule, as long as the favor of votes are more than half, motion and the decision is adopted; according to the proportion of majority rule, in favor of votes must be more than half of the specified proportions' motion (such as 2/3, 3/4, 3/5 or 4/5, etc.), which may be passed, so that majority rule is very important to ensure that elections the democratic rights of the authorized person. Indirect election is not directly by voters vote, but by the next level of national representative body, or by the elected representatives (or electoral) election on a national representative body representative and personnel of national public office election, refers to the electors by elected representatives to exercise their voting rights of the election [8, 9].

66.2 Research Method and Establish Model

66.2.1 Research Object

Elections have been adopted the two methods that are direct election and indirect election.

66.2.2 Research Method

Using image analysis, mathematical modeling, literature and other research methods, to study the elections that have been adopted the similarities and differences of direct election, indirect election in two ways. On this basis, to analyze the pros and cons of two kinds election method, providing a reference for the development of democracy.

66.2.3 Modeling and Analysis

Election model is one of the many statistical physics model. In the election model, it can be considered all points that are occupied by all held different views people in configuration space $\Omega = \{0,1\}^{Zd}$. Namely, elector holding a view is "O", otherwise it is "1". Making $\{\xi(S), S \geq 0\}$ is the electoral collection of supporting "1" view, $|\xi(S)|$ means that the number of collection, $\{\xi^C(S), S \geq 0\}$ is the electoral collection of supporting "0" view. In the election model, voters can always change their view, they will change their view based on the view of neighbors, to change in a random time. Assumption People's thoughts are very simple in this model, the number people of their neighbors holding different view to change their view. In order to build models, marking $\{T_n^x, n \geq 1\}$ $\left(x \in Z^d\right)$ rate is 1, $\{y_n^x, n \geq 1\}$ is independent and identically distributed random variables sequence. In order to image of the instructions this process, it can adopt the method of the icon: drawing an arrow for each x and x from $x + y_n^x$ to (x, T_n^x), and noting a δ in the (x, T_n^x) process. In this model, it can be imagined that the water inflow is from the bottom of δ_0 to the whole structure. It is as to impede the dams of water flow through; the arrow as to allow the water flow through channels, making the flow direction in a certain direction. Electoral model diagram is shown in Fig. 66.1.

The above calculation leads to the duality process of the election model, this process is that all arrows of Fig. 66.1 are in the opposite direction, and mapping to define the path:

$$\xi_t^B = \{x : \forall y \in B, a(x,0)\text{path}\} \tag{66.1}$$

Obviously:

$$\{\xi_t^A \cap B \neq \phi\} = \{\xi_t^B \cap A \neq \phi\} \tag{66.2}$$

Thus, to obtain the election path.

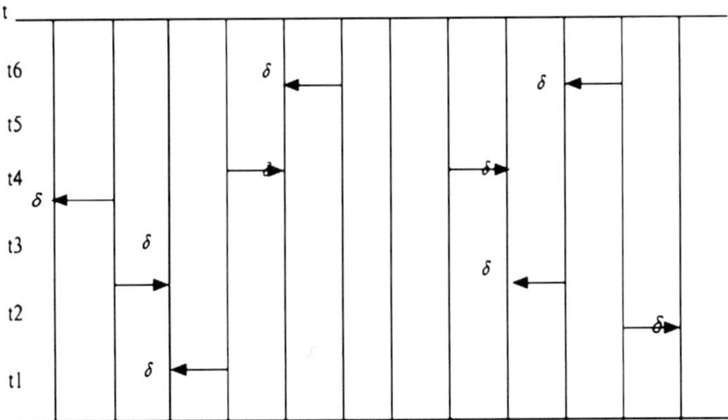

Fig. 66.1 Electoral model diagram

66.3 Model Analysis

66.3.1 Modeling Analysis of Direct Elections

In the model of the construction direct elections, according to Fig. 66.2, assuming that a total of nine candidates, only one can be eventually elected; assuming the election process is in the ideal state, and the election is at least five votes. In accordance with the qualified majority principle, those elected will be at least six votes; it is shown in Fig. 66.3. Direct election shows the democratic superiority, which can be obtained to intuitively grasp from the model diagram.

66.3.2 Model Construction and Analysis of Indirect Elections

According to the Fig. 66.4 model, first nine are divided into three groups, which is first group, second group and third group. To set A can be smoothly elected in the first group, through an election, the two of the first group agree to vote A ballot, the two of the second group agree to vote A ballot, and the third of the third group do not agree to vote A ballot; Secondly, through two elections, two votes are in favor and one vote is against, A is smoothly elected. However, according to the Fig. 66.4 election process can be seen the true result, which has 4 people that agree to choose A, and has 5 people that do not agree to choose A. If taking the direct election, A clearly cannot be elected. Similarly, if B has an assent in the first group, the second group is also a assent, however, there were three people agree in the third group, B should be elected in accordance with the democracy principles. If using the indirect election, B has not elected.

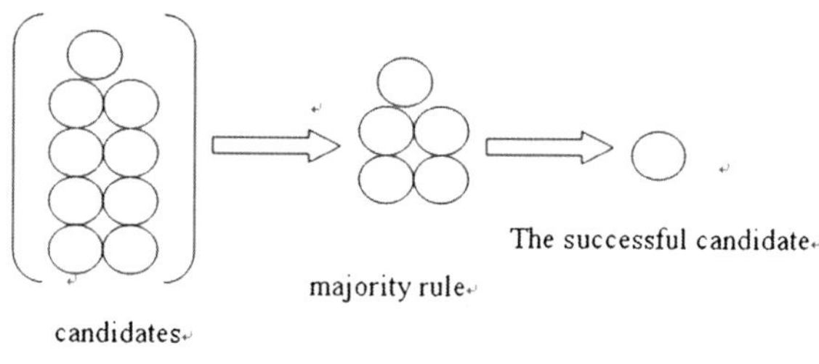

Fig. 66.2 Adopt the direct election model under the simple majority principle

Fig. 66.3 Adopt the direct election model under the qualified majority principle

majority rule

The successful candidate

candidates

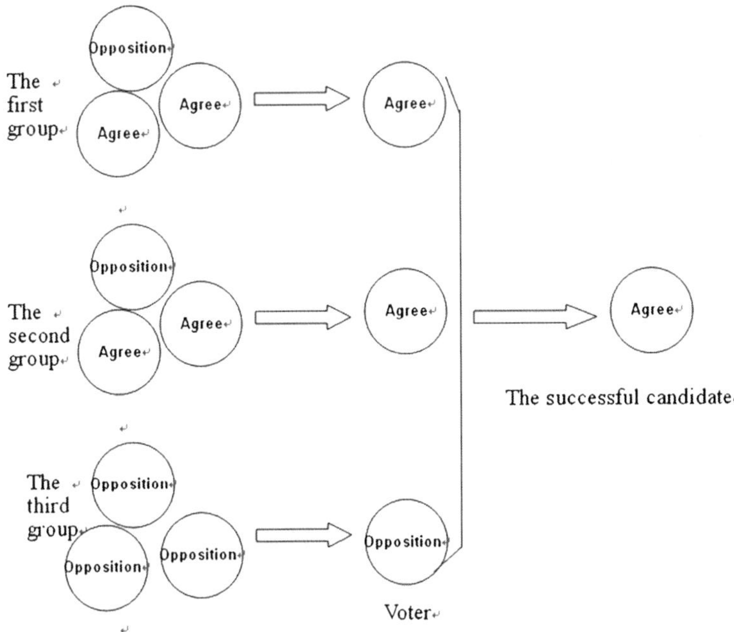

Fig. 66.4 Indirect election model of multi-level election

Above is hypothetical extreme cases, of course, it can also design a more extreme case. From the Fig. 66.5 model, it can be seen that there are a selected, and the first group and the second group are all agree; against group are in accordance with the majority principle or the majority opposed principle, that is, the first group and the second group have three assent, the third group is also a assent, so the number of consent is 7. There may be more extreme, Fig. 66.5 shows that it is all agreed through the two elections, which is 9 people agree Fig. 66.6.

Three cases can be seen that the candidates have been elected. From the process of model, this is difference in the amount of democracy and the loss of the voter democratic rights, especially through the indirect election of more than two, the class of indirect election is increasingly more, and it is more likely to cause the election results deviate from the wish of the electors. In the indirect election, increasing election level may make to increase efficiency. However, the voters and the connotation of democracy are further.

66.4 Results Analysis

The established election model can be seen:

1. Direct election can be more accurate, more direct to express voters' will; voters will achieve, to achieve the will of the voters. To participate elections is

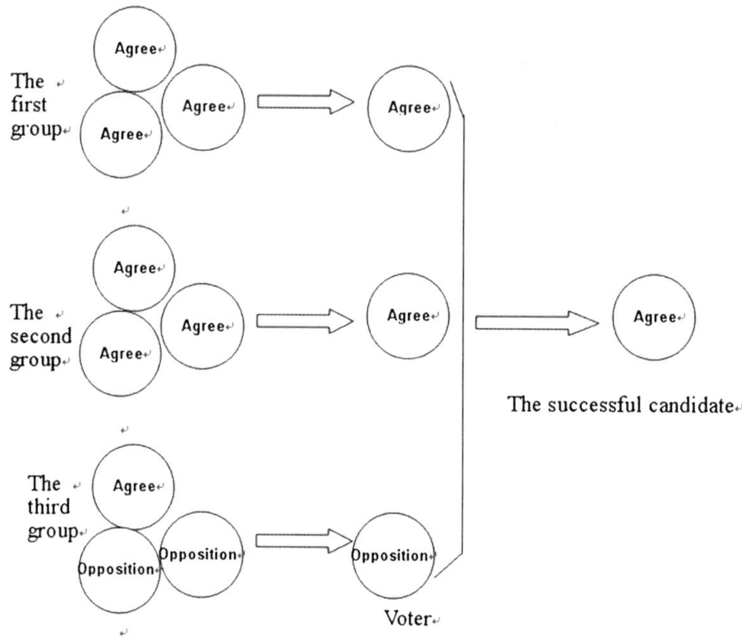

Fig. 66.5 Indirect election model of multi-level election

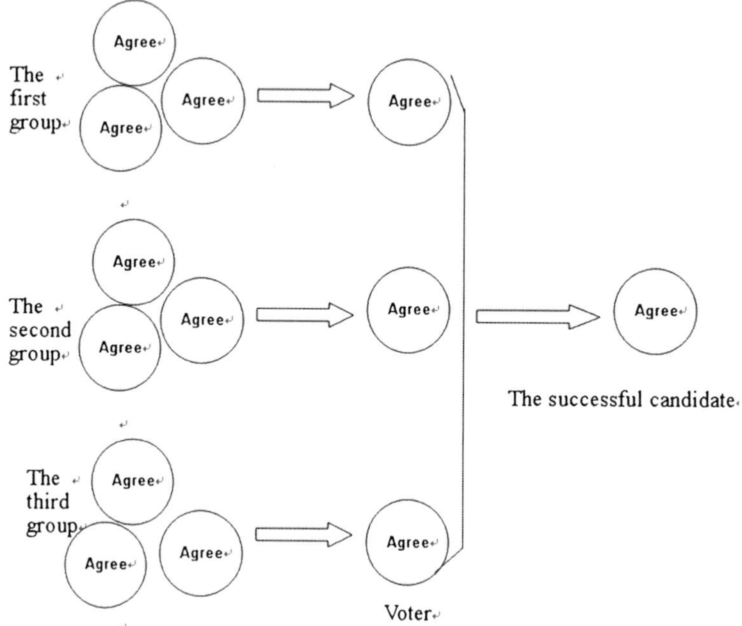

Fig. 66.6 Indirect election model of multi-level election

an important form of citizens' political participation, which is also free, real and independent to express their political will, to exercise one of the important ways of their political rights. Indirect election reflects the people indirectly authorized, direct election reflects the people directly authorized. Direct election allows the will of voters and the interests to get expression of full and direct, without going through the intermediary role of the "check people" to express. According to their own views of candidate personality charm and personal political views, to make direct evaluation and selection, direct vote chooses the true sense of the agent of their approval, according to their willingness to exercise public power and safeguard their own interests. This election makes the relationship between voters and elected that is more close, voters will are not be distorted, the individual choice of voters and social choice has small distance. From the effective realization real to comprehensive democratic sense, direct election system is the best choice; it is undoubtedly higher democracy than indirect election.

2. Direct election is the sharp weapon of opposing bureaucracy. In a direct election, voters is directly by voting to determine the candidate elected or not, by the elected representatives, its power comes from voters, it must directly responsible for voters. At the same time, the voters elected and recall are superimposed, in authorized direction is convergent. Voters have direct supervision and restriction power for voters, which can choose you, also recall. This will make the person elected who are really realized, their power is derived from the majority of the voters, and not from a few people, if the majority of voters powers could not very good exercise, voters will recover their power, so they dare not slacken our efforts, always eyes down, but did not dare do what one wishes without restraint, because of millions of pairs of eyes staring at you. Directly elected people also may be an unknown person, but is not a bad man, he must be very hard to lead to big mistakes.
3. Direct elections can be better reflecting the fair and equitable. Indirect election is higher level; it will have many disadvantages, which have more easily happen things that are bribery and intimidation in election. The number of the final election is greatly reduced than the primary election, bribery and intimidation naturally become easier. In the direct election, because voters are numerous, every vote value is also reduced, candidate bribed voters power weakening, buy ability decreases at the same time. Buy off dozens of people is not difficult, but to buy tens of thousands of voters is quite difficult, it can be said that the bribery natural killer of direct election is itself.
4. Direct election has powerful political education function.
5. When people take the direct election, they will be out of the occupation, industry or life and other organizations limitations, to understand social whole interest and long-term interests, to learn to self-esteem and respecting other people, respecting for the majority. In the importance of "personal wishes" and treat correctly "the public will", learning through empathy and consensus to reach, and to take collective political action. Such a process, to voters, it was a real and vivid ideological and political training. It allows voters

consciously as d a member of sense of responsibility and sense of justice and in society, the intelligence level and moral capability of voters has been greatly improved, it contributes to the voters to form a sense of belonging, identity, not free from social groups.

66.5 Conclusion

As can be seen from the above model, direct election is better than indirect election. It can be said that direct elections can get all the benefits of indirect elections. If direct elections are not the benefits, indirect elections also are not.

References

1. Huang W (2010) Soviet communist party decades Ji, vol 8. Jiangxi University Press, Nanchang, pp 78–79
2. Shu A et al (2009) Existing problems and their reasons in the inner party democratic elections. Hubei Soc Sci 3:67–69
3. Jianzhong L (2010) From respecting the principal position of party members set out to promote inner-party competitive election. Inner Mongolia Soc Sci 4:90–96
4. Zhang S (2009) The final breach of China's democratic political development is party secretary difference elections. Chongqing Soc Sci 10:12–18
5. Wang X (2010) Study of Chinese county and township people's congress direct election, vol 8. China Financial and Economic Press, Beijing, pp 45–48
6. Ren X (2010) The superiority of the direct election. Mianyang Econ Technol Coll 12:90–97
7. Gan L (2011) Liaoning: Feasibility study of the current expansion of people's congress direct election range, vol 8. Northeastern University, Liaoning, pp 78–80
8. Chen L (2010) Study of Chinese communist party inner-party election, vol 8. Jiangsu University, Jiangsu, pp 78–79
9. He X (2011) Competitive selection system and improve our electoral system, vol 7. Nanjing Normal University, Nanjing, pp 12-17

Chapter 67
A New Method of Inland River Overloaded Ship Identification Using Digital Image Processing

Lei Xie, Jing Chen, Zhongzhen Yan and Zheyue Wang

Abstract With the development of low-carbon economy, inland river transportation has been attracting more and more attention in China. At the same time, driven by some economic benefits, the ship overload phenomenon continues to occur. Therefore, overloaded ship detection has been a key factor for reducing marine traffic accidents. This paper presents a robust method for detecting overloaded ship and the proposed algorithm includes three stages: ship detection, ship tracking, and overloaded ship identification. Ship detection is a key step and the concept of ship tracking is built on the ship-segmentation method. According to the segmented ship shape, we propose a predict method based on Kalman filter to track each ship. The data of ship length and ship speed will be used to identify overloaded ship. The proposed method has been tested on a number of monocular ship image sequences and the experimental results show that the algorithm is robust and real-time.

Keywords Inland river • Kalman filter • Overloaded ship identification

67.1 Introduction

Recently, countries and groups have increasingly expressed concern about environmental protection and low-carbon life. The trend of social development not only promotes the great development of inland river transportation, but also causes growing overloaded phenomenon of inland river ship.

With the development of information technology, video camera and high definition camera are considered as suitable sensor devices for capturing and

L. Xie (✉) · J. Chen · Z. Yan · Z. Wang
Engineering Research Center for Transportation Safety (Ministry of Education), Wuhan University of Technology Wuhan 430063, Hubei, China
e-mail: xl_for_paper@163.com

L. Xie · J. Chen · Z. Yan · Z. Wang
Intelligent Transport System Research Center, Wuhan University of Technology
Wuhan 430063, Hubei, China

Z. Zhong (ed.), *Proceedings of the International Conference on Information Engineering and Applications (IEA) 2012*, Lecture Notes in Electrical Engineering 220, DOI: 10.1007/978-1-4471-4844-9_67,

recognizing spatio-temporal aspects of inland river structures and traffic situations, and have been introduced in many social systems for various purposes, especially transportation. It is well recognized that vision-based surveillance systems are more versatile than others for traffic parameter estimation and overloaded phenomenon identification [1].

Ship detection and ship tracking can offer us a continuous description of the vessel traffic flow. Therefore, it has been an important and challenging issue in video-based overloaded ship identification. However, there are several problems that remain unsolved in this process. On the one hand, the tracking result strongly depends on the quality of ship detection. Apparently, it augments the instability of tracking system. On the other hand, overloaded ship identification is [2] to be processed in real-time and it needs a simple and efficient approach to extract the ship feature.

In the last several years, extensive research work has been done and many traffic monitoring systems have been exploited that include road transportation[2] and water transportation [3]. Researchers have developed various algorithms to extract object feature and track moving objects but most of the water transportation monitoring systems are focused on infrared cameras [4], automatic identification system (AIS) [5] or radar [6, 7], while little work has been done for optical image 8. Considering the complexity of the Kalman filter, [8] many researchers presented their own algorithms to construct imaging filters and achieve real-time operation [9, 10].

This paper presents a robust and real-time method to identify overloaded ship through a video camera and high definition camera. The proposed algorithm includes two stages: ship detection, ship tracking, and overloaded ship identification.

The remainder of the paper is organized as follows. We first introduce the process of ship detection in Sect. 67.2. The algorithm related to ship tracking is given in detail inSect. 67.3. Overloaded ship identification and its experimental results are presented in Sect. 67.4. Finally, the conclusion are drawn in Sect. 67.5.

67.2 Ship Detection

It is a very important step to extract the ship shape out of the river background. The ship detection method requests to automatically segment every ship so that there can be a unique tracking associated with the ship. In this phase, we will solve several problems as follows:

Extract the background image automatically from a sequence of river traffic images and update the background continually according to the change in ambient lighting, weather, etc.

Select an adaptive filter to eliminate abnormal moving object in the binary background subtraction image so that the system can be more robust.

Detect ship from the binary background subtraction image.

Apparently, it is desirable to extract the initial background image automatically from a sequence of road traffic images before background subtraction. Therefore, we propose a background extraction method based on moving object pixel detection, in which each pixel of image will be identified whether its intensity had an obvious change or not. Moving object pixels could be extracted from current input image by performing a difference on three consecutive inter-frames. First, it is needed to calculate inter-frame subtraction image of two pairs of images, i.e., the first subtraction image between $(k-2)$ frame image and the $(k-1)$ frame image, and the second subtraction image between $(k-1)$ frame and (k) frame. Second, two binary images could be transformed from the two subtraction images via a dynamic subtraction threshold. Then, we apply the bitwise logical and operation to the two binary subtraction images to clarify the moving object pixels. Finally, the original background image could be obtained through patching up non-moving object pixels from a sequence of input images. The same method could be used to update the background image.

After the update process, we calculate the difference between background image and input image for each pixel, and then apply binary filter to background subtraction image to clarify the moving object region. While the binary difference image of moving ships is obtained, we apply a block filter based on statistics to filter the noises and then adopt a simple seed-growing arithmetic to detect ship. The detailed results of our proposed scheme are represented in Figs. 67.1 and 67.2.

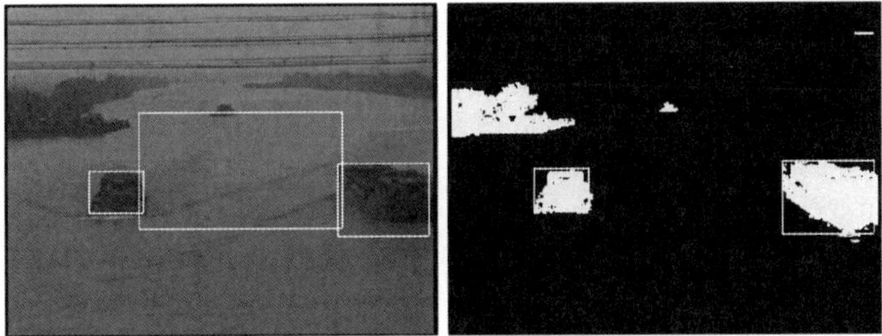

Fig. 67.1 An example of background estimation and updating

Fig. 67.2 An example of ship detection

67.3 Ship Tracking

Based on the segmented ship shape, which can be represented by a simple square model, we propose a Kalman filter method to track ship.

The Kalman filter is the minimum-variance state estimator for linear dynamic systems with Gaussian noise. In addition, the Kalman filter is the minimum-variance linear state estimator for linear dynamic systems with non-Gaussian noise.

Consider the system model as follows:

$$xk + 1 = \cdots Fxk \cdots + \cdots wk \tag{67.1}$$

$$yk \cdots = Hxk \cdots + \cdots vk \tag{67.2}$$

where k is the time step, xk is the state, yk is the measurement, wk and vk are the zero-mean process noise and measurement noise with covariances Q and R respectively, and F and H are the state transition and measurement matrices. The Kalman filter equations are given as follows:

$$Pk = F\xi K - 1 \cdot FT + Q \tag{67.3}$$

$$Gk = Pk \cdot HT\,(HPk \cdot HT + R) - 1 \tag{67.4}$$

$$x\,(k|k) = F \cdot x\,(k-1|k-1) + Gk\,(yk - HF \cdot x\,(k-1|k-1)) \tag{67.5}$$

$$\xi k = (I - GkH) \cdot Pk \tag{67.6}$$

For $k = 1, 2\ldots$ where I am the identity matrix $x\,(k|k)$ is the priori estimate of the state xk given measurements up to and including time $k - 1\,x\,(k-1|k-1)$ is the posteriori estimate of the state xk given measurements up to and including time k. Gk is the Kalman gain, Pk is the covariance of the priori estimation error $xk - x\,(k-1|k-1)$, and $\xi\,K$ is the covariance of the posteriori estimation error $xk\,x\,(k|k)$

When the noise sequences $\{wk\}$ and $\{vk\}$ Gaussian, uncorrelated, and white, the Kalman filter is the minimum-variance filter and minimizes the trace of the estimation error covariance at each time step. When $\{wk\}$ and $\{vk\}$ are non-Gaussian, the Kalman filter is the minimum-variance linear filter, although there might be nonlinear filters that perform better. When $\{wk\}$ and $\{vk\}$ are correlated or colored, (3–6) can be modified to obtain the minimum-variance filter. Our system schematic diagram is shown in Fig. 67.3

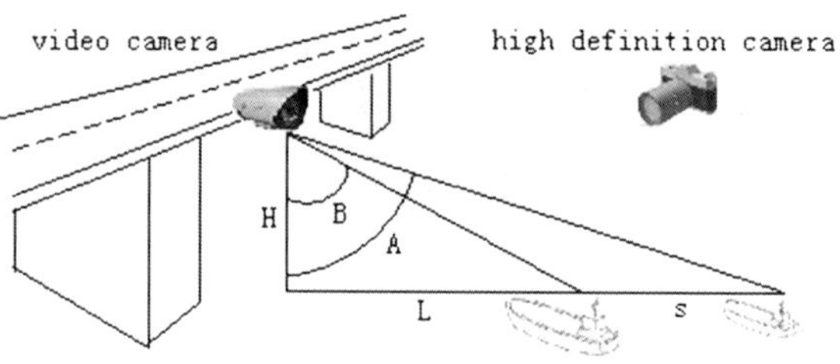

Fig. 67.3 The system schematic diagram

As we know, ship feature extraction plays an important role in ship tracking. The extracted information must be robust and essential to the accurate visual interpretation of the image so that the tracking result is not dependent on parameters controlling thresholds, which is often established empirically to achieve acceptable performance. Our arithmetic extracts many ship features to characterize each ship, such as ship center of mass, average intensity, and ship speed, etc.

Now suppose it satisfies the equality constraints. Let Xk be the ship feature vector of the time k. Then:

$$Xk = \cdots [xi\,(k) \cdots xvi\,(k) \cdots xj\,(k) \cdots xvj\,(k) \cdots xD\,(k)] \tag{67.7}$$

$$Yk = \cdots [yi\,(k) \cdots yj\,(k)] \tag{67.8}$$

where $xi\,(k)$ and $xj\,(k)$ are the state values of ship center's coordinates in the input image, $xvi\,(k)$ and $xvj\,(k)$ are their speed values, $xD\,(k)$ is the average intensity of the whole ship, $yi\,(k)$ and $yi\,(k)$ are the measurement values of $xi(k)$ and $xj\,(k)$ $Xk = \cdots [xi\,(k) \cdots xvi\,(k) \cdots xj\,(k) \cdots xvj\,(k) \cdots xD\,(k)]$.

Based on the above definition and the system schematic diagram, we define the Kalman system model as:

$$Xk + 1 = FXk \cdots + \cdots wk \tag{67.9}$$

The final result may cause the Kalman filter updates the Kalman gain Gk and the ship feature vector is recalculated. Our system tracks all these ships by implementing the above processes recursively until they have left the field of view.

$$F = \begin{bmatrix} \cos A & T\cos A & 0 & 0 & 0 \\ 0 & 1 & 0 & 0 & 0 \\ 0 & 0 & 1 & T & 0 \\ 0 & 0 & 0 & 1 & 0 \\ 0 & 0 & 0 & 0 & 1 \end{bmatrix} \tag{67.10}$$

$$Yk \cdots = HXk \cdots + \cdots vk \tag{67.11}$$

$$H = \begin{bmatrix} 1 & 0 & 0 & 0 & 0 \\ 0 & 0 & 1 & 0 & 0 \end{bmatrix} \tag{67.12}$$

67.4 Overloaded Ship Identification

The research in this step is primarily how to get the position of the waterline by digital image processing technology. In order to ensure the detection accuracy of waterline, another high definition camera will be used to shoot the ship photograph. Taking into account the difference of coordinate systems between the video camera and high definition camera, it is necessary to convert the ship center's coordinates and obtain the data about ship length, ship speed, and navigation direction through the ship tracking result.

Our paper takes two steps to detect the waterline edge.

The first step is the edge detection approach based on the navigation direction. Compared to other operations, the canny operator gets the better results in detection. In order to find the exact waterline, the false edges are removed by navigation direction projection. Finally, the waterline line is fitted by the least square method.

The second step is the edge pick-up approach based on the Hough transform. This paper adopts a voting method to choose the possibly true edges, in which the weight values of the following attributes are large: the hue grads, the saturation grads, the intensity grads, and the times of searching edges. According to the difference in the average values between the upside area and the downside, the method selects the true edge of waterline. The horizontal line is fitted by the least square method.

Finally, the distance between the top of ship and the waterline will identify whether the ship is overloaded or not. The detailed results of our proposed scheme are represented in Figs. 67.4 and 67.5.

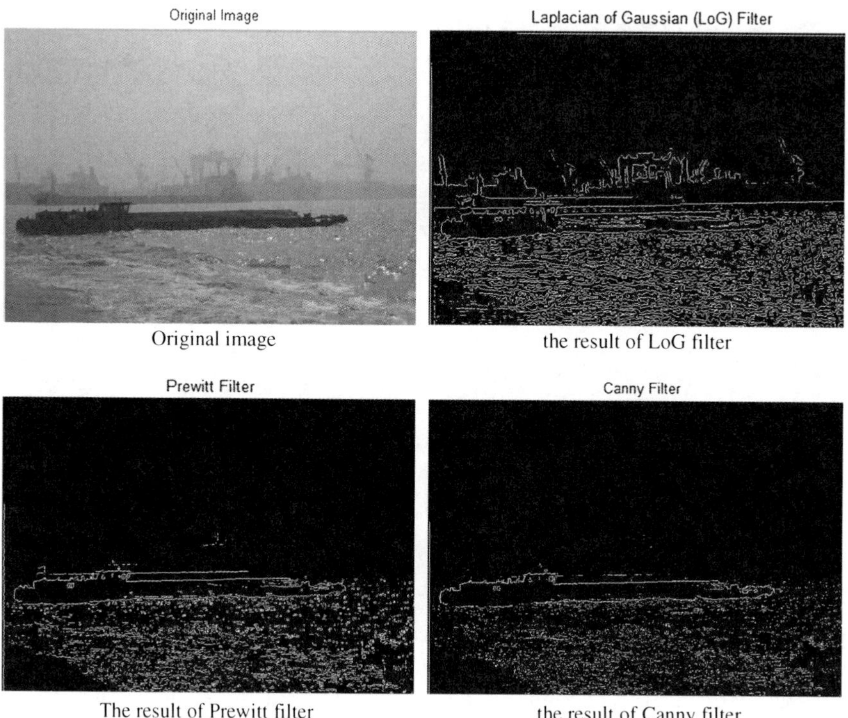

Fig. 67.4 An example of edge detection

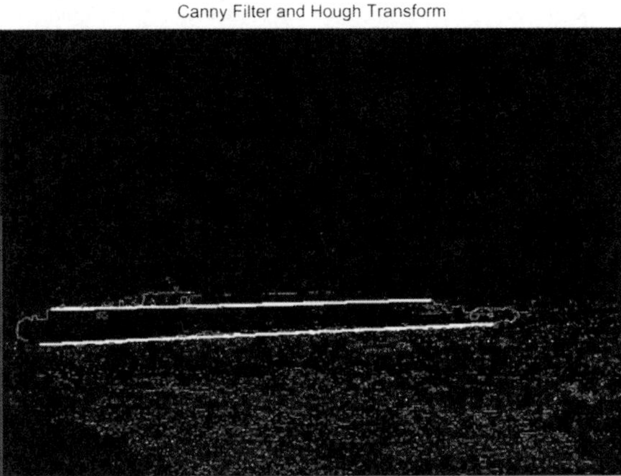

Fig. 67.5 The edge approach about the top of ship and the waterline

67.5 Conclusion

In this paper, we present a new algorithm for ship tracking in a video-based ITS. The experiment results on real-world videos show that the algorithm is effective and real-time. The correct rate of ship tracking is higher than 85 %, independent of environmental conditions.

Acknowledgments The author thanks his colleagues for their influence. Thanks to the referees for their suggestions which have greatly improved the presentation of the paper. This work was supported by Transportation Construction Technology Project (201132820190) and Department of Transportation Industry tackling Project (2009353460640).

References

1. Lee E-K, Ho Y-S (2010) Generation of multi-view video using a fusion camera system for 3D displays. IEEE Trans Consum Electron 56(11):2797–2805
2. Ha DM, Lee JM, Kim YD (2004) Neural-edge-based ship detection and traffic parameter extraction. Image Vis Comput 22(6): 899–907
3. Hu H, Tian J, Dai G, Wang M, Peng Y (2011) A new method of ship detection for SAR images. Int J Advancements Comput Technol 3(9):64–71
4. Wu J, Mao S, Wang X, Zhang T (2011) Ship target detection and tracking in cluttered infrared imagery. Opt Eng 50(5):234–247
5. Obad D, Bošnjak-Cihlar Z (2004) Benefits of automatic identification system within framework of river information services. In: Proceedings Elmar—International Symposium Electronics in Marine Proceedings, vol 46(23), pp 143–147
6. Vicen-Bueno R, Carrasco-álvarez R, Jarabo-Amores MP, Nieto-Borge JC, Rosa-Zurera M (2011) Ship detection by different data selection templates and multilayer perceptrons from incoherent maritime radar data. IET Radar Sonar Navig 5(2):144–154
7. Ruiz ARJ, Granja FS (2009) A short-range ship navigation system based on ladar imaging and target tracking for improved safety and efficiency. IEEE Trans Intell Transp Syst 10(1):186–197

8. Zhu C, Zhou H, Wang R, Guo J (2010) A novel hierarchical method of ship detection from spaceborne optical image based on shape and texture features. IEEE Trans Geosci Remote Sens 48(9):3446–3456
9. Li P, Zhang T, Ma B (2004) Unscented kalman filter for visual curve tracking. Image Vis Comput 22(2):157–164
10. Piovoso MP, Laplante PA (2003) Kalman filter recipes for real-time image processing. Real-Time Imag 9(6):433–439

Chapter 68
An Improved Image Segmentation Approach Based on Pulse Coupled Neural Network

Yongxing Lin and Xiaoyan Xu

Abstract In this paper, we introduce a new image auto-segmentation algorithm based on PCNN and fuzzy mutual information (FMI). The image was first segmented by PCNN, and then the FMI was used as the optimization criterion to automatically stop the segmentation with the optimal result. The experimental results demonstrated that the CT and ultrasound images could be well segmented by the proposed algorithm with strong robustness against noise. The results suggest that the proposed algorithm can be used for medical image segmentation.

Keywords Image segmentation • Pulse coupled neural network • Fuzzy mutual information

68.1 Introduction

Image segmentation is important in the field of medical imaging team. To accurately segment the image has been a hot topic. All these half-or fully automatic algorithms can be divided into five types mainly depending on the strategy of region of interest (ROI) division and edge detection, texture and characteristic analysis, deformation and positive mode; the algorithm is a mixed methods and multi-scale-based method [1, 2]. The model method often uses a prior knowledge and active contour model and the statistical model, or does not use any deformation prior information about the interested region return-on-investment (ROI). The foundation of active contour model, including snake model and the level set, is very popular, usually by semi-automatic division. The mix method is based on the combination of different algorithms to optimize the result [2, 3].

Y. Lin (✉)
Zhejiang Science and Technology University, Hangzhou 311121, Zhejiang, China
e-mail: linyongxing@cssci.info

X. Xu
Zhejiang SUPCON Technology Co., Ltd, Hangzhou 310053, Zhejiang, China

Z. Zhong (ed.), *Proceedings of the International Conference on Information Engineering and Applications (IEA) 2012*, Lecture Notes in Electrical Engineering 220, DOI: 10.1007/978-1-4471-4844-9_68, © Springer-Verlag London 2013

68.2 Method

68.2.1 The Principle of PCNN Image Segmentation

Figure 68.1 shows the neurons in mathematical model. It consists of three parts: the accept area, a field, and pulse generator. Accept areas the role of the other neurons is affected by input and from outside of two channels that connect material F and channel L. The feeding input F_{ij} receives the external stimulus I_{ij} and the pulse Y from the neighboring neurons. The linking input L_{ij} receives the pulses from the neighboring neurons and output signals. In the modulation field, F_{ij} and L_{ij} are input and modulated. The modulation result U_{ij} is then sent to the pulse generator, which is composed of a pulse generator and a comparator. The U_{ij} is compared with the dynamic threshold θ_{ij} to decide whether the neuron fires or not. If the U_{ij} is greater than threshold θ_{ij}, the pulse generator will output one and the dynamic threshold will be enlarged accordingly. When θ_{ij} exceeds U_{ij}, the pulse generator will output zero. Then a pulse burst will be generated. The corresponding mathematical model is expressed as follows:

$$F_{ij}[n] = \exp(\alpha_F) F_{ij}[n-1] + V_F \sum M_{ijkl} Y_{kl}[n-1] + I_{ij} \tag{68.1}$$

$$L_{ij}[n] = \exp(\alpha_L) L_{ij}[n-1] + V_L \sum W_{ijkl} Y_{kl}[n-1] \tag{68.2}$$

$$U_{ij}[n] = F_{ij}\left(1 + \beta L_{ij}[n]\right) \tag{68.3}$$

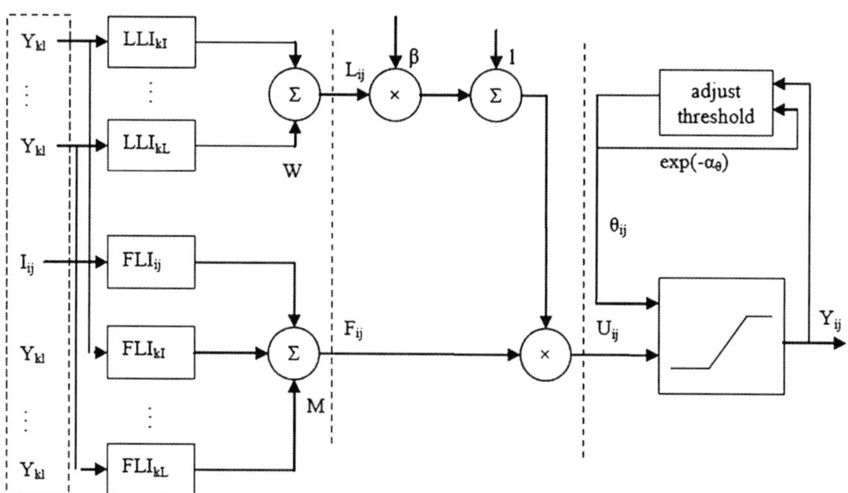

Fig. 68.1 Model of PCNN neuron

$$Y_{ij}[n] = \begin{cases} 1 & U_{ij} > \theta_{ij} \\ 0 & U_{ij} \le \theta_{ij} \end{cases} \tag{68.4}$$

$$\theta_{ij}[n] = \exp(-\alpha_\theta)\,\theta_{ij}[n-1] + V_\theta Y_{ij}[n-1] \tag{68.5}$$

where i and j refer to the pixel positions in image, k and l are the dislocations in a symmetric neighborhood around a pixel, and n denotes the current iteration (discrete time step). The M_{ijkl}, W_{ijkl} are the constant synaptic weights, and V_F, V_L and V_T are the magnitude scaling terms. The α_F, α_L and α_θ are the time delay constants of the PCNN neuron, and β is the linking strength.

From the point of view of image processing, the model still has some limitations in practical applications. There are many parameters required to be adjusted in the model, which is time-consuming and also difficult. In order to further reduce the computational complexity, the improved PCNN model [4] as follows:

$$F_{ij}[n] = I_{ij} \tag{68.6}$$

$$L_{ij}[n] = V_L \sum W_{ijkl} Y_{kl}[n-1] \tag{68.7}$$

$$\theta_{ij}[n] = \begin{cases} \exp(-a/n)\,\theta_0, & Y_{ij}[n-1] = 1 \\ \theta_0, & Y_{ij}[n-1] = 0 \end{cases} \tag{68.8}$$

Production neurons pulse will lead to their launch around the fire of the interaction between neurons, which will cause nearby neurons to be used in the same way. Hence it can produce a pulse spread outside the activity of the field. In a neuron, fire will fire any group of neurons the whole group. So the image segmentation can quickly realize the use of synchronization characteristics.

68.2.2 Fuzzy Mutual Information

Mutual information (MI) is a kind of similarity measure because of its proven versatility, and has been widely used in image processing [5]. This information-theoretic is not dependent on any hypothetical data, and does not assume particular relationships in different forms of strength. As regards image processing, it is assumed that the largest reliance is on the shades of gray image between them to the correct aligned. The Max-MI standard has been used for image segmentation. However, it does not always get the best segmentation effect, because it will be affected by the transformation of missile value for the overlapping area image. The fuzzy theory applied to image segmentation puts forward the fuzzy mutual information (FMI). This paper introduces the MI based on the correlation coefficient as follows:

Given image A and B, the MI is defined as

$$MI(A,B) = \sum_{a,b} p_{AB}(a,b) \log \frac{p_{AB}(a,b)}{p_A(a) \cdot p_B(b)} \tag{68.9}$$

where $p_{AB}(a, b)$ is the joint probability distribution of two images, and $p_A(a)$ and $p_B(b)$ are marginal distribution of image A and B, respectively. Then the FMI is given by

$$FMI(A,B) = \sum\sum (\rho(a,b))^{\alpha}\, p_{AB}(a,b) \log \frac{p_{AB}(a,b)}{p_A(a) \cdot p_B(b)} \tag{68.10}$$

where α is an adjustable factor which is greater than 0, and $\rho(a, b)$ is the correlation coefficient of image A and B. FMI has the following properties:

1. Symmetry: FMI $(A, B) =$ FMI (B, A).
2. Conversion: If the adjustable factor $\alpha = 0$, then the FMI is the same as MI.

68.2.3 Auto-Segmentation Algorithm

The algorithm can be implemented with the following steps:

1. Setting the parameters: $VL = 0.5$, $\alpha = 10$, $\theta 0 = 255$, $\beta = 0.1$, and iteration number $n = 1$.
2. Inputting the normalized gray image to PCNN network as the external stimulus signal I_{ij}.
3. Iterating $n = n + 1$.
4. Segmenting the image by PCNN.
5. Computing the value of FMI. If FMI $<$ FMImax, go to Step 3, otherwise stop segmentation and the final result is obtained.

68.3 Results and Discussions

Figure 68.2 shows the segmentation results of tire images by different algorithms. The images from left to right in the first line are the original tire image, image with Gauss noise, image with salt and pepper noise, and image with multiplicative noise, respectively.

Figure 68.3 shows the segmentation results of medical cerebral CT image. Figure 68.3a is the original CT image, and Figs. 68.3b–d are the images segmented by Otsu, PCNN with max-entropy, and PCNN with max-FMI algorithm, respectively. Figure 68.4 shows the segmentation results of breast tumor ultrasound image with the same sequence as Fig. 68.3. The PCNN with max-FMI

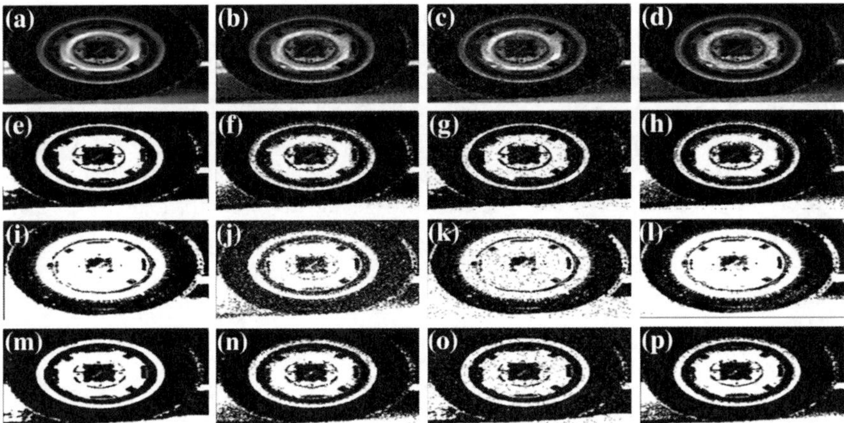

Fig. 68.2 Segmentation results of tire images

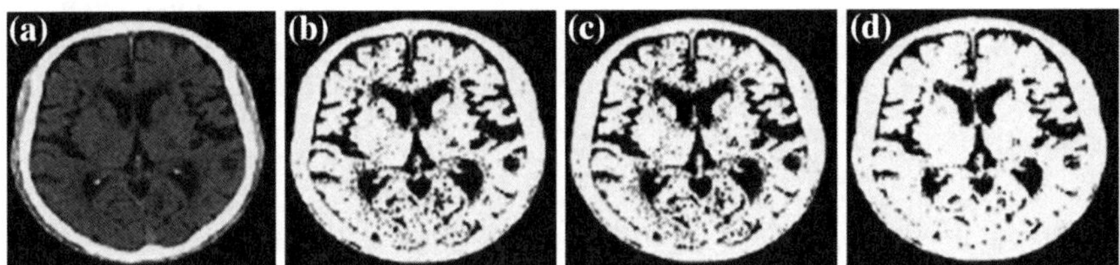

Fig. 68.3 Segmentation result of CT image. **a** Original CT image. **b** Segmentation image with Otsu. **c** Segmentation image with max-entropy PCNN. **d** Segmentation image with max-FMI PCNN

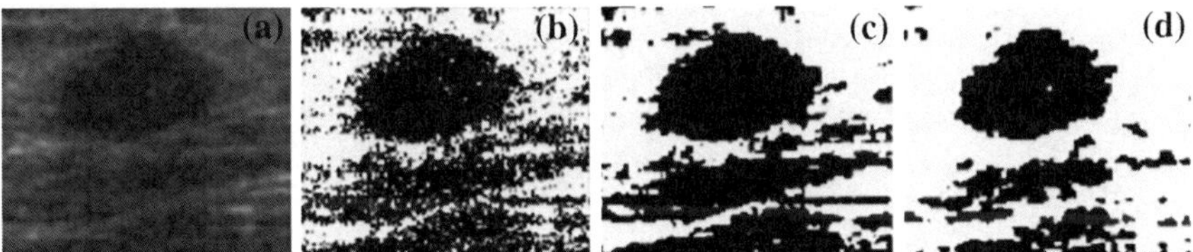

Fig. 68.4 Segmentation result of breast ultrasound image. **a** Original ultrasound image. **b** Segmentation image with Otsu. **c** Segmentation image with max-entropy PCNN. **d** Segmentation image with max-FMI PCNN

algorithm again illuminated the well segmentation effect than other two algorithms, especially in the boundaries of region of interest. Despite the volume effect in CT, the CT image is clear enough without pre-processing. The intracranial regions in cerebral CT, such as the cerebrospinal fluid and the brain matter, are well segmented by PCNN with max-FMI, while there is too much noise in the images segmented by Otsu and PCNN with max-entropy. For ultrasound image, there are too many small spots in Fig. 68.4b, which means that Otsu algorithm suffered from the speckle noise. The boundary in Fig. 68.4c is obviously smooth, and the details are lost for PCNN with max-entropy algorithm. However, the boundary of breast boundary could be accurately segmented in ultrasound image, although the speckle noise was inherent in ultrasound image. Furthermore, our algorithm can segment the ultrasound image without the pre-processing of denoising or enhancement, which will reduce the running time. So the proposed algorithm has strong robustness against noise and high performance efficiency (Fig. 68.4).

68.4 Conclusion

In conclusion, we introduce a new image auto-segmentation algorithm based on PCNN and FMI. The proposed algorithm was able to effectively segment the CT and ultrasound images, which was confirmed by the experiments. The results suggest that the proposed algorithm has the potential application in medical images.

References

1. Noble JA, Boukerroui D (2006) Ultrasound image segmentation: a survey. IEEE Trans Med Imaging 25:987–1010
2. Tsantis S, Dimitropoulos N, Cavouras D et al (2006) A hybrid multi-scale model for thyroid nodule boundary detection on ultrasound images. Comput Methods Programs Biomed 84:86–98
3. Archip N, Rohling R, Cooperberg P (2005) Ultrasound image segmentation using spectral clustering. Ultrasound Med Biol 31:1485–1497
4. Eckhorn R, Reitboeck HJ, Arndt M (1990) Feature linking via synchronization among distributed assemblies: simulation of results from cat cortex. Neural Comput 2(3):293–307
5. Madabhushi A, Metaxas DN (2003) Combining low-, high-level and empirical domain knowledge for automated segmentation of ultrasonic breast lesions. IEEE Trans Med Imaging 22:255–269

Chapter 69
Video Self-Adaptive Noise Reduction Algorithm

Jing Li

Abstract In digital video processing systems, studying an efficient noise reduction algorithm has become an important issue. As the noise is uncertain, so I hope to develop an algorithm that can adapt to different levels of noise. The temporal filter has the advantage of retaining the edge and details of a picture while the spatial filter has better abilities of reducing noise. In order to take into account the noise filtering and the protection of details and edges of video and pictures, I approached the joint temporal and spatial filtering algorithm. As different methods have various effects to noises of different levels, I adopted the self-adaptive system to achieve the best result.

Keywords Joint temporal and spatial filtering algorithm • Adaptive • Images and video • Recursive • Mean filter • Median filter

69.1 Introduction

The algorithm in this paper is targeted at YUV video streaming. Y component is brightness, so it is necessary to process the Y component and convert it with U, V components to RGB format to be displayed together [1]. Research has shown that the result from processing the three channels of RGB, respectively, will encounter distortion in color. Therefore, to avoid such a distortion, it is necessary to directly process the Y.

J. Li (✉)
Department of Information Science and Electronic Engineering, Zhejiang University, Hangzhou 310000, Zhejiang, China
e-mail: lijing@cssci.info

Z. Zhong (ed.), *Proceedings of the International Conference on Information Engineering and Applications (IEA) 2012*, Lecture Notes in Electrical Engineering 220, DOI: 10.1007/978-1-4471-4844-9_69, © Springer-Verlag London 2013

69.1.1 Pre-filtering

The gray value of most salt and pepper noise points is concentrated at 255 or 0, while most signal points are not concentrated at this area [2]. Hence, the author proposes this algorithm: (1) assume the pixel value is judged as the salt and pepper noise point when the gray range is between [0, 16] or [235, 255] and other areas are regarded as signal points; (2) the noise point (Z_{ij}) can be judged from the above and the peripheral pixels of the 3*3 window centered at Z_{ij} are marked as (Z_1, Z_2, ..., Z_8), from which the points whose gray is not in the salt and pepper noise gray range can be selected and also marked as $\{X_1, X_2, \ldots, X_l\}$, $0 \leq 1 \leq 8$; Z_{ij} is the gray value of the noise point ij, and Z_1, Z_2, ..., Z_8 and X_1, X_2, ..., X_l are the gray value of the point around the pixel ij; (3) when there is $l = 0$, there are signal points around the pixel ij, and S_{ij} will not be processed and its value is still equal to itself; (4) when there is $l \neq 0$, there are some signal points around the pixel ij, and now the gray mid-value of the surrounding signal points can be used as the

Table 69.1 The noise reduction effect of pre-filtering on the forman sequence added with salt and pepper or Gaussian noise

The *n*th PSNR (dB)									
	1	2	3	4	5	6	7	8	9
Gaussian noise after pre-filtering	30.353	30.087	29.960	30.164	30.399	30.139	30.263	30.087	30.218
Salt and pepper noise after pre-filtering and processing	30.888	30.496	30.727	30.572	30.496	30.970	30.496	30.806	30.572

Fig. 69.1 The seventh frame image of forman: **a** image after salt and pepper noise experienced pre-filtering; **b** image with 24 dB salt and pepper noise; **c** image after Gaussian noise experienced pre-filtering; **d** image with 32 dB Gaussian noise; **e** original image

gray value of this point; order Zij = median $\{X_1, X_2, \ldots, X_l\}$, in which median $\{X_1, X_2, \ldots, X_l\}$ is the mid-value operator to solve the set $\{X_1, X_2, \ldots, X_l\}$, and this is called as the noise mid-value algorithm as $\{X_1, X_2, \ldots, X_l\}$ is possible signal points. For the Forman sequence, when the Gaussian noise (variance: 36; PSNR: 32.6 dB) or the salt and pepper noise (PSNR: 32.6 dB) is added into it, the result in Table 69.1 can be gained after the pre-filtering [3].

From the above, it can be known that pre-filtering has a large elimination effect on the salt and pepper noise, but it is not obvious for the sequence disturbed by Gaussian noises. After an objective evaluation by PSNR is made, the subjective evaluation can be seen as shown in Fig. 69.1.

69.1.2 Time Domain Filtering

The time domain filtering method in this paper is based on the weighted recursive filtering algorithm of front and back frames. In this algorithm, when the front frame and the frame before the front frame are the images after experiencing pre-filtering, their salt and pepper noise have been removed, but have do not proceed by the spatial domain smoothing method. In selecting the threshold, the video sequence Football within movement can be used for detection for the sake of solving the critical threshold. When it is larger than this threshold ($T1$), it is unnecessary to be processed. Gaussian noise (variance: 36; PSNR: 32.6 dB) can be used for detection to test the sequence input result when the threshold is between 5 and 100. The details can be seen in Table 69.2.

From the above, PSNR value decreases when threshold is too small; the difference between two frames is larger than the set thresholds most until the result of the front frame is output; PSNR value also decreases when threshold is too big, but results of the front frame and the frame before the front frame are still output by the recursive filtering within the threshold range; when the difference between two frames is larger than 18, W is 1 and is weighed with the front frame.

Table 69.2 The time domain filtering effect under different threshold ($T1$)

PSNR after noise reduction	PSNR (dB) after the nth frame noise reduction								Average of 8 frames
	9	10	11	12	13	14	15	16	
$T1 = 5$	31.496	31.349	31.277	31.572	31.970	31.422	31.349	31.206	31.455
$T1 = 10$	32.648	32.422	32.422	32.648	32.648	32.648	32.572	32.137	32.518
$T1 = 15$	33.349	33.572	33.349	33.572	33.727	33.496	33.422	33.277	33.470
$T1 = 18$	34.572	34.349	34.277	34.727	34.807	34.496	34.277	34.206	34.464
$T1 = 20$	33.496	33.277	33.496	33.572	33.888	33.496	33.422	33.206	33.482
$T1 = 100$	27.496	27.277	27.349	27.572	27.727	27.572	27.422	27.349	27.470

Table 69.3 The filtering effect of diffident threshold ($T2$) on gaussian noise

PSNR (dB) after the nth frame noise reduction									Average of 8 frames
	1	2	3	4	5	6	7	8	
$T2 = 1$	32.786	32.737	32.993	32.7856	32.736	32.837	32.837	32.736	32.806
$T2 = 5$	34.637	34.736	34.888	34.686	34.786	34.686	34.786	34.786	34.749
$T2 = 10$	32.686	32.837	32.686	32.737	32.589	32.637	32.540	32.498	32.650
$52 = 15$	31.637	31.492	31.637	31.637	31.637	31.540	31.540	31.493	31.577

Table 69.4 The filtering effect of diffident weighted value ($W1$) on gaussian noise

PSNR (dB) after the nth frame noise reduction								Average of 8 frames
1	2	3	4	5	6	7	8	
$W1 = 0.632.588$	32.540	32.786	32.588	32.686	32.686	32.588	32.588	32.631
$W1 = 0.733.736$	33.940	33.686	33.786	33.67	33.686	33.686	33.837	33.749
$W1 = 0.834.637$	34.736	34.888	34.686	34.786	34.686	34.786	34.786	34.749
$W1 = 0.933.736$	33.888	33.888	33.786	33.837	33.736	33.736	33.837	33.805

Table 69.5 The filtering effect of diffident weighted value ($W1$) on Gaussian noise

PSNR (dB) after the nth frame noise reduction								Average of 8 frames
1	2	3	4	5	6	7	8	
$W1 = 0.632.488$	32.540	32.786	32.588	32.686	32.686	32.588	32.588	32.631
$W1 = 0.532.437$	32.366	32.978	32.856	32.646	32.786	32.826	32.689	32.749
$W1 = 0.433.645$	33.823	33.543	33.886	33.332	33.324	33.243	33.543	33.529
$W1 = 0.331.736$	31.888	31.888	31.786	31.837	31.736	31.736	31.837	33.805

The movement intensity is large when the difference between two frames is larger than threshold, but is small when the difference is smaller than threshold. Test the second threshold from 1 to 17, and assume W is 0.8 (larger than $T2$ but smaller than $T1$) or 0.5 (smaller than $T2$). The domain filtering has a good effect on the stationary sequence, so the stationary sequence News is selected to be added with Gaussian noise (variance 36, PSNR: 32.6 dB) for testing $T2$ so as to solve the best $T2$ value, as shown in Table 69.3.

From the above, the threshold (T2) is 5. Determine the value of $W1$ first, namely, the weighted value when the difference in the pixel brightness values related to two frames is between [5, 15]. To test $W1$, the value range [0.6, 0.9] can be considered. The details are shown in Table 69.4.

The value of $W2$ can be determined from Table 69.5.

Thus, the best value for $W1$ is 0.4. Based on Tables 69.2, 69.3, 69.4, 69.5, the thresholds $T1$, $T2$, $W1$ and $W2$ are determined, and the equation of the first movement detection of the domain weighed recursive filtering can be gained as follows:

$$\begin{aligned} W &= 1, \quad |F_1(i,j) - F_2(i,j)| > 18 \\ W &= 0.8, \quad 5 < |F_1(i,j) - F_2(i,j)| < 18 \\ W &= 0.4, \quad |F_1(i,j) - F_2(i,j)| > 5 \end{aligned} \tag{69.1}$$

In Eq. (69.1), $F_1(i, j)$ is the pixel value of the last frame, and $F_2(i, j)$ is the pixel value of the current frame. Next, the second weighed recursive filtering is on, and the equation is as follows:

$$F(i,j) = (1 - W) * F_1(i,j) + W * F_2(i,j) \tag{69.2}$$

In Eq. (69.2), $F\ (i, j)$ is the result of the domain filtering, namely, the gray value of this point.

69.1.3 Spatial Domain Filtering

69.1.3.1 Edge Detection

First, edge detection can be made. Gauss-Laplacian is also called LOG operator. This operator will carry out a smoothing processing on the original image with Gaussian algorithm before solving edge, to realize the maximum suppression on noise. Two-dimension Gaussian function is used to smooth the image as follows:

$$G(x,y) = e^{-\frac{x^2+y^2}{2\delta^2}} \Big/ \left(2\pi\delta^2\right) \tag{69.3}$$

Its shape is like the clock inverted in the two-dimension space. For this function, Laplace transform can be used, namely, the second-order directional derivative can be taken for Gaussian function, so the function equation of the 2-D LOG operator can be gained as follows:

$$\nabla^2 G(x,y) = \frac{\partial^2 G}{\partial x^2} + \frac{\partial^2 G}{\partial y^2} = \frac{1}{2\pi\delta^4}\left(\frac{x^2+y^2}{\delta^2} - 2\right) e^{-(x^2+y^2)/(2\delta^2)} \tag{69.4}$$

Use this LOG operator to do convolution operation on input image $I(x, y)$ and the output image is as follows:

$$\begin{aligned} F(x,y) &= \int\limits_{\alpha\equiv-\infty}^{+\infty}\int\limits_{\beta\equiv-\infty}^{+\infty} \nabla^2 G(\alpha,\beta) I(x-\alpha, y-\beta) d\alpha d\beta \\ &= \frac{1}{2\pi\delta^4}\int\limits_{\alpha\equiv-\infty}^{+\infty}\int\limits_{\beta\equiv-\infty}^{+\infty} \left(\frac{x^2+y^2}{\delta^2} e^{-\frac{x^2+y^2}{2\delta^2}}\right) * \\ & I(x-\alpha, y-\beta) - 2e^{-\frac{x^2+y^2}{2\delta^2}} I(x-\alpha, y-\beta) d\alpha d\beta \end{aligned} \tag{69.5}$$

Fig. 69.2 LOG operator

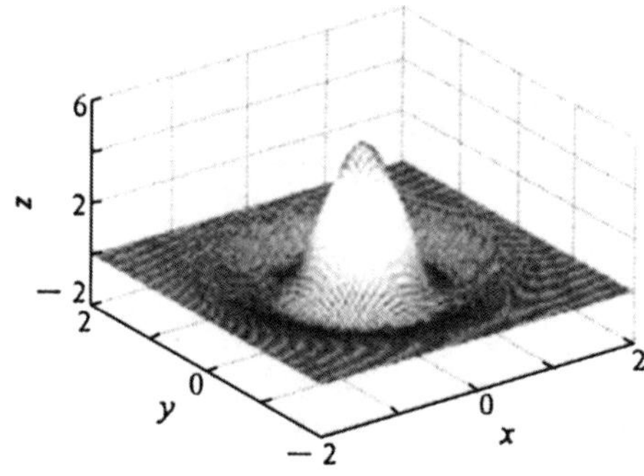

LOG operator is a low pass filtering process, which is used to eliminate the image intensity change when the spatial size is much less than Gaussian space coefficient to reduce noise. Then, use Laplace (∇^2) to gain the second-order directional derivative image $F(x, y)$ of smooth image $I(x, y)$. From the shape of LOG operator in Fig. 69.2, LOG operator is about symmetry of round, and is the same in sizes of edge detection in all directions.

The 5*5 template of the LOG edge detection operator in this algorithm is as follows:

$$\begin{matrix} -2 & -4 & -4 & -4 & -2 \\ -4 & 0 & 8 & 0 & -4 \\ -4 & 8 & 24 & 9 & -4 \\ -4 & 0 & 8 & 0 & -4 \\ -2 & -4 & -4 & -4 & -2 \end{matrix} \tag{69.6}$$

69.1.3.2 Noise Level Detection and Self-Adaptive Filtering

The noise level detection method is based on statistical method. The image without edges should be smooth under no noises. Under no noises, Forman, Football video sequences are detected, and noise detection is increased gradually, gaining cumulative probabilities of gray differences between [20, 256]. PSNR values are in Table 69.6 after sequences experience straight passing, 3*3 center-weighted filtering, 3*3 mean filtering, 5*5 mean filtering, and 7*7 mean filtering.

From the above, the divided four thresholds are $T_1 = 0.7$, $T_2 = 0.5$, $T_3 = 0.2$ and $T_4 = 0.1$. Order P is the cumulative probability of adjacent point brightness difference between [20, 256]. If there is $P > T_1$, 7*7 mean filtering is used; if $T_1 > P > T_2$, 5*5 mean filtering is used; if $T_2 > P > T_3$, 3*3 mean filtering is used; if $T_3 > P > T_4$, 3*3 center-weighted filtering is used; If $P < T_4$, straight output is used, which is based on Weber's Law, and now the noise is in the acceptable range. The algorithms of mean filtering, mid-value filtering and center-weighted filtering have been introduced above. Template of 3*3 center-weighted is as follows:

$$\begin{matrix} 0.08 & 0.08 & 0.08 \\ 0.08 & 0.36 & 0.08 \\ 0.08 & 0.08 & 0.08 \end{matrix} \tag{69.7}$$

Table 69.6 PSNR values after forman and football sequences cumulative probabilities reduce noise when entering different filters or straight under different Gaussian noise

	Gaussian noise variance/PSNR (dB)	Cumulative probabilities	Straight output (dB)
Forman	9/38	0.071	38
	37/32	0.11	32
	81/29	0.21	29
	144/26.5	0.32	26.5
	324/23	0.48	23
	900/19	0.65	19
Football	9/38	0.07	38
	36/32	0.11	32
	81/29	0.22	29
	144/26.5	0.32	26.5
	324/23	0.48	23
	900/19	0.65	19
3*3 canter weighted filtering (dB)	3*3 mean filtering (dB)	5*5 mean filtering (dB)	7*7 mean filtering (dB)
29.4	29	27.6	27
30	31	29.6	28.5
26.3	29.8	28.5	27.3
25.6	27.9	27.5	25.8
25.3	25.8	26.2	25.9
29.4	29	26.3	25.6
29	30.8	29.5	27.8
27	29	28.8	28.5
24.2	25.8	26.2	24
21	22.9	23.3	23.2

69.1.4 Weighted Output

The space-time combined filter is used in this algorithm, and its results are shown in Table 69.7; the Wz is the weighted value of the spatial domain.

From the above, it can be known: $Wz = 0.9$, if $P > T1$; $Wz = 0.8$, if $T1 > P > T2$; $Wz = 0.55$, if $T2 > P > T3$; $Wz = 0$, if $P < T3 > P > T4$; $Wz = 0$, if $P < \mathrm{T4}$. The weighted equation is as follows:

$$Y(x,y) = Y_1(x,y) * (1 - Wz) + Y_2(x,y) * Wz \tag{69.8}$$

In Eq. (69.8), $Y(x, y)$ is the brightness value of space-time outputs; $Y1(x, y)$ is the brightness value of domain filtering; the $Y2(x, y)$ is the brightness value of spatial domain filtering. The final result is outputted with the form of RGB converted by Y component after processed and the U and V component. The conversion method is introduced above.

Table 69.7 Filtering effect of different space–time weighted value *Wz* on different level Gaussian noise

Movement detection in domain filtering(*Wz*/PSNR): $\|F_1(i,j)-F_2(i,j)\| < 18$ (dB)					$\|F_1(i,j)-F_2(i,j)\| > 18$
$P > T_1$ /17	$T_1 > P > T_2$ /21.7	$T_2 > P > T_3$ /28	$T_2 > P > T_4$/ 31	$P < T_4$ /33	
$Wz = 0.9$ 23.6	$Wz = 0.8$ 26.2	$Wz = 0.7$ 32	$Wz = 0.5$ 31.4	$Wz = 0.1$ 33.8	$Wz = 1$
$Wz = 0.8$ 23.5	$Wz = 0.7$ 26.1	$Wz = 0.6$ 32.2	$Wz = 0.4$ 31.8	$Wz = 0$ 33.8	
$Wz = 0.6$ 22.4	$Wz = 0.6$ 25.8	$Wz = 0.5$ 32.2	$Wz = 0.3$ 32.2		
–	–	$Wz = 0.4$ 31.9	$Wz = 0.2$ 32.5		
–	–		$Wz = 0$ 32.7		

69.2 Conclusion

The time domain in this algorithm can keep the edges and details of each frame, and the spatial domain has a good effect on smooth noise reduction. For different speed and sequences with different characteristics, the noise reduction effect is excellent. Subjective evaluation on it is good; the PSNR objective evaluation also suggests it has a good noise reduction effect.

References

1. Xie Z, Bao Z, Xu C, Zhang G, Zhang S, Yang Y (2009) Video pre-processing algorithm based on motion-compensation. Comput Eng 4(19):521–528
2. Li Y, Qiao Y, Gao F, Gao Y, Sun Z (2007) Adaptive temporal filter based on motion compensation for video noise reduction. Chin J Electron Devices 3(05):288–297
3. Song H, Xun Y (2009) Study on the whole variation image noise reduction. J Chengdu Electromech Coll 34(04):379–386

Chapter 70
An Image Stereo Match Algorithm Based on Multi-Area Match

Jiangyan Sun

Abstract Micro computer three-dimensional matching stereovision is an important research topic which opened new research contents of three-dimensional object information to measure small. In this paper, an image stereo match algorithm is proposed based on multi-area match. The proposed match algorithm uses the matching strategy point calculation similarity and matches between pixels and characteristics of pixels. Finally, pixel is chosen as the biggest similarity of matching pixels. Detailed numerical results show the effectiveness of the proposed scheme.

Keywords Color image • Image match • Multi-area match

70.1 Introduction

At present, the research micro assembly operation is very popular. Microcomputer stereovision provides to stereo images operation and micro assembly and is the operation of the eyes and the assembly [1]. Based on three-dimensional matching on the stereo image it is an important step in information acquisition, three-dimensional data reconstruction three-dimensional information, and high precision measurement. Stereo precision of matching the determination and image data reconstruction precision measurement precision plays an important role in stereovision microcomputer [2].

J. Sun (✉)
Modern Education Technology Center of Xi'an International University, Xi'an 710077, Shaanxi, China
e-mail: sunjiangyan@guigu.org

Z. Zhong (ed.), *Proceedings of the International Conference on Information Engineering and Applications (IEA) 2012*, Lecture Notes in Electrical Engineering 220, DOI: 10.1007/978-1-4471-4844-9_70,

70.2 Similar Measurement

In this paper, methods of using the matching strategy point calculation similarity and candidate of match between pixels and characteristics of pixels are used. Finally, pixel is chosen as the biggest similarity of matching pixels. Similar algorithm is a kind of important matching algorithm through calculation of the real matches the selected pixels and the correlation between the feature candidates pixel matching pixel point by point of matching area [3, 4]. In the literature work, single color channels of color images are used to assist scalar type of similarity principle, the establishment of a scalar type stereo matching algorithms. The advantage of the presented method is rapid processing speed, defect is, this algorithm of ignoring the correlation between the color channels, so more false color image matching point points, not matching are deduced. This paper chooses 1-2-standards and clear correlation coefficient. We the exact name of the left and right image in pixels images in the left is called the pixel, pixels characteristics on the pixels is called the candidate game. In order to expand the accuracy of the three-dimensional matching, center of adjacent pixels pixel consists of a pixel set, and the correct pictures, a similar pixel set is choice [5]. The goal of similar algorithm is to establish the correlativity between corresponding pixel sets. If the feature pixel set is $\{C_{lk}\}$, $k = 1, 2, 3\ldots$,M, a candidate match pixel set in the matching area is $\{C_{rk}\}$, they correspond of average color vector respectively is

$$\bar{C_l} = \frac{1}{M}\sum_{k=1}^{M} C_{lk} \tag{70.1}$$

$$\bar{C_r} = \frac{1}{M}\sum_{k=1}^{M} C_{rk} \tag{70.2}$$

The corresponding vector criterion standard respectively is

$$\lambda_{lk}(C_{lk} - \bar{C_l}) \tag{70.3}$$

$$\lambda_{rk}(C_{rk} - \bar{C_r}) \tag{70.4}$$

$$\lambda_{lk}(C_{lk} - \bar{C_l}) \tag{70.5}$$

$$\lambda_{rk}(C_{rk} - \bar{C_r}) \tag{70.6}$$

The average vectors are

$$\bar{\lambda}_l(C_{lk} - \bar{C_l}) \tag{70.7}$$

$$\bar{\lambda}_r(C_{rk} - \bar{C_r}) \tag{70.8}$$

So, we definite the correlation coefficient as

$$\rho = \frac{\sum_{k=1}^{M} |\lambda_{lk}(C_{lk} - \bar{C}_l) - \bar{\lambda}_{lk}(C_{lk} - \bar{C}_l)||\lambda_{rk}(C_{rk} - \bar{C}_r) - \bar{\lambda}_{rk}(C_{rk} - \bar{C}_r)|}{\sum_{k=1}^{M} |\lambda_{lk}(C_{lk} - \bar{C}_l) - \bar{\lambda}_{lk}(C_{lk} - \bar{C}_l)|^2 \sum_{k=1}^{M} |\lambda_{rk}(C_{rk} - \bar{C}_r) - \bar{\lambda}_{rk}(C_{rk} - \bar{C}_r)|^2} \tag{70.9}$$

70.3 The Influence of Matching Areas and Similarity Measurements to the Stereo Matching

The corresponding objects depth stereo matching point points behind is different, and its range is different, unknown. Microscopic surface depth, the greater the fluctuating ranges of large gap, and size distribution of the dynamic changes of the gap between the matching of the area size changes as the gap size. If the size of the area is too small, matching feature points of the matching point reporter can't be in matching area, and then select a the most similar to the matching point pixel matching pixels from the other candidates, cause false match. From actual reconstruction experiment, matching that area is too small, reconstruction surface become plane and greater practical surface of different shapes, because a lot of false match appearance, size, in Fig. 70.1 matching area is too small, matching feature points point a corresponding index four is located outside the matching area feature points, the matching point reporters a can choose match point set in the candidate is given, and the two indices false match. The worst way to miss someone is to describe matching Fig. 70.2 large matching area, a feature points corresponding to the matching point actually four, for matching area, put forward the super large consumer are very similar to the matching point matching points, many similar five with possible noise effect, also there may be similarities of matching point matching point four match points candidate, it will lead to less than 5 worse matching.

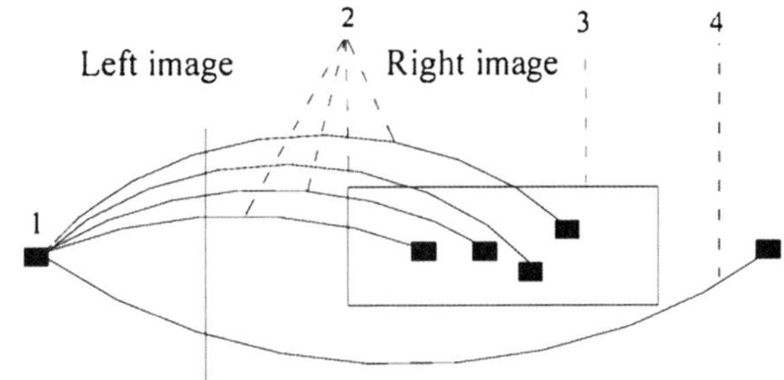

Fig. 70.1 Size of matching area is small, and matching point is outside the matching area

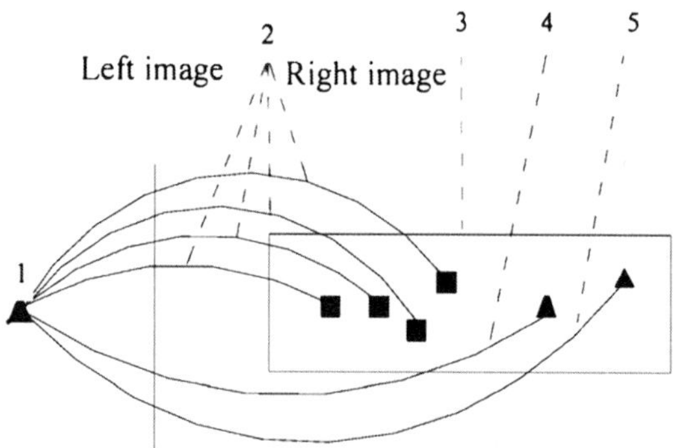

Fig. 70.2 Size of matching is big, and there are more similar matching points

Based on the arithmetic of multi-matching the initial matching areas get samples by increasing the number of matching area, and gave the best matching output of the mean square error of the statistics. This algorithm and the other algorithm is good at single matching area in the performance and stability. But the multi-matching algorithm is used in a similar measure looking for matching point. In fact, different similar may lead to a measurement matching different output the effect of need. A similar measure is an important aspect of influence of three-dimensional matching accuracy and stability. We tried to export matching algorithm based on a similar measure.

The judge in the similarity between the vector of color, people will take similar measures, summarizes the formula of more than 10. Different types of similarity measure method is applicable to different types of image, and adopt corresponding measures, the accuracy of the sensitive to noise in weak field noise conditions, high anti-interference measures, but the game is actually noise low precision, so take similar measures have different stereo matching different performance characteristics. Select and use a similar measure undeserved can also cause the false match results and not matching phenomenon.

70.4 The Algorithm Based on Multi Similarity Measurements

Multi similarity measurement algorithm can combined many kinds of measurements, such as I-norm, 2-norm, 3-norm, infinite-norm, etc. al. If a series of matching output samples are given, X_k expresses the output sample with the *l*, *k*th similar measure. $X_k = \{X_{ki}, Y_{ki}\}$, $i = 1,2,3,\ldots,m$, $k = 1,2,3,\ldots,n$, *i* is the index series number of samples. The optimum matching output can be acquired by three steps: Firstly, samples are input into a filter, and the false matching data and the wrong matching data is

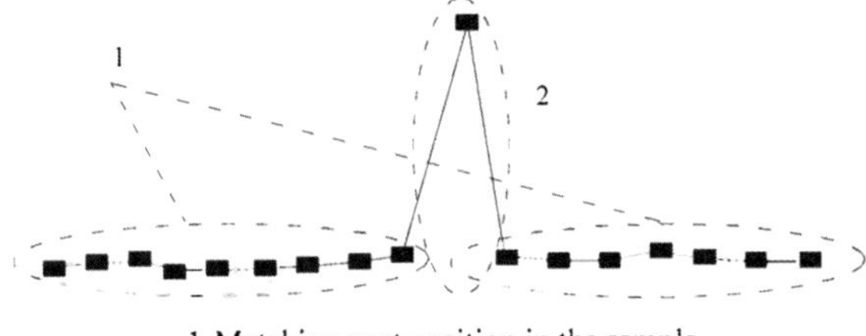

Fig. 70.3 The similar measure causes the impulse noise

filtered. Because size of matching area can be adjusted by software, false matching data is existent, and there is remarkable difference between false matching data and other element data. So false matching data can be seen noises, as shown in Fig. 70.3.

Standard deviation of each element relative to other elements of the sample are calculated.

$$\sigma_{ki} = \left(\sum_{j=1, j\neq 1}^{m} \left[\left(x_{ki} - x_{kj}\right)^2 + \left(y_{ki} - y_{kj}\right)^2 \right] \Big/ (m-1) \right)^{\frac{1}{2}} \tag{70.10}$$

Based on the above function, the corresponding mean-square error samples is obtained, $\sigma_k = \{\sigma_{ki}\}$, $k = 1,2,3,\ldots,n$, $i = 1,2,3,\ldots,m$. And the minimum value of standard deviation set is found, $\sigma_{k\min} = \min(\sigma_k)$, if σ_{ki} satisfies the condition ($\sigma_{ki} \leq T\sigma_{k\min}$ the corresponding data of σ_{ki} is the actual matching data. Then new sample can be obtained, $X_k = \{X_{ki}, Y_{ki}\}$, $k = 1,2,3,\ldots,n$, $i = 1,2,3,\ldots,m_k$. Secondly, different sample X_k represents different level, and under different similar measurement conditions, emerge probability of wrong matching data is also different. It is possible for new sample including bigger different matching data with different similar measurement. By analyzing the difference between levels, filtering bad level and calculating the average value of remaining levels, wrong matching data can be corrected furthermore. The average value is calculated by the function

$$\bar{x}_k = \left(\sum_{i=1}^{m_k} x_{ki} \right) \Big/ m_k \tag{70.11}$$

$$\bar{y}_k = \left(\sum_{i=1}^{m_k} y_{ki} \right) \Big/ m_k \tag{70.12}$$

Based on the above function, the average sample can be obtained as

$$\bar{X}_k = \{\bar{x}_k, \bar{y}_k\} \tag{70.13}$$

If bigger different elements appeared in the sample, the standard square error is used to find the best matching output. Mean square-error of one element relative to other elements for $\bar{X}_k$ is

$$\bar{\sigma}_k = \left(\sum_{j=1,j\neq1}^{m} \left[\left(\bar{x}_k - \bar{x}_j\right)^2 + \left(\bar{y}_k - \bar{y}_j\right)^2 \right] \Big/ (n-1) \right)^{\frac{1}{2}} \tag{70.14}$$

A standard deviation sample is constituted by $\bar{\sigma}_k$, and the sample is

$$\sigma = \{\bar{\sigma}_k\} \tag{70.15}$$

$k = 1, 2, 3,\ldots, n$. The minimum value of the above sample is

$$\bar{\sigma}_{k\min} = \min(\bar{\sigma}) \tag{70.16}$$

If σ_{ki} satisfies the condition $\sigma_{ki} \leq T\sigma_{k\min}$, there is little difference among corresponding data, and data are valid to be kept. Via the above two steps, the matching output sample can be finally obtained as

$$\breve{X} = \left\{\breve{x}_k, \breve{y}_k\right\} \tag{70.17}$$

$k = 1,2,3,\ldots,p$. And the average value of above sample elements is as the final matching result.

The output of stereo matching algorithm is the disparity signal indicated with $d(m)$, $m = 1,2,\ldots,$M. $\hat{\mathrm{d}}\,(m)$ is the estimation of $d(m)$ if noises are considered. We use CMSE to evaluate the performance of a stereo matching algorithm. CMSE is

$$CMSE = \frac{1}{M}\sqrt{\sum_{m=1}^{M}\left[\mathrm{d}_t\,(m) - \hat{\mathrm{d}}\,(m)\right]^2} \tag{70.18}$$

Only use of CMSE to evaluate the performance is not enough, and other mathematical quantities are used too, such as matching effective ratio (S), matching rate of accuracy (PO) and single pixel matching rate of accuracy (PI). Many kinds of threshold are used in a stereo matching algorithm, such as correlation threshold, and these thresholds can filter some candidate matching points. So it is possible for a matching algorithm to owning high precision, but more small amounts of matching points.

70.5 Conclusion

A match algorithm based on multi-similarity measurements and multi-matching areas also have the deficiency, it increasingly consumes the computer resources, is more time-consuming more, and is not suitable for the real-time movement environment.

References

1. Nasser MN (1992) A stereovision technique using curve-segments and relaxation matching. IEEE Trans Pattern Anal Matching Intell 14(5):566–572
2. Fua P (1993) Aparallel algorithm that produces dense depth maps and preserves image feature. Mach Vis Appl 6:35–49
3. Wang Y (1998) Principles and applications of structural image matching. ISPRS J Photogram Remote Sens 53:154–165
4. Quam LH (1999) Hierarchical warp stereo. In: Image understanding workshop, vol 12(3). Sciences Applications International Corporation, New Orleans, pp 149–155
5. Cox IJ, Hingorani SL, Rao SB, Maggs BM (1996) A maximum likelihood stereo algorithm. Comput Vis Image Underst 63(3):542–567

Part VII
Mathematics and Computation

Chapter 71
Research of the Electric Power Universal Service Cost Accounting Methods Based on DET Econometric Model

Tiandeng Chen

Abstract With the continuous development of China's economy launching its well-off society, universal electric power service is becoming increasingly important. However, this process will increase the power enterprises infrastructure construction costs and a decline in revenues. This paper analyzes the practicality of the method of the DEA cost of universal service cost compensation model and uses the model to determine the evaluation and then makes an empirical analysis of selected performance lever, and puts forward appropriate policy recommendations.

Keywords Electric power universal service costs • Cost accounting methods • Policy recommendations

71.1 Introduction

In order to guarantee people's daily life and work on electricity in order that all people can enjoy the country's electricity supply, the government is often forced to intervene in electric power enterprises, prompting the promotion of the universal service [1, 2]. However, now power companies do not have good power generally costing the government a lot in targeted subsidies, universal service, the cost to only rely on themselves to bear, undermining the competitive conditions of the electricity market and thus losing a lot of income [3, 4].

T. Chen (✉)
Taizhou Institute of Economics Taizhou, Taizhou 318000, China
e-mail: tiandengchen@126.com

Z. Zhong (ed.), *Proceedings of the International Conference on Information Engineering and Applications (IEA) 2012*, Lecture Notes in Electrical Engineering 220, DOI: 10.1007/978-1-4471-4844-9_71, © Springer-Verlag London 2013

71.2 Electric Power Universal Services Overview

71.2.1 The Concept of Electric Power Universal Service

The universal electric power service refers to unified power enterprises, where everyone has an acceptable price to provide the same quality of electricity service. The main emphasis in this concept is there are three points [5]. First of all, to ensure uniform quality of service, we charge a uniform price for electricity in universal service without discrimination or differential pricing. Second, the uniform price of service if the vast majority of users are acceptable to recognized, will not be set too high or random prices. Finally, in the conditions of the previous price, to ensure that many people can enjoy the national government to force to promote the universal service discount, especially in remote and economically backward areas.

After the power system reform, the interests of the separation of power enterprises themselves have been unable to support the universal service, the implementation of the main by the enterprise to a combination of business and government. We can see from the definition of universal service, the service is aimed at the object is some of the power system or policy is not in very developed sites and areas.

71.2.2 The Status of the Electric Power Universal Service

In big cities, the grid coverage is always good. In recent years, largely through our government's continuous development and efforts to remote and impoverished areas of the power problem has been resolved, many provinces and cities have achieved the "Power". China has a vast territory, want to completely solve the supply problem, take some time and the accumulation of financial resources. In China, there are still many people without access to electricity, the need for the state to further the implementation of the Electric Universal Service.

In order to achieve this goal, it is necessary for the state and power companies to expand the coverage of the grid, the original grid upgrade, extension and expansion, or local conditions in a number of electricity-less areas of new energy power generation to ensure that every household smooth access to electricity. In the process of infrastructure, the electricity businesses to invest in hardware costs and management costs, and the government should give appropriate compensation, which is also the purpose of universal service costing.

71.2.3 The Importance of Electric Universal Service

In this modern society, the use of electricity is closely related with people's daily lives. No electricity, it means that people not take electric lights, and not take TVs, rice cookers and other household appliances, does not take computers, mobile phones this communication tool, not network, completely isolated from modern life. And the universal service not only extends the power grid, but also controls electricity; provide basic power protection for the residents of those remote and poor areas, to improve their standard of living and quality [6].

In addition to life, the power is with the production operations close contact. Now produce more and pay attention to automation and information to those with power are inseparable from the implementation of the universal service of electricity, but also to promote change in the way of local productivity, and accelerate their socio-economic development and improve local living conditions.

On the political side, the current social crowd phase need attention, such as low-income groups, persons with disabilities, as well as the population located in remote and poor areas, they also need the universal service to improve its production status and improve their quality of life [7]. If can not, you may will social inequality has further widened the social contradictions are growing wider. Especially today's information society, the power of universal service is particularly important for our country to build a harmonious socialist society.

71.3 The Electric Power Universal Service Cost Accounting Methods

71.3.1 Basic Concepts of the Electric Power Universal Service Cost Accounting Methods

To ensure that universal service can be carried out smoothly, it must be accurate accounting of the net cost of electricity due to the introduction of the universal service, which the Government may have targeted compensation for business losses. The difference between the income and input costs is the implementation of the net cost of universal service. Mainly to the cost of the concept of the following two [8]:

1. Historical cost: This concept is mainly focus on the actual cost of the operators, data from the corporate accounting statements, which is also the regulator's cost analysis of the main data sources.
2. forward-looking costs: the cost of application of this concept in order to achieve efficiency pricing, that is, to achieve the maximization of consumer and producer surplus, emphasizing the variability of the cost, marginal and opportunistic.

Different accounting methods correspond to the intensity of the excitation. China's electric power universal service costing methodology should be based on forward-looking cost calculation method. This accounting method excludes the impact of non-efficiency factors on the cost of universal service high-cost areas of the enterprise to reduce costs restraint mechanism. In the current situation, electricity regulatory agencies commissioned by the power grid companies provide to universal service areas that are remote and undeveloped and are not yet the power of the region, hence calculating the cost of electricity to these areas can only be forward-looking cost. Although the accuracy of this approach is not high, taking into account the particularity of the universal service, this method is the most practical. The advantages of this approach are that we can promote grid adoption of new technology to improve the operational efficiency.

71.3.2 The Establishment of Data Envelopment Analysis Model

Data Envelopment Analysis (DEA) analysis method is a most commonly used evaluation model. This model is used to evaluate the overall evaluation of DMU between multiple inputs and outputs.

Assumptions of this model in the same type of s departments or enterprises (DMU); every DMU has m types of input (the cost of resources), and n types of input (the effectiveness of the amount of information), each j-th decision making unit DMUj ($j \in I = \{1,2,\ldots, n\}$) corresponds to the input and output indicators (X_j, Y_j); The vector is expressed as $X_j = (X1_j, X2_j,\ldots, Xn_j)$ T > 0; the output vector is expressed as $Y_j = (Y1_j, Y2_j,\ldots, Yn_j)$T > 0; xI_j is the DMUj ith input, yI_j is DMUj the rth output ($j = 1, 2,\ldots, n$; $i = 1, 2,\ldots, m$; $r = 1,2,\ldots, s$).

Weight means $V = (v1,v2,\ldots, vm)T$, $U = (u1,u2,\ldots, um)T$ and then[9]

$$h_j = \frac{u^T y_1}{v^T x_1} = \frac{\sum_{k=a}^{a} u_k y_k}{\sum_{j=1}^{m} v_j x_j} \tag{71.1}$$

And we establish the model [10]:

$$\begin{aligned} &\max h_j = \frac{u^T y_1}{v^T x_1} = \frac{\sum_{k=a}^{a} u_k y_k}{\sum_{j=1}^{m} v_j x_j} = V_p \\ &st \frac{\sum_{k=a}^{a} u_k y_k}{\sum_{j=1}^{m} v_j x_j} \le 1, j = 1, 2, \ldots, n \\ &v = (v_1, v_2, \ldots, v_m)^T \ge 0 \\ &u = (u_1, u_2, \ldots, u_m)^T \ge 0 \end{aligned} \tag{71.2}$$

It can be concluded that online function of the model is the optimal solution that reflects the relative effectiveness of different types of decision-making units on the model of optimal solution u and v, the objective function $vp = 1$; this decision-making unit develops methods for the DEA to be effective.

71.3.3 Electric Power Universal Service Cost Calculation

We use the DEA C2R model and the MaxDEA5.0 calculation, according to the universal service evaluation indicators, to estimate the universal service performance evaluation results as shown in Tables 71.1 and 71.2.

Table 71.1 Analysis results of the universal service DEA evaluation index value and the relative performance value

DMU	Relative performance values	Input index ($)			
		Gross domestic product (GDP) per capita	Rural residents' annual income	Investment cost of rural power network	Rural power network operation and maintenance costs
City 1	1	22,732	4,832	453,466	32,456
City 2	0.97	24,256	4,945	512,455	46,433
City 3	0.87	31,452	5,536	643,345	78,432

Table 71.2 Analysis results of the universal service DEA evaluation index value and the relative performance value

DMU	Relative performance value	Output indicators (%)				
		Village power rate	House hold power rate	Per capita electricity consumption growth rate	Power supply reliable rate	Voltage qualification rate
City 1	1	100	99.6	8	99.7	98.1
City 2	0.97	100	99.2	6.4	99.6	95.4
City 3	0.87	100	99.3	7.7	99.7	95.3

By analyzing the results we can see that the lowest city relative performance index value, the input costs of electric power universal service, on the contrary, in the power rate and the consumption level is relatively high, so in the implementation of the universal service the effect of high, but its relative performance is the lowest.

By analyzing the results we can see that the lowest city relative performance index value, the input costs of electric power universal service, on the contrary, in the power rate and the consumption level is relatively high, so in the implementation of the universal service the effect of high, but its relative performance is the lowest.

71.4 The Related Policy Proposals of the Electric Power Universal Service Costs

The government executive order to achieve universal service is already in line with China's economic reform, resulting in the level of universal service not high, uneven regional development, and lack of funds, especially in the less developed western regions. These issues have become very prominent. Therefore, the government-led transitional measure is that of establishment of long-term mechanism of universal service fund in the electric power industry reform to provide universal service.

71.4.1 Subsidies Objects and Content of Explicit Universal Service Fund

Power universal service is dynamic; therefore, the object of the universal service may be time changes and changes. However, at a particular time, the universal service also has the relative stability of the strategic objectives of a country at a certain period of time [11]. As a developing country, for China's electric power industry, electric power universal service fund subsidies are mainly targeted at two categories; first, the power of high-cost areas to provide universal service operators, and second the low-income electricity users. Electric Universal Service Fund subsidies can be divided into two phases. The first stage is the compensation cost areas, the cost of electricity operators to provide universal service, to maintain the development potential of the power operator and to achieve the goal of universal service; the second stage, when mature, is to direct subsidies to low-income users to realize the effectiveness of the universal service objectives. Power universal service cost compensation is calculated as [12]:

$$F_i = C_0 \times \theta \tag{71.1}$$

In the equation, F_i represents the amount of compensation in i parts of the power supply enterprise in the implementation of the universal service; C_0 is the selected benchmark area of the universal service net cost; θ indicates performance factor.

71.4.2 Clear the Source of the Universal Service Fund

National fiscal transfer payment In addition to providing a universal service policy, the Government needs to provide financial support. It was mainly through financial subsidies directly to give businesses the universal service fund mechanism established by the financial transfer payment funds into the universal service fund accounts. The proportion of the financial transfer payment can be determined according to the socio-economic development and the state's financial ability to pay.

Power ventures to pay the universal service fund In order to create an open, fair and impartial competitive environment, all power enterprises should fulfill the universal service obligations to pay the universal service fund, according to their income by a certain percentage. To take into account in the development of payment of the proportion of the difference in profitability between different electricity enterprises and sustainable development capacity, so that enterprises can afford.

Social sources Electricity services are necessities of the masses through propaganda, affluent population, and foreign groups to donate funds to the development of China's electric power universal service. Although there is some uncertainty in the way of this to raise funds, but along with economic development and progress of civilization, for the Electric Universal Service Fund to provide support to people concerned about the problem of poor electricity will more and more willing to universal service for electricity Fund support will be more and more.

71.4.3 Improve the Operational Efficiency of the Universal Service Fund

First of all, we can introduce a competition mechanism to improve the efficiency of the operation of the Fund. The distribution phase of the universal service fund introduces franchise bidding, auction format, and the number of enterprises to compete on the electric power universal service entry date of a franchise.

Secondly, the concept of performance management into the daily operation of the Electric Universal Service Fund to assess the effect of the use of the universal service fund, supervise the performance of enterprises to use the funds to improve the operational efficiency of the universal service fund. In addition, the government and the public in accordance with the results of performance evaluation of enterprises run supervision.

71.5 Conclusion

In the course of the development of our socialist society, electricity is a huge problem for people's livelihood. The government should implement universal service as soon as possible, and the power industry should be established as soon as possible with good power universal service cost accounting methods. In accordance with this request the government should give out reasonable subsidies to achieve the maximum results at minimum cost.

References

1. Aliey AE, Kholmanov IN, Whabibullaey PK et al (1999) Study of the thermoelectric power in amorphous and single crystalline lithium tetraborate-Li2O + 2B2O3. Solid State Ionics 118(1/2):111–116
2. Leibensperger RL, Stover JD (1999) Inclusions, rolling contact fatigue and power dense transmission design. Iron & amp; Steel Rev 42(11):90–96
3. Xue G (2010) Selection of clean power technologies for Chinese power industry under the new policy and new power market environment. Power Syst Technol 36(20):1–6
4. Peng C, Wang Z (2010) The origin and development of the telecommunications universal service. Enterp Manag Posts Telecommun 7(11):7–9
5. Xin G (2008) Telecommunications universal service international comparison and inspiration. Commun Bus Manag 435(8):13–15
6. Luo G, Liu Z (2009) China rural electric power universal service analysis. Yunnan Electric Power 46(6):3–4
7. He M (2011) The theoretical model and the behavior of the network consumption, vol 66(3). Heilongjiang People's Publishing House, Heilongjiang, pp 84–115
8. Liu W (2011) Public goods supply of wind power market operating strategy. J North China Electric Power Univ (Beijing) 87(08):11–13
9. Benhuo Y (2009) Power sector reforms in China economics of regulation. Shenyang Inst Eng 24(08):11–13

10. Tian J (2009) Power company performance based on entropy weight fuzzy comprehensive evaluation. Chinese Manag Inf 25(6):55–58
11. He C (2006) Business performance data envelopment model to evaluate the method. Mord Bus 73(4):43–47
12. Wang C (2009) Jiangsu province electricity market segmentation and its application. J SE Univ 13(11):225–227

Chapter 72
Research on Foreign Folk Sports and Cultural Characteristics Based on Fuzzy Mathematical Theory

Shen Yang, Jing Jia and Ying Huang

Abstract Needless to say, folk sports culture is the root of a country's traditional sports, which is particularly important, though folk sports culture from ancient to modern times is not an easy matter. This paper is based on this issue, citing foreign traditional folk sports and their characteristics, and analyzing the course of development of foreign folk sports. Finally, the paper uses the theory of fuzzy mathematics, folk sports, and cultural characteristics with evaluation results showing the development at the leading level.

Keywords Foreign folk sports • Folk sports culture • Fuzzy evaluation method

72.1 Introduction

Folk sports wealth is a treasure house of human culture and the bridge connecting human thoughts and feelings. Whenever on this bridge, different colors, languages, ideas, and personalities of people will be rapid in spiritual intimacy, together amazing, impressing, and blending into the soul. They were generated to motivate the women who struggled with each other to develop them.

Folk sports are one of the most active in folk culture, the most active and most extensive influence, and the most far-reaching of practical activities [1]. Folk sports also gave birth to the main source of all sports. Folk sports are a national culture, and is like a bright pearl in world history. Today, according to sports folk on the rightful place in the sports science and folklore of the two fields of science, people have good reasons to develop and make him shine and glossy glorious for the height of the new re-met.

S. Yang (✉) · Y. Huang
Department of Sports Art, Hebei Institute of Physical Education, Shijiazhuang, China
e-mail: shen_yang12@126.com

J. Jia
Department of English, Hebei Institute of Physical Education, Shijiazhuang, China

Z. Zhong (ed.), *Proceedings of the International Conference on Information Engineering and Applications (IEA) 2012*, Lecture Notes in Electrical Engineering 220, DOI: 10.1007/978-1-4471-4844-9_72,

Here, the author thinks it is artificially essential to distinguish between the terms folk sports and social sports. Folk sports is an integral part of social sports. Widespread among the people, it has a distinctive national style and local characteristics of traditional physical exercise activities. Rich in content, forms, and entertaining, the majority of projects are not subject to restrictions of time, place, and equipment. There is entertainment, folklore, games, performances, and festivals of the features. Folk sports is created by certain people, heritage, and enjoyment, and is integrated and attached to the customs of everyday life (such as festivals, rituals, etc.) into a collective, pattern, traditional life of sports activities, both as a sports culture and as a living culture. Folk sports is an important part of folklore, which is an important part of folk culture. Sports in the folk life is a living culture relying on the customs of people's daily lives (such as festivals, rituals, etc.), passing on a special life and culture, folklore, and sports people daily life organic component.

72.2 Analysis of Foreign Folk Sports Events

This paper lists four folk sports: basketball, table tennis, track and field, and marathon. The origin and history of these folk movements are as follows:

Basketball was introduced in the United States early in December 1891 in the City YMCA International Training School in Springfield (Springfield,), Massachusetts (Dr Springfield College) by James Naismith, the physical education teacher of the school in order to solve the problem of outdoor sports in winter, since football and baseball cannot be played outdoors. He drew on the rules of netball to develop the rules of basketball. There are only 13 basketball rules.

Table tennis is derived from the English in this country; its official English name is "table tennis" and means "table tennis". In 1890, several (navy) officers stationed in India of the British Navy accidentally discovered table tennis which was quite exciting. Later, they switched to the small hollow ball instead of the elasticity of the small solid ball, and the use of timber instead of a racquet, a new kind of "tennis" on the table, which is named after the origin of table tennis. Of table tennis soon became a rage of popular sport [2]. The early twentieth century, the United States began to complete its production of the game of table tennis appliances.

Track and field events were created by the Greeks. According to records, the oldest track and field competition on the first ancient Olympic Games was held in 776 BC in Greece, the Olympic Village. The project was only one—the sprint, the runway for a straight length of 192.27 m. In the 10th Olympic Games in 708 BC, it was formally included in the long jump, discus, and javelin field events. Only men were allowed to participate, woman could not even watch and offenders were sentenced to death.

Marathon, known to ancient folk sports came from the Greek. In 492 BC, in the Persian aggression against Greece, Greek marathons with fewer forces defeated Persia. The Athenian generals in Asia were too eager to pass on the news of their victory to the anxiously waiting people of Athens, so they selected a long-distance running expert Fei Euripides to transmit messages. The long-distance expert, in order to relate to fellow citizens as soon as possible the news of the victory ran hard, and when he went to the central plaza of the city of Athens he was out of breath and cried excitedly: "joy, the people of Athens, we are victorious!". After that his cries faded, he pitched down in the presence of all and did not wake up [3]. To commemorate the victory of the battle and in recognition of the diligence of hero Fei Euripides' achievements, in 1896 the Athenians in the first Olympic Games provided a new competition—Marathon. The distance is the distance of the Marathon to Athens to determine when Fei Euripides through the line for the entire 40 km after 200 m. In 1920, after careful determination, the distance was changed to 42 km after 192 m.

72.3 The Course of Foreign Folk Sports Culture Development

Based on the historical data, we accessed data from 1979 to recent years, and make analysis of more than 30 years; the process of foreign sports folk culture in their respective countries, through this study, in order to further the characteristics of folk culture of sports, the analysis will play a crucial role.

First, we make the analysis of the data from 1978 to 1988 shown in Table 72.1.

Table 72.1 shows in accordance with the geographical location, a summary of the research projects of foreign folk sports and cultural. The table displays European countries and their own folk sports culture which is particularly of great importance, compared to North America which is lagging behind. But the folk sports culture has entered a period of development.

Next is the data statistics after 10 years shown in Table 72.2.

Table 72.1 Foreign folk sports and cultural development during 1979–1988

Years	North America	Europe	Others	Total
1979	7	11	0	18
1980	9	15	5	25
1981	8	11	3	26
1982	10	8	3	21
1983	4	17	4	25
1984	10	14	4	28
1985	6	16	3	25
1986	8	14	5	27
1987	6	11	3	20
1988	5	14	2	21
Total	73	131	32	236

Table 72.2 Foreign folk sports and cultural development during 1989–1998

Years	North America	Europe	Others	Total
1989	5	6	1	12
1990	7	10	1	18
1991	8	11	2	21
1992	9	10	1	20
1993	7	15	2	24
1994	3	19	1	23
1995	6	16	1	23
1996	4	14	5	23
1997	5	13	5	23
1998	8	10	6	24
Total	62	124	25	211

Table 72.3 Foreign folk sports and cultural development during 1999–2008

Years	North America	Europe	Others	Total
1999	6	10	7	23
2000	9	9	6	24
2001	3	12	7	22
2002	10	9	2	21
2003	5	14	4	23
2004	4	19	2	25
2005	9	11	3	23
2006	8	16	4	28
2007	5	15	2	22
2008	10	12	8	30
Total	66	117	45	231

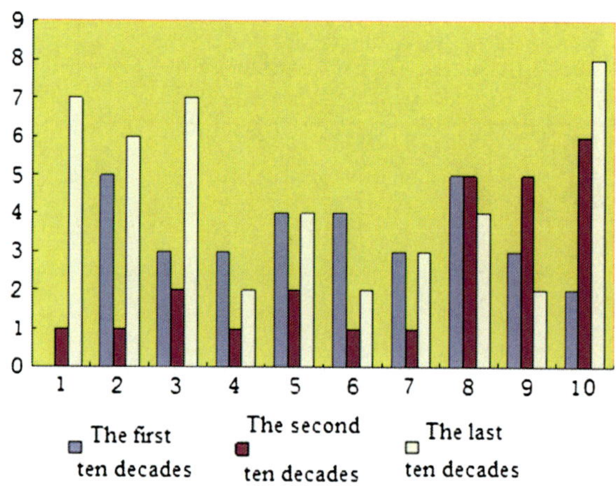

Fig. 72.1 Sports folk developing data in three decades in other countries

The table shows a significant downward trend, foreign sports folklore and cultural development is not so active before, but it's still in good condition (Fig. 72.1, Table 72.3).

The cultural development into the heyday, in all aspects of the development trend of the more prominent, compared to the folk culture of sports in North America and European countries remain unchanged, basically there are no major the magnitude of the change. The following figure shows the development of three decades in other countries sports folk.

72.4 Sports Folk Cultural Characteristics Based on Fuzzy Mathematics Theory

We use the theory of fuzzy mathematics and make sports folk culture features of conversion to digital, through the establishment of the mathematical model to analyze the characteristics of the folk culture of the foreign sports good or bad, the model is as follows:

We order $X^{(0)}$ as he modelling sequence of GM (1, 1),

$$X^{(0)} = (x^{(0)}(1), x^{(0)}(2), \ldots, x^{(0)}(n)) \tag{72.1}$$

$X^{(1)}$ is $X^{(0)}$'s 1-AGO (Accumulated Generating Operation) sequence [4],

$$X^{(1)} = (x^{(1)}(1), x^{(1)}(2), \ldots, x^{(1)}(n)) \tag{72.2}$$

We order $Z^{(1)}$ is $X^{(1)}$'s mean (MEAN) generating sequence [5],

$$Z^{(1)} = (z^{(1)}(2), z^{(1)}(3), \ldots, z^{(1)}(n)) \tag{72.3}$$

$$z^{(1)}(k) = 0.5 \cdot x^{(1)}(k) + 0.5 \cdot x^{(1)}(k-1) \tag{72.4}$$

We define: GM (1,1) 's gray differential equation model as [6]

$$x^{(0)}(k) + az^{(1)}(k) = b \tag{72.5}$$

We set $\hat{\alpha}$ as the parameter vector to be estimated, that is, $\hat{\alpha} = (a, b)^T$, and the column differential equation of the least squares estimated parameters can meet $\hat{\alpha} = (B^T B)^{-1} B^T Y_n$ [7].

In the Eq. [8]

$$B = \begin{bmatrix} -z^{(1)}(2) & 1 \\ -z^{(1)}(3) & 1 \\ \cdots & \cdots \\ -z^{(1)}(n) & 1 \end{bmatrix}, Y_n = \begin{bmatrix} x^{(0)}(2) \\ x^{(0)}(3) \\ \cdots \\ x^{(0)}(n) \end{bmatrix} \tag{72.6}$$

Based on the characteristics of the folk culture of foreign sports, we construct data matrix B and data vector Y_n [9].

$$B = \begin{bmatrix} -\frac{1}{2}\left[x^{(1)}(1) + x^{(1)}(2)\right] & 1 \\ -\frac{1}{2}\left[x^{(1)}(2) + x^{(1)}(3)\right] & 1 \\ -\frac{1}{2}\left[x^{(1)}(3) + x^{(1)}(4)\right] & 1 \\ -\frac{1}{2}\left[x^{(1)}(4) + x^{(1)}(5)\right] & 1 \\ -\frac{1}{2}\left[x^{(1)}(5) + x^{(1)}(6)\right] & 1 \end{bmatrix} = \begin{bmatrix} -4.235 & 1 \\ -7.425 & 1 \\ -10.73 & 1 \\ -14.19 & 1 \\ -17.83 & 1 \end{bmatrix} \tag{72.7}$$

$$Y_n = \begin{bmatrix} x^{(0)}(2) \\ x^{(0)}(3) \\ x^{(0)}(4) \\ x^{(0)}(5) \\ x^{(0)}(6) \end{bmatrix} = \begin{bmatrix} 3.13 \\ 3.25 \\ 3.36 \\ 3.56 \\ 3.72 \end{bmatrix} \tag{72.8}$$

We calculate [10] $\hat{\alpha} = \begin{bmatrix} a \\ b \end{bmatrix} = (B^T B)^{-1} B^T Y_n$

$$B^T B = \begin{bmatrix} 707.46375 & -54.41 \\ -54.41 & 5 \end{bmatrix} \tag{72.9}$$

$$(B^T B)^{-1} = \begin{bmatrix} 0.008667 & 0.094319 \\ 0.094319 & 1.226382 \end{bmatrix} \tag{72.10}$$

$$\hat{\alpha} = (B^T B)^{-1} B^T Y_n = \begin{bmatrix} -0.043879 \\ 2.925663 \end{bmatrix} \tag{72.11}$$

Finally, we get the model [11]

$$\frac{dx^{(1)}}{dt} - 0.043879x^{(1)} = 2.925663 \tag{72.12}$$

$$\hat{x}^{(1)}(k+1) = 69.3457e^{0.043879k} - 66.6757 \tag{72.13}$$

$$(x^{(0)}(1) = 2.67; \frac{b}{a} = -66.6757) \tag{72.14}$$

Finally, the author gives the error variance of each grade range of foreign folk culture, and the results of its features are shown in Table 72.4.

Table 72.4 The results of the features

Small error probability p value	Variance ratio c value	Rating
>0.95	<0.35	Good
>0.80	<0.5	Qualified
>0.70	<0.65	Barely qualified
≤0.70	≥0.65	Failure

72.5 Conclusion

Finally, we compared foreign sports folk culture and sports folk culture, and summed up the characteristics of the folk culture of the foreign sports. From the table we can conclude that foreign sports folk culture is mainly characterized by heaven as opposed to individual-oriented, seeking value, heavy aggressive, chasing executive one, the trend extreme, heavy rational analysis, abstract thinking and risk-taking, and so on.

References

1. Zhang Z (2005) Reflections on the integration of the Western sports culture. Datong Vocat Tech Coll 24(03):22–24
2. Liang M, Yan S (2003) Mixed innovation, integration is progress—from the history of our sports culture changes and the sports culture of the East-West blend of China's future sports development. Educ Pap (HEAD) 52(11):10–13
3. Wang Y, Huang C (2010) Western sports culture preliminary study. Suzhou Vocat Univ 62(01):5–7
4. Wang J (2008) Western sports culture differences and complementarity. China Electric Power Educ 13(18):3–4
5. Liu Z (2010) Conflict and fusion of Eastern and Western cultures in the development of traditional sports analysis. Movement 34(08):17–19
6. Zhang T (2004) Diverse blend of sports culture and the development of traditional Chinese sports culture. Sci Technol Inf 74(06):20–23
7. Wang C, Gan T, Tan L, Huang Z (2010) Parkour sports cultural features and the integration of college sports culture. Inner Mongolia Sports Technol 131(04):66–68
8. Zhou M (2008) Globalization and development of physical culture philosophical reflection. Sci Technol Inf 223(31):34–36
9. Ge Q, Fan C, Gong D (2006) Sports and cultural differences in global perspective and diversification. Sports Sci 42(06):14–16
10. Zhang X (2012) Spring and autumn period, Wu De thinking. Goods Qual 62(S2):57–60
11. Li P (2012) Martial arts education for preschool aggression discussion. Movement 63(04):67–69

Chapter 73
Research on Aesthetic Education in School Physical Education Based on Multiple Linear Regression Method

Shen Yang, Ying Huang and Qiang Luan

Abstract Based on the school physical education situation of the aesthetic education, physical beauty, sports beauty, clothing beauty, language beauty, behaviour beauty, relationship beauty and environmental beauty are researched through 261 college students of 5 universities in the survey. Then according to the different genders and different disciplines, the school aesthetic education in physical education is counted, and differences testing experiment is given. Finally, the above factors are calculated by using multiple linear regression method.

Keywords Physical education • Aesthetic education • School • Multiple linear regression method

73.1 Introduction

Aesthetics is not affiliated to any other subjects; and the "natural beauty", "art beauty", "society beauty" are the "things" that give "aesthetic" concept; beauty in things are ahead of human thinking and existence. Beautiful feelings do not need philosophical analysis and thinking activities, and people want to express this feeling. They will start to think of procedure by selecting expression, vocabulary and so on. It appears to be conceptual issues, so that "beauty" does not "want to" come out.

Sports aesthetics in physical education is very important. Because of different ideologies, all sports aesthetics research object, method and result have bigger difference. In our country, aesthetic education in school physical education research is a developing stage [1]. There are many shortcomings; there are many areas in need of improvement. Based on the following physical beauty,

S. Yang (✉) · Y. Huang
Department of Sports Art, Hebei Institute of Physical Education, Shijiazhuang, China
e-mail: shen_yang12@126.com

Q. Luan
Physical Education Office, Langfang No.12 Middle School, Langfang, China

Z. Zhong (ed.), *Proceedings of the International Conference on Information Engineering and Applications (IEA) 2012*, Lecture Notes in Electrical Engineering 220,
DOI: 10.1007/978-1-4471-4844-9_73, © Springer-Verlag London 2013

sports beauty, clothing beauty, language beauty, behaviour beauty, relationship beauty and environmental beauty, a survey of the students and some advice are given.

73.2 Aesthetic Education in School Physical Education Research

This is achieved by the following 261 university students of five universities. At present, all countries have to do aesthetic education reform and strengthen aesthetic education in physical education in various solutions in the world. In the twentieth century, beautiful psychology has a new theory: a perfect human thinking should be the summation of perceptual thinking and rational thinking. Aesthetic education makes emotional liberation and sublimation of people. Because it is the rationality and human life communication; and it can create a sound personality [2]. Investigation of specific is shown in Tables 73.1 and 73.2:

For the school sports in aesthetic education research, the author thinks that physical beauty, sports beauty, clothing beauty, language beauty, behaviour beauty, relationship beauty and environmental beauty will play a vital role in aesthetic education. Therefore, in the process of PE teaching, teachers should strengthen aesthetic education in physical education in the following aspects [3]:

(1) Strengthening of the physical appearance of practicality and aesthetics. Any physical appearance should be practical, such as building makes life for the people living in it comfortable and convenient, practical crafts

Table 73.1 The distribution of students

University	Hubei university	Northeastern university	Shandong university	Central China normal university	Anhui university	Total number	Percent (%)
Male	14	24	40	30	27	135	51.72
Female	32	23	17	31	23	126	48.28
Total number	46	47	57	61	50	261	100

Table 73.2 Student subjects distribution

University	Hubei university	Northeastern university	Shandong university	Central China normal university	Anhui university	Total number	Percent (%)
Art	25	0	25	60	19	129	49.43
Science	21	47	32	1	31	132	50.57
Total number	46	47	57	61	50	261	100

should make people feel as one wish. Utility is the basis of aesthetic, aesthetics can also enhance the usefulness, and the two things promote each other.

(2) Outstanding physical performance and the beauty form. Performance is an important feature of the physical appearance, also can be made different with drama, fiction, film and other reproduction art. The representation and the beauty forms are inseparable (Table 73.3).

Then, the author makes a difference test; the results are shown as follows: (Table 73.4)

(3) Strengthens our country of nationality and era. A practical sport shows the combination of nationality and the modern time. The physical aesthetics shows a strong national flavour and ethnic characteristics.
(4) Attention to sports science rhythm and rhyme beauty. Rhythm is the most basic and important means of sports action fluctuation change expression. It is not only can express a certain content and emotion, but also can form dance sports beauty.

Then, the author through a survey of 261 college students in 5 universities, male and female students, art and science students are compared in this chapter [4, 5]. Details are shown as follows:

Where, $P < 0.05$, P is significant, $P < 0.01$, is highly significant.

Then, different subject college students' information is investigated, the results are shown as follows: (Tables 73.5, 73.6)

Where, $P < 0.05$, P is significant, $P < 0.01$, is highly significant.

Table 73.3 Male and female college students in the aesthetic education need

Gender	Physical beauty	Sports beauty	Clothing beauty	Language beauty	Behaviour beauty	Relationship beauty	Environmental beauty
Male	45.86 ±22.07	55.09 ±12.98	41.76 ±22.98	49.14 ± 16.27	59.85 ±15.24	54.68 ±18.66	51.11 ±18.23
Female	51.88 ±24.64	63.86 ±16.87	43.16 ±24.18	41.57 ±17.75	53.77 ±17.47	57.14 ±18.17	53.05 ±20.67

Table 73.4 Male and female college students in the aesthetic education difference test

Factor	MD	F	Sig
Physical beauty	2360.04	4.33	0.04
Sports beauty	2587.34	10.30	0.001
Clothing beauty	162.87	0.23	0.64
Language beauty	4857.24	17.30	0.000
Behaviour beauty	1791.84	6.48	0.01
Relationship beauty	517.08	1.52	0.22
Environmental beauty	240.71	0.63	0.14

Table 73.5 Effect of different disciplines of sports aesthetic education

Gender	Physical beauty	Sports beauty	Clothing beauty	Language beauty	Behaviour beauty	Relationship beauty	Environmental beauty
Science	46.72 ±25.04	57.89 ±17.47	45.13 ±23.57	42.47 ±16.59	54.24 ±16.31	57.67 ±17.34	54.63 ±20.77
Art	50.81 ±21.79	60.35 ±14.64	38.17 ±23.27	48.55 ±17.39	58.746 ±17.11	53.85 ±19.33	49.45 ±18.47

Table 73.6 Male and female college students in the aesthetic education difference test

Factor	MD	F	Sig
Physical beauty	1069.05	3.46	0.16
Sports beauty	469.32	1.94	0.18
Clothing beauty	1904.24	1.78	0.07
Language beauty	2578.57	9.26	0.003
Behaviour beauty	1067.14	3.78	0.05
Relationship beauty	896.74	2.65	0.11
Environmental beauty	1849.37	4.73	0.03

73.3 Aesthetic Education Based on the Multivariate Linear Regression Method

Multiple linear regression and linear regression can use statistical test with the significance of regression equation, can also use P value method (P Value) as a test. F Statistic is:

$$F = \frac{\text{MSR}}{\text{MSE}} = \frac{\text{SSR}/p}{\text{SSE}/(n-p-1)} \tag{73.1}$$

When H_0 is true, $F \sim F(p, n-p-1)$. The significance level α is given. Check the F distribution table of critical value $F_\alpha(p, n-p-1)$, F_0 of F is calculated. If $F_0 \leq F_\alpha(p, n-p-1)$, accept H_0. Namely, in the significant level α, we think linear relationship of y and $x_1, x_2, \ldots, x_p$ is not significant; if $F_0 \geq F_\alpha(p, n-p-1)$, this linear relationship is obvious. Use P value method for significance test is very convenient: this P value is $P(F > F_0)$. Use of computer is easy to calculate the probability. Many statistical softwares (such as SPSS) are given a test value P, which eliminate the censored distribution table trouble, for a given significance level α, if $p < \alpha$, H_0 is refused, instead, H_0 is accepted [6].

On linear correlation conditions, two or more independent variables change on a dependent variable quantity. It shows a relationship among numbers of mathematical formulas that is called the multivariate linear regression model. This chapter is based on multiple linear regression method to study aesthetic education. Established model is described as below:

Let y be an observable random variables, it is impacted by P non random factors $x_1, x_2, \ldots, x_{\mathrm{p}}$. y and $x_1, x_2, \ldots, x_{\mathrm{p}}$ has a linear relationship:

$$E(y) = \beta_0 + \beta_1 x_1 + \cdots + \beta_p x_p \tag{73.2}$$

where, $\beta_0, \beta_1, \cdots, \beta_p$ is $p+1$ unknown parameters,ε is unmeasured random error, and it is usually assumed as $N(0, \sigma^2)$. We call formula (73.2) is multiple linear regression model. y is known as explanatory variable (the dependent variable), $x_i (i = 1, 2, \ldots, p)$ is explanatory variable.

$$E(y) = \beta_0 + \beta_1 x_1 + \cdots + \beta_p x_p \tag{73.3}$$

According to the sports aesthetic education research, the multiple regression equation is established. First, the unknown parameters $\varepsilon_1, \varepsilon_2, \ldots, \varepsilon_n$ are estimated. For which, we have n independent observations for getting n group of sample data $(x_{i1}, x_{i2}, x_{i3}; y_i)$. They meet formula (73.2).

$$\begin{cases} y_1 = \beta_0 + \beta_1 x_{11} + \beta_2 x_{12} + \cdots + \beta_p x_{1p} + \varepsilon_1 \\ y_1 = \beta_0 + \beta_1 x_{21} + \beta_2 x_{22} + \cdots + \beta_p x_{2p} + \varepsilon_2 \\ \cdots\cdots \\ y_1 = \beta_0 + \beta_1 x_{n1} + \beta_2 x_{n2} + \cdots + \beta_p x_{np} + \varepsilon_n \end{cases} \tag{73.4}$$

where, $\varepsilon_1, \varepsilon_2, \ldots, \varepsilon_n$ are independent of each other and are subject to $N(0, \sigma^2)$.

Formula (73.4) can be expressed in matrix form:

$$Y = X\beta + \varepsilon \tag{73.5}$$

where, $Y = (y_1, y_2, \ldots, y_n)^T, \beta = (\beta_0, \beta_1, \ldots, \beta_p)^T, \varepsilon = (\varepsilon_1, \varepsilon_2, \ldots, \varepsilon_n)^T$, $\varepsilon \sim N_n(0, \sigma^2 I_n)$, I_n is n order unit matrix.

$$X = \begin{bmatrix} 1 & x_{11} & x_{12} & \ldots & x_{1p} \\ 1 & x_{21} & x_{22} & \ldots & x_{2p} \\ \ldots & & & & \\ 1 & x_{n1} & x_{n2} & \ldots & x_{np} \end{bmatrix} \tag{73.6}$$

By the formula (73.4) and the properties of the multivariate normal distribution, Y obeys dimensional n normal distribution; its expectation vector is V, variance and covariance matrix is $\sigma^2 I_n$, i.e. $Y \sim N_n(X\beta, \sigma^2 I_n)$.

Finally, by using of the Kendall coefficient matrix, physical beauty, sports beauty, clothing beauty, language beauty, behaviour beauty, relationship beauty and environmental beauty of the 7 factors of value are shown in the following formula (73.7):

Table 73.7 Results

Factor	Physical beauty	Sports beauty	Clothing beauty	Language beauty	Behaviour beauty	Relationship beauty	Environmental beauty
Physical beauty	/	31	6	14	21	27	12
Sports beauty	1	/	5	7	14	18	17
Clothing beauty	26	27	/	27	6	15	29
Language beauty	18	25	5	/	24	26	32
Behaviour beauty	11	18	16	8	/	7	15
Relationship beauty	5	14	17	6	25	/	23
Environmental beauty	20	15	3	0	17	9	/

$$
\begin{aligned}
\frac{\partial Q(\hat{\beta})}{\partial \beta_0} &= -2\sum_{i=1}^{n}(y_i - \hat{\beta}_0 - \hat{\beta}_1 x_{i1} - \hat{\beta}_2 x_{i2} - \hat{\beta}_p x_{ip}) = 0 \\
\frac{\partial Q(\hat{\beta})}{\partial \beta_1} &= -2\sum_{i=1}^{n}(y_i - \hat{\beta}_0 - \hat{\beta}_1 x_{i1} - \hat{\beta}_2 x_{i2} - \hat{\beta}_p x_{ip})x_{i1} = 0 \\
\frac{\partial Q(\hat{\beta})}{\partial \beta_k} &= -2\sum_{i=1}^{n}(y_i - \hat{\beta}_0 - \hat{\beta}_1 x_{i1} - \hat{\beta}_2 x_{i2} - \hat{\beta}_p x_{ip})x_{ik} = 0 \\
\frac{\partial Q(\hat{\beta})}{\partial \beta_p} &= -2\sum_{i=1}^{n}(y_i - \hat{\beta}_0 - \hat{\beta}_1 x_{i1} - \hat{\beta}_2 x_{i2} - \hat{\beta}_p x_{ip})x_{ip} = 0
\end{aligned}
\tag{73.7}
$$

The final calculated results show science students of sports aesthetic education teaching environment is not as good as art students. Between male and female, the 7 factors—physical beauty, sports beauty, clothing beauty, language beauty, behaviour beauty, relationship beauty and environmental beauty—has little difference. It shows the sports aesthetic education teaching environment can be basically regardless of gender issues [7] (Table 73.7).

73.4 Conclusion

Investigation is given based on the 261 college students of the 5 universities in this chapter, and then multiple linear regression method of the results is analysed. Gender does not influence sports aesthetic education teaching environment; and the subject will influent on sports aesthetic education teaching environment. Thus, this is sure to affect the education, the school physical education teacher should distinguish science and art students.

References

1. Yang Z (2011) Are sports an art—basic discussion of late twentieth century western sports aesthetics. Sports Sci 24(2):47–49
2. Wang S (2011) Physical concerns: sports aesthetics and contemporary mission. J Phys Educ 2(2):57–60
3. Yan W (2010) Sports aesthetics origin overview and development. Private Sci Technol 42(09):32–35
4. Jiang C (2006) Analysis on modern sports aesthetics aesthetic connotation of cultural philosophy. J Cult 56(10):7–10
5. Hu X (2008) Review of sports aesthetics study. J Phys Educ 11(10):17–20
6. Jin W (2008) Sports aesthetics in school physical education. J Cult 46(12):35–38
7. Li Y (2007) The social demand and the development of sports aesthetics. J Chengdu Sport Univ 36(04):63–66

Chapter 74
Analysis on Indigenous Psychology Based on *T*-test Statistical Regularity

Meidan Liu and Yujuan Sun

Abstract Indigenous psychology is getting more and more attention by the modern society, but indigenous psychology research from all walks of life are limited in natural science point of view, and many of them are the analysis of the theoretical importance of indigenous psychology, significance, and direction. This paper is based on psychology culture turn from the perspective of the indigenous psychology to the analysis of development trend, the cultural turn is not only attention to the nature but also to the people's psychology and behavior, making it more profound understanding of and through, based on a review of the related literatures, the cultural turn to indigenous psychological effect on the development of research hypotheses, combining to the quantitative data analysis and coming to the conclusion, in order to instruct the development of indigenous psychology, making the indigenous psychology can combine to local culture and Humanistic culture and get better and more sustainable development.

Keywords Cultural turn • Indigenous psychology • Quantitative • Development tendency

74.1 Introduction

With the continuous development of society, social deep indigenous psychology research attracted all walks of life's attention in the society, but to the research of local psychology, it also is the theoretical analysis of the importance and significance in the majority, and much of the research on the basis of western psychology, or is only for the local psychological analysis through the psychology of ancient Chinese, not really internalize the western psychology methods to use or the comparative analysis, the real practical characteristics of native to the analysis

M. Liu (✉) · Y. Sun
School of Humanities and Sciences, Northeast Agricultural University, Harbin, China
e-mail: meidan_liu@126.com

Z. Zhong (ed.), *Proceedings of the International Conference on Information Engineering and Applications (IEA) 2012*, Lecture Notes in Electrical Engineering 220, DOI: 10.1007/978-1-4471-4844-9_74,

of cultures are rarely, research of localization are marching for, in the research topic, theoretical guidance and methods lacking the original analysis [1, 2].

Cultural turn will fill cultural factors into indigenous psychology investigation, not only provide a novel and broader view, make the indigenous psychology research more in-depth and comprehensive, be helpful for you to have more intuitive understanding to objects, nature, and main body of the psychology research, also let everybody to internalize the essence of the culture in psychology category, be helpful for people more close to reality to know about the mutual intrinsic relationship of cultural features and indigenous psychology, be helpful for the indigenous psychology research on cultural diversity and methods of the integration and effective comprehensive development [3]. This paper is a psychology of the orientation of culture in perspective, putting the local traditional cultural factors into the psychology of the research, combined with proper effective quantitative analysis way, further establishing indigenous psychology development paradigm, the construction of indigenous psychology development system of culture orientation, and analyzing the development trend of the indigenous psychology [4].

74.2 Theory and Research Method

Indigenous psychology is based on the traditional culture of each country and get the comprehensive system psychology, but also is the change and use of the concept, mode, and method of western psychology. Psychology and culture will be closer together, and further to apply to the local culture, to supplement each other. Indigenous psychology is making up of theoretical issues, the methodology and discipline system of the localization of the content. And they need more issues area shows that embodies the way of the local people, most local humanism features and psychology behavior [5]. The localization of the theory requires from the local century, which will be able to reflect the situation of Chinese characteristics and traditional culture and unique psychology got considered. The localization of method not only refers to the model and method of western psychology and joins special Chinese native way combining two for one. Subject system will be planned and arranged combining the local education way to curriculum content, time, training plan and methods, such as academic activities, organization form.

Variable data access method of the psychology have evaluation, experimental psychology, and cognitive neuroscience paradigm (see chart 1), this paper is to use measurement, combined with the form of questionnaire measuring, for a regional personnel divided into several sample group that is the control group and the investigation on the data obtained for the quantitative analysis of the data by statistical software (Fig. 74.1) [6].

This paper is from the perspective of indigenous psychology of cultural orientation to study the development trend of psychology culture, which need the quantitative evaluation analysis for influence of indigenous psychology of culture orientation, according to the literature material of psychology set the

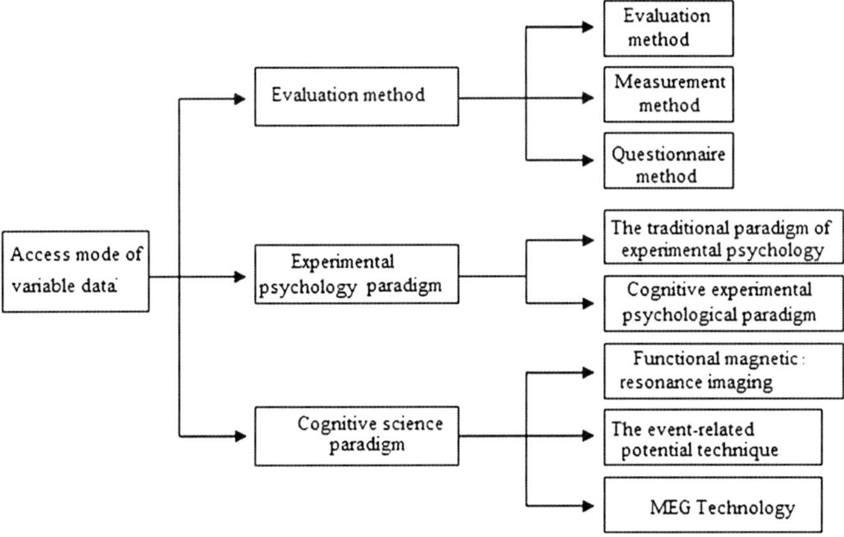

Fig. 74.1 Access method figure of psychology variable data

representative factors index such as motivation, psychological change degree and behavior degree to five individual table, cultural orientation variable selected six index factors to exploratory factor analysis, through the investigation and get the data score, and statistics influence degree from the local orientation for culture psychology. Thus, we can take effect quantity (with *d* said) to determine [7].

$$d = (\overline{\chi}_1 - \overline{\chi}_2)/s_2,\ S = \sqrt{\frac{1}{N}\sum_{i=1}^{N}(\chi_i - \mu)^2},\ \text{S for standard deviation.}$$

According to correlation statistical analysis method of two samples variable has four (see Table 74.1). Use coefficient square of point two column related (show by r^2_{pb}) [8].

$r^2_{pb} = \frac{(t)^2}{(t)^2 + df}$, df is two samples variable degrees of freedom that is for $n_1 + n_2 - 2$.

Table 74.1 Related statistical testing method of sample variables

Sample variable type	Statistical analysis
Continuous	Pearson product-moment correlation, regression analysis
Type (two) continuous	*T* test, two column related, point two column related
Type (three above) continuous	F inspection, the series related (rank)
Type	The $\chi 2$ test, and all relevant contingency table

74.3 The Present Situation of Indigenous Psychology and Data Analysis

Indigenous psychology is humanistic psychology on the basis of the psychology and gets the local behavior model, richer and enriches the content of the psychology and areas. It was originally the study of Chinese local people from the perspective of the psychology in the west, in order to analyze the local personage psychological guidance and the development characteristics and behavior situation; next is combined to China's actual culture, economic and social background, the thorough analysis of Chinese people's psychological condition and behavioral development situation on the reform. It now enters into the new study stage of the comprehensive theory, method, and idea of the localization. We must overcome the concept confusing, completely devoid method of western psychology, etc., which only pay attention to the object and neglect the content of the indigenous psychological research.

Use exploratory factor analysis to analyze factors indicators of the sample data of cultural orientation (see Table 74.2), from six index factors of the cultural orientation on two aspects of exploratory factor analysis. Through the analysis of SPSS which is statistical software, when freedom DF for 300, p value is less than 0.01, under the condition of the factors index overall Chi square spherical inspection is 773.872, this shows that the index of culture orientation in the range of 0.01 is significant relationships, so the index factors of culture orientation can offer factor analysis. And the total variance level which culture orientation factors can explain was 73.051 %.

Based on the sample survey data from the Table 74.3 and Fig. 74.2, it is known that, the changes of the data except for Mean deviation are 5 in the different

Table 74.2 Exploratory factor analysis of psychological cultural orientation

	The first factor	The second factor
The first factor: relational orientation		
1. More like a atmosphere of harmonious relationship to work	0.765	
2. Like	0.598	
3. Look, harmony in all things	0.668	
Second factor: the family orientation		
1. Have a close relationship with his family		0.871
2. Have a strong sense of responsibility to the family		0.613
3. Try hard for personal ideal		0.785

Table 74.3 Capacity change data tables of T sample inspection

Samples	Mean deviation MS	Standard deviation S	*T*	DF	*P*
14	5	0.81	1.38	29	0.18
34	5	0.81	2.08	67	0.07
134	5	0.81	4.10	267	0.002

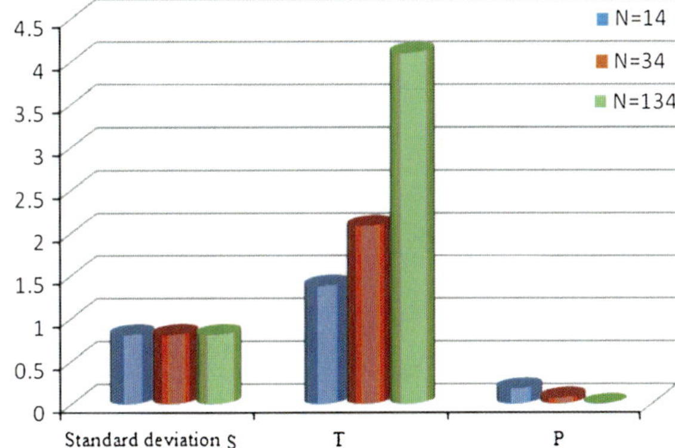

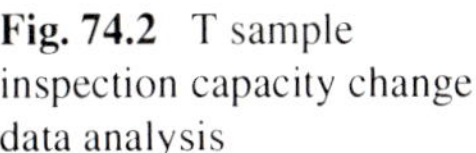
Fig. 74.2 T sample inspection capacity change data analysis

sample size, while the standard deviation having no change is 0.81, the rest have change. With the increase of sample size, numerical value of T value and DF value increase, but the increase amplitude is different; the growth of DF is very big. But along with the increase of the sample p value is decreasing. This fully explain the sample selection should be as much as possible, and representative, the more real, accurate to get the influence degree of cultural orientation for the development of local psychology.

According to the Table 74.4, the correlation coefficient of the development of cultural orientation and indigenous psychology is 0.76, when t value is for 11. *P* value is less than 0.01 level, which analyzed data can build the structure model between variables.

According to the Table 74.5 and Figs. 74.3, 74.4, the analysis results of psychology culture orientation in indigenous psychology influence show that the orientation of culture is 32.6, the mean error is only 2.6, the effect size *d* is 0.828 when *p* value is 0.015, saying psychology culture orientation can explain the nearly 83 % of indigenous psychology development variance, also reflect that the cultural orientation have a big influence to the development of the indigenous

Table 74.4 Analysis of related coefficient development of cultural orientation and indigenous psychology

	Mean	Standard deviation	
Cultural orientation	32.6	0.81	(0.76)
The development of local psychology	29.5	0.83	

Table 74.5 The results of analysis of psychology culture orientation in indigenous psychology influence

	SS	DF	MS	F	P	D
Cultural orientation	33.5	320	32.6	8.95	0.015	0.828
Error	30.0	9	2.6			
Sum	45.2	329				

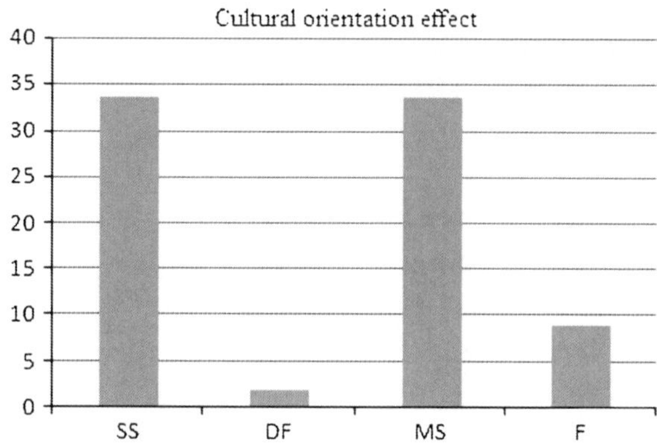

Fig. 74.3 Analysis of psychology culture orientation to indigenous psychological impact

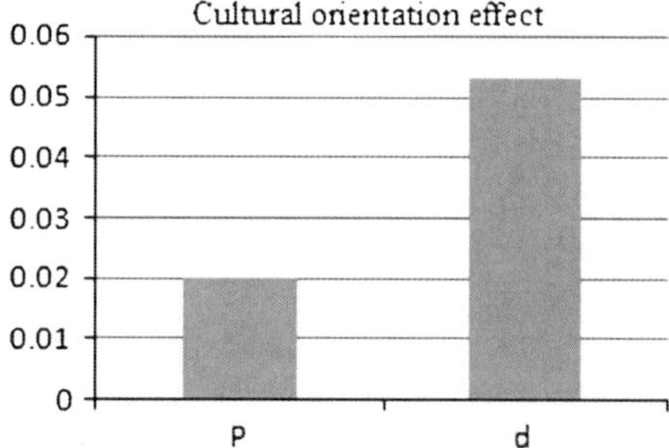

Fig. 74.4 Analysis of psychology culture orientation to indigenous psychological impact

psychology, for the local psychology research it has practical significance, so need to take psychology culture orientation seriously.

74.4 Conclusion

Research on culture orientation for indigenous psychology can be small to analysis of local humanistic psychology and compared with the western and have a contact with the international. The development trend of the indigenous psychological need further strengthens combination of Chinese and western cultures from the theory, the concept and method provide the development way of indigenous psychology.

The first is the core as the research content, using western psychology to thick live essence, seeking common ground while putting aside differences, fully the advanced guidance to the volatile, creating effective way belongs to local.

The second is to fully combine with own conditions; join local traditional national culture ideas, to constantly be enriched, perfect and more Chinese characteristics. China has a long history of culture, the broad and profound thought knowledge; this is also the advantage of Chinese indigenous psychology. Having the local culture, it is not only inheritance and innovation for their culture, but also rich with the ancient culture and ideology, make its continuous to carry forward.

The third is that apply comprehensively concept and principle of dialectical materialist to undertake thought and method of guidance. The localization of the psychology requires not only its actual conditions, and also need to fill foreign outstanding cultural principle into the traditional national culture, to get the good method, using to the study of the local psychology and behavior development, which will comprehensive fully from local humanities, economic, political, and other environmental factors to solve the problem of the indigenous psychology and realize the characteristics of the localization psychology.

Acknowledgments The research was supported by the scientific research initiation fund project of Northeast Agricultural University with the name revising the Indigenous Psychology under the Trend of Psychology Cultural Turn, and the project number QNW2011.

References

1. Haiying J (2009) The development trend of culture to psychology. Soc Scientists 46(11):37–39
2. Ma Y (2010) Methodology of contemporary psychology culture turn. Psychology agent new no 24(2):49–51
3. Ma C (2011) The reflection and ahead of China local psychology research. J Hexi Inst 342(3):76–77
4. Jin Z (2011) Contributing factors analysis of South Korea foundation education achievement-based on South Korea indigenous psychology research. Mod Educ Sci Gen Res 467(6):32–34
5. Li X, Jiang Y (2010) Concerning meaning of the Chinese culture psychology and localization. J Jixi Univ 12(3):53–55
6. Kuang X, Duan B (2011) Scientific psychology research thinking. J Gansu Radio TV Univ 223(12):31–33
7. Mo L, Wang R, Caiqi C (2010) System analysis and system reconstruction of the psychology research method. Psychol Sci 29(5):102–103
8. Xiao X, Yu J, Jiang Y (2011) Understanding models from psychology data. Comput Sci Explor 96(3):68–71

Chapter 75
Solution of Algebraic Theory Model

Shaorong Liu

Abstract Usually, we use the arithmetic to build correlation equations for solutions when we encounter problems. This paper builds problem-solving model through the use of theory model of higher algebra, which are the research methods for problems based on algebra, to establish model matrix of relevant algebraic and response relationship of logic problems, and to formulate a concept of mathematical symbols system (MSS), to interpret problems in the process of the teaching, and to bring target of problems into the specific teaching of management; it lets students become supervisor of problem solving, concretely analyze and establish algebra theory model of related problems for problem-solving method.

Keywords: Algebraic model • Mathematics teaching • Equation • Solution method

75.1 Introduction

In the late 1980s of the twentieth century, Eugenio de filloy developed a kind of methodology and for mathematics education research. In this method and theoretical framework which played a central role in mathematics teaching, it is the study of one local theory model (LTM) describing the idea of organization. The model describes the fact that the content of phenomenon produced in the process of some special mathematics teaching and learning will involve some special students, the purpose of the model is adequate for the observed phenomena in the special environment.

In fact, this model is derived from a thing. Not only that, if it occurs, this phenomenon will be observed on the basis of the characteristics of the model. Therefore, the model is descriptive, explanatory, and predictive, but do not exclude

S. Liu (✉)
Liuzhou City Vocational College, Liuzhou 545002, China
e-mail: liu_shaorong@163.com

Z. Zhong (ed.), *Proceedings of the International Conference on Information Engineering and Applications (IEA) 2012*, Lecture Notes in Electrical Engineering 220, DOI: 10.1007/978-1-4471-4844-9_75, © Springer-Verlag London 2013

that the observed phenomena can be a means of explanation, prediction description in different ways: different model. One of the main reasons of this theory is the phenomenon explained which cannot be observed from the design expected and not establish the appropriate analysis program in mathematics education or related disciplines on the basis of the general theory, such as psychology, pedagogy, sociology, history, epistemology, or linguistics.

75.2 The Symbols of Mathematical System

In consideration of the phenomenon above, we found that it is also very important that it should be considered as a symbolic school algebra sign system, learning algebra and the symbol system before or the relationship of an intermediate language level, thereby incorporating a concept of system of mathematical symbols (MSS).

The phenomenon used in mathematics is not the nature of all languages, which makes it not better to use as markers of a term or concept, it belongs to linguistics, therefore speak referred is casual, instead of using the term "expression". Relatively it is also very convenient, because it is used to using "algebraic expression" or "arithmetic" to refer to the corresponding written form in mathematics.

However, in the phenomenon of some important facts, there are no isolated signs stressing (either mathematics or in any text) it may be concealed. This is a very common language description for the mathematical text to differentiate, the two events as a composition of strict mathematical signs and some other vernacular language phenomenon. As a whole, Symbol system must be a mathematical description of system, rather than the description of the phenomenon, because the system is responsible for the meaning of words. Therefore, "system of mathematical symbols" which must understand the term is regarded as symbols of mathematical signs or mathematical systems, which is the system about the nature of mathematics, and not just the individual phenomenon.

In addition, Filloy described the concept of mathematical symbol system in 1990, which was widely used [1]. It is a service tool to analyze the text produced by student, when they teach in the school system; these texts of mathematics are regarded as the results of production process and the analysis of history mathematics texts. When they take these math texts as the object of study, rather than the so-called ideal text as performance or text of "the language of mathematics", they measured symbol system of concept mathematic and text which must open in all directions. Therefore, when we must say that system of mathematical symbols or its corresponding code, the possibility of social convention, it has generated symbol function and has established relevant teaching function, and intend to use in their preaching artifacts i. But also it must be considered that the notation system or a class of learners produced in order to give the significance put forward of their teaching mode, although they could be subjected to influence of social phenomena, it has yet established a corresponding quality system.

Mathematicians and students in the field repeated the experiment process in the read/convert text of mathematics with space for text. In particular, from this teaching model perspective, it creates meaningful text sequence as TS to read/conversion into other TSS for learners. In addition, combined with a wide range concept of MSS and the concept of generalized, part of text, mathematical text is the so-called texts written of mathematical language (symbol). In contrast, mathematical text the stratification of production using means of system of mathematical symbols, the substance expressed has the difference.

75.3 The Problems of Arithmetic Algebraic

Any procedure of indication, usually in teaching or solving word problems, it puts forward translation algebra MSS in textbooks. To some extent, this is also what we call the Cartesian method. In Descartes' method, algebra MSS makes it possible to leave the meaning referred to in the statement of their natural language problem, so as to achieve the transformation, under algebraic equations of the way of the solution or equation (S) description. It uses the expression level a natural language. Let us put the informal methods aside, such as trial and error, conduct arithmetic with method which students like very much, such as continuous analytical reasoning (MSAI) and (AMSE) continuous exploration of analysis method. These methods are formally described as follows: sequential analysis inference method (MSAI) is a classic analysis of method to solve the problem. In this method, the statement of the problem was regarded as "reality" or "possible state of description of the world", so this text transformation, through the analysis of the sentence, i.e., Using "fact" is effective in any possible world. Describing actions of these sentences constitute the logic solver; "likely", recognized until the conversion used as solutions. In use of continuous exploration analysis method (AMSE) solver, especially in data exploration for analysis of the problem, we can found the relation between the quantities in the movement and thereby solve the problems.

Mathematical discovery in the book of the "Cartesian mode" rewrites the rules of Descartes, in such a manner, which can be seen as the principle about using algebra MSS to solve problem. Descartes' rules used by Polya are as follows [2]:

(1) First of all, to have good determination of understanding of problems and reducing some unknown quantity;
(2) Use the natural way to survey questions, in the appropriate order to regard it as the relationship of between problem solving and visualization, according to the conditions, it must hold the unknown data;
(3) Separate part of condition, you can express under two different ways in the same number, and therefore obtains unknown number of the equation. It finally should be divided into conditions of many parts, and thus to obtain the equation system as much as possible;
(4) To reduce an equation or equation system.

Problem of text reading/transformation is the core in this process, using MSS algebra in natural language expression, namely an equation expression. Text reading/conversion of the process of problem requires a special behavior, which extracts from its number and their relationship. Extract semantic statement of this group number and exploration in the field of a set of number and relationship, which is enough expressed as search equation of algebraic SMS (or as many non equivalent equation, named for the number events). This text reading/conversion reduction and enlargement: it reduces the amount of text and relations, expands the text volume of text read and not mentioned explicitly in a statement of relationship issues. We call the "analysis of reading", which is a special reading/conversion. Therefore, through converting text of the analysis of problems in reading into a group of number and relationship, that is to say, we need to translate the algebra (SMS phenomenon, in which only system of the quantity and relation.) In form, CM is divided into the following steps [3]:

(1) Analysis of the problem, converting it to the list quantity and relation quantity.
(2) Choose one letter (or several different letters) and specifical number (or several number).
(3) Write algebraic expressions to specify other number, used the letter in the second step, found the relationship of the analysis of reading in the first step.
(4) Write an equation (or launch independent equations of event number in the second stage), written algebraic expression of the same number in the third step (non equivalent).
(5) Equations are transformed into standard form.
(6) Application formula and algorithm solution, in order to standardize the form of the equations.
(7) Explain question result in the statement.

The selected CM split into various steps, described the ideal subject behavior, the elements of step method for solving word problems (algebraic) ability, in this sense, it is a formal competency model. In fact, students mainly administrate the methods to solve the problems of users, they first through a brief reflection phase, assessing whether they can foresee the method of the steps, namely write a logical symbol frame with their situation. In this logic, we will determine how to use the outline of semiotics to problem-solving strategies in the process of students using arithmetic in MSAI or in CM algebraic.

75.4 The Equivalence Classification Method of Algebraic Classification

The equivalence classification method of Algebraic Classification is an approach of system classification to algebraic problems of t using algebraic properties which have very big effect on in-depth understanding of arithmetic system structure.

First, it should construct the equivalent matrix to question: equivalence relation of problem-solving method, which can be divided into n + 1 type in accordance with equal in rank all, kinds of representative element is [4]:

$$\left[\frac{Ir}{0}\right], \text{(r = 0, 1, ..., n)} \tag{75.1}$$

Using equivalence relation to classify to space, which represents element is F^n, the type is [5]:

$$1/2(n+1)(n+2) \tag{75.2}$$

So, represent element is [6]:

$$x_p^2 - x_{p+1}^2 - \cdots - x_r^2 (0 \leq r \leq n, 0 \leq p \leq r) \tag{75.3}$$

At the same time, the establishment of algebraic problem solves one-dimensional equation [7]:

$Ca + Cr = Cp, Dp \times Cp = T, Da \times Ca = Cda, Cda = T.$

Construct structure diagram for Fig. 75.1:

Figure 75.1 shows that using the specified Network relationship, any solution is that start through the analysis of the reading, in which have unknown number, because there is no edge, it does not necessarily mean that the algebraic relation type is unknown. This means leading to the algebraic solution the analysis of reading. However, this does not mean that this problem is algebra in essence. In the fifth section, we will find that in two process of the analysis of problem solving, in order to solve the problem, students have different analytical reading for the same problem and produce different networks relationship, allowing arithmetic operation to get the solution, only having one unknown number, because there are edges and path passing from these edges problem. In this case, analysis of reading allows the arithmetic solution. Arithmetic or algebraic qualification, therefore, the network relation is by the analysis of reading, not the problem.

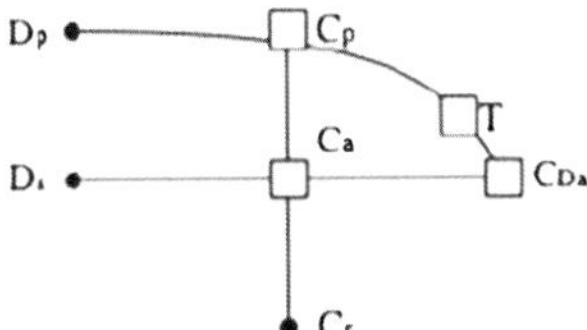

Fig. 75.1 The diagram of algebraic problem structure solving

75.5 Optimization Solution of Algebraic Model

Whether feasible and expected solution would use algebra of the MSS or not, it determines, to a great extent, the remaining stages of algebraic or arithmetic properties and the whole solution process. The above is specific management they know and control practices (such as the implementation of solving, for example, the use of expressions and movements of review of the systematic in each step).

Therefore structural formula the optimization model for [8]:

Dp + Dm = Da, Ca + Cr = Cp, CDp + CDm = CDa, Dp × Cp = T, Dp × Ca = CDp, Dp × Cr = St, Da × Ca = CDa, Dm × Ca = CDm, CDm = St, CDa = T.

This volume and the network of the relationship can be shown in Fig. 75.2, which explain the number and relations put forward in the process of the whole solution from the solution in space.

The correct solution in the finals, they only used quantitative relation which contributes to the establishment of equation [9]:

$$x = \left(\frac{x}{57} - 113\right) \tag{75.4}$$

But the logical formula always follows the logic symbol outline, from the beginning, operating the unknown and known quantities they expected in the same way. This can be described as an version of algebraic model method, which consists of three steps and their action plan is as follows: (a) The establishment of the unknown x; (b) Exanimate appearing number in the statement of questions and in the narrative stories, and the calculate numerical value or write them down, refers that the algebraic expression is unknown, until an equation sum. (c) Solution of equations, Origin and the lack of analysis or analysis of quantitative relationship may be bad students, causing the error calculation or write the number of algebraic expressions of the proper relationship which is failure respond, but responding incorrect quantity relation.

It can be seen from Figs. 75.3 and 75.4 that although finding the error on their own, using a J was from the beginning of the algebra in the short-term solutions, and they do not leave this arithmetic or trial and the error strategy from this process. As they are process management, so you can find and correct the error possibility. The students have been reviewing the instruction given and controlling their

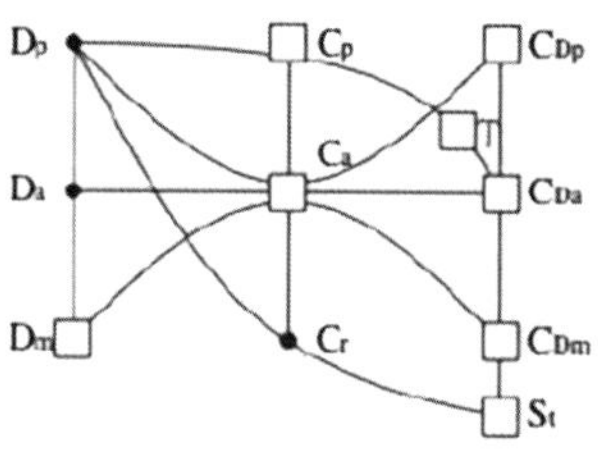

Fig. 75.2 Network graph of the optimization quantity and relation

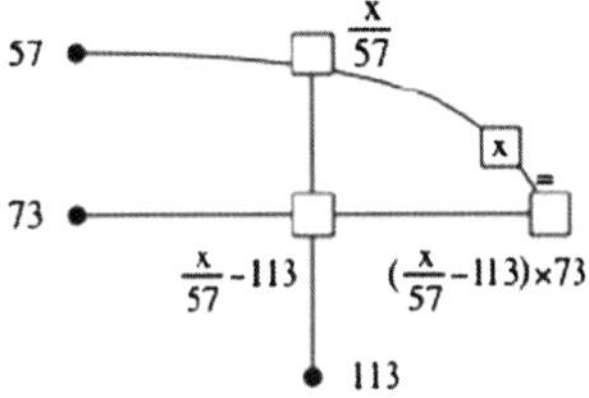

Fig. 75.3 Relationship diagram of logic equation number

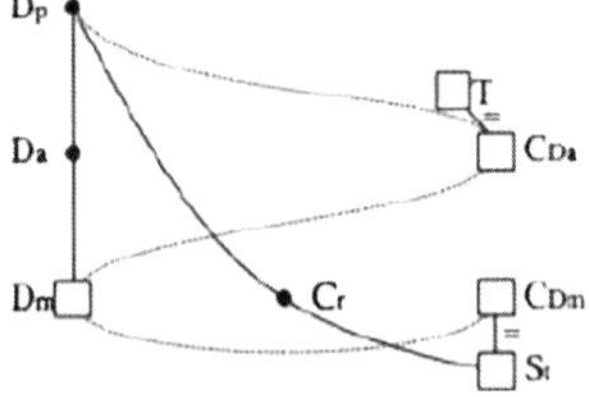

Fig. 75.4 Relationship diagram of algebraic problem model

expression. By resorting to the very examination to control meaning expression and description, the use of the vernacular name express this meaning. Through the joint efforts and cooperation, students are forced to reach a shared meaning, thus preventing them from unnecessary work in grammar.

75.6 Conclusion

The target brought into a teaching mode of problem solving and specific management, control, make students find their own practical action, in order to develop the steps of problem solving which is required is, when let students become competent of problem solving, and even change their solution strategy or method, in the whole process they control, it shows their efforts has not met success. Although a systematic review and correct his steps, all the students still select the solution path on each from the beginning, even though they had been looking instructions of a different solution program. The use of the algebra MSS and analysis of graph corresponding to plot, we can see the algebraic properties of methods of problems solving. That is to say, the algebraic properties of the solution is not just from their choice of MSS which represent relationships of logical symbol contour, but also from these relationships, operation of the unknown amount also appeared the network character.

However, new number and relationship are added to their first analysis of reading in a series of trial and error modification, these new relationships generated network, which includes possibilities of arithmetic solution, but its initial logic, semiotic outline is new possibilities opened up and maintains the algebraic path. In this case, the algebraic properties of logic symbol throughout the entire solution path. However, at any time, they calculate the unknown number. Only reflection in the verification process, they can be finally corrected, but does not leave the arithmetic reasoning behind.

References

1. Wang Z (2009) Methodology of mathematical thinking and teaching, vol 35(2). Higher Education Press, Beijing, pp 445–447
2. Bao X, Liu H (2010) Research on the theoretical system of linear algebra. Eng Math 64(12):73–75
3. Wang J (2009) Theoretical framework of Fuzzy algebra. Pure Math Appl Math 24(9):22–25
4. Wei Y, Lin C, Tian L (2011) Description of adaptive workflow model method in Process Algebra. Chin J Electron 11(11):79–82
5. Tuo H (2010) Formation and cognitive function of mathematical symbol system. J Northeast Norm Univ 173(6):11–13
6. Liu J (2010) Analysis of linear algebra problems and model, vol 53(09). China Agricultural Press, Beijing, pp 622–624
7. Jia X (2011) Analysis of students' learning problems and solutions. Reform and opening up in 123(2):67–69
8. Tang J, Sun Z (2009) Exploration and practice in teaching reform of higher Algebra. J Fuyang Norm Coll (Nat Sci Ed) 242(02):31–34
9. Xue Y (2009) Urriculum construction and practice of "higher algebra and analytic geometry". J Higher Educ 134(04):101–103

Chapter 76
College Students Education Based on Multivariate Statistical Analysis

Guo Dongyan and Jia Gao

Abstract Party members council system is an important part of Party members to continue education in college students, it relates to the play of the Party vanguard and exemplary role. By analyzing the current situation of the education of Party members council of university students, Build the Education Convergence index system, and analyzed the result of Education Convergence by using multivariate statistical analysis, strengthen the scientific and reasonable of Party member education evaluation, In order to improve the Party member quality of college students.

Keywords: The principal component analysis • The quality of Party members • Education convergence

76.1 Introduction

Party members of college students are the outstanding elements. It is an important issue of College Students' Party about how to do the education of Party members and appraisal work, improving their awareness constantly and maintain the Party member's advanced. Construction of the education of Party members appraisal index system on the analysis of the present situation education evaluation based on Party members in this paper, and introduce multivariate statistical analysis to the Education Convergence Explored more scientific and reasonable method of the Party member education evaluation, improving their awareness, enhance the vanguard and exemplary role of Party members, promote the continuing education of Party members [1].

D. Guo (✉) · G. Jia
Hebei United University, Tangshan, Hebei, China
e-mail: 43698059@qq.com

Z. Zhong (ed.), *Proceedings of the International Conference on Information Engineering and Applications (IEA) 2012*, Lecture Notes in Electrical Engineering 220,
DOI: 10.1007/978-1-4471-4844-9_76,

76.2 Education Convergence Status of College Students Party Member

Currently, universities increasingly attach importance to the work of Party construction, especially in the carried out to maintain the advanced education. After thorough study and practice of the scientific concept of development, more clearly recognizes the leading role of Party-building work, See the achievements of College Students' Party, and should also be a sober analysis of the shortcomings, continuing education of Party members, especially education of Party members and appraisal work is relatively weak affected the improvement of student Party Members.

76.2.1 Focus on the Development of Party Members and the Expense of Education Still Exists

The education of Party members reviewed the degree of attention is not enough. In the college students' Party construction, Party member development has always been concerned and attention, it is very strict specification from the requirements to the procedures of the development, layers of screening, step-by-step checks, to ensure the quality of Party members. However, after joining the Party, it has been relaxed that continuing education students deficiencies in the supervision and evaluation of the student Party members. For example, the probationary period of a probationary member examines only a formality. Theoretical study of the Party members catch was also in-depth. For party members and exemplary function requirement of play is not high enough and so on makes lack rigid requirements on the management, supervision, and evaluation of the education council of the Party members, Weakened Party spirit training, the theoretical level, and exemplary role.

76.2.2 The Relative Lack of a Reasonable Evaluation Mechanism and Lack of Scientific Research on the Education of Party Members Reviewed

In the current Party members continuing education, often confined to the theoretical study of the branch or participating in the theme of practical activities that the arrangement of the next higher Party organization, Not only limit the Party members to carry out continuing education, but also constraints more widely of the play of the Party member function, this led to the council idea of the education of Party members is not enough open. Education of Party members reviewed scientific research was not deep enough. Emergence of the universal education of Party members consultative way is a personal summary of the branch council or simple classification assessment. There is no good taking into account the inside and outside environmental factors of the entire Party branches and individual factors of Party members, There is no effective grasp and the relationship between the new era of Party members specific requirements of the Party and the law of growth of college students Party members. There is no scientific integration of the traditional ideological and political education model and modern management methods and so on.

76.2.3 The Relative Importance of Qualitative Evaluation in the Deliberations, There is Lack of Qualitative and Quantitative Research in the Education of Party Members

Due to the impact of the traditional education of Party members review mode, the education evaluation to Party members and individual result is often generate through the personal summary, criticism and self-criticism. Listen to the masses of Party members feedback and recommendations or other forms. So in the traditional education evaluation on Party members in given is qualitative evaluation and feedback. In the development of Party members appraisal index system of education, it is necessary to make full use of scientific theory of modern management and statistical knowledge, promote the education of Party members review the validity and reliability, Play an exemplary role of Party members in learning, life and work, and other fields.

76.3 The Initial Building of System of Education Convergence Indicators of Party Member in College Students

The education council of the Party members involved in a wide range of content is a relatively complex system [2]. It is necessary to consider all aspects of the Party members in the Education of Party members reviewed the indicators, and with high requirements in considering the reality operability. We solicit the views of some Party building experts also to do research on special topics students of Party members and ordinary students, on the basis of considering a variety of opinions and suggestions. Construct the evaluation index system indicators of education of Party members. According to the Party to the requirements of the Party members, combined with the actual characteristics of the student Party members, mainly to consider the thinking quality, professional quality, play the role of three aspects, in each of the main content consists of specific appraisal indicators, comprehensive appraisal the overall quality of Party members (Table 76.1).

By setting up three types of large targets and 15 small indicators. Clear the specific requirements of Party members in the continue education sector. Not only the requirements of the Party's ideological and political Party members to resolve student Party members see tangible indicators, always has the general direction of the ideals and beliefs, but also quality requirements of the Party on Party members to resolve the specific content and specific objectives of university personnel training. Finally, in the embodiment of aspects of the Party member to tease out the role of the four aspects, making the student Party members have a clear value orientation in the life, learning and work, and implement the council of the education of Party members, to become a benchmark in the hearts of every Party member [3].

In the formal education of Party members reviewed, departure from the comprehensive consideration of Party branches inside and outside environment. Select a Party member self-assessment, peer assessment, teacher evaluation, student council, and other forms. In order to further sort out the weight of indexes, the greatest degree of assurance of science and objectivity of the appraisal results, introduced the multivariate statistical analysis on the appraisal results, using principal component analysis for data processing, and in accordance with the overall factor score for quantitative evaluation; re-use system clustering method for qualitative analysis.

Table 76.1 Party education convergence indicators

Thinking quality	1. Consciously strengthen the theoretical study and constantly improve the political awareness and theoretical training. 2. Firm political direction, adhere to the principle conscious and the CPC central committee consistent 3. Consciously abide by the rules and regulations, law abiding. 4. Honest and trustworthy, self-discipline and lenient toward others. 5. Actively participate in branch activities, to fulfill its duties and obligations. 6. Unite students, helpful, good civilized way.
Professional quality	1. Clear learning objectives, attitude upright, attend class on time and learning diligent 2. Smoothly through the computer, english test, a good academic performance. 3. Good professional basis, actively involved in research contest, and continue to broaden the professional perspective. 4. Good habits, and actively create a civilization bedroom. 5. Actively participate in collective activities, focusing on the improving the overall quality of the individual.
Effect play	1. Have a holistic perspective, and plays an active role in the school and stable work. 2. Actively complete arranged work. 3. Actively play party influence, driving around students with learning and common progress. 4. Actively play the exemplary role of Party members, and enthusiastic service for students.

Through the building of index system, the definition of review mode analyzes the results using the multivariate statistical analysis, to achieve a combination of qualitative and quantitative, to avoid the man–made interference of subjective which determines the index weight, so that Party members Education Convergence more targeted, operability, council more comprehensive, scientific, and rational. At the same time, with consultative indicators, the direction of the efforts of student Party members clear, targeted, and plays an active role of exemplary vanguard role of Party members play. Through studying the appraisal system of education of Party members, it is a useful attempt for the objective management of the education of Party members. Good practice and exploration for College Students Party Member of the Education Convergence, which provides a new way of thinking to strengthen the Party members continue education.

76.4 Case Study of College Students Party Member Education Convergence

Extract a branch of the appraisal results as a sample, there are 10 Party members of council participated in the branch, in which the classes have 29 students and 3 of the faculty involved in evaluation. Evaluation is divided into four grades: excellent, good, in general, and poor, which were assigned points 9.5, 7.5, 6, and 3, calculate the index average score of the consultative Party members, In the appraisal calculations, It is necessary to have selection process of evaluation form, removal the incomplete, arbitrary evaluation form of the evaluation, make the results more realistic response of the actual performance of the Party members (Data are summarized in Table 76.2).

Principal component analysis is a method which multiple indicators into one or several of the composite indicator. Basically, retained the vast majority of information that reflected in the number of indicators, and then through the new indicators to achieve the purpose of analysis [4, 5].

In this paper, take advantage of the DPS software for standardized processing of the raw data. First calculate the correlation coefficient matrix, and then calculate the characteristic roots and eigenvectors, variance contribution rate (Table 76.3) and the principal component loading matrix (Table 76.4), and so on. Then variance contribution rate of each principal component as the weights. Construct comprehensive evaluation function of the main ingredients, and come to the main components, integrated factor score, and ranking (Table 76.5).

The variance contribution rate is a measure of the relative importance of each factor indicators. The size of the variance contribution rate shows all of the main components of relative importance [6]. Generally, think of main components of accumulation contribution rate above 85 %, and then few principal components can represent the original more than most of the indicators. Table 76.3 shows cumulative contribution is 95.782 % of the first two principal components of the statistical data.

Table 76.2 Party education convergence subindicator data sheet

Appraisal index	XS01	XS02	XS03	XS04	XS05	XS06	XS07	XS08	XS09
Index B1	9.15	7.05	8.03	8.65	7.5	8.54	8.08	7.17	8.36
Index B2	9.2	6.9	7.65	8.5	6.68	9.2	8.79	6.82	8.52
Index B3	9.22	7	8.36	9.05	6.88	9.35	8.5	7.65	8.45
Index B4	8.88	7.1	8.15	8.9	7.75	9.2	8.4	6.98	8.85
Index B5	8.85	7.05	8.2	9	7.85	9.15	8.3	7.1	8.9
Index B6	8.76	6.96	8.03	8.6	6.9	9.35	8.25	7.06	8.52
Index B7	9.08	7.45	7.71	8.65	6.75	8.76	8.8	7.77	8.5
Index B8	8.78	7.25	7.55	8.35	6.88	8.65	8.1	7.45	9.08
Index B9	9.02	6.85	7.95	8.45	7.25	9.15	8.25	6.84	8.15
Index B10	9.05	7.16	7.61	8.8	6.65	8.45	8.2	7.29	8.3
Index B11	8.98	6.75	7.82	8.4	7.15	8.73	8.35	6.71	8.35
Index B12	8.8	6.9	7.9	8.35	7.06	8.78	8.2	6.89	8.32
Index B13	8.92	7.36	8.13	8.65	7.35	8.84	8.42	7.08	8.48
Index B14	8.86	7.25	8.05	8.55	7.08	8.78	8.38	7.27	8.5
Index B15	8.95	7.18	8.15	8.68	7.26	8.92	8.45	7.32	8.57

That is, keeping most of the information of the original indicators and has significantly with representative.

Table 76.4 shows, all variables have a considerable positive load in the first principal component that changes in each indicator equal effect on the first principal component, It can clearly be seen a composite indicator, and the second principal components and index B7 have larger positive related that is the attitude and Party members have bigger index association, seen the second principal component is a learning attitude factor.

Table 76.3 Principal component eigenvalue, contribution rate and the cumulative contribution rate

Number	Eigenvalue	Contribution rate (%)	Cumulative contribution rate (%)	Number	Eigenvalue	Contribution rate (%)	Cumulative contribution rate (%)
1	13.835	92.233	92.233	6	0.060	0.397	99.563
2	0.532	3.550	95.782	7	0.033	0.222	99.785
3	0.246	1.639	97.421	8	0.020	0.132	99.917
4	0.156	1.039	98.460	9	0.012	0.083	100.000
5	0.106	0.707	99.167				

From the Table 76.5, the first principal component and factor rankings are consistent, large difference with the second principal component sorting. According to the level of the factor score can determine the performance of the pros and cons of student Party members. Therefore, the overall performance optimal is student XS01, student XS06 followed by, and students XS02 is worst performance. According to the second principal component scores, students XS08 learning attitude is the best, students XS05 attitude is the worst.

Table 76.4 The first and second principal components loading matrix

Serial number	F1	F2
Index B1	0.2567	−0.211
Index B2	0.2634	0.1056
Index B3	0.258	0.1034
Index B4	0.2538	−0.409
Index B5	0.2503	−0.4496
Index B6	0.2617	0.0157
Index B7	0.2386	0.6137
Index B8	0.2446	0.2224
Index B9	0.2616	−0.165
Index B10	0.2541	0.2607
Index B11	0.265	−0.1244
Index B12	0.2654	−0.0298
Index B13	0.2657	−0.0732
Index B14	0.2656	0.1381
Index B15	0.2665	0.0437

Table 76.5 The principal component of the review of student party members, comprehensive factor score and ranking

XS01	4.4963	1	0.4383	4	4.162616	1
XS02	−5.4795	10	0.6104	3	−5.03222	10
XS03	−0.8459	7	−0.5026	8	−0.79804	7
XS04	2.7506	3	−0.0273	6	2.535984	3
XS05	−5.1896	9	−1.5053	10	−4.83994	9
XS06	4.2082	2	−0.2179	7	3.873602	2
XS07	1.2766	6	0.6773	2	1.201483	6
XS08	−5.0942	8	1.1255	1	−4.65857	8
XS09	2.1468	4	0.0804	5	1.982905	4
XS010	1.7307	5	−0.6787	9	1.572181	5

76.5 The Thinking of Education Review in Party Members of College Students

First, the Party members' education review is the effective exploration of strengthening the Party members' objectives management. In this paper, on the basis of extensive research to build the system of Party members' education evaluation, and to be verified by the examples, then establish the sound education review mechanism for the student Party members and take effective measures for Party members to education review. The education review of the Party members to strengthen objectives management of Party members, and to play an active role in giving full play to the advanced nature of the student Party members is useful to explore and try.

Second, Party education review index system needs to improve constantly. Education of Party members in the review system in existence, the proportion of qualitative and quantitative need to be adjusted, the review index need to further refine. Reality operability needs to keep on strengthening. The set of the review index needs the scientific concept of development as guidance and adhere to people oriented, integrated. In practice, according to the requirements of the Party members on Party, the personnel training in university is to adjust constantly. Further enhance the scientific rationality of education review system and the true meaning of objectives management, and keep Party alive.

References

1. Ping Z (2005) Emphasis on the continuing education of the student party members School Party building and ideological education 8:26–28
2. Chen S, Tu C, Zhao W (2006) Exploration of the graduate education of party members' management evaluation index system. High Educ Explor 5:85–87
3. Wang X (2004) The application of multivariate analysis, vol 1. Shanghai University of Finance and Economics Press, Shanghai, pp 192–196
4. Kaitai F (1999) Utility of multivariate statistical analysis, vol 2. East China Normal University Press, Shanghai, 291–292
5. Huang H, Bao H, Zhao Y (2009) Uniform application of multivariate statistical analysis method to evaluate teaching quality. Ningbo Univ (Educ Sci Ed) 4:56–58
6. Tang Q, Feng M (2002) Practical statistics and DPS data Processing System, vol 1. Science Press, Beijing, pp 333–339

Chapter 77
Mathematics Education and Employment Quality Cultivation

Dong-mei Song and Xiao-qian Zhang

Abstract The paper expounds the key role that mathematics education plays in employment quality cultivation from three aspects. They are advantages of mathematics education in employment, existing problems in mathematics education, and methods to solve problems and promote employment.

Keywords Mathematics education • Employment quality • Cultivation

77.1 Introduction

Since our higher education expands enrollment, institution of higher learning have more and more diversified graduates. However, college graduates are faced with sever employment situation, as employers and the society are not only satisfied with graduates' excellent scores and class cadre, but also lay stress on graduates' overall quality and competence. Therefore, training students with high overall quality and strong overall competence is the fundamental goal of higher education and an issue badly in demand of solving. As an indispensable part of higher education, mathematics education plays a key role in training students' employment quality and competence. Mathematics teachers with working experience should fully be aware of the heavy burden they shoulder, try hard to improve mathematics education, improve students' overall quality, and help students lay a solid foundation for successful employment [1].

In order to succeed in employment, students should be equipped with good employment quality. What is employment quality? It refers to good scientific culture, humanistic culture and morality, strong social adaptability, continuous learning ability, organization and management ability, teamwork spirit, responsibility, creativity, pattern, behavioral norm, value orientation, ideal pursue, and so on in man's mathematics interpersonal skills and good psychological quality

D. Song (✉) · X. Zhang
Hebei Women's Vocational College, Shijiazhuang 050091, Hebei, China
e-mail: songdongmei@hrsk.net

Z. Zhang (ed.), *Proceedings of the International Conference on Information Engineering and Applications (IEA) 2012*, Lecture Notes in Electrical Engineering 220,
DOI: 10.1007/978-1-4471-4844-9_77,

in addition to solid professional and basic knowledge. According to research, employers lay stress on the following employment qualities: professional dedication (73 %), teamwork spirit (70.1 %), and problem-solving ability and adaptability (68.4 %) [2]. However, these problems generally exist in college graduates, which result in hard employment. Therefore, institutions of higher learning should comprehensively improve students' employment quality to dominate competitive advantages in employment.

77.2 Advantages of Mathematics Education in Students' Employment Quality Training

First, mathematics education is universal in basic education of colleges and universities. Currently, either in district with hi-tech information or remote area, institutions of higher learning is able to open college mathematics course as long as they have basic infrastructure. Mathematics teachers can train students' geometrical intuition ability, judging and choosing ability, exploring ability, problem-solving ability from different aspects. Undoubtedly, college mathematics education is important for training students' abstract thinking, logic reasoning ability, rational thinking and flexibility. These qualities and abilities are required by modern high technology talents and managers. The above abilities play a key role in training students' employment quality.

Second, college mathematics education can train students' mathematics spirit. Mathematics spirit is the concentrated reflection of human beings' psychological intention, such as thinking activities. Meanwhile, college mathematics education is the product of human beings' continuous summarization and internalization of mathematics experience, knowledge, methods, ideas, awareness, and valid values.

Third, college mathematics can train students' psychological quality of social intercourse and teamwork spirit. Their strong mathematics problem-solving ability can stimulate people's courage to pursue the truth and enhance people's self-confidence. Meanwhile, it can train student's ability to explore the truth and solve problems independently and further lay a solid foundation for their own employment qualities [3].

Fourth, college mathematics can promote students' overall development. Mathematics is closely related to many subjects. It provides basis for other subjects' learning. College mathematics is the main course for students to master mathematics, an important carrier to train rational thinking and improve learning abilities, and also a way for students to receive exposure to esthetics. People who truly understand mathematics know that mathematics can promote man's perfect development and improve his overall quality. As people step into knowledge economy age, the popularity of computers facilitates the universal application of mathematics. Mathematics and its application play a role different from its previous essence. College graduates who studied college mathematics have advantages in employment.

77.3 Existed Problems

First, with the expansion of college enrollment in recent years, the number of students enrolled in institutions of higher learning increases sharply. Because of differences of regional teaching resources and teaching levels, there are relatively big differences existing in most of students' mathematics bases, learning attitudes, abilities, and initiatives. But syllabus and textbooks are unified and lessons are given in large class and lectures, so teaching results are bad. Meanwhile, final tests are universally closed book examinations, which bring great pressure for students with low starting point and weak foundation.

Second, in terms of teaching contents, textbooks are severely disjoint with college development. In order to meet market and social demands, colleges and universities generally add professional and actual course contents and cut down class hours of basic courses. Too many course contents and few class hours have become a critical paradox. In basic courses teaching like mathematics, in order for one-sided emphasis on "enough usage" and professional service, mathematics class hours are cut down and teaching contents are compressed. Teachers accelerate teaching process for realization of teaching tasks. They no longer introduce deducting and proving process. Instead, they directly instill application-oriented formula, theorem, and conclusion from textbooks into students. Thus, students lack logical deduction and analytical abilities in mathematics learning and cannot understand application background and requirements for formula, theorem, and conclusion. They mechanically memorize formula, theorem, and conclusion, but do not know how to flexibly put them into analyzing and solving actual problems. Students put all time and energy to pure mathematics computing skills training. As a result, some students are fed up with mathematics learning, and some even give up mathematics [4].

Third, mathematics education is exam oriented. Mathematics education does not truly reflect its application orientation. In current high vocational colleges, schools and students emphasize on teaching quality, but schools measure and assess teaching quality by seeing whether students can understand lessons in classrooms, whether they can do homework without review, and whether they can pass exams. Thus, teachers teach for exams while students learn for exams. For example, in class teaching of advanced mathematics, teachers only emphasize on teaching of mathematics concept, theorem, formula, and examples, and those contents that will be tested.

Finally, educational core in colleges and universities is to cultivate students' practical ability and creativity to cultivate managers and technical talents in production line for the society. The goal of talents cultivation is practical-oriented cultivation instead of academic or theoretical-oriented education. Therefore, we should change general requirements of logical and thinking stringency by colleges and universities and take application-oriented contents, thinking openness, and problem-solving initiative as an important task in mathematics education reform in colleges and universities.

77.4 Proposed Scheme

First, reform of mathematics educational contents. Teaching contents of mathematics in colleges and universities should be decided according to overall goal of education in institutions of higher learning. Teaching contents should be centered on students, provide necessary basic services of mathematics for students' further learning and take improvement of students' mathematical quality and practical problem-solving abilities in mathematical method as the fundamental goal. In the first place, we should flexibly arrange teaching contents based on actual and professional needs, make applicable transformation of past-theorized teaching contents and recombine teaching contents according to different degrees and subjects to realize the integration of mathematics course and professional courses. In the second place, mathematics teaching contents should fully reflect the basic principle of "aim at application, measure by enough use, take concept-mastering and application-strengthening as the key point". They should stress the application and using value of the knowledge, try to avoid profound logical deducting process of theorem and explain using meaning in actual problem solving, for example, computing in architectural design, technical analysis in stock, storage and processing in information, statistics in national economy, design in different financing and insurance, weather forecast, key science decisions, and so on. As long as we transform these actual problems into mathematical model, we can compute and get results from computers. Therefore, quantification and mathematization are the inevitable trend in social and economic development. As a thinking method and an expressive language, mathematics has penetrated into different fields. It should be guided by teachers in lesson teaching. In the third place, fun, popularity, and application of mathematics teaching contents should be emphasized to stimulate students' learning interest and cultivate their mathematical character. In the fourth place, mathematical experiments and model should be enhanced and application of mathematics in actual life should be introduced through modern educational technology. The setting of teaching contents should be transformed from "take school courses as the main part and emphasize scientificity" to "emphasize construction in professional fields and pay attention to the consistency of college learning and working experience, value employment instead of knowledge in book". In the fourth place, for classes with big differences, teaching according to different levels can be applied to solve the problem of complicated graduates [5].

Second, reform of teaching methods should transform from taking classrooms, laboratories, and libraries as main learning places and basing book learning to integrated design of classrooms and intern place. We should emphasize that the combination of intern and learning as employment is an important drive for learning. Teaching methods should reflect the application of mathematics (cultivate students' ability to solve actual problems in mathematical method), fun (help students to come over fear of learning mathematics and to appreciate the fun and beauty of mathematics), and experiments (stimulate students' initiative to participate and facilitate them to emphasize on direct experience). The following principles should be complied by. The first one is the principle of creativity. Teaching methods of college mathematics should be continuously innovated while old teaching methods should be abandoned to create new teaching methods which can help students fast master knowledge. The second one is an easily understandable principle. Teaching aims to let most students acquire much and accurate knowledge in the rapidest speed. The third one is an open principle. Comprehensive application of different teaching methods is an effective way to improve teaching quality. The fourth one is to improve teachers' professional quality. Teachers of basic courses should have certain professional knowledge to provide better service.

Finally, through the nurture of learning spirit of mathematics, students' employment quality and professional characters are shaped. The key point is to explore mathematical thinking pattern and thoughts, such as direct thinking, logical deduction, simple computing, and accurate conclusion, and so on in mathematical contents. Through it, students can develop the habit of intellectual activities, plan their work, find and select rational way to accomplish their work, and criticize and judge results.

References

1. Li MW (2006) College maths and the development of students' employment quality. J Chongqing Univ Sci Technol (Soc Sci Ed) 5:29–35
2. Zhong YB (2006) Analysis of college students' employment quality and employment abilities cultivation. Chin Sociol 38(4):438–445
3. Courant R, Robbins H (2005) What is mathematics. Fudan University Press 2(4):380–387 (American)
4. Zhang Y (2009) Discussion of strengths and characteristics in Chinese mathematics education. Edu Chin After-school 2(S3):438–445
5. Zhang CB (2007) Discussion of mathematical spirit and professional character. Vocat Edu Res 9:67–74

Chapter 78
Efficient Scheme of Mathematical Modeling Contest

Yan-fang Wu, Yan-chao Yang, Hong-qiang Cai and Yuan Gao

Abstract This paper has analyzed the importance of the mathematical contest in modeling. The mathematical contest in modeling is able to improve the comprehensive abilities of the students. The paper has analyzed the students in the northwest universities during 2008–2010 and the circumstances of the participation in the mathematical context in modeling. At the same time, the paper has made analysis on the reasons. On the basis of these, several efficient ways have been put forward. The efficient ways aim to improve the enthusiasm of the participation degree of the mathematical context in modeling for the students in the northwest universities.

Keywords Mathematical contest in modeling • University students • Participation • Activity

78.1 Introduction

The mathematical contest in modeling for national university students are held by the coordination between the high education sector in the ministry of education and the National Council for Industrial and Applied Mathematics [1, 2]. The mathematical contest in modeling is a kind of scientific innovation activity that is of mass participation. It is facing the national university students [3, 4]. The mathematical contest in modeling has already become the extracurricular science activity that is of the largest scale among the universities all over the country and has influential power both inside the country and outside the country. In addition, the mathematical modeling is no longer what it used to be. It does not require the

Y. Wu (✉) · H. Cai
School of Chemical Engineering, Qinghai University, Xining 810016, Qinghai, China
e-mail: wuyanfang@cssci.info

Y. Yang · Y. Gao
School of Civil Engineering, Qinghai University, Xining 810016, Qinghai, China

Z. Zhang (ed.), *Proceedings of the International Conference on Information Engineering and Applications (IEA) 2012*, Lecture Notes in Electrical Engineering 220,
DOI: 10.1007/978-1-4471-4844-9_78,

Table 78.1 The statistics and analysis of participation situation of the mathematical contest in modeling for northwest university students

Northwest provinces	Shanxi	Gansu	Qinghai	Ningxia	Xinjiang
Years	2008				
Participating schools	44	21	1	2	14
Participating teams	592	239	4	17	130
Proportion in five northwest schools	0.61	0.23	0.004	0.016	0.14
Proportion of per year increase of participating teams	982				
Years	2009				
Participating schools	47	23	1	3	16
Participating teams	689	279	7	28	149
Proportion in five northwest schools	0.6	0.24	0.005	0.025	0.13
Proportion of per year increase of participating teams	1,152				
Years	2010				
Participating schools	54	25	1	4	16
Participating teams	897	304	6	42	196
Proportion in five northwest schools	0.62	0.21	0.004	0.029	0.137
Proportion of per year increase of participating teams	1,445				

students to recite a few mathematical formulas and solve a few application subjects. It has changed and no longer rigid any more. The mathematical contest in modeling has a very strong application. There are very wide application fields. The subjects that the mathematical contest in modeling has involved include chemistry, biology, economics, finance and information, and so on. The mathematical contest in modeling not only requires the students to transform the practical problems into the mathematic problems, but also requires the students to apply mathematics in a flexible manner. It requires the students to solve the problems with the knowledge of computer and other subjects. Moreover, the participation method is the team made up of three persons. It is to be accomplished making use of the open libraries and such resources as the network. In the end, it needs to hand in a piece of paper. In such learning and participation of context, the students are not only able to improve their learning ability, application ability, and innovation ability, but also able to improve their communication ability, teamwork cooperation ability, and paper writing ability.

78.2 Statistics and Analysis of Participation Situation of the Mathematical Contest

78.2.1 Data Statistics

The author has made analysis on the participation statistics on each playing area all over the country during 2008–2009. In addition to this, the author has summarized the data of part of the playing areas and their participation situation. The data being summarized has been formulated into Table 78.1.

It can be easily seen from Table 78.1 that among the playing areas in five provinces in the northwest, the proportions of the Shanxi playing area participation team is very large. They are basically on the level of more than 60 %. Proceeding to the next, it is the Gansu playing area. The proportions in three years have exceeded 20 %. As for the smallest proportion, it is the Qinghai playing area. The proportion of Qinghai playing area is basically at the level of 0.4 %. It can be seen from Table 78.2 that the participation teams in the west region of the country have been on the increasing trend. Although it is the case, the proportion of the west region participating teams is still lower than the proportion of the nationally increased participating teams per year in the total participating teams in the playing areas all over the country. Under these circumstances, we can make the conclusion that the participation degree of the mathematical contest in modeling is quite low for students from northwest universities.

78.2.2 Reason Analysis

There are a great amount of reasons.

The students lack deserved activities. This characteristic has certain relationship with the learning abilities of the students themselves. Compared to the university students in the other parts of the country, the university students in the western part of the country have very bad foundation. They do not have profound foundation on

Table 78.2 Comparison statistic tables of northwest playing area total teams with national playing area total teams

Years		2008	2009	2010
Area statistics	Total participating teams in northwest schools	982	1,152	1,445
	National teams	12,846	15,050	17,317
Proportion of west participating teams		7.6 %	7.7 %	8.3 %
Proportion of per year increase of participating teams		–	14.6 %	13.1 %

the professional theories. In addition to these, their ability to do things is relatively bad. As for the mathematical context of modeling, it has brought out relatively high requirements for the students. The mathematical contest in modeling not only requires the students to transform the practical problems into the mathematic problems, but also requires the students to apply mathematics in a flexible manner. It requires the students to solve the problems with the knowledge of computer, and other subjects. Therefore, there may be some students who have shown great interests in the participation of the mathematical contest in modeling and they have signed their name for the mathematical contest in modeling. However, they fail to have a comprehensive idea about the mathematical contest in modeling. During the participation process, they have to be out of the mathematical contest in modeling because of the restrictions of the knowledge structure and level. In addition to these, there may be the restrictions of other subjective conditions.

The schools do not pay enough attention to the mathematical contest in modeling. The schools do not have enough propaganda for the mathematical contest in modeling. They do not have enough promotion and organization intensity. Take Qinghai University as an example, the participating teams for the recent 3 years in Qinghai University are quite a few. In addition, the mathematical association in modeling has been established only in recent years. The development of mathematical contest in modeling lacks more important support. Therefore, students do not show great interest in it.

The narrow participation area of the students has affected the enthusiasm for the students to take part in the mathematical contest in modeling and relevant activities as well. The current guidance work of the mathematical modeling construction relies on the teachers from the college of mathematics to a large extent. In addition, teachers from the other majors have little understanding and cognition about the mathematic modeling construction. Teachers do not have a great participation on the mathematical contest in modeling. Their guidance is quite limited as well. For a great amount of schools, they do not have the concentrated trainings under 1 month before the mathematical contest in modeling. However, the mathematical contest in modeling is a kind of systematic engineering. In this case, this kind of guidance does not have much effect on the students and it should be changed.

78.3 Proposed Methods

78.3.1 Mathematical Modeling Construction

As for a great amount of universities in northwest, they set the course of mathematical modeling construction as an elective course. In addition, they have a very easy examination on the course. Therefore, the author suggests that the universities should take the mathematical modeling construction as a kind of compulsory course. In this case, the university students can have a chance to get to learn about the mathematical modeling construction in advance. They are able to grasp a few theories based on mathematical modeling construction. At the same time, they should operate mathematical experimental course that require the students to grasp multiple mathematical software. Students should pay attention to the cultivation of the ability comprehensively and actively take part in the mathematically modeling construction contest.

78.3.2 Improved Mathematical Knowledge

During the ordinary learning process, they should enrich the knowledge on mathematics, computer, and engineering. The participation method is the team made up of three persons. It is to be accomplished making use of the open libraries and such resources as the network. In the end, it needs to hand in a piece of paper. In such learning and participation of context, the students are not only able to improve their learning ability, application ability, and innovation ability, but also able to improve their communication ability, teamwork cooperation ability, and paper writing ability.

78.3.3 Enhance the Enthusiasm

The mathematical contest in modeling not only requires the students to have relatively high abilities, but also requires the teachers to have higher abilities. Therefore, the teachers should keep on complementing their knowledge and take part in the teaching activity in a creative manner. In the guidance for mathematical contest in modeling, the teachers and the students should learn together. The teachers should explore the matters with the students actively and they should promote each other so as to improve their mutual abilities. Conclusion As the mathematical modeling construction context in the universities in the western part of the country develops slowly, and the schools do not pay enough attention to the mathematical modeling construction, it has been a very long-term and hard task to cultivate the mathematical modeling construction abilities of the students. Therefore, we should be insistent so as to better cultivate the ability in the mathematical context of modeling.

78.4 Conclusions

The universities in the western part of the country fail to pay much attention to the mathematical contest of modeling. They have a bad foundation and each aspect has not gained much progress considering the actual conditions of the schools. We need to realize that it is a long-term and hard task. Therefore, we should be insistent.

References

1. Aislla K (2010) Excellent organization work. National university student mathematical contest on modeling 25:115–117
2. Dai H (2010) Exploration on mathematical contest of modeling promotion in districts. New Curriculum Res 3(207):89–92
3. Wei C, Jiang HY (2009) Survey analysis of mathematical model construction for Guangxi. Math Res 18(3):51–54
4. Li X, Zhang D (2006) An exploration to increase students activities to study mathematics of vocational academies. J Guilin Coll Aerosp Technol 3(3):103–105

Chapter 79
Fault Diagnosis of Non-Gaussian Nonlinear Stochastic Systems Based on Rational Square-Root Approximation Model

Lina Yao, Jifeng Qin and Hong Wang

Abstract In this paper, the rational square-root B-spline model is used to represent the dynamics between the output probability density function (PDF) and the input. This is then followed by the novel design of a nonlinear neural network observer-based fault diagnosis algorithm so as to diagnose the fault in the dynamic part of such systems. A simulated example is given to illustrate the efficiency of the proposed algorithms.

Keywords Stochastic distribution systems • Fault diagnosis • Rational square-root B-spline approximation

79.1 Introduction

To improve the reliability and security of practical control systems, fault detection and diagnosis (FDD) has long been regarded as an important and integrated part in control system design [1]. For stochastic systems, when faults and other disturbance inputs are random processes (variables), usually, two kinds of approaches can be used to deal with the related FDD problems [2]. The first group of methods is originated from the statistic theory, where the ratio of likelihood and Bayesian methods are used to estimate the abrupt changes of the states or parameters of the concerned systems. In this context, the FDD algorithms are obtained by using some numeral computation methods such as the Monte Carlo methods or the particle filtering methods [3]. As for the second group of methods, observer and filtering design theory are employed where the min–max optimization techniques have been applied to the estimation error systems in order to guarantee some of the required performances [4].

L. Yao (✉) · J. Qin
School of Electrical Engineering, Zhengzhou University, Zhengzhou, China
e-mail: yaoln@zzu.edu.cn

H. Wang
Control System Centre, Manchester University, Manchester, UK

Z. Zhong (ed.), *Proceedings of the International Conference on Information Engineering and Applications (IEA) 2012*, Lecture Notes in Electrical Engineering 220, DOI: 10.1007/978-1-4471-4844-9_79,

In most cases, the current literatures on the FDD algorithms for stochastic systems aim at those stochastic systems subjected to Gaussian distribution. However, faults, inputs, and outputs of general stochastic systems can be of non-Gaussian type. As such, there is a need to further develop fault diagnosis methods that can be applied to the stochastic systems subjected to arbitrary random parameters. In addition, there are many systems in practice where the output concerned is the PDF of the system output, rather than the actual output values [5]. For such a group of systems, the output that can be used for the feedback control that is the measured output PDFs. Such types of stochastic systems are called stochastic distribution systems which were defined in Ref. [6]. While the control objective of SDC systems is to control the shape of the output PDF, the aim of FDD in SDC systems is to use the measured input and output PDFs to obtain information of the fault aim at using the measured faults. An observer-based algorithm was previously developed in Ref. [5] to detect the fault in the non-Gaussian SDC system based on the linear B-spline approximation model. However, there are several disadvantages in using the linear B-spline model. For example, when the number of the basis functions is not big enough, the calculated output PDF may become negative due to its weak numerical robustness. Subsequently, the rational B-spline model, the square-root B-spline model, and the rational square-root model are developed to enhance the numerical robustness of the controller design of SDC systems [7]. It has been shown that the rational square-root B-spline model can effectively combine the advantages of the rational and the square-root B-spline models, where its feasible domain of weights covers almost all the entire space except for the origin. As such, it would be ideal if one can combine the FDD with the rational square-root B-spline model for SDC systems [8, 9].

In Ref. [9], an observer-based FDD method has been proposed for the stochastic distribution control system based on the rational square-root approximation. However, in that paper, the system dynamic part is about the dynamic change of the weight, which has no physical meaning in practice. This forms the main purpose of this paper; the stochastic distribution control system is composed of two parts, one part is the dynamic part of a practical nonlinear system and there is a relationship between the weight vector and the system state. The other part is the static part, where the rational square-root B-spline model is used to represent the dynamics between the output PDF and the input. For this SDC system, a nonlinear adaptive neural network observer-based fault diagnosis algorithm is developed to diagnose the fault in the dynamic part of the rational square-root B-spline model.

79.2 Rational Square-Root B-Spline Model Description

Based on the well-known B-spline neural networks, the following rational square-root B-spline model can be used to approximate PDF $\gamma(y, u(t))$.

$$\sqrt{\gamma\left(y,u\left(t\right)\right)}=\frac{\sum_{i=1}^{n}\omega_i B_i\left(y\right)}{\sqrt{\sum_{i,j=1}^{n}\omega_i\omega_j\int_a^b B_i(y)B_j(y)d}}=\frac{C(y)V}{\sqrt{V^TEV}}\forall y\in[a,b] \quad (79.1)$$

where $B_i(y)\geq 0$ is the prespecified basis function, $\omega_i(i=1,2,\ldots,n)$ is the approximation weight which is only related to $u(t)$ and n is the number of basis functions. In Eq. (79.1), it has been denoted that

$$\begin{aligned} C(y) &= [B_1\left(y\right) B_2\left(y\right),\ldots,B_n(y)], \\ E &= \int_a^b C\left(y\right)^T C\left(y\right)\mathrm{d}t,\ V=[w_1,w_2,\ldots,w_n]^T\ (V\neq 0) \end{aligned} \quad (79.2)$$

Assuming that the dynamic part described by the general system states and the inputs can be expressed as a nonlinear system, then the whole dynamic system, that uses the rational square-root approximation model for the output PDFs, is described as follows

$$\begin{cases} \dot{x}(t)=Ax(t)+Gg\left(x\left(t\right)\right)+Hu\left(t\right)+\rho\left(x,u\right) \\ V\left(t\right)=Dx\left(t\right) \\ \sqrt{\gamma\left(y,u\left(t\right)\right)}=C\left(y\right)V\left(t\right)/\sqrt{V\left(t\right)^T EV\left(t\right)} \end{cases} \quad (79.3)$$

where $x\in R^n$ is the state vector, $V(t)\in R^m$ is the output weight vector, $u(t)\in R^n$ is the control input vector and $\rho(x,u)$ is the fault vector. $A\in R^{n\times n}$, $H\in R^{n\times n}$, $D\in R^{m\times n}$ and $G\in R^{n\times n}$ are known system parameter matrices.

Assumption 1 $g(x(t))$ is supposed to satisfy $g(0)=0$ and the following norm condition:

$$\left\|g\left(x_1\left(t\right)\right)-g\left(x_2\left(t\right)\right)\right\|\leq m_x\left\|\left(x_1\left(t\right)-x_2\left(t\right)\right)\right\| \quad (79.4)$$

for any $x_1(t)$ and $x_2(t)$, where m_x is a known Lipschitz constant.

79.3 Fault Diagnosis

When a fault has been detected, the fault diagnosis needs to be carried out in order to estimate the size of the fault.

Denote $X=[x^T,u^T]^T\in R^N(N=n+m)$, where $X\in A_d\subset R^N$ and A_d is a compact set. $\hat{X}=[\hat{x}^T,u^T]^T$, where $\rho(\hat{X})=B_0\hat{W}S(\hat{X})$.

The approximating property for the nonlinear model depends on the center vector and width vector of 'Gaussian' function. It is difficult to choose center vectors and width vectors of two 'Gaussian' functions by trial and error. Hence, an adaptive RBF network is used to approximate the system fault so that the parameter vectors d and σ are on-line updated.

The fault can be modeled as

$$\rho(X) = B_0 W^* S(X, d^*, \sigma^*) + \in (X) \tag{79.5}$$

where W^*, d^*, σ^* are the ideal weight matrix, the center vector and the width vector. $\in (X)$ denotes the approximation error of RBF networks, $S(X, d^*, \sigma^*) = s_i, s_i = s_i(||X - d_i^*||, \sigma_i^*)$, $W^* \in R^{n\times q}$, $S(X, d^*, \sigma^*) \in R^{q\times 1}$, $d_i^* \in R^{n\times 1}$, $d^* = [d_1^*, \ldots d_q^{*T}] \in R^{k_0 \times 1}$, $\sigma^* \in R^q$

and $k_0 = n \times q$, where q is the neuron number of the neural network approximating the system fault.

Assumption 2 There exist an ideal weight matrix W^* and ideal vectors d^*, σ^* such that $|| \in (X)|| < \bar{\in}(\bar{\in} > 0)$ for all $X \in A_d$. Assuming that there exist a constant matrix $\bar{W}$ and constant vectors $\bar{d}, \bar{\sigma}$, satisfying $||W^*|| \leq ||\bar{W}||$, $||d^*|| \leq ||\bar{d}||$ and $||\sigma^*|| \leq ||\bar{\sigma}|$. Equation (79.3) can further be written as

$$\begin{aligned} \dot{x}(t) &= Ax(t) + Gg(x(t)) + Hu(t) + B_0 W^* S\left(X, d^*, \sigma^*\right) + \in (X) \\ V(t) &= Dx(t) \end{aligned} \tag{79.6}$$

The fault diagnosis observer is constructed as follows:

$$\begin{aligned} \dot{x}(t) &= A\hat{x}(t) + Gg\left(\hat{x}(t)\right) + Hu(t) + B_0 \hat{W} S\left(\hat{X}, \hat{d}, \hat{\sigma}\right) + K(t)\varepsilon(t) \\ V(t) &= D\hat{x}(t) \\ \varepsilon(t) &= \int_a^b C(y)\left(\frac{\hat{V}(t)}{\sqrt{\hat{V}^T(t) E \hat{V}(t)}} - \frac{V}{\sqrt{V^T E V}}\right) \mathrm{d}y \end{aligned} \tag{79.7}$$

where $\hat{x}(t)$ is the state vector of the fault diagnosis observer, $K(t)$ is the observer gain to be determined, and $\varepsilon(t)$ is the residual vector.

Lemma 1 *There exists a* $\lambda(T_1 \leq |\lambda| \leq T_2), T_1 = \lambda_{\min}(E)/\lambda_{\max}(E))$, $T_2 = \lambda_{\max}(E)/\lambda_{\min}(E))$ *such that the following equation*

$$\sqrt{V^T E V} - \sqrt{\hat{V}^T \hat{E} \hat{V}} = \lambda(\sqrt{\hat{V}^T V} - \sqrt{\hat{V}^T \hat{V}}) = \lambda(||V|| - ||\hat{V}||) \tag{79.8}$$

holds.

It is obvious that the initial condition should be chosen as $\hat{W}(0) = 0$, in order to make the output of fault approximator being zero prior to fault occurrence, that is to say, $B_0 \hat{W} S(\hat{X}, \hat{d}, \hat{\sigma}) = 0$.

Denote the observation error vector as $e(t) = \hat{x}(t) - x(t)$.

Let $K(t) = L\sqrt{\hat{V}^T(t)E\hat{V}(t)}$, where L is a gain vector to be determined. Using Lemma 1, Eqs. (79.6) and (79.7), the observation error dynamic system can be written as follows:

$$\begin{aligned}\dot{e}(t) = (A + L\Sigma D)\, e(t) + G\tilde{g}(t) + L\Sigma V\frac{\lambda(\|V\| - \|\hat{V}\|)}{\sqrt{V^T EV}} \\ + B_0\hat{W}S\left(\hat{X},\hat{d},\hat{\sigma}\right) - B_0W^*S\left(\hat{X},d^*,\sigma^*\right) + B_0W^*\tilde{S} - \in(X)\end{aligned} \tag{79.9}$$

where $\tilde{g} = g(\hat{x}) - g(x)$, $\tilde{S} = S(\hat{X},d^*,\sigma^*) - S(X,d^*,\sigma^*)$.

The Taylor's series of $S(\hat{X},d^*,\sigma^*)$ is expanded at $(\hat{d},\hat{\sigma})$, which is shown as follows: $S(\hat{X},d^*,\sigma^*) = S(\hat{X},\hat{d},\hat{\sigma}) - S'_{\hat{d}}\tilde{d} - S'_{\hat{\sigma}}\tilde{\sigma} + O(\cdot)$
where $\tilde{d} = \hat{d} - d^*$, $\tilde{\sigma} = \hat{\sigma} - \sigma^*$, $S'_{\hat{d}} = \frac{\partial S}{\partial d}|_{d=\hat{d}} \in R^{q\times k_0}$, $S'_{\hat{\sigma}} = \frac{\partial S}{\partial \sigma}|_{\sigma=\hat{\sigma}} \in R^{q\times q}$.
Then, it can be further formulated that

$$\begin{aligned}B_0W^*S\left(\hat{X},d^*,\sigma^*\right) &= B_0W^*S\left(\hat{X},\hat{d},\hat{\sigma}\right) - B_0W^*S'_{\hat{d}}\tilde{d} - B_0W^*S'_{\hat{\sigma}}\tilde{\sigma} + B_0W^*O(\cdot) \\ &= B_0\hat{W}S\left(\hat{X},\hat{d},\hat{\sigma}\right) - B_0\tilde{W}S\left(\hat{X},\hat{d},\hat{\sigma}\right) - B_0\hat{W}S'_{\hat{d}}\tilde{d} + B_0\tilde{W}S'_{\hat{d}}\tilde{d} \\ &\quad - B_0\hat{W}S'_{\hat{\sigma}}\tilde{\sigma} + B_0\tilde{W}S'_{\hat{\sigma}}\tilde{\sigma} + B_0W^*O(\cdot)\end{aligned} \tag{79.10}$$

where $\tilde{W} = \hat{W} - W^*$.

Substituting Eq. (79.10) into Eq. (79.9), the error dynamic equation can be further obtained as

$$\begin{aligned}\dot{e} = (A + L\Sigma D)\, e + G\tilde{g} + L\Sigma V\frac{\lambda\left(\|V\| - \|\hat{V}\|\right)}{\sqrt{V^T EV}} \\ + B_0\tilde{W}S\left(\hat{X},\hat{d},\hat{\sigma}\right) + B_0\hat{W}S'_{\hat{d}}\tilde{d} + B_0\hat{W}S'_{\hat{\sigma}}\tilde{\sigma} + v\end{aligned} \tag{79.11}$$

where $v = B_0W^*\tilde{S} - \in(X) + (B_0\tilde{W}S'_{\hat{d}}\tilde{d} + B_0\tilde{W}S'_{\hat{\sigma}}\tilde{\sigma} + B_0W^*O(\cdot))$.

Lemma 2 *If the following parameter-updating laws* (79.12)–(79.14) *are given, then v is bounded in the set A_d.*

$$\dot{\hat{W}} = \begin{cases} -\Upsilon\Phi + \Upsilon\frac{(\|\hat{W}\|^2-\beta)tr(\Phi^T\hat{W})}{\|\hat{W}\|^2}\hat{W} & \|\hat{W}\|^2 > \beta \text{ and } tr(\Phi^T\hat{W}) < 0 \\ -\Upsilon\Phi & \text{otherwise}\end{cases} \tag{79.12}$$

$$\dot{\hat{d}} = \begin{cases} -\Upsilon_d\Phi_d + \Upsilon_d\frac{(\|\hat{d}\|^2-\beta_d)\Phi_d^T\hat{d}}{\|\hat{d}\|^2}\hat{d} & \|\hat{d}\|^2 > \beta_d \text{ and } \Phi_d^T\hat{d} < 0 \\ -\Upsilon_d\Phi_d & \text{otherwise}\end{cases} \tag{79.13}$$

$$\dot{\hat{\sigma}}_i = \begin{cases} -\Upsilon_\sigma\Phi_{\sigma_i}(-1 + \hat{\sigma}_i - a_i), & \hat{\sigma}_i < a_i \text{ and } \Phi_{\sigma_i} < 0 \\ -\Upsilon_\sigma\Phi_{\sigma_i}(-1 - \hat{\sigma}_i + b_i), & \hat{\sigma}_i > b_i \text{ and } \Phi_{\sigma_i} > 0 \\ -\Upsilon_\sigma\Phi_{\sigma_i} & \text{otherwise}\end{cases} \quad i = 1, 2, \ldots, q \tag{79.14}$$

where $\Phi = M^T \sqrt{\hat{V}^T E \hat{V}} \varepsilon(t) S^T(\hat{X}, \hat{d}, \hat{\sigma}) \in$ $\Phi_d = S_d'^T \ddot{W}^T M^T \sqrt{\ddot{V}^T E \ddot{V}} \varepsilon(t) \in$ $\Phi_\sigma = S_d'^T \hat{W}^T M^T \sqrt{\hat{V}^T E \hat{V}} \varepsilon^T \in R^q$, $M \in R^{1 \times n}$ and $\beta, \beta_d, a_i, b_i \Upsilon, \Upsilon_d, \Upsilon_\sigma$ are the parameters to be determined.

Lemma 3 *Parameter-updating laws* (79.12)–(79.14) *guarantee that the following three inequalities are satisfied*

$$tr\left[\tilde{W}^T \left(\frac{1}{\Upsilon}\dot{\hat{W}} + \Phi\right)\right] \le 0, \tilde{d}^T \left(\frac{1}{\Upsilon_d}\dot{\hat{d}} + \Phi_d\right) \le 0, \tilde{\sigma}^T \left(\frac{1}{\Upsilon_\sigma}\dot{\hat{\sigma}} + \Phi_\sigma\right) \le 0 \tag{79.15}$$

Theorem 2 *For the nonlinear fault system* (79.1), *state observer* (79.7) *and error equation* (79.11), *it is supposed that there exists a set A_d large enough to satisfy Assumptions* 1. *The following equation should be satisfied*

where P and Q are two positive definite matrices.

$$\left(A + L\sum D\right)^T P + P\left(A + L\sum D\right) = -Q \tag{79.16}$$

If parameter-updating laws are selected as (79.12)–(79.14) and B_0, M satisfy the condition $PB_0 = (\Sigma D)^T M$, $e(t)$ is uniformly ultimately bounded. That is to say $e(t) \in L_\infty$.

79.4 Simulation Examples

To illustrate the effectiveness of the proposed fault diagnosis, a stochastic system is considered as follows:

$$A = \begin{pmatrix} -0.5 & 0.3 \\ 0 & -1.3 \end{pmatrix}, G = \begin{pmatrix} 0.1 & 0 \\ 0 & 0.1 \end{pmatrix}, H = \begin{pmatrix} 0.2 & 0 \\ 0 & -0.3 \end{pmatrix}, D = \begin{pmatrix} 1 & 0 \\ 0 & 1 \\ 0 & 1 \end{pmatrix} \tag{79.17}$$

To simulate the algorithm, it is assumed that the fault is constructed as $\rho(x, u) = [\cos(0.8x(1)), 0]^T t \ge 5s$.

The response of the residual signal is shown in Fig. 79.1 and the fault diagnosis result is shown in Fig. 79.2. From these two figures, it can be concluded that desired fault diagnosis results have been obtained.

79.5 Conclusions

In this paper, the stochastic distribution control system is composed of two parts, one part is the dynamic part of a practical nonlinear system and there is a relationship between the weight vector and the system state. The other part is the

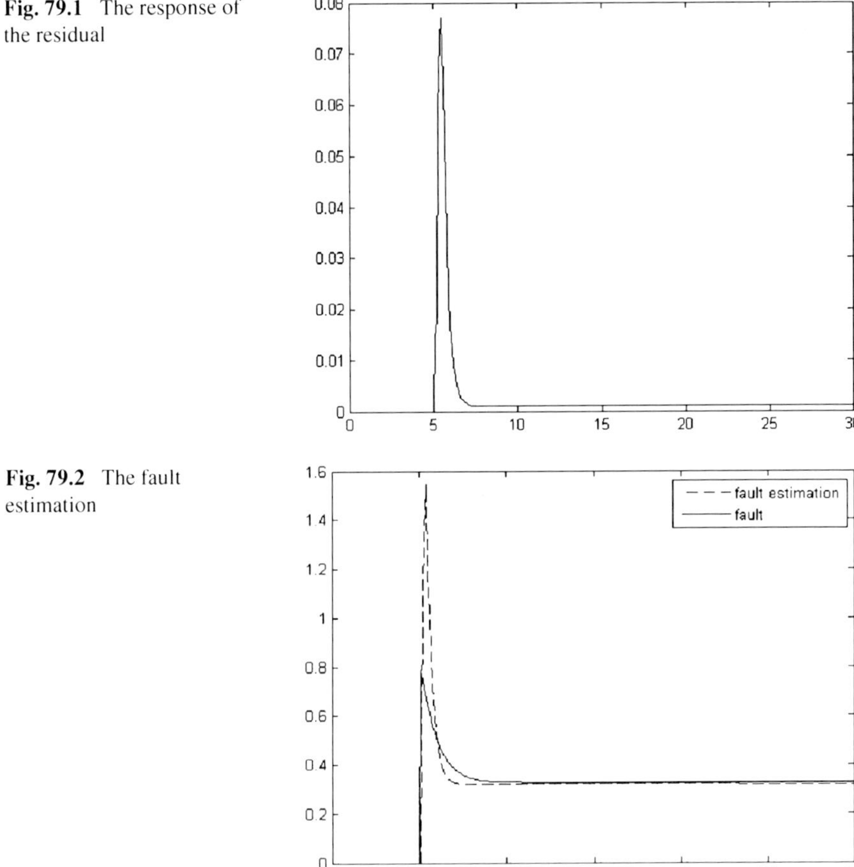

Fig. 79.1 The response of the residual

Fig. 79.2 The fault estimation

static part, where the rational square-root B-spline model is used to represent the dynamics between the output PDF and the input. A nonlinear neural network observer is constructed to detect and diagnose the fault in the dynamic part of this system, where a Lyapunov function based technique is used to formulate the required observer gain and adaptive tuning rule for the fault diagnosis.

Acknowledgments This paper was supported by Chinese NSFC 61104022 and 10971202.

References

1. Basseville M (1998) On-board component fault detection and isolation using the statistical local approach. Automatica 34:1391–1395
2. Hibey JL, Charalambous CD (1999) Conditional densities for continuous-time nonlinear hybrid systems with applications to fault detection. IEEE Trans Automat Control 44:2164–2170

3. Zhou DH (1995) A new approach to sensor fault detection and diagnosis of nonlinear systems. ACTA Automatica Sinica 21(3):362–365
4. Chen RH, Mingori DL, Speyer JL (2003) Optimal stochastic fault detection filter. Automatica 39:337–340
5. Wang H, Lin W (2000) Applying observer based FDI techniques to detect faults in dynamic and bounded stochastic distributions. Int J Control 73:1424–1426
6. Wang H (2000) Bounded dynamic stochastic systems: modeling and control, vol 3. Springer, London, pp 127–129
7. Zhou JL, Yue H, Wang H (2005) Shaping of output PDFs based on the rational square-root B-spline model. Acta Automatic Sinica 31(3):343–348
8. Girosi F, Poggio T (1990) Networks and the best approximation property. Biol Cybern 63:169–176
9. Yao LN, Wang H, Yue H, Zhou JL (2006) Fault detection and diagnosis for stochastic distribution systems using a rational square-root approximation model. Proceedings of the 45th IEEE conference on decision control, vol. 1 (USA) pp 4163–4168

Part VIII
Information Management Systems and Software Engineering

Chapter 80
Housing Design Based on SIP System

Wenshan Lian, Chenhui Tang and Youpo Su

Abstract IP (insulated panel) is a better material for interior wall and outer wall feature. SIPs (structural insulated panels) are one of the best insulated wall panels and insulated roof panels for commercial and residential use. They are an environment -friendly, dimensionally stable, modular building products. With high insulation value, excellent water resistance, light weight, and low cost, the versatile insulated wall and roof panels are ideal for use in all types of building plans. It has the features of fire resistance, sound insulation, heat insulation, and it is easy to install.

Keywords SIP panel • House system • Environment • Structure

80.1 Introduction

This paper is about exploration on the building science, especially the one that uses new materials, such as SIPs, which are short for structural insulating panels.

The traditional block buildings, which piled up either by stones or by bricks, are widely used in the ancient temples or historical buildings. Little by little, this structure has disappeared to be replaced by framing structure, due to its reasonable mechanical properties, more stable wall system, and less load. Since frame is the main stress components, wall becomes an enclosure, purely for protecting privacy and insulating heat, noise, and moisture. As long as the skin could carry its own weight or part of the structure loads; the framing system could be eliminated completely. This theory has been tested and developed by scientists and engineers. In 1947, a complete structure using corrugated paperboard was

W. Lian · C. Tang (✉) · Y. Su
College of Civil and Architectural Engineering, Hebei United University,
Tangshan 063000, China
e-mail: Chenhui_tang2008@yahoo.cn

W. Lian
e-mail: lwsttt@126.com

Z. Zhong (ed.), *Proceedings of the International Conference on Information Engineering and Applications (IEA) 2012*, Lecture Notes in Electrical Engineering 220,
DOI: 10.1007/978-1-4471-4844-9_80,

built in Wisconsin. The structure had been disassembled continuously for testing. By observing the panel stiffness, the structure was analyzed and developed over 31 years. In 1969, foam core was introduced into the structure, and thus the insulated panels formed [1].

The SIP neither limited to weather nor to construction site; not only does it save lots of time and labor, but also makes it possible for constructors to solve the problem prior to construction efficiently. Since the panels are manufactured and fabricated under factor restriction and standard, the accuracy is guaranteed; furthermore, the speed of on-site assembling is reassured.

Nowadays, the SIP is mostly made up of four parts: structural facings, rigid foam insulation, structural adhesive, and connectors. In which, the insulation foam core material uses usually expanded polystyrene (EPS), extruded polystyrene (XPS), polyurethane, or polyisocyanurate. The structural facing is the enclosure system both for the foam core and the interior space; normally, it is referred as shell construction, no matter it is straight or curved, whose point loads are dispersed in all directions.

80.1.1 The Analysis of the House System

The house system is composed of three main parts: the environment, the housing, and the people while every elements of the system are interrelated and interact all the time. In 'The Architecture of the Well-Tempered Environment', Banham investigated the effects of modern environmental control systems on architectural form. The primary theme of this work is that technology should be used by the designer to enhance human comfort, but it should not be allowed to overpower the natural processes of climate, culture, and architectural precedent [2]. Moreover, the system has also been divided into subsystems. For example, building enclosure is a subsystem of roof assembly, wall assembly, and foundation assembly. Wall assembly is a subsystem of paint, insulation, and certain other layers; whereas inside the enclosure is a subsystem of electrical power, plumbing, heating and cooling, and so on, while mechanical system is an initial element in wall system. Hereby comes the next point, structure [3].

80.1.2 The Structure Analysis of the Panel

The mechanical properties of SIP are that the upper part of the panel stands the compression, when the load is applied, while at the same time the bottom part of the panel suffer tension. Therefore, considering the mechanical properties of the material, the choice of the SIP material could be concluded as the thickness of the core, the tensile strength of the panel, and the compressive strength of the panel. Salvadori once investigates building structure as an internal part of architectural form and the design process. Much of what must be learned about structures by architects can become mired in abstract mathematical analyses and formulae. Salvadori elevates this technical dimension of the architectural process to show how structural form can and should be inseparable from the more subjective facets of design. As Nina Rapport notes: "The use of overt patterning in structural system blurs the line between what is structure and what is decorative, and results in a third thing, a deep decoration- a decoration that is both below and in the surface, that creates new spatial effects and comprises holistic atmospheres" [4].

80.2 The Insulation Property of SIP Material

Furthermore, the insulation is also an important consideration in the choice of the SIP material. Besides the opening of the elevation design, there are certainly other thermal bridges in buildings; the joint of the panel is one of them. The external environment and internal environment disclose many potential dangers to the SIP wall assembly, such as rain, moisture, vapor, air, and heat. Actually, water is the main concern in building design. Water comes to lots of different forms, since rain and ground water is the liquid form of water, wall assembly should be effectively prevented from getting wet from the exterior and interior; the second form of water is the solid part as snow, which demands SIP wall for the strength carrying the load of snow, in case it is not melted or been shoveled in time; at last, it is the vapor phase, such as vapor and moisture, which are easily being ignored, not like controlling water, could be drained easily. The difficult situation of SIP wall is how to protect the SIP wall from moisture outside, foundation from both upside and soil downside. Once the moisture passes through SIP walls, the issue would be how to avoid moisture accumulating and remove moisture. In general, the process of preventing water and moisture from entering is the same; and the strategy of keeping the moisture out is also the reason of trapping vapor inside.

XPS foam board is a good roof insulation board. XPS is insulation manufactured through a plastic extrusion process. The resulting boards are almost 100 % closed cell, strong, highly moisture resistant, and easy to cut and shape.

XPS is particularly indicated for places where a high mechanic resistance is required. Its resistance to water and compression makes it the ideal insulation solution for extreme conditions. Moreover, it is very easy to install and offers a high-performance thermal insulation for your building. XPS foam board widely used in wall insulation, low-temperature storage facilities, parking platform, the airport runway, construction of concrete roof, and the structure of the roof, highways.

80.2.1 Technical Data Sheet of the XPS Foam Board

Technical data of the XPS form board

Density	Kg/m³	36–40
Thermal conductivity, 90 days, 10 °C	W/mK	0.027–0.03
Compressive strength at 10 % deflection or yield, (vertical)	kPa	≥300
Tensile strength	kPa	≥300
Water absorption	Vol %	≤1.00 %
Capillarity	Nil	Nil
Coefficient of linear thermal expansion	mm/mK	0.07
Temperature limits	°C	−50 °C, +75 °C

80.2.2 Main Characteristics of XPS Foam Board

First, the performance is stable and anti-aging. It has been experimented that the XPS insulation boards can be used for at least 35–50 years.

Second, it is resistant to compression, compared with other thermal insulation material, such as EPS 200 kPa, it has better compressive strength of around 300 kPa, which makes it the ideal solution for heavy domestic or industrial loads.

Third, it has the potential of water resistance due to the reason that polyfoam is almost 100 % closed cell and as such is unaffected by moisture.

Fourth, the material is lightweight, volume optimized, and easy to handle. Although the volume optimization is not difficult, it is resistant to deformation and outer impact.

Fifth, the closed cell structure and density of polyfoam allow specific edge details and surface finishes to be cut into the boards to make them as fit as possible.

Fig. 80.1 The picture of SIPs

Moreover, polyfoam can be cut into almost any shape, so that the versatile façade of the house enriches the image of the building. Moreover, polyfoam is recyclable, and it is 100 % ozone friendly. Generally speaking, there are two kinds of SIP structure which is widely used.

1. OSB + XPS + OSB.
2. Wood + MDF + XPS + OSB.

Compared to EPS core, the SIPs are made with XPS foam core, have higher compressive strength, lower water absorption, lower thermal conductivity, longer service life, and so on. Normally, the size of SIPs: length: 2400 mm, width: 600 mm, thickness: 30–140 mm (optional), packages: pack in cartons (Fig 80.1).

MOQ: 1*40'HQ.
The features of the panel:
Thermal conductivity: Less than 0.030 W/(m·k);
Compressive strength: more than 250 kPa;
Tensile strength: more than 300 kPa;
Bond strength: more than 200 kPa.

80.3 Conclusion

After all, SIP-manufacturing system is time-reducing, money-saving, and recycled material. The design of SIP system is such that it reduces the amount of air leaking into and out of the house, thus reducing your energy bills. As a matter of fact, the detail of the system, especially the joint is quite important. SIP system emits less greenhouse gasses than traditional stick frame construction. It is suitable for various climates, including long, cold, wet winters.

References

1. Michael M (2000) Building with structural insulated panels (SIPs): Strength and Energy Efficiency Through Structural Panel Construction, 1st edn. Taunton Press, Newtown, pp 12–17
2. Reyner B (1984) The architecture of the well-tempered environment, 2nd edn. University of Chicago Press, Chicago
3. Marlo S (1980) Why buildings stand up: the strength of architecture, New York, ww Norton & Co
4. Rappaport N (2006) "Deep Decoration", in 30/60/90. Archit J 10:10–13

Chapter 81
Research on Real-Time Pedestrian Counting System Based on Computer Technology

Liming Song and Jinghui Zhang

Abstract In order to propose a real-time pedestrian counting system, which can be used to estimate the number of pedestrians in the video pedestrians, the video sequence must be carried out according to the following procedures: First, the candidate region is divided into water droplets, background subtraction, extracted from each BLOB corresponding to the estimated number of a set of functions and neural networks, and each set of pedestrian. In order to achieve real-time processing, we use a simple but effective feature, and the use of the Parzen window density estimation, the input image is fast with accurate target detection and adaptive background modeling.

Keywords Visual monitoring • Pedestrian detection • Real-time count

81.1 Introduction

At present, surveillance cameras for various purposes are used more widely. Under normal circumstances, observers collected images from these cameras for monitoring and control [1]. In practice, this manual monitoring includes the following problems: even if there are only a few observers, monitoring costs are very high. Increase of their working hours increases the fatigue of the observers, leading to their lower alertness. The increase in the number of cameras will lead to increased fatigue monitoring costs and observers. Especially when the image of the observed real-time analysis of the security measures required, the

L. Song (✉) · J. Zhang
Tangshan College, Tangshan 063000, China
e-mail: song_liming89@126.com

Z. Zhong (ed.), *Proceedings of the International Conference on Information Engineering and Applications (IEA) 2012*, Lecture Notes in Electrical Engineering 220,
DOI: 10.1007/978-1-4471-4844-9_81,

fatigue of the observer will become a serious problem [2]. Therefore, the automatic analysis applications, such as image processing technology, are promising methods. To this end, we develop the system for real-time video sequences to be monitored to estimate the number of pedestrians. This study will enhance the pedestrian system to count real-time technology for future pedestrian technology leaps and bounds.

81.2 The Principles and Methods

The use of image processing is to count the estimated number of pedestrians in the input image. From the development of pedestrian counting system based on image processing, it is expected to reduce monitoring costs and observer fatigue. Pedestrian, you can use in a variety range of applications [3]. They can be divided into three methods [4]: (1) Individual pedestrian detection; (2) the trajectory clustering of visual function; (3) Based on the characteristics of the return.

First, it is the individual pedestrian detection algorithm, it is estimated that all pedestrian detections input image. Viola and others proposed to improve the appearance and movement of people counting method based on function. Zhao et al. proposed a segmentation model based on Bayesian methods. However, these methods cannot be applied in a very crowded and significantly occluded scene, because they need to detect and segment all pedestrians.

Visual function trajectory clustering method, that is, over time functions, people are constantly tracking and identification of visual computing. As the pedestrian estimated by this method within a specific time, their real-time processing is difficult. The third method is based on the characteristics of regression method to estimate the characteristics of the pedestrian regression function, extracted from the input image, use. These are usually the working methods [5]: (1) Background subtraction; (2) The various functions of the extracted foreground region; (3) The regression function to extract the eigenvalues, such as pedestrian liner, segmented liner, or a neural network. It was also based on the input image edge detection of a regression function neural network is used to calculate back, such as edge orientation and application of background subtraction to obtain the histogram function of the size of the BLOB. Chan et al. proposed a method in the regression function is a Gaussian process retrogression 28 feature extraction crowd paragraph dynamic texture blending. However, these methods cannot estimate the location of the input image, pedestrians, or they cannot be of real-time execution. For example, some people proposed method based on the input image cannot be estimated pedestrian, because it has only one input image to extract the feature vector. Chan et al. proposed method cannot achieve the principle of real-time processing, because of the mixed dynamic texture segmentation using a series of successive images, from past to future. In addition, it may take a lot of time to extract 28 features such as Cliff's kiwi degree, homogeneity, entropy, complexity, etc.

Feature-based regression analysis method is most accurate, because it is able to cope with occlusions and background changes. However, as mentioned above, the previous several remaining issues. In this file, we recommend that real-time counting estimates "the number of pedestrians in the video sequence".

The methods include [6]: (1) Background subtraction and shadow elimination; (2) The function to estimate the prospects spots; (3) in each BLOB by the neural network of pedestrian number. We achieved real-time processing through the following strategies [7]: (1) Simplify the function: only simple and effective use of feature extraction faster; (2) Segmentation and background modeling of the foreground object detection based on fast Parzen window density estimation.

81.3 The Counting Method Using the BLOB Function

The outline of the proposed system and the input image is divided into mobile spot object, using background subtraction and shadow elimination. They are the normalization according to each BLOB, the size of the approximation of the real scene, to extract a variety of functions. However, it must be to train the neural network to create a pedestrian in the training data BLOB relationship between the eigenvalues.

81.3.1 The Walk Segmentation

We use background subtraction to detect the non-background pixels in the input image. We only subtract a background image from the observed image, without the need for prior information about object in the target areas. In this method, we use the fast algorithm for adaptive background model construction. Parzen window density estimation provides fast and accurate detection of small intensity or color change of the target areas, such as the impact of the light changes. The algorithm estimates the Parzen window density estimation, which is not using the background model, the observed pixel value from a video sequence of parametric density estimation. However, the shaded area casts the object to be detected as the object of its size and location of the change over time. Because the shaded area may affect the counting system, we need to obtain prospects from the background subtraction. Therefore, we use a shadow detection method, using the YUV color in order to eliminate the shaded area. This method is based on the observation on the surface of cast shadow that will reduce the saturation of the intensity values and a similar small proportion of shadow tonal values is similar to the original background. In other words, this method is based on the observation of the surface to cast a shadow over the same attenuation of the three components of the YUV color values. Finally, we can only detect objects in the region, not including the shadow of the shadow of the elimination of the region to use background subtraction.

81.3.2 The Extraction of Feature

Therefore, we adopt the method based on neural networks, using six functions to extract the BLOB of each detected foreground regions for these nonlinear modeling.

BLOB functions capture the shape and size of BLOB foreground segmentation. Perimeter is the total number of pixels in the BLOB peripheral. BLOB perimeters in perimeter area than measuring the complexity of a drop shape. A high percentage of spots included in its surrounding bumps, which may be the number of indicators included inside length—the total number of edge pixels in the BLOB. The edge of the BLOB is a powerful clue in the BLOB. Canny edge detector is applied to the entire image, hide the edge of the image of segmented regions, and calculate the side length.

Rectangular functions capture the scale and location of the partition rectangle that contains the BLOB prospects. Aspect Ratio measures pedestrian rectangular configuration of the ratio between the width and height in the BLOB. Protruding state of the BLOB pedestrians in the entire organization is the internal state of associated input image or not. On the side of the rectangle that contains the BLOB coincides with the upper side of the input image, we determine the pedestrians on certain parts of the internal image and status set to "no upper body". The other two sides are in the same way [8]. Part of the body is not within the image, extract the image feature value changes, and the prominence of presentation in order to deal with this situation. Highlight the relationship between the position and image features is also neural network training.

81.3.3 Functional Normalization

We must take into account the impact of the projection angle of the camera system, because in most cases, in the pedestrian moves, the camera projection plane is not parallel to the ground plane. Therefore, the region of the same object, its position on the input image and extract features from BLOB different large-scale objects. To solve this problem, we have investigated the effectiveness of functional normalization, and the use of two methods: explicit normalization and implicit normalization.

In this method, we first change the input image; the pedestrian area size is an independent position by a plane angle is called the homograph transform. Planar perspective transformation point (u, v), the correspondence between the input image with a ground plane describes a matrix H3 $\times$ 3 (x, y) as follows:

$$\begin{aligned} u(h_{31}x + h_{32}y + h_{33}) &= h_{11}x + h_{12}y + h_{13} \\ v(h_{31}x + h_{32}y + h_{33}) &= h_{21}x + h_{22}y + h_{23} \end{aligned} \tag{81.1}$$

The four-point correspondence is sufficient to solve these equations, we can convert the image input image for H., the same area approximately equal to the object in an object at some point/area/person at another point. However, if the pedestrian tilts left or conversion rights image, as they are standing perpendicular the pedestrian and the ground plane are not coplanar. Under these circumstances, such as BLOB peripheral or marginal features, it is still dependent on the point in the transformed image. Therefore, the normalization function using planar perspective transformation changed between the area ratio of the original area of the BLOB based on the ratio of the areas (such as BLOB area) function, applicable to them. For linear features (such as BLOB periphery or edge), normalization of the square root of the ratio of perimeter, area, the ratio of BLOB area calculation, and BLOB circumference in this way. Based on the function of the rectangle that contains the BLOB extracted from the original input image, because they are independent of the scale, that is its own independent position.

In this method, we extracted from each BLOB seven characteristics: six features described in Sect. 81.3.2 and location parameter, the normalization of our functional training a neural network with the implicit training data, including implicit normalization: scale parameter. The input area of the same object changes its position and image; we can use the BLOB implicit characteristic value of the normalization of the location information. Then, in the coordinates of a rectangle that contains the BLOB value as the location information in the BLOB. Therefore, we can estimate the image of the pedestrian network of the number of pedestrian scale transition in the BLOB to do a nerve.

81.3.4 The Number of Pedestrians to Estimate the Neural Network

In order to capture the nonlinear relationship between the characteristics of a BLOB and pedestrians training data, we use a neural network. In our system, the neural network model has two hidden layers, shown in Fig. 81.1. Therefore, n is either 6 or 7, which depends on the normalization. The output layer has only one unit, they correspond to the number of pedestrian in the BLOB. Sigmoid activation function is used for all nodes in the network. In order to make the training data, we use the automatic spot detection and extraction spots. "Manual gives the number

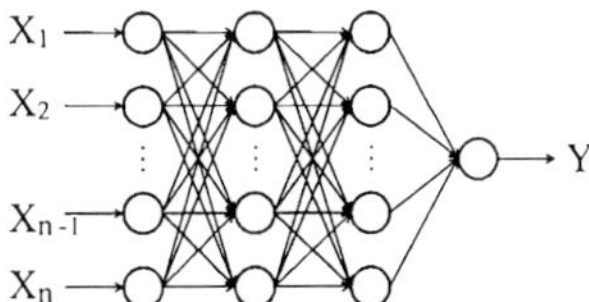

Fig. 81.1 Neural network used in the proposed method

of pedestrian in each spot. This training the neural network training data, we can use the network to estimate pedestrian Online.

81.4 The Experiments

In this section, we describe the following results: (1) to evaluate the accuracy; (2) calculate the cost of the assessment; (3) further validation with other data sets 4.1 to evaluate the accuracy.

In this experiment, we have used the resolution of the image data set PETS2006 (PETS2006 reduced to 320 × 240 pixels. The data set consists of 3,020 images, we use images from the 2,035th out of 3,020 images to train the neural network framework, then, choose a test framework for each of the seven images in the data set, we use 290 assessment test frame, nonoverlapping training frame accuracy and errors in Table 81.1, how do we correctly evaluate our proposed method can estimate the number of pedestrian in a BLOB, for example, how many spots contain a pedestrian was identified as "a" is incorrect and really believe that "two" in the case of spots, containing less than two pedestrian Table 81.1 shows that the method can identify the correct number of contained pedestrian. This is because the neural network can learn a variety of information about these spots, because they continue to observe the training frame. On the other hand, Table 81.1 also shows that they cannot be estimated contains the correct number of water droplets in the pedestrian, which contains three pedestrians.

In Table 81.2, we have assessed value is estimated to ground truth the accuracy of the proposed approach, namely a clear normalization of Sect. 81.3.3 as described in two different normalization. Figure 81.2 shows the people count; the

Table 81.1 The confusion matrix of a BLOB in the pedestrian quantity (PETS2006)

		To assess the value					Total	Accuracy(%)
		0	1	2	3	4		
Ground truth	0	3,399	37	0	0	0	3,436	98.9
	1	41	851	36	0	0	928	91.7
	2	0	38	94	0	0	132	71.2
	3	0	0	5	10	4	19	52.6

Table 81.2 Method accuracy

	Clear the standardization	Implicit standardization
Confirm	1,249	1,249
False positive	77	199
False negative	84	78
Error	161	277
Accuracy (%)	87.1	77.8

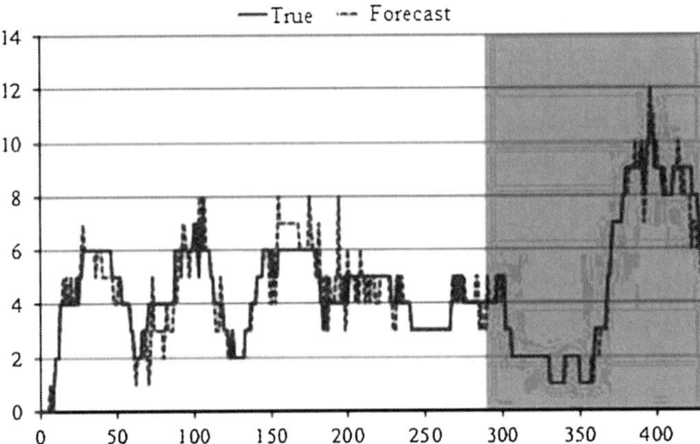

Fig. 81.2 Clearly the normalization

scope of the shadow is used to frame the results of training. In Table 81.2, the false positives and false negatives were defined as follows: false positives—counting errors caused by the nonpedestrian objects for pedestrians, and count the number of pedestrians over the value of the ground truth. False negative—resulting in missing pedestrians and count the number of errors caused by pedestrians is less than the value of the ground truth. Table 81.2 shows the false-negative values are not much different, regardless of their use of the normalization. On the other hand, Table 81.2 also shows that false positives implicit normalization of explicit normalization. It also shows that this is more explicit normalization of the number of errors caused by incorrect understanding spots containing two pedestrians where one pedestrian in the larger implicit normalization. Implicit normalization, the images provided by the characteristics of neural network location, but cannot get

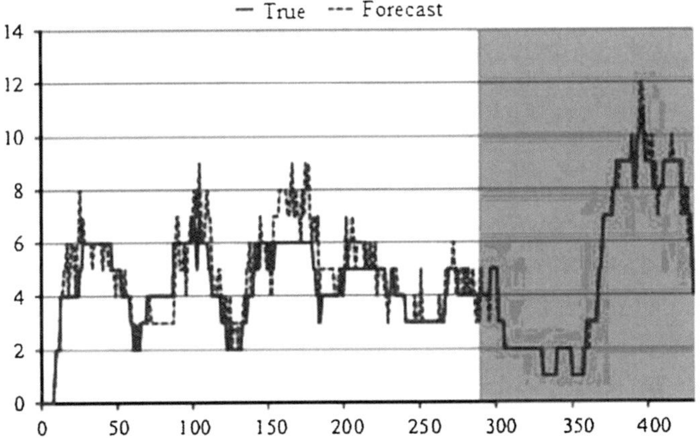

Fig. 81.3 Implicit standardization

the location of the neural networks rely on the image with very accurate. This is partly because the implicit normalization of image features included in the remote areas, high and clear normalization of extraction in small water droplets and the accuracy of the eigenvalue. As shown in Fig. 81.3.

81.5 Conclusion

In this paper, we propose a method to estimate the pedestrians' number by real-time video sequence. We have shown that the method of execution speed of more than 10 FPS and its accuracy is higher than 80 %. The remaining works are as follows: background subtraction improved: In the proposed method, background subtraction is used to split. For background subtraction, we have used the background model, using the Parzen window density estimation. Because this model is the observed pixel value based on the latest N-frame sequence, pedestrians living in the same location into the background model, they cannot extract correctly. Therefore, in these cases, neural networks often output incorrect estimates, such as pedestrians and cannot correct the error as the label of nonpedestrian areas. The detection of spots adaptive modeling, fixed spots around the background model update problem will be solved. Function: better than those who use this method may be a better function. Therefore, we must try a variety of feature sets in order to obtain better performance.

References

1. Zhang Y (2001) Image segmentation, vol 09. Science Press, Beijing, pp 201–203
2. Liu C (2010) The pedestrian counting method based on gradient orientation histogram. Bei Hang Univ 08:67–69
3. Wang Q, Feng Y (2010) The rapid statistics method based on color and shape information. Comput Measur Control 18(9):2157–2159
4. Fu S, Zhang X (2004) The moving targets real-time detection method based on image sequences. Opt Technol 30:103–105
5. Zhu M, Lu J (2003) The real-time digital image processing system based on TMS320C6202 DSP. Opt Precis Eng 05:135–138
6. Shu R (2010) Research on human detection method based on gradient orientation histogram. Henan Polytech Univ 09:2157–2159
7. Zhao M (2009) Research on the technology of automatic passenger counting based on the recognition of multiple moving targets. Chongqing Univ 05:45–48
8. Li M (2006) The applied research in the APC of video moving target detection and tracking. Chongqing Univ 08:67–70

Chapter 82
Information Security Evaluation System

Hong-lei Lu and Erxi Zhu

Abstract The level of analysis, study of information security evaluation index system, identified the main index and weight coefficients, and was judged on consistency to test the weight is reasonable, so that the weight was more in line with objective reality, and easy-to-quantified, and thus improved the comprehensive evaluation of the reliability and accuracy. For quantitative analysis of network information security provides an effective method.

Keywords Information security • Evaluation method • Analytic hierarchy process • Weight coefficient

82.1 Introduction

With the rapid development of information technology and extensive application, one of the growing dependence of information, information security has become an important component of national security [1, 2]; effective protection of national critical infrastructure and important safety information systems to enhance information security risk management has increased as a matter of national political stability, social stability, and orderly operation of the global economic problems. Information security risk management is a systematic project, the need for technical and nontechnical means, related to safety management, education, training, legislation, emergency recovery, and other aspects of cooperation [3, 4].

But how to build on the information security system to measure success or the construction of the information security system effective? How can the effectiveness of this quantitative assessment, etc., which are urgently needed set of

H. Lu (✉) · E. Zhu
Dept of Computer Engineering, Jiangsu College of Information Technology, Wuxi 214153, China
e-mail: sszdha@yeah.net

Z. Zhong (ed.), *Proceedings of the International Conference on Information Engineering and Applications (IEA) 2012*, Lecture Notes in Electrical Engineering 220,
DOI: 10.1007/978-1-4471-4844-9_82,

evaluation criteria, methods, and tools. Evaluation of traditional information security is often single qualitative or quantitative indicator of the lack of systematic, targeted, non-authoritative assessment of the mutual recognition of evaluation results are also out of the question. So, how to regulate the security of information systems security system, how to conduct a security assessment of information systems, which is in the process of information facing enormous challenges—we need to establish a scientific, quantifiable, and actionable information security evaluation system.

Comprehensive evaluation of information security, the core is to establish an information security evaluation index system. This requires from three considerations. First is to establish a good indicator system framework; the second is what kind of evaluation system for the comprehensive evaluation methods and the third is indicator data collection methods and standards. At present, the evaluation of network information security, there are still many problems, such as the lack of scientific evaluation of the quantitative basis, mostly based on experience or simple arithmetic average to determine; evaluation results more abstract, difficult to show that the evaluation of the actual situation was. The above problems, in this paper, Analytic of Hierarchy Process (AHP) network information security evaluation system were studied.

82.2 Main Index

AHP, the first of many complex factors from the selection of the most important indicators of critical evaluation, and according to the constraints of relations between them constitute a multilevel index system. The quality indicator system to evaluate the structural prerequisite for success the number of index system-level complexity of the issues from the inspection and evaluation of the accuracy to be achieved by request. To this end, the establishment of the network information security evaluation index system.

Information Security Evaluation and Certification in the application and run the standard type, the country has developed a number of evaluation criteria.

We will evaluate the indicators. They are divided into three indicators; level indicators, including physical environment, information security products, systems, and networks, which is an integrated indicator of the nature, and no specific form of expression; second indicator is the specific target projects, including covering the entire process of network information security; three indicators in the secondary index is more specific under some of the projects, due to the large number, only an example. Evaluation indicators can be added according to actual situation, delete changes, but any kind of evaluation methods are aimed at evaluators can be effective monitoring and control, in order to improve certain aspects of the evaluators were in the process should be adjusted in some of the specific content evaluation.

Specific evaluation index system as shown in Table 82.1

Table 82.1 Network security evaluation system

The main factors		Secondary factor		
Level indicators	Weight	Secondary indicators	Weight	Level indicators
Physical environmental indicators	A1	Computer security site	B11	Completion, complete quality and progress etc.
		Low supply voltage fluctuation	B12	
		EMC limits	B13	
		EMC test and measurement	B14	
		The security of information technology equipment	B15	
		Anti-around information technology equipment and measuring limits	B16	
		Low-voltage electrical current limit	B17	
Information security index	A2	Network proxy server security technology	B21	The number of personnel, equipment etc.
		Router security technology	B22	
		Information technology security technology	B23	
		Safety filtering firewall	B24	
		Application-level firewall security technology	B25	
		Information technology development	B26	
		Information technology production	B27	
		Information technology evaluation	B28	
		IT procurement	B29	
Systems and network indicator	A3	Information technology open	B31	Number, type, strategy etc.
		Interconnection of information technology	B32	
		Information technology interconnection network layer security	B33	
		Information technology systems management	B34	
		Information technology systems security	B35	
		Police information technology	B36	
		Information technology report	B37	

82.3 Comparison Matrix and Determine the Weight

We take the AHP to carry out; this method can effectively solve multiobjective evaluation system, multijudge situation.

This paper will have two levels, that is the main factor layer and the subelement level. Under each of the main factors contributing to containing several subfactors, so, first through the various subfactors under the main factors and the relative importance of pairwise comparisons between and with reference to the hierarchy chart, comparison matrix can be constructed. Physical factors of environmental indicators to determine the submatrix shown in Table 82.2.

To solve the above matrix, obtained by feature vectors $w_1 = (0.333, 0.228, 0.121, 0.163, 0.058, 0.038, 0.060)^T$, the largest eigenvalue $\lambda max_1 = 7.049$. Information security products subfactors that determine the matrix of indicators are shown in Table 82.3.

To solve the above matrix, obtained by feature vectors $w_2 = (0.019, 0.036, 0.045, 0.086, 0.161, 0.116, 0.231, 0.122, 0.184)^T$, the largest eigenvalue $\lambda max_2 = 9.012$. Systems and networks secondary factor indicators are shown in Table 82.4 Comparison Matrix:

To solve the above matrix, obtained by feature vector $w = (0.340, 0.257, 0.082, 0.102, 0.123, 0.038, 0.058)^T$, the largest eigenvalue $\lambda max_3 = 7.129$. Comparison of the main factors that determine the matrix is shown in Table 82.5.

Table 82.2 Matrix of physical environmental indicators

	B11	B12	B13	B14	B15	B16	B17
B11	1	5	3	2	7	4	1
B12	1/5	1	7	4	7	3	6
B13	1/3	1/7	1	2	3	5	7
B14	1/2	1/4	1/2	1	7	7	9
B15	1/7	1/7	1/3	1/7	1	5	3
B16	1/4	1/3	1/5	1/7	1/5	1	2
B17	1	1/6	1/7	1/9	1/3	1/2	1

Table 82.3 Information security index matrix

	B21	B22	B23	B24	B25	B26	B27	B28	B29
B21	1	1/5	1/3	1/7	1/9	1/5	1/6	1/4	1/2
B22	5	1	1/2	1/5	1/7	1/6	1/2	1/4	1/3
B23	3	2	1	1/7	1/9	1/5	1/3	1/2	1/3
B24	7	5	7	1	1/7	1/3	1/4	1/2	1/3
B25	9	7	9	7	1	1/2	1/3	1	1/2
B26	5	6	5	3	2	1	1/7	1	1/2
B27	6	2	3	4	3	7	1	1	1
B28	4	4	2	2	1	1	1	1	1/3
B29	2	3	3	3	2	2	1	3	1

Table 82.4 Systems and network index matrix

	B31	B32	B33	B34	B35	B36	B37
B31	1	5	7	5	3	5	2
B32	1/5	1	9	4	7	5	2
B33	1/7	1/9	1	1	2	2	3
B34	1/5	1/4	1	1	2	3	4
B35	1/3	1/7	1/2	1/2	1	5	7
B36	1/5	1/5	1/2	1/3	1/5	1	1
B37	1/2	1/2	1/3	1/4	1/7	1	1

Table 82.5 Comparison of the main factors that determine the matrix

	A1	A2	A3
A1	1	3	2
A2	1/3	1	2
A3	1/2	1/2	1

To solve the above matrix, obtained by feature vector $w = (0.537, 0.268, 0.195)^T$, the largest eigenvalue $\lambda max_3 = 3.006$. Therefore, the main factors relative to the level indicators, several factors should be sorted according to weight: the importance of security to sort the order should be: (1) physical environmental indicators; (2) information security products index; (3) system network indicators.

82.4 Consistencies

Determine the consistency test is to identify the weight coefficient matrix an obtained is reasonable. A fully consistent with the time, $\lambda max = n$, and in addition λmax, the other characteristic roots are 0. When the comparison matrix has satisfactory consistency, it is slightly larger than the largest characteristic root matrix of order n, and the remaining eigenvalues close to 0. This level of analysis based on the conclusion is basically reasonable.

Testing process is as follows: the matrix obtained λmax of the judge on behalf of the $C.I. = (\lambda max - n)/(n-1)$ Consistency index C.I. derived values; based on the average random consistency index R.I. numerical tables found R.I. values; Finally, $C.R. = C.I./R.I.$ relative consistency index obtained C.R. values. When $C.R. < 0.1$, the comparison matrix is reasonable, appropriate weight coefficient obtained, or to determine the matrix of adjustment, according to the above steps to re-find the weight coefficient matrix.

Judgment on the consistency of F–S to determine:

$C.I._1 = (\lambda max_1 - n_1)/(n_1 - 1) = (7.049-7)/(7-1) = 0.008$, properly check the following table matrix order $N = 7$, $RI = 1.32$, this time, $CR_1 = CI_1/RI_1 = 0.008/1.32 = 0.006 < 0.1$. So that the matrix has satisfactory consistency. That is determined by AHP in the cells of the weight factor is objective and acceptable.

Table 82.6 Matrix order and the corresponding RI value

Matrix order	1	2	3	4	5	6	7	8	9	10	11	12	13	14
RI values	0.00	0.00	0.58	0.90	0.12	1.24	1.32	1.41	1.45	1.49	1.51	1.48	1.56	1.57

C.I.$_2$ = ($\lambda max_2 - n_2$)/($n_2 - 1$) = (9.012−9)/(9−1) = 0.002, Table 82.6 matrix by properly check order N = 9, RI = 1.45, this time, $CR_2 = CI_2/RI_2 = 0.002/1.45 = 0.001 < 0.1$. So that the matrix has satisfactory consistency. That is determined by AHP in the cells of the weight factor is objective and acceptable.

C.I.$_3$ = ($\lambda max_3 - n_3$)/($n_3 - 1$) = (7.129−7)/(7−1) = 0.021, Table 82.6 matrix by properly check order N = 7, RI = 1.32, this time, $CR_3 = CI_3/RI_3 = 0.021/1.32 = 0.016 < 0.1$. So that the matrix has satisfactory consistency. That is determined by AHP in the cells of the weight factor is objective and acceptable.

C.I.$_4$ = ($\lambda max_4 - n$)/($n - 1$) = (3.006−3)/(3−1) = 0.003, Table 82.6 matrix by properly check order N = 3, RI = 0.58, this time, CR = CI/RI = 0.003/0.58 = 0.005 < 0.1. So that the matrix has satisfactory consistency. That is determined by AHP in the cells of the weight factor is objective and acceptable.

82.5 Weight of Each Index to Determine the Integrated

Obtained in the same hierarchical level between the various elements of the relative importance, you can top–down factors on the calculation of the overall integrated at all levels of importance. In this article, there are three main factors. A layer of A_1, A_2, A_3, its goal has been the importance of the total respectively, W_1, W_2, W_3, therefore, A layer of the main factors combined weight: $W_k = \Sigma W_{ij} \times W_{jk} = (0.537, 0.268, 0.195)$, shown in Table 82.6

82.6 Conclusions

This chapter studies how to establish information security evaluation index system, according to certain analytical framework, evaluation criteria and methodology, the use of AHP to determine weights, the use of rigorous mathematical methods to delete as much as possible subjective and judged whether the matrix has satisfactory to test the consistency of a reasonable weight, making the weight more easily quantified objective reality and to improve the comprehensive evaluation of the reliability, accuracy and objective impartiality. For quantitative analysis and determination of information security degree of realization of the development process or provide a try.

References

1. Vaughn RB Jr, Henning R, Fox Kevin (2002) An empirical study of industrial security-engineering practices. Syst Softw J 61(53):225–232
2. Lee JS (2002) Uncertainty analysis in dam safety risk assessment, in civil and environmental engineering. Utah State Univ 53(32):6–13
3. Mohan J, Razali R, Yaacob R (2004) The malaysian telehealth flagship application: a national approach to health data protection and utilisation and consumer rights. Med Inform 73(35):217–227
4. Pack S, Choi Y (2004) Fast handoff scheme based on mobility prediction in public wireless LAN systems. IEEE Proc Commun 151(5):489–495

Chapter 83
Real Estate Management Information System

Yanzhen Sun

Abstract Real estate management information system for database design problem use Microsoft SQL Server 2000 as a database development tool, real estate management information system database for the detailed design, and gives the database connection login and key code, proved. The design can ensure the security of real estate management and efficient operation of information systems to adapt to the real estate information management needs.

Keywords Real estate • Management information systems • Databases • B/S

83.1 Introduction

In recent years, China's hot real estate market continued to show momentum, all the real estate company have developed a large number of commercials, offices, and so on. The housing market includes not only houses, but also second-hand housing market, so the classification of real estate data becomes increasingly important. In the absence of the formation of a mature management system, data management is chaotic, query, statistics are not comprehensive [1]. The management information system for real estate sales in the scientific, standardized, network requirements, the sale of real estate rental information, buy Quiz information, real estate until a comprehensive management information, both to improve efficiency, and reduce the workload and also faster and more accurate, can the level of enterprise management to achieve higher level of real estate management systems and other management information systems, whether using C/S mode or B/S mode, the database needs to have the background support. The system is running the real estate information management foundation and source of the data. James Martin, an American scholar pointed out: in the enterprise's data processing

Y. Sun (✉)
Qinghai University, Xining 810016, QingHai, China
e-mail: whejxiong@126.com

Z. Zhong (ed.), *Proceedings of the International Conference on Information Engineering and Applications (IEA) 2012*, Lecture Notes in Electrical Engineering 220,
DOI: 10.1007/978-1-4471-4844-9_83,

work in the "data is stable; treatment is changing the data in modern data processing center". Enterprise data should be the overall planning and organization, and establish a unified database management [2].

General background database development tool that is Oracle, DB2, and other development tools for large databases, there are Microsoft SQL Server, Visual FoxPro, Access, and other database development tools for small and medium sized [3]. The design is based on B/S model, the use of Microsoft SQL Server 2000 as a database development tool.

83.2 The Basic Structure of the System

For real estate management information system, the overall objective is to provide users and administrators rapid and efficient information services, manage the real estate information service rationalization. The demand for real estate companies is to manage the system users, management, and real estate information. Real estate information management information management is mainly the sale of rental, buy Quiz information management, real estate management, quality housing management, quality real estate management, real estate information inquiries. The tenants are mainly residential real estate information query [4].

System design, a basic requirement is the ability to input, query, modify the various real estate related information, such as revising the sale of rental information, buy Quiz information, update real estate related information, such as real estate information, update and delete and so on, the system's modules structure shown in Fig. 83.1. Real estate information management system by the powerful query capabilities, real estate information management companies in the information shared on the basis of height, effectively improve its management tools, standards and adjust their business processes and improve business processes of various functional departments of reliability and operational efficiency, and then on the various management systems organize and analyze data, provide decision support for the leadership of the basis for the rational allocation of enterprise resources [5, 6].

83.3 Database Design

The main database of information processing solves three problems: first, the effective organization of the data here mainly refers to the rational design of data to computer access; second, easily input the data into the computer; third, according to the requirements of the user data are extracted from the computer [5].

Real estate management information system database design involved five steps: database requirements analysis, conceptual design, logical design, physical design, and load testing. Information system's main task is to obtain large amounts

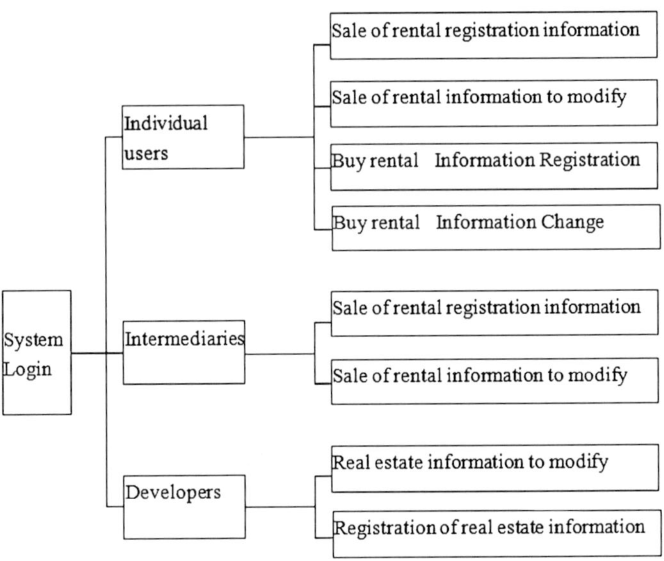

Fig. 83.1 System block diagram

of data through the information management needs, which must store and manage large amounts of data. Thus, a good data structure and database, so that the whole system can quickly and easily and accurately call and manage the data required is a measure of information systems development the key indicators of good or bad.

We often use the database concepts including tables, fields, views, indexes, synonyms; some systems will be used in the system tables, database design, logical database, mainly for the design, about the classification of data according to some, group organized system and logical level is user oriented. Database design needs to integrate each department of the archived data and data needs, analysis of the relationship between the various data, in accordance with the description of DBMS features and tools provided to design an appropriate scale, correctly reflect the data relations, data redundancy less access efficiency high, to meet the requirements of a variety of query data model.

83.3.1 Conceptual Design Database

Conceptual design task is to precede from reality, to draw out the system entity–relationship diagram, and list the relationship between the various entities and the corresponding table. Designers from the user perspective on data and processing requirements and constraints, to produce a conceptual model reflecting the user point of view. Conceptual model and then converted into the logical model.

Conceptual design will be independent from the design process open, so that the task of each stage is relatively uniform, the design complexity significantly reduced, without a specific DBMS limitations.

This system is based on "entity–relationship model" the concept of database design concepts. "Entity–relationship model" is a semantic model, which tries to the relationship between the world of things has its own characteristics, awakened the abstract and that they also describe the linkages and interactions. In the "entity–relationship model", the basic semantic unit is the entities and relationships, which entity is the information can be identified by its own abstract representation of anything. From specific people, objects, events, to other abstract concepts and status, can be expressed with abstract questions. Relationship is something internal, or things between the abstract representations of semantic relationships. Links between entities in different soybean oil, we can accord the association between entities linked to classification.

System entity relationship ER diagram shown in Fig. 83.2, real estate entities ER diagram shown in Fig. 83.3.

83.3.2 The Design and Operation of the Data Table

As two-dimensional relational database tables can clearly describe the link between the data in the ER diagram design, the use of database normalization design, the design of the database tables should also follow the paradigm of the design, which is the basis for database design, to normalize the database design, to minimize data redundancy and duplication, a reasonable design of the database table should be the premise of meeting the demand, so that the minimum amount of data duplication, as much as possible in accordance with the relational database system theory approach to designing a database reduce data redundancy.

In the design of the database structure, sometimes in order to achieve the simple programming and ideas clear, often deliberately increase the number of

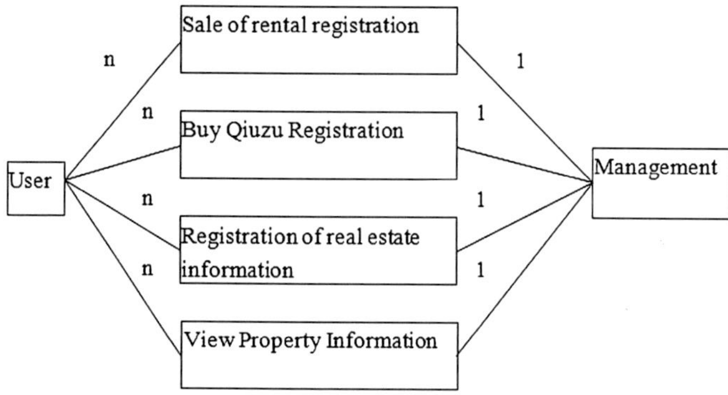

Fig. 83.2 E-R diagram entity relationship

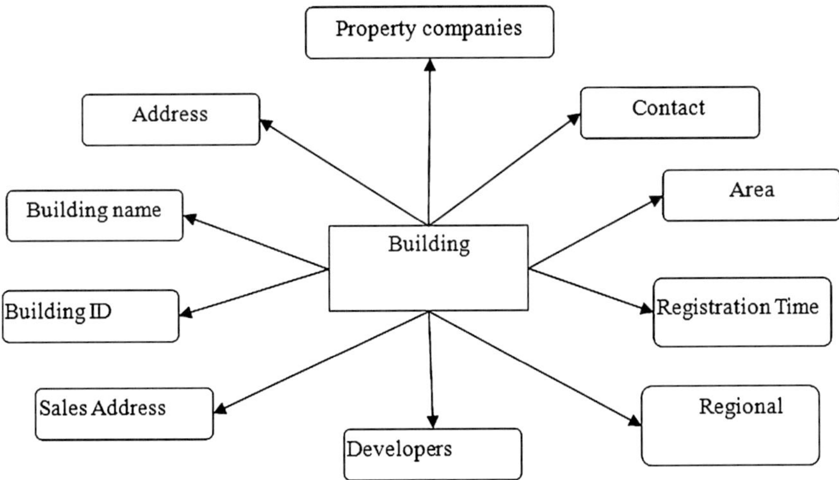

Fig. 83.3 E-R diagram real estate entities

redundant data. Although this is contrary to the traditional relational database theory, but given the choice of machine type and grade of other factors, if the increase will not significantly increase redundancy and reduce storage space efficiency, and the great help of programming, often using this method.

Table defined in the system is strictly in accordance with the ideas and requirements paradigm to complete, database of all forms have reached the requirements of the three paradigms. Based on the above three design principles, according to the system ER diagram for the system characteristics, standardization of the data collected.

The basic operation of the data table is the basic property information management system functions; database management system software is the most basic operations. The basic operation of the datasheet information, including real estate add, delete, modify, and other operations. Real estate Information added feature: the user manually increase the real estate information.

Increasing function of real estate information: the user manually increases the real estate information.

Real estate information delete function: to be deleted by the user to select real estate information, click the Delete, delete the property information, first remove the buffer, then submitted to the database if not, you may revoke the removal, but if submitted to the database, and delete will not restore the information.

Real estate information editing: the user selects to modify the real estate information; you can re-modify the property registration information.

83.3.3 Database Login and Connection

Database connectivity is mainly used for database connection stored in the system all the information, do not have to log in again to set the database, but data in the database changes, the user can configure the input to the unit or the data on the LAN server name and database name. If the application and the database connection are successful, run the program the next time this configuration will no longer have pop-up window, unless there are changes to the database server that will pop up.

Database connection in the entire system need to log in several times to run, is the realization of the most important function of the system, and database connectivity is not, the system will not be able to run. Log in achieving the key code database is as follows:

```
protected void Button4_Click (object sender)
{
Connect String = "Data Source = FB9304AE5EB04F5; Initial Catalog = fangchan;
  Integrated Security = True";
Sql Connection conn = new Sql Connection (connect String);
if (DropDownList5.SelectedItem.Text.ToString () == "personal" )
  {
  string sql String = "select * from members of the table where the user
     name = '"+ this.TextBox3.Text +"' and password = '"+ this.TextBox4.
     Text +"' and user type = 'personal'";
  conn. Open ();
  Sql Command cmd = new Sql Command (sql String, conn);
  Sql Data Reader reader = cmd. Execute Reader ();
  if (reader. Read () == true)
  {
  Session ["yonghuming"] = this.TextBox3.Text;
  Session ["yonghuleixing"] = "personal";
  Response. Write (" <Script language = JavaScript > alert ('Login successful !!');
     </Script >");
  }
  else Response. Write ("< Script language = JavaScript > alert ('Login failed,
     please check the username or password is correct, then log !!'); </Script >");
  reader. Close ();
  conn. Close ();
}
```

Button4 which is the "Login" button, click "Login" button to be connected with the database to achieve the corresponding functions. Should be mentioned that the code in the background, connect the data source need to use part of the connection code is:

String connection String = System. Configuration. Configuration Manager. App Settings ["weirong"]. To String ();

Add "using System. Data. SqlClient"; in the Web. Config file and then add the following code:

```
<configuration>
<app Settings>
```

<add key = “weirong” value = “Data Source = FB9304AE5EB04F5; Initial Catalog = weirong; Integrated Security = True”/>
</App Settings>
<connection Strings>
<add name = “weirong” connection String = “Data Source = FB9304AE5EB04F5, Initial Catalog = weirong; Integrated Security = True” provider Name = “System. Data. Sql Client”/>
</Connection Strings>

83.4 Conclusion

Currently, in most of the real estate agencies, real estate information management section, or the manual management, efficiency is very low and cannot keep abreast with the various types of property in the hands of information and real estate market, compare needs, develop information management system for real estate is very necessary, but in the process of real estate management information systems, database design is very important, the database design is the core of management information systems and infrastructure, it puts a lot of the information system model of data organized according to some, to provide storage, maintenance, the function of retrieving data, so that information systems can be convenient, timely and accurate from the database to obtain the necessary information. Scientific and effective during the database design, to ensure efficient operation of the system.

References

1. Shan W, Sa SX (2006) Database system, vol 34(13). Higher Education Press, Beijing, pp 103–286
2. Davidson L (2003) SQL Server 2000 database design the definitive guide, vol 7(2). Electric Power Press, Beijing, pp 45–108
3. Zhao YJ (2008) Database access in VB application. Bus Technol 18(5):20–21
4. Jin H, Wong K-F (2002) A Chinese Dictionary constitution algorithm for information retrieval. In: ACM transactions on Asian language information processing, vol 9(5). ACM Press, USA: New York, pp 281–296
5. Emmerich W (2006) Engineering distributed objects, vol 13(5). Wiley, USA: New York, pp 23–28
6. Yang X (2006) EJB Architecture Based On Hierarchical Model of Structure. In: The Changbai, the regular Juan Tianjin polytechnic university, vol 7(2), pp 6–8

Chapter 84
Research on Data Transmission

Yuqing Mo

Abstract This paper mainly introduces the data transmission and datagram sending and receiving procedures. The datagram sending and receiving are realized through the two classes of VC++ MFC. Currently, the frequently used network protocols include TCP, RTP, and UDP, among which the most complex one is TCP with the highest reliability but poor efficiency. Generally, RTP protocol is applied to real-time streaming media transmission with less reliability than TCP, and also relies on UDP but with poorer efficiency than UDP. UDP is the simplest transport layer protocol without controls on errors, traffics, and congestions. Its simplicity brings high efficiency, but it is only responsible for data transmission, leaving the traffic and error controls to the high layer to solve, so by such a way it can shorten data transmission time to the minimum and improve the efficiency of data transmission system.

Keywords UDP protocol • Data transmission • Receive message • Sending message

84.1 Introduction

The sending-end host sends message to the receiving-end host with the receiving-end front-end processor responsible for the receiving of message [1, 2]. After receiving the message, the front-end processor sends a broadcasting of UDP in LAN to inform related receiving-end hosts of processing message [3, 4].

Y. Mo (✉)
Hunan College of Information, Changsha 410200, Hunan, China
e-mail: moyuqing@hrsk.net

Z. Zhong (ed.), *Proceedings of the International Conference on Information Engineering and Applications (IEA) 2012*, Lecture Notes in Electrical Engineering 220,
DOI: 10.1007/978-1-4471-4844-9_84, © Springer-Verlag London 2013

84.1.1 Arithmetic Thoughts

A file is divided into several data blocks; sender first sends "Request to Send" message, and applies for a certain number of data blocks [5, 6]. The "Request to Send" message carries the MD5 message digest of the file to be sent; and receiver can retrieve buffer pool based on "Message Digest". If no buffer area of this file is in buffer pool, it is necessary to allocate buffer to this file and also initialize the block list [7].

Receiver can search unallocated data blocks in block lists, and then send "Clear to Send" message to sender to tell the data block able to be continuously sent at a time. After these data blocks are sent completely, sender sends "Clear to Send" message again to apply for the data block to be sent next time.

If multiple senders simultaneously send files to the same receiver, he can decide whether they are sending the same file (with the same content but different files name) according to the message digest in "Request to Send" message. If the files all senders are sending are the same, the receiver can allocate the data blocks of file to multiple senders according to the network bandwidth. At this moment, although each sender does not know where the senders are, the sent files can be shared with other senders. Namely, the data block sent by a sender is unnecessary to be re-sent by other senders.

84.1.2 Analysis on Cases

Figure 84.1 describes that multiple hosts send message to one receiving end (communication front-end processor). The sender has four message sending hosts which can be allocated to different LAN, among which host 1, 2, 3 send datagram "A", while host 4 sends datagram "B".

It is assumed that host 1 first sends datagram "A". It sends "Request to Send" message by which sender can apply for the block number and quantity of the data blocks able to be continuously sent next time. This message carries the MD5 "Message Digest" and file length of datagram "A". The message digest stands for the features of datagram "A", applicable to differentiate other datagram, while the message length is used by receiver to allocate block lists and decide their sizes in buffer. For example, each data block is 5 K (the size of data block is a constant value agreed by both receiver and sender); if the length of datagram "A" is 20 K, then "A" should be divided into four blocks and needs four items in block list with 8 bit of each list standing for three status (Received, Allocated and Unallocated) of data block. The size of block list $= 4 + 4 + 4 + 4 * 8 / 8 = 16$ bytes. By retrieving buffer, receiver can find a new buffer allocated to datagram "A" in buffer pool, which was not received by datagram "A" in the past time; and also creates initialization list for datagram "A" in buffer based on datagram length. At the moment, the status of all data blocks in block lists is "Unallocated". After the receiver is in

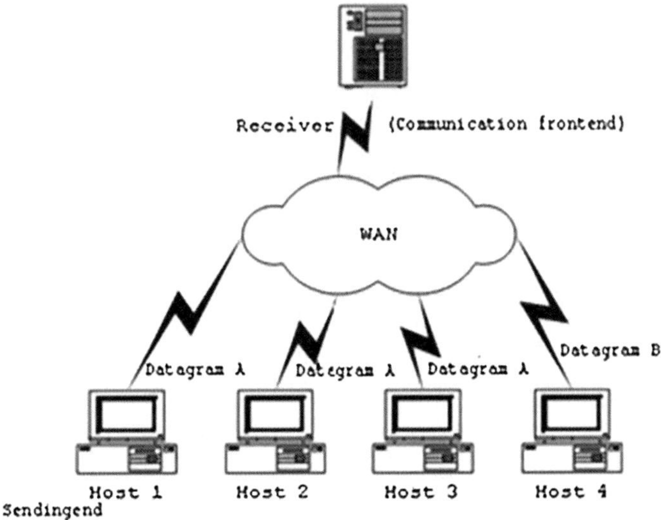

Fig. 84.1 Send datagram from multiple points

initialization buffer, it sends "Clear to Send" message to host 1, which points out the block number and quantity of the data block able to be continuously sent by sender at a time, in which the process is called as "Load Sharing". After data block is allocated, the status of this data block in block list is altered into "Allocated".

After a period of time, it is assumed that datagram "A" successfully sends 20 data blocks whose status has altered into "Received" in block lists of receiver's buffer. At this moment, host 2 participates in the sending row, which also sent datagram "A" to the same receiver (communication front-end processor). Host 2 sends "Request to send" datagram to receiver (communication front-end processor) which relies on the "Message Digest" to find that the message sent by host 2 is the same message being sent by host 1, and hence receiver allocates several left data blocks to host 2 and also alters the status of these data blocks into "Allocated" in block lists.

Host 1 and 2 simultaneously send datagram "A", and both hosts will not repeatedly send the same data block. Moreover, the data block sent by one host can be shared with the other host. Multiple senders simultaneously sending the same file to one receiver can accelerate the file transfer.

Through the common sending of host 1, 2, the communication front-end processor completely receives datagram "A". At this moment, host 3 requests datagram "A" from receiver (communication front-end processor) which finds the status of all data blocks is "Received" by retrieving buffer pool and then sends "Clear to Send" message to host 3, and sets 0 for the quantity of the allocated data blocks, indicating this message had been completely received. At this moment, as for host 3, it actually does not care about whether datagram "A" is sent by other processors and only feels the sending of datagram "A" is completed instantly, promoting the file transmission speed to be lifted greatly.

When receiver (communication front-end processor) receives datagram “A”, it can also receive the message sent by other processors (e.g. simultaneously receiving the datagram “B” of host 4).

84.2 The Sending Datagram and Receiving Datagram

Sending datagram and receiving datagram include five formats totally, all of which are sent in UDP. As UDP datagram header length field only has 2 bytes but it should take up 8 bytes, the length of the largest datagram is “65,535 – 8 = 65,527 bytes”.

84.2.1 The Sending Datagram

84.2.1.1 The Sending Message

Messages sent by sender include two formats of “Request to Send” and “Datagram”.

“Request to Send” is that sender sends signals to receiver. “Datagram” is the data information sent by sender. Moreover, only one data block can be sent at a time.

84.2.1.2 The Procedure of the Sending Message

Sender (one or multiple senders) simultaneously sends files to receiver, and all senders are unnecessary to coordinate with each other. Each sender can join to send the files anytime and also can exit anytime. The “Load Sharing” is completed by receiver. What the senders mainly need to solve is the control on traffics, and they can adjust the sending rate of message based on the successful sending rate, applying the binary exponents to retreat the algorithm to adjust the sending rate. Also, it is necessary for senders to carry on timeout retransmission for the “Request to Send” message.

In order to improve the efficiency of the program execution, it is necessary to create two threads. One is to send message and the other is to listen in message. The two threads communicate with each other through messages and semaphores.

84.2.2 The Receiving Message

84.2.2.1 The Messages Sent by Receiver Include “Enabling Signal”, “Suspended Signal”, and “Ended Signals”

“Enabling signal” is the response to the “Request to Send” message sent by sender, and informs the block number and quantity of the data block able to be continuously sent by sender next time. “Suspended Signal” is to control traffics and informs sender of suspending to send datagram. "Ending Signals" is to inform sender that all data blocks have been received and the data receiving was successfully completed.

84.2.2.2 The Procedure of the Receiving Message

Receiver mainly considers the identification of data source, the management of buffer pool, the dynamic sharing of load, etc. The procedure of message transmission is as shown in Fig. 84.2.

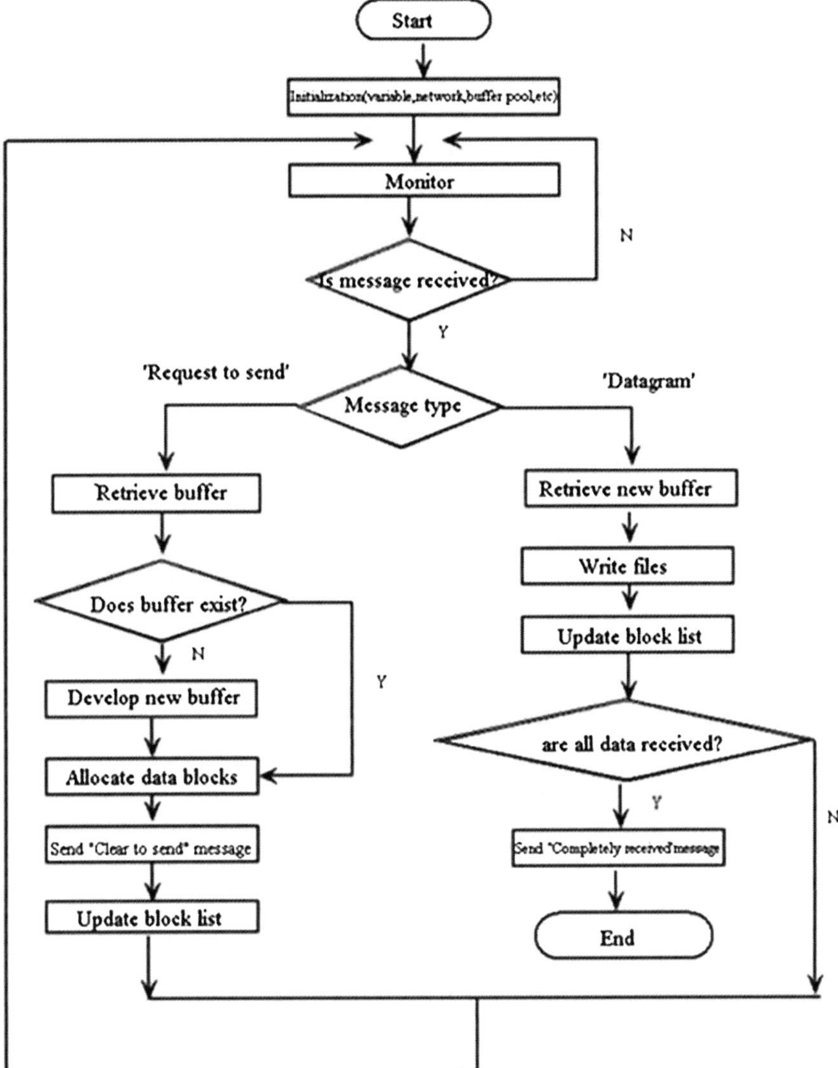

Fig. 84.2 The procedure of the receiving message

84.2.3 Class CSocketSend and Class CSocketReceive

VC++ MFC provides high supports for network, and is easy to create the network-based application. It mainly includes class CAsyncSocket and its derived class CSocket. Class CAsyncSocket encapsulates Windows Socket API, and has the advanced features of asynchronous I/O, able to provide flexible network communication functions and also convenient to process the occurring network events. Class CSocket is derived based on class CAsyncSocket, providing class CAchive similar to file operation and also high-layer supports for Winsock-based programs.

Class CSocket works in the congestion mode, and is easy to use but with poor flexibility. However, CAsyncSocket encapsulates the lower layer of Windows Socket API, and is able to flexibly control on network program by overriding CAsyncSocket network events. By such a way, the network communication program can acquire good performance.

By inheriting class CAsyncSocket, the two basic classes applicable to data communication are realized: Class CSocketSend (Data sending) and Class CSocketReceive (Data receiving), which will be simply introduced in the following:

84.2.3.1 Class CSocketSend

It is a derived class of CAsyncSocket, and realizes the sending function of data. In order to be easily understood, a great number of details are eliminated and only the basic sending procedure is described in the following:

```
Class CSocketSend : public CAsyncSocket
{
Private:
CObList m_DataList; //Linked list to be sent
UINT m_nRemotePort; //Remote Port
CCriticalSection m_cslock; //Key area
CWinThread *m_pThread; //Thread object
Private:
UINT static DataSendProc (LPVOID pParam); //Thread processing function
Public:
CSocketSend();
Virtual~CSocketSend();
Public:
BOOL Create (UINT nRemotePort =720); //Initialization & create the sending
  thread
SendData (LPVOID lpBuf, int nBufLen); //add data into list and wake up thread
};
```

In the method of "Create", the interface parameter defines the UDP port of remote host. It creates a socket and hooks object, and binds it to the local port, and creates a data sending processing thread. The data buffer of the object monitored by the thread will suspend if it is null, in order to save CPU resources.

```
BOOL CSocketSend::Create (UINT nRemotePort)
{
m_nRemotePort=nRemotePort;
BOOL bRes=CAsyncSocket::Create (0, SOCK_DGRAM);
m_pThread=AfxBeginThread (DataSendProc, this, THREAD_PRIORITY_
  NORMAL, 0, CREATE_SUSPENDED);
Return bRes;
}
```

In the method of "SendData", data are added to buffer linked list and immediately returns after waking up the data sending processing thread. The design extremely improves the response speed of program, which makes the program be able to quickly return and needless to wait for the completion of the data sending.

```
CSocketSend::SendData(CByteArray myByteArray)
{
m_csLock.Lock(); //lock key area
m_DataList.AddTail (myByteArray);
m_csLock.Unlock ();
m_pThread-> ResumeThread ();
Return;
}
```

The data sending processing thread function sends data to network and will suspend until the buffer is null.

```
UINT CSocketSend::DataSendProc(LPVOID pRaram)
{
CSocketSend *pcsSend=(CSocketSend*)pParam;
While (true)
{
If (pcsSend-> m_DataList.isEmpty())
pcsSend-> m_pThread-> SuspendThread();
Else
{
pcsSend-> m_csLock.Lock();
CByteArray mySend=pcsSend-> m_DataList.RemoveHead();
pcsSend-> m_csLock.Unlock();
Sleep (nDelay) //delay
pcsSend-> SendTo(mySend.GetData(),
mySend.GetSize(), pcsSend-> m_nRemotePort);
}}
Return 0;
}
```

84.2.3.2 Class CSocketReceive

It is also a derived class of CAsyncSocket, and realized data receiving function. Similarly, the following list will only show demonstrative description.

```
Class CSocketReceive: public CAsyncSocket
{
Private:
CObList m_DataList; //The receiving buffer
UINT m_nLocalPort; //local port
CCriticalSection m_csLock; //key area
CWinThread* m_pThread; //thread object pointer
Private:
UINT static DataReceiveProc(LPVOID pParam); //data receiving thread function
Virtual~CSocketReceive();
Public:
BOOL Create (CWnd* pWnd, UINT nLocalPort=1204); //Initialization & create
   the receiving thread
CString GetData(); //get data
Private:
CWnd* m_pWnd; //message sending window
}
```

The tasks of the data receiving class CSocketReceive are receiving data and informing the primary thread of the receiving of the data. The "Create" function of the class includes two parameters (pWnd and nReceivePort). The pWnd is the pointer to the window to receive the user-defined messages sent by the receiving data thread. The nReceivePort specifies the port to receive data. The function creates the local socket and binds it to the local port nReceivePort, and also creates a data receiving processing thread, informing the function "OnReceive" of waking up it when there is data to come.

```
BOOL CSocketReceive::Create (CWnd* pWnd, UINT nLocalPort)
{
m_pWnd=pWnd;
m_pLocalPort=nLocalPort;
BOOL bRes =CAsyncSocket:Create(m_nLocalPort, SOCK_DGRAM);
m_pThread=AfxBeginThread(DataReceiveProc,this,THREAD_PRIORITY_
   NORMAL,0,
CREATE_SUSPENDED);
Return bRes;
}
```

When UDP data packets come, class CSocketReceive will auto-call the function "OnReceive" which will wake up the data receiving thread.

```
Void CSocketReceive::OnReceive (int nErrorCode)
{
m_pThread-> ResumeThread ();
}
```

Data receiving processing function receives data and stores then in the buffer linked list, and informs the primary thread of data receiving, and then it will suspend.

```
UINT CSocketReceive: DataReceiveProc (LPVOID pParam)
{
CSocketReceive *pcsReceive=(CSocketReceive*) pParam;
CByteArray myRecv;
myRecv.SetSize (UDP_SIZE);
CString strRemoteAddress;
UINT nRemotePort;
int nCount;
While (true)
{
nCount=pcsReceive-> ReceiveFrom (myRecv.GetData (), UDP_SIZE, strRemote-
  Address, nRemotePort);
pcsReceive-> m_csLock.Lock();
pcsReceive-> m_DataList.AddTail (myRecv):
pcsReceive-> m_csLock.Unlock ();
pcsReceive-> m_pWnd-> PostMessage (WM_USER_DATAREADY);
pcsReceive-> m_pThread- > SuspendThread ();
}
Return nCount;
}
```

The task of GetData method is getting a UDP data packet from the data buffer linked list to provide it for the primary thread to process.

```
CByteArray CSocketReceive::GetData ()
{
CByteArray strData;
m_csLock.Lock();
strData=m_DataList.RemoveTail();
m_csLock.Unlock();
Return strData;
}
```

USER-DEFINED MESSAGE WM_USER_DATAREADY is informing the primary thread that there is data in the receiving buffer.

```
#define WM_USER_DATAREADY (WM_USER + 0 × 100)
```

The primary thread has message processing function "OnDataReady (WPARAM wParam, LPARAM lParam)" to take charge of getting data and processing them.

References

1. Jamsa K, Cope K (1996) Internet programming, vol 5(1). Electronic Industry Press, Beijing, pp 123–127
2. Yao L (2006) Be proficient in MFC program design, vol 18(13). Post and Telecom Press, Beijing, pp 84–88
3. Shepherd G, Scot W (2003) MFC internals, vol 24(16). (trans: Zhao J et al.). Electric Power Press. China, pp 238–145
4. Yufeng W (1995) Apply UDP to realize reliable transmission system and the program design. Radio Commun Technol 22(1):29–33
5. Charles P (1999) Programming windows, vol 16(10), 5 edn. Microsoft Press, USA: Washington, pp 1119–1158
6. Anthony J, Jim O (1999) Network programming for microsoft windows, vol 25(21). Microsoft Press, USA: Washington, pp 117–207
7. David JK (2000) Inside Visual C ++, vol 7(3), 3rd edn. Microsoft Press, USA: Washington, pp 95–151

Chapter 85
Identity Authentication Based on Campus Network Design

Qian Xu

Abstract In order to ensure the safety of the campus network with RAIM technology as the core, we designed a network for the campus identity management system, and the system's detailed description of the key technologies through practical application shows that the system can better adapt unified authentication and access control, in order to achieve a more secure campus network system authentication.

Keywords Campus network • Identity authentication • RAIM system • SSO

85.1 Introduction

In 1994, the China Education Research Network was started to build [1]; our college campuses have set up their own broadband networks, and web-based database applications to establish various types of networks [2]. From the current development, the use of more authentication in three ways: (1) Maintain a separate user database application system for identity authentication. (2) Adopt a unified identity authentication technology, the same user database authentication services for a number of applications. This approach not only solves the unified management of multiple applications as the problem, but also reduces user management costs. (3) Use single sign-on technology to address multiple applications issue multiple logins, users simply log in once, you can use the specified number of systems. However, the existing campus network application development lacks a unified standard, applications may belong to multiple departments, the system did not follow the establishment of uniform data standards, data formats vary, the system cannot be achieved between effective data sharing, so they formed network environment information silos. Need to use several different applications for users

---

Q. Xu (✉)
Electrical and Information Engineering College, Shaanxi University of Science and Technology, Xi'an, Shaanxi 710021, China
e-mail: xuqian2@foxmail.com

Z. Zhong (ed.), *Proceedings of the International Conference on Information Engineering and Applications (IEA) 2012*, Lecture Notes in Electrical Engineering 220,
DOI: 10.1007/978-1-4471-4844-9_85,

of the system, if each of the storage management system, a different authentication method, and users need to remember several different passwords and identity, and the user access to different applications in the need for multiple logins. This is not only to the user but also to the system management it has brought great inconvenience.

Design a uniform, high-security, and high-reliability identity and access management system to complete the entire campus network user's identity and rights management [3] to ensure that the application system based on a unified model, to focus on development and upgrading the environment, and to reduce the overall system operation and maintenance costs; on the other hand it ensures that the entire school system, use and management of security, so as to enhance the overall quality of digital campus and level.

85.2 The Analysis of Identity Authentication System Problems

Currently, the university's authentication system mainly in the following questions:

1. The application system has a separate authentication and access control policy; as a user to enter different applications, you need to input each different account number and password. Users may find their need to remember that the application system account and password is also increasing. This is easy to generate the password that is forgotten, it could cause some users account and password for the convenience of the memory, in a number of applications use the same account and password. Hackers can crack the security level from low to get the system account and password, and account number and password to access with the high level of system security, the application of the entire campus network system security has posed a great threat.
2. The user roles and access control policies are difficult to maintain consistency. Of each campus has its own application user management system, they are by their own access control policy to assign user rights. When each application for their users to add, delete, and modify the permissions, in order to maintain different user roles and permissions application consistency and integrity, and only on the relevant application system interoperability, this operation is complex and error-prone, easily leads to different application systems that are having inconsistent user roles and permissions.
3. Additional applications, as useless unified authentication and access control policy specification constraints, resulting in authentication and access control code duplication and development, and independent, is not conducive to cross-system integration. With the "Digital Campus" in-depth, integrating existing applications and increasing the system for the new unified standard have become inevitable trends.

In order to solve the current campus network, the application of existing authentication systems and access control issues, the establishment of a unified service for all applications authentication and access control platform, unified identity authentication, user rights, and reasonable access control is necessary.

85.3 RAIM Architecture

Campus network users may use two types of Web applications, one is the Internet, Web applications, such as Sina-mail, 163 mail, Gmail mailbox, etc. [4]; the other is the internal campus Web applications. Web applications on the Internet are independent; these applications cannot be modified to fit the unified authentication and access control. Campus online Web applications, is within the campus network applications, can be appropriately modified in order to better adapt to a unified authentication and access control. Campus-based Web applications RAIM implementation is more complex than the first category, security requirements are also higher. Unified authentication and access control architecture are shown in Fig. 85.1:

Teachers, administrators, students, and parents through the client (browser or other application), access to Internet or Web applications on campus, first log in a unified platform for identity authentication and access control, through the platform provides single sign-on portal, further access to Internet or campus network Web applications.

RAIM system includes five parts, namely, credentials, SSO client, a unified authentication and access control services, identity and access control database, and Web applications.

85.3.1 System Flow

When users access Web application systems, Web blocking agent will intercept access requests. If the user has been authenticated and has permission to access

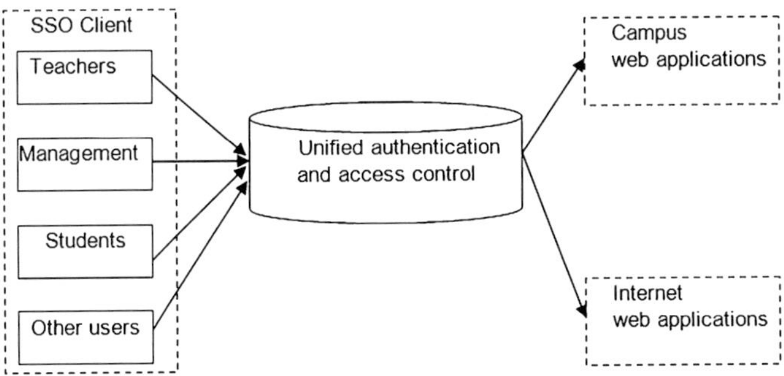

Fig. 85.1 Uniform identity authentications and access control architecture

the requested resource, it allows users to access just the request. If the user is not authenticated, then redirect the basic login screen. Although if the user authenticated, but no permission to access the resource, display an error page. If the user is not logged, the user needs to log in using the credentials, including username, smart card, USB Key, and so on.

After the user credentials entered by the authentication services verify that the user is the legitimate user. Identity authentication services from the database identifying information removed, and user login credentials to use when compared if the same, then the user is correct. Otherwise, redirect to SSO login page. If the authentication service authenticates the user correctly, the resulting user identity token.

Basic Access Control service will be the basic access control policy database to determine the user can access the corresponding Web systems and resources. After user authentication, not only to obtain the basic user access control, also called extended access control services will be made RAIM platform requested by the user of the Web application-level resource access control URL.

85.3.2 Extended Access Policy

Extended Access Control service access policies using the extended library, resulting in the expansion of the user control of services, If the user is logged directly from RAIM unified platform system, it redirect to the SSO portal page. This page list all the user has access to Web applications. If the user is to access the Web application, as redirected identity has not been certified unified platform for authentication, you can skip.

Extended Access Control Service has extended the control of the user's session information. And the previously generated and the user identity token together constitute a session authentication and access control information. Web application system for authentication and access control agent calls.

85.3.3 System Login

Users from the SSO portal page, select what you want to log in the system. When the user selects the Web application to use later, will call the second login credentials. Secondary user can store login credentials to access the Web application account and password. If the second login is successful, the user can use the Web application system. RAIM platform users can use the global write-off, global logout; the user continues to access the Web application requires re-login.

85.4 RAIM System

RAIM system includes a credential, SSO client, a unified authentication and control services of the question, identity and access control database and Web application system agents five parts, each part also includes several modules.

85.4.1 Design Credentials

Is the identification of different credentials that uniquely identifies the user's identity? The most common is the user name/password pair. Can also be a campus card or a USB Key and so on? Usually, credential applications generally use the campus identity database as a more authoritative reference.

85.4.2 SSO Client Design

SSO client RAIM platform is mainly used for user login and use the Web application can access. It is the view layer of RAIM. Relationship between the three modules is shown in Fig. 85.2:

When a user makes a request to the login module, if you need to actually login with a Web application URL request parameters, then call the second login module directly, log on to the need to use Web applications. If not with the parameters, call the SSO gateway module, the user can select from the SSO login required login portal Web application and calls the second login module, log on to the selected Web applications.

85.4.3 Identity and Access Control Database

Save the identity information, access control information database. You can use LDAP directory server, you can also use the relational database.

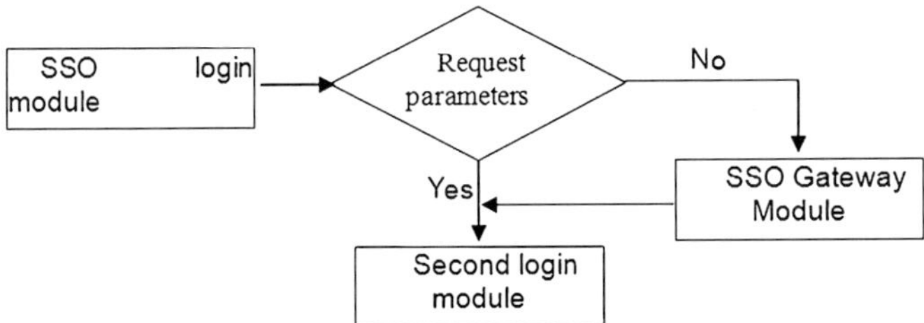

Fig. 85.2 Login module relations

85.4.4 Web Application System

Legacy Web applications, Web blocking agents and need for additional authentication and access control agents, they are responsible for the application authentication and access control of the core. RAIM platform through the agent and the exchange of information, access the user's identity tokens and access control information.

85.4.5 Unified Authentication and Access Control Service

Unified authentication and access control services are the core of the whole system is responsible for identity authentication and access control implementation.

User registration service and Web application system up service center platform is essential to achieve complete part. User registration service for the user's identity registration, Web application system registration services, Web applications that required the self-signed digital certificate through to the center platform provided by the center platform is responsible for vetting, if approved, the Web application to can join the center platform.

85.5 Conclusion

Construction of the traditional campus network is isolated in each application system design; identification of the function is repeated in the design and implementation, distributed in various application systems, resulting in substantial waste of time and money.

In this paper, a unified authentication system to achieve the specific technology used, the key question is described in detail. Proposed a lightweight authentication approach RAIM, which overcomes the traditional remote script to https and authentication; there are some disadvantages to achieve effective authentication and to ensure the reliability of authentication, security, and efficiency. And the actual situation of different applications, providing a different method and different mandates call control methods, taking into account the system's security. Finally, the actual show programs and technologies used are feasible.

Acknowledgments This work is partially supported by Shaanxi Provincial Science and Technology Department: Computer Software and Security Technology (2009JM8003).

References

1. Crochon E, Gomez B, Caucheteux D, Vacherand F (2005) New technologies for contactless Air Interfaces. In: Proceedings of the E-Smart 20(5):873–879

2. Abrial A, Bouvier J, Renaudin M, Senn P, Vivet P (2001) A new contactless smart card IC using an on-chipantenna and an asynchronous microcontroller. IEEE J Solid-State Circuits 36(7):1101–1107
3. Best RM (1980) Preventing software piracy with crypto-microprocessors. Proc. IEEE Spring Compcon 80(53):466–469
4. Clark PC, Hoffman LJ (1994) BITS: a smartcard-protected operating system. Comm. ACM 12(9):66–70

Chapter 86
EJB Technology-Based Enterprises EAM System

Hua Sun

Abstract The internal repair and maintenance of enterprise asset management system, using components based on EJB technology, design, enterprise asset management system gives the design method and system structure; these are discussed in detail EJB implementation method, and Entity Bean's design was described. The system is based on EJB component method development of distributed multilayer structure, increasing the EAM system's scalability, flexibility, and good maintainability.

Keywords EJB • J2EE • Entity bean • Asset management

86.1 Introduction

Enterprise asset management is an important part of enterprise management, enterprise asset management system is mainly for internal repairs and maintenance of asset management, auxiliary enterprises use computer systems to manage physical assets, protection of the safe operation of equipment, maintenance of maximum efficiency, reduce maintenance costs. According to Gartner Group of companies who have already been implemented before the survey EAM, EAM system successfully applied in several side businesses to obtain the following economic benefits: lower maintenance costs and reduce the number of unnecessary maintenance [1]; increase of the effective working time by 10–20 %; reduction by 10–25 % of inventory costs; reduction of equipment downtime by 10–20 %; increase of the device efficiency by 20–30 %; extension of equipment life cycle by about 10 %; and increase of the inventory accuracy rate of 95 %.

EAM system development and computer technology, network technology, database technology, and the development of repair methods are inseparable, in the

H. Sun (✉)
Yancheng Institute of Technology, Yancheng 224001, Jiangsu, China
e-mail: whejxaid@yeah.net

Z. Zhong (ed.), *Proceedings of the International Conference on Information Engineering and Applications (IEA) 2012*, Lecture Notes in Electrical Engineering 220,
DOI: 10.1007/978-1-4471-4844-9_86,

20 years from the early DOS versions, the stand-alone mode and a pure equipment maintenance management software developed to be cross-platform, Web-based and a variety of maintenance by way of an asset management system. Many well-known manufacturers have launched their own EAM products, such as the United States MAXIMO, DATASTREAM, Sweden, IFS, Germany's Siemens BFS++, are the EAM market, one of the best systems, Oracle has launched its own EAM products. But there are all kinds of enterprise asset management software, the logic layer is not independent, software reusability is poor, maintenance and updating of inconvenience to the shortcomings of the system's maintenance costs are high, an excessive burden on the client, the system resource utilization is low.

This combination of various characteristics of enterprise asset management systems, through advanced J2EE specification, using its core EJB (Enterprise JavaBeans) technical terms [2], the establishment of enterprise asset management system to support EJB, J2EE based on the four-tier architecture, the WEB layer using JSF (Java Server Faces) technology, the presentation layer and business logic separation; business logic layer using EJB component technology to improve the system scalability and maintainability; for the J2EE Platform the diversity of technology and enterprise application development EJB performance will reduce the complexity of the situation; in the EJB design, design and application of design patterns for EJB integration and optimization program to optimize performance, system performance has improved significantly. It solves the enterprise asset management and platform security issues, scalability issues, achieve the standardization of enterprise asset management and information technology.

86.2 Introduction J2EE and EJB Technology

86.2.1 J2EE

Sun, J2EE [3] is the definition of a distributed enterprise application development specification. It provides multilevel models and a range of distributed application development technology. Multilevel distributed application model is based on feature to the application logic into multiple levels, each level of support for the appropriate server and components, components in a distributed server components run in the container, the container through the relevant agreements between the communication to achieve intercomponent calls.

J2EE is a Java2 platform to simplify the use of multienterprise solutions development, deployment, and management of complex issues related to architecture. J2EE-based technology is the core Java platform or Java2 Platform Standard Edition, J2EE is not only the standard version to consolidate a number of advantages, such as "write once, run anywhere" features, easy to access the database JDBC API, CORBA technology and the ability to Internet applications in the security mode to protect data.

86.2.2 EJB

At present, the component model can be divided into three categories. The first is COM/DCOM technology (Microsoft's Component Object Model and Distributed Component Object Model); the second is the Object Management Group (OMG) Common Object Request Broker Architecture (CORBA); and the third category is the enterprise level Java Bean (EJB) architecture. March 1998 Sun, officially announced the industry version of the long-awaited EJB1.0 [4], EJB specification defines a Java server-side component model. EJB specification defines a Java client built model of JavaBeans, to support the server-side application development. In the EJB specification, the server from the EJB container provides many features; developers can focus on developing the core application functionality. Technology can greatly reduce the use of EJB application development server workload. Now EJB2.0, 3.0 versions have been released, while the corresponding EJB component products have been a lot. This technology to the current situation in the world IT industry is still a cutting-edge technology can open up a wide area. EJB component technology can be good to eliminate the existing shortcomings in the asset management business, which represents a cutting-edge technology, making the complex multilayer structure of application development easier.

86.3 EAM Systems

Enterprise assets, enterprise asset management, equipment accounting as the basis, work order submission, approval, and implementation of the main line, in accordance with the defect treatment, planned maintenance, preventive maintenance, predictive maintenance of several possible models to increase maintenance efficiency and reduce overall maintenance costs as the goal, purchasing management, inventory management, human resource management data in a fully integrated information sharing system, while achieving information integration and combination of standardized management.

86.3.1 Development Environment

Operating system: server operating system is Windows2000 Server Edition, the client operating system versions can be Windows98/2000/xp series and above IE5.0 browser.

Database System: MS SQL Server2000 Enterprise Edition.

Development Tools: The system uses Borland's JBuilder2006 development. JBuilder is a powerful Java development tool, which provides a visual integrated development environment that supports the latest JDK version; you can facilitate the development of Java applications, Java Bean, EJB, and so on. Can be efficient and easy to complete system development work, and it can quickly develop and deploy J2EE applications and application server products, and many mainstream well integrated, including Jboss, Web logic, Tomcat, and so on.

Application Platform: the system uses Jboss. Jboss is the mainstream of the Java application server for building large, multilayered, secure, and distributed WEB applications. It provides the development and utilization of the server business logic of the basic architecture to support distributed programming model. Jboss J2EE services can implement distributed applications for enterprise development and provide an excellent operating environment.

86.3.2 System Structure

System according to business needs EAM, EAM system includes the following modules: inventory management, purchasing management, work order management, operation management, document management, preventive maintenance, equipment management, workflow, resource management, planning, and rights management.

EAM System is B/S mode. Distributed multitier application architecture to take WEB server used for the JBOSS. J2EE technologies used include EJB, Servlet, JSP, JNDI, and so on. Figure 86.1 is based on J2EE multitier distribution model EAM system. Client layer is the application of the display part, customers view the page through a browser and associated operations, in this system mainly through

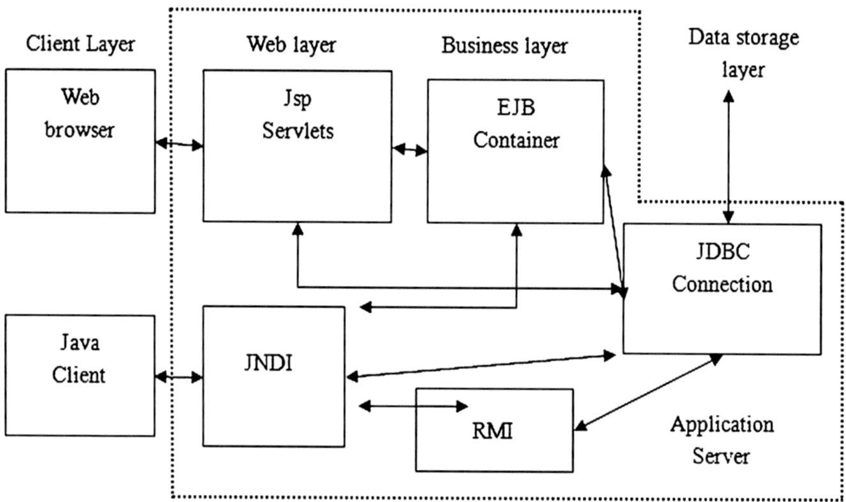

Fig. 86.1 EAM system structure

the EJB business logic packaged applications, such as data processing, control, and management, it can encapsulate the business logic changes in business logic to reduce the impact on the system data storage layer including the database system.

86.4 Implementation of EJB

Business layer of the system mainly used EJB Entity Bean and Session Bean's implementation. PM Organization designed to rights management, for example, the system used by the container-managed entity Bean persistence (CMP), and support for the database insert, delete, search, and other operations. Entity bean in order to avoid direct exposure to the Web tier, the use of Session Facade model designed PM Organization Session Facade, the entity Bean (PM Organization Bean) "packaging" in the PM Organization Session Facade. Thus, separating the client and the entity that hides its client's complex interaction among participants.

In the application of data between all levels of delivery PM Organization Session Bean in the design, organization of all relevant data is encapsulated in a one-time transfer PM Organization DTO class, greatly improving system performance. PM Organization DTO is a simple class that implements the Fertilizable interface, this class of each representative of a private member variable in an information organization, usually a private variable for each data table corresponds to a field, the other is for each a private variable, have designed a pair of get and set methods to read and this variable set.

PMO organization DTO sample code is as follows:

```
public class PMO organization DTO implements Serializable
{
private String organization ID;//Organization ID
private String unification No;//Organization Code
private String organization Name;//organization name
private String description;//Description
private PMUser DTO [] PMUsers;//user organization
//Get the organization name
public String get Organization Name ()
{return organization Name;}
//Set the organization name
public void set Organization Name (String organization Name)
{this. organization Name=organization Name;}
……
}
```

When business organizations need a method ID check organization's other information, they can use public PM Organization DTO PM Organization Find By Primary Key (String organization ID) method returns PM Organization DTO, to obtain all the information of the organization.

86.5 Entity Bean Design

Entity Bean is a database and user cache data between applications; it is in memory WEB container, it is not like the manifestation of a row in the database, but an object instance. Entity Bean does not perform complex tasks or workflow logic, because they are representative of the data, are persistent state of the object.

To rights management, for example, with the database table corresponding to the following main entity bean: PM Organization corresponds to the data sheet PM Organization; PMUser correspondence and data sheets PM User; PM Resp Employee correspondence and data sheets PM Employee; PM Resp Department corresponds to the data sheet PM Department; PM Navigation corresponding Data Sheet PM Navigation; PM Role corresponding data table PM Role; PM Right corresponding data table PM Right; PM Group corresponding data table PM Group and so on.

Figure 86.2 shows PM Organization Bean, PM User Bean, PM Navigation Bean relationship, and other EJB same, PM Organization Bean also by the remote interface class (or local interface class), Home interface class (or the local Home interface class), and the Bean implementation classes. Their corresponding documents were PM Organization, PM Organization Home, PM Organization Bean, PM Organization, PM Organization Home are two interface files, PM

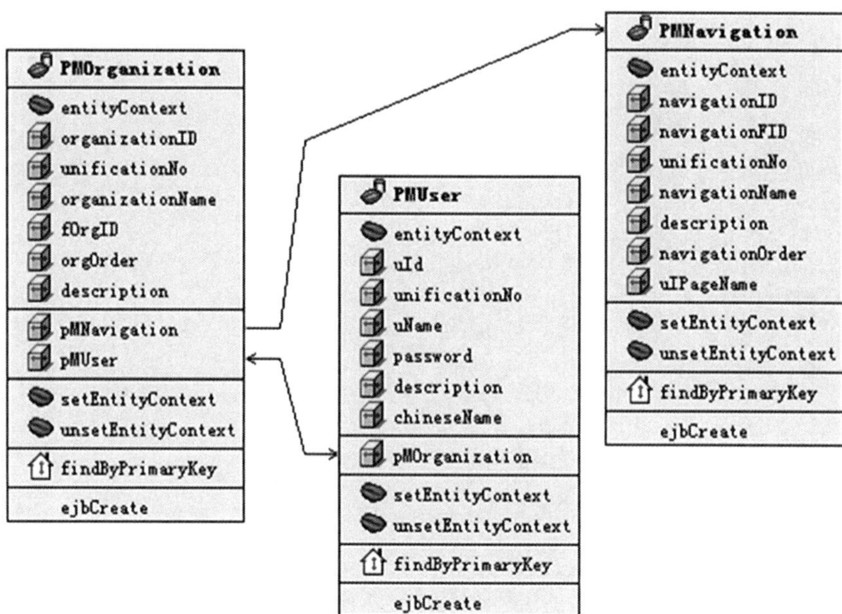

Fig. 86.2 Relationships among PM Organization Bean, PM User Bean, PM Navigation Bean

Organization Bean is the EJB class file itself. PM Organization Bean implementation of the code is as follows:

Home interface is responsible for finding and creating PM Organization Bean, return PM Organization Bean reference to the customer, the customer calls PM Organization Bean components using the reference method, get results, the last object removed PM Organization Bean object. In the Home interface, the user can add own finder, sleet other logical method.

86.6 Conclusions

This chapter describes the J2EE- and EJB-related technical standards, design EJB technology-based enterprise asset management system, given the system design of the structure and function modules, EJB are discussed in detail the specific implementation method, and the design of an entity described Bean. The system uses advanced EJB component technology, component-based EJB method development of distributed multilayer structure. EAM system increases the scalability and flexibility. Server-side component applications can be modified quickly and the location in the network components and application independent. The EAM system in system maintenance, expansion, and security has certain advantages.

References

1. Xiaoxia Z, Ying B, Liwen D (2003) Distributed implementation based on EJB middleware technologies. Kunming Univ Technol (Sci Technol) 28(5):90–92
2. Yong L (2006) EJB-based e-commerce system design. Chongqing Univ 14(8):143–146
3. Qingyuan B, Cheng-zhong LI (2005) Based on EJB component framework for distributed e-business applications. Univ Info Technol 20(16):26–28
4. Xiaohui I, Chi shift W (2003) J2EE architecture and middleware layered structure of the design. Railway Comput Appl 12(7):13–14

Chapter 87
System Dynamics Simulation for Reservoir Management

Zhihui Fan and Yang Li

Abstract Reservoir management of oilfield (also usually called assets management) is an effective method to improve the performance of reservoir development. A new method to study reservoir management was carried out with system engineering. Reservoir management is looked as a system and it is a sort of system engineering to be established according to the practice of oilfield with Chinese features. Based on system analysis and system structure study, the proceeding of information exchanging and feedback cycles between artificial production subsystem and natural reservoir subsystem are described to support a system dynamics simulation model of compound production–reservoir system. Real data from SL oilfield were run on the model to demonstrate an example. Key index of oilfield development were selected for history fit. Established model was used to predict the production system trend and corresponding advice for reservoir management was put forward based on prediction.

Keywords: Reservoir management • System engineering • System dynamics • Simulation research

87.1 Introduction to Reservoir Management Study

Reservoir management is a practical science that uses elements of geology and petroleum engineering to predict the behavior of oil and natural gas within subsurface rock formations. Thakur takes reservoir management as all sorts of cute methods for maximizing profit. Satter thinks reservoir management looking reservoir as natural resource and applying all measures to know and develop it to maximizing economic profit and social benefits. Wiggins and Startzman defined reservoir

Z. Fan (✉)
College of Petroleum Engineering, China University of Petroleum, Beijing, China
e-mail: gaojc_17@163.com

Y. Li
Oilfield E & P Department, China Petroleum & Chemical Corporation, Beijing, China

Z. Zhong (ed.), *Proceedings of the International Conference on Information Engineering and Applications (IEA) 2012*, Lecture Notes in Electrical Engineering 220,
DOI: 10.1007/978-1-4471-4844-9_87,

management from the view of operation: from discovered to disposed reservoir management is a series of operation of recognition, measure, producing, development, monitoring, evaluation, and decision. Woods and Abib considered reservoir management as an optimal dynamic strategy as reservoir development.

Reservoir management is used throughout the full life cycle of oil and natural gas reservoirs. It is used to determine the most cost-effective way of managing the development of a new field or to bring new life to an older field with, for example, enhanced oil recovery measures such as steam flooding or steam injection. Through the use of a suite of technologies, including remote sensors and simulation modeling, reservoir management can improve production rates and increase the total amount of oil and natural gas that is ultimately recovered from a field.

Could we review reservoir management from a new prospect closer to management? If the answer is YES, which management study tool is the best for this?

87.2 Reservoir Management System Analysis

87.2.1 Conception of Reservoir Management

Reservoir management is a sort of complex system engineering, which includes such features described as the following. Reservoir is always buried under more than 1,000 m rocks. Even advanced reservoir description technique being applied, there is no doubt that description of reservoir could be detailed to finer level as reservoir development going ahead. Reservoir management, as most authors said, is a basket ball team, involving many descriptions such as geology, reservoir engineering, artificial pumping engineering, storage and transport engineering, and financial evaluation. All subjects are bound to be considered in the same time for higher reservoir production performance. Reservoir management has two aspects, technique character and management character. Any management activity must be affected by the surrounding social, economical, and even geographical factors. All these features have endowed reservoir management with nonlinear solution. So, any linear optimization tool is hard to be applied in reservoir management performance improvement. Reservoir management is destined for a complex system. Reservoir management system (RMS) is defined: A technical and economical compound system includes nonrecyclable natural reservoir and effective integration all resources of human, assets, cash, and information, and is a reservoir operation and management process beyond the precise prediction of reservoir behavior, and with an objective maximizing the recovery and profit.

87.2.2 Reservoir System Structure Analysis

Based on RMS definition, a structure chart of RMS is demonstrated as Fig. 87.1. RMS is divided into four subsystem, management organization subsystem,

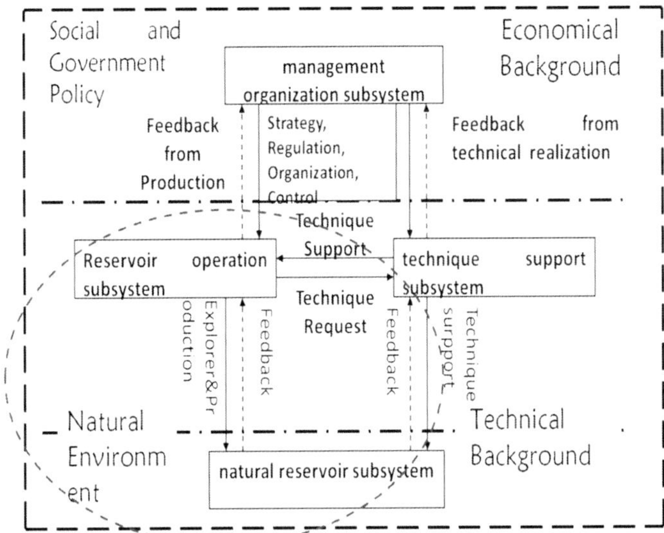

Fig. 87.1 Reservoir system structure

reservoir production/operation subsystem, technique support subsystem, and natural reservoir subsystem. There are three artificial subsystems except natural reservoir resource subsystem. Between any two artificial subsystems there is an immediate contact, with which information and energy are exchanged, too. Reservoir operation subsystem and technique support subsystem both have an immediate effect on natural reservoir subsystem, but management organization subsystem has to control the lowest subsystem through the middle two subsystems with policy optimization.

As most reservoir management authors have noted that natural reservoir resource is the final object which accepting the exploitation and giving out corresponding feedback. All artificial subsystems could be rearranged and redefined except natural reservoir. Natural reservoir subsystem just lies there underground no matter being found, being described correctly, or being operated suitably. Reservoir operation subsystem takes dual roles, one is to carry out action on the lowest subsystem, and the other is to accept the regulation and support from the upper and right subsystems. It also has most assets of reservoir management, costs much investment and overhead, certainly produces profits. Reservoir operation subsystem is the system which directly operating reservoir. There are much information and energy exchanged between these two subsystems. Action and reaction processes between them are complex. Natural reservoir subsystem and reservoir operation subsystem are first selected to complete a model which can simulate the basic process of the reservoir management. The boundary of study is circled by an eclipse in the Fig. 87.1.

87.3 Reservoir Management System Dynamics Study

87.3.1 System Dynamics

Mathematically, the basic structure of a formal system dynamics computer simulation model is a system of coupled, nonlinear, first-order differential (or integral) equations. Simulation of such systems is easily accomplished by partitioning simulated time into discrete intervals of length delta (t) and stepping the system through time one delta (t) at a time.

Is system dynamics suitable to RMS? Requests for a complex system which is modeled by system dynamics simulation are subjective, self-controlled, and nonlinear. Subjective of RMS lies in the decision character of management. Development decision has to consider many aspects arranging from policy to technique, and also is subjective to multiobjective. Standing on one point only one decision is made and followed. Self-control lies in the cooperation between elements constituting system. Reservoir management process involves many phrases of multidisciplinary cooperation. All operation will be routed to the final objective automatically; even the performance is usually not as so-called the best one. For the process not only relies on self-control, it also on the phrase people standing on. There is no result beyond time. Geological information seems never certain and developing; reservoir gives feedbacks on all measures or operation on reservoir with a so-called time-delayed, RMS is obviously nonlinear. Three necessary conditions are met, the next is to observe the simulation itself.

87.3.2 Feedback Loop of Reservoir Management System

Feedback is a mechanism, process or signal that is looped back to control a system within itself. In more general terms, a control system has input from an external signal source and output to an external load; this defines a natural sense or path of propagation of signal. Such a loop is called a feedback loop. In systems containing an input and output, feeding back part of the output was to increase the input is positive feedback; feeding back part of the output in such a way as to partially oppose the input is negative feedback.

RMS contains many feedback loops which may be positive or negative. It is more common for a RMS there being a positive feedback loop, where a positive-going wave (i.e. increased investment) on the input (i.e. production) leads to a positive-going change on the output (i.e. increased production), will amplify the input signal, leading to more modification (i.e. investment to be increased). Negative feedback is necessary for whole RMS, which controls RMS in a balanced cycle. Figure 87.2 demonstrates the feedback loops among natural reservoir subsystem and reservoir operation subsystem.

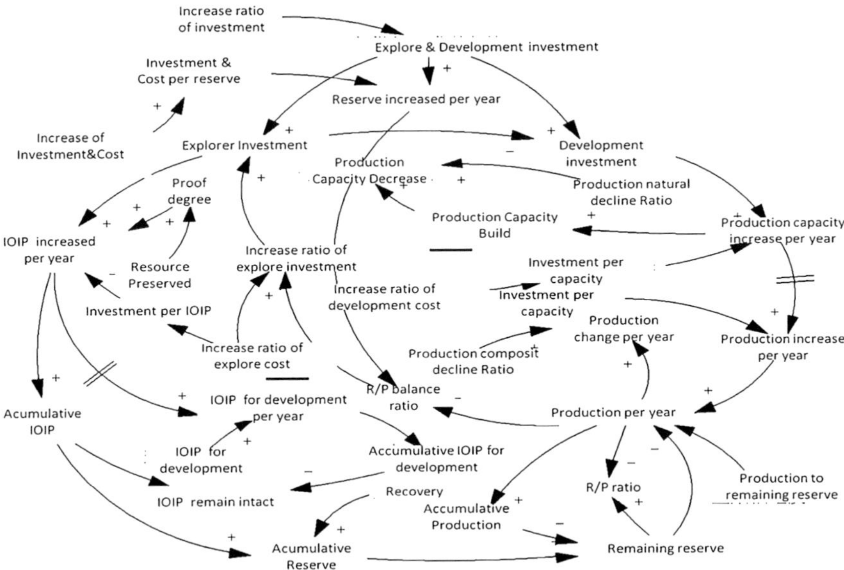

Fig. 87.2 Feedback loops among natural reservoir and reservoir operation compound subsystem

Internal feedback loops in natural reservoir subsystem has a main loop that is explore and development investment → production capacity increase production → production reserve balanced ratio → explore and development investment. This loop is a negative feedback loop, which depends on the number of negative symbols contained in the loop, that is, odd number says negative feedback, mean number says positive feedback. By control theory, the core negative feedback loop gives RMS a stable cycle, which is in compliance with the practice of reservoir management.

Internal feedback loops in reservoir operation subsystem play the role of reservoir predict, production, and financial evaluation. This subsystem consists of many positive feedback loops which lead to an unstable state that in return increase the performance of reservoir management. Only does this kind of increase overload the RMS balanced level so far, it leads a system crash. In fact, one of the objectives of study on RMS or reservoir management is to keep the balance by prompting and control. In general, reservoir operation subsystem is still a stable system with odds number feedbacks in the main route. This core loop is detailed as: Investment realized as oil facility, oil facility improved production capacity, capacity realized as production with a delta (t), production calculated the P/R ratio, P/R ratio was an important index to make up investment, investment was considered two parts, explore and development, the proportion of two parts was further made according to balance reason. It is not hard to recognize this loop not only is full of positive feedback, but also leads the system back to a stable point as the next cycle start.

When natural reservoir and reservoir operation subsystems is compound, the feedback loop becomes more complex for reservoir response to human action being both considered in the same time. But it is easy to find it is still a negative feedback loop in the core loop. That is easy understood by thinking any natural resource has inertia to react to any force to change it. Any overloaded effect will reduce it marginal performance when it keeps going up.

87.4 System Dynamics Simulation of RMS

87.4.1 System Dynamics Simulation Software

Commercial software is packed by many companies or institutes, such as Model Maker, STELLA, and Vensim. Vensim PLE is the main tool for run system dynamics model of RMS in this paper. Vensim is used for developing, analyzing, and packaging high-quality dynamic feedback models. Models are constructed graphically or in a text editor. Features include dynamic functions, subscripting (arrays), Monte Carlo sensitivity analysis, optimization, data handling, and application.

87.4.2 System Dynamics Simulation of SL Oilfield

Feedback loops of natural reservoir and reservoir operation compound subsystems were built for further validation test which make sure that model suitable for practical running. History data from SL were put into the model for history matching. Three key index, P/R ratio, production, reserve found per year, which are important to evaluate performance, even the survival, of a reservoir management team, are selected for history matching. History data of 2001–2010 were checked in the model. Model output and history data are compared in Table 87.1.

Comparing to history data, calculated three key indexes have relative errors ranging from 0.24 to 4.88, which indicates confidence in the prediction of that model of statistical sense. RMS dynamics model took reservoir characters and operation indexes into consideration. Compared to traditional system engineering methods, it includes more conditions that a successful reservoir management has to rely on. The model not only considers time sequence like Gray model, also establish this nonlinear relationship on feedback loops among the system elements.

Data of SL oilfield in 2010 was set as the start point; prediction of the following 5 years was carried out by the model. The output was calculated by the model on input to a RMS, that is, output is a sort of response to input. So, it supplies reservoir management with a new sensitivity analysis tool rather than prediction method.

Table 87.1 History matching of key index of RMS (SL oilfield)

Item	Unit	Year				
		2006	2007	2008	2009	2010
Production	104t	2695	2742	2770	2774	2785
Production (calculated)	104t	2599	2732	2695	2702	−710
Relative error	%	−3.56	−0.33	−2.73	−2.60	−2.69
Reserve new found per year	104t	10621	10339	10677	10408	1087
Reserve new found per	104t	1087	10542	10442	10346	10250
Relative error	%	−0.95	1.96	−2.19	−0.60	−2.58
P/R ratio	%	161.8	119.6	104.0	−04.0	103.4
P/r ratio (calculated)	%	155.7	116.1	105.4	103.0	100.6
Relative error	%	−3.90	−3.01	1.28	−0.95	−2.78

Table 87.2 Key indexes changes corresponding to changes in investment

Item	Changing ratio (%)							
Investment	−10	−8	−5	−2	2	5	8	10
Reserve new found per year	−4.83	−3.88	−2.64	−1.17	1.28	2.60	4.33	5.48
Production	−3.22	−2.52	−1.64	−0.80	0.91	1.71	2.55	3.12
P/R ratio	−0.81	−0.75	−0.40	−0.22	0.24	0.57	0.91	1.16

One of the concerned questions of reservoir management team is the performance of investment. Table 87.2 shows three key indexes' corresponding output/response to changes in investment.

There are positive correlations between investment and reserve new found, production, and P/R ratio. The sensitivities of three key indexes are different; from higher to lower are reserve new found, production, and P/R ratio. Since investment seems has an exclusive effect on production and reserve new found, the changes in these two ones are notable. P/R ratio is partly affected by investment; in fact, with more factors, such as reserve new found and time delay, P/R ratio changed minimum.

87.5 Conclusion

Natural reservoir subsystem was taken as a conceptually compound entity. One side, by adjusting operation parameters and reservoir character setting, the system dynamics model fits various types of reservoir and different phrase of every type. On the other side, human activity and subjective were accepted to help finding the best policy.

A suitable production objective will be the most important index that not only affecting the performance of reservoir management, but also the sustainable development. Production per year is sustained at a level of 2700 ± 30t.

Investment has two parts, development and explore. Explore portion settles the possible reserve to be found. So it is advised to enlarge the explore portion from present 35 to 50 % in a long range.

References

1. Thakur GC (1990) Reservoir management: a synergistic approach. SPE 20(13):8–21
2. Satter A, Varnon JE, Hoang MT (1992) Reservoir management: technical perspective, vol 27(22). Presented at the SPE International Meeting on Petroleum Engineering held in Beijing, China, pp 350–364
3. Abib O, Moretti FJ, Cen M, Yang Y (1991) Application of geological modeling and reservoir simulation to the west Saertu area of the Daqing oil field SPE reservoir engineering 6(1):36–43
4. Michael J. Radzicki and Robert A. Taylor (2008) Origin of system dynamics: Forrester and the history of system dynamics. In: Jay W (ed) U.S. Department of Energy's Introduction to System Dynamics, U.S., 23(14):20–28
5. Forrester J (1971) Counterintuitive behavior of social systems. Technol Rev 73(3):52–68
6. Sterman JD (2001) System dynamics modeling: tools for learning in a complex world. California Manag Rev 43(4):8–25
7. Repenning NP (2001) Understanding fire fighting in new product development. J Prod Innovation Manag 18(12):285–300. doi:10.1016/S0737-6782(01)00099-6
8. Repenning NP (1999) Resource dependence in product development improvement efforts, Massachusetts Institute of Technology Sloan School of Management Department of Operations Management/System Dynamics Group 35(25):435–440
9. Radzicki MJ, Taylor RA (2008) Feedback. In: U.S. Department of Energy's Introduction to System Dynamics 23(10):200–214

Chapter 88
Information Security Emergency Response Procedures and Disposal System

Fan Wang

Abstract The key of building smart platform in information security is the emergency response and handle events of the realization mechanism of the system intelligence response and the control center. It is equal to the role of the human brain, the basis of knowledge base of problem-solving strategy based on system security incident, case base, and network security expert knowledge base. It consists of an emergency report subsystem and emergency response system, providing emergency response and processing services to assist the emergency response database and emergency response of the expert system. Information security emergency response and processing system focus on information security of the emergency response technology, they are: intrusion detection technology, accident diagnostic technology, against source separation technology and rapid recovery technology, network attack technology, and computer forensics tracking technology.

Keywords Information security • Information security emergency response • Computer forensics

88.1 Introduction

The number of cases of classification happened to be more than those in the network in the traditional form. In fact, under the influence of information warfare against, hacking, code, and natural disasters in attack, an attack on an information system, or a small defect may bring disaster, large information of a major disaster information will be inevitable. Due to a wide range of information security in the process of information, the feature information of the attack, sudden emergency response measures must emphasize processing information security. In case of network and information security incident, which may affect the information security interests, timely and effective measures must be taken to control the development of the crisis and reduce the loss [1] and [2].

F. Wang (✉)
Guizhou College of Finance and Economic, Guiyang 550004, Guizhou, China
e-mail: wangfan@cssci.info

Z. Zhong (ed.), *Proceedings of the International Conference on Information Engineering and Applications (IEA) 2012*, Lecture Notes in Electrical Engineering 220,
DOI: 10.1007/978-1-4471-4844-9_88, © Springer-Verlag London 2013

88.2 Construction of Information Security Emergency Response

The platform consists of the following four core components.

88.2.1 Incident Control Management Center

This part is responsible for the control and management work of the entire emergency response and disposal platform, including system security incidents case base, the expert knowledge base, and model base [3].

88.2.2 Network Status Monitoring Subsystem

This part is responsible for real-time monitoring network risk source and collecting evidence from computer, including intrusion detection subsystem, log auditing system, and computer forensics system.

88.2.3 Attack Response System

This part mainly takes emergency response and disposal to network risks, including automatic protection systems, attack classification and positioning systems and attack analysis system.

88.2.4 System Backup and Rapid Recovery System

This part is mainly about the disaster recovery processing of network attacks, including storage backup system, adaptive recovery system, and the incident management system [4].

The architecture of Information security emergency response and disposal system is as follows in Fig. 88.1:

88.3 Key Technologies of Information Security Emergency Response

88.3.1 Intrusion Detection Technology Using SVM Feature Weighting Classification Methods

In the study of intrusion detection system (IDS), based on the traditional machine learning method, the premise is, the number of samples is not enough, so only

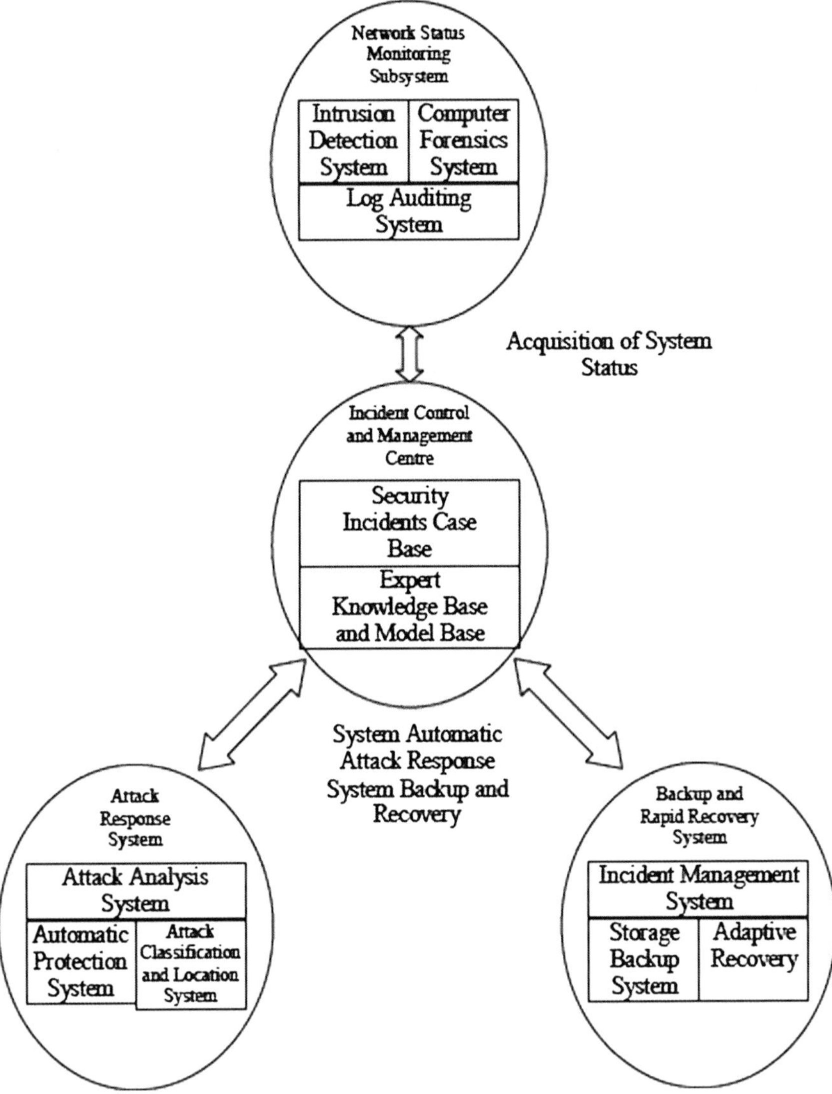

Fig. 88.1 The architecture of Information security emergency response

when sample size tends to infinity, can act guarantee theory. In the modern network attack behavior of the species, constantly updated, therefore, training data on machine learning will inevitably encounter problems [5]; by contrast, small sample of the inspection data in real application, the use of traditional method and machine learning will meet problem, under-learning, over-learning, the local minimum points, etc.

The support vector machine (SVM) is a new kind of machine learning method based on statistical learning theory, and on the basis of realization structure risk minimization principle.

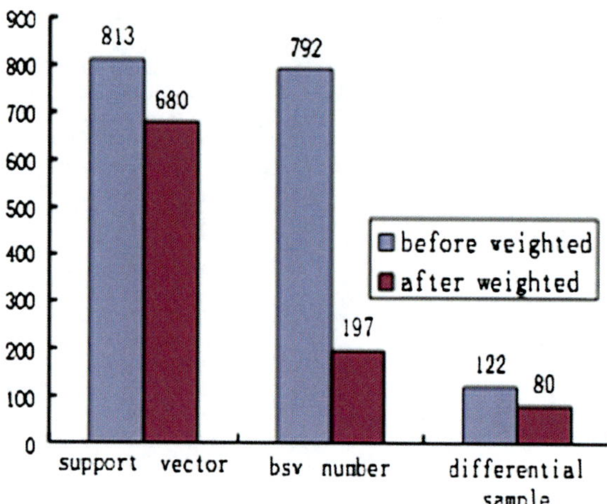

Fig. 88.2 Support vector set before and after feature weighting

On the basis of the statistical analysis of the existence of a large number of network data, the method of the generation of support vector machine classifier, use of generating invasion to determine the visit or connection function, make the attack event classification. It can ensure a higher accuracy of classification of support vector machine in the relative lack of knowledge [6].

Because of the support vector characteristics fully describe the training data set SV set into equivalent of the training data set. To test the effectiveness of the method, and the comparison of the support vector suit before and after weighted feature, the experimental results shown in Fig. 88.2 are as follows.

Seen from the chart, after weighting, the total number of support vector dropped to 680 from 813, indicating that the improved generalization of new training. The number of BSV (bounded support vector) dropped to 197 from 792, a decrease of 75.1263 %, indicating that there is a large number of samples at the classification boundary before weighting, and after that, the samples at the classification boundary are significantly reduced, indicating the effectiveness of this feature weighting classification method.

88.3.2 A Multidimensional Model of Computer Forensics

The traditional computer forensics model has two characteristics: first, the forensics process, therefore inevitable fractionation process of the legal supervision and computer forensics technology process, and, in some cases, the legal supervision results in concerning the process of vacuum, cannot afford strict authentication. Concerning the process and supervision process should be at the same time and cannot be separated, go. Second, ignored changes over time, the criminal act of criminal means are constantly changing. A forensics model should be able to meet the forensics in different periods of time demand. This platform application of computer forensics model is shown in Fig. 88.2 .

Any evidence of the course should be from the crime feeling that is a three-dimensional space curve of the origin of the evidence. This curve is by concerning the process, legal supervision process and requirements. These three aspects are indispensable, from the origin of the forensics, especially the process of legal supervision. Cannot we use the vacuum model that have the following features?

This model is based on the conclusion of existing model, including periods such as preparation of forensics, physical forensics, digital forensics, full monitoring of forensics, and presentation and summary of evidence, etc. It is the traditional forensics model when the time and need of the multidimensional model contract to a point, therefore, the shrinkage of the model is better.

As a result of constant changing and updating criminal means, different forensics strategies are needed at different times even for the same forensics demand; therefore, this model adds time constraints, putting forward the new theory that forensics strategy should be changed with time.

This model monitors the whole forensics process and generates complete monitoring data. In case of being challenged in court, the forensics process can be played back to prove the legitimacy and fairness of the process.

This model is a multidimensional framework; therefore, forensics staff is required to accomplish the task from various angles, thus avoiding some mistakes and shortcomings during the course (Fig. 88.3).

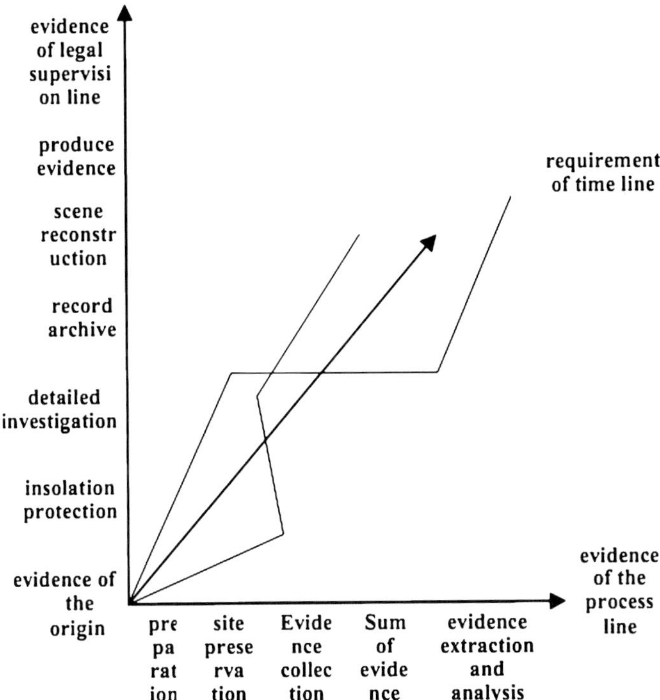

Fig. 88.3 A multidimensional model of computer forensics

88.4 Application of Information Security Emergency Response

Taking the construction of emergency information system in a certain city as an example, unique working processes of this system are analyzed and expounded dealing with unexpected information security incidents. Information security and emergency response is divided into six stages: preparation—detection—inhibition—eradication—recovery—tracking.

88.4.1 Preparation

Preparation for emergency response should precede the happening of network information security incidents.

This stage focuses on prevention, including:
Establish a group of reasonable defense control measures bases on threats.
Establish a set of incident processing procedures as effective as possible, such as which procedures to take, priority, division of labor, acceptable risks, etc.
Achieve necessary resources, equipments, and personnel for problem processing.

88.4.2 Detection

Detection is to determine whether there are malicious codes, files, and whether directories have been tampered with, etc. Methods of detection are as follows:

Rely on relevant software.
Determine through a number of signs.
Deep investigation of windows.

88.5 Conclusion

Some technologies, such as the accident diagnosis technology will fail to meet demand, the national standards, the information counter; therefore, further research should be made about how to make the use of all kinds of technology of existing problems and solve the network to attack and defense technology upgrade from a single point cluster. Tracking network attack is a challenging work, especially for distributed denial of service attacks. The invaders can be a lot of host, through some of the offensive use fake address. Based on the combination of the current TCP/IP network infrastructure construction, network tracking is a very challenging task technical difficulty.

References

1. Lee W (2009) A data mining framework for constructing features and models for intrusion detection systems: [PhD Dissertation], vol 3(3). Columbia University, pp 438–443
2. Brownlee N, Guttman E (2010) Expectations for computer security incident response RFC2350. Ref Netw Work Group 4(6):32–38
3. Fraser B (1997) Site security handbook. RFC2196. Ref Netw Work Group 9(2):399–407
4. Malkin G (1996) Internet user glossary. RFC1983. Ref Netw Work Group 8:64–69
5. Kahn C, Porras PA, Staniford-Chen S, Tung B (2009) A common intrusion detection framework. Internet Secur 8(7):73–78
6. Denning D (1987) An intrusion detection model. IEEE Trans Softw Eng 13(2):222–232

Part IX
Multimedia Aechnology and Applications

Chapter 89
Flash-Based Multimedia Courseware's Production and Implementation

Huixin Zhang

Abstract As the computer in college teaching is gradually in-depth, multimedia courseware for teaching has become more popular modern teaching methods. Multimedia courseware is not only rich in resources and colorful, but also able to fully mobilize the enthusiasm of the students, broaden students' horizons, and add more auxiliary learning tools such as multimedia player. It not only helps students to absorb knowledge, but also greatly improves the teaching efficiency of teachers. There are many ways and means of multimedia courseware's production, including the more powerful Flash, which is an extensive way of production of courseware. This paper mainly introduces the concept of multimedia courseware and steps of producing multimedia courseware based on Flash.

Keywords Flash • Multimedia courseware

89.1 Introduction

The computer has been deeply into all walks of life. The computer brings us ever-changing working life and, at the same time, it gradually moves into the teaching activities of teachers [1, 2]. Multimedia courseware for teaching realizes books in many science-fiction movies can play a video scene, stimulates the enthusiasm of the students, makes students more interested in the course, and promotes the teaching activities of teachers [3, 4]. Multimedia courseware not only provides very intuitive presentation of a variety of multimedia video and audio, but also can simulate the micro even macro things and simplify the reproduction of things responsible for the process [5, 6, 7]. This paper selects Flash as courseware authoring software. Flash can be imported to a variety of pictures, audio, animation, video and other media, and be linked with teaching contents that can be presented vividly, which is unmatched by traditional teaching.

H. Zhang (✉)
College of art and design, Mudanjiang Normal University, Mudanjiang, China
e-mail: X7421@163.com

Z. Zhong (ed.), *Proceedings of the International Conference on Information Engineering and Applications (IEA) 2012*, Lecture Notes in Electrical Engineering 220,
DOI: 10.1007/978-1-4471-4844-9_89, © Springer-Verlag London 2013

89.2 Multimedia Courseware Overview

Media, the general concept refers to the carrier of transmission of information. Multimedia is a variety of media elements collectively. That is to say, it is a new integrated media which comes from an organic combination of kinds of media. Generally speaking, multimedia is the scope of the computer. Multimedia technology, a new type of computer technology, can integrate a variety of text, graphics, images, audio, 2D and 3D animation and video, and other multimedia data effectively and then be integrated into the interaction of an IT. In Traditional teaching, the blackboard, textbooks, and other static resources are owned by the media [8, 9]. However, computer multimedia includes texts, static, dynamic graphic images, sound, and video.

Multimedia courseware is tailor-made for the course content of secondary teaching software. It utilizes all kinds of multimedia technology, combines with the course content, and enriches the teaching presentation. It combines the relevant text, graphics, images, sound, animation, and video with knowledge points, makes the combination interact with each other, integrates various elements, and then shows a colorful multimedia application courseware. Students can deepen their understandings of knowledge points through images, audio, and video showed in multimedia courseware, and greatly improve their interests to acquire knowledge. What is the big feature of Multimedia courseware is that if students do not understand some of the teaching content, we can play the video in the multimedia courseware repeatedly until the students understand it.

89.3 General Steps of Multimedia Courseware

89.3.1 Clearing Teaching Objectives

First, Multimedia courseware contents should be in accordance with the requirements of syllabus to stress focused points, showing students more basic contents that must be mastered rather than less important contents. Second, we must break through the difficulties. A major advantage of the multimedia courseware is that the difficulty in the course which is not easy to be understood can be showed visibly by more popular pictures, animations, photo albums, and other effects. The first step of Multimedia courseware is to clear teaching purposes, hope what kind of effects that the courseware will eventually reach, and how to make that to help students deepen understandings of knowledge, assist memory, deepen impression, and how to help students to broaden their knowledge, inspire imagination, and practice creativity. These are all worth considering before producing and determining the problem. Only when teaching purposes are to be determined, the production of courseware can proceed objectively. As we make a multimedia courseware according to the teaching requirements, it can better assist teaching (Fig. 89.1).

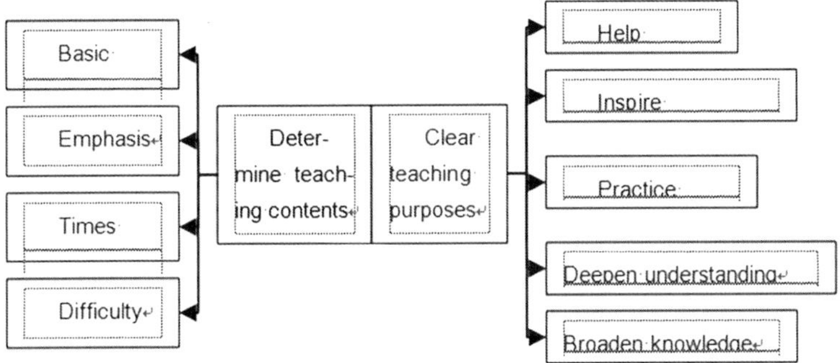

Fig. 89.1 The figure of the primary task of multimedia courseware

89.3.2 To Prepare Material

After determining the teaching contents and teaching purposes, we begin to produce Flash courseware. We start to search for relevant materials. The selection of multimedia courseware materials should be based on teaching materials, should follow the teaching objectives and teaching tasks, and enrich courseware sufficiently. A good courseware needs to increase the number of vivid and intuitive multimedia materials rather than to play the material contents just like the slide, so the preparation of the material in the production of multimedia courseware is crucial. During selecting Material, We can start it in the multimedia material library to find the right one, and then find the richer materials on the Internet-related courses' websites. Under normal circumstances, the format of the material we prepared is jpg, gif, RMVB, AVI, and so on. If we are downloading a format that the Flash software does not support, then we can convert it into the format we can use based on a variety of formats conversion software.

89.3.3 The Design of Creative Script

The production of multimedia courseware is also a creative project. The so-called script refers to the idea of multimedia courseware need to just like the script showed on films and television, and we need some of the script like that. The role of these scripts is how to combine the teaching contents and preparation of material, and then to describe the presentation orders and ways of the text, images, video, and so on. These scripts also describe a variety of guiding significant contents like the layout of the courseware, text structure, and manifestation of the interface. The conception of these scripts needs to be revised and improved repeatedly.

89.3.4 To Achieve the Script

After having determined the courseware contents, collected enough materials, designed script, we can begin to create multimedia courseware. There are a variety of ways to produce and achieve multimedia courseware. The enabling technologies are: Flash basic animation techniques, such as techniques of frame by frame animation, mask, action, guide, forms; Photoshop processing picture technology. Multimedia courseware process requires being familiar with basic computer operation and the usage of related software.

Under normal circumstances, production of courseware will not only need an authoring tool, but a variety of software, the higher requirement of mastery of computer software. As to graphics and image processing, we need to be able to use Photoshop and other software. As to the sound, we need to use the audio clipping software. If it still has animation, Flash and other software are needed. Only using kinds of software flexibly, having a good esthetic ability, you can produce an excellent courseware.

89.3.5 Feedback and Modify

To really apply produced Flash courseware to the teaching process, we still need to modify and improve it repeatedly. We need to modify student's feedbacks more. If the multimedia courseware cannot arouse students' interests or aid teaching better, we need to complete and modify it based on the views of teachers and students. I believe, after repeated effective modifications, a good courseware will be born.

89.4 Use of the Flash in Multimedia Courseware Production

The main interface of the Flash software consists of the menu bar, toolbar, timeline, stage, actions, and attributes. Timeline includes scenes, layers, frames, and playhead. Frame is equivalent to a picture in films; layers consist of many frames, which are equivalent to a film; scenes consist of many layers, which are equivalent to a story.

First, you need to build a new film document in the Flash, then set the attribute background as dark green, the size of 800 × 600 pixels. The document can be set to full-screen size which makes the picture fill the entire screen to achieve the best viewing experience. On the initial frames of the main scene setting "fscommand (\" fullscreen \ ", \" ture \ ")", we can fulfill the full-screen size settings. It should be noted that you should better set up a layer to write the scripting language at the top layer of the main scenes. Before writing a scripting language, you must select the frames to enter. All script input can only be written on the button, frames, and movie clips.

Second, to input the relevant materials to the library.

Finally, to compile the courseware program. Open the File menu, select Insert "scenes", and enter "Scene1", "Scene2", and "Scene3" or according to the teaching content to name the scene different names.

(1) "Scene1", mainly introducing the production of picture's playback

To set up three new video clip components in the library, name them "Picture1", "Picture2", "Picture3" or name them associated with the image content or teaching content, here with Pictures 1, 2, 3. The following names can be changed according to Flash elements. Another thing to be noted is that the size of the three images should be exactly the same, if not; you need to deal with by Photoshop and other mapping software.

To make these pictures scroll animation. Create a new layer in the scene named "Picture Play";

To drag and drop video clip component—"Picture1" to the first frame of "Scene1", "Picture2" to the 15th frame, "Picture3" to the 30th frame.

Add a stop action to the "Picture Play" layer. The method to stop adding is to add the "Stop" instruction to the 14th frame and the 30th frame.

Then we can achieve the effects of playback picture.

(2) "Scene2", mainly introducing display methods of the image playback on the base of "Scene 1".

In "Scene2", set up a new layer and name it "Effects Show";

To input "Show 1" to the first frame in the layer and save it as the components, and set the action "Sliding into"; To input "Show2" to the 15th frame in the layer and save it as the components, and set the action" getting big at middle"; To input "Show3" to the 30th frame in the layer and save it as the components, and set the action "Rotating bottom to fly into". The action effects can be set based on the design situation.

To drag and drop the components of "Show1", "Show2" and "Show3" to the layer of "Effects Show" and adjust the display position.

Finally, to add the action "Stop" to the 14th frame and the 30th frame.

After the above steps, we can set different text effects in Flash.

(3) "Scene3", mainly introducing adding the video control buttons.

To produce buttons in the component library. To add a button component to the library, and establish a "Scene3" layer, and draw a rectangular box, size of 100 × 80, black color (the button to lift the color), in the first frame of the layer, and then draw a blue rectangle, size 90 × 70. This will make the button a 3D image. Alignment: center. To make the second frame and the third frame a key frame. The method is: to select the second frame and the third frame, right-click to insert a key frame; In the key frame in the second frame, change the blue rectangle's color to green (mouse pointer over the button color), the third frame's color to dark green (the button is pressed). To copy four times "the video playback button 1", which makes the size and the color of the buttons consistent, and the four buttons are named "Scene2", "Scene3", "Back", and "Next".

In the first frame in the "Scene1" into the "scene", to put into "Scene1", "Scene2", and "Scene3" three buttons, and adjust the position. To select the button "Scene1", click "action—button/action/movie control (select standard mode)/ double-click the" on ", and select "press"/double-click the "goto", and then select "Go and play". "Scene1" needs to fill in the scene in order, then type frame label. To add to " Scene2" and "Scene3", it needs to mark the first frame of the two scenes so that this mark can be found. The marking is: to right-click to add a frame label in the scene, and input the frame label name. So when adding the button to jump action, you can choose the frame label name and make it possible to jump to the first frame at a scene. The button's action set of "Step" and "Next" is similar to the above. Do not describe it here.

(4) Courseware publishing

First, to save all the files above, and then open "the menu bar" in the file menu, and to choose "Export Movie" or "publish" to save the file as SWF or.exe format. Thus, you can operate Flash courseware program independently away from the editing page of the Flash development tools.

89.5 The Broad Prospect of Flash Courseware Applications

With the rapid dissemination of multimedia courseware, Flash-based multimedia courseware is also developing rapidly. From the stand-alone version to the network version, from simple text effects to a rich display of word, picture, sound, and video, from the automatic demonstration to interactive type, and from the secondary teaching to secondary self-study. Flash animation in courseware can easily realize auto-scaling, discoloration, distortion, deformation, rotation, and interaction design with other material. Gradual improvement of the design of multimedia courseware makes teachers get rid of manual writing on the chalkboard of traditional teaching methods, can turn some traditional boring conceptual knowledge into action, vivid screen image displayed; students can get rid of the boring abstract books; a variety of Flash animation, material will firmly seize the eyes of students providing students with a pleasant image and video experience, invisibly, they can master knowledge imperceptible. These are the things that traditional textbooks cannot match.

The design of Flash multimedia courseware, will give full consideration to the age, characteristics, and psychological factors of students, fully stimulate students' interest in learning, appropriate use of multimedia material, and mobilize the enthusiasm of students to appreciate music, develop students' imagination and creativity, and create a strong learning and teaching study atmosphere. Flash multimedia courseware has a variety of characteristics, such as rich content, innovative technology, times material, vivid images, simple operation, and friendly interface.

These features are constantly improving, which will make Flash multimedia courseware more loved by teachers and students.

89.6 Conclusion

In the process of producing Flash multimedia courseware, key points of multimedia courseware production are language, logicality of code, and practicality. At the same time, we synthetically apply Flash, Photoshop, and other software. In order to enrich the courseware content, we also apply a variety of components, script, animation, and other elements. With the continuous application of multimedia courseware in teaching, especially the Flash-based multimedia courseware is used more widely. Because of its strong performance and compatibility, it becomes an increasingly important proportion of the teaching activities. The Flash-based multimedia courseware helps students getting rid of the boring theoretical knowledge, greatly enhances students' intellectual curiosity, activates classroom atmosphere, and helps to master the knowledge easily. With the continuous development of computer technology, Flash-based multimedia courseware is becoming increasingly popular and mature.

References

1. Li Y (2009) The classical tutorial module template Jingjiang on flash multimedia courseware production, vol 10. Tsinghua University Press, Beijing, pp 22–27
2. Wu W (2009) The tutorial on Flash CS3 basis and examples, vol 8. Chongqing University Press, Chongqing, pp 93–97
3. Fang Q (2002) Flash multimedia CAI courseware design tutorial examples, vol 6. Tsinghua University Press, Beijing, pp 22–25
4. Guoxiong X (1997) Computer-aided teaching principles and courseware design, vol 4. University of Electronic Science and Technology Press, Chengdu, pp 12–19
5. Zheng B, Zhilong Z (2005) Of flash MX action script syntax reference dictionary, vol 4. China Railway Publishing House, Beijing, pp 4–25
6. Zhang M (2005) On the multimedia courseware. Electron Comput News 12:19–20
7. Wang J (2009) Based on the flash 8 platform the medical network multimedia courseware design and development. Contemp Chin Med 15:88–90
8. Qin Y (2010) The following principles on FLASH courseware production. Occupational 6:45–49
9. Zhai ZQ (2010) Analysis of few errors and countermeasures flash courseware produced. Occupational 3:11–14

Chapter 90
A Novel Algorithm for Embedding Watermarks into Audio Signal Based on DCT

Chao Yin and Shujuan Yuan

Abstract A new algorithm for embedding watermarks into DCT domain of digital audio signal is proposed in this paper. First, the original audio signal is divided in several segments, and the number of segment is two times as long as the number of watermark bits; Second, perform discrete cosine transform on each segment; By comparing middle-frequency coefficients of the two segments chosen by us, the larger one is chosen to be embedded watermark. Thus, it can not only guarantee the robustness, but also settle the synchronization problem. The experimental results show that this algorithm is robust to many operations to audio signal, such as low-pass filter, white noise, and so on, and the extracted watermark is more visible. But it not a blind detection algorithm.

Keywords DCT • Audio signal • Watermark

90.1 Introduction

With the manufacture and issuance of the digital AV products in quantity, the copyright protection of the audio data becomes more and more important. The copy restriction, use track, and the peculation confirmation can be realized by embedding watermark into audio carrier [1]. In recent years, the research of the audio digital watermark technology is developed rapidly, especially the technology of embedding watermark in the transform domain of digital audio signal. It makes the research more practical, because it can embed information in sensitive area of the audio carrier.

Similar to image document, the audio document can be made some modifications to embed information. These modifications cannot be removed if the attackers do not destroy the original audio. The audio watermark technology is used for copyright protection, contents authentication, audio search, secret communication, etc.

C. Yin (✉)
Department of Physics and Electronic, Langfang Teacher's College, Langfang, China
e-mail: liuxw_16@163.com

S. Yuan
Department of Qinggong College, Hebei United University, Tangshan, China

Z. Zhong (ed.), *Proceedings of the International Conference on Information Engineering and Applications (IEA) 2012*, Lecture Notes in Electrical Engineering 220, DOI: 10.1007/978-1-4471-4844-9_90, © Springer-Verlag London 2013

At an earlier time, there are four algorithms in time domain [2], LSB, echo hiding, phase coding, and spread-spectrum method. But also, we embed watermark into transform domain of the audio signal. They are DWT, DCT, DFT, and so on. In this paper, we embed watermark bits into discrete cosine transform domain of the audio signal, because it has several advantages [3]: (1) Discrete cosine transform (DCT) and IDCT are a pair of orthogonal transforms, and in comparison with FFT, the coefficients of transform domain only have real part that makes us embed watermark into audio signal conveniently. Also it has fast algorithm. (2) Some popular audio compression algorithms have used discrete cosine transform, such as MPEG-2, MP3, and so on.

For digital audio watermark, somebody (like Cox) [4] says that watermark should be embedded into the best important parts of auditory system (the most important parts are the low-frequency coefficients [5]) in order to improve the robustness, because the most important parts carry more signal energy. Thus, it can still remain the main parts.

In the presence of audio signal distortion, it will have better robustness if we embed watermark into the best important parts of auditory system. At the same time, the most important parts should be modified carefully in order to guarantee invisibility [6]. However, others say that watermark should be embedded into the middle-frequency coefficients [7, 8].

In this paper [8], the digital audio signal is divided into several segments and they perform (DCT) on each segment. The watermarks are embedded into audio signal by modifying its middle-frequency coefficients of DCT domain.

This paper presents an improved algorithm on the basis of Ref. [8].

90.2 The Embedding and Extracting Process

Let a represents the original digital audio signal, whose length is L. W represents the watermark to be embedded. W is a binary image whose number of lines are M_1 and number of rows are M_2. W is shown in the formula (90.1):

$$W = \{w(i,j), 0 \le i < M1, 0 \le j < M2\}, w(i,j) \in \{0,1\} \qquad (90.1)$$

Watermark is embedded into original audio signal as noise, so we must not influence its service quality. Let N represents the number what we need to embed one bit. Tests show that L shall conform to the requirement in formula (90.2).

$$L \ge M1 \times M2 \times N \qquad (90.2)$$

90.2.1 The Embedding Process

The embedding process is shown in Fig. 90.1

Convert the watermark into a one-dimension sequence. This paper uses a planar image as watermark and the image is represented by W. The number of lines of W

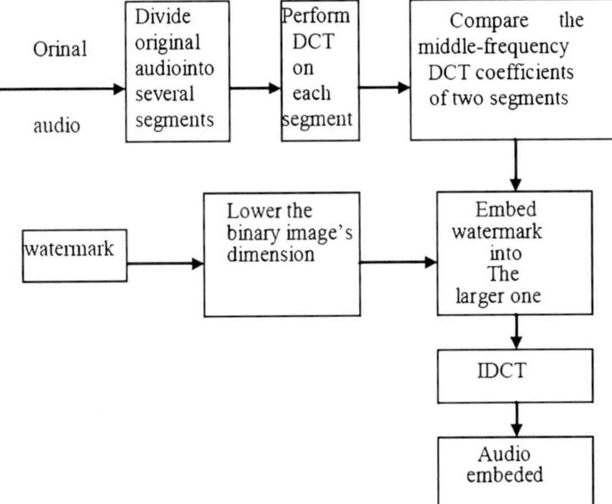

Fig. 90.1 The embedding process

is $M_1$and the number of rows is M_2 Thus, the total number of bits of the image is N ($N = M_1 \times M_2$). The planar watermarking image must be converted into a one-dimension sequence in order to embed watermark in audio in mono. We assume V is the sequence, viz.

$$V = \{v(k) = w\,(i, j)\,, 0 \le i < M1, 0 \le j < M2, k = i \times M2 + j\} \quad (90.3)$$

Divide the audio signal into several segments: we choose a section of the audio to embed watermark and divide the audio into several segments. The total number of these segments is represented by P, and P shall conform to the requirement in formula (90.4).

$$P = M1 \times M2 \times 2 \quad (90.4)$$

Perform discrete cosine transform on each segment.

Embed watermarks: Through comparing the second coefficient of two segments chosen in my way, the larger one is chosen to embed one watermark bit. Thus, it not only can guarantee the robustness, but also the synchronization.

We embed watermark like this [9]:

$$d'_k\,(2) = d_k\,(2)\,[1 + \alpha \times v\,(k)] \quad (90.5)$$

α Is a parameter that controls the strength of modification? If the value of α is too small, the robustness of the watermark is poor; If the value of α Is too high, it will decrease the audio's use value. Thus, we should choose α appropriately $d_k(2)$. Represents the second coefficient before modified, and $d'_k(2)$ represents the second coefficient after modified.

Through performing discrete cosine inverse-transform on the coefficients; we get the audio sinal embedded.

90.2.2 The Extracting Process

The extracting process is shown in Fig. 90.2:

Divide the audio signal embedded audio into several segments.
Perform discrete cosine transform on each segment.
Extract watermarks, as in Eqs. (90.6) and (90.7):

$$v_s(k) = \frac{1}{a \times d_k(2)}\left[d_k''(2) - d_k(2)\right] \tag{90.6}$$

$$v_s(k) = \frac{1}{a \times d_k(2)}\left[d_k''(2) - d_k(2)\right] \tag{90.7}$$

$v_s(k)$ is a sequence extracted from audio embedded? The value of threshold is set as needed. $d_k''(2)$ Is the larger coefficient obtained from the audio embedded?

Change the sequence to a planar image, and the image is represented by W_s.

$$W_S = \left\{w_s(i,j) = v'(k), 0 \le i < M1, 0 \le j < M2, k = i \times M2 + j\right\} \tag{90.8}$$

To estimate similarity of the original watermark and the extracted watermark, we set a definition of ρ as follows:

$$\rho = \frac{\sum_{i=1}^{M1}\sum_{j=1}^{M2} w(i,j)\, w_s(i,j)}{\sqrt{\sum_{i=1}^{M1}\sum_{j=1}^{M2} w(i,j)^2} \times \sqrt{\sum_{i=1}^{M1}\sum_{j=1}^{M2} w_s(i,j)^2}} \tag{90.9}$$

90.3 Simulation Experiments

In this part, we give the experimental results based on MATLAB.

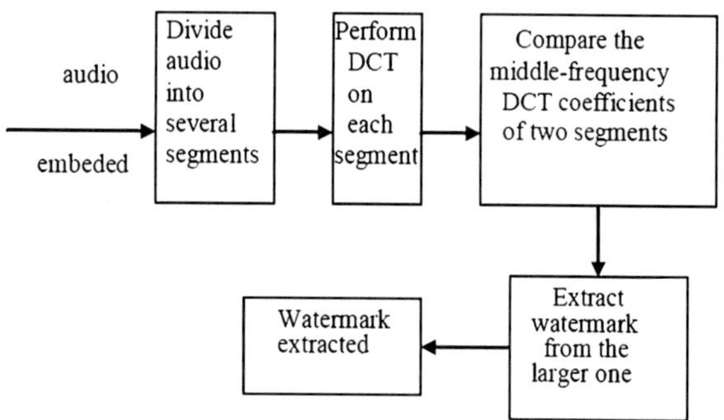

Fig. 90.2 The extracting process

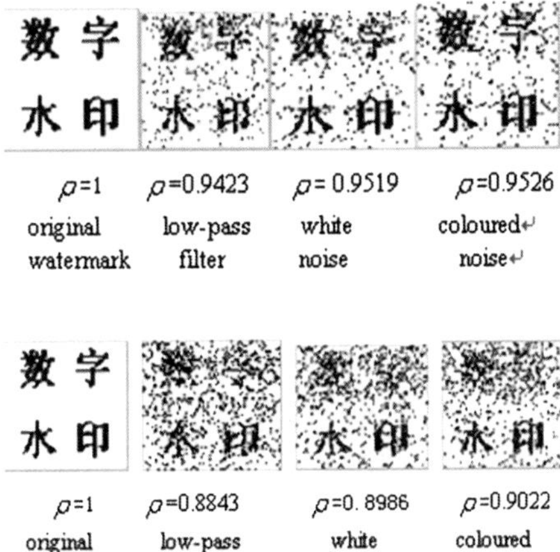

Fig. 90.3 The extracted watermark in this paper

Fig. 90.4 The extracted watermark in paper [8]

We choose 12 s long pop music in the simulation experiment, whose sample frequency is 44.1 kHz and its resolution is 16 bits. The watermark is a binary image whose number of lines is 64 bits and number of rows is 64 bits too.

To test the robustness of the watermark, we attack the audio as follows [10]:

Add white Gaussian noise whose mean is zero and variance is 0.01 to the audio signal.

Add colored noise which is obtained through filtering white noise with a band pass filter whose band width from 300 to 3400 Hz.;

Filter the audio with a low-pass filter, whose cutoff frequency is 11.025 kHz;

We listen to the pop music before embedded and the pop music after embedded, we cannot distinguish them. And we calculate the audio signal's SNR (The ratio of the signal to the random noise in a set of data). The audio signal's SNR in this paper is 25.3570 dB. The extracted watermark is shown in Fig. 90.3.

The audio signal's SNR is 25.2537 dB according to the algorithm in Ref. [8]. The extracted watermark is shown in Fig. 90.4.

90.4 Conclusions

This paper presents a novel algorithm for embedding watermark into digital audio signal based on DCT. The original audio signal is divided in several segments, and the number of segment is two times as long as the number of watermark bits; Second, perform discrete cosine transform on each segment; By comparing the middle-frequency coefficients of the two segments, the larger one is chosen to be embedded watermark. Thus, it can not only guarantee the robustness, but also settle the synchronization problem.

By comparing the algorithm with algorithm in Ref. [8], the experiments show that the watermark extracted in this paper is more visible, and the algorithm is robust to many attacks to the music, such as low-pass filter, noise, and so on. The drawback is that it needs original audio while extracting, so it is not a blind detection algorithm.

Acknowledgments The project was completed according to the scientific study project of langfang teacher's college (Number: LSZY201101). I would like to express my gratitude to my department leader and colleagues. I was thankful for their help and supports.

References

1. Wang B, Chen Q, Deng F (2003) Digital watermarking technology. Xidian University publishing company, Xian
2. Chang J (2005) "An investigation of digital watermarking technique based on wavelet analysis", master paper. Guangdong Univ Technol 12(2):123–128
3. Huang J (2004) "The research of audio digital watermark algorithm", master paper. Wuhan Univ Technol 4(12):332–337
4. Cox IJ, Kilian J, Leighton T, Shamoon T (1997) Secure spread spectrum watermarking for multimedia. J IEEE Trans Image Process 6(12):1673–1687
5. Tew fik H, Swanson M (1997) Data hiding for multimedia personalization, interaction, and protection. J IEEE Signal Process Mag 14(4):41–44
6. Kirovski D, Malvar HS (2003) Spread-spectrum watermarking of audio signals. IEEE Trans Signal Process 51(4):1020–1033
7. Hsu CT, Wu JL (1999) Hidden digital watermarks in images. J IEEE Trans Image Process 8(1):58–68
8. Wang Q, Sun S (2001) A novel algorithm for embedding watermarks into digital audio signals. J Acta Acoustica 26(5):464–467
9. Jiang T (2007) "Research on digital audio watermarking algorithm", master paper. Wuhan Univ Technol 32(11):432–436
10. Zhang M (2004) "Research on technology of digital audio watermarking", master paper. Harbin Eng Univ 5(16):21–25

Chapter 91
A Scalable Video Coding Algorithm Based on Human Visual System and HSI Color Space

Han Dan Cen, Wenhui Zhang and Song Yang

Abstract On the basis of theory and practice, the sensitivity of human eye is unbalanced to different color objects. Traditional video coding mainly focused on considering the statistical property and structure of video image as well as compression on the source material itself, while ignoring the effective use of the human visual characteristics. In the view of this, a new scalable video coding algorithm considering color visual sensitivity based on human visual system (HVS) was proposed. The experimental results show that this new algorithm can effectively reduce the distortion of sensitive color objects, so as to improve the color quality of the video image.

Keywords Scalable video coding • Sensitivity color • HVS • Video compression

91.1 Introduction

With the rapid development of techniques in video communication, the study of scalable video coding received increasing attention [1]. As the scalable coded video could be decoded and displayed at variety of temporal or spatial resolutions as well as quality levels, it has been widely used in HDTV/SDTV [2] compatible transmission, video on demand (VOD) [3], multipoint video conferencing and many other fields. In the MPEG-2 standard for video compression, there are several tools for bit stream scalability.

H. D. Cen (✉) · W. Zhang · S. Yang
Information Engineering School, Communication University of China, Beijing, China
e-mail: ipoqwzw@yeah.net

Z. Zhong (ed.), *Proceedings of the International Conference on Information Engineering and Applications (IEA) 2012*, Lecture Notes in Electrical Engineering 220,
DOI: 10.1007/978-1-4471-4844-9_91,

91.2 Color Visual Characteristics of Human Visual System

On human retina, there are some certain kinds of cells which contain pigments and could perceive lights [4]. There are two types of these cells. One type is the rod, which is numerous and spread all over the retina, but only respond to light in darkness. The other type is the cone, which is located in one small area of the retina but sensitive to colors. The cone is also divided into three types: the red, the green, and the blue. Red cones are sensitive to the red light. Green cones are sensitive to the green light which is a little more than the red ones. Blue cones are sensitive to the blue light, but their sensitivity is only about 1/30 that of the red or green ones. Figure 91.1 describes the sensitivity of these three types of cones relative to different wavelength of lights.

On the other hand, human eye can distinguish different colors which have different saturation. But they have different threshold of hue. Figure 91.2 shows the relationship between the sensibility (saturation level can be identified, from 0 to 100 %) and wavelength.

From Fig. 91.2, we concluded that human eye (about 580 nm) has the lowest sensibility to yellow region which could only distinguish less than 10 levels of saturation variation. But in red region, on the contrary, one could distinguish more than 25 levels of saturation variation.

It is human visual characteristic that causes the different distortion perceived by human eyes. A scalable codec model proposed as follows will solve this problem.

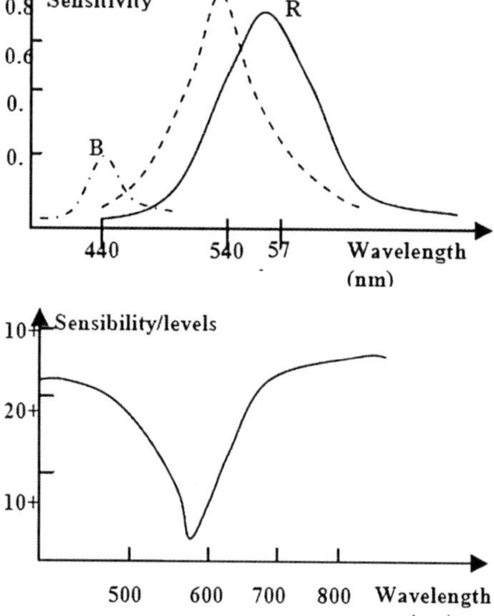

Fig. 91.1 Relative sensitivity of the cones

Fig. 91.2 Relationship of resolution threshold of saturation and wavelength

91.3 Improving Quality of Image Scalable Codec

91.3.1 HSI Color Space

In order to achieve the scalable codec in sensitive region, specific macro block should be marked in each frame firstly. Unfortunately, most typical color space like RGB and YUV has some drawbacks: components are highly correlate, lack of human interpretation, nonuniformity, and so on. So they are unsuitable for color identify. To allow image processing to be performed in easily understood terms. Another coordinate system based upon hue, saturation, and intensity has been defined. The transformation from the RGB into the HIS system is described as formula (91.1):

$$\begin{cases} H = \begin{cases} \dfrac{\theta}{360} \\ \dfrac{360-\theta}{360} \end{cases} \\ \theta = \cos^{-1}\dfrac{0.5[(R-G)+(R-B)]}{\sqrt{(R-G)^2+(R-B)(G-B)}} \\ S = 1 - \dfrac{3*\min(R,G,B)}{R+B+G} \\ I = \frac{1}{3}(R+G+B) \end{cases} \quad (91.1)$$

The saturation (S) corresponds to relative color purity, $0 < S<1$. The hue (H) refer to the domain wavelength of the color stimulus and it is a function of the angle, $0 < H<1$. In this paper, H and S are used to identify human sensitive colors. Finally, the intensity (I) represents the distance along the axis perpendicular to the polar plane, $0 < I<1$.

Figures 91.3 and 91.4 show the result using HSI color space to identity sensitive color. In Fig. 91.3, Result of calculating indicates that average value of hue $\overline{H}$

Fig. 91.3 Image of water lily

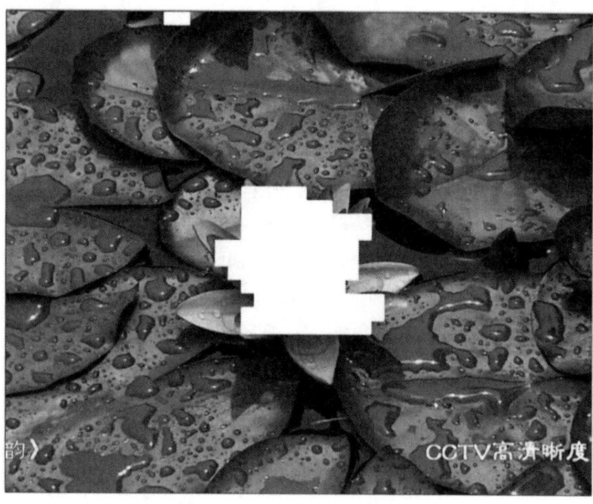

Fig. 91.4 Image of distinguished sensitive block

and saturation $\overline{S}$ in red region is respectively 0.849 and 0.842. In Fig. 91.4, corresponding region whose $\overline{H}$ value located in coordination (0.843, 0.856) was shown in white blocks.

91.3.2 Scalable Codec Design

In image/video codec, error is mainly caused by process of quantizing DCT coefficients. Generally speaking, the higher bit rate, the finer quantization step, the better the quality of image is, and the more vivid color is, vice versa. According to the human visual characteristic mentioned in Sect. 91.2, a scalable codec designed as Fig. 91.5 is used to improve the color quality of sensitive regions.

Encoder as shown in Fig. 91.5a, before DCT transform, bit stream access to the Color Distinguish (CD) module in which hue and saturation value (H and S) of every pixel was calculated by using formula (91.1). In each macro block, if the number of specific pixel (whose H and S value located in the sensitive region) exceeds threshold of 128, half of the total pixels in macro block. This macro block would be recognized as a block sensitive to color. Then the base layer is formed firstly by quantizing the 8×8 DCT coefficient block using a coarse quantization step size Qb which is then coded by using VLC to form the base bit stream. On the other hand, the reconstructed (inverse quantized) base block is then subtracted from the original DCT coefficients in order to derive the quantization error (QE) introduced by the base.

The decoder shown in Fig. 91.5b, two separate inverse quantization and inverse transformation processes are used. These are required because the use of the base layer picture alone for inter frame prediction at the encoder. To keep track with the encoder, data from the enhancement layer were not accumulated in the predictor. In this case, even if the large packed loss from the enhancement layer, the base quality of the video could be insured. In decoder, two different quality outputs

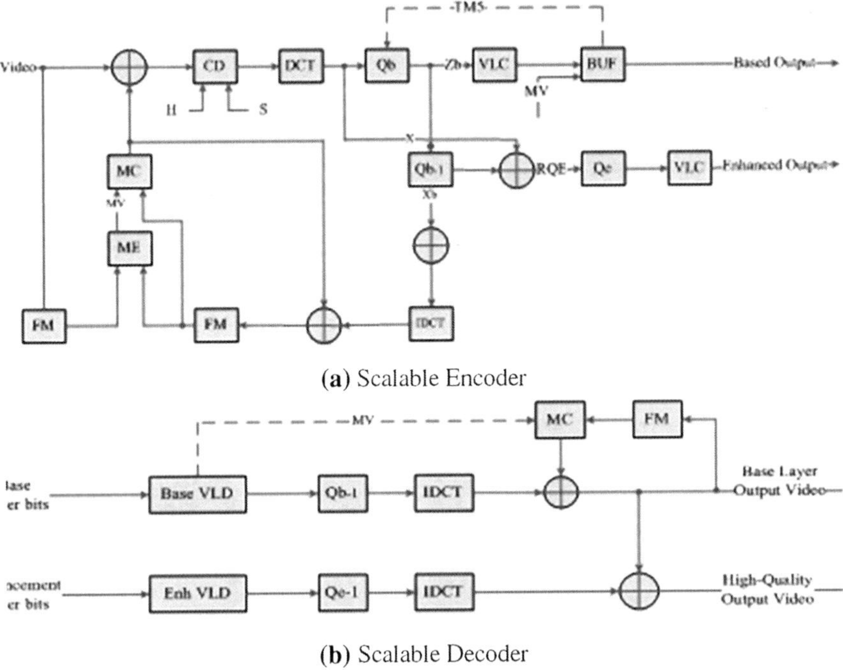

(**a**) Scalable Encoder

(**b**) Scalable Decoder

Fig. 91.5 Block diagram of the proposed scheme (**a**) Scalable Encoder (**b**) Scalable Decoder

have been afforded. Base quality provided only by base layer. Whereas as high quality output based on the basic level, at the same time, it compensates the quantization error of sensitive region, the represented images are more vivid and living.

91.4 Results and Discussions

To evaluate the performance of the New Algorithm, both objective and subjective assessments are adopted. The test has been conducted on HD (1,920 × 1,080) sequences RED ROSE which has a very slow pan, bright colors, and high contrast. Sensitive area is set between H = 0.8534 and H = 0.9948 (including all red regions). The MPEG GOP length is N = 12 and 2 B-frames (M = 3) between P-frames is used in this test.

91.4.1 Subjective Quality

The base and enhanced sequences were displayed on a plasma screen for visual assessment. We set base layer bit rate as 17 Mbit/s, while enhanced layer quantization step Qe is 8. Figure 91.6 shows the original and (region of sensitive) ROS image.

Fig. 91.6 Original and ROS image at bit rate 17 Mbit/s RED ROSE sequence from number: 95

Fig. 91.7 Decoding without (*left*) and with (*right*) enhancement layer. Part of frame 95 of RED ROSE

Then, enhancement layer conducted in the sensitive region. We work on base layer and enhancement layer respectively in the decoder. Figure 91.7 shows a comparison of a portion of a frame from the sequence without and with enhancement layer.

91.4.2 PSNR Quality

In this paper we use PSNR described as formula (91.2) as the objective criteria of image quality.

$$PSNR = 10\lg\left[\frac{255^2}{\left(\frac{1}{N\times M}\right)\sum_{j=0}^{N-1}\sum_{i=0}^{M-1}(y(i,j)-x(i,j))^2}\right] \tag{91.2}$$

Table 91.1 Value of PSNR in each bit rate

Base layer bit rate PSNR(dB)	Qe	Total bit rates (Mbit/s)	PSNR (dB)	PSNR Gain/dB (dB)
15 MBit/s 30.24(dB)	8	26.4	32.52	2.28
	10	23.1	31.15	1.91
	12	20.7	31.33	1.09
	14	18.9	30.89	0.65
18 MBit/s 34.21(dB)	8	27.6	36.46	2.16
	10	24.4	35.85	1.64
	12	22.6	35.03	0.82
	14	21.5	34.87	0.66
20 MBit/s 36.03(dB)	8	28.8	38.05	2.02
	10	26.4	37.48	1.45
	12	23.8	36.74	0.71
	14	22.1	36.54	0.51

In the formula, $M \times N$ indicates to image size, x(i,j) and y(i,j) represents the gray values of original image and reconstructed image, respectively.

The base layer bit rate of HD sequences is set at 15, 18, and 20 M bit/s. Enhancement layer bit rate is determined by its quantization step Qe. First, only the base layer is used in experiments and average PSNR is measured on each bit rate. Then, both base and enhancement layer are decoded, PSNR gain in each bit rate is calculated and results are presented in Table 91.1.

From Table 91.1, we can find that with the increasing of bit rate in base layer, the PSNR rise accordingly. To the enhancement layer, the finer quantization step Qe, the more bit rate increased. Consequently, the quality of the reconstruct image is better.

91.4.3 Compare with SNR Scalable Coding

As we know, the main drawback of SNR scalable codec is the increasing of total bit rate. While the advantage of New Algorithm is that it redistributes the coding bits in the enhancement layer based on human visual characteristics. Therefore, at a specific bit rate of base layer when distortion of the image mainly from sensitive region in subjective feelings, the reconstruction image in decoder could have higher subjective quality with a lower bit rate increasing.

In order to prove this, subjective test has been conducted as follows. Test sequences and other parameters are selected as before. Base layer has been coded at eight different coding rates: 15, 16, 17, 18, 19, 20, 21, and 22 Mbits/s. While four different quantization steps: 8, 10, 12, and 14 are used in enhancement layer, with two different scalable ways: new algorithm and SNR scalable codec. According to ITU-R BT.710, Double Stimulus Continuous Quality Scale

Table 91.2 Specifications of the monitor

Resolution	1,680 × 1,050
Dot pitch	0.282 nm
Peak luminance	150–250 cd/m2
Contrast ratio	1,000:1
Viewing angles	1,600 horizontal, 1,600 vertical
Response time	5 ms
Gamut	92 %

(DSCQS) is adopted as the procedures of the subjective quality evaluation. The original and test sequences are shown to the user twice in alternating fashion, the order chosen randomly. They rate the material on a scale ranging from bad to excellent, and the rating has an equivalent numerical scale from 0 to 100.

Viewing conditions comply with ITU-T P.910 for home environment. Viewing distance was set at about three to four times the height of the video image. The specifications of used screen were shown in Table 91.2.

Experiment results are shown in Fig. 91.8. When the base and enhancement layer are both available at the receiver, the New Algorithm and the SNR scalable codec could achieve a notable higher subjective quality. When at lower bit rate, such as 15, 16 Mbits/s, the SNR scalable codec performs better than New Algorithm since both sensitive and nonsensitive regions have a large distortion. However, the 17 Mbits/s is a turning point for New Algorithm: the MOS increases

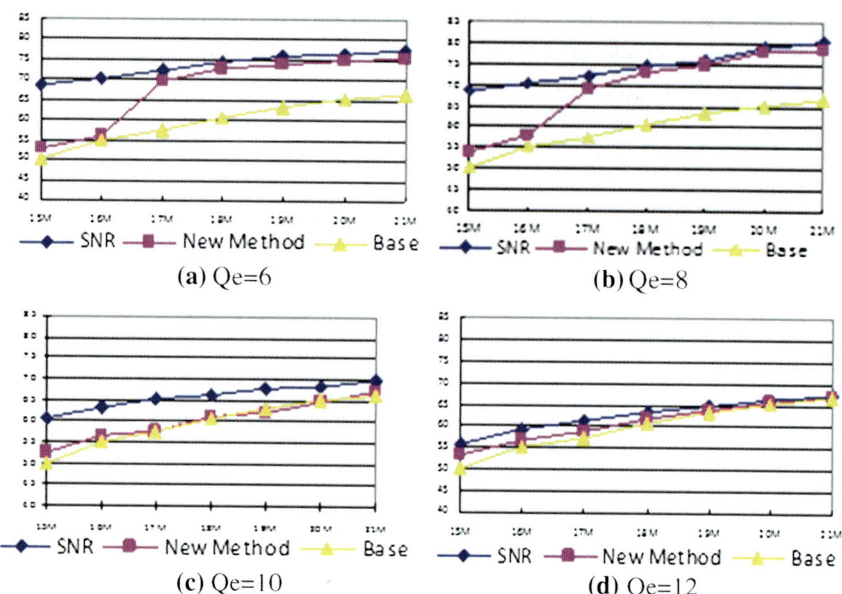

Fig. 91.8 MOS results for given bit rate **a** Qe = 6. **b** Qe = 8. **c** Qe = 10. **d** Qe = 12

significantly at this point. That is because, in a higher bit rate, image quality deterioration is mainly from sensitive regions which are encoded in the New Algorithm using the enhancement layer. Compare with SNR scalable codec, New Algorithm only needs to encode specific macro blocks instead of all blocks (in this sequence, encode blocks only account for 30 %). That is to say, in a higher encode bit rate, the New Algorithm could achieve the same subjective evaluation as the SNR but with less bit rate increasing, which is about 30 % of the SNR.

91.5 Conclusions

This paper gives a new algorithm of scalable coding system considering the color visual sensitivity based on human visual system (HVS). Experiment results show that this new algorithm could improve image quality on objective and subjective evaluation. Especially, compare with SNR scalable codec, it could provide nearly the same subjective feelings with less bits rate increased at a higher bit rate of base layer. The cogitation of this scalable codec based on HVS, it is suitable for most image and video compression stand nowadays, such as JPEG, MPEG-2, and H.264.

Acknowledgments This work was supported by the "211 project" of Communication University of China.

References

1. Mi X (1991) MPEG (ISO/IEC JTCI/SC29/WG11) coding of moving picture and associated audio for digital storage media at up to 1.5 Mbit/s. 14(3):11172–11174
2. Robers MA (1997) "SNR scalable video coder using progressive transmission of DCT coefficients," Master's thesis, Northwestern University, Department of Electrical and Computer Engineering 36:262–264
3. Wilson D, Ghanhari M (1996) Transmission of SNR scalable two-layer MPEG-2 coded video through ATM networks'. In: 7th international workshop on packet video, vol 19(7), pp 37–42
4. ITU-T (1999) Recommendation P.910, subjective video quality assessment methods for multimedia applications. Stand Descr 28(3):267–269

Chapter 92
Research on the Spirits of Teaching Teamwork Based on Interactive Multimedia Courseware

Ling Zhang

Abstract To have a strong grasp of English is to ensure that a critical success factor in the teaching of English. This paper attempts to investigate the English-learning students to understand the relationship between teamwork and interactive multimedia courseware (CDiCL). The experimental design carried out before and after the test, these experiments involved four complete groups of students, the experimental control group, the only colpublicorative learning (CL) group, and the only courseware (CD) group and CDiCL-groups. This paper also carried out interviews and participant observation of a combination of used to collect qualitative data. The results of the quantitative data analysis showed that these two groups of the CD better than the other groups did not. Qualitative data analysis revealed that the team used between users of the CD is more effective when the team smaller (3–4).

Keywords Teamwork • CDiCL • Colpublicorative learning • Quantitative analysis

92.1 Introduction

In such a competitive society, semi-skilled translator and English writer such as English-learning assistants need to have a certain level of math skills often need to solve the problem in its basic learning work formed the fixed analysis tools. For English students, how to become the future pillars of society, by the widespread concern of the community. New ways of teaching are determined to ensure that these students understanding of English culture. Some scholars said that the process of learning English can not only enhance learning, but also to promote team learning [1]. A unified teacher-training program combined with the views of all the professional standards to a new conceptualization of general and special education courses for their respective programs. In order to more effectively

L. Zhang (✉)
School of Foreign Languages, Huanghe Science and Technology College,
Zhengzhou 450063, China
e-mail: lingzhang450063@163.com

Z. Zhong (ed.), *Proceedings of the International Conference on Information Engineering and Applications (IEA) 2012*, Lecture Notes in Electrical Engineering 220,
DOI: 10.1007/978-1-4471-4844-9_92,

cultivate a colpublicorative and inclusive, pre-service teachers should develop a program of pre-service teacher education courses and professional practice and the concept of shared vision, establish a comprehensive program to provide opportunities for special education and general education to work together. Not only in learning, including many teams other cooperative efforts to work harder and get a multiplier effect because of the division of the team spirit. Complete any work effort and the effect of doubling is subject to the trust and entrust the motivation and improve their attitude and commitment to adapt to the culture of these technical staff to work. However, this research level also depends on the learning environment for students. For example, the English was used in the classroom teaching a significant difference and to assess students' satisfaction, learning, and teaching. Learning is from the Internet, access to a wide range of advantages of the world.

The purpose of this research includes the following: (1) the use of interactive multimedia courseware (CDiCL) whether a significant impact on learning English and learning teamwork, to promote the progress of the colpublicorative learning team members; (2) What factors promote the importance of team cooperation enhance the use of computers and the Internet on the mastery of English and calculus.

92.2 The Relevant Problems

Teamwork is essential in today's world, teamwork and a variety of ability to work together to become more important. In English teaching, traditional teaching methods are to take the "spoon-fed" taught in the teaching process, limiting the students explore English problems and timely questions. Therefore, teaching is to provide courses to encourage cooperation between people and different communities [2].

The students not only in the classroom to research English, but also in the public is trying to use their own learning, this form of behavior in the students' growth stage can be crucial. The Internet has become every one of us in the life indispensable important information platform, how to bring into full play its positive role, make it better for our country campus safety culture construction, we must now have a problem to be solved. Internet-based campus network is the important infrastructure in school teaching, scientific research, shouldering the important task of management and foreign exchange, network information safety is directly related to the teaching, and research activities such as safety [3]. Therefore, various departments in the school to further research the relevant laws and regulations, and actively improve and implement the relevant rules and regulations, fully aware of the importance of computer network security, improve the work of network security awareness, strengthen security concept, and cogent safeguard school in network information security. The school authorities to network safety into the school security focus positions, with the school security work together, together to implement the deployment check, ensure the school network information flow, safe, effective, and ultimately makes the safety culture construction is fast finish. Learning process, teachers must master a few aspects of students' motivation, affect the effectiveness and level of commitment for the students to solve the problem, because it does not control any kind of learning is very difficult to quantify. In particular, teachers' limited energy, leading to a teaching is not the time to more fully thinking, cognitive, to select and plan how teaching students the relative lack of knowledge exploration and reflection.

Computer-based learning (CBL) for personalized learning. This is because, it provides the opportunity to give individuals a basic math skills and concepts of training and practice making their own to get to meet its development. CBL designed to meet different learning tendencies as the goal [4]. However, the design of English learning software and multimedia English in teaching is a challenge. At present, computer supported colpublicorative learning (CSCL), the very popular because it helps students to do self-learning, new English also allows the teacher to monitor the progress of many students in many classes anytime, anywhere. This paper is CDiCL based on the research to promote teamwork.

92.3 Research Methods

This paper is intended for students of the Universities of English research, and selected 135 students, the method used in the CL group, that is, each team member needs to play a particular role, such as team leader, deputy leader, record keepers, managers, and timekeepers. And each time we carry out the role of rotation. Team members are the highest level of choice by teachers based on the English level of the SPM, the team leader. It is mainly characterized by the special needs according to the characteristics of students and other content, including the general education environment and teaching classroom, strategies for these children. However, in addition to special education content teachers are for nondisabled people in the general classroom preparation. The immersion model of teaching is characterized by general and special education disciplines and Union College to supervise the education of a mode of experience. From the disciplines of the two schools into general and special education content with the existing curriculum to meet the diverse needs of the student's regular teachers.

Independent variable is learning CDiCL, CD and CL, and a traditional teaching method, the dependent variable is the score before and after the test. Test for each group of members of the index factors, mainly to analyze several dimensions of learning ability, efficiency, the arrangement of the screen, graphics, user satisfaction, and then come to the overall performance of each group, as shown that CD indicators of the group, in Table 92.1.

From Table 92.1 and Fig. 92.1 can get score greater than 70 % indicate that the courseware in the category at an acceptable level. The avaipublicility of the CD group, the learning ability of the CD group accounted for 86 % of the rest,

Table 92.1 CD interactive avaipublicility

Respect	Percentage
Ability to learn	86
Efficiency	75
Screen arrangement	75
Graph	76
Customer satisfaction	78
The overall performance	87

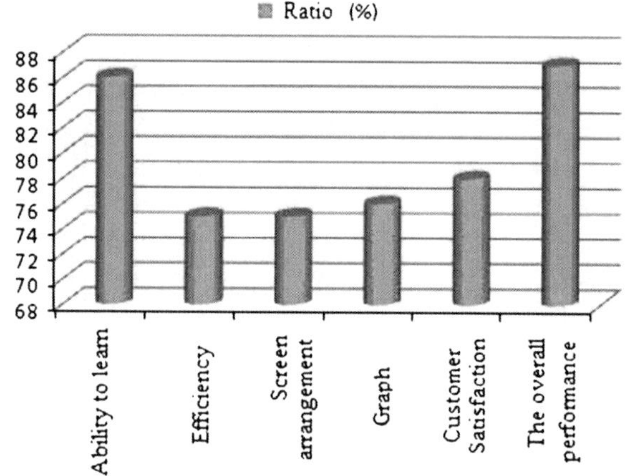

Fig. 92.1 The avaipublicility of a percentage of all aspects

such as efficiency, the screen arrangement, graphics, and user satisfaction in about 76 %, but also in the acceptable range. And overall performance is quite satisfactory, accounting for 87 %.

The CD layout, content and interface design and development is built on the basis of the theoretical model of Herman. Master of education required for the design of the CD using the psychomotor, cognitive, and emotional elements. In addition, the role of the courseware will reduce the overload of students' memory. Handout for each member of the main is grade 0 (none) to 10 (highest) classification, and the interaction for the culture of commitment among team members, accountability. Before and after the test, each test has 12 questions. It covers figures, replacement, decomposition, simplified linear equations, signed numbers, fractions, and word problems. Also combined with the form of peer review, audio, video recording and semi-structured interview to triangulate, and using SPSS for the quantitative analysis of the results of the ATLAS/TI was used to analyze the dialog. Therefore, to strengthen the research of network knowledge, you put the network knowledge skilfully applied to work research and life, you is the campus safety culture construction is an important part of. Therefore, the school should be in the setting of network curriculum based on knowledge, must also often hold some network knowledge, to give students systematic curriculum. Students in the universal network knowledge at the same time, the teacher should improve the network knowledge, thus better to teach students, management of students. In the efforts of restructuring teacher education programs, it is important to discuss the problem of teacher education, and agreed to the concept of a unified, aware of its content and process in an atmosphere of openness, and trust. Traditional education is often the method of division, differentiation, and experts on the work, and now education is no longer relevant, this is a special education opportunities are clear values, hopes to promote the education of students' diverse and inclusive.

Try to concentrate on solving the problem of the CL mode a week, to encourage group discussion. Most transactions between the daily life of the businessman and his clients, they need foreign exchange. Others are within the scope of works, English and technical personnel need to be converted into the right thing, in order to take the right paint, the right road, and clearly marked.

Here is 60 min per week to accept a different picture of the learning process:

From Fig. 92.2 of the CL group team, can help their teachers to solve the problem of words. At the same time, the group of CDiCL ongoing research group at the beginning of a 20 min CD, 40 min from the move here.

The basis of the principles is compulsory, the school should provide education for all children, regardless of the student has any perception of difference, disability or other social, cultural, and language differences [5]. For the diverse needs of these learners and the pursuit to make the school more the need for regular and special education teachers and learners as well as family and community to consult and cooperate with each other, strategically effective teaching and learning. In China, independent of teacher education courses, regular and special education do not have the desired effect, a comprehensive knowledge of the functions and responsibilities in line with the diversity of learning needs in the classroom. The purpose of this paper is to establish a new mode of education requires teachers with the attitudes, knowledge and ability of the various factors to be effective in the classroom to meet the diverse learning needs. The paper will focus on the need for regular restructuring and special education, teacher education, and education programs to develop the overall teacher in order to promote an inclusive pre-service teacher education courses to students of all disciplines.

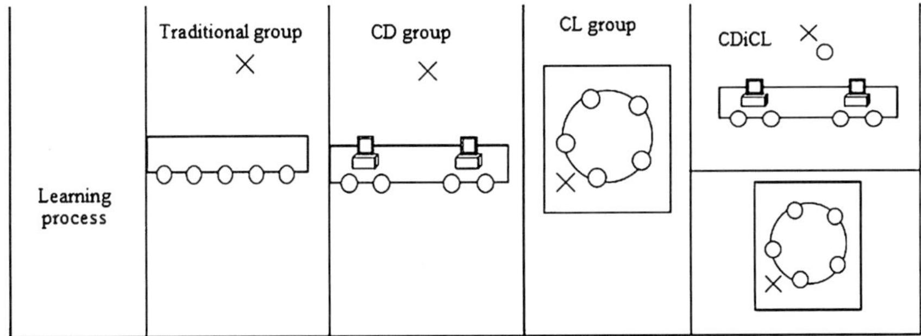

Fig. 92.2 Different methods used in each learning process

Table 92.2 Descriptive statistics of each participating group

Group	Before the test		Test	
	The mean	Standard deviation	The mean	Standard deviation
The traditional group	5.87	4.3	9.46	6.3
CL group	6.43	5.7	8.41	6.4
CD group	8.45	4.5	14.07	5.6
CDiCL	5.98	5.5	11.95	8.6

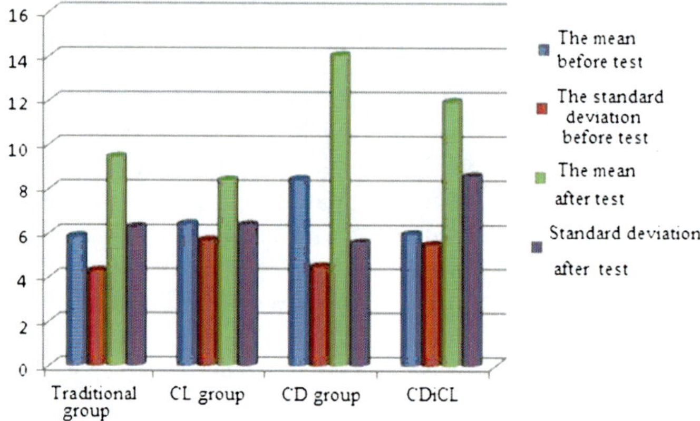

Fig. 92.3 The comparison chart of statistical analysis in each group descriptive

92.4 The Data Results

The qualitative and quantitative analysis, descriptive statistical analysis is of the score of each group, after in-depth data analysis in accordance with the SPSS software.

Obviously, from Table 92.2 and Fig. 92.3, the test before and after the test, the mean of each group are showing an upward trend, but CDiCL group the largest increase, followed by the CD group, then there are the traditional group, and finally the CL group, but before the test, the mean maximum for the CD group, the second is the CL group, the third is CDiCL group, and the traditional group ranked last. During the tests, the mean of the largest CD group, CDiCL group moved up one row of the second, the third is the CL group, and finally the traditional group. These show that in teaching activities, the traditional teaching methods ineffective.

92.5 Conclusion

This research to determine whether interactive multimedia can be seen in the research of English learning teamwork plays an influence produced an effective role, the Herman advantages of application of the model for positive interaction between members and teachers. However, the level of interaction between team members is less than 5 %, the future needs for more positive interaction with team size to generate better English learning. Team building must be properly organized, team spirit needed more time, the limited capacity of the teaching and experience, to fully understand the attitude of learning, motivation, learning methods and ways to effectively control and to guide full use of interactive multimedia learning, and fully mobilize the enthusiasm of students, enhance their interest in learning, and enhance teamwork and promote common progress.

References

1. Wu T (2010) Multimedia teaching courseware design in the English classroom. Examination Wkly 12:3–5
2. Li Y (2009) Multimedia-assisted English teaching to solve a good problem. New Curriculum In 08:36–39
3. Wu F, Wang W (2010) IT context of the use of multimedia chemical experiment teaching thinking. Xinjiang Norm Univ (Nat Sci) 25(3):55–57
4. Liu L, Chen X (2010) Multimedia information English in the College physical education. Era Educ (Educ Teach Ed) 07:24–27
5. King K (2008) Multimedia applications of information English in the modern education reform significance. Radio TV Univ 13(3):37–38

Chapter 93
Analysis on Language Skills-Centered English Teaching

Hua Fan and Wenjuan Xiao

Abstract Through a questionnaire survey of English teaching programs on college students, this paper makes the data analysis of the survey results from basic language skills, structure and details of the various elements and contents. It also uses multi-level evaluation techniques to conduct a comprehensive evaluation of the English teaching program, drawing that the current English teaching program is not appropriate to students' requirements. To this end, a number of measures are proposed to improve the English curriculum to address specific problems, which can provide reference to the English teaching program improvement.

Keywords Language skills • English teaching programs • Comprehensive evaluation

93.1 Introduction

English language is the English teaching information carrier, and it has become the main tool for the teacher to complete the task of teaching courses. Language skills are a teacher with the correct speech, semantics, intonation, in line with the logical structure of spoken language, to make narrative and elaborate description of the teaching program, and student problems [1]. It is a personal behavior. Teachers' teaching of English language skill has direct impact on the enthusiasm of an important element of learning English, and can provide guidelines for students to learn English and improve the learning of English thought, as well as can provide them with important guidance. However, the use of language skills is a combination of students' learning activities like observation capability, experimental ability, and discussion activities.

1. Basic language skills: Everyone must have the language skills. They consist of five elements [2]: voice and articulation, as the physical materials of language,

H. Fan · W. Xiao (✉)
School of Foreign Languages, Hebei University of Technology, Tianjin 300130, China
e-mail: wenjuanxiao@yeah.net

Z. Zhong (ed.), *Proceedings of the International Conference on Information Engineering and Applications (IEA) 2012*, Lecture Notes in Electrical Engineering 220,
DOI: 10.1007/978-1-4471-4844-9_93,

it is the form of sound show and hear expressed in the language of information; the volume and speed, which is the speaking volume and speed. The ears of everyone have some endurance, generally normal speech rate of 210–260 words/minute word; the tone and rhythm, which are various changes of the voices; vocabulary, which requires standardized, accurate, and vivid expression; syntax, which is the words of logical rules.

2. Special language skills: language skills in a special form of communication.

The three elements of language skills (phase): introduction, intervention, and assessment.

Therefore, in order to analyze the English teaching program, it will be based on the structure and elements of the language skills of English language teaching programs for multi-system evaluation and analysis of the feasibility of the English teaching program [3] Table 93.1.

93.2 The Survey of English Teaching Program

This survey for the basic language skills is mainly around the college graduates to investigate the satisfaction of the English teaching program, which provides seven options, and they are as following: very dissatisfied, dissatisfied, slightly

Table 93.1 Language skills in a special form of communication

Elements	Details of elements	Details
Introduction	Boundaries mark	Introduce new topics, and new requirements
	Bring out the theme and focus	Achieve new topics and new requirements
	Name	Name specified students to carry out the answer
Intervention	Tip	Tip knowledge and behavior as the basis
	Repeat	Repeat the answer, thought provoking
	Ask	Ask based on answers to a question, thought provoking
Assessment	Evaluation	Evaluation and analysis on the answer to comment
	Repeat	Repeat the same answers to get attention
	Corrections	Corrections to be analyzed and corrected
	Ask	Continue to ask questions, causing the depth and breadth of thinking
	Expansion	Expand to other relevant information to get better understanding

dissatisfied, average, slightly satisfied, satisfied, and very satisfied [4]. The selection of this distribution is mainly according to the current situation of university graduates in English teaching, and it is more closely tied to reality, and they can be accurate and intuitive responses to the English teaching program.

The figure shows, in terms of language skills, students are not satisfied with the English teaching program. The vast majority believe that the current English curriculum does not meet the requirements of language skills, and cannot reach the satisfaction. This also shows that the English teaching program is not very satisfactory, and the effect is very bad Fig. 93.1

93.3 The Evaluation Model of English Teaching Program

The fuzzy evaluation is to quantify the fuzzy information, with a reasonable choice of factors threshold, to make a multi-factor quantitative evaluation. The following is a multi-level fuzzy comprehensive evaluation of the language skills of English teaching programs. With more elements of language skills, we consider only the five aspects of English teaching program, as is shown in Table 93.2. Weight value is to determine the students' overall satisfaction of English teaching.

1. Comprehensive evaluation of the beginning level of English teaching program.

Comprehensive evaluation of the elements of language skills:

First determine the evaluation factors. There are three elements which have the impact of language skills, consisting of the domain [5, 6]:

U = {introduce (n1), intervention (n2), and assessment (n3)};

There are four evaluation sets, the resulting composition of the evaluation on the field is V− = {dissatisfied (x1), general (x2), satisfied (3), and very satisfied (4)}.

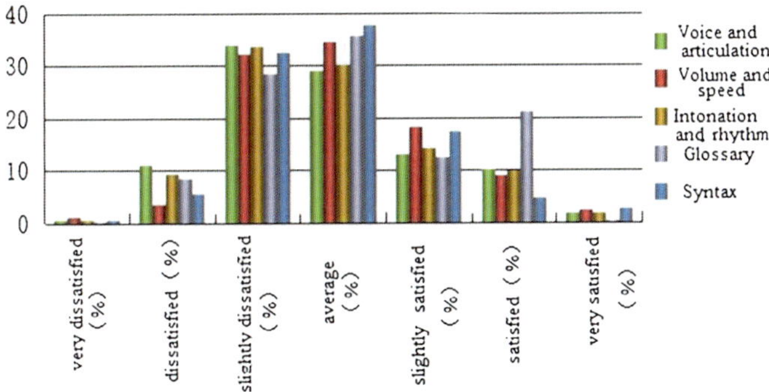

Fig. 93.1 Proportion of basic language skill and english teaching program

Table 93.2 The table of basic language skills and english teaching program

	Very dissatisfied (%)	Dissatisfied (%)	Slightly dissatisfied (%)	Average (%)	Slightly satisfied (%)	Satisfied (%)	Very satisfied (%)
Voice and articulation	0.7092	10.9929	34.0426	29.0780	13.1206	10.2837	1.7730
Volume and speed	1.2433	3.4483	32.1839	34.4828	18.3908	9.1954	2.2989
Intonation and rhythm	0.5420	9.2141	33.6043	30.3523	14.3631	10.0271	1.8970
Glossary	0.2510	8.3621	28.5829	35.8492	12.5234	21.1241	0.2627
Syntax	0.6440	5.4371	32.4673	37.6837	17.3683	4.5737	2.6478

Second, determine the degree of membership of various factors: 16 % are not satisfied; average accounting for 52.5 %; satisfaction accounting for 22.5 %; very satisfied is 10 %.

The introduction of membership: a1 (0.16, 0.525, 0.225, and 0.10)
Intervention: a2 (0.25, 0.65, 0.225, 0.2)
Assessment: a3 = (0.35, 0.375, 0.42, 0.12)
Thus, the evaluation matrix is

$$R_1 = \begin{bmatrix} 0.16 & 0.525 & 0.225 & 0.10 \\ 0.25 & 0.65 & 0.225 & 0.2 \\ 0.35 & 0.375 & 0.42 & 0.12 \end{bmatrix} \tag{93.1}$$

The number of weights normalized is: 0.4 + 0.2 + 0.4 = 1
Thus constituting the vector of a fuzzy factor set V, V = (0.4 0.2 0.4)
And the availability of comprehensive evaluation is

2. Comprehensive evaluation of the two level of English language teaching programs comprehensive evaluation of the two-level elements [7]:

$$A_1 = X_1 Y_1 = \begin{pmatrix} 0.4 & 0.2 & 0.4 \end{pmatrix} \begin{bmatrix} 0.16 & 0.525 & 0.225 & 0.10 \\ 0.25 & 0.65 & 0.225 & 0.2 \\ 0.35 & 0.375 & 0.42 & 0.12 \end{bmatrix} = \begin{bmatrix} 0.163 & 0.75 & 0.087 & 0 \end{bmatrix} \tag{93.2}$$

By the sign of the elements' details, we can get marks, focus concentration, name, tips, repeat questioning, evaluation, repeat, correct, and asked, expansion, extending the two comprehensive evaluation system that make up the English teaching program. Among them, take the limits of signs and expansion, and extends as important the right weight. "Boundary mark" expands and extends the number of weights constitute a fuzzy vector $B_1* = (0.3\ 0.7)$. Comprehensive evaluation of the obtained result is

$$B_1^* = A_1^* \begin{bmatrix} \frac{B_1}{B_2} \end{bmatrix} = \begin{pmatrix} 0.3 & 0.7 \end{pmatrix} \begin{bmatrix} 0.263 & 0.75 & 0.077 & 0 \\ 0.4 & 0.425 & 0.085 & 0 \end{bmatrix} = \begin{bmatrix} 0.218 & 0.7 & 0.082 & 0 \end{bmatrix} \tag{93.3}$$

The "introduction" of comprehensive evaluation result is

$$B_2^* = \left(0.2 \quad 0.8 \right) \begin{bmatrix} 0.36 & 0.55 & 0.19 & 0 \\ 0.18 & 0.57 & 0.37 & 0 \end{bmatrix} = \begin{bmatrix} 0.4 & 0.522 & 0.278 & 0 \end{bmatrix} \tag{93.4}$$

Similarly, evaluation result of "intervention" is

$$B_3^* = \left(0.4 \quad 0.6 \right) \begin{bmatrix} 0.3 & 0.2425 & 0.4345 & 0 \\ 0.5875 & 0.145 & 0.4345 & 0 \end{bmatrix} = \begin{bmatrix} 0.245 & 0.5375 & 0.3475 & 0 \end{bmatrix} \tag{93.5}$$

Similarly, the comprehensive evaluation of the assessment result is

$$B_4^* = \left(0.3 \quad 0.7 \right) \begin{bmatrix} 0.3 & 0.456 & 0.34 & 0 \\ 0.5 & 0.22 & 0.35 & 0 \end{bmatrix} = \begin{bmatrix} 0.351 & 0.3542 & 0.537 & 0 \end{bmatrix} \tag{93.6}$$

3. Comprehensive evaluation of the three level of English language teaching program

We use the details of the elements of language skills as an English teaching program evaluation fuzzy vector

$$B^* = \begin{bmatrix} 0.2 & 0.2 & 0.1 & 0.3 & 0.2 \end{bmatrix} \tag{93.7}$$

Thus, the comprehensive evaluation result is

$$B^* = A^* \begin{bmatrix} B_1 \\ B_2 \\ B_3 \\ B_4 \\ B_5 \end{bmatrix} = \begin{bmatrix} 0.2 & 0.2 & 0.1 & 0.3 & 0.2 \end{bmatrix} \begin{bmatrix} 0.238 & 0.5 & 0.282 & 0.12 \\ 0.442 & 0.51 & 0.278 & 0.32 \\ 0.255 & 0.665 & 0.275 & 0.32 \\ 0.411 & 0.642 & 0.237 & 0.46 \\ 0.159 & 0.306 & 0.459 & 0.06 \end{bmatrix}$$
$$= \begin{bmatrix} 0.255 & 0.548 & 0.21 & 0.307 \end{bmatrix} \tag{93.8}$$

The above three comprehensive evaluation results show that the evaluation of students' English language teaching program is still not satisfying. To this end, we must strengthen the English teaching curriculum reform in order to improve the standard of English teaching.

93.4 Measures of Improving English Teaching Language Skills

English teaching language skills reforms include import teachers teach, writing on the blackboard and modern means of application, students observe, and practice experimental teaching activities. In these activities, language skills are very important, and must therefore be in the English teaching program which has strict requirements:

Import the required subject, namely the need to import well-designed language, including: the creation of context, the introduction of materials, and arousing the desire for knowledge; the visual demonstration, which is the instance to mobilize students to explore new knowledge; the analog and contact to improve the quality of teaching to get the best results; using suspense to arouse enthusiasm.

The required language of lectures, mainly in the logic and is inspired. Penetrating, containing: the logic, logical reasoning, easy to understand; the penetrating, i.e. elucidation of the accurate analysis of solid teaching content; the inspiration: To stimulate students' interest and inspire their thinking ability.

Induction and summary of the language requirement, reflected in a concise, plain, and extended performance.

93.5 Conclusion

Through research on college students in the English teaching program, and from the structure and various elements of the language skills and elements of the details and content, this paper makes an analysis of survey results. We use multi-level evaluation techniques to conduct a comprehensive evaluation of the English teaching program, drawing the conclusion that the English teaching program is not appropriate to students' requirements; to this end, a number of measures are used to improve the English curriculum and to address specific problems, which can be used as references for the English teaching program improvement.

References

1. Weihua L, Sheqiang Z, Hongtao C (2007) Students' satisfaction evaluation system survey of the ideological and political theory teaching. Elite Cult 05:123–125
2. Yayi S (2011) Evaluation of linear structural equation modeling. Chin Hosp Stat 04:33–35
3. Hansheng S (2004) Language intuition of the phrase in english teaching. High Educ Stud Sci Technol 23(1):118–119
4. Ruifang Z (2002) Sense of language training in english teaching, vol 24(2). Jining Teachers College, Jining, pp 68–71
5. Yueping W (2002) Analects sense characteristics, vol 05. Fuyang Teachers College: Social Science, Fuyang, pp 60–62
6. Jian Z (2003) Second language teaching should be to cultivate the sense of language-oriented study. Lang Trans 01:53–57
7. Chunsheng L (1989) Safety evaluation on the weighted average multi-level fuzzy comprehensive evaluation parties supporting. Industr Saf Dust 02:16–20

Chapter 94
Analysis of Physical Exercise Mode on College Students

Shuyou Wu and Fusong Hu

Abstract China's national physique monitoring results show that the physical fitness of college students is continuing to decline slowly. This paper, based on the current situation, analyzes several students' suitable physical exercises, and records the impact of university students' physical exercises. Finally, the paper uses the last elements calculation, which is multiplied by the weight of the various physical exercise mode indicators and the scores of their income, and then adding them together to get the conclusion that these types of physical exercise model significantly improve college students' physical fitness.

Keywords Physical exercise mode • College students • Physical fitness • Factor scoring method

94.1 Introduction

With the continuous improvement of our living standards, people are constantly to improve their living standards, including amateur hours of physical exercise [1]. In 2011, China's national physique monitoring results showed that our national constitution overall pass rate has continued to rise in primary school physical fitness for 20 years, the downward trend is under preliminary control, but at the same time, China's university students physical fitness has continued to decline slowly [2]. The twenty-first century is the century of talent, but to face the physical fitness of our students is declining, a serious breach of the physically round development of the original intention of our national leaders. Of college students physical decline in physical fitness of college students continued to show a slow decline.

S. Wu
Hainan College of Software Technology, Qionghai 571400, China

F. Hu (✉)
Physical Education College, Qiongzhou University, Sanya 572022, China
e-mail: wu_shuyou1@163.com

Z. Zhong (ed.), *Proceedings of the International Conference on Information Engineering and Applications (IEA) 2012*, Lecture Notes in Electrical Engineering 220, DOI: 10.1007/978-1-4471-4844-9_94, © Springer-Verlag London 2013

With them in middle school, high school stage to ignore the physical exercise has a direct relationship [3]. The pressure of college entrance examination for high school students has virtually no exercise and physical decline. To college, they continued the inertia of the high school, still emphasis on exercise. On the other hand, a lot of the University of Physical Education have taken the club system, many projects are elective, students participate in physical exercise depends on interests and hobbies, is bound to make that they have special interests to only one subject and their physical fitness is not comprehensive, balanced, and enhanced.

At present, all over the country attention and implementation of school sports is not balanced, and even there are significant differences, in many places, the target of an hour of exercise a day merely to meet the inspection, a mere formality. In fact, exercise beneficial to the health of college students. Students participate in regular exercise will make students good health, and energy to complete the study and work and tasks [4].

94.2 The College Students' Physical Exercise Mode

Generally are based on college students staying at school, so we think that the main mode of university students' physical exercise: exercise of the bedroom, classroom exercises, stair corridor exercise, and sports complex exercise. Methods of exercise in the bedroom, such as dumbbell curl, lift the chest, squats, chest exercise, inverted, tai chi, or take a deep breath. Dumbbells are generally held flat, lateral raise, alternating grip lift and other 12 kinds of actions to choose [5]. Should not be too much or too little exercise good idea to regularly practice the next day once every 30–40 min is appropriate, the exercise arm muscle, could also contribute to the body muscle development for the degree of fever, swollen. Chest movements include common steel spring system, the practice generally has the arms to 13 kinds of action on the drop-down, flat, and Latin America. Exercise the latissimus dorsi, pectoralis major, and brachial four muscles, which are very useful to young students and bust their development. Teachers exercise less, and the other students to match. In the exercise of the staircase Road step jumps, use the stairs and steps, which can improve the students' leg strength. Finally, the exercise of the sports venues includes mainly badminton, basketball, boxing, football, gymnastics, handball, judo, swimming, taekwondo, tennis, table tennis, and volleyball.

And then, we divide the college students participating in physical exercise in the form of physical exercise mode into standalone, together, group, and school organization. Through college investigation and analysis, we get the following results in Table 94.1.

Table 94.1 Proportion of students' participating in physical exercise modes

Gender	Alone	Together	Group	Organized
Male	35.8 %	20.3 %	25.4 %	18.2 %
Female	12.6 %	42.5 %	9.5 %	5.2 %

Table 94.2 Students' understanding of this model

Type	Very clear	Basically clear	Not very clear	Do not understand
Traditional	5.8 %	25.3 %	60.2 %	8.7 %
Modern	13.2 %	61.8 %	18.7 %	6.3 %

It can be found from the table, the boys prefer to go alone to participate in physical exercise, but the girls are afraid of loneliness, and care less to choose to exercise alone, so more girls banded together to participate in physical exercise; for groups type physical activity, girls are less participate in, fewer girls will choose to participate in physical exercise organized by the school, schools should strive to improve in this area ratio.

We can also divide physical exercise mode into traditional patterns and modern mode. The traditional mode of exercise sports activities including archery, wrestling, qigong, meditation, Ba Duan Jin, Yi Jin Jing, and other health guidance class [6]. While the modern mode of exercise, such as badminton, basketball, boxing, football, gymnastics, handball, judo, swimming, taekwondo, tennis, table tennis, and volleyball. The following is a survey for college students the level of understanding of both physical exercise mode which can be seen from Table 94.2.

From the above table, we can see that most college students are not very understanding of the ancient Chinese way of physical exercise, keeping fit but those are our ancients, the school should vigorously improve the publicity of the ancient traditional exercise mode.

94.3 The Physical Exercise Modes' Impact on Constitution

From these physical exercise modes we extract a few college students most commonly used sports items, the results of the investigation then focused on the student early, and then of other test analysis of this model to the students physical changes [7].

94.3.1 The Outdoor Football

Football fitness level is very high, which needs high-speed running ability, the ability to control body weight, flexible pace, step point, the confrontation of power quality, and good endurance, and flexibility qualities. Football training can promote the body's metabolism, improving the function of the respiratory system, heart and other organs, increase appetite, and improve the absorption capacity. Good exercise, to improve young people's muscles, ligaments, and strength. The results show that: the growth of the athletes' body weight, considerably more than the college students of the same age weight gain index. The muscles are strong,

Table 94.3 Football's impact numerical on constitution

Group	N	Football	
		Freshman	Sophomore
50 M	966	8.97 ± 5.47	7.35 ± 3.98
Standing long jump	966	221.25 ± 165.73	250.68 ± 18.21
Pull-up	966	8.34 ± 5.02	5.97 ± 4.35
1,000 M	966	260.03 ± 40.16	270.97 ± 196.64

Table 94.4 Track and field's impact numerical on constitution

Group	N	Track and field	
		Freshman	Sophomore
50 M	199	7.57 ± 4.39	7.87 ± 4.68
Standing long jump	199	242.16 ± 55.23	240.27 ± 14.59
Pull-up	199	6.34 ± 2.98	6.77 ± 5.37
1,000 M	199	232.83 ± 32.42	268.21 ± 40.54

developed, and have an important role on the fitness of your body, improve athletic ability, and a variety of sports injury prevention. College students in the following table after a year of physical fitness exercise (Tables 94.3, 94.4).

94.3.2 Track and Field

Students systematically engaged in athletics, can promote the body's metabolism, to coordinate the link between the nervous system and motor organs, improve the functioning of the cardiovascular system, respiratory system, and other internal organs; the full development of strength, speed, endurance, sensitivity, coordination physical fitness, to promote normal development, and improve health; but also to promote walking, running, jumping, and throwing skills more rationally and effectively, thereby maintaining and improving human life and learning, the ability to adapt to work; can delay human aging process. College students in the following table after a year of physical fitness exercise.

94.3.3 The Indoor Basketball

Basketball the sport can improve the vitality of life: basketball activities cover the running, jumping, cast a variety of forms of physical exercise, and exercise intensity, therefore, it can be comprehensive, effective and comprehensive manner to promote physical fitness and all-round development of the human body functions

Table 94.5 Indoor volleyball's impact numerical on constitution

Group	N	Indoor volleyball	
		Freshman	Sophomore
50 M	200	7.27 ± 6.774	7.18 ± 5.76
Standing long jump	200	238.82 ± 16.19	240.62 ± 15.57
Pull-up	200	7.48 ± 4.38	6.97 ± 3.87
1,000 M	200	245.27 ± 17.34	237.53 ± 28.84

to enhance and maintain vitality, all man's activities lay a solid physical basis, thereby improving the quality of life [8]. College students in the following table after a year of physical fitness exercise (Table 94.5).

94.3.4 The Indoor Volleyball

Participating in volleyball can not only improve people's strength, speed, flexibility, endurance, jumping, reaction physique and athletic ability, and improve the functional status of the body organs and systems, but also to cultivate the wit, determination, and calm the psychological quality of. College students in the following table after a year of physical fitness exercise.

94.3.5 Traditional Martial Arts

Martial arts practice effects on the body is very large, and the cardiovascular system can be random is in continually receiving such a large stimulus process, and gradually improved. The requirements of practicing martial arts on the respiratory systemare very high, often practicing the martial arts will enhance the function of the respiratory system. Practicing martial arts when parts of the body muscle coordination, exercise the functions of the nervous system. College students in the following table after a year of physical fitness exercise. Table 94.6 numerical martial arts on Constitution.

Table 94.6 Traditional martial arts' impact numerical on constitution

Group	N	Traditional martial arts	
		Freshman	Sophomore
50 M	100	7.39 ± 3.94	7.31 ± 3.76
Standing long jump	100	236.52 ± 20.09	233.62 ± 18.57
Pull-up	100	8.08 ± 4.52	7.57 ± 4.63
1,000 M	100	244.47 ± 21.44	247.83 ± 23.79

Table 94.7 Modern aerobics' impact numerical on constitution

Group	N	Modern aerobics	
		Freshman	Sophomore
50 M	300	8.59 ± 11.75	8.51 ± 11.04
Standing long jump	300	179.32 ± 40.89	179.52 ± 48.96
Pull-up	300	8.81 ± 5.72	8.67 ± 4.93
1,000 M	300	232.81 ± 18.33	230.79 ± 20.67

Table 94.8 Moran I index value

Project	Moran I	Moran I Expectations	Standard deviation	Normality statistic	Probability *p* value
Football	0.256	−0.035	0.108	2.703	0.005
Track and Field	0.249	−0.028	0.112	2.629	0.015
Basketball	0.231	−0.030	0.114	2.405	0.025
Volleyball	0.215	−0.024	0.107	2.576	0.015
Martial arts	0.210	−0.029	0.116	2.129	0.020
Aerobics	0.227	−0.024	0.117	2.783	0.027

94.3.6 Modern Aerobics

For the six test items, the paper related to the test results are as follows: (Tables 94.7, 94.8).

The bodybuilding operating as an aerobic exercise, the fitness function of all aerobic exercise, such as full to improve their physique and improve cardiovascular fitness and muscular endurance, and promote the coordinated operation of various tissues and organs of the body, the human body to achieve the best functional state [9]. In addition, different from other aerobic exercise aerobics is that it is an easy, beautiful sport in fitness at the same time, bringing people to enjoy the arts, people happy to revel in the exercise fun, to reduce the psychological pressure, and promote physical and mental health, which greatly enhances the effect of fitness. College students in the following table after a year of physical fitness exercise.

94.4 The Factor Scoring Method

In this paper, we mainly use the element method of calculation, the weights of the various physical exercise mode indicators multiplied by the score of their respective income, then the sum of the main model is as follows.

Table 94.9 A_{ij} assignment reference table

Importance aij	Definition	Explanation
1	Equally important	i and j are equally important
3	Slightly important	i is slightly important than j
5	More important	i is more important than j
7	Obviously important	i is obviously important than j
9	Absolutely important	i is absolutely important than j
2,4,6,8	Between the two adjacent	Between the two adjacent

$$F = \sum_{i=1}^{m} P_i \omega_i, P_i = \sum_{j=1}^{n} x_{\text{ij}} p_j$$

$$\sum_{i=1}^{m} a_i \omega_i, \sum_{i=1}^{m} a_i w_2^i, \ldots\ldots, \sum_{i=1}^{m} a_i w_n^i, W^k = B_k, B_{k-1}, \ldots\ldots, B_2, B_1 \quad (94.1)$$

$$C_{.1} = \frac{\lambda\text{max} - n}{n - 1}$$

In the above formulas, F is the final overall evaluation scores; w_i is a single weighted value of the training elements (indicators); P_i is expressed as a fraction of experts to a single training elements (indicators); X_{ij} represents the i-th training management indicators, evaluation elements from the final score, P_j is the j evaluation elements of the final proportion. A_{ij} values can refer to Table 94.9. An index weights A_i = i index the frequency divided by the number of samples, (i=1–4); two index weights $a_{\text{ij}} = a_{\text{ij}}$ appear frequency divided by the number ($a_i 1 + a_i 2 + a_i 3 + \ldots$ aij of frequency) a_{ij} is a subsidiary of i-level indicators index of the jth two indicators.

94.5 Conclusion

This paper analyzes several models of physical exercise and the obtained physical exercise mode has a more pronounced role in the improvement of physical fitness of students through four sports. Therefore, in the coming days, we need to pay attention to the following points: First, governments at all levels, especially the education of the sports management department should further implement the spirit of the Central Document No. 7, to increase investment in school sports; university schools should conscientiously implement the national curriculum standards, quality, and quantity of athletic class, there is no day of physical education, and schools should be one hour in the afternoon after school organized students to collective exercise; is to increase the school sports ground, sports equipment improve and equip, and strengthen the full-time physical education teachers with the existing physical education teachers the ability to enhance the work, and provide protection for the effective implementation of school physical education and extracurricular sports activities; strengthen propaganda to appeal to the full range of family and social concern to the health of young people problems, and also to establish healthy talent and growth concept.

Acknowledgments The research was supported by the education scientific research project of Hainan Province Department (Grant No. Hj 2009-161). I would like to thank all the people who have helped me in my research.

References

1. Jianmin L (2011) Physical education teaching and mental health education, vol 86(2). Trade Union Management Institute, Shandong Province, pp 58–60
2. Luo Yi H, Xin W (2010) College physical education in how to cultivate students' sports ability. Liaoning Inst Publ Adm 6(05):25–27
3. Fuchun S, Yijun S (2009) Students' psychological barriers and school physical education. Coal Manage Inst 73(02):12–14
4. Lin J (2009) On the college physical education cultivates students' creative ability, vol 13(02). Heilongjiang Province Law, 78–80
5. Xiaoli Z (2009) Implementation of university physical education teaching harmony education, vol 131(01). Beijing Traffic Management Institute, pp 54–57
6. Xuelin Z (2011) Psychological factors of female students to participate in sports analysis, vol 21(02). Liaoning Police and Justice Management Institute, pp 34–36
7. Jiayong Z (2006) Moral education in physical education. Polit Sci Law 24(04):44–46
8. Yingjun Z (2009) Multimedia technology in college physical education, vol 4(01) Trade union forum Shandong Province union Management Institute, pp 12–14
9. Junrong Z (2004) Analysis of university students in management by objectives-a case study of Chongqing university of posts and telecommunications, vol 63(02) Chongqing Institute of Socialism pp 52–55

Chapter 95
English Writing Teaching Scheme Based on Multimedia

Wanyu Liu

Abstract With its visualization and intuition in the teaching mode, rich teaching resources and good interactive interface, multimedia is of great importance in promoting and assisting teaching. It will be of great significance both in theory and in practice to introduce it into English writing classes. This essay will analyze and discuss this related issue.

Keywords Multimedia • English writing • Network platform • Vividness

95.1 Introduction

Teaching English writing is an important link in English teaching. It is also a comprehensive challenge for the English teaching elements of grammar, vocabulary memorizing, sentences forming, and the ability for English reading [1, 2]. Due to the reason above, teaching of English writing is one of the difficulties of English teaching in the long run. Many of the students do not know how to grasp the essentials, writing English compositions that are made up of simply by words without any significance [3, 4]. These compositions do not rise to the degree of standards, which can be traced back to two aspects. On the one hand, English writing belongs to the comprehensive project, which is of great difficulty. On the other hand, it has great connection with our teaching methods.

Nowadays, teaching of English writing still stays in the stage of writing a composition basing on a topic. In this stage, students are hard to overcome the psychological fear of difficulties and form the vicious circle of the more writing and more afraid, while more afraid and worse. In that case, pursuing innovation of teaching in English writing, motivating the interest of students in learning as well as enhancing the effects on teaching in writing have become the focus of teaching in English writing. And in this regard, multimedia can throw us a lot of inspiration.

W. Liu (✉)
Wuhan Polytechnic University, Wuhan 430000, Hubei, China
e-mail: Liuwanyu@chinaqikan.net

Z. Zhong (ed.), *Proceedings of the International Conference on Information Engineering and Applications (IEA) 2012*, Lecture Notes in Electrical Engineering 220, DOI: 10.1007/978-1-4471-4844-9_95, © Springer-Verlag London 2013

Multimedia is also called CAI teaching system which was put into teaching in 1980s. It has played a good role in teaching in the past 20 years. With its visualization and intuition in the teaching mode, rich teaching resources and good interactive interface, multimedia is of great importance in promoting and assisting teaching. Introducing it into the teaching of English writing is very meaningful both in practice and in theory. In relation to the use of multimedia in teaching English Writing, we can analyze from the following aspects.

95.2 Establish and Improve the System of Multimedia Teaching

Establishing and improving the system of multimedia teaching is the base of accomplishing the target of adopting multimedia in teaching English writing. Its core is to make resources, tasks and the arrangement of staffs needed in teaching writing with multimedia in details. Generally speaking, the method of filling the module can be adopted to fill different parts and make them in details, which leaves significant influence on the guides and standard in the follow-up work. And the refinement of functions also helps to discover the blind areas of management, which at the same time makes it possible to be properly adjusted according to the outer conditions. It has strong and normative guidance. Usually, the current system of teaching English writing with multimedia should take the following elements into consideration.

95.2.1 Teaching Purposes

The plan proposed of purposes on teaching English writing with multimedia should include both general teaching objectives and stage teaching objectives.

95.2.2 Equipment Management

It refers to the management of multimedia teaching equipment such as network maintenance, information collection equipment maintenance and processing, and transmission equipment maintenance.

95.2.3 Hardware Equipment

It requires the consideration of the requirements on hardware equipment during the whole teaching process. The focus should be on the computers, networks, practical software related to information technology, and so on. We should also consider whether there are any deficiencies or gaps in related funding and hardware building in schools and determine timely solutions.

95.2.4 External Environment

We should plan and manage the external environment of teaching such as classroom discipline and the approach of room management to make sure that there is a scientific and good external environment for teaching.

95.2.5 Teaching Staff

In order to achieve related teaching objectives, excellent teaching staff is very essential. And the teaching staff mainly consists of two levels of factors. The first one is that whether the teaching staff can satisfy the needs of teaching in writing. If the needs can be satisfied, teaching plans should be adjusted or resources of teaching staff should be further fulfilled. The second factor mainly includes the re-education of the teaching thinking and skills and job training for the teaching staff to make sure that their teaching levels can meet the current requirements.

95.2.6 Teaching Test

There are two aspects on testing the teaching effects. On the one hand, the focus is to decide which method to test the teaching effects through examination or the development of small issues. On the other hand, the evaluation of the test effects and the feedback from students are both main contents in teaching test.

95.3 Stimulate Students' Interest in Writing with Multimedia

Interest is the best teacher. To achieve the improvement in teaching English writing, we should firstly stimulate students' interest and passion in learning writing. Allowing for this aspect, CAI multimedia teaching system has good introductory influence. First, as for the contents of writing, we can make full use of the multimedia to make them vivid and visualized with the aid of screens, pictures, and so on which can provide students with an intuitive impression on the topic they are going to dig into. For example, when teaching the text of Disney, teachers can play some introduction of the theme park in English, advertisements or some classic fairy tales such as the sleeping beauty to give students a general idea. By taking advantage of the functions of capture and tips in multimedia and abstracting some key words from the material, teachers can lead students to express their own idea toward Disney according to the vivid multimedia and keywords abstracted. The writing mentioned here should be open, allowing students to refer to various kinds of material. After everyone finishes their work, teachers can arrange for the part of voting the best work in class. When the result comes out, teachers should guide students to think the reasons why they think the work is the best. Is there any usage of words, sentences, and phrases worth learning? After conclusion is done, we can go back to the text to see how these aspects are expressed. In this way, we come across the process of comparison. On the one hand, we can stimulate students' interest in learning. On the other hand, students can have great impression on the key words, vocabulary, related grammars, and writing skills.

95.4 Enrich the Writing Level with Multimedia

In the past, English writing was based on the topics that had been given in advance. However, this kind of thinking and teaching mode has been out of the date. It can not be accustomed to the current education in teaching English writing. Under this circumstance, the system of multimedia teaching can help to enrich our level of teaching writing. First, we can display a variety of scenes such as campus, supermarkets, companies, and amusement parks. Many kinds of writing topics can be proposed in these scenes. For example, we can make up a traveling scene with the system of multimedia and play some related traveling English videos and advertisements, guiding students to play roles of visitors, travel agents, guides, and so on. On the other hand, students can write about related contents, advertisement designs, tourism projects, and commentary according to the role assigned. Provided with examples and all kinds of material, students can enrich their writing levels to a great extent. They can overcome the traditional limitation of boring mode. What is more, their writing topics and formats are also enriched, which can be seen in Fig. 95.1.

Guided with this kind of thinking, we can take advantage of the scenes created by the system of multimedia and assign roles to students to play. This enables students to think over the contents they should express and make the contents into writing. And online translation is also a very good source. Students translate Chinese into English, while multimedia reverts their translation into Chinese again. By comparing the two versions, students can find out the differences and pay more attention to the expression details in writing. This kind of training can start from simple sentences to a paragraph of words.

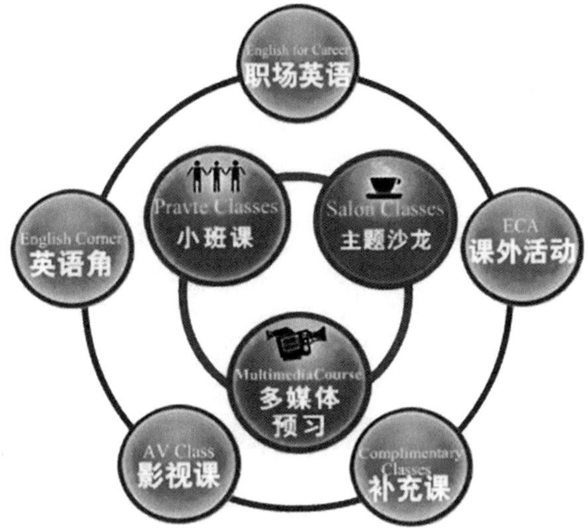

Fig. 95.1 Diagram of enriching levels of english writing

95.5 Enrich Writing Channels with Multimedia

In traditional teaching modes, the improvement of writing was supposed to write unceasingly. However, according to the viewpoint of modern English teaching, English listening, speaking, reading, and writing are closely related. These four elements have close influences on each other and promote each other. With the aid of multimedia, we can effectively integrate the four elements and enrich their writing channels.

For example, we can make full use of teaching method of role playing and transposition. We can assign students to make a five-minute speech before every English class begins. As for the contents, they can be chosen by the students themselves. Sometimes, students can be encouraged to tell a story in English or share some divergent thinking of the texts learnt previously. By adopting the multimedia, we can record the whole process of the speech. After that, we can encourage students to write a composition according to their own records. When the composition is finished, we should lead students to discover their weaknesses and shortcomings of the composition. Amended step by step, the composition is becoming more and more plentiful and polished. After the composition has been amended completely, we can ask students to come in front of the class telling the same contents. At this time, the system of multimedia can play back the student's previous speech. Then we should ask the rest of the students to compare the two speeches and find out the differences. In this way, students can have better memory and understanding of the influences that grammars, sentences frames, and paragraph convergence have on a composition with the help of the functions of re-comparison and real-time capture.

Let us make another example. In the writing classes, teachers will explain a lot of sentence patterns, example sentences as well as model essays. If they are having classes with multimedia, there will be more material such as pictures, flash, and videos. The traditional blackboard is difficult to have such a huge carrying capacity. As a result, students often forget to take notes when listening to the explanation. Even if some of them could take some notes, the notes are often not coherent. However, PPT is often adopted in the teaching mode with multimedia. With its help, teachers can be relieved from the traditionally heavy writing on the blackboard and devote more passion to the explanation and assistance for the students. That this kind of courseware can be copied easily makes it convenient for students to download or copy through the school information platform after class. This enables students to listen to the teachers wholeheartedly in classes. What is more significant, they can review the courseware after class to have deeper impression. This kind of teaching effect is prominent.

95.6 Enhance the Construction of the Network Platform

Another advantage of teaching with multimedia lies in its interaction with network education. It can set up the network education conveniently. The establishment and improvement of the network platform can greatly reduce the limitation of time and space. On the one hand, students can study related knowledge through the self-learning network platform anytime they want besides the basics in class learning and discussion. They can also find out the FAQ's self-diagnosis, which greatly improve the time dimension of learning. On the other hand, teachers can also upload and publish some writing exercises, sources as well as model essays regularly through the network platform for students to download and refer to. They can also release the teaching arrangements in the near future or some writing tasks for students to choose freely, which makes it possible for students to arrange their time more reasonably. What is more, students can take advantage of the formats of online communication, message leaving and e-mails to break the limitation of time and space. They can interact with their team members and teachers effectively. The construction of the network platform makes the connection between students and teachers closer and closer. At the same time, the network operation is beneficial for the data analysis with the system of multimedia. Therefore, resource library can be built based on that, making it possible for teachers to analyze the related problems, so that more targeted follow-up teaching can be achieved.

It has been the main trend that teaching English writing with multimedia. Its promotion and in-depth usage lead the revolution of teaching. Multimedia can stimulate students' interest in study, making their learning attitude from passiveness to activeness. It also turns the stage learning into lifetime learning to a great extent. With the development of the multimedia technology, the ability of teaching English writing can step onto a higher development platform without doubt.

References

1. Zhou W (2008) Some universities in beijing to the united states expedition, study of the teaching management in the U.S. colleges. Educ Res 7(2):53–57
2. Liu L (2006) How to improve the teaching standard of english writing. Hubei Educ 7(2):45–48
3. Ma MS (2008) English writing teaching mechanism layer. Educ Res 5(1):24–29
4. Zou LL (2009) Study of the digital and information technology in teaching of english writing. Educ Res 8(4):65–69

Chapter 96
Efficient Speeding Up Scheme of AutoCAD Drawing

Weihong Luo

Abstract The development of machinery automation enables AutoCAD graphics software to practice more universally. Depending on its exclusive advantages, this software has played a very good role in drawing mechanical graphics. Compared to previous traditional manual drawing, CAD drawing owns many advantages, operating speed, dimensional accuracy, graphic scale, and so on in drawing can all be controlled strictly according to the machinery industry standard. After a period of using, technical staff found that AutoCAD graphics software has much room for development, and many places are needed for improvement.

Keywords AutoCAD software • Drawing • Speed • Technological upgrading

96.1 Introduction

AutoCAD is automatic aided design computer software which is mainly used in two-dimensional drawing, detailed mapping, document designing, basic three-dimensional design, and so on. As CAD has a wide range of design performances, it is more universally used in machinery manufacturing industry. Although it is automatic drawing, the use of CAD software still relies on operators. Having a research into related methods of speeding up CAD drawing is very crucial to effective drawing.

W. Luo (✉)
Information Department of Jiangxi Vocational and Technical College, NanChang 330013, JiangXi, People's Republic of China
e-mail: Luoweihong@hrsk.net

Z. Zhong (ed.), *Proceedings of the International Conference on Information Engineering and Applications (IEA) 2012*, Lecture Notes in Electrical Engineering 220,
DOI: 10.1007/978-1-4471-4844-9_96,

96.2 Deficiencies in Manual Drawing

Being restricted to socioeconomic and scientific technologies previously, most of machinery drawing in our national industries depended on "manual drawing". Although in specific period, manual drawing played an effective role in machinery industry's designing. While with the fast development of market economy, machinery manufacturing industries began to realize the problems of "manual drawing".

96.2.1 Increases the Difficulty

"Manual drawing" previously needed an in-depth anatomy of the whole part structure, and then it analyzed the location of graphics from various angles, only after the different understanding of parts, can people began to draw [1]. As to some simple mechanical components, drawing operators can sketch according to the parts' external contour, while facing difficult components, the difficulties of drawing became much more.

96.2.2 Being Easy to Make Error

According to actual experiences of machinery manufacturing drawing, product drawing in machinery industries' manufacturing involves various contents, such as parts structure, circle "R" size, contour design and so on, these paper contents are needed to be verify, record and calculate one by one by drawing operators. Long-term status of this kind of complex work, makes operators inevitably make errors and lead to unqualified producing and processing.

96.2.3 Having Low Speed

Mechanical drawing is the main basis for industries' daily production, and also the producing basis of sustainable production in machinery industry. Although manual drawing can deal with some drawing issues, once it faces imported machinery, manual drawing is hardly competent for drawing task of machinery industry. Therefore, graphics of industries cannot keep up with the steps, which cause a lot of obstacles of post-production arrangement.

96.2.4 Having Substandard Product

It is undeniable that manual drawing owns its individual advantages, such as: fast mapping, simple drawing, drawing anywhere, and so on. While manual drawing is operated by people and cannot totally meet mechanical standard. Size is the key in mechanical drawing, if size is not ensured, graphics will not meet production standard, manual drawing is still difficult to meet standard in dimensional accuracy.

96.3 The Advantaged of AutoCAD Drawing

At present, wide use of computer technologies provides favorable conditions for modernized producing in machinery industry and promotes the improvement of enterprises' economic benefits. Mechanical drawing has also experienced higher level of transformation, "CAD drawing" gradually replaces manual drawing and is widely used in mechanical industry. Since the introduction of AutoCAD drawing technologies, its advantages in using become more evident.

96.3.1 Integration

Replacing the manual drawing, CAD drawing realizes the integrated control of part size and greatly reduces the difficulty of drawing staff [2]. From part data, artifact design, size (see picture one), to graphics draft, determine drawing, print drawing and so on, these procedures can all be finished in computer software. Therefore, we say that the practice of AutoCAD software can help mechanical drawing realize the integration (Fig. 96.1).

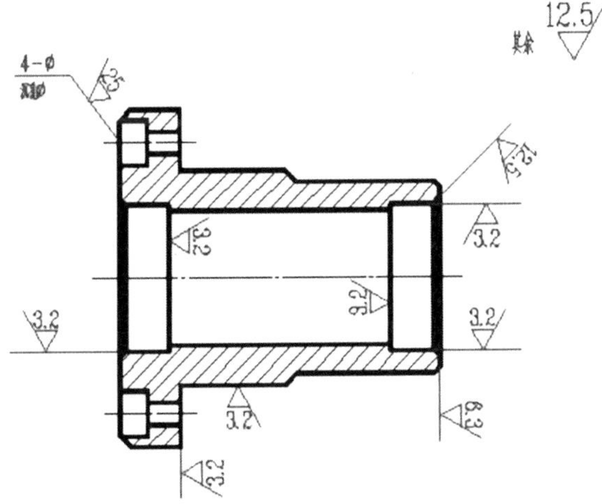

Fig. 96.1 Roughness of the part

96.3.2 Put into Disk

After the drawing staff finishing the drawing tasks by computers, the original drawing draft cannot be stored by disks; information of all the other papers related to mechanical production can be stored in computer disks. If papers are needed to be checked by mechanical engineers and business leaders, we can print the graphics stored in disks on standard papers; therefore, we can get intuitive understanding of production design.

96.3.3 Automation

Automation of mechanical manufacturing must brings the automation of mechanical drawing, AutoCAD software drawing is the typical signs of automation. Modifying, adjusting, imitating, and other operation and drawing staff' operation of the paper can be integrated with software to realize automation. Computers can not only finish the drawing according to original designs and display two-dimensional design, but also analyze graphics from three-dimensional angle.

96.3.4 Science and Technologies

There is no doubt that scientific technology is the first productivity, the nature of enterprises' "CAD drawing" promotion is the practice of computers. With the adjustment of management model for our national small and middle-sized enterprises, demand of technologies used in economic development is also needed to improve continuously [3]. We can realize economic scientific and technological development by dealing with mechanical and processing issues with computer technologies.

96.4 Upgrading Technologies of AutoCAD Software Drawing

Considering the universal use of AutoCAD software drawing, drawing staff should improve drawing technologies continuously, they should speed up drawing as fast as possible on the condition that paper's quality is ensured, so can we promote the development of mechanical drawing technologies. Facing the technological road of industrial technology, we should improve the present drawing so as to ensure the processed mechanical products' good quality, high standard, and lofty precision.

96.4.1 Preparation

"If a workman wants to finish his job wonderfully, he must improve his tools firstly." This sentence can best be fit for AutoCAD software drawing, whatever CAD drawing or manual drawing; we must do the preparation well. After the drawing staff getting manual drafts or actual part, they must analyze the paper and part carefully. Then they choose the appropriate layer, line, dimension style, target capture, unit format and graphic boundaries, and so on.

96.4.2 Establish Gallery

The so-called "gallery" means the common designs such as round, square, triangle, and so on; drawing staff can conserve these designs in a folder, people can take them directly if they are required, then only some size changes are needed. Some small common parts can be collected in gallery, such as screw, bump lot and so on, not every symbol is needed to modify, only the corresponding designs are needed to modify.

96.4.3 Keyboard Shortcuts

"Shortcuts" are the fastest method in computer drawing, skilled cartographers can finish drawing on the keyboard. Quick order of AutoCAD software can modify according to individual hobbies. The speed of quick drawing is 40 % higher than common method. As usual, frequently used shortcut keys are only about 15, as long as people remember operation skillfully, quick drawing is not very difficult.

96.4.4 Make Normative Base Map

Base maps are very important to mechanical paper and geological graphics, if drawing staff can ensure the normative paper, they can avoid the adjustment in the future, and this is also a method to improve drawing speed [4]. For example, we must ensure base map accurate and meet the industry's standard when we copy graphics by scanners; digital graphics board entry must implement effective control of the border and reduce drawing time greatly.

96.4.5 Round Handle

"Round" is very common in CAD drawing; the time needed in round handle will influence the whole drawing time. In order to speed up drawing, it is necessary to improve treatment of irregular round. Irregular smooth curve can be scattered into multiple lines, the control points of curves are the peak of the multiple lines. Then people connect, cut or extend the multiple lines, smooth chooses FIT or SPLINE [5].

96.4.6 Optimization

Since computer serves as a main platform for AutoCAD software drawing, we should ensure the high performance of the computer's internal configuration, thus we can get results as fast as possible in dealing with the data. Optimization of the operation systems can enlarge storage space, make CAD software having a larger operating space, and reduce the frequency of changing data with hard disk when AutoCAD works, therefore AutoCAD drawing becomes faster.

96.5 Conclusion

The development of scientific technologies promotes the update of drawing technologies, which have a great role in promoting enterprises of modern industrial production, geological exploration, engineering and construction, and so on. As common drawing software, AutoCAD software improves largely compared with manual drawing. In order to improve drawing efficiency, we should master drawing techniques skillfully and make drawing more smoothly.

References

1. Zhou J (2009) The update of modern mechanical manufacturing product drawing technologies. J Nanjing Univ Sci 32(13):45–47
2. Shang J (2008) Consider "CAD drawing" replacing "manual drawing" from the perspective of production efficiency. Manage Obs 24(8):71–73
3. Jin S (2009) Updating machinery industries' drawing technologies in economic globalization. Manage Small Middle-sized Enterp 40(18):28–30
4. Li X (2009) Elementary discussion of effective ways of speeding up auto CAD software. Res Chin Enterp 16(11):54–56
5. Zhu Y (2008) Study the influence of software on machinery industry producing. Mach Manuf Autom 25(12):72–74

Chapter 97
Implementation Bump Map by GPU Based on HLSL

Dong Xu and Hai Ning Qin

Abstract Bump map is the algorithm by disturbed surface normal direction to generate a bump or rough objects. This paper studies how to use HLSL to achieve bump map rendering, full use of GPU computing power and programmability to improve rendering efficiency. The five steps of the principle and implementation are described, and a summary of this realization.

Keywords Bump map • Texture mapping • HLSL

97.1 Introduction

In the generation of real time and realistic graphics, the simulation of object surface texture details is called texture mapping technology [1]. The general texture mapping technology deals with the texture of color, whereas the surface attribute of an object in the real world not solely includes color but also possesses a sense of roughness, such as the bumpy textural details presented on the surfaces of orange, tree trunk, rock and mountain, and so on. To present such a sense of roughness, bump map technology is then needed to be applied [2, 3]. Bump map is a calculation method which is used to generate bumpy or rough object by perturbing the surface normal of object. At present, the rendering speed of GPU is far faster than CPU along with the development of the programmable GPU technique, providing much more room for the improvement of rendering efficiency of three-dimensional scene. HLSL is a sort of high level shading language developed by Microsoft, which is implemented on the GPU of graphics card. This paper studies

D. Xu (✉)
School of Information Engineering, Nanning College for Vocational Technology, Nanning 530008, Guangxi, China
e-mail: liulz_12@126.com

H. N. Qin
Computer Engineering, Guangxi Economic & Trade Polytechnic, Nanning 530021, Guangxi, China

Z. Zhong (ed.), *Proceedings of the International Conference on Information Engineering and Applications (IEA) 2012*, Lecture Notes in Electrical Engineering 220, DOI: 10.1007/978-1-4471-4844-9_97,

how to apply HLSL to implement the graphing of bump map and how to make full use of GPU computing ability and programmability, fulfilling the bumpy texture effect and improving the mapping performance.

97.2 Process of Implementing Bump Map

Main steps:

1. Load the normal map of the surface of object.
2. Generate tangent-plane space of matrices through vertex normal, tangent line, and binormal.
3. Incorporate the vertex, Light vector, and normal vector into space of tangent plane.
4. Obtain the perturbing normal vector by sampling the normal mapping.
5. Calculate diffuse reflection and specular reflection, and ultimately completing the illumination calculation.

From the above steps, object with the sense of bumpy texture details is accomplished. In this paper, the perturbation to object surface normal is completed through the imported normal map.

97.3 Implementation of HLSL-Based Bump Map

97.3.1 Obtaining and Loading of Normal Map

To implement the sense of bumps and dents is to fundamentally alter the normal vector of the surface of object, and then implement its roughness sense by using illumination model. Normal map is a sort of graphic file which exists in the format of *.tga or *.bmp, in which the value saved for each vertex is different from general color texture graphics. Also, this normal map saves the normal vector of the curved surface vertex which has been perturbed. 3D modeling tools such as 3DMAX or Maya can be applied to model and make normal map or generated through height map. Through C++, the following codes can load bumps and dents textures (i.e. normal map), and then can be imported into the pixel shader (PS) of HLSL, and finally it is submitted to CPU to be processed.

```
//create bumps and dents texture
V_RETURN (D3DXCreateTextureFromFile (pd3dDevice, L"*.tga", &g_p Bump
   Map));
//Import the two textures into PS
g_p Effect → Set Texture ("Bump Map", g_p Bump Map);
```

97.3.2 Constructing the Space of Tangent Plane

During the calculation of illumination, it is required to incorporate the object's vertex coordinate, vertex normal vector, vertex light vector into one coordinate; but three vectors are defined in their own local coordinate system, which are required to be transferred to the world coordinate system. However, the calculation needed for transferring every vertex coordinate of object and its corresponding normal vector into the world coordinate system is tremendous and will cost too excessive system resources, which are almost infeasible under the condition of demanding high-standard real-time operation. The simplified method adopted here is to transfer the three vectors to the coordinate system of the space of tangent plane which is the space where the imported normal map is, namely the texture space (Fig. 97.1). Therefore, it is required to calculate the basis matrix of the coordinate system of the space of tangent plane, and then the transformation of the light vector can go on.

Three mutual-vertical vectors are required to construct the basis matrix of a coordinate system of the space of tangent plane, which here are marked as "normal vector" (n), "tangent vector" (t) and "binormal vector" (b). Postulate there is a vertex "p" on the surface of object; the normal vector of "p" can be the tangent vectors "t" of both "n" and "p"; and then cross "n" and "t", from which the obtained vector is perpendicular to "n" and "t", which can be regarded as the binormal vector "b" of "p" (i.e. $b = b = n \times t$). Based upon such three vectors, a basis matrix can be gained for conversion. The codes of vertex shader (VS) of HLSL are showed as follows (Fig. 97.1) :

```
//Construct a Matrix of World to Tangent Space (texture Space), which is used to
   incorporate Light vector into the coordinate system
//only the coordinate system is incorporated can the illumination calculation go on
Float3 × 3 worlds To Tangent Space;
World to Tangent Space [0] = mul (Tangent, mat World);
//cross tangent vector and normal vector to gain the binormal vector which is per-
   pendicular to tangent vector and normal vector
   World to Tangent Space [1] = mul (cross (Tangent, Normal), mat World);
```

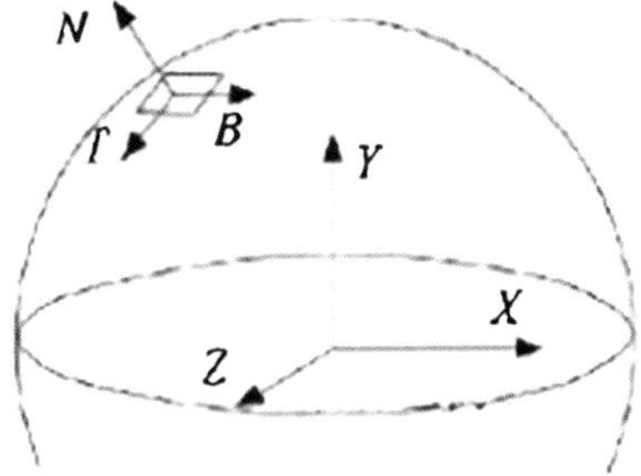

Fig. 97.1 Vertex transferring from object coordinate system to tangent space coordinate system

World to Tangent Space [2] = mul (Normal, mat World);

$$L\begin{bmatrix} T_x & B_x & N_x \\ T_y & B_y & N_y \\ T_z & B_z & N_z \end{bmatrix} \tag{97.1}$$

97.3.3 Incorporate Vertex, Light Vector and the Normal Vector of Texture Space into Space of Tangent Plane

Subsequently, the vertex light vector and the normal vector of texture space are transferred to the coordinate system of the space of tangent plane. What needs to be noted here is that the line-of-sight direction of point "p" should be used as normal vector of texture space after it has been transferred. The relevant VS codes are listed below:

//Transfer the vertex to the world space coordinate system
Float3 Pos World = normalize (mul (Pos, mat World));
//the light vector of the vertex in the world space coordinate system
Float3 Light = vec Light Dir—Pos World;
//calculate the vertex light vector in the texture space (space of tangent plane)
Out. Light. *xyz* = mul(world To Tangent Space, Light);
//The line-of-sight vector of the vertex in world space coordinate system
Float3 Viewer = vec Eye—Pos World;
//calculate the vertex line-of-sight vector in the texture space (space of tangent plane), and this vector is understood as the normal vector of the normal texture space.
Out. View = mul (world To Tangent Space, Viewer);

97.3.4 Obtaining the Normal Vector by Sampling the Normal Maps

In this step, samples are taken from the normal maps, or it can be said that the normal maps determined the orientation of object's surface. Normal vector is a 3D vector which is made up of three detached vectors X, Y, and Z. In computer, however, the normal vector stored by normal map is actually stored in the format of no-symbol texture, namely R, G, and B three formats whose values all are between [0 and 1]. This indicates that the three detached vectors X, Y, and Z of the normal vector in normal map are respectively saved in R, G, and B color channel, whereas the value extent of normal vector of vertex in practice is between [−1 and 1]. The relevant HLSL codes in pixel shader (PS) are as follows:

//Bump and dent texture (normal map) sampling obtains the normal vector which is used to substitute the vertex normal to implement the calculation of illumination.

Float3 bump Normal = 2.0 * (tex2D (Bump Map Sampler, Tex) −0.5).

97.3.5 Calculation of Illumination

Calculation of illumination is the process to calculate the illumination and determine the color of pixel. In bump texture mapping, Phong illumination model is adopted, which is based upon pixel. In other words, the illumination will be calculated in PS. Phong shader processing method applies the normal vector in the triangle's vertex to calculate the normal vector of every point in the interior of triangle through interpolation, and then re-calculate the color of pixel according to the normal vector of all pixels. Phong illumination model is consisted of three parts: Ambient light, diffuse reflection light, and specular highlight. Ambient light will be given a specific value in application. The difficulties here lie in the latter two items.

The processing of diffuse reflection illumination model complies with the Lambert theorem which concludes that the reflected light for object surfaces with ideal diffuse reflection (totally unsmooth and without luster) is defined by the cosine value of the included angle formed by the object surface normal "N" and the light vector "L" (which points from the surface point of object to the light source) (See Fig. 97.2).

$$I_{\text{diffuse}} = S_{\text{diffuse}} \times \cos\alpha \tag{97.2}$$

$$I_{\text{diffuse}} = S_{\text{diffuse}} \times (\mathrm{L} \cdot \mathrm{N}) \tag{97.3}$$

The implementation codes in the PS of HLSL are as follows:

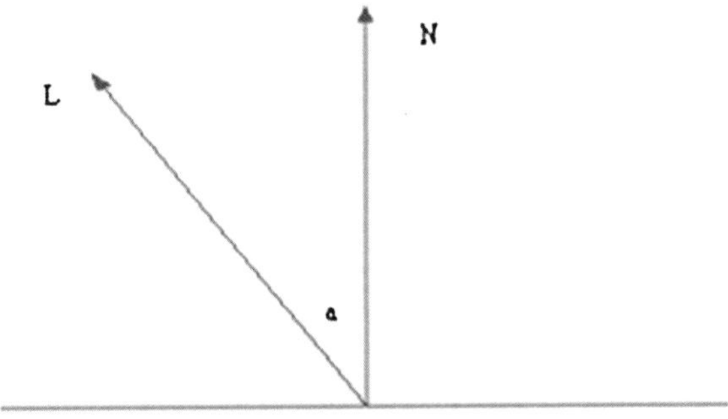

Fig. 97.2 Diffuse reflection light strength calculation

```
//light vector "L" in standardized texture space
Float3 Light Dir = normalize(Light);
//the cosine value of the included angle and the contribution of diffuse reflection
   illumination can be obtained in accordance to the dot product made by multiply-
   ing the normal line "N" of texture space and the light vector "L"
Float4 diff = saturate (dot (bump Normal, Light Dir));
```

The next step is the implementation of the calculation of specular reflection. Take the surface of metalwork or glasswork (e.g. a glass) for example. The luminance of the glass surface changes while the observer moves his/her position. Such a phenomenon often happens in the highlight reflection of the surface of a smooth object, which is called as Specular Reflection. "R" represents vector of reflection light which can be acquired by multiplying the light vector "L" and normal vector "N" (See Fig. 97.3).

The implementation codes in the PS of HLSL are as follows:

```
//calculate the strength of reflection light, corresponding to R = 2(N.L) N-L, N.L
   is diff
Float3 Reflect = normalize (2 * diff * bump Normal—Light Dir).
```

Next, the strength of specular reflection light which is viewed from the angle of observer is calculated through the strength of reflection light (See Fig. 97.4). The point "p" in the equation represents the strength of specular reflection light of object surface. The closer "V" and "R" are, the brighter the reflection ray is.

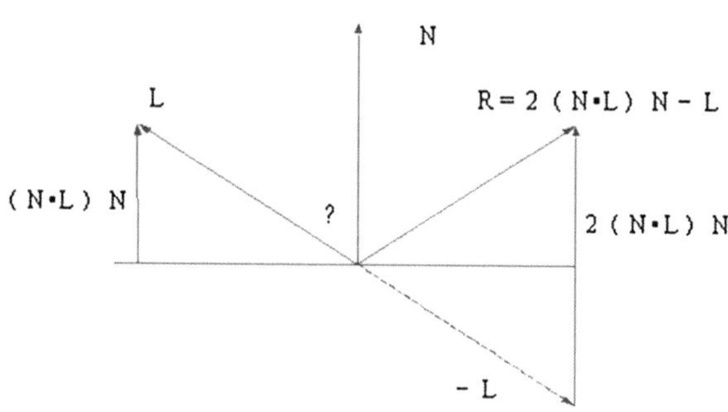

Fig. 97.3 Calculation of the vector of reflection light

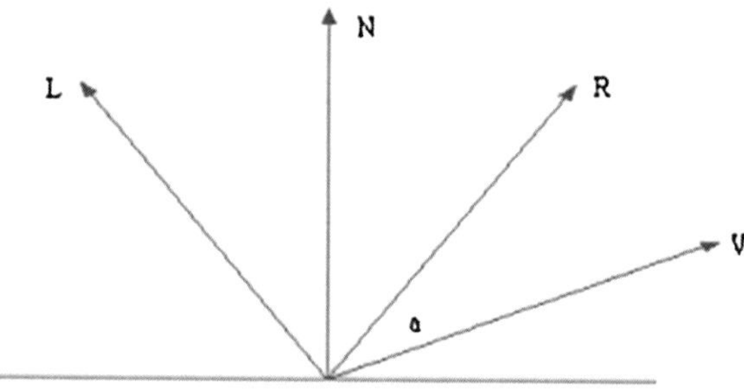

Fig. 97.4 Calculation of strength of specular reflection light

The calculative equation of specular reflection is shown below:

$$I_{\text{specular}} = S_{\text{specular}} \times (\cos\alpha)^{p} = S_{\text{specular}} \times (\mathrm{V} \cdot \mathrm{R})^{p} \tag{97.4}$$

The implementation codes in the PS of HLSL are as follows:

```
//calculate the strength of specular reflection in accordance to the equation, here p
  is 15
Float4 spec = min (pow (saturate(dot(Reflect, View Dir)), 15), color.a);
```

After the calculation of illumination, the luminance value of point "p" is obtained to implement the bump and dent sense of the point "p", but in the real-world object surface not only owns the bump and dent sense and also has the corresponding color attribute to show object details. Therefore, the luminance value of point "p" should be modulated with the color value obtained by general color texture mapping or be added together or multiply them. The value obtained in the end will be exported. The summing equation of ambient light, diffuse reflection and specular reflection is the following:

$$I = I_{\text{ambient}} + I_{\text{diffuse}} + I_{\text{specular}} \tag{97.5}$$

The implementation codes in the PS of HLSL are as follows:

```
//calculate itself's obscured shadow item
Float shadow = saturate (40 * diff);
//obtain the final color of each pixel by blending texture color with illumination
  color
Out_ps.Color = 0.2 * color + (shadow * (color * diff + spec)).
```

97.4 Conclusion

In this paper, the bump map alters the original normal vector of the object surface through the imported normal map, and then the bumpy and dent sense of object surface is presented through the calculation of illumination, and the specular high detached vectors are added as well. The object mapped by bump map has a strong bumpy sense. The normal value stored by normal map substitutes the normal vector of each vertex of object surface without using a series of transformation of height maps, hence saving the calculative time. Compared with other bump map algorithms, the implementation methods in this paper are not only the detached vector of diffuse reflection and also include the detached vector of specular highlights, by which the rendering effect is more vivid. This effect observed under the condition of fixed viewpoint has done a fairly good job. Moreover, the mapping speed of HLSL has been improved further by implementing it through GPU programming.

However, this method has its own limitations. It works under the precondition that the rendering effect will not vary in accordance to the change of observing direction, namely regardless of your observation from whichever viewpoint you can see the same surface at all the time as long as the object is mapped by this technique. This phenomenon differs from the effect showed on the surface of real-world object, thus this will result in distortion of the image. Otherwise, the sense of reality will be weakened when you closely observing the object surface due to the fact that this technique does not practically alter the geometry shape of the object.

References

1. Qunsheng P, Hujun B, Xiaogang J (2002) The arithmetic fundamentals of 3D computer graphics, vol 4(2). Science Press, Beijing, pp.65–69
2. Luna FD (2003) Introduction to 3D game programming with DirectX 9.0, vol 16(11). Word Ware Publishing, Inc, pp 85–91
3. Wang L (2006) DirectX Documentation for C++. Microsoft 25(18):36–43

Chapter 98
Color Art Research and Creativity in Exhibition Space

Honglei Xu

Abstract Colour displays a vital role in the exhibition environment especially in Exhibition space. Illumination, show of the exhibits' colour and details, and atmosphere creation can't work without colour. But both natural colour and artificial colour do harm to the exhibits, which will reduce the strength of the paint and material, even because the paint fading and paper curl. The damage can't be recovered, and the exhibits will not exist anymore if the colour design is ignored. So the relationship between the exhibits protection and the art visual will become one of the core issues of exhibition space design research.

Keywords Color art • Exhibition space • Creativity

98.1 Introduction

Color is the first element during exhibition space visual design, good color design will not only attract consumers, but also show the exhibition space designing art, express the designer's feeling, enhance the impression of exhibition space's brand in customers. One investigation of American popular color research centre shows that, there is a "seven seconds rule" when people choose exhibition spaces: facing kinds of exhibition spaces, people need only 7 s the make show if they are interest in them. And during the 7 s, color will influence 67 %, which has become a key element in people's judgment. From this, it can be seen the importance of exhibition spaces' color design. The focus concerning of exhibition space design is how to improve the exhibition space creativity from color creative design. More and more scholars and enterprises have done sufficient color research and creativity designing. This article discusses color research of modern exhibition space and the trends of creativity design development based on the current situation of exhibition

H. Xu (✉)
Xi'an University of Architecture and Technology, Xi'an 710055, Shaanxi, China
e-mail: xuhonglei@cssci.info

Z. Zhong (ed.), *Proceedings of the International Conference on Information Engineering and Applications (IEA) 2012*, Lecture Notes in Electrical Engineering 220, DOI: 10.1007/978-1-4471-4844-9_98,

spaces' color designing research. Exhibition space's color creative design relates to operating method, sales strategy, market efficiency, technical implement and whole exhibition space's system. The design process owns the characteristic of scientific, phasing, systemize, classifies, which needs to make a strict program to aid the exhibition space's creative design. The process can be divided to color research, concept born and brand image establishing, management, etc. [1]. Color research includes gathering related information, market positioning and requirement research and working technology, etc. concern the relation of the exhibition space's using crowd, using method, using environment and the using places to make reliable research foundation of exhibition space's concept born and image establishment [1]. Color design management is to make people investigation according to exhibition space's image, collect the result and make modification to the color program and manage systematically the final program to insure the exhibition space and the exhibition space's image. Color coordination is a important element of showing exhibition space's art. Color research and exhibition space's color creative design is closely related, which is an indispensable segment.

98.2 Significance and Contents of Exhibition Space Color Research

98.2.1 Significance of Exhibition Space Color Research

"The most valuable, the most operative research often occurs before creative work."(Scott Young), designing research has become the technology support and guide during the whole exhibition space's designing process to realize the commercializing of exhibition space's development [2]. The essence of designing research is to succeed in exhibition space's designing, exhibition space's color research choose a kind of color plan which could properly show the relative information and brand new specifics from the research of color attributes, culture, psychology and color technology and popularity research, etc. Consider from art, the significance of color research is to improve the exhibition space's artistic quality, enhance the visual appreciation value; consider from business, design research is the origin of enterprise competition advantages, sufficient color research is the insurance of launching color sales and improving exhibition space's value.

98.2.2 Contents of Exhibition Space Color Research

98.2.2.1 Exhibition Space Color's Attributes

Exhibition space's color design relates closely to exhibition space's attributes [6, 7]. For example, china postal service has its "postal green" as postal brand color, their working garments and vehicle take this color too. So, exhibition space's color attributes research can not only make basic research on the "color three attribute": hues, bright, purity, the more important thing is to make clear the exhibition space's exhibition space color and the normalized limit, to jump out of the limit will face certain risk. In the research and analyze of current exhibition space's tradition or popular color, then conclude from the result, to make the color design trend for the exhibition space's color creativity. "Color three Attribute" is the foundation of all kinds of color design; use the three attributes reasonably to create kinds of exhibition spaces design. Exhibition space's tradition color, exhibition spaces exhibition space color limit will be stable in a certain period, but will gradually change with the time, technology and fashion [8]. Just like the

"Imperial yellow" used only in imperial family has now popular in even alley.

98.2.2.2 Exhibition Space Color's Culture and Psychology

Different ways of living during different culture decide the different color effects [3]. Exhibition space's color culture research should make detailed investigation on place, culture, tradition and fashion, and then supply the exhibition space's color design by exhibition space's function and market requirement. Culture difference should be the basic element in the effects of exhibition space's color design [4]. When designer designs, exhibition space's.

Color culture research has become more important in order to make the exhibition space's color harmony with the consumers' culture background. E.g.: blue shows sensitive in England, and intoxication in German [4]. To exhibition space its own, color feel is the first place, this is a process which is from feeling to rationality. Color's influence in psychology gradually occurs in people's society active. People living in different culture background react different on color. For the different experience, education, society status, sex, age, quality, hobby, people feels, senses, thinks differently. For example Chinese feels red great difference from western people. What's more, color research should consider not only the inherent color of exhibition space, but also should meet the environment color and the color source color which could influence the exhibition space's color.

98.2.2.3 Exhibition Space Color's Technology and Popularity

As the example raise before, yellow was once the imperial color in ancient China, it is not for beautiful, but for the difficulty of producing technology. The nature resource material has now replaced by chemical exhibition spaces. Sufficient and low-priced color producing technology has made the exhibition space's color richer. Relative research of exhibition space's color technology mainly focus on the safety, reliability, and matching between color function and exhibition space's attributes and function, color technology and the exhibition space material and surface handling.

Consider exhibition space's color technology can avoid meaningless cost, increase profit, and improve market sales and occupation rate by the power of fashion color. This is a great tool for enterprise to avoid meaningless producing and investment, reduce society resource waste. Fashion color has great influence on economy prosperity and enterprise producing management. Hand exhibition space fashion color could forecast exhibition space producing and sales status. Research on the exhibition space's fashion color is judge the future exhibition space fashion trend and color plan according to scientific regularity to mostly satisfy consumer's requirement trends.

98.3 Exhibition Space Color Creative Design

Color creative designing process concerns not only the artistic and seasonality, but also the functional. Color research has become one of the creative design develops foundation. Designer conform the restriction relation between exhibition space's attributes, functional form and color design according to the research on foundation of color research, realize the exhibition space creativity through color collocation which make exhibition space owns visual beautiful and fresh. To counter the color research's main contents, color creativity now mainly develops in artistic aesthetics, engineer, management; it covers cognition, psychology, artistic, engineering, computer science, material science, management, etc.

Color art design starts from aesthetics, designer's experience or color trend will be integrated into the exhibition space by the way of art processing, color will be chosen by specific design requirement and elements as material, function, environment, characteristic, technology, cost, etc. and these colors will be used by kinds of methods as contrast and reconcile, balanced and stability, ratio of partition and rhythm to high color the aesthetic of exhibition space's art, establish exhibition space image with its brand. Color plan is high coloring the art processing—art creativity based on the foundation of choosing color reasonable in which matching colors and exhibition space's color style establishing is the main respect of handling color art creativity [5]. The second is the consistency of color and content.

Exhibition space color design has to be consistent with the exhibition space to make sense. Establishing color image is the important element of exhibition space style establishing. Always, "color's three elements" is still a powerful weapon to realize greatly influencing people's emotional feeling. Among the three elements, hue is the most important one, even for the same exhibition space, when the hue changes, the exhibition space will show completely different style. Bright and purity express more directly to establishing exhibition space's style establishing. For example, using high purity, especially the three RGB, will high color children exhibition space's designing features, and purple, no matter heavy or thin, will express female exhibition spaces style. What's more, lots of experience show that increasing contrast will enhance vitality and decreasing it will show steadily. "Contrast" can be seen as an art processing method to exhibition space color and style design, it can be comprehensive used of bright contrast, purity contrast, hue contrast, area contrast, to form the union of exhibition space's color style and image.

98.4 The Problems in Color Environment Design of Domestic Exhibition Space

The name of "Exhibition space" is originated from Japan. It should be called Art museum more accurately [2], because Exhibition space is one kind of museum in western countries. Exhibition space and Museum follow the same design standards. Museum illumination design standard. In order to establish this standard, The establishment group, from Architectural Physics Institute of the China Construction Science Research Institute, who has spent 4 years investigating the color environment of hundreds of representative Museum in Beijing, Shanghai, Hangzhou, Wuhan and other places, and summarized existing main problem in domestic museum color environment design [3]: In terms of illumination levels, the showroom that place natural coloring predominantly has too high luminance fluctuations. The showroom that natural coloring incorporates with artificial illumination has high luminance value. The showroom that artificial illumination is used alone has maximum luminance value of 200–300 lx, minimum value reaches 40 lx.

In terms of illumination facilities, the color source is mainly the fluorescent lamp and incandescent lamp, the fluorescent lamp accounts for a large proportion. Lamps used in the museum are general utility lamp sold in market. The color source mask in most showcases is not good, one or two reflections are quite serious. In terms of exhibits protection the vast majority of showroom take no against ultraviolet measure, the conditions for the storehouse are relatively poor. While construction funds were relatively scarce, it is difficult to radically improve the illumination conditions. Besides problems concluded by establishment groups.

"Recently, the exhibits display in domestic museum formed a kind of obvious trend: All exhibits used full-closed, artificial illumination techniques" [5]. As concluded from the above problems, the color environment level in museum is backward as a whole, such as single illumination technique, uncontrolled luminance, single illumination facilities, without any effective measures in exhibits protection, ignorance of visitors' visual feeling.

98.5 The Exhibits Protection Methods of Natural Coloring

The reasonable use of natural color can not only display the appearance of exhibits with energy saving, but also satisfy visitors' physiological and psychological need of closing to nature. But natural color is composed of various wavelengths, harsh ultraviolet among sun color will make photosensitivity exhibits fade, deteriorate even ruin. So more ideal natural color source is soft, even, bright diffuse color through special radiation-proof sky color or side-window towards the north. Ultraviolet can be controlled by the special glass, the blind, the curtain and the window hole of radiation-proof sky color or side-window towards the north. For example, the lead glass or normal glass coating may filter ultraviolet in sun

color, photosensitivity exhibits are protected by radiation-proof glass cube, and different exhibits can be protected by different retractable curtain. The ancient Roman civilization showroom of National Museum of China is an excellent natural coloring design case. Huge side-window is used to introduce natural color for Non-photosensitivity exhibits area, which can express the ancient Roman art. Photosensitivity exhibits areas afford three kinds of curtain group according to exhibits characteristic and illumination level, as following: white curtain and opaque curtain is used in colored mural painting area, which is most sensitive, white curtain and ultraviolet-isolation curtain is used in general exhibition area, which is more sensitive, ultraviolet-isolation curtain is used separately in sculpture area, which is normal sensitive.

References

1. (U.S.) M David Egan Victor Olgyai (2006) Architectural illumination (Foreign Coloring Design Series), 2nd edn. vol 3(3). China Building Exhibition space Press, Beijing, pp 499–507
2. (1998) The people's republic of china exhibition space standard–museum coloring design standards. National Heritage Board, Beijing
3. Zhang X, Zhan QX (2005) The history and current situation of China's museum color environment–national gallery of China as an example. Architectural J 2:62–65
4. Zhang X, Zhan QX, Wang L (2005) Low-tech + low cost-Rome exhibition transformation of color environment of the national museum. Lamps Coloring 2:4–6
5. Zhu X, Zhu ZB, Wang J (2006) On the function of color in museum interior design for exhibition. Lamps and Coloring. Hundred Schools in Arts 41:77–86
6. Ai J (2007) How to select proper color sources to protect heritage in exhibition design. Chin Nat Mus 3:380–386
7. Tang L (1992) The research and application techniques on natural coloring in museum showroom. Huazhong Archit 2(4):391–399
8. Ding ZJ (2003) Museum illumination and art works exhibition. Coloring Eng 1(14):5–16 (Beijing)

Chapter 99
Tourism English Education Based on Multimedia Technology

Hong Quan

Abstract Because of recent movement toward contemporary digital learning, this research provides learners with tourist English resources for practice. The material is conducted by video on demand (VOD) or through the Internet. Specialty of digital material is that it integrates the world's top 10 most popular tourist spots and conversations for travel with the basic skills of English learning. Included are vocabulary usage, pronunciation practice, speaking opportunities, translation practice, listening practice and tests, and short answer writing exercises. These skills are used frequently in the tourism and business field. Through the additional links, the learner has access to rich resources for greater knowledge. This promotes self-motivation and interest. This material can be used through CD-ROM, webcam, or e-learning platforms to augment further interest. In addition to courses and teachers as guides, a modern learning environment is reached. This research is aimed at increasing a multimedia digital learning material for practical learning. This design can be expanded to other professional fields in the future. Through the way of digital learning, it can develop the learner's professional skills and achieve the goal of using what the learner have learned.

Keywords Tourism english • Multimedia technology • Education

99.1 Introduction

According to the estimates of the world travel and tourism council (WTTC), tourism will reach 8.6 trillion dollars and people seeking tourism related jobs would increase to 25 million in the next 10 years [1, 2]. Tourism will play an important role in the worldwide economy. As Taiwan enters into the prestigious world trade organization (WTO), increased opportunities of contact with foreign

H. Quan (✉)
Jiangxi Science and Technology Normal University, Nanchang 330013, Jiangxi, China
e-mail: quanhong@guigu.org

Z. Zhong (ed.), *Proceedings of the International Conference on Information Engineering and Applications (IEA) 2012*, Lecture Notes in Electrical Engineering 220,
DOI: 10.1007/978-1-4471-4844-9_99,

visitors in business or pleasure will have to be met [3, 4]. Hence, upgrading the public cognition regarding tourism English for the increased understanding of the world is significant for the tourism industry's development.

Through the rapid development of the Internet and multimedia technology, e-learning has become the most important learning system. To achieve higher learning effectiveness, the Taiwan Government is promoting "Challenge 2008" as a national project that includes a program, "Dual Stars," to stimulate the industry of digital contents, which will be the main objective of developing knowledge and economic progress and get involved in entertainment, life, education, culture, and publication.

99.2 Development Approach

Designing e-learning materials is a multi-disciplinary task that emphasizes coordination and integration of different fields of study. Take this digital material, for example. It incorporates tourism English, digital technology, multimedia functions, and other related technology to integrate all aspects of resources into practical teaching system for learner convenience. This material focuses on intermediate-level students and people from all walks of life.

For the section of "Introduction to Famous Tourist Spots", the 10 famous tourist spots are represented and are made into teaching materials for learners, including Niagara Falls, The Great Wall, the Mausoleum of the First Quin Emperor and Terra-Cotta warriors, the Pyramids of Giza, Taj Mahal, Angkor Wat, Moais in Easter Island, the Leaning Tower, the Great Canyon, and the Great Barrier Reef.

For the section of "Conversations for Travel", the most encountered conversations when traveling are used for learners to practice on. Topics include: dinner situations, hotel accommodations and bookings, public transportation, booking of airplane and train tickets, airport check-ins, department store, and customs. The eight short conversations pertaining to the topics are shown in videos to allow the learner to view the conversations on the screen. Learners can imitate the conversations or listen to the actors' pronunciation.

In order to facilitate learning in a user-friendly environment, the e-learning material discussed here is designed based on the following learner's needs [5, 6].

1. Design a basic and consistent e-learning material to suit learners' needs.
2. Create a user-friendly environment based on learner's qualities and experiences in hopes of helping learners are familiar with the learning material.

99.3 Results and Discussions

The material includes two sections: Introduction to Famous Tourist Spots and Conversations for Travel. Both selections are prominently shown on the interface when learners access the website. By clicking on the selection. The following topics are explained:

After clicking the Introduction to Famous Tourist Spots on the main page of the website, a new interface, shown in Fig. 99.1, will show up. The selection of the tourist spots is on the left hand side of the screen so learners can click on the spot they are interested in. In the middle of the page is the text introduction to the tourist spot. On the right hand side, there are selections related to English learning including vocabulary, pronunciation, speaking, listening, and test questions. The function of the selections is discussed below:

1. Vocabulary: The important vocabulary words in the text are highlighted. The translation, definition, and sentence usage of the word are provided when the cursor is above the word and a frame pops up as shown in Fig. 99.2.
2. Pronunciation: The color of the text is black, and when clicked on the pronunciation choice, the color will change as the sentence is being spoken like karaoke style. This guides the learner's pronunciation improvement, reading focus, and listening skills. It also increases learning interest.
3. Speaking Practice: After clicking on the Speaking Practice selection, the word color will change. The recorded voice becomes mute, allowing learners to practice their pronunciation and reading. The material functions as an interface on the screen. The learner can record his or her voice through a recording program. As the color on the text moves, the learner can record his or her voice into the system and listen by playing it back.

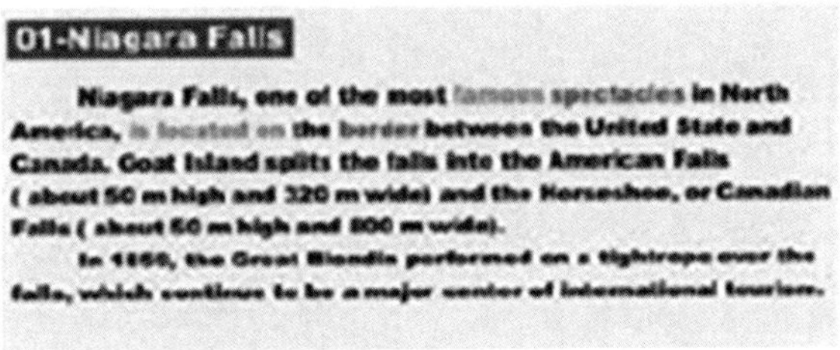

Fig. 99.1 The main page of the selection of introduction to famous tourist spots

01-Niagara Falls

Niagara Falls, one of the most famous spectacles in North America, is located on the border between the United State and Canada. Goat Island splits the falls into the American Falls (about 50 m high and 320 m wide) and the Horseshoe, or Canadian Falls (about 60 m high and 800 m wide).

In 1860, the Great Blondin performed on a tightrope over the falls, which continues to be a major center of international tourism.

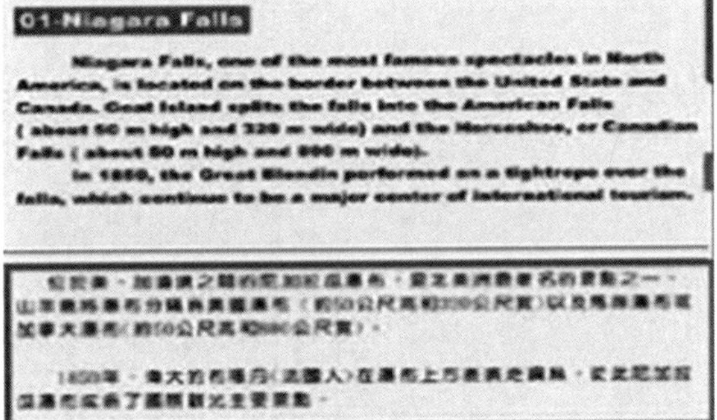

01-Niagara Falls

Fig. 99.2 The selection of translation and hyperlink

After selecting the conversations for travel part of the section (as shown in Fig. 99.3), the link takes learners to an interface as shown in Fig. 99.4. In order to achieve consistency and convenience, the frame for the conversations for travel is the same as the frame for the famous tourist spots. The left side of the window contains the themes (conversation situations). In the middle are the texts. In the right side are English Learning selections including vocabulary, pronunciation practice, speaking, translation, listening practice and conversation videos. Most of functions of the selections in this section are the same as those in the section of introduction to famous tourist spots. A few minor changes are present in this section. The explanation for the sections is provided in the following:

1. Pronunciation: Along with color change as the sentence is being read in karaoke style, there are different highlights for different actors (two speakers in dialog) to guide the learner's pronunciation, reading, and listening skills. This is shown in Fig. 99.5.

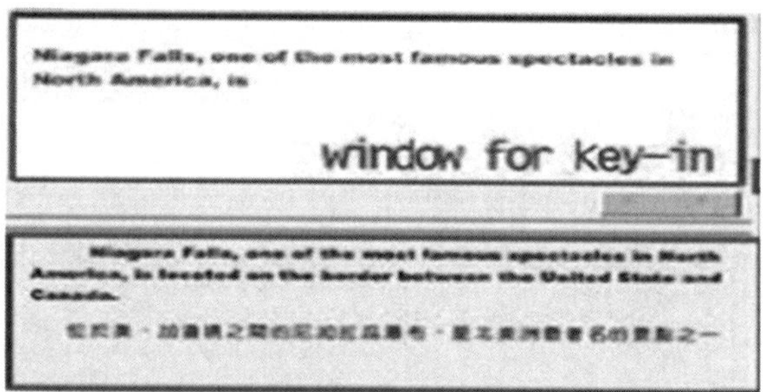

Fig. 99.3 The selection of listening test

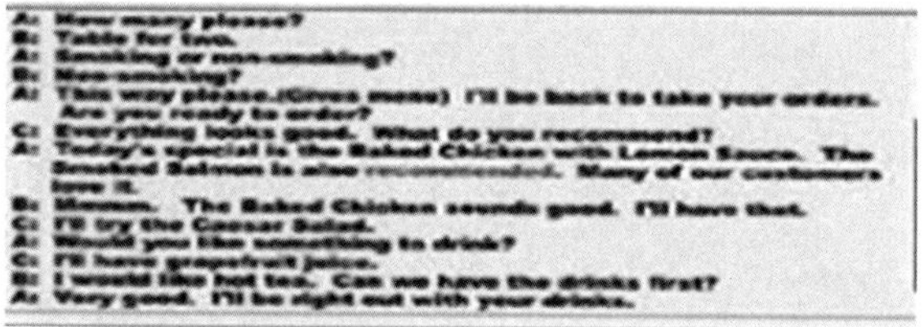

Fig. 99.4 The main page of the selection of conversation for travel

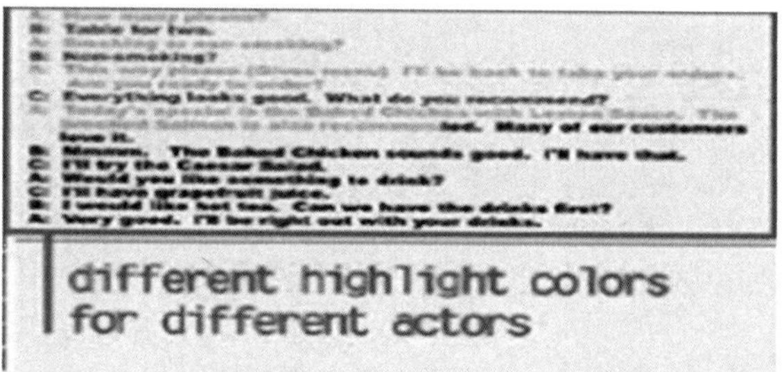

Fig. 99.5 The selection of pronunciation

2. Conversation Videos: Conversation videos are made according to the theme to raise learning effectiveness through visual stimulation. This creates an interactive learning environment.

The materials are designed for tourism English; however, for other professionals in similar fields, this can also be useful. Tourism English is one aspect in the various materials that could be used with this system. This digital system is designed using Flash software to achieve an interactive and convenient learning platform. Various materials can be inputted into the system so modification and updates are not difficult for programmers. To achieve a good and contemporary learning atmosphere, up-to-date information and pictures are used. Due to copyrights, pictures unable to be used can be accessed by users by simply clicking on the Links option. The Links function as a supplemental resource for learners to increase further interest and knowledge.

99.4 Conclusion

The materials itself play a key role to interact with the learner. Evaluating the learner's needs incorporates suitable teaching methods with strategy in digital material production. The English learning effectiveness can thus be achieved. In order to raise the learner's motivation and interest in learning consistently, the following elements should be considered:

1. In order to decrease the pressure of learning, increase cognitive ability, develop better learning attitude and characteristics, and nurture frequent and practical usage of multimedia materials.
2. Utilizing technology to create an interactive and diverse learning environment will be the trend of the new generation. Multimedia usage and design should not distract user's attention or lose his or her interest.
3. Topic of materials should be based on daily life. It should not only help to incorporate learner's prior experience but also acquire new knowledge through experience and usage.

References

1. Li P (2010) Computer, e-learning and creative thinking, vol 12. pp 39–46. http://www.fhjh.tp.edu.tw/erc/ Hyper-Media.htm
2. Lin GM (1999) Principle of design in constructing CAI on internet, In: The 8th international CAI conference, vol 4. pp 480–487. http://paper.nt1.isst.edu.tw/data01/acbe/iccai8/84/84.htm
3. Tsai SC, Lin JD (2003) Study on development of a multimedia digital material for semiconductor english. In: Proceeding of seminar on humanity, technology and human resource development, vol 9. Kaohsiung, pp 535–540

4. Tsai SC, Li KJ, Lin JD (2004) Theories and practices of interactive interface design, vol 6. Industrial Design Association, Taichung, pp 429–438
5. Fang YM (2003) Study on development of a multimedia digital material for conversation of business english. In: Proceeding of the 21st international conference on english teaching and learning, vol 5. Taipei, pp 676–683
6. Alessi SM, Trollip SR (2001) Multimedia for learning: methods and development, vol 1. Allyn and Bacon, Boston, pp 348–354

Part X
Database and Data Mining

Chapter 100
Research on Distributed and Heterogeneous Database Integration Based on Web Service

Wei-hua Wang and Wei-qing Wang

Abstract To resolve the problem of uniform access of distributed and heterogeneous networks database resource in modern enterprise, the Web service technology was used to construct distributed and heterogeneous database integrated service system. The communication principle and the characteristics of Web service technology, and the advantages of its application to distributed and heterogeneous were analyzed. A distributed and heterogeneous database system integrated structure and a dynamic database increased or decreased model based on Web service technology were established, and a distributed and heterogeneous database service system was designed.

Keywords Web service • Distributing • Database

100.1 Introduction

Today, every corporation and inside the enterprise have their own information system, the format of the data in these system is not compatible each other. With the development of the traffic, the distribution of the work space of the departments of the enterprise expand continually, sometimes, it has been spans the national boundaries. To integrate the information systems in the distributed work space of the departments of the enterprise is extremely urgent, and to integrate the heterogeneous distributed databases is the key technology of the information integration, the aim of the information integration is to exchange the information

W. Wang (✉)
School of Software Engineering, Key Laboratory of Machine Vision and Intelligent Information System, Chongqing University of Arts and Sciences, Yongchuan 402160, China
e-mail: y2002ww@163.com

W. Wang
Department of Information Management, Southwest University, Rongchang Campus, Rongchang, Chongqing 402460, China

Z. Zhong (ed.), *Proceedings of the International Conference on Information Engineering and Applications (IEA) 2012*, Lecture Notes in Electrical Engineering 220, DOI: 10.1007/978-1-4471-4844-9_100, © Springer-Verlag London 2013

without transforming each other, and problems in the integration are the database distributing and the heterogeneous information format.

The research activity in the information integration domain has resulted in survey papers being published in the last decade. A method of Agent-based information shared and individual service policy is proposed in [1, 2], [3, 4] proposed a way of information integration and shared based on extensible markup language (XML), Xiao and Wang [5–9] etc., introduced multi-source marine environmental information and enterprise information integration. Most of these methods have the same implement mechanism based on a stand document, it has the same problem which cannot use the information straightway each other because of its data format, and it needs a program to transform the data format before use. In order to settle this problem, we propose a method of distributed and heterogeneous database integration based on Web service in this paper.

100.2 Heterogeneous Database System Integrate

100.2.1 Web Service Technology Advantage

To settle the problem of distribute heterogeneous database system integration using the Web service technology has many advantages, this method is not only independent of data type in heterogeneous database system when exchanging each other, but also the data can be computed in distributed system environment using this technology which boosting the computing speed of the system. Xiao [10] introduces an application using this technology in file system, and Xu [11] proposed an application using this technology in cooperating with assistant design.

Hence, integrating the heterogeneous computing system using the technology of Web service has the great advantage [12].

100.2.2 The Principle of Database Integration

Using Web service to integrate the heterogeneous database system is the basic principle of database integration; in the other words, a Web service can be used to access the database of a department. There are two methods to access the database; they are shown in Fig. 100.1, in Fig. 100.1a, the local data accessing separated from remote data accessing, in order to advance the access speed, the local data do not need Web service, for the Web service need register which need other resource. In Fig. 100.1b, the local and remote data accessing are all need Web service.

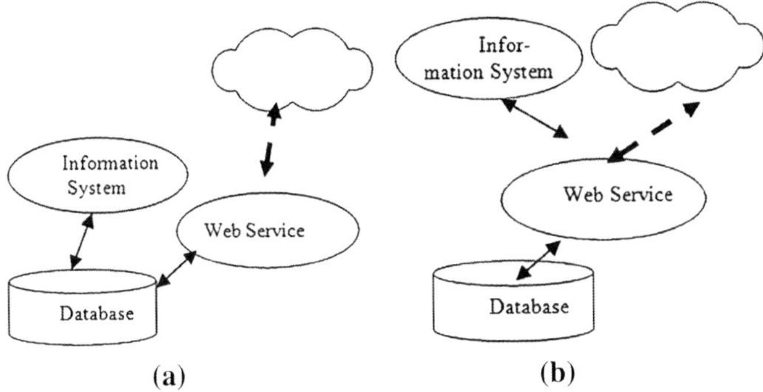

Fig. 100.1 The principle of database integration. **a** Local accessing separate from remote access. **b** Integrate accessing

100.2.3 Integration Structure

The integration structure of distribute heterogeneous database is shown in Fig. 100.2. All the remote database are accessed by Web service interface, the data accessing interface of the Web service are distributed to network after registering, and then the clients in the network can accessing the database from network as the same from the local database.

100.2.4 The Principle of the Web Service Communication

As an application program, Web service permits a computer using the functions in an other computer by an universal data format and protocol, such as XML can be

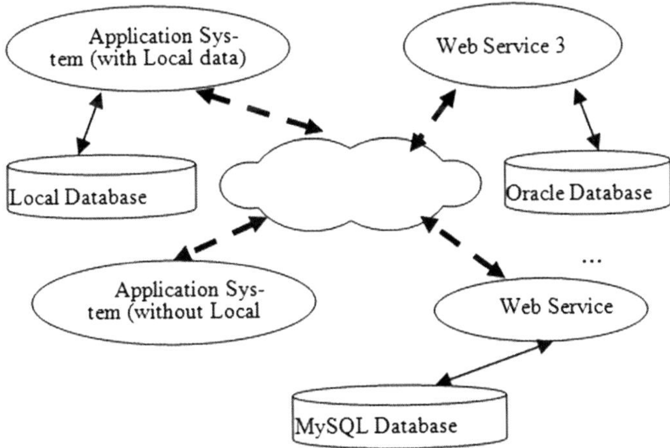

Fig. 100.2 Integration structure of distribute heterogeneous database

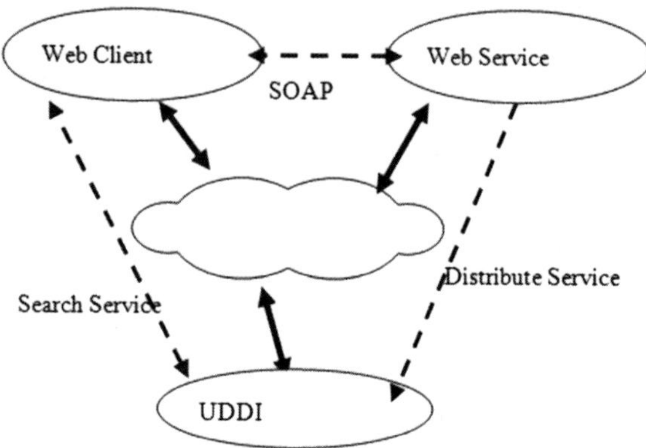

Fig. 100.3 Web service communication principle

used to describe the request and the answer of the data in the distribute computing. Web service is made of three components include a set of standard communication protocol, interface specification Web services description language (WSDL), and Web service manager. The structure and principle of the Web service communication is shown in Fig. 100.3.

Universal description, discovery and integration (UDDI) is used to create and distribute Web service in the network [5]. The simple object access protocol (SOAP) message mechanism is,

To create and distribute Web service in the network is used by UDDI which implements by SOAP message mechanism [5], and SOAP is an standard XML/HTTP protocol. Hence, the key technology and the advantage of this method is that Web service depends on universal standard protocol.

100.3 Database Integration Implement

Accessing data can be implement by using the object oriented (OO) technology, it including querying, inserting, updating and deleting, and so on. Following is an example of data accessing of SQL service database that is implemented by C#, the function of this example is getting the data from the datable Books of the database MyBookShop according to giving condition BookId.

```
[WebMethod(Description = “Getting data from table Books according condition
  BookId”)]
public DataSet SelectDB (int BookId)
{
string connStr = “server = localhost; database = MyBookShop; uid = sa;
  pwd = sa”;
```

```
string sqls = "select Id, Title, Author from dbo.Books where Id < " + BookId;
SqlConnection conn = new SqlConnection (connStr);
SqlDataAdapter ad = new SqlDataAdapter();
ad.SelectCommand = new SqlCommand (sqls, conn);
DataSet ds = new DataSet ();
ad.Fill(ds);
conn.Close();
return ds;
}
```

Following is the request example of SOAP 1.2 of the above example.

```
POST/WebService/Service.asmx HTTP/1.1
Host: localhost
Content-Type: text/xml; charset = utf-8
Content-Length: length
SOAPAction: "http://tempuri.org/SelectDB"
<?xml version = "1.0" encoding = "utf-8"?>
<soap:Envelope xmlns:xsi = "http://www.w3.org/2001/XMLSchema-instance"
  xmlns:xsd = "http://www.w3.org/2001/XMLSchema" xmlns:soap = "http://
  schemas.xmlsoap.org/soap/envelope/">
<soap:Body>
<SelectDB xmlns = "http://tempuri.org/">
<BookId > int </BookId>
</SelectDB>
</soap:Body>
</soap:Envelope>
```

Following is the answer example of SOAP 1.2 of the above example.

```
HTTP/1.1 200 OK
Content-Type: text/xml; charset = utf-8
Content-Length: length
<?xml version = "1.0" encoding = "utf-8"?>
<soap:Envelope xmlns:xsi = "http://www.w3.org/2001/XMLSchema-instance"
  xmlns:xsd = "http://www.w3.org/2001/XMLSchema" xmlns:soap = "http://
  schemas.xmlsoap.org/soap/envelope/">
<soap:Body>
<SelectDBResponse xmlns = "http://tempuri.org/">
<SelectDBResult>
<xsd:schema > schema </xsd:schema > xml  </SelectDBResult>
</SelectDBResponse>
</soap:Body>
</soap:Envelope > .
```

100.4 Conclusions

Integrating database system helps to settle the integrating problem of heterogeneous information system in a enterprise, it promotes the emperies to unifies programming in the departments of the enterprise and to unionize assort with each other in the departments of the enterprise, from this way, the cost of the enterprise can be reduced effectively. A new method based on Web service is proposed to resolve the problem of uniform access of distributed and heterogeneous networks database resource in modern enterprise in this paper, the communication principle and the characteristics of Web service technology, and the advantages of its application to distributed and heterogeneous were analyzed. A distributed and heterogeneous database system integrated structure and a dynamic database increased or decreased model based on Web service technology were established, and a distributed and heterogeneous database service system was designed.

Acknowledgments This research was supported by the science foundation of the Chongqing State Education Commission under grant No. KJ111220, the higher education innovation foundation of Chongqing State Education Commission under grant No. 09-3-181, the science foundation of Chongqing University of Arts and Science under grant No. Z2011JS13 and No. Y2009JS56, the education innovation foundation of Chongqing University of Arts and Science under grant No. 100240.

References

1. Zhu J, Hu Y, Gao Y, Gou F (2011) Design and implementation of mobile agent-based on spatial information sharing service. J Southwest Jiaotong Univ 46(3):427–433
2. An S, Cui N, Yu H (2010) Personalized service strategy of travel information based on multi-agent negotiation. J Southwest Jiaotong Univ 45(4):627–634
3. Shi Y, Cai Y (2009) Visualization and mapping for information system integration based on xml. Comput Eng 35(8):65–67
4. Li J, Deng X, Xu A, Li T, Gao Z (2010) Sharing and integration of gear manufacturing information based on xml and web services. Trans CSAE 26(7):169–173
5. Xiao R, Du Y, Su F (2009) Multi-source marine environmental information grid platform and its implement. Geomatics Inf Sci Wuhan Univ 34(8):932–935
6. Wang Y, Yi S, Long Y, Zhang L, Tang P (2010) Study of car corporate technology data integration based on SOA. J Hunan Univ Nat Sci 37(5):35–39
7. Du Y, Feng W, He Y, Xiao R (2010) Geographic information services integration with web services. Geomatics Inf Sci Wuhan Univ 35(3):347–349, 364
8. Shen L, Lu F, Li F (2011) Web service recommendation model oriented enterprise information system integration. Comput Integr Manufact Syst 17(1):186–190
9. Yin S, Ying C, Liu F, Guo K (2009) Integrated service mechanism of networked collaborative product development resources. Comput Integr Manufact Syst 15(11):2233–2240
10. Xiong H, Liu Y, Guo D (2008) Design and implementation of distributed heterogeneous database migration system. Comput Eng 34(4):57–79
11. Xu H, Zhu X, Tong J, Yang F, Zhao J (2004) Key technology on web service based distributed file service system. Comput Eng 30(24):43–44
12. He Z,Wu L, Zhang H, Li H, Lai H (2011) Research on semantic security supply-and-demand policy for web service. J Sichuan Univ Eng Sci Ed 43(1):116–122

Chapter 101 Product Quality Risk Information Collecting System

Huali Cai, Wanjin Tang, Yuexiang Yang, Lizhi Wang and Xia Liu

Abstract To build product quality risk information collecting system in China, the definition and classification of product quality risk information is firstly given. In addition, the architecture of risk information collecting system is built, the data flow diagram is drawn, the metadata of the risky case are listed and the system is implemented. At last, several supporting measures are established.

Keywords Risk • Collecting system • Product quality

101.1 Introduction

During these years, there are lots of accidents caused by products threatening the human being or environment. Not only the government, but also the manufacturers are paying more attention to the products quality. The product quality risk information, which located at lots of places, will bring benefits to human's health and the society's safety if being deeply mined. However, rare institutes have built the comprehensive system or made study on the risk information. Several authors have done much work in the field of food [1], hospital [2], or specific product [3].

There exists several countries have built small-scale risk information collecting systems such as American CPSC [4] and European Rapex [5]. They play very important role in reporting the poor quality products. But the information they collect only cover little channels and have hardly been deeply mined for more use.

This paper aims to build product quality risk information collecting system in China to collect more risk information, so that more accidents can be avoided and more feedback information can be given to the manufacturers.

H. Cai (✉) · W. Tang · Y. Yang · L. Wang · X. Liu
Quality Management Branch, China National Institute of Standardization, Beijing 100088, China
e-mail: caihuali101@126.com

Z. Zhong (ed.), *Proceedings of the International Conference on Information Engineering and Applications (IEA) 2012*, Lecture Notes in Electrical Engineering 220, DOI: 10.1007/978-1-4471-4844-9_101,

The rest of the paper is organized as follows: the definition and classification is introduced in Sect. 101.2. Section 101.3 builds the risk information collecting system. Related supporting measures are established in Sect. 101.4.

101.2 The Definition and Classification of Product Quality Risk Information

101.2.1 The Definition

All kinds of information on product's quality, including those like threatening life and asset, poorly influencing import and export trade, related industry and other social aspects, bringing worse effect on safety, health, environment and fraud, are all called product quality risk information. They all need to be analyzed, judged, and disposed in time.

101.2.2 The Classification

According to the information source, risk information can be classified into seven categories, they are:

1. Complaints information, which includes the information from 12,365 platforms, emergency call hotline, online messages, letters, and other related aspects of risk information.
2. Law enforcement and regulatory information, which includes the information of production license, registration and filing, mandatory certification, defective product recalling, law enforcement investigation, supervision and inspection, market access, port inspection and quarantining, and revision of standards.
3. Active monitoring information, which includes the information regularly obtained by relevant business divisions, local councils, and other risk monitoring institutes.
4. Laboratory testing information, which includes the information tested by the relevant inspection bodies and laboratories in the routine inspection.
5. Media and network information, which includes the information on radio, television, newspapers, Internet, advertising, and other promotional media.
6. Other countries reporting information, which includes the information being reported when international organizations, foreign government departments, and other research institutions notifying Chinese products.
7. National institutions and organizations reporting information, which includes the information published by relevant government, trade associations, intermediary organizations, research institutions, and enterprises.

101.3 Risk Information Collecting System

101.3.1 The Architecture of Risk Information Collecting System

The architecture of risk information collecting system can be drawn from two aspects. One is from the aspect of government region shown in Fig. 101.1, from which the risk information can be transmitted from the county to its superior hierarchical level until the nation. The other is from technical aspect: the data layer consists of several databases, which are the base for the whole system; the business layer mainly process and make full use of the basic data; the service layer is oriented to the users. The three layers are supported by the information safety system and information standard system, which is shown by Fig. 101.2.

101.3.2 The Data Flow Diagram

From the detailed business flow, the data flow diagram is drawn as Fig 101.3.

101.3.3 The Metadata of the Risky Case

According to all aspects of the risky case, the metadata should consist of the information on case itself, product, victim, causes, and harm. The detailed metadata are listed in Table 101.1.

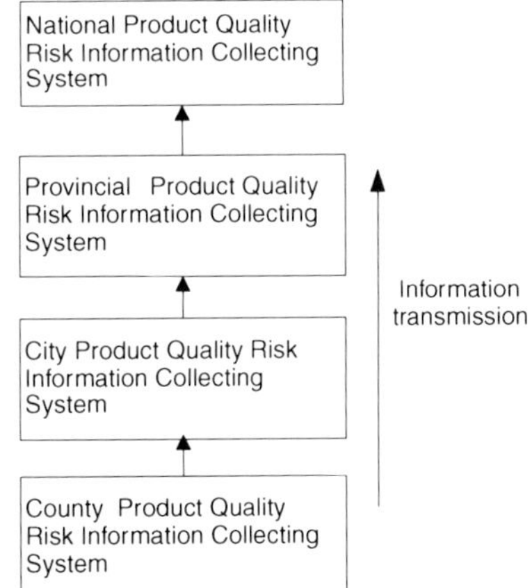

Fig. 101.1 The architecture from government region

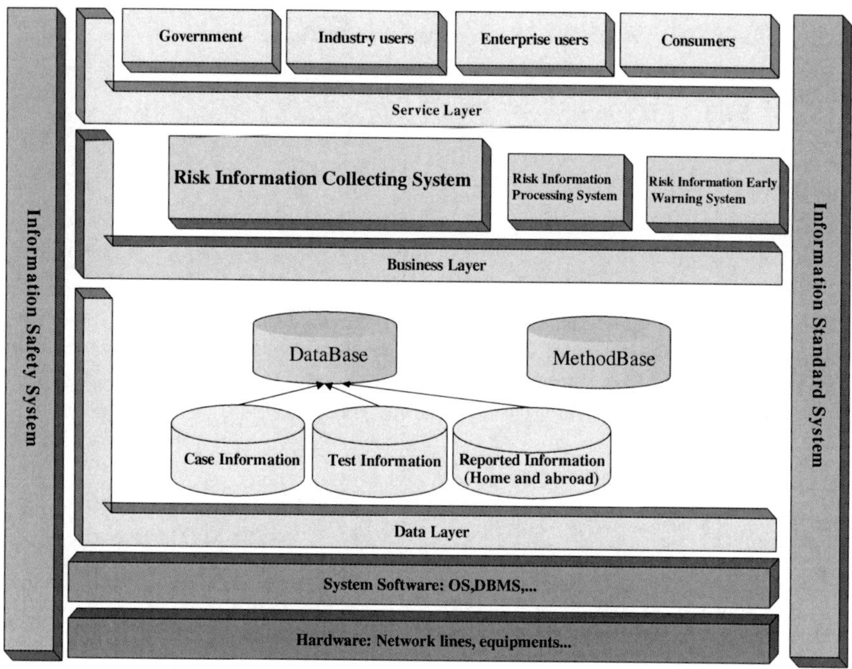

Fig. 101.2 The architecture from technical aspect

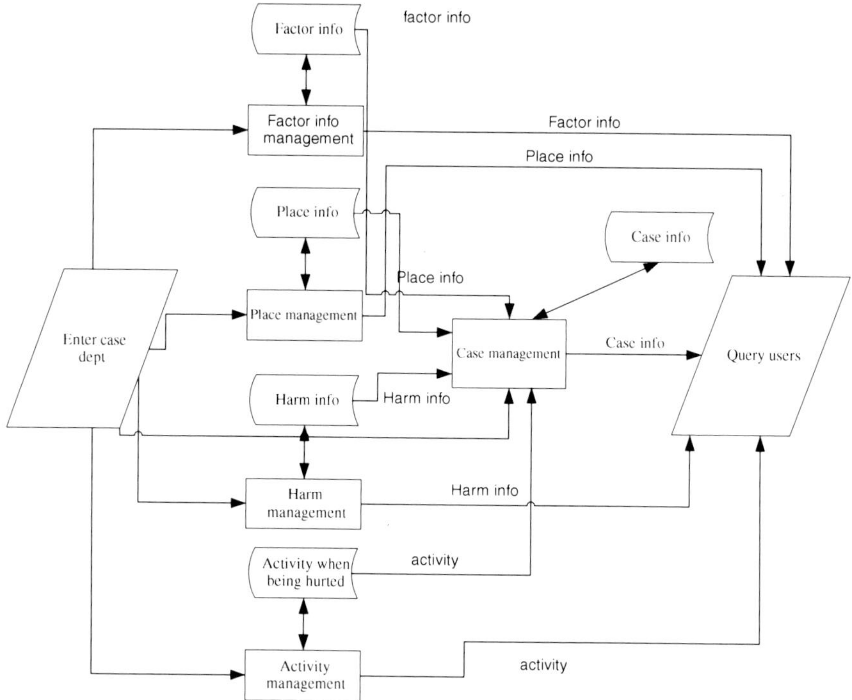

Fig. 101.3 The data flow diagram

Table 101.1 The detailed metadata of the risky case

Item name	Description
REGISTER_ID	Register coding
NF_TITLE	Title of the case
CASE_KEY	Keywords of the case
CASE_SOURCE	The source of the case
NF_COUNTRY	The reported country
HARM_PLACE	The part where the case brings to the people
HC_HUMAN	The harm to people
HC_DEGREE	The degree of the harm
HARM_EVENT	The activity when the case happen
OCC_TIME	The happen time
OCC_PLACE	The happen place
CUS_NAME	The name of the consumer
CUS_GENDER	Sex of the consumer
CUS_AGE	Age of the consumer
CUS_ID	ID card no.
PRD_NAME	The name of the product
PRD_ID5	The classification of the product
PRD_BRAND	The brand of the product
PRD_MODEL	The style of the product
PRD_ORGIN	The origin of the product
PRD_OBJECT	The users
PRD_DATE	Producing time
PRD_IMPORTER	The importer/wholesaler
PRD_SALES	The sales situation
PRD_AMOUNT	The amount of the product
PRD_PHOTO	The photo of the product
HC_ENVIRONMENT	The harm to environment
FAC_PHYDESC	The physical factors
FAC_CHMDESC	The chemical factors
FAC_BIODESC	The biological factors
FAC_VIOLAWS	The violated of laws or regulations
FAC_COMPARATIVE	The violated items
FAC_MEASURES	The measures taken
USED_PERIOD	Service life of the product
RESULTS_DESC	The result of the case
NOTES	Other information

101.4 The Establishment of Risk Information Collecting Supporting Measures

Some supporting measures must be established to support the implement of the risk information collecting system. They are:

1. To strengthen the leadership and make clear the responsibility. All the quality safety related departments and units should attach great importance to product quality and enhance information security risk evaluation.

2. To put the institutions and personnel into operation. The quality inspection departments at all levels should designate a special body or person responsible for the job of collecting, reporting, judging, and disposal of risk information.
3. To refine the programs of the work. Every responsible body should refine the program, so that the scheme can be implemented.
4. To stand out the work emphases. We can concentrate on the food safety and form an effective food quality and safety risk information management system.
5. To strengthen the organization and coordination. Product quality and safety risk management involves a wide range of information. Comprehensive, quality inspection departments at all levels should clarify their respective responsibilities to strengthen the organization and coordination of information on product quality and safety risk management.

Acknowledgments The authors gratefully acknowledge the support of the Chinese Quality Inspection Commonweal Projects (Grant No. 201110013, 201010013, 200910088, 200910279 and 201010268) and the project by CNIS's Chairman (Study and development of the system of Network public opinion monitoring service (I)).

References

1. Verbeke W, Frewer LJ, Scholderer J, et al (2007) Why consumers behave as they do with respect to food safety and risk information. 5th international symposium on hormone and veterinary drug residue analysis 11:2–7
2. Tsumoto S, Matsuoka K, Yokoyama S (2006) Risk mining in hospital information systems. 6th IEEE international conference on data mining 5:699–704
3. Hagel BE, Pless IB, Goulet C et al (2004) Quality of information on risk factors reported by ski patrols. Inj Prev 13:275–279
4. Martin, Lowell F (2001) The CPSC product safety circle. Annual quality congress transactions 5:602–606
5. Tirendi S (2009) Chemical emissions from toys-the case of stink blasters. Int J Environ Anal Chem 21(06):929–938

Chapter 102
Data Mining in Load Forecasting of Power System

Guang Yu Zhao, Yan Yan, Chun Zhou Zhao, Chao Wang and Hao Zhang

Abstract This project applies Data Mining technology to the prediction of electric power system load forecast. It proposes a mining program of electric power load forecasting data based on the similarity of time series research, effectively overcoming the negative effects on the prediction results caused by the limited and incomplete data. It also illustrates a list of examples to prove that the conducted method is effective and efficient.

Keywords Load forecasting • Data mining • Time series

102.1 Introduction

With the development of computer, network, and communication of information technology and data storage technology, especially the appearance of data warehouse, an increasing number of data is stored. For these massive data, not only do the traditional data analysis methods and data retrieval mechanism consume a lot of time, but also they completely depend on the advance of the data on the relationship between assumptions and estimates of data, which have been unable to satisfy people's eager on implicit knowledge in these data. Data mining technology emerges as the times' require, provide powerful means putting automatically and intelligently massive data into useful information and knowledge.

Collecting large number of historical data is the key to power load forecast, establishing a scientific and effective forecast model, adopting effective algorithm, based on historical data, conducting a large number of study and analysis, continuously correcting the model and the algorithm, in order to reflect the real load variation. According to the load forecasting theory, using the method

G. Y. Zhao (✉) · C. Z. Zhao · C. Wang · H. Zhang
Qing gong College, Hebei United University, Tangshan, Hebei, China
e-mail: zhaoguangyu@126.com

Y. Yan
College of Sciences, Hebei United University, Tangshan, Hebei, China

Z. Zhong (ed.), *Proceedings of the International Conference on Information Engineering and Applications (IEA) 2012*, Lecture Notes in Electrical Engineering 220,
DOI: 10.1007/978-1-4471-4844-9_102,

of Data Mining based on historical data of electric power load data, power load forecast is one of the power prediction methods. The result of Data Mining is based on a large amount of historical data being analyzed, it can be objectively given by the data implied useful information and knowledge law, have higher credibility.

This article based on the Students Science and Technology Innovation of Qing gong College, Hebe United University "The research of load forecasting of power system".

In power system, various data sources are mainly come as follows: real-time data, file data, and simulation data. At the same time, each data source also contains many different kinds of data; all these data constitute an extremely large information storage system. Therefore, data are deeply analyzed, from which getting more valuable information, for the power system prediction problem has important significance [1].

102.2 Data Mining Theory and Research Status

Data Mining refers to the data extracted from the implicit in which people prior to the unknown, but potentially useful information and knowledge, and expresses them as understandable patterns of senior process. Data Mining is also called knowledge discovery in database, data analysis, data fusion, and decision support. The starting point of Data Mining is to reduce the experts getting hidden knowledge from large amounts of Data Mining; it makes the data storage technology into a more advanced stage. It not only makes the use of database storage functions, the historical data query and traverse, to answer the question "what", but also can be able to find out potential links among historical data, digging out many hidden important information, which can well support people 's decision, to answer the question "why". Data Mining can discover that knowledge can be divided into the following several kinds of knowledge models:

a. Classify model: based on the known classes of individuals were summarized, extracted to represent common features of groups, classification model.
b. Regression model: analyze historical data of object attribute to predict future trends.
c. Time series model: can use the existing data to predict the future. Compared with the regression model, time series model emphasizes the consideration of time characteristic, especially considering the time period of the level, such as days, weeks, months and so on, sometimes also consider calendar effects, such as holidays.
d. Clustering model: it is a group which is divided into a number of categories, so that the same individual as similar as possible and in homogeneity individual differences as large as possible. And it is different from classification model; the clustering model belongs to unsupervised learning process.

e. Association model: reflecting the knowledge of dependence or association between things, known as association rules. Association rules are the general form: if A happens, B has c % can happen, C called the confidence of the association rules.
f. Sequence model: it is similar to associated model, but the difference is that the objects of sequence model are distributed in the time, and the discovered rules also relate to order.

All the above knowledge can be found in different conceptual levels, and with the concept of tree improvement, from microscopic to macroscopic, to meet the different need of users and levels of decision making.

The subjects of Data Mining are data accumulated from a professional field, and it is a man–machine interaction, repeated process, and the results of mining to be used for the professional. Therefore, the whole process of Data Mining is inseparable from the application domain knowledge. It belongs to the application demand of interdisciplinary development product [2], (Fig. 102.1).

Data mining in load forecasting of power system is mainly used as following several points:

102.2.1 Outlier Data Analysis

Before load forecasting requires pretreatment of the acquisition data refers to recognize and correct bad and adverse data, including the natural complement and impact data stripping. Cluster analysis divides the similarity data into a category, used in data pretreatment and outlier Data Mining, effective searching for abnormal data, accurately position them, and get the load pattern of each time.

102.2.2 Load Pattern Analysis

Power load has complexity, randomness, time varying, dispersion, diversity, discontinuity and other characteristics, to determine the exact load model category

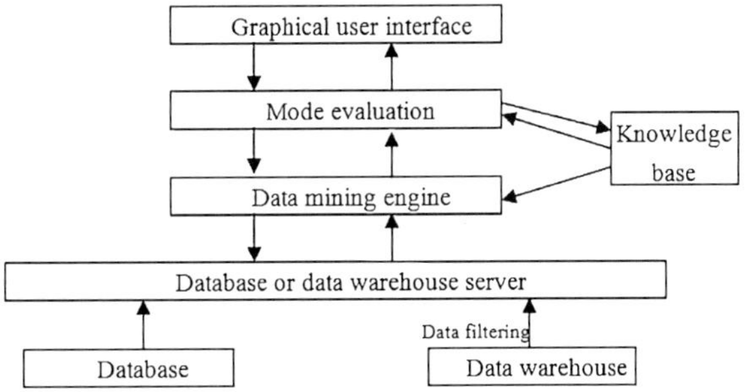

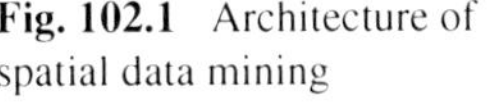
Fig. 102.1 Architecture of spatial data mining

is a difficult problem to solve. But the general trend of load, its cycle has a certain similarity, the similarity of Data Mining to find different conditions of load state, found out different operation conditions, the load similarity of weather conditions, and establish different load operation mode.

102.2.3 Load Forecasting Data Mining

Load forecasting Data Mining process includes load forecasting mining definition, data collection and pretreatment, execution of load forecasting mining, mining result interpretation, and evaluation [3].

102.3 Based on Time Series Similarity Research in Data Mining

Time series refers to the chronological observation value set, such as electric power load. Time series similarity search (also known as the similarity query) is to find out the similar sequence to the given sequence pattern in the time series database. For example, search the similarity rule of today and yesterday, this month and last month is helpful to predict future load [4].

102.3.1 Data Collection and Pretreatment

The first collection is historical data related with all load forecasting, including weather data, historical load data, holiday type data, transforming the data format, removing the noise in the data, and cleaning incomplete and inconsistent data, after the full exploration of load forecasting related with history data, selecting strong prediction capacity variable establishing load forecasting mining model [5].

After knowing the characteristics of dataset preliminary, we should choose a suitable algorithm to establish load forecasting mining model. Therefore, the part integrates a variety of algorithms, including neural network algorithm, multiple regression method, the regression tree and support vector machine algorithm. Through the comparison of various mining methods for the effect load forecasting, discovering limited samples learning supporting vector machine algorithm in the test dataset shows better effect, the regression function by the following equation:

$$f(x) = \sum_{i=1}^{\rho} \left(a_i - a_i^*\right) K(x_i, x) + b \tag{102.1}$$

Where: $(x_1, y_1), \ldots, (x_\rho, y_\rho)$ is training set, $x_i \in R^n, \rho$ is sample number, n is input dimension, $y_i \in R, K\ (x_i, x)$ is Kernel function, b is a constant, a_i, a_i^* is Lagrange multiplier.

In order to strengthen the mining model prediction ability, combining with boosting method based on support vector machine model is improved, to get the final support vector machine propulsion model:

$$\text{Model}\ (x) = \sum_{k=1}^{Q} \varepsilon_K f_k(x) \tag{102.2}$$

where ε_k is based on method to establish the k load forecasting regression model $f_k(x), (k = 1, \cdots, Q)$.

102.3.2 Feature Extraction

Load time series data are usually incomplete, inconsistent, deficiencies, noise, and redundancy of data. Data pretreatment technology can improve the quality of the data, which are helpful to improve the precision and performance of the subsequent mining process, Data Mining process is the indispensable important step. Pretreatment time consuming of usually accounts for the entire Data Mining task 70–80 % of workload. Load time series data preprocessing contains data integration, data selection, data cleaning, data conversion, and so on [6].

Piecewise Average Approximation (PAA) acts as the load time series dimension reduction method, the time series is divided into K of equal segments, then forming a new series to express approximately the original sequence with each segment average value. The average of PAA series dimension is lower than the original time series dimension, and making the effective use of multidimensional index, simple, intuitive, and efficient.

Divide time-interval rules length n into N sections $(N \ll n)$, secondly select the average of each partition to it forms a new average value sequence. Thus, it transforms the time series X on time domain into the average sequence $\overline{X}$ (the Dimension is N) on domain [7].

Divide the sequence x length n into N sections. For convenient, if n can be divisible by N; otherwise, it can meet the conditions by means of zero-padding through the time sequence at the end. n/N says each segment contains data points.

Then use the following formula to calculate average column of the ***i***ts $(i = 1, 2, \ldots, N)$ segmented column, which means the ***i***ts data points of the sequence

$$\overline{X}\ [i] = \frac{N}{n} \sum_{j=\frac{n}{N}(i-1+1)}^{\frac{n}{N}i} X[j]. \tag{102.3}$$

As shown in Fig. 102.2 is the comparison between the PAA features (mean value) with the original time series, the time-series length 160, transformed into arrange series $\overline{X}$ length eight by PPA.

The PAA method of two kinds of extreme circumstances: (1) when N = n, average sequence is the original sequence $\overline{X} = X$; (2) when N = 1, simplified for the entire sequence of half the mean X = mean (X).

102.3.3 Index Structure

By means R^+ - tree memory index Minimum Bounding (hyper)-Rectangle (MBR), to realize the similarity query. R^+- Tree is R- tree's improved structure; R- tree is B- tree in the multi-dimensional objects stored on the expansion, it is a balanced tree, and ensures 50 % of the space utilization rate. In order to achieve subsequence matching, introduce the sliding window and MBR technology. Sliding window technology can be a subsequence matching into isometric whole sequence matching, which is one of the effective technologies for realizing subsequence matching.

Combine the sliding window with segmented average technology, the sequences are obtained by the sliding window, then make each subsequence segmentation do sectional average transformation, from the same original sequences correspond to the N—dimensional feature space on the $n - \omega + 1$ points, forming a little point chain (the pre and post is near). Through the method of MBR, the first point chain is divided into several sub chains. Each sub chain is represented by MBR with containing each point, and storage MBR by R^+- tree in the end [8].

102.3.4 Similarity Search

Isometric sequence perfectly matched is relatively simple. But in reality, there are often the most unequal-length sequences in the query. Usually, the query sequence length is less than the sequence length, also known as a subsequence matching. In

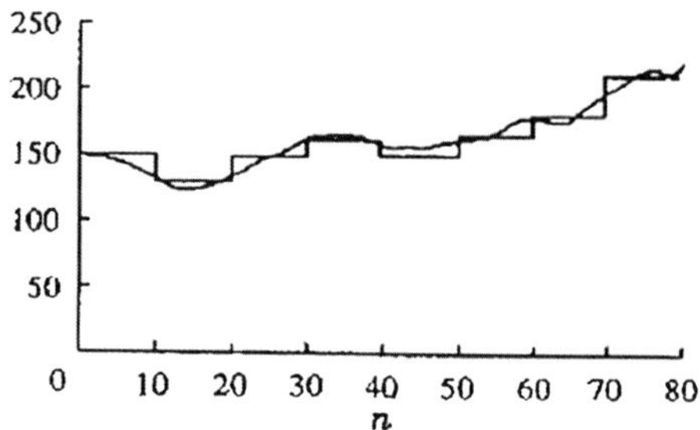

Fig. 102.2 The comparison between PAA characters reconstructed series and primary time series

order to achieve a subsequence matching, combine the basic PAA dimensionality reduction method with sliding window and MBR technology, and take R-tree as a memory index, to save the occupied space and improves the query efficiency.

The query sequence Q by feature extraction is mapped by a point to the feature space, known as query point, the point and similarity threshold ε define the query area together. Such as two-dimensional spatial search region which center is a query point, with ε as the radius of the circular area? In the *R*-tree retrieves all the query area MBR, known as the candidate MBR [9]. The following is the specific algorithm:

Algorithm: $Search\_MBR(Q, \varepsilon)$
In put: The query sequence Q, Similarity threshold ε
Out put: Similar to Q and promoter sequence
Steps:

1. The Q through the PAA mapping for the query object Q, as a feature on the domain of the query point;
2. According to the position of query point and threshold ε, determine the query area, checking all the intersection of MBR in the MBR index, as a candidate MBR.
3. Each of the candidate MBR, one is the retrieval of the MBR corresponding sub chain, the other one is the retrieval of the chains corresponding to multiple sequences, as a candidate subsequence. In the time domain, calculate the distance between candidate sequences and Q.

102.4 Error Analysis of Load Forecasting

Load forecasting model is a generalization of statistical data, which reflects the general characteristics of the experience data internal structure, not completely consistent to the data about the specific structure, so inevitably produces errors. Study on the reasons of the errors, establish a proper model of error analysis, calculate and analyze of the size of the error, can realize the degree of accuracy of the prediction results, has important reference value to make decisions in the use of predictive information [10].

102.4.1 The Causes of Error

The causes of error mainly have following several aspects:

1. Mathematical model used in the prediction mostly includes only research on some of the main factors of phenomenon, which is just a simple load condition reflection, and there is a gap between with the actual load, use it to predict will inevitably produce errors;

2. The selection of the methods for predicting the improper; inappropriate prediction method is inevitable to generate a prediction error;
3. The load forecasting is used all the data which may not be completely accurate and reliable;
4. The accident happened and the sudden changes in the event or situation.

Over a variety of causes of errors are blended together to show, therefore, when the error is large, the commonly used method is to predict the outcome of severe loss of real time, we need to examine these problems one by one, the roots, to be improved.

102.4.2 Load Forecasting is commonly used in Calculation, Analysis of Prediction Error Indicator, and Method

Set x represents the actual value, $\bar{x}$ represents predictive value, $x - \bar{x}$ is known as the absolute error, $\frac{x-\bar{x}}{x}$ is called relative error.

1. The mean absolute error analysis method: avoid predicting errors of the position and negative offset.

$$MAE = \frac{1}{n}\sum_{i=1}^{n}|E_i| = \frac{1}{n}\sum_{i=1}^{n}|x_i - \bar{x}_i| \tag{102.4}$$

2. Analysis of mean square error method: Do not avoid the positive and negative error of additive problems.

$$MSE = \frac{1}{n}\sum_{i=1}^{n}E_i^2 = \frac{1}{n}\sum_{i=1}^{n}(x_i - \bar{x}_i)^2 \tag{102.5}$$

3. Root mean square error analysis method: strengthen the numerical large errors in the target function, thereby improving the index sensitivity.

$$RMSE = \sqrt{\frac{1}{n}\sum_{i=1}^{n}E_i^2} = \sqrt{\frac{1}{n}\sum_{i=1}^{n}(x_i - \bar{x}_i)^2} \tag{102.6}$$

4. Standard error analysis method.

$$S_x = \sqrt{\frac{\sum(x_i - \bar{x}_i)^2}{n - m}} \quad (i = 1, 2, \cdots n) \tag{102.7}$$

N: The number of the historical load data;

M: Degrees of freedom, which is the number of variables; namely, the total number of the independent variable and dependent variable [11].

102.5 Conclusion

There are many worth of studying and discussing the problem. This paper describes Data Mining technology and its application in power system, and applies the similarity of Data Mining to analyze load forecasting. This is only preliminary application of Data Mining technology in load forecasting [12]. In the future, multidimensional time series similarity technology and decision tree technology in Data Mining will be applied in the user's load clustering and classification for power load forecasting, and combining Data Mining technology with artificial neural network to predict, further study relationship of load and time, the climate and the users, providing more advanced and accurate prediction method for load forecasting.

Acknowledgments This paper is based on the students science and technology innovation of Qing gong college, Hebei United University "The research of load forecasting of power system".

References

1. Jawed H, Michelin K (2000) Data mining concepts and techniques simon fraser University, vol 1(3). Morgan Kaufmann Publishers, California pp 34–38
2. World-wide competition within the EUNITE network EUNITE competition report, company behind: East-Slovakia power distribution company
3. Wei W, Cheek PL, Kook KP (2003) Predicting drug dissolution profiles with an ensemble of boosted neural network: a time series approach. Ire Trans on neural networks 3(2):459–463
4. Espy D (2000) Adaptive logic networks for East Slovakian electrical load forecasting EUNITE competition report, company behind: East-Slovakia power distribution company 12:398–409
5. Zhou Q, Pu H, Zhai YJ (2010) Data processing and experimental research on load forecasting. Comput Eng Appl 5(15):193–195
6. Khan MR, Abraham A (2003) Short term load forecasting models in Czech Republic using soft computing paradigms, Int J Knowl-Based Intell Eng Syst, vol 6(17). IOS Press, Netherlands pp 179–183
7. Zhang JH, Qiu W, Liu N, Zhang XF (2010) Grey model based on orthogonal design and its application in annual load forecasting. Power Syst Clean Energ 7(02):181–184
8. Ding J (2010) Technique of power system load forecasting. Heilongjiang Sci Technol Inf 8(3):213–217
9. Liu J (2010) Research on peak load forecasting using RPROP wavelet neural network, Herbin institute of technology, master degree paper 9(12):256–259
10. Du XH, Zhang LF, Li Q, Liu SX (2010) Research on the VB-Mat lab interface method in load forecasting software of power system. Power Syst Prot Control 10(19):121–125
11. Matteson DS, McLean MW, Woodard DB, Shane G (2011) Henderson, Forecasting emergency medical service call arrival rates. Ann Appl Stat 11(13):79–85
12. Chen X (2010), Research of power load forecasting for power based on neural network and software developing, Central South University master degree paper 12(3):34–38

Chapter 103
A Complex Learning System for Behavior Factor Based Data Analysis

Wei Guan

Abstract In this paper, we propose a complex learning system for data analysis (DP) that is based on behavior factor. This system combines features of both rule-based systems (RBS) and rule-based DP framework. We have identified and analyzed the properties that distinguish behavior factor and rules from data for better determining the most components of the proposed system. For representing domain behavior factor, we investigated a uniform and unified rule representation form based on the creation of the environment component that stored information especially for the management of rule and its quality. Besides dealing with some limitations of currently RBS and BFS cited above, the system through the case study allows us to observe many advantages.

Keywords Data analysis • Behavior factor • Complex learning system • Components

103.1 Introduction

Data quality (DQ) has always been an important issue and is even more the case today. The research works look at the role of Data analysis (DP) tools in helping improve DQ and clarify the need to take an enterprise wide approach to DQ management.

There is a wide variety of DP tools. Their functionality can be classified as follows: Declarative DP and Rule-Based approaches for DP (RBDP). Whatever tools are chosen, a systematic, the rule-based approach yields better results than an unstructured approach [1–3]. Although the rule-based system (RBS) that encodes behavior factor as rules and used to process complicated tasks have been firmly established for many years, they have not been well formally and adequately addressed for the DP tasks. Hence, there is a need to an appropriate RBS for DP.

W. Guan (✉)
Information Center, Xi'an University of Posts and Telecommunications,
Xian 710121, China
e-mail: guanw@xupt.edu.cn

Z. Zhong (ed.), *Proceedings of the International Conference on Information Engineering and Applications (IEA) 2012*, Lecture Notes in Electrical Engineering 220,
DOI: 10.1007/978-1-4471-4844-9_103,

In this paper, we assume that a RBS can play yet another role; it can be an agent of change to improve the DQ. Our primary objective is to improve the functional capability of a DQ management by embedding it with behavior factor of the problem area. When the behavior factor acquisition is performed, the system states the behavior factor in the form of rules and uses these embedded rules to provide advice. As the currently RBS allows only the manipulation of production rules without precise understanding of rules theoretical foundations and without considering rule quality, our system provides an uniform rule representation based on first-order structure (FOS) theory of logic that allows the representation of all the types of rules (derivation, production, integrity, transformation, and reaction) and their quality characteristics. Hence, the proposed RBS for DP contains two subsystems: the first is related to the rules management and the second is focused on the rules quality management. In this system, we have incorporated two semi-automatic processes for the behavior factor acquisition and transformation.

Another feature of our proposal is that the control and the improvement of rule (resp. behavior factor) quality are performed online, incrementally, during the design of the system. It permits an early clean of behavior factor and rules from anomalies and inconsistencies.

103.2 Background and Some Related Works

As our objective is to enhance the DQ by applying a rule based approach in DP, it is necessary then to represent some works related to behavior factor based system, RBS and rules-based data analysis.

Behavior factor-based systems and a BFS is defined as an interactive software system that uses stored behavior factor of a specific problem to assist and to provide support for decision-making activities bound to the specific problem context. BFS has been developed and used for a variety of applications [4, 5].

Behavior factor acquisition is the most important element in the development of BFS. Behavior factor could be obtained from domain experts, raw data, documents, personal behavior factor, business models, and/or learning by experience.

The BFS is vague, and that it is generally hard to specify in advance a BFS completely because of the incremental nature of the behavior factor elicitation process.

Interviews are considered as the most popular and widely used form of expert behavior factor acquisition. One serious drawback of interviews is the fact that they are time consuming.

A review of the literature on BFSs highlights the following major many advantages:

Practical behavior factor made applicable: Systems can assist experts in decision making even if they have that behavior factor in hand; this improves the accuracy and timeliness of the decision made.

Consistency: Results produced by a behavior factor system are consistent throughout its operational life span.

Very often people express behavior factor as natural language or using letters or symbolic terms. There exist several methods to extract human behavior factor.

RBS proves to constitute one of the most substantial technologies in the area of applied artificial intelligence (AI). They have used in diverse intelligent systems. Some interesting recent applications of RBS are the ones that are related to business and the Semantic Web. Both current research works and applications go far beyond the traditional approach.

In the RBS, behavior factor is represented in the form of (If condition Then conclusion < action >) production rules. A rule describes the action that should be taken if a rule is violated. The empirical association between conditions and conclusions in the rule base (BS) is their main characteristic.

Various rule representation techniques are available. However, the rules for DP would require an appropriate rule representation technique that allows the management of rule and its quality, and the presentation of all type of rule.

103.3 Rule-Based Data Analysis System

An important advantage of the proposed RBDP system is that its design is based on strong rule-based DP meta model. In addition, the method is used as a basis for the development and specification of the uniform rule representation.

103.3.1 Rule-Based Data Analysis Meta-model

The generic method in Fig. 103.1 which is drawn with the UML notations illustrates the most important information which is worth tracking on RBS for DP. The

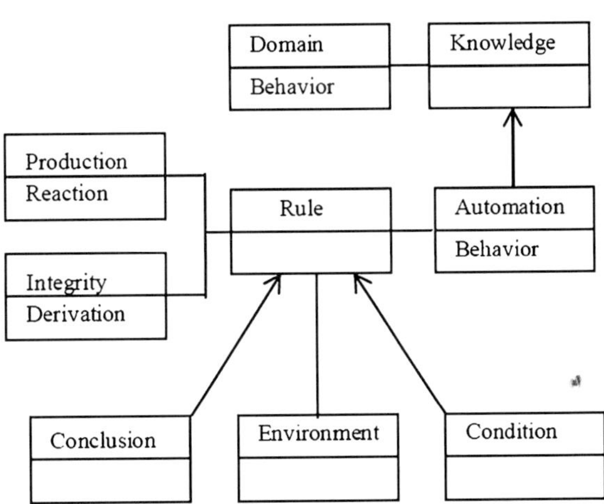

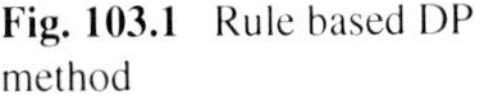
Fig. 103.1 Rule based DP method

information is captured through the life cycle of the rule-based DP systems and their supporting data.

The role of the method is to provide the most components of the system and their relationships. Hence, it offers a good basis to design the RBDP system with the understanding of its functionality and to define a uniform rule representation. The method is divided into two parts, one that presents the behavior factor/RBS and the other that contains the complex learning system and rule quality management system. As depicted in Fig. 103.1, the main types of rules considered in this meta model are: reaction, production, integrity, derivation, and transformation.

The concept of component aims at realizing a unified rule form. As the condition and conclusion are the common components of each type of rules, the idea behind the use of component concept is to represent the uncommon components (for example, the component: post-condition is only used for a reaction rule).

103.3.2 Rule Representation

The expressive power of our rule representation constitutes a crucial feature with respect to the ability of behavior factor specification. Hence, it is of primary importance to properly find a uniform rule form that deals with the limitations described above.

Although there are several rule representations built on top of first-order logic, the concept of structure that gives a context for determining the value (true, false) of a given formulae has not been taken into account. In this work, we provide rules with the concept of environment which is an adoption and an adaptation of the structure in order to create a uniform rule form.

103.3.3 Overview of the RBDP System Components

Figure 103.2 shows the design of the proposed RBDP system. The system is a set of behavior factor sources, process, tasks, structures, and subsystems. It is based on the BFS and RBS technologies. The proposed RBDP system consists of 10 elements.

The main behavior factor sources used by the RBDP system are:

Domain behavior factor: it is the set of behavior factor sources from which behavior factor can be obtained. In our work, the legacy system, programs, and databases referred also to as domain behavior factor parts.

Behavior factor/Rule bases: the behavior factor base (KB) is temporary area where the behavior factor transformations are applied. The BS is the physical area where rules are stored in order to be used for DP. Once the behavior factor is transformed into a rule, it will be stored in the RB and deleted from the KB.

The most typically tasks and processes of the system are:

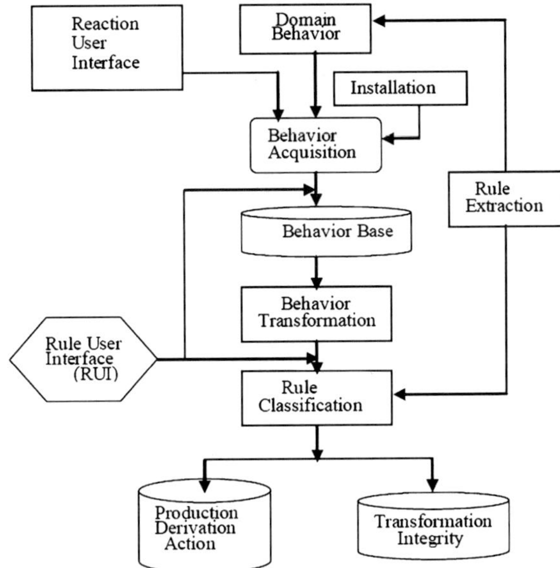

Fig. 103.2 Rule based DP system

Installation: as the design of RBDP system is independent of the databases to be cleaned; then, it becomes necessary to configure the environment required for performing the RBDP.

Behavior Factor Acquisition: This process focuses on extracting behavior factor from expert behavior factor, books, documents, and so on by using manual form text or questionnaires.

Behavior Factor Transformation: it provides a set of operations necessary to transform behavior factor onto rules.

Rule Classification: the goal of this task is to affect each rule to its type (Production, Reaction, Derivation, Integrity and Transformation). For each type, we have defined the corresponding rule template, with which a given rule will be compared.

Mapping: it provides a set of transformations rules that permit the expression of each rule of temporary RB with respect to the uniform rule syntax.

Rule extraction: it uses the techniques of extraction of rule automatically from computers (Databases, legacy systems, and programs).

103.4 Description of the System

As we have indicated above, the proposed system provides two subsystems for the management of rules and their quality. These subsystems which are fully integrated with the RBS for DP use the Environment component as basis for complex learning.

Complex Learning System: A Complex learning System is a software package that performs and controls the creation, modification, suppression, and the archiving of rules. The rules are stored in relational tables. When the rule is not yet applied on data, it can be deleted or archived. The user can be involved to manipulate the rule through the RUI.

Let us notice that the quality dimension can be measured by combining multiple metrics and also a metric can be used by multiple dimensions. So in our system each metric is implemented as atomic function (code).

Quality Management System As the quality of DP results is determined by the quality of rule, our system integrates a module for rule quality management which can be performed automatically or by the user through the QUI. The QMS follows the continuous life cycle of Total Data Quality Management: Define, Measure, Analyze, and Improve rules as essential activities to ensure high quality of data.

Differently to data where the DQ improvement can change the value of data, the QMS can only deactivate or archive rule when it is evaluated poor. Therefore, the user defines the quality dimensions and their quality metrics which allow measuring the quality of rule through the QUI and after the QMS analyzes the quality of rule in order to decide how to improve it. Let us notice that the use of Environment is to allow an incremental define of quality dimensions and their metrics.

103.5 RBDP Functions

Installation: The installation as we have described above is an important task, because it allows to the user to define the information and requirements necessary to the RBDP. Among the most typically information, we cited:

Location of databases, users authorized to use the rule-based DP system,

List of sources that are parts of the domain behavior factor, a set of ontologically based DQ dimensions and metrics used in the literature and also are important to rules.

Keywords of the domain behavior factor will be used to regrouping the behavior factor.

Behavior Factor Acquisition: This process follows two steps: Collection and Storage. The collection step aims at collecting the information extracted from sources of DB using questionnaires in natural language. The memorization of the information is made during the storage step according to structured language sentences through the BUI in the KB. Note that the user can store directly this information through the BUI.

Rule Extraction: As we have indicated above, the proposed phase allows the extraction of rules from programs, legacy systems, and so on by using the data mining and programs slicing techniques. In general, these techniques extract behavior factor as association rules with confidence and support metrics.

103.6 Case Study and Some Experimental Results

In this section, we present some experimental results of the validation of our proposal in the health sector. We study the impact of embedded DB in DP for DQ performances. These experiments show that the proposed RMS and QMS subsystems deal well with some of the limitations of the existing RBDP works.

The atomic conditions and conclusions of rule in this experimentation are based in the use of attributive logic, which uses attributes for denoting some proprieties (A) of objects (o) and atomic values of attributes (d).

The results of this case study provide several important observations: Quality of rules, Expected rules, Uncertainty Rule, Rule and Quality Dimensions Groups Validation, and Autonomous.

Although the design of the architecture allows one to use the notions of parallelism and pipeline, the time consuming is the major challenge of this proposition especially during the initial building of the RBDP system.

103.7 Conclusions and Perspectives

This work confirms that conventional software engineering and behavior factor engineering are complementary and both essential for developing the increasingly larger systems of today. In this context, this paper describes an approach that advocates the use of behavior factor RBSs techniques for DP systems. In particular, we aim to make domain behavior factor explicit and separate from the system where will be used for data analysis. For representing domain behavior factor, we investigated a uniform and unified rule representation form based on the creation of the environment component that stored information especially for the management of rule and its quality. Besides dealing with some limitations of currently RBS and BFS cited above, the system through the case study allows us to observe many advantages. The major challenge of our proposal is the time consuming. Then, it makes its implementation tedious and complex. This is related to the problem of automation of behavior factor acquisition that is not yet resolved.

Finally, once this system is designed, we intend to make it independent from specific database. Therefore, it would be interesting to perform this system with other possible databases. The QMS will also include a feedback loop to enhance the rule quality.

References

1. May LJ (2002) Major causes of software project failures. http://stsc.hill.af.mil/crosstalk/1998/jul/causes.asp
2. Taylor (2000) IT projects: sink or swim. Comput Bull 1(24):24–26. doi:10.1093/combul/42

3. Ropponen J, Lyytinen K (2000) Components of software development risk: how to address them? IEEE Trans Softw Eng 26(2):98–111. doi:10.1109/32.841112
4. Jones C (1997) Applied software measurement: assuring, productivity and quality, 2nd edn, vol 23(7). McGraw-Hill, New York, pp 242–249
5. Standish group, extreme chaos, 2001, Standish group international. http://www. standishgroup.com

Chapter 104
Concepts Mining of Non-level Relationships from Teaching Management Network Information

Shaobin Huang

Abstract Concept learning in teaching management looks for identifying concept elements like nonlevel relationships from information sources. This paper described a procession to mine nonlevel relationships from English network information in teaching management. The process is semi-automatic once it presents to the specialist, a list of probable relationships that will be selected manually. The process uses Unified Relational Language-URL techniques to mine candidate relationships. It aims at mining pairs of concept in a sentence with the verb that probably link them.

Keywords Teaching management • Concept learning • Nonlevel relationships • Unified relational language

104.1 Introduction

A concept is a formal and explicit specification of a conceptualization of a domain of interest [1] that defines concepts and relationships between those concepts to represent knowledge in that domain. Concepts have a great importance for modern knowledge systems since they provide a form for structuring knowledge bases and their reusing and sharing.

There are two fundamental aspects in concept learning. The first is the availability of prior knowledge, which may be in the form of concept to be extended or to be transformed into the first version of concept. This version is then automatically extended by learning procedures or manually by the knowledge engineer. The other aspect is the format of data sources from which you want to mine knowledge. There are three different types of data sources: unstructured sources (documents in relational language like traditional webpages), semi-structured sources, and structured sources (database schemas).

Some approaches for concept learning from structured [2] and semi-structured sources [3] were proposed and showed good results. However, even considering that

S. Huang (✉)
Shantou University, Shantou 515063, China
e-mail: hsb@stu.edu.cn

Z. Zhong (ed.), *Proceedings of the International Conference on Information Engineering and Applications (IEA) 2012*, Lecture Notes in Electrical Engineering 220,
DOI: 10.1007/978-1-4471-4844-9_104, © Springer-Verlag London 2013

these approaches provide support for the development of concepts, most of the available knowledge, especially on the Web, is in the form of teaching management network information in teaching management using relational language [4].

This paper proposes a process for automating the mining of nonlevel relationships between concepts of concepts from textual sources. These relationships correspond to slots in a frame-based concept. For example, in the field of Family Law, we expect to mine relationships such as "represents" between the frames "school" and "client" and "study" between the frames "teacher" and "action".

104.2 A Concept Definition

Concepts are formal specifications of concepts in a domain of interest. Their classes, relationships, constraints, and axioms define a common vocabulary to share knowledge [5]. Following, a concept definition and a simple example in the domain of family relationships are presented.

Formally, a concept can be defined as the group (104.1):

$$O = (C, H, I, R, P, A) \tag{104.1}$$

where, $C = CC \cup CI$ is the set of entities of the concept. They are designated by one or more terms in relational language. The set CC consists of classes, i.e., concepts that represent entities that describe a set of objects (for example, "mother" $\in$ CC), while the set CI is constituted by instances, (for example, "Anne Smith" $\in$ CI).

$H = \{kind\_of (c1, c2) \mid c1 \in CC, c2 \in CC\}$ is the set of level relationships between concepts, which define a concept hierarchy and are denoted by "kind of (c1, c2)", meaning that c1 is a subclass of c2, for instance, "kind of (Mother, Person)".

$I = \{is\_a (c1, c2) \mid c1 \in CI \wedge c2 \in CC\} \cup \{propK (ci, value) \mid ci \in CI\} \cup \{relK (c1, c2, \ldots, cn) \mid \forall i, ci \in CI\}$ is the set of relationships between concept elements and its instances. For example, "is_a (Anne, Mother)", birth (Anne Smith, 02/12/1980)" and "mother_of (Anne Smith, Clara Smith)" are relationships between classes, relationships, and properties with its instances.

$R = \{relk (c1, c2, \ldots, cn) \mid \forall i, ci \in CC\}$ is the set of concept relationships that are neither "kind_of" nor "is_a". For example, "mother_of (Mother, Daughter)". $P = \{propK (ci, datatype) \mid ci \in CC\}$ is the set of properties of concept classes. For instance, "date_of_ birth (Mother, mm/dd/yyyy)".

$A = \{condition\ x \Rightarrow conclusion\ y (c1, c2, \ldots, cn) \mid \forall j, cj \in CC\}$ is a set of axioms, rules that allow checking the consistency of a concept and infer new knowledge through some inference mechanism. The term condition x is given by $condition = \{(cond1, cond2, \ldots, condn) \mid \forall z, condz \cup H \cup I \cup R\}$. For example, "∀Mother, Daughter1, Daughter2, mother_of (Mother, Daughter1), mother_of (Mother, Daughter2) ⇒ sister_of (Daughter1, Daughter2)" is a rule that indicates that if two daughters have the same mother then, they are sisters.

104.3 Nonlevel Relationships

Nonlevel relationships can be classified in domain independent and domain dependent. The domain independent relationships are of two subtypes: ownership and aggregation. Aggregation is the "whole-part" relationship. For example, in the sentence "The horse is out of order". We have a nonlevel relationship of aggregation between "horse" and "legs". The linguistic realization of the relationship of aggregation occurs in two forms: the possessive form of English (apostrophe) and the verb "have" in any conjugation.

However, the converse is not true, i.e., the occurrence of such linguistic accomplishments does not imply a relationship of aggregation as will be explained in the next case. Ownership relationships are held as in the example: "Parents will wait for the teacher's decision" in which there is a relationship of possession between "teacher" and "decision". The linguistic realization of this kind of relationship occurs in two forms: the possessive form of English (apostrophe) and the verb "have" in any conjugation. However, the converse is not true, i.e., the occurrence of such linguistic accomplishments does not imply a relationship of possession. Domain dependent relationships are expressed by particular terms of an area of interest. For example, the sentence "The school will enhance the study in three days" shows the relationship "enhance" between the terms "school" and "study" and is characteristic of the legal field. Table 104.1 summarizes the subtypes of nonlevel relationships and their dependence/independence of the domain.

104.4 The Proposed Process

The proposed process makes use of URL [1] and data mining techniques [3] to mine, from textual sources in English, nonlevel binary relationships between two concept classes. The technique retrieves the relationships indicated by verbs in a sentence, and suggests the possible best level in the concept hierarchy where the relationship should be added. The process is composed of three phases: mining of candidate relationships, analysis of the level, and manual selection of relationships. These phases are detailed in the following sections.

104.4.1 Mining of Candidate Relationships

The mining of candidate relationships phase makes use of URL techniques to mine from network information in teaching management an initial set of relationships. Initially, the network information in teaching management is split into sentences since relationships are identified only between terms in the same sentence.

Table 104.1 Nonlevel relationships and its dependency/independence with a domain

Kind	Sub-kind	Linguistic realization	Mining	Example
Domain independent	Aggregation	Possessive and the verb "have"	RLP techniques	"A horse has four legs"
	Ownership	Possessive and the verb "have"	RLP techniques	"Father and mother will wait for the teacher's decision."
Domain dependent	-	Domain verbs	Statistical methods	"The school will enhance the study in three days."

Then, a search is done in the sentences to find those that have at least two terms that can represent concepts from the class hierarchy of the concept. For that, the class concepts of the concept hierarchy are expanded with their synonyms and possibly with their hyponyms and heteronyms in a generalization/specialization level defined by the user. For example, beyond the term "knowledge" we can consider one level higher in the hierarchy of drink concepts and then include "student study" in the search. These two parameters are intended to increase the recall of the search. Next, a lexical analysis is performed only on the sentences retrieved in the previous step.

The goal is to find the verb forms as indicative of nonlevel relationships.

The last step consists on the generation of group composed of two concepts and a verb relating them from the sentences considered in the previous activity. Two situations are possible. First, there can be sentences having terms that represent concepts that are at a maximum distance of D terms (being D a nonnegative integer) and have a verbal form among them. For this situation, a group in the form <concept 1, verb form, concept 2> is generated. Second, there can be sentences that have the contract form "'", as in "teacher' decision". In this case, a group is generated with the format <concept 1, has, concept 2>. It is also generated an alert to the user that he/she needs to take a decision about the label of this relationship, since it may not be an aggregation, and so "has" cannot be the best label. For example, in the sentence "Parents will wait for the teacher's decision", the best label for the relationship between "teacher" and "decision" might be "take".

104.4.2 Analysis of the Appropriate Level

To suggest the most appropriate level in the concept hierarchy where to insert the relationship as a class slot, the algorithm for discovering generalized association rules is used. One popular application of this algorithm is to find associations between products that are sold in a supermarket and describe them in a more appropriate level. The basic algorithm for mining association rules uses a set of transactions T = {ti | i = 1, …, n}, each transaction ti consists of a set of items, ti = {ai, j | j = 1, …, mi, ai, j ∈ C} and each item ai, j is an element of a set of concepts C. The algorithm computes association rules Xk ⇒ Yk (Xk, Yk ⊂ C, Xk ∩ Yk = {}) which have values for the measures of support and confidence above a given threshold. The support of a rule Xk ⇒ Yk represents the percentage of transactions that have Xk ∩ Yk as a subset; the confidence is defined as the percentage of transactions that have Yk as consequent when Xk is the precedent of the rule. Formally, support and confidence are given by the formulas: Support (XK ⇒ YK) = | {ti | XK ∪ YK ⊂ ti} |/n and Confidence (XK ⇒ YK) = | {ti | XK ∪ YK ⊂ ti} |/| {ti | XK ⊂ ti}|. To mine associations between concepts in the correct level of a hierarchy, every transaction ti is extended to include the ancestors of each item ai, j, for example, ti': = ti ∪ {ai, l | (ai, j, ai, l) ∈H}. Then, support and confidence are computed for all possible association rules Xk $\Rightarrow$ Yk, such that Yk does not have an ancestor of Xk since this would be a trivial association. Finally, we exclude all association rules Xk $\Rightarrow$ Yk that have lower values for support and confidence than an ancestor rule Xk $\Rightarrow$ Yk. Item sets Xk and Yk contain only ancestors or items found in item sets of the rule Xk $\Rightarrow$ Yk.

104.5 Evaluation

Since the mining of nonlevel relationships can be seen as an activity of information retrieval, measures to evaluate such systems can be used in this context. The usual measures of effectiveness of retrieval systems are precision and recall. Recall measures the system's ability to retrieve relevant information from all relevant information available, which in this context corresponds to the mining from the corpus as many groups corresponding to real relationships.

To this end we define the following sets: "T+" corresponds to all relevant groups present in the corpus, i.e., those that represent relationships. "T−" corresponds to all irrelevant groups present in the corpus, i.e., those that do not represent relationships and "r" is the retrieved set of groups (r = A ∪ B). "A" is the set of retrieved groups that represent relationships and "B" corresponds to those that do not represent relationships.

Usually, mechanisms to improve recall reduce the precision and vice versa. Thus, it is not desirable to provide good values only for recall (R) or precision (P). It is important to have a good combination of both. A measure frequently used to reflect this combination in a single value is the F-measure, a harmonic average of both that is given by (104.2).

$$\text{F-measure} = (2 \times R \times P)/(R + P) \qquad (104.2)$$

104.6 Related Work

Table 104.2 shows a comparison of some approaches for the mining of nonlevel relationships with the one proposed in this work. Most of them combine both URL and AL techniques. The [3] proposed an approach based on the premise that nonlevel relationships are usually expressed by verbs that relate pairs of concepts (elements of the C set in the definition of Sect. 2). First, sets of synonyms of concepts from concept are created, using Word net. The use of synonymous increases the recall of concepts mined from a corpus. The corpus is then searched to identify pairs of concepts that occur in the same sentence with verbs that relate them. Thus, each occurrence of a relationship has the form of a group <concept 1, concept 2, and verb> which make up the set of candidate relationships. It is then applied over this set a mining algorithm of association rules, which in this

Table 104.2 Non-level relationships from network information

	Employed techniques	Automated learning technique	Relationships are indicated by verbs	Suggests the level
Zhang peng	RLP and data mining	Mining of association rules	Yes	No
Wang dong	RLP and data mining	Mining of generalized association rules	No	Yes
Zhao jun	RLP and web search engine	-	Yes	No
Guan yuan	RLP and data mining	Mining of generalized association rules	Yes	Yes

case will be of the form {<ci, cj > ⇒ v | ci, cj ∈ C and v is a verb}. As a result in [3], rules are mined that, according to the statistical evidence measures, support and confidence, represent good suggestions of nonlevel relationships. For example, given the concepts "parent" and "child" and the verb "has", the group <parent, child, has> represents the co-occurrence of these three terms in a sentence of at least one of the documents. If the rule <parent, child> ⇒ <has> has a support greater than the minimum, the strength of association between these two concepts, linked by this verb, is given by the confidence of the rule. The recommendation of an association rule is ultimately made based on the measure of his confidence, which depends on its format. Thus, it would be necessary an evaluation about the consequences of the rule format in the final result. For example, the rule <ci, cj> ⇒ v could be recommended, whereas v ⇒ <ci, cj> could not be. The experts [4] propose a similar process to that of other experts. Foster et al. [3], with the difference that it uses an algorithm of generalized association rules to suggest the possible most appropriate level for the relationship. This approach works with network information in teaching management, and proposes an automatic to the authors, the redundancy of information in an environment as vast as the Web is a measure of its relevance and veracity. The first stage is the mining and selection of verbs that express typical relationships of the area. Based on morphological and syntactic analysis, verbs that have a relationship with the domain-keyword are mined. To avoid the relational language complexity, some constraints are used, for example, verb phrases containing modifiers in the form of adverbs are rejected. Then, it measures the degree of relationship between each verb and the domain. To do so, statistical measures are made about the term distribution on the Web. The values obtained are used to rank the list of candidate verbs. This let one choose the labels of nonlevel relationships that are closely related to the domain. The domain dependent verbs are used to discover concepts that are nonlevel related. To do so, the system queries a Web search engine with these patterns: "verb domain-keyword" or "domain-keyword verb" that returns a corpus related to the specified queries.

104.7 Concluding Remarks

Most efforts on concept development are required to identify and specify its nonlevel relationships and there is still a lack of effective techniques and tools to automate and even provide an appropriate help to these tasks.

This paper described a process to mine nonlevel relationships from network information in teaching management. The process is semi-automatic once it presents to the specialist a list of probable relationships that will be selected manually. The process uses URL techniques to mine candidate relationships. It aims at mining pairs of concepts in a sentence with the verb that probably link them. For this purpose, the specialist is asked to interactively adjust a parameter that indicates the maximum distance, in words, between two concepts in a sentence for them to be considered related by a verb located in between. The technique also includes a phase, which can be optionally performed, for the identification of the best level of the concept hierarchy where a nonlevel relationship should be included. For this purpose, an AL technique for the mining of generalized association rules is used. A tool is been developed to automate and better evaluate the proposed process. For the evaluation, a corpus of 450 documents in the Family Law doctrine will be searched for nonlevel relationships and the result will be compared against Family Law, reference concept in the same domain. The effectiveness of results will be measured using the traditional information retrieval measures (recall, precision, and f-measure).

References

1. Foster I, Kesselman C (1999) The grid: blueprint for a unified relational language, vol 14(13). Morgan Kaufmann, San Francisco, pp 144–148
2. Czajkowski K, Fitzgerald S, Foster I, Kesselman C (2001) Grid information services for distributed resource sharing. In: 10th IEEE international symposium on high performance distributed computing, vol 6(16). IEEE Press, New York, pp 164–169
3. Foster I, Kesselman C, Nick J, Tuecke S (2002) The physiology of the grid: an open grid services architecture for distributed systems integration. Tech Rep Glob Grid Forum 15(11):546–549
4. Smith TF, Waterman MS (1981) Identification of common molecular subsequences. J Mol Biol 147(15):195–197
5. May P, Ehrlich HC, Steinke T (2006) ZIB structure prediction pipeline: composing a complex biological workflow through web services. In: Nagel WE, Walter WV, Lehner W (eds) Euro-Par 2006, vol 4128(24). LNCS, Springer, Heidelberg, pp 1148–1158

Chapter 105
Research on the Web Data Preprocessing

Chaodong Lu and Xin Xiong

Abstract In order to Web data mining and data preprocessing of text classification problems, the Gini index on the Web data mining preprocessing, through in-depth analysis of principles and the text features of the Gini index, the Gini index constructed a new measure function, and in the original feature space for feature selection, using the Gini index of the purity of principle. The results show that the classification accuracy of the method has high computational complexity, smaller, and can improve the classification performance for Web data preprocessing.

Keywords Web data mining • Data preprocessing • Gini index • Text classification

105.1 Introduction

Data mining technology continues to improve and applications for Web Mining production and extensive application of foundation. Web information as users use the basic content of the data based on the Web, a new data derived Web log. It contains a variety of objects, including users; click the query words and the Web page, etc. These objects include not only its own nature, and also with other existing between different kinds of objects inter-related relationship, take advantage of these information can effectively improve the user access to information on the Web satisfaction, improve the utilization of information [1].

Web Mining can help users to query relevant information quickly and accurately, from the Web data found in the unknown information potentially useful to

C. Lu (✉)
Wuhan University of Technology, Wuhan 430070, Hubei, China
e-mail: luchaodong@hrsk.net

C. Lu · X. Xiong
Henan Institute of Engineering, Zhengzhou 451191, Henan, China

Z. Zhong (ed.), *Proceedings of the International Conference on Information Engineering and Applications (IEA) 2012*, Lecture Notes in Electrical Engineering 220,
DOI: 10.1007/978-1-4471-4844-9_105, © Springer-Verlag London 2013

understand the client's interests, customized for specific users, and other personalized information.

But the Web log data collected may have redundant data, may also be missing some data, it is necessary to in the excavation before the processing of these data to obtain the appropriate data format for pattern discovery [2]. The process normally includes data cleaning (Data Cleaning), user identification (User Identification), session identification (Session Identification) and the path to add (Path Completion) and so on.

The task of data cleaning data mining process is to delete the registry entries for the log does not need; user identification is to refer to the same page, different users, even those who use the same IP address of the user, the process associated with it; session identification to a given All page references the user to a log together and to classify them into a user session; path added to fill the browser cache and proxy server causing the missing page reference.

105.2 Web Log Data Source and Pretreatment Processes

Web users access the information needed before mining the data pretreatment and data preprocessing will directly affect the merits of sequential pattern mining accuracy.

105.2.1 Web Log Data Source

Web usage mining data from the user and the Web server, the user interaction, when the user requests the application server to go through the middle of the client browser, and may go through one or more intermediate proxy server, and finally reach the Web server through a firewall, again by the Web server to interact with the application server, the response data are also subject to these intermediate processes. Thus, Web usage mining the distribution of the data used in the server side, client and proxy server, as shown in Fig. 105.1.

105.2.2 Web Log Preprocessing Procedure

The main task of preprocessing stage is removed from the original log file is selected the user's browsing, discovery algorithm using standardized data, the results will directly affect the accuracy of the algorithm processing the results and confidence, including: data cleaning, user identification, and page view identification. Data cleaning is to remove unnecessary data mining process. User identification will be requested by the user and his/her associate pages, of which there are major problems, when multiple users through a firewall or proxy server to access the site, multiple users have the same IP address. In this case, to identify with the same IP address, multiple users, server log files are not only known site topology, but also the need to use some heuristic rules, carried out on the Web Log Data Preprocessing. The process shown in Fig. 105.1.

105.3 Web-Based SVM for Text Preprocessing

For no structural or semi-structured text documents, how to become a computer can understand and the mathematical model can handle text categorization, text clustering, or other text mining the premise and the key steps. Feel that the current model of the text is: the Boolean model, vector space model, cluster model, and probability model.

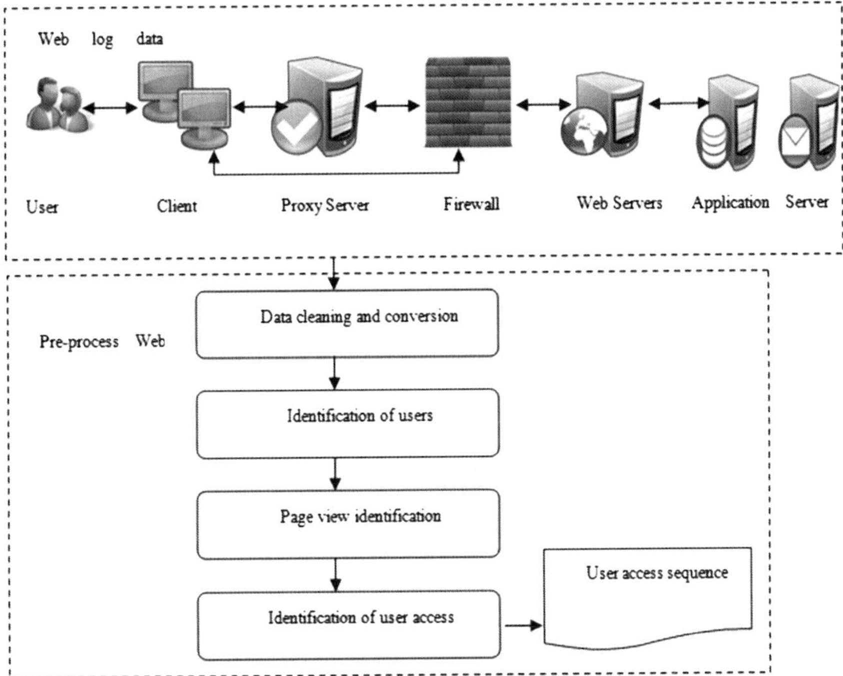

Fig. 105.1 Web log data sources and pre-process

105.3.1 SVM Model

The several models, vector space model has a strong computability and feasibility, especially with the great wealth of online information, its application is not limited to text retrieval, automatic abstracting, automatic extraction of keywords and other traditional issues, but is widely applied to the search engine, personal information agent, online press releases and other information retrieval, and have achieved good results [3].

Web page will document all the features of the item as a document of coordinates to vector form of the document is represented as a point in multidimensional space, vector space by a group composed of linearly independent basic vectors and a particular way for each a feature item weight given to different factors, vector dimension and the same vector space dimension, and can be described by a vector space, so space will be used in Web page documents a feature vector to be described. As a result of the definition of the text in the form of a vector to the real number field, making the data mining process can be achieved in the calculation method to improve the natural language text of the computability and maneuverability. Therefore, the model is widely used in Web text mining text preprocessing.

105.3.2 Feature Selection

For text classification, the most common method is to use the feature selection function to further reduce the dimension of feature space to improve the classifier performance. Feature selection refers to the characteristics from the original N-select t ($t < N$) characteristic, and this characteristic can be more concise and more effective than the contents of the text. The general idea of feature selection is to construct an evaluation function, the characteristic features of each subset of independent assessment, the assessment of each feature to get a sub, and then all the characteristics of points according to their assessment of the size of the sort, select to meet the threshold predetermined number of characteristics of the formation of a feature subset.

Many of the traditional feature selection method, commonly used feature selection functions are: Information Gain, Expected Cross Entropy, Document Frequency, and so on.

105.4 Improved Classification Algorithm Gini Index

Shankar and Karypis application of the principle of the Gini index of the characteristics of text feature selection weighting issues, Charu and Aggarwal, who studied the text with a Gini index of feature selection principle, but they used the

Gini index of miscellaneous degree theory [4, 5]. Gini index used in this paper, preprocessing of data mining to obtain a better classification accuracy, through in-depth analysis of principles and the text features of the Gini index, the Gini index constructed a new measure function, and in the original feature space for feature selection, the use of Gini index is the purity of principle for the classification of Web text, and its computational complexity smaller, faster.

Breiman et al. by the Gini index was proposed in 1984, is widely used in CART algorithms, SLIQ Intelligent Miner decision tree algorithm and the algorithm, the algorithm described as follows [6]:

If the Q is a collection of s data samples,

$$Q = \{c_1, c_2, \cdots, c_s\} \quad s \in R \tag{105.1}$$

Class Q label attribute has different values m, defining m different classes:

$$\{c_i, i = 1, 2, \cdots, m\} \tag{105.2}$$

where $|C|$ is the total number of categories, according to the different class label attribute value can be divided into Q m subset:

$$\{Q_i, i = 1, 2, \cdots, m\} \tag{105.3}$$

Let Q_i is a sample belonging to the class c_i collection, s_i is a collection of sample size Q_i, the set of Q_i's Gini Index as:

$$Gini(Q) = 1 - \sum_{i=1}^{|C|} P_i^2 \tag{105.4}$$

where P_i the probability of any sample is c_i and s_i/s used to estimate.

When the $Gini(Q)$ minimum is 0, set all the records that belong to the same category, that can get the maximum useful information; when the $Gini(Q)$ is maximum, the collection of all the samples in terms of uniform distribution for the type of field, it means to be the smallest useful information.

If you do, according to a property division, the dataset is divided into k sub-set of Q: $(Q_j, j = 1, 2, \cdots, k)$, $Gini_{split}$ after the split index:

$$Gini_{split}(Q) = \sum_{j=1}^{h} \frac{s_j}{s} Gini(Q_j) \tag{105.5}$$

where h the number of child nodes is, s_j is a child node the number of records at j, s is the number of records at node q. $Gini_{split}$ provided the property was chosen as the smallest division of property of the node.

The basic idea is that the Gini index for each property to be through all possible segmentation method, if the Gini index provides minimal, it was selected as the node splitting criteria, both for the root node or child nodes.

In order to apply the theory of the Gini index weighted characteristics of the text, the purity of the Gini index formula can be unified described as follows:

$$GiniTxt(w) = \sum_{i=1}^{|C|} P_i(w)^{\delta} \tag{105.6}$$

where δ is a nonzero value, w is a term, $p_i(w)$ is the probability of term w. Type can be turned on:

$$GiniTxt(w) = \sum_{i=1}^{|C|} \frac{\sqrt{P(w|c_i)}}{P(w|c_i)} \tag{105.7}$$

Construction of a new classification decision-making formula is as follows:

$$Gini_j(d) = \frac{sim(c_j|d)}{n_{cj}} \times \sum_{i=1}^{n} weight(w_t) \times Gini(w_t) \tag{105.8}$$

Algorithm using TF-IDF weighting coefficient adjustment $Gini(w_i) = \sqrt{P(w_i|c_i)}$ is the word w_i obtained by Gini Index probability formula. Gini Index here is to take the $\delta = 1/2$ situation during the cross-validation experiments show that when the $\delta = 1/2$, the classification of the best, so when Gini Index as a classifier, take $\delta = 1/2$. The decision rules for the $Gini_j(d) = \max_i Gini_j(d)$, decision making as $d \in c_j$.

First of all, this forms more suitable for text classification, because for text classification, the information contained in the more favorable classification of text; Second, the calculation of the similarity of the words in the full use of the weight of information: $sim(c_j|d)$, make full use of the word training set weights; again, due to the use of the TF-IDF weighted this probability further, this form of the probability of the word fully considered the impact on the classification of information to improve the performance of Gini Index classification.

105.5 Conclusions

Faced with such vast amounts of Web data, how to deal with text classification where to begin a partial solution to this problem. For text classification, the text preprocessing is the most difficult part has a direct impact on the classification performance and efficiency. Based on the study, while text categorization text preprocessing algorithm of study, preprocessing algorithms in the analysis of the existing text of the advantages and disadvantages, based on an improved text feature selection algorithm, the Gini index improved, making the original application the properties of tree selection method can be applied to text feature selection, on this basis, could be used for feature selection method for the Gini index of further improvement, the algorithm improved to some extent improves the classification performance for Web data preprocessing.

References

1. Paliouras G, Papatheodorou C, Karkaletsis V, et al (2000) Clustering the users of large web sites into communities. In: Danyluk, A (ed) Proceedings of the 17th international conference on machine learning, vol 24(15). Morgan Kaufmann Publishers, San Francisco, pp 719–726
2. Nanopoulos A, Manolopoulos Y (2001) Mining patterns from graph traversals. Data Knowl Eng 37(3):243–266
3. Sha F., Saul L. K. (2005) Analysis and extension of spectral methods for nonlinear dimensionality reduction. In: Proceeding of the 22nd international conference on machine learning (ICML-05), vols 32(16). pp. 785–792
4. Shankar S, Karypis G (2005) A feature weight adjustment algorithm for document categorization. http//www.cs.umm.edu/~karypis
5. Aggarwal CC, Gates SC, Yu PS (1999) On the merits of building categorization systems by supervised clustering. In: KDD'99 vol 31(26). San Diego, pp 352–356
6. Breiman L, Friedman J, Olshen R et al (1984) Classification and regression trees, vol 7(3). Wadsworth International Group, Monterey, pp 683–694

Chapter 106
Des Acclimatization Development and Mining of Multimedia Databases

Zhimei Zhao and Chaodong Lu

Abstract To solve the des acclimatization research data management problem, a form of a multimedia database, data management, indications are given off the plateau overall design of multimedia database and the database engine for its development, gives a plateau associated multimedia data off indications description method, proved that the design can be adapted to the development of multimedia database des acclimatization and general data mining requirements.

Keywords Des acclimatization • Multimedia • Database • Data mining

106.1 Introduction

Altitude sickness is acute mountain sickness acute high altitude disease (AHAD) is after people reach a certain altitude (usually around 2,700 m), the body to adapt to the altitude caused by the natural physiological responses [1, 2]. The low-altitude sickness is the original response corresponds with: When people return from the highland plateau to the plain, from the low pressure, low oxygen environment, into the normal atmospheric pressure, the normal oxygen content of the environment, to adapt to the new environment, re-adjustment of physiological functions, their heart rate, stroke volume increased, the disappearance of the phenomenon of hyperventilation, called off the low of the original response or adaptation (des acclimatization) [3, 4].

After lengthy study, the des acclimatization has accumulated a lot of research data, but these data are still in relatively scattered state, a large number of images, video, sound, and other information of mutual cross-penetration of large amount of data, data types and more, processing time is long. For these multimedia data management, the traditional database technology cannot teach them, with the power of

Z. Zhao (✉)
Henan Institute of Engineering, Zhengzhou 451191, Henan, China
e-mail: zhaozhimei@hrsk.net

C. Lu
Wuhan University of Technology, Wuhan 430070, Hubei, China

Z. Zhong (ed.), *Proceedings of the International Conference on Information Engineering and Applications (IEA) 2012*, Lecture Notes in Electrical Engineering 220, DOI: 10.1007/978-1-4471-4844-9_106, © Springer-Verlag London 2013

computers growing multimedia presentations, multimedia, database development, and research has become a hot spot, and off the plateau indication for the development and multimedia database mining provides a scientific and effective management methods for the effective control and prevention of high altitude off indication.

106.2 Des Acclimatization the Overall Design of Multimedia Database Indications

Multimedia (multimedia) is the integration of two or more kinds of media man–machine interactive information exchange and communication media. Multimedia– hypermedia system is a subset, should include text, images, video, audio, and other multimedia data. Multimedia data types and data features with the traditional data in the database are completely different, although the multimedia DBMS, such as the traditional DBMS to store data, query and transaction processing, but in the input, storage, retrieval, output, and other aspects of dealing with the substance were much more complex.

We will plateau off the indication of data, in accordance with the contents of the study were to establish three multi-media databases: "The investigation object database", "symptom database", "Technology Indicators Database". In the multimedia environment, object-oriented database technology, each entity is modeled as objects, each instance of a class is to support abstract data types. Multimedia data types used to expand its semantic classification, aggregation, association, summarized, and consolidated data abstraction, the use of multimedia query language (MSQL) for management, maintenance, and provides search services.

System structure shown in Fig. 106.1, is divided into three layers: the first layer is the user interface layer to complete the system and the exchange of information between users, interactive behavior; second is the multimedia data management

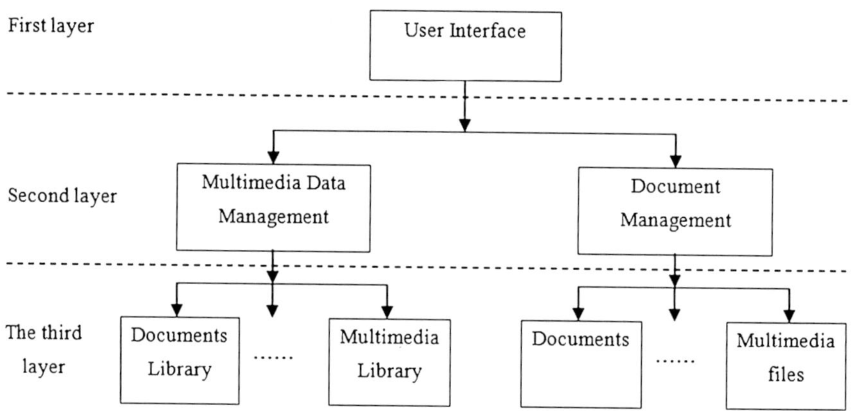

Fig. 106.1 Multimedia database system structure of des acclimatization

and file management system is to achieve the core of one hand is used to format the data, the data on various media operation and maintenance of database information, on the other hand, the actual file operation and maintenance, and accordingly the multimedia database system and its corresponding file system connections; Third is the physical data layer, the database in a variety of data and the composition of different types of documents, including text library databases, graphics, image library, sound library, video library, and multimedia libraries.

Des acclimatization multimedia database management system, in addition to traditional database management system functions, but also with the management of images, sounds, and text formatting data, video and other nonfunctional, while effective in a variety of media in the database records with their corresponding correspondence files up, which is characterized by a strong database in the background, the support of various data organization and management of an orderly, effective in improving the efficiency of the query statistics; front by means of other applications to the real file contents at a glance shown to make the database more intuitive simple, easy to understand.

106.3 Database Engine Design

Multimedia database engine is a relational database management system, the expansion, which is to provide database support for multimedia applications, to simplify multimedia application development complexity. Multimedia database engine takes full advantage of relational database management system for these extended features, multimedia data processing capability will be tightly embedded into the database management system to go.

Database engine as shown in Fig. 106.2, which works as follows:

First, the definition in the database, a new more user-defined functions, and the establishment of multimedia data within the management table. After that, users can create in the database field that contains the custom list types, all kinds of information will be stored in the field of multimedia to multimedia data management table. Upon completion of multi-media database engine, users can access through a variety of multimedia data of the database interface. When the fields of multimedia add, delete, change, and other operations, defined in a table that contains the field of multimedia data is triggered on the one hand, multimedia data management directly modify data in the table; on the other hand, user-defined function calls through the OLE automation components to complete the multimedia service data processing and maintenance of all internal data. Finally, this operation will trigger the information stored in the multimedia data management table. This multimedia database engine to separate from the database system, can be developed, installed and used in all the independent release of relational database products. Engine to support different media independent, the user can according to actual needs of the multimedia database engine to extend and improve the development and application capabilities.

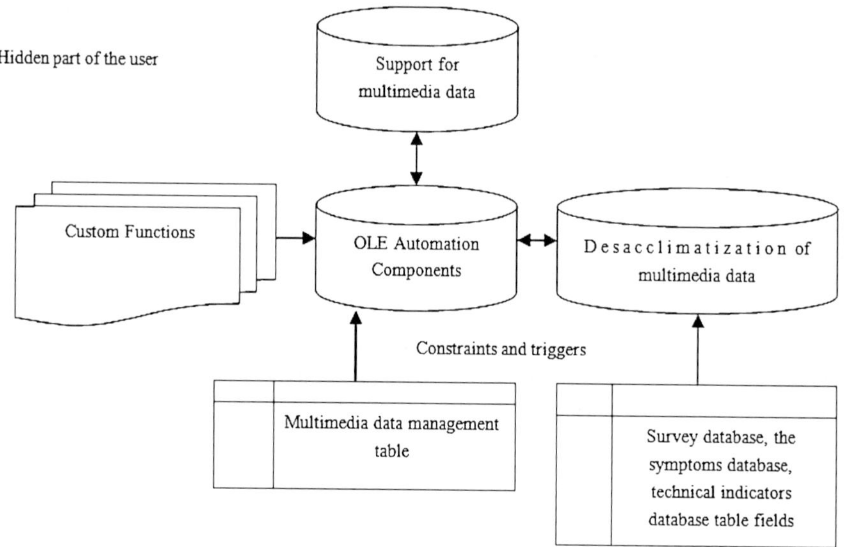

Fig. 106.2 Schematic diagram of multimedia data engine work

106.4 Description of Multimedia Data Associated

Often needs a variety of multimedia information integration are described. Such as a multimedia. Both images. And a voice. There are text, which would change the traditional form of database operations and database interface, in particular the establishment of the illustrated and query the database, to be described in a unified semantic association.

Need to find out association rules between items in a given dataset of useful links. Association rules multimedia data object or characteristic are frequent between the models. Therefore, a given dataset must be on the same subject, or association rule mining will lose practical significance. Associated multimedia data are generally divided into two steps, the first step to identify all the frequent description of sets; second, derived from the frequent association rules describe the focus.

Define any type of multimedia data as follows:

D set of multimedia datasets, association rules are:

$$M_1 \wedge M_2 \wedge \ldots \wedge M_i \rightarrow N_1 \wedge N_2 \wedge \ldots \wedge N_j(c\%) \qquad (106.1)$$

where $M_1, M_2, M_i, N_1, N_2, \ldots, N_j$ is the data set D of data description,

For des acclimatization in the "investigation object database" can be listed according to gender, ethnicity, age, and physical condition, such as residence time in the highlands, the data used to describe the characteristics of the survey.

For des acclimatization in the "symptom database", you can list all of the des acclimatization symptoms: dizziness, lethargy, unable to focus, insomnia, dizziness, fatigue, chest tightness, palpitation, cough, constipation, anorexia, and edema in 12 common symptoms of high altitude adapted species off. To have mild symptoms but does not affect the daily work; moderate needs treatment for the symptoms but does not affect their work; severe treatment for the severe symptoms needs to rest. At the same time, and the symptoms were divided into mild, moderate, severe, and so described.

For des acclimatization in the "target database", can be described by the following indicators: including heart rate, blood pressure, muscle strength, reaction time, lung capacity, maximal oxygen uptake, oxygen saturation, blood lactate, hemoglobin and red blood cell levels, serum creatinine kinase (CK), serum malondialdehyde (MDA), superoxide dismutase (SOD), and nitric oxide (NO) levels, and so on.

$c\%$ is the confidence of the rules;

The rules can be described as follows: occurs when $M_1, M_2, \ldots, M_i$, $N_1, N_2, \ldots, N_j$ also occur, namely:

$$M \rightarrow N(c\%) \tag{106.2}$$

If you define $\{M_1, M_2, \ldots, M_i\}$ description for each support to describe the whole image that the concentration of the probability, denoted by $\delta(Mi/D)$, describes the set of support within the collection of all descriptions that while the probability, denoted by

$$\delta(M_1 \wedge M_2 \wedge \ldots \wedge M_i/D) \tag{106.3}$$

Because only those who care about the pattern of high frequency, so for those with low probability is not necessary to consider the description, where the actual application can set a background to the minimum support, only to meet the minimum support description or description set before considering, if the description of a group focused on a subset of a description does not meet the minimum support, which describes the set would not have considered. Meet the description of the minimum support set is called frequent description set.

Concentration can be derived from the frequent description of the rules, frequently describe the set of rules derived $M_1, M_2, \ldots, M_i$:

$$M_1 \rightarrow M_2 \wedge M_3 \ldots \wedge M_i \tag{106.4}$$

$$M_1 \wedge M_2 \rightarrow M_3 \wedge M_4 \rightarrow \cdots \rightarrow M_{i-1} \wedge M_i \tag{106.5}$$

$$M_1 \wedge M_2 \wedge \ldots \wedge M_{i-1} \rightarrow M_i \tag{106.6}$$

$$\vdots$$

Frequently described by a set of rules derived are useful not only is those who meet the requirements of the rules of credibility what we care about. The credibility of a rule can be obtained by conditional probability formula.

$$\text{confidence} = M\left(M_1 \middle| M_2 \wedge M_3 \wedge \ldots \wedge M_i\right) = \frac{\delta(M_1 \wedge M_2 \wedge \ldots \wedge M_i)}{\delta(P_1)} \tag{106.7}$$

There may also need a pre-given minimum confidence min_support, to meet the minimum confidence of the rule as strong association rules, strong association rules that are the heart of the association rules.

106.5 Summary

With the medical standards and the development of computer technology, people have accumulated on the des acclimatization a lot of research data, all relevant information have gradually developed deep seated, all kinds of documents, data, support the growing diversity, in particular, there the coexistence of a variety of media file storage form. The multimedia database management and development of these data provides a platform used. This paper off indications for the plateau is given multimedia database system architecture design ideas, and the key technology of the database engine design is discussed, in the multimedia database design and the integration of data mining, association gives a description of multimedia data method, which can effectively de-indications for the data associated with the plateau described, to meet the general multimedia data mining.

References

1. Oracle 9i Replication Management API Reference Releasel (9.0.1) Part Number A87502-01[EB/OL] (2002). http://otn.oracle.com/docs/products/
2. Subrahmanian VS (1998) Principles of multimedia database system. Morgan Kaufmann 3(1):56–59
3. Chen W (2003) Compulsion, banyan Multimedia database engine design and implementation of computer application. 15(11):104-109
4. Ning G (1994) MEDB multimedia database management system design. Multi-Media World 9(7):36–41

Chapter 107
On the Construction of Internet Data Center in Colleges

Xiangya Tao and Cheng Yang

Abstract With the evolution of the communication through network and the storage technology, the construction of Internet data center (IDC) is also accelerating its development. Based on the analysis of the current status of IDC construction, the paper also studies the construction of its network, sever, memory system, application system, self-service system, and computer room to provide macroscopically with the reference to setting up IDC in the colleges.

Keywords IDC • Current status • Construction • Study

107.1 Introduction

The application of Networking and Information Technology in education has different forms in different times. From ISP/ICP to .COM, and then to Internet Data Center (IDC), they all reflect innovation and changes of Networking and Information Technology. IDC is the coalition of traditional data center and Internet. It not only has the features of traditional data center like "reliable host, concentrated data", etc., but also has such features as quick response, comprehensive and systemic data service, and flexible access method, and so on. IDC is a platform of data resource cluster and resource service. It pays special attention not only to service capacity construction such as data storage room environment, safety assurance, network bandwidth, software environment and data storage host performance, etc., but also to the construction of IDC service concept and mechanism. IDC university construction is to centralize and share good quality teaching resources, so that optimization of teaching can be realized and education equity can be promoted.

X. Tao (✉)
Information Resources Center of Huaihai Institute of Technology,
Lianyungang 222005, Jiangsu, China
e-mail: terials@sina.cn

C. Yang
Information Dissemination College in Xuzhou Normal University,
Xuzhou 221009,
Jiangsu, China

Z. Zhong (ed.), *Proceedings of the International Conference on Information Engineering and Applications (IEA) 2012*, Lecture Notes in Electrical Engineering 220,
DOI: 10.1007/978-1-4471-4844-9_107,

107.2 Current Situation and Problem Analysis of IDC University Construction

107.2.1 Current Situation of IDC University Construction

IDC university construction is a gradual and progressive process. The initial data center is a server and a data storage center of teaching resources; with the increase of teaching applications, an ERP system is needed, which is a management and application center of teaching data; then the data center also assumes the function of calculating, and it gradually becomes a teaching data processing center; next, simple calculation cannot meet data requirement. The data center has to face teaching resources, face digital applications of teaching. It has to be developed into a service center which bases on the Internet, has integrated data processing and management abilities, and a teaching resource type database.

In recent years, the integration of teaching resources in universities has been accelerating. The consequent requirements of data center further grow. The data center has gradually become independent forces in the running of college education and administrative resources and the IT physical carrier and the cardinal number of code in a university. In the construction wave of IDC, universities are also an important physical carrier for the IDC development. Higher requirements are put forward on its availability and operation efficiency.

Currently, IDC in Chinese universities is mainly at the level of education data application center, education data operation service center. In the future, we will see more and more IDC becoming an independent business agency in universities. In the university education information construction, its construction criterion not longer rigidly sticks to the following specific areas: providing storage for education, providing computing, and providing processing, and so on. It should expand to areas of resource service [1]. This requires IDC to continuously strengthen soft environment, so that it can provide better and more human teaching information data for education.

107.2.2 Problem Analysis of IDC University Construction

For IDC, high availability of the whole information systems is the most important, while high availability of the data section which is the core of information systems is of top priority. The construction of IDC currently has the following problems: (1) infrastructure is not enough, including the availability of network application, data backup and disaster recovery, anti-virus system and the overall system needs to be further strengthened; (2) the data processing of memory system implements distributed, serial processing, adding the differences of interconnection rights of server systems and data outlet, which greatly affects the storage and extraction rate of data; (3) the data of IDC are less secure, storage data among application systems mutual licenses and there is no unified management of authority which is likely to cause undue privileges and directly affects the security of teaching resources. IDC in many universities often suffers the attack of virus and hacker [2]; (4) IDC netted texture is difficult to ensure the uniformity in the exchange of data storage. At the same time it increases the relevance of application system. As a slight move in one part may affect the whole situation, local problems easily affects the overall efficiency of IDC, which affects the quality of information services.

107.3 Study of University IDC Construction

According to the IDC construction situation and the analysis of the problems in construction, the author thinks university IDC construction must strengthen its work and improve construction quality and level in the following aspects.

107.3.1 Network Construction

The efficient operation of university IDC mainly counts on a high-performance network to provide data resources service to education. This requires continuously optimizing the network structure, updating and upgrading the core equipment of IDC network in order to improve the operational efficiency of the network. This high-performance network should include: (1) –AN construction of IDC: including -AN infrastructure building, -AN hierarchical division; and –AN performance design, etc., which should make the built overall network the best; (2) IDC data interconnection platform construction, that is, each independent teaching resource library of IDC should realize the high quality of interconnection and data interchange. (3) IDC teaching subscriber access system construction: the system should be able to ensure that teachers and students store and quote IDC data in a safe and reliable way, or maintain equipments of teachers and students stored in IDC, which requires IDC provides correspondent Internet access methods such as dial-up access, dedicated access and VPN, and so on. [3]. (4) IDC network management construction: because IDC network structure is very large and complex, the network construction is to ensure that the external services of network are uninterrupted and are of high performance. A high-performance network management system should be constructed.

107.3.2 Server Construction

Server constructions of university IDC can be divided into mangy aspects, generally into basic service system server and application service system server, mainly as follows: (1) basic system servers: For the purposes of ensuring IDC provides services to education, this kind of server should include Network Management Server, directory server, firewall server, system performance monitor server, all security servers, DNS server, etc.; (2) Database server: For the purpose of ensuring IDC provides basis of various application services for teaching subscribers, IDC database server construction should be able to support for high-capacity access and ensure the variety of database; (3) Application server: this is a server which provides related application services for teaching subscribers by IDC. To ensure business expansion of IDC, application server should have good ductility to ensure the variety of the supportiveness of application software; (4) Data backup server: to ensure the safety and reliability of teaching subscribers, as IDC servers have a great variety and many databases, data backup should support multiple models, various data formats and have a large capacity; (5) Load balance construction of servers: for many servers to work coordinately, a load balance administration mechanism is needed, which is one of the important technical supports for IDC to provide data service of high performance and high reliability to teaching subscribers. The load balance of server can be accomplished by hardware (like network switching equipment) or software.

107.3.3 Memory System Construction

Memory system construction is one of the key construction content in university IDC. It has to ensure that its data storage capacity is large enough. With the widely application of various multimedia application technologies, teaching information resources data grows rapidly like a geometric set. The capacity of data has developed from Level GB to Level TB. Such a large data require a safer and more reliable storage system in university IDC. In addition, as the dense crowd at university, the centralized amount accesses IDC is very large, so the storage system construction should consider that it must have high data processing efficiency. Also, storage system should have very good expansibility to meet the requirements of IDC developments.

107.3.4 Software System Construction

Software system construction requires substantial investment of university IDC. It is the means for IDC to carry out external services on the base of Internet, server, and storage system building. IDC software system construction should mainly include: (1) Web system: it is one of the content for IDC to carry out Web-Hosting services. Web system software should have Web system function that provides data services to multi-teaching subjects; (2) E-mail system: E-mail system should support multiple e-mail protocols, such as SMTP, POP3, IMAP4,

Web-Mail and Voice-Mail, and so on. Meanwhile e-mail system should also have very good expansibility; (3) database system: IDC should build multi-vendor database systems, such as databases of Oracle, Informix, SQ-Server, SyBas, etc., to meet the requirements of various teaching subscribers. (4) Security system: such as firewall software (Except hardware firewall), anti-hacking software, anti-virus software and so on. It is a premise to ensure that IOC provides data security services to teaching subscribers; (5) Data backup software: backup software should support multiple backup devices, machinery of a variety of manufactures, multiple databases, which can ensure the variety of IDC data construction; (6) Application development system: IDC should provide development system platform corresponding development tools to meet the human requirements of teaching subscribers.

107.3.5 IDC Self-Service System Construction

University IDC receives great attention by universities and welcome by teachers and students because of its good quality teaching data service. IDC should strengthen its self-service system construction. It includes: (1) Customer Relationship Management System (CRM): CRM is a basic service system on which IDC and teaching subscribers establish a good relationship. Its application in university IDC can precisely provide information such as teachers and students' developments and new requirements; (2) Network and server management system: IDC has a large network and server systems. To manage these systems, a powerful network, server, and application management system is needed to ensure IDC teaching service quality [4]; (3) IDC internal management system: it can ensure that each data system within IDC can work uniformly and coordinately and complete teaching service of high quality; (4) Charging system: The charging system is a guarantee of IDC income, which is beneficial for the healthy development of IDC construction.

In addition, university IDC construction should also strengthen central machine room construction. Machine room construction is the largest part of IDC preconstruction investment. As IDC subscribers store their important data and application in IDC storage equipments, the requirements for IDC room environment where storage equipments are placed is very high. IDC room environment mainly includes power supply system, security system, wiring system, communications system and air-conditioning system, and so on. Among these, power supply system is one of the most important, which is the key of computer room construction. As a large numbers of devices of IDC need great electric power, reliability and expansibility of power supply system is of great importance. Loss of data resources caused instantaneous power cut or power failure sometimes can be huge. The power supply system mainly includes service power, UPS construction, distribution cabinet, electric wire, socket, illuminating system, grounding system, lighting protection system and self-generating system, and so on.

107.4 Conclusions

This paper starts research in the application of IDC in university teaching resource construction. With the thorough construction of university digital campus and further development of information construction, data of office system, student management system, education administration system, and logistics system, etc., will also be brought into IDC and then IDC will truly complete large-scale construction. Currently, university IDC, as the IT base that provides teaching resources data service, it provides platforms for the construction and application of all teaching resources in universities; it also provides virtual space for various modes of practice teaching; it is a new means of which a brand new educational technology applies to teaching. IDC construction is still in its growing period of construction and has no unified strict standards. Due to the differences of IT application in different universities, the meanings of IDC differ from each other. Under these circumstances, the research on IDC construction is of great importance. In this paper, several important dimension of IDC construction is studied, which regulates a certain extent of IDC construction areas and solves the problems of how to construct. All in all, IDC energizes the reform of university digital teaching. It now has been a trend of information technology construction and will be developed faster and better.

Acknowledgments Remote Multimedia Teaching Data Transmission System based on IP Multicast and JMF technology (Grant No.: 2010150044); Higher Education Management Research Subject of Huaihai Institute of Technology: Network Teaching Resources Establishment Research of Huaihai Institute of Technology (Grant No. GJ2010-18); Teaching Reform Subject of Huaihai Institute of Technology: Research on the Network Teaching Resources Storage and Mobilized Utilization Model of Huaihai Institute of Technology in the Information Integration and Sharing Stage (Project Approval No.: 2010-12); the host of the three subjects: Tao Xiangya.

References

1. Jian H (2004) Construction on network data center. China Road Transport 2:48–49
2. Zhensong Z, Wei Z (2003) Security Design of a Large Network Data Center. J Pla Univ Sci Technol Nat Sci Edn 5:40–43
3. Guohua G (2001) Integration of techniques at the data center for the crustal movement observation network of China. Earthquake 4:12–18
4. Qislyuw J (2008) On the construction of data center in digital campus [EB/OL]. http://www.kankan.cn/SuperLibtary/freearticle.asp?AID=28077.2008-4-9 vol 24, pp 145–148

Chapter 108
Web Data Mining on Intelligent Network Course System

Xiao-zhong Zong, Zai-tie Chen and Hong-jun Zhang

Abstract Based on Web mining theory and technology, introduced the process of the Web mining. The article proposed a model of intelligent network course system adopts B/S model, mainly includes two problems: the first is intelligent forecast of students visited the curriculum resources; second is dig out the hot course; finally in the form of visualization relayed to the students. System of intelligent core is intelligent forecast recommend system. This model used offline part of mining and the on-line partial mining phase separation mentality, discusses the intelligent network based on Web mining system structure of curriculum system model and its algorithm, and verified algorithm by analysis.

Keywords Web mining • Web usage • Association rule • Data preprocessing

108.1 Introduction

Data mining is to abstract or "mine" knowledge from a great deal of data [1]. Web data mining is to use data mining technology to find and abstract knowledge from Web documents and visit data. Apply Web mining to Web education can find the learning interests, learning key points, learning directions, and visit habits of students from a great amount of Web visit data. It is able to find out the learning interests of the students and recommend appropriate course content and learning materials to students according to this. The application of Web data mining can offer important methods for the intelligence and personality for the Web course system [2]. Dig the potential models of the users' visit behaviors, predict the possible results of users visit and select and recommend Web information related to the interests of users.

X. Zong (✉) · Z. Chen
Shazhou Polytechnical Institute of Technology, Zhangjiagang 215600, China
e-mail: zongxiaozhong@cssci.info

H. Zhang
Educational Technology Donghua University, Shanghai 201620, China

Z. Zhong (ed.), *Proceedings of the International Conference on Information Engineering and Applications (IEA) 2012*, Lecture Notes in Electrical Engineering 220,
DOI: 10.1007/978-1-4471-4844-9_108,

108.2 Web Mining

108.2.1 Basic Concepts

Web mining is a kind of activity that indentifies potential model p from a great amount of document collections C. The relationship between C and p can be represented with the mapping: $\xi : C \to p$ [3].

The difference of Web information results in the diversifications of Web mining. According to the difference of information source, Web mining can be divided into Web content mining, Web structure mining, and Web usage mining. In addition, the Web structure can be considered as a part of Web content mining as well. In this way, Web mining can be simply divided into Web content mining and Web usage mining. It is shown as Fig. 108.1:

108.2.2 The Process of Web Data Mining

Compared to the traditional database and data warehouse, information on the Web is unstructured, semi-structured, and dynamic. In addition, it is easy to confuse. Therefore, it is hard to do mining directly to the data on the website. It has gone through necessary data treatment. The treatment process of typical Web mining is shown as Fig. 108.2.

108.2.2.1 Data Collection Selection

According to the principles related to themes and the requirements of clients, a group of data related to the target is abstracted from a great amount of primitive data so as to offer relatively clean data.

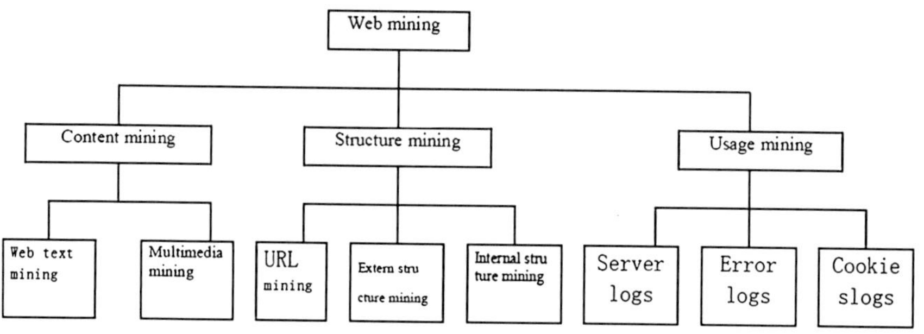

Fig. 108.1 Web mining classification structure

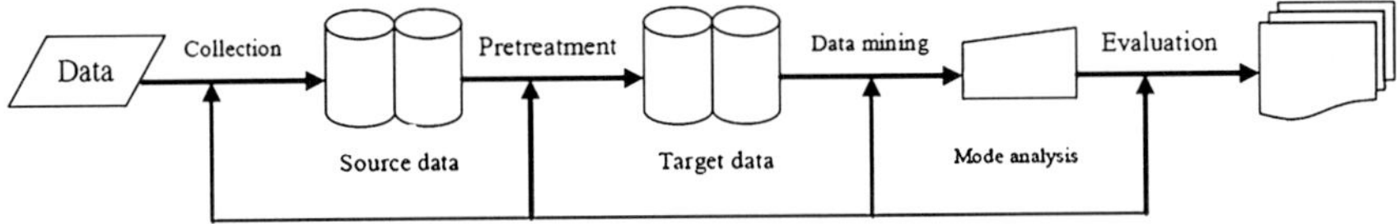

Fig. 108.2 The process of Web data mining

108.2.2.2 Data Pretreatment

It refers to delete the data irrelevant to mining task in Web data. It is to carry out processing and organization reconstruction on source data. It is to transfer useful Web data to appropriate data form and get good preparation for the front stage of data mining.

108.2.2.3 Data Mining

Through data mining methods, information that is meaningful to support and decision making can be found. The procedure is the implementation process of data mining task. Efficient, potential and useful information, and knowledge can be gotten from data.

108.2.3 Explanation and Evaluation

The process requires the clear understanding of clients' requirements as well as the reasonable explanation and presentation of mining results. The stage is to testify and explain the model produced by the previous procedure. The work is automatically done through machine. It is as well done by the interactions of analysis people.

108.3 Intelligent Course System Model Design Based on Web Mining

108.3.1 System Structure Function

The intelligence system mainly realizes two problems: intelligence prediction on students' visit to course resource and mining favorite course. It then transfers the information to students with visible manner. The system adopts B/S model. The intelligence core for the system is intelligence prediction recommendation system, shown as Fig. 108.3.

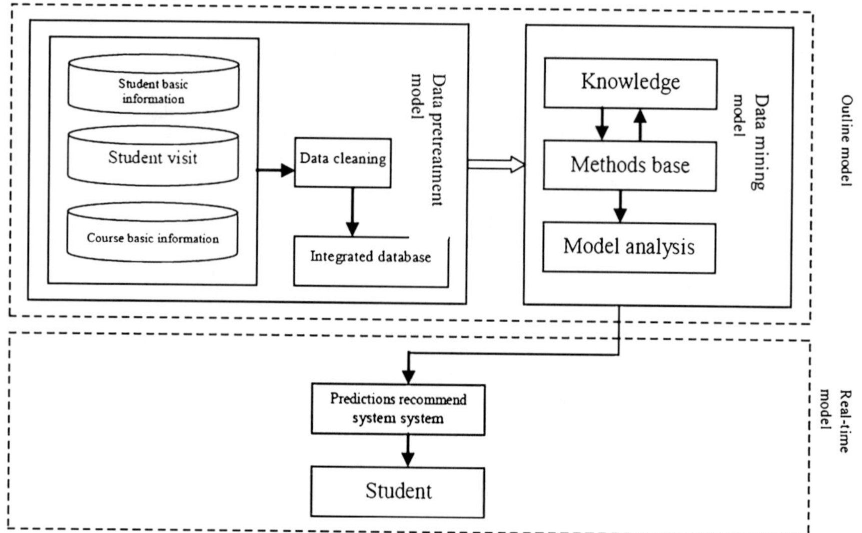

Fig. 108.3 Intelligence course system model based on web mining

There are two kinds of intelligence course system types: video and HTML form. There are about 3,000 courses and there are about 8,000 students in school.

It can be seen from Fig. 108.3 that the application of mining technology can improve the intelligence of the system, making it equip with characteristics of independence, irritability and cooperation, and so on [4].

The entire framework of the model is divided into outline and online [5]. The outline part bears functions such as data pretreatment and knowledge mining while the online part bears functions such as offering real-time service to clients [6]. The relationship between the two is that: the outline part offers foundation support for the online part while the online part offers recommendation service on the basic of the outline part [7].

108.3.2 Realization Way

To calculate the visit rate for someone to a certain course

f_i Represents the visit rate, that is to say, the first visit $f_i = 1$, the second visit $f_i = 2 \ldots x_i$ represents the duration for the visit, the visit rate for a course can be calculated through weighted average method:

$$\bar{x} = \frac{f_1 x_1 + f_2 x_2 + \cdots + f_k x_k}{f_1 + f_2 + \cdots + f_k} = \frac{\sum_{i=1}^{k} f_i x_i}{\sum_{i=1}^{k} f_i} = \frac{\sum fx}{\sum f} \tag{108.1}$$

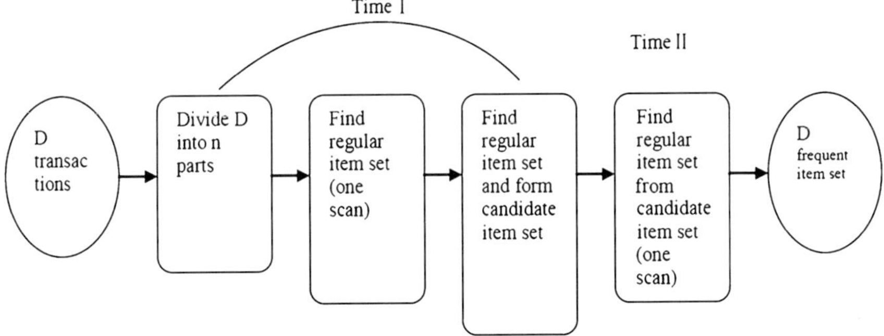

Fig. 108.4 Principle picture dividing technology

Average visit rate can be calculated, with N represents the visit amount:

$$\bar{X} = \frac{\bar{x}_1 + \bar{x}_2 + \cdots + \bar{x}_N}{N} \tag{108.2}$$

Which courses are the most popular ones? Conclusion can be drawn from the average visit rate.

From the beginning of the second moth, conclusions can be drawn from the ranking.

Deviation from the mean, that is $(X - \bar{X})$ can represent the attribute and degree for a observation value to the average value. The larger the value is, the more popular the course is Fig. 108.4.

108.3.3 Key Technology

In Web data mining, associate minging technology is to mine the connection of visited page or documents from the server in a session [8]. The most famous correlation rule finding methods is the a priori methods brought out by Agrawal.

An Apriori method is the most influential boolean association rules. With the iteration of layer by layer, k- is used to search $(k + 1)$-item set. First, find frequent 1– item set, recording as L_1 and so on. Finding each L_K requires a database scanning [1].

108.3.4 User Interface Model

Visible technology, It is mainly divided into basing on points and basing on sequence. Visibility based on sequences pays attention to the sequence characteristics of users' behavior while basing on points show various values.

Table 108.1 The predicted success rate is 56.80 %

Course name	Visit rate (%)	Duration time (m)	New visit rate (%)	Pape turn rate (%)	Success rate (%)
Recommend course 1	40.71	47	47.23	47.36	34.44
Recommend course 2	23.56	34	27.86	44.93	14.42
Recommend course 3	16.57	18.52	21.37	55.91	5.52
Recommend course 4	9.32	11.29	13.82	67.83	1.89
Recommend course 5	3.26	8.87	6.21	84.93	0.52

Knowledge inquiry technology, search-related rules, and model and other knowledge automatically. It can help to analyze target and answer inquires intelligently.

108.4 Simulation Results and Analysis

From the mining results, we have predicted five courses for students. In order to testify the feasibility of the model, we set the weight of the system as h = 1.8. The results are shown in Table 108.1.

108.5 Conclusions

The problem researched in the paper is to select appropriate courses for the leading users to utilize mining dynamic based on Web [9]. It will immediately recommend appropriate courses for the next time based on the previous visit records. Practice has shown that the application of mining technology based on Web in feature course system has improved the personal service level of feature course system. It has offered intelligent assistant methods for the decision analysis of system.

Acknowledgments This work was supported by the National Natural Science Fund (51079045) and supported by the Natural Science Fund in Jiangsu Province (BK2010264).

References

1. Han J, Kamber M (2006) data mining: concepts and techniques. Translated by Ming F, Xiaofeng M, vol 12(4). China Machine Press , Beijiing, pp 498–508
2. Zhi-guo Z (2010) Design of architecture of web usage pattern mining system. Inf Syst 33(4):97–101

3. Li-jun S, Fan-rong M (2007) Research and design of XML-based web text mining model. Comput Eng Design 28(10):2230–2287
4. Wen-lan F, Guo-lin Y (2009) Research and application of web data mining on personalized search engine. J Inner Mongolia Agric Univ (Social Sci Edn) 30(4):223–226
5. Hun Y, Bo M (2003) A kind of WED data mining method based on XML. Comput Appl 23(6):160–161
6. Xue-zhi W, Jing Z, Jun-huai L, Xiao-li Z (2006) A self-adaptive cache schedule optimization algorithm based on data mining of web log. Comput Appl 32(11):116–118
7. Qiu-ping G, Quan-lan W (2010) One recommendation-system model of digital library and realization based on web mining. Library Work Study 29(6):53–54
8. Xing-wen L, You-biao Y, Hai-bin C (2007) Study of personalized modern distance education system based on web mining. Comput Eng Design 28(12):3016–3022
9. Xiao-dong X, Ke L, Shi-rui Z (2010) Research on the improvement of a priori methods used by Web. Comput Eng Design 31(3):539–541

Author Index

Z. Zhong (ed.), *Proceedings of the International Conference on Information Engineering and Applications (IEA) 2012*, Lecture Notes in Electrical Engineering 220, DOI: 10.1007/978-1-4471-4844-9, © Springer-Verlag London 2013

X (*cont.*)

Y

Z

Printed by Publishers' Graphics LLC
DBT140313.15.17.21